Wandering Significance

Wandering Significance

An Essay on Conceptual Behavior

MARK WILSON

CLARENDON PRESS · OXFORD

OXFORD

UNIVERSITY PRESS

Great Clarendon Street, Oxford OX2 6DP

Oxford University Press is a department of the University of Oxford.
It furthers the University's objective of excellence in research, scholarship,
and education by publishing worldwide in

Oxford New York

Auckland Cape Town Dar es Salaam Hong Kong Karachi
Kuala Lumpur Madrid Melbourne Mexico City Nairobi
New Delhi Shanghai Taipei Toronto

With offices in

Argentina Austria Brazil Chile Czech Republic France Greece
Guatemala Hungary Italy Japan Poland Portugal Singapore
South Korea Switzerland Thailand Turkey Ukraine Vietnam

Oxford is a registered trade mark of Oxford University Press
in the UK and in certain other countries

Published in the United States
by Oxford University Press Inc., New York

British Library Cataloguing in Publication Data

Data available

Library of Congress Cataloging in Publication Data
Wilson, Mark.
Wandering significance : an essay on conceptual behavior / Mark Wilson.
p. cm.
Includes bibliographical references and indexes.
1. Concepts.   2. Cognition.   3. Philosophy of mind.   4. Thought and thinking.
5. Psycholinguistics.   I. Title.
BD418.3.W53 2006   121'.4—dc22   2005023339

Typeset by Newgen Imaging Systems (P) Ltd., Chennai, India
Printed in Great Britain
on acid-free paper by
Antony Rowe Ltd, Chippenham, Wiltshire

ISBN 0–19–926925–4   978–0–19–926925–9

1 3 5 7 9 10 8 6 4 2

To the memory of Geof Joseph and Tamara Horowitz

Of all the comrades that I've had, there's none that's left to boast
And I'm left alone in my misery like some poor rambling ghost.
And as I travel from town to town, they call me the wandering sign:
"There goes Tom Moore, that bummer shore, from the days of 'forty-nine."

American folk song, apparently adapted from
a music hall original by Charles Rhodes

Adam names the animals

SUMMARY CONTENTS

CONTENTS

PREFACE AND ACKNOWLEDGMENTS

Any work this prolonged demands both an apology and a map for navigating its expanses efficiently. As to the former, although my fondness for digressive curios contributes its share of extraneous pages here, the largest blame for the book's verbal exuberance can be laid at the door of the prevailing state of philosophy, which has long depended, without seeming adequately aware of the reliance, upon a collection of innocuous-looking evaluative notions: *concept, property, theory, possibility*. Within their proper compass, these words serve us as useful assistants, toiling busily within the humble rounds of everyday application. However, they are also surprisingly complicated in their ministrations, for their descriptive successes typically depend upon a complicated patchwork of diverse strategies that easily pass unrecognized by their employers. That surface simplicity often trades upon hidden complexity is not an unfamiliar phenomenon: the effective operations of a hand tool such as a screwdriver demand the confluence of quite subtle supportive factors to work properly. Despite this de facto complexity, we are naturally, but falsely, inclined to look upon "concepts" as rather simple in their inherent constitution. This innocent faith then tempts us to presume that the shifting soils arrayed under the heading of "concept" provide firm and fixed ground upon which great projects can be confidently founded. Trusting to this ill-situated confidence, we frame blueprints of our intellectual capacities that, although flattering to our vanity, prove misguided in their execution and, on occasion, trick us into truly unfortunate decisions when our real life buttresses and piers begin to shift inevitably within the sands in which they have been posted. In my opening chapters, I attempt to supply some sense of the harm wrought, for "concept" and "theory"-induced misapprehensions adversely affect many walks of life, even those far removed from the realms of the overtly "philosophical."

Ideally, the counsel of academic philosophy should temper these missteps somewhat but, in fact, my profession has more often served as an avid cheerleader to excess. The essential background to this odd situation is this: near the start of the twentieth century a host of quite substantive concerns, including some troubling practicalities arising within the mathematical and physical practices of the day, became quite critical. A number of important thinkers suggested that a certain blend of themes, drawn from both ordinary life and longstanding philosophical tradition, might provide exactly the tonic required to cure these woes. Although many of the tenets they emphasized continue to quietly dominate current thinking in philosophy, no scientific worker approaching their original array of practical concerns would recommend the same remedies today, anymore than we still entrust our health to Carter's Little Liver Pills or Harness' Electropathic Belts (undoubtedly, many of the nostrums we presently cherish will appear equally ridiculous a hundred years hence). For reasons I will outline next, much of this old conceptual consensus congealed into tacit dogma and has lumbered on more or less intact ever since. These persisting doctrines, in their sundry varieties, will be labeled the *classical view of concepts* in the book.

Given that this respected "cure" no longer answers to any real life malady, why do we still consume great gobs of the stuff avidly? Some of this appetite undoubtedly derives simply from the inertia that keeps old doctrines aloft even after they have become detached from the bow from which they were originally sprung. The fragmentation of intellectual tasks typical of our modern era often supplies the partial vacuum that abets these low friction flights. When the classical view was first distilled, the great thinkers who blended the concoction together were astonishingly knowledgeable about the physics, philosophy, psychology and mathematics of their day. But the pressures of increasing specialization since have led philosophy as an academic subject to become largely detached from the pragmatic urgencies that brought the classical portrait of *concept* and *theory* into prominence and, accordingly, there are fewer folks around able to survey its wide variety of interlocking topics suitably. For we philosophers, this disciplinary myopia has proved particularly unfortunate, because it insures that we rarely profit from the rich veins of efficacious wisdom that have been slowly uncovered over the past century with respect to the scientific specifics that were originally tangled up in those founding musings.

In fact, if presented entirely in abstraction from concrete application, the turn of the century consensus with respect to attributes and the nature of science appears entirely reasonable and innocuous. Its subtle problems emerge only when its fundamental tenets are once again set in engagement with some form of demanding problematic. In many respects, it demonstrates a kind of *perversity* on Nature's part that she has decided that, in the final analysis, she will not submit to our a priori classical expectations. But if we ask little of Nature, she is unlikely to criticize our misapprehensions much.

A important side effect of classical thinking is that it inherently elevates philosophy's dominion to airy levels beyond the slings and arrows of inconvenient fact and methodological complication. This conception of what "philosophy should be about" is no doubt soothing, even if not very realistically founded, and likely explains why many contemporary philosophers cling devotedly to classical assumption (even if they fail to recognize that they do so). Having grown accustomed to our unmerited disciplinary autonomy, our conceptions of "concept" and "theory" are likely to become congruently vague and pliant in a manner that prevents their sprockets from engaging firmly, as formerly they did, with the machinery of practical concern. This progressive disengagement deftly isolates the kingdom of philosophy from external criticism, but it is no surprise that such well-protected arrangements are apt to leave the unbiased observer with the impression of a great contraption grinding away in aerial irrelevance ("irrelevance" may be too mild a criticism, for such a levitated display is likely to harm passing motorists if they bump into one another while gawking at the damned thing). In truth, what might be properly considered as "philosophical thinking" constitutes a vital aspect of everyday life, but we must continually ensure that it retains a linkage to genuine instrumentalities through ascertainable belts, gears and rods. Unfortunately, I believe that the drift towards scholastic aloofness has increased in recent years and many of my contemporaries now pursue projects that strike me as functionally pointless, often under the self-styled banner of analytic metaphysics. To me, much of this bountiful activity merely represents the foliage that naturally blooms when the grounds of classical presumption are no longer tended by a gardener who keeps their practical ramifications firmly in view.

Many critics have likewise sensed something deeply amiss in the basic classical picture and have offered various diagnoses of its underlying problems (amongst these authorities W. V. Quine will prove most salient here, in a manner to be outlined in Chapter 5). Unfortunately, many of these anti-classical accounts suffer from the same eagerness for excessive generality as betrays the analytic metaphysician and their proposals usually run to implausible contraries as a result. There are a number of important ways in which the original classical story manages to capture important aspects of everyday practical decision correctly, but this worthy germ is often discarded along with its accompanying chaff.

To maintain a firmer grip on the tiller of practicality, as well as benefitting from the capable insights subsequently won by an army of advancing intellects, we could do worse than simply revisiting the scientific dilemmas of the founding era and observing how its concrete concerns are addressed today. Classical mechanics, after all, has never really gone away: its myriad methods still embody our best strategies for discussing the behaviors of macroscopic materials profitably (responsibility for their maintenance has now shifted to the supervision of departments of engineering and applied mathematics, rather than physics, however). No comparable study could provide, I think, a richer illustration of the lesson that philosophical dilemmas are best approached with commonsensical caution and an eye for subtle detail, rather than by hastily raising the flag of grandiose hypothesis. In fact, many of the theses advanced here were suggested to me in the course of working on a project entitled "Classical Mechanics: One Hundred Years After" sponsored by the National Science Foundation (would that I had been able to complete this reexamination in the full detail it deserves). In this regard, I have found the writings of the celebrated electrical engineer Oliver Heaviside to be particularly inspirational.

But to pursue such a course exclusively would quickly engulf the book in arcana beyond the ken or patience of my intended audience and so I have confined my discussion of affiliated issues largely to the later parts of Chapter 4, as well as a few supplements scattered as insular sections here and there and marked with an asterisk. However, if I am right in my diagnostic surmise, processes of linguistic development similar to those common in applied mathematics can be expected to arise within entirely domestic settings as well. Accordingly, I have attempted to prosecute my argument mainly through the consideration of homespun notions such as "rainbow," "weighs five pounds" and "filbert." To be sure, I often sketch some variant of the scientific circumstances that inspired my analysis alongside, for it is usually within a context of technical urgency that the *strategic wisdom* of the gambit under discussion becomes most evident (my humbler, "everyday dress" illustrations of allied processes may seem merely ephemeral or whimsical, if examined in isolation). Indeed, I very much hope, if nothing further is achieved, that my readers will gain a warmer appreciation of the clever and unexpected thinking that a good engineer regularly brings to problems that, upon cursory inspection, may seem routine or unimportant. Nonetheless, my little passages of popular science can be easily skimmed or skipped without losing the essential thread of our discussion and if some disquisition upon millwork seems excessive, it is probably time to advance to the next section heading. Some supplementary remarks have been set in finer print simply because accuracy requires that certain technical issues be canvassed in greater detail; the uninterested reader may certainly ignore these (as such, they comprise the stuff of which footnotes are generally made, but I have reserved the latter largely for the citation of sources).

Even profiting from a liberty to glide past technicalities, a mighty thicket of pages remains to be negotiated in this book. The basic structuring of my argument is as follows. In Chapter 1 I delineate the book's main themes as best I briefly can, especially in section (iv). Chapter 2 surveys the manner in which worries about concepts typically insinuate themselves into everyday practical concerns, in spite of our earnest efforts to "avoid philosophy." Chapter 3 outlines the classical picture of concepts in greater detail, whereas Chapter 5 sketches the manner in which its tenets have been opposed by a loosely defined school of pre-pragmatist thought (my own suggestions represent a blend between these two positions). Chapter 4 is the most overtly science-focused in its emphases, outlining the odd legacy of "theoretical content" that greatly hinders clear thinking about concepts, as well as developing the positive portrait of *facades* that remain central throughout the rest of the book. This chapter's discussion, unfortunately, involves somewhat nitty gritty considerations that will not prove to everyone's taste and so the entire topic of facades is reopened from a fresh point of view in Chapters 6 and 7, which are less technical and can be regarded as the most central to our entire discussion. Finally, the remaining chapters take up the crucial topic of how we should rationally deal with a language prone to behave in the unruly ways that facade-like behavior represents. It is here that we will finally appreciate the good works that everyday appeals to "concepts" et al. perform on our behalf, as well as understanding the mechanisms whereby they occasionally lead us astray.

Given the abundance of typeface before them, those readers most avidly interested in contemporary philosophy of language may find it profitable, after perusing the overview of Chapter 1, to jump directly to Chapters 6 and 7, where certain unexpected patterns of linguistic development are outlined in some detail (the appendix to Chapter 3 should supply an adequate sense of what I intend under the heading of *classical theory*). These studies directly illustrate the behaviors with which the book is centrally concerned and may provide more robust motivation for revisiting the venerable themes surveyed in the prior chapters. In truth, I regard the earlier discussion as crucial to my overall argument, for these pages highlight various developmental stages within the philosophical careers of "concept" and "theory" that are commonly forgotten or left neglected within contemporary discussions. This inattention often leaves the omnibus of contemporary philosophy of language rumbling vigorously onward, although it seems to have forgotten to take along its fare-paying passengers. In addition, a large body of accumulated folklore about science presently impedes progress in philosophy: beguiling caricatures of "what science is about" that are wrong in their fundamentals and readily tempt credulous souls into unfortunate alleyways as a result. Chapters 3 to 5 attempt to survey these entangled details from an essentially historical point of view and many lay readers may find these materials the most engaging in the book, for they show that, at base, analytic philosophy does not represent a disengaged topic of no practical import but is originally founded in robust issues of substantial concern (even if that legacy is often forgotten today).

Nonetheless, my picaresque recounting of "concept" and "theory" 's misadventures is rather lengthy. Since many contemporary philosophers of language do not view their preferred topics as grimly as I do—as ill-motivated and inextricably encrusted with layers of "scientific folklore"—, they may reasonably elect to skip beyond my initial discursive chapters, agreeing to return only if they find robust reason to do so within my later cache of examples. Such leapfrogging readers may well wonder, however, as they confront these localized illustrations, "Gee, couldn't this case be handled by X's theory as well?", where reference is made to some proposal that falls more squarely within the ambit of classical tradition. No doubt, a fair answer will usually be, "Yes, it can." Nonetheless, I have not engaged in a good deal of the usual comparative tit-for-tat here. My unhappiness with the classical point of view lies in the fact that it paints an implausible portrait of human intellectual capacity and practicality, not that its some-what hazy descriptive vocabulary can't be adapted to any situation that comes down the pike. After all, any substantive and well-established creed finds ample ways to provide its practitioners with a conviction of coherence and I do not believe that the houses of "analytical metaphysics" and the like can be easily toppled by discovering intrinsic flaws in their construction, just as few apostates can be expected to abandon the Church of Latter Day Saints simply because of inconsistencies within The Book of Mormon (however strange its contents may seem to the rest of us). Accordingly, I find it more important to return to the wells of original motivation than laboring mightily to prove alternative accounts unacceptable. And it is exactly this basic doctrinal reappraisal that my opening chapters attempt to provide.

To be sure, many contemporary philosophers regard it as virtually axiomatic that the nature of philosophy requires that doubts assume the forms of the internal contra-vention that I largely abjure ("Philosophy deals exclusively with the realm of conceptual possibility," they contend, "and if a view is wrong, it can be refuted entirely by armchair reflection"). But such expectations are founded squarely in the views of "concept," "theory" and "possibility" under critical review here. Indeed, philosophy's favored methodology of interior confutation would scarcely be accepted as credible within any other branch of learning and its requirements have seemed plausible within our ranks only because unexamined assumptions with respect to "conceptual grasp" have made them so; more exactly, the inherited traditions of classical thinking establish an a priori portrait of philosophy's prerogatives that stems directly from the manners in which we commonly misunderstand the evaluative utilities of everyday talk of concepts and the like. Or, at least, that is the theme this book proposes to argue, if considered in its wider entirety.

But I recognize that much of my prospective audience will not initially share my misgivings with respect to the stalwart trustworthiness of our intuitions with respect to "concept" and allied topics. I appreciate that such readers may not sympathize with my decision to emphasize motivational fundamentals over current debates and may therefore lose patience with the rather elaborate sifting of themes that transpires within my opening chapters. For this readership, perhaps a good jolt of unusual examples provides a better incentive for reopening old issues, whose hidden difficulties, after all, prove rather delicate in their details. For myself, I am very much of the opinion—shared by the most admirable portions of the older Anglo-American tradition in philosophy—that we should rarely trust the sweeping Thesis taken in its own terms and should always endeavor to tag its putative contents to real life motivation and application. This venerable brand of skeptical inquiry anticipates that, when firm connections with the concrete are eventually forged, the doctrine that once seemed obvious and transparent on unexampled reflection will often prove to be tacitly laden with a large schedule of small, but nonetheless vital, misapprehensions with respect to human capacity (the devil and the Good Lord both reside within the details). But philosophical caution of this stripe seems to have lately faded from the academic landscape and I have found that recent audiences are sometimes perplexed by the roving and apparently unconstrained forms of examination practiced here. In this preface I have tried to explain why I believe our rambles are obligated by the vast territory in which our chosen topics naturally distribute themselves. We can properly trim our travel docket only when we are pretty certain that everything we seek lies within proscribed bounds. Accordingly, the specific examples and proposals provided in Chapters 6 and 7 are not independent of the rest of the book, nor are they even constitutive of its main themes. However, the focused oddities they embody may motivate a wider search in which we become more willing to turn over some of the neglected and apparently unprepossessing rocks that lie scattered here and there upon the sprawling moors of "concept" and "theory."

Accordingly, I hope the book as a whole persuades its readers that the circumspect approach it outlines better accords with a plausible appraisal of human intellectual capacity than does current orthodoxy (I will be flattered if the work is regarded as a worthy continuation of the school of tempered common sense pioneered by Thomas Reid and J. L. Austin). In any event, its lamentable massiveness represents the only way I have discovered to advance its brief persuasively, at which point I can only echo my muse Heaviside, who wrote of his efforts to introduce some quite peculiar methods for solving differential equations:

> The above may help others on the way. But perhaps, like the fishes who were preached to by the saint: "Much edified were they, but preferred the old way."[1]

---

[1] Oliver Heaviside, Electromagnetic Theory, iii (New York: Chelsea Publishing, 1971), 291.

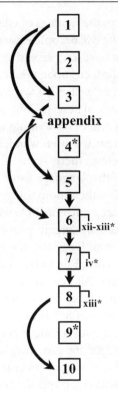

Reading plan: optional material is marked with an asterisk.

According to his biographer Paul Nahin, Heaviside's reference is to Antony of Padua who once proclaimed in a celebrated sermon:

Hear the word of God, oh ye fish of the sea and the river, for the infidel heretics despise it.[2]

Presumably Heaviside was amused by Antony's assumption that his substitute audience was likely to find much of value in his fulminations.

However that may prove, I can honestly promise the apprehensive reader that this book is filled more with curious example than grand architectonic and that it has accumulated its bulk in the fashion of The Pickwick Papers rather than The Brothers Karamazov. In any case, although I would ideally prefer that my argument be followed straight through, I have supplied a chart that marks out several shorter programs of study. I trust I will be pardoned for the occasional redundancies that make these alternative routes feasible.

Many of the suggestions I advance were originally prompted by methodological remarks offered by applied mathematicians and other scientific investigators: in reading these, I have often thought, "Gee, that's a very sensible policy which would have never occurred to me a priori; I wonder if such strategies might be applicable elsewhere." In this

---

[2] Paul Nahin, Oliver Heaviside (Baltimore: Johns Hopkins Press, 2002), 239.

regard, I am particularly indebted to the writings of Oliver Heaviside, Jacques Hadamard and Franz Reuleaux, for reasons that will become evident later. On the philosophical side, Bertrand Russell and W. V. Quine have long served as the pylons between which I have endeavored to steer and my specific focus upon predicates and concepts grew out of my thesis work under Hilary Putnam as well as his writings of the time. Several reviewers have characterized the opinions offered here as "Wittgensteinian" and perhaps they are. When I was young, I read a good deal of his writings under the able tutelage of Charles Marks and I find it quite striking that we often wander onto similar topics in our philosophical peregrinations. Nonetheless, there seem to be many persistent themes in Wittgenstein—some mystic belief that language game archetypes will show themselves to philosophers in the manner of Goethe's morphology of plants[3]—that utterly elude the compass of my own thinking and seem incompatible with its formative tenor. Insofar as I can see, our topical resemblances may largely prove a function of the territory: our transients look much the same, but his long term trajectory is attracted to a far different corner of the phase space than my own. But, in fact, I don't know, because I don't really understand his overriding ambitions. All I can do is acknowledge the eerie "Kilroy was here" quality that I often experience when my own lines of thought push me into yet another neighborhood that Wittgenstein has already visited.

As this project has been a-borning for a longer period than I'd care to think about, more people should be acknowledged for their helpful suggestions than I can actually manage, having outlined parts of this material over the years in a number of talks and seminars. To all the useful comments I received, thanks. And thanks, in pride of place, to my family, Winston and Kathleen, for putting up with it all and for serving as guinea pigs in mysterious "linguistic experiments." To my brother George for not only getting me into philosophy, but, more importantly, getting me *through* it. To three especial friendships formed when we were all at Chicago Circle together: Penelope Maddy, Michael Friedman and Anil Gupta. Their conjoined philosophical influences, different as they all are, riffle quietly through all the pages here. To Bob Batterman, Jeremy Butterfield, Joe Camp, Bill Demopoulos, Jeremy Heis, Jeff King, Michael Liston, Bob Schwartz, Lionel Shapiro and Sheldon Smith for much help on specific topics. To my editor, Peter Momtchiloff, for urging me up and over the last hill with good humor and for arranging for several exceptional referee reports.

Finally, I'd like to remember once again the two friends to whom this book is dedicated: to Tamara Horowitz, whose invariable common sense shines through in her posthumous The Backtracking Fallacy,[4] and to Geof Joseph, who taught me that, in philosophy, a bit of whimsy can be worth a thousand words. Would that I could have better benefitted from his help in shortening the pages here.

Mark Wilson

[3] My opinions in these matters have been much influenced by David G. Stern, Wittgenstein on Mind and Language (Oxford: Oxford University Press, 1995) and John Koethe, The Continuity of Wittgenstein's Later Thought (Ithaca, NY: Cornell University Press, 1996). For my own uncertain speculations on these matters, see Mark Wilson, "Wittgenstein: Physica Sunt, Non Leguntur," Philosophical Topics (1999).

[4] Tamara Horowitz, The Backtracking Fallacy (Oxford: Oxford University Press, forthcoming).

1

WIDE SCREEN

Since I got my lens, I'm feeling so glad;
I fit any kind of screen that come to Trinidad.

The Duke of Iron[1]

(i)

Our topics introduced. To be honest, the central concerns of this book—issues relating to the status of *concepts, notions, properties, attributes, traits, characteristics* and other notions of that ilk—have acquired a hard-won reputation for dullness, such that otherwise ardent students of philosophy frequently shun the subject as irrelevant to the normal run of human concerns. And the usual literature on the topic often confirms this somewhat leaden impression. I once received a new philosophical text on properties[2] from a publisher that came accompanied by a fulsome blurb extolling its educational virtues: "Here is just the work," some scribe from the Grub Street of textbook advertising wrote, "to fire the imaginations of all your undergraduates in your next philosophy class." Inside I found a little box with the word "the" inscribed several times inside. *"How many 'the' 's do you think are in the box?,"* the text asks and this query provides the sole motivation for the investigation of a lengthy sequence of rather bizarre (to my thinking) "theories of universals." The enthusiast from the publicity department evidently believed that, in a classroom situation, some clever pupil will suggest the answer "One" and this startling proposal will ignite such heated debate that the entire class will

---

[1] The Duke of Iron (Cecil Anderson), "Wide Screen," Monogram Record M-934. I worry about this accreditation because Anderson often covered the compositions of other calypsonians. Indeed, W. V. Quine made the mistake of attributing his title From a Logical Point of View to Harry Belafonte, when the originating source ("Ugly Woman") was composed by the Mighty Lion who never received adequate credit for his work (and made superior records to boot).

[2] David M. Armstrong, Universals: An Opinionated Introduction (Boulder, Colo.: Westview Press, 1989). A similar example is provided in Nicholas Wolterstorff, "On the Nature of Universals," in Michael J. Loux, ed., Universals and Particulars (Garden City, NY: Doubleday, 1970). Peirce employs "the" as an illustration of his type/token distinction; perhaps this tradition traces to him: Charles Saunders Peirce, The Essential Peirce, ii (Bloomington: Indiana University Press, 1998), 480.

the the

the the

sit in transfixed attention throughout an entire semester. For myself, I would not trust my pedagogy to such a slender motivational reed.

In any case, I propose to investigate the problems of concepts and attributes in a different spirit. To me the most salient fact about such notions is that they frame the basic vocabulary through which we *justify and criticize* a wide range of human activities. As the celebrated Ludwig Wittgenstein writes:

> *Concepts lead us to make investigations; are the expression of our interests, and direct our interests.*[3]

For example, with respect to the appraisal of mathematical performance, we might variously declare: "Archie has never fully grasped the concepts of the calculus, so of course he can't work the problems" or "Betty, on the other hand, has looked more deeply into its central notions and believes she has discovered a better way to work with these notions" or "Veronica maintains that Betty's ways of reasoning cannot be justified according to the characteristics she has so far been able to articulate." And so on, through many possible variations. Through such appeal to the proper content of sundry concepts we correct and steer onward our own projects and those of others.

I will call words like "concept," "attribute," "notion," "property" and so forth *terms of conceptual evaluation*, for the simple reason that these provide the phrases we employ in everyday life to evaluate the degree to which we believe ourselves "conceptually prepared" to execute some prospective task or other (later I shall add "truth" and "validity" to the heap we consider, but for the time being the first faction will keep us busy enough).

The rub is that, in critical cases, the *exact guidance* supplied by a purported "concept" can prove less than clear—where do our judgments of "what concepts tell us" come from? On what grounds should we condemn Archie for not having "fully grasped the concepts of the calculus"? What little bird informs Betty that she has successfully "looked more deeply into the central notions of the calculus" than others? How should Veronica justify her claim that "Betty's ways of reasoning cannot be justified according to the concepts she has been able to articulate thus far"? From what sources do these sundry judgments with respect to correct and incorrect application spring? We can easily imagine circumstances where any of our claims might prove controversial. *What is it* to "grasp a concept" anyhow?

---

[3] Ludwig Wittgenstein, <u>Philosophical Investigations</u>, G. E. M. Anscombe, trans. (New York: MacMillan, 1953), §570.

Indeed, from the history of science alone, we can readily provide examples where confident appeals to "conceptual authority" have subsequently proved detrimental and unwarranted. Often the chariot of scientific progress might have rolled more swiftly onward if such specious forms of conceptual friction had not impeded its advance (indeed, my Archie, Betty and Veronica claims correlate neatly with certain unfortunate episodes in mathematical history to be surveyed in Chapter 8). Our basic human nature often seeks perches of unearned advantage from which we can lustily applaud our own endeavors while dismissing the divaricate proposals of rivals. Spurious appeal to the "proper content" of a concept can readily provide a dandy picket from which such lofty forms of intellectual sniping can be executed. The complaint, "Oh, you're not using that concept quite right," has so frequently served as a pretext for unearned privilege that we might easily succumb to cynicism with respect to *all* judgments of this nature.

Indeed, quite sweeping disparagements of the claims of "conceptual authority" have invaded the academic humanities in recent years, to generally deleterious effect (we shall examine a case in point in 2,v). Within this strain of self-styled post-modernist critique, most appeals to "conceptual content" are dismissed as rigorist shams, representing scarcely more than polite variants upon schoolyard bullying. Run-of-the-mill appeals to "conceptual authority" tacitly masquerade prejudiced predilection in the form of falsely constructed universals which, in turn, covertly shelter the most oppressive codes of Western society. But such sweeping doubts, if rigorously implemented, would render daily life patently unworkable, for we steer our way through the humblest affairs by making conceptual evaluations as we go. In what alternative vocabulary, for example, might we appraise our teenager's failings with respect to his calculus homeworks? Forced to choose between exaggerated mistrust and blind acceptance of every passing claim of conceptual authority (even those issuing from transparent charlatans), we should plainly select gullibility as the wiser course, for the naïve explorer who trusts her somewhat inadequate map generally fares better than the doubter who accepts nothing. We will have told the story of concepts wrongly if it doesn't turn out to be one where our usual forms of conceptual evaluation emerge as appropriate and well founded *most of the time*.

Of a milder, but allied, nature are the presumptions of the school of Thomas Kuhn, which contends that scientists under the unavoidable spell of different *paradigms* often "talk past one another" through their failure to share common conceptual resources, in a manner that renders scientific argumentation more a matter of brute conversion than discourse. We shall discuss these views later as well.

Although their various generating origins can prove quite complex, most popular academic movements that promote radical conceptual debunking of these types draw deeply upon inadequate philosophies of "concepts and attributes." Such doctrines often sin against the cardinal rule of philosophy: *first, do no harm*, for such self-appointed critics of "ideological tyranny" rarely prove paragons of intellectual toleration themselves.

(ii)

The classical picture of concepts. In contrast to these injurious critiques of conceptual authority, the analytic tradition in philosophy (a heritage to which this book largely belongs) has generally painted a rosier portrait of human capacity wherein the internal contents of traits are assumed to be both comparatively sharp and objectively assessable. "If they would only scrutinize their concepts rightly," the analytical school contends, "Archie, Betty and Veronica should be able to sort out their squabbles definitively, for conceptual clarity is a sure path to unquestionable correctness." As we shall see, such sentiments represent the natural development of the attitudes we manifest within the resolution of everyday conceptual problems.

To be sure, the optimistic and commonsensical assumptions of the analytical school are often articulated in terms that can startle the unprepared reader. For example, the nineteenth century German philosopher Gottlob Frege (a predecessor greatly cultivated within the analytical tradition) frequently evokes a hypothetical "third realm of existence" (that is, neither mental nor physical in nature) wherein the full slate of possible concepts and thoughts is supposed to dwell:

> [Concepts] *are neither things in the external world nor ideas. A third realm must be recognized. Anything belonging to this realm has in common with ideas that it cannot be perceived by the senses, but has in common with things that it does not need an owner so as to belong to the contents of his consciousness.*[4]

Such passages, to put it gently, may strike the sober minded as odd or occult. Some of us, in nominalist reflex, may feel roused to the office of becoming Robert Ingersols of metaphysical excess, seeking to cleanse our intellectual landscape of the blight of mystical universals. Others may discern a converse duty to defend Frege's redoubt of abstraction from attack by the excessively hardheaded (such are the crusades to which the man with the "the"'s in a box hopes to summon his audience).

However, in this book I suggest we resist such calls to ontological battle. Frege, in fact, was a professional mathematician greatly concerned with advancing his subject to a state of such perfect rigor that all of its results could stand as permanently unimpeachable. In the passage cited, shorn of Platonic metaphor, Frege simply articulates his strong conviction that (i) we can determinatively compare different agents with respect to the degree to which they share "conceptual contents"; (ii) that initially unclear "concepts" can be successively refined by "clear thinking" until their "contents" emerge as impeccably clear and well defined; (iii) that the truth-values of claims involving such clarified notions can be regarded as fixed irrespective of our limited abilities to check them. His peculiar talk of unearthly kingdoms, parsed sympathetically, represents little more than an appeal to our everyday faith that most conceptual disagreements can be definitively and crisply resolved through a diligent program of clear thinking. And, in the same tolerant spirit, every important thesis that Frege advances in "third realm"

---

[4] Gottlob Frege, "Thoughts" in Collected Papers on Mathematics, Logic and Philosophy, Peter Geach and R. H. Stoothoff, trans. (Oxford: Basil Blackwell, 1984), 363.

guise can be easily restated within the homely vernacular of commonplace intellectual evaluation.

Such tempered replacements stand near the heart of what I shall call *the classical picture of concepts* in the sequel; it represents the general run of doctrines with respect to concepts that have proved the most widely shared across the historical spectrum of formally articulated forms of philosophical thinking. In truth, the most problematic aspects of this classical picture trace, not to its "wild ontology," but rather to the *manner in which we grasp concepts* is there described: that Archie, Betty and Veronica differ simply in relating to the common concepts of the calculus according to different degrees of contemplative engagement. Purged of metaphysical metaphor, such assumptions should seem entirely plausible, bordering on the tautological and embodying scarcely more than the commonsensical attitudes we evince in our everyday weighing of conceptual authority. Has Archie truly mastered the calculus concepts? Is Betty's claim of deeper insight sound? Is Veronica right to fault Betty's appeals?

Indeed, within the most dominant portions of the analytic tradition, classical assumptions like (i)–(iii) seem so obvious that the prospective student of concepts quickly imagines that there is little to adjudicate beyond determining in what ontological dominion these gizmos properly sit. Since this task, as we've noted, can seem less than enthralling, many philosophers abandon this metaphysical chore to the specialists and pursue more gratifying forms of investigation.

I might indicate that, although I frequently cite Gottlob Frege in this book, I nevertheless regard the early twentieth century philosopher Bertrand Russell as a more perfect representative of the classical picture (Frege maintains an appreciable range of eccentric opinions that we needn't explore here). Later, in an appendix to Chapter 3, I shall codify a lengthy list of the theses that I consider to be most characteristic of a classical point of view. Here Russell's evocative Problems of Philosophy[5] of 1912 provides our basic frame, although I have freely added some other popular claims not articulated in Russell when they help fill out the picture in natural directions (e.g., with respect to notions of possibility and possible world, about which Russell would have been personally dubious). However, I intend to cast the mesh of "classical picture" rather widely in this book and so allow our list to embrace popular opinions that differ from Russell's own in some respects (he was much prone to changing his mind on some of our lesser topics in any case). We'll be mainly concerned with the general *tenor* of the classical picture (whose foundations lay firmly planted in the soil of everyday, nonphilosophical thinking), rather than fussing extensively with every tenet in the compendium of classical themes that I provide in the appendix to Chapter 3. I formulate the doctrine in such lengthy terms mainly so that my intentions won't seem intolerably vague when I write of the "classical picture." At first glance, many of its contents should appear vapid truisms. In truth, they're not; materials capable of tempting us into great foolishness (or worse) lie sheltered here. But the sum total, good and bad, derives entirely from the fabric of ordinary life. Why this happens is the primary subject of our book.

---

[5] Bertrand Russell, The Problems of Philosophy (Oxford: Oxford University Press, 1912).

(iii)

Conceptual evaluation. Few modern philosophers in the analytic tradition—and certainly no post-structuralists or Kuhnians!—will consider themselves advocates of such a classical picture (to be "classical" hardly sounds like being up-to-date). In some ways, such demurrals are correctly indicated; in others, rather confused. Let me therefore outline why we concentrate largely upon classical themes in this book, rather than turning forthwith to more revisionary accounts of these matters. It is easiest, I think, if I simply outline my overall appraisal of the intellectual circumstances in which we presently find ourselves, leaving the details to be filled in later.

(1) We utilize terms like "concept" and "attribute" to profitably appraise and redirect the classifications, inferences, inventions and other projects we pursue in the course of everyday life.

(2) In the course of so doing, we tend to form rough pictures of these evaluations that are too simplified to be entirely correct. However, for many relatively undemanding purposes, these faulty portraits do not impede the practical work we achieve speaking of "concepts" and "attributes." A good analogy to this happenstance can be found in Isaac Newton's experiments on the composition of light, where, with his prism, he believed he had decomposed daylight into its ingredient strains:

> And to the sage-instructed eye unfold
> The various twine of light, by thee disclosed
> From the white, mingling blaze.[6]

Although the underlying difficulties were not clearly recognized until the 1880s, this natural portrayal of what occurs in Newton's investigations is quite misleading, for, in a very real sense, the light's "components" are actually created within the prism or diffraction grating. That daylight has a preexistent spectrum is, nonetheless, a correct claim, but one that needs to be justified according to the rather surprising and elaborate statistical treatment initiated in the early twentieth century (this situation will be discussed again in 9,iii). Newton's simpler picture approaches correctness closely enough that it can guide us adequately through many varieties of optical phenomena, to the extent that a neophyte may advance fairly far in her studies before she hears any whisper of the complex revised story. But eventually the day comes when she must plunge into more sophisticated waters.

The doctrines dubbed as the classical picture of concepts in this book largely represent the explicit codification of these sketchy pictures from ordinary life as explicit philosophical or methodological theses. For many purposes, they guide us ably, but, in delicate circumstances, we are easily led astray.

(3) Accordingly, the unprepossessing term "concept" can sometimes play tricks upon any of us, even the most determinatively "unphilosophical." In virtually every subject matter, seemingly plausible assumptions about the working basis of innocent-looking

---

[6] James Thompson, "Spring," quoted in Marjorie Hope Nicolson, <u>Newton Demands the Muse</u> (Princeton: Princeton University Press, 1966), 31.

words are capable of sending able investigators scampering away on the most quixotic of projects; folks who otherwise appear as if they haven't a trace of ontological hankering in their bones. These misadventures do not trace to errant academic thinking; instead, there lie seeds deeply planted within the humblest forms of everyday thought that stand ready to sprout great globs of undesirable foliage if supplied the least encouragement. No husbandry from formal philosophy is required at all; misguided forms of conceptual appeal will readily blossom of their own accord. Like it or not, all of us must tacitly turn "philosopher" at certain stages in our endeavors and this is very much part of the story I wish to tell in this book.

(4) In the main, our familiar vocabularies of "concept," "idea" and "trait" are nicely adapted to the sleepier lanes of everyday usage where pressures to innovate or explore unexpected pathways are not rudely demanded. But, as our everyday descriptive terms become pressed to higher standards of accuracy or performance, as commonly occurs within industry or science, a finer and more perplexing grain of conflicting opinion begins to display itself within our applications of "hardness," "force" and even "red." In truth, this same texture usually lies delicately embossed upon our more nonchalant patterns of classification as well, but the filagree is there more subtle and easier to miss. However, once this hidden weave is foregrounded, anomalies in reasoning become evident and questions of how we should proceed with our classifications become oddly perplexing.

In Chapters 6 and 7, I shall present a variety of related models (to be called *facades* or *atlases*) that attempt to articulate the pattern latent in some of these tacitly evolving patterns, as well as articulating theoretical reasons why they should be expected to emerge as a descriptive practice gains increasing practical success.

(5) Indeed, along a wide frontier, the late nineteenth century witnessed unexpected blossomings of descriptive disharmonies within both mathematics and the physical sciences that baffled traditional preconceptions with respect to methodology. It is common in popular histories to bundle these sources of puzzlement together under the heading "problems in the rigorization of science," but this familiar categorization does not adequately recognize that many of these difficulties represent the emergence of the resistive grain I have just sketched.

(6) A general program for addressing these methodological concerns was then hammered out, based centrally upon the simplified pictures of conceptual behavior that were earmarked under (2), but now rendered explicit and formally "philosophized." It is this family of articulated doctrines I call the *classical view of concepts* here (whereas the more diffuse everyday attitudes from which they emerge will be labeled as *ur-philosophy*). These classical proposals for making corrections in our intellectual course were quite optimistic in character, maintaining that any diligent thinker can, if she only sets her mind to the task, permanently avoid the strange conceptual snares into which scientific topics otherwise fall. It is within this nineteenth century context of response to methodological crisis that what I call the classical picture really comes to life and supplies a context where we can truly appreciate the *practical work* the approach intends to accomplish.

I should hastily add that most of the doctrine packaged into the classical picture is of a venerable philosophical vintage (much of it lies latent in Descartes or Locke, for example), but I consider that an important recrystallization occurred in and around 1900.

(7) By any standard, this classical synthesis should be regarded as a great <u>tour de force</u>. Although many nineteenth and early twentieth century authors participated in its development, I believe the Russell who wrote <u>The Problems of Philosophy</u> deserves much credit for articulating the nicest epitomization of the philosophical core of what constitutes classical thinking about concepts. And across of the wide swath of his other intellectual projects (e.g., within the philosophy of language or the foundations of mathematics), we witness a vivid expression of the range of *tasks* with which the classical portrayal was expected to engage.

(8) However, more than one hundred years of subsequent effort in mechanics and other fields have demonstrated that such dilemmas are not so easily or permanently resolved as the classicists believed. As noted in the Preface, classical mechanics has never died, but has instead marched robustly forward to our times in the genial custody of engineers and applied mathematicians, for it remains our best linguistic vehicle for auguring the behaviors of everyday macroscopic materials successfully. Through the probing of later investigators, some of us now appreciate that the nineteenth century's characteristic problems with "force" et al. were not adequately resolved by the classical "cures"—that the problems of those times did not trace simply to conceptual sloppiness or non-rigorous articulation, but flow instead from deeper mathematical issues connected to the basic intractability of many forms of physical description. Any practical term of macroscopic classification, it turns out, is confronted with the formidable task of trimming a vast amount of underlying complexity to humanly manageable standards and such considerations supply the real causes of why peculiar textures naturally spring up within our successful employments of "hardness," "force" and "causation." In later chapters I shall articulate several basic models (my *facades*) that indicate how such underlying strains sometimes induce a complex fragmentation in surface syntactic structure.

In other words, the nineteenth century's characteristic "methodological" problems turn out, from the perspective of a century later, to reflect the generally cantankerous proclivity of the physical world to force our ongoing employments of language to evolve along curious and sometimes mystifying pathways. Scientific worries that once seemed as if they merely required a dash of heightened rigor now turn out to trace to less remediable aspects of human circumstance. For the problems that plagued the Victorians cannot be adequately cured by simply correcting a bit of sloppy thinking on X or Y's part, as the optimistic reformers of the era hoped, but instead mandate the acceptance of quite unusual strategies in the prosecution of successful descriptive policy. It is a pity that these revised lessons are not familiar to a greater audience, for it is too often assumed in general intellectual circles that the old classical cures *did* work, thereby perpetuating a very unhelpful mythology of faulty methodological anecdotes that continue to plague philosophical thinking to this day (in the form, "The Victorians were once troubled by symptoms X, which were then cured by tonic Y"). It is then commonly presumed that Nature's uncooperative tendencies with respect to descriptive acquiescence emerge mainly with the rise of relativity and quantum mechanics, but this is not true; allied difficulties glower sullenly even at the core of what we may mistakenly regard as the most stolid and respectable corners of engineering (we will obtain a better chance of dealing with the quantum oddities if we first do a better job with respect to classical mechanics' peculiarities). Likewise, the old struggles over rigor within mathematics should not be regarded as merely minor, and now fully remedied, niceties with respect to the appropriate definition for "limit" or "derivative," but as tracing to valiant classical attempts to control the bizarre conceptual domains into which mathematical thought seems, almost against its will, ever forced to migrate. We moderns, unfortunately, have lost much of our appreciation of the *strangeness* of these developments, thereby leading to what I regard as a rather sterile era within the philosophy of mathematics.

(9) Back in the brighter days of the Edwardian era, however, the prospects for achieving permanent rigor looked less bleak, for it seemed as if, in classical thinking, the tools had finally been forged to end conceptual wars forever. As secondary spoils of this apparent conquest over confusion, two major themes enter our modern intellectual heritage:

First, the novelties introduced by new forms of scientific terminology can be adequately controlled by setting their presumptions within an articulated web of explicit theory, which can, in some sense, implicitly define the core behaviors of the terms in question. This innocent-looking and cheery supposition forms the germ of many dubious assumptions about "theory" that flower more fully later. I will canvas how much of this has unfolded in Chapter 4.

Secondly, as noted in the Preface, a pleasant niche for philosophy as a distinctive subject matter gets carved out within the ambit of classical thinking, wherein the village philosopher (often dismissed as a dreamy layabout in less appreciative times) is now assigned a trade as briskly delineated in its obligations as "blacksmith." This new calling is that of *custodian of the conceptual domain*, a supposed vocational entitlement that now leads many of us to look upon the problems of "concept" and "theory" in an altogether skewed fashion. Better, I think, that the philosopher accept a less clearly marked portfolio, for that better suits the fashion in which life bequeaths its problems to us.

(10) For a considerable period—say, circa 1880 to 1950—, this classical legacy remains largely dominant, at least within Anglo-American and European philosophical and scientific circles. Because so many folks falsely presume that the problems of rigor highlighted under (5) have been successfully tamed by classical methods, it will greatly assist our speculations if we can make the old problems of rigor come alive again, rather than falsely continuing to regard them as happily vanquished.

(11) Despite the many worthy projects that have been pursued under its aegis, the full classical synthesis, when fully and baldly assembled as a "philosophy," incorporates a range of assumptions about human conceptual capacity that look plainly implausible

and even supernatural taken all together, although any exact pinpointing of where the distortions lie proves elusive (which isn't surprising, because most of the classical picture is simply cobbled together from the intuitive strands of everyday thinking). Accordingly, a wide variety of contemporary philosophers, whether of an analytic or alternative cast, have wished to reject the full classical story in some way or other. Certainly, a seminal event within classicism's declining fortunes can be dated to the 1952 publication of Wittgenstein's Philosophical Investigations, which is plainly anti-classical in its tenor even as its other objectives remain obscure. However, earlier thinkers like John Dewey or roughly contemporaneous figures such as W. V. Quine are clearly troubled by the full classical melange quite independently of any Wittgensteinian influence. Indeed, the present book reflects many of the neo-pragmatic themes that have been emphasized by these authors, although I hope its specific concerns are more tempered by a commonsensical scientific realism than is often the case.

However, the "correctives" to classical thinking offered by its critics are often worse, in their sum effect, than the ills they seek to "cure." This is particularly true with respect to the so-called *holism* that is often central within these critiques, as I shall outline more fully in 5, xii. Our later investigation of the factors that cause theory facades to form (which represents a distinctly non-holist phenomenon) should help to steer us past these unfortunate anti-classical proclivities.

(12) It often happens that, when some intellectual project that has promised too much finally exits the stage, some fossilized residue of assumptions as to "what most needs to be done" is left behind. I daresay, by way of parallel illustration, that the unhappy heritage of Freudian thought unwittingly shapes our ongoing assessments in this way. To an extent that we are probably unable to appreciate fully, we are still driven to suppose, "Something important needs to be said about those creepy dreams we sometimes have; surely they *must* mean something hidden." The story of dreams remains an intriguing scientific question, but our conviction of the continuing urgency of the topic is likely a remnant of the preposterous hopes that psychiatry once invested in their interpretation.

I believe that similar intellectual inertia affects many of our modern musings about concepts, even within the realm of relatively straightforward empirical researches within psychology. We are still inclined to pursue will-o'-the-wisp goals without adequate motivation simply because such projects once held pride of place within the classical picture. I believe this is especially true of the halcyon ambitions described under (9) with respect to permanent rigor and clearly delineated philosophical mission. As noted in the Preface, I will often depart from prevailing standards of philosophical method in this book simply because I believe those very requirements are grounded within the dubious conceptions of concept under review here.

(13) If so, then what is to be done? Three primary tasks need to be addressed. First, we should revisit the original patterns of everyday descriptive practice and study more carefully the finer grain that can be found there. Here we will learn that its latent complexities often supply evidence of underlying forms of sophisticated descriptive *strategy* whose employment we have probably not recognized. Leaning upon the

wisdom of the engineers, I shall attempt to delineate the basic sinews of several of these strategic gambits in Chapters 6 and 7. The unnoticed emergence of these unexpected descriptive complexities often create crises in linguistic management: how do we control words that have wandered unexpectedly in their strategic underpinnings?

It is in this regard that words like "concept," "attribute" and "theory" emerge as the central vocabulary we employ when the need to resettle language upon less confusing rails arises. The only problem is that we are naturally inclined, without benefit of any philosophical indoctrination, to picture "concept"'s corrective functions in simple and overly schematic terms, rather as we invariably picture "friction" as a simple physical process when, in fact, an astonishing variety of processes congregate together under this heading. It is from this native semantic naïvety that the classical picture of concepts emerges, as natural inclination is eventually converted into explicit philosophical doctrine. So, secondly, we need to recognize that evaluative notions such as "concept" and "theory" do not hew to a fixed function, but instead trace shifting and contextually sensitive diagnostic paths, adapting to the idiosyncratic personalities of the bothersome primary words ("force," "red", "hardness") they seek to appraise. That is, "concept" and "attribute" do not behave in totally regular ways simply because *it is their job* to monitor materials that do not behave regularly either.

If these conclusions are just, then we have plainly invested excessive philosophical hope in the expectation that the contents of our concepts can be held firmly fixed, if only we remain sufficiently vigilant. We need to frame, I think, a far more mitigated appraisal of our capacities to anticipate our linguistic futures. Once again, I think the hard won lessons of twentieth century applied mathematics can assist in framing a more tempered view of our actual capabilities.

(14) The main consideration that drives the entire argument of the book is the thesis that the often quirky behaviors of ordinary descriptive predicates derive, not merely from controllable human inattention or carelessness, but from a basic unwillingness of the physical universe to sit still while we frame its descriptive picture. Like a photographer dealing with a rambunctious child, we must resort to odd and roundabout strategies if we hope to capture even a glimpse of our flighty universe upon our linguistic film. In this regard wisdom gradually accumulated within applied mathematics can help us understand the difficulties involved, for they've evolved some very effective methods for dealing with recalcitrant subjects.

This view of our subject dictates that the bulk of the book will largely be concerned with a range of revealing and somewhat unusual examples, all designed to bring forth the finer grain I have described. From their puzzling behaviors we can gain a deeper appreciation of the substantive practical goals that the original classical picture sets itself, as well as pondering how we should proceed if we no longer believe its story. Generally speaking, I won't attempt to reproduce the true arcana of the original history, but instead frame simpler cases that can still supply an appropriate sense of the kinds of troubles displayed within the nineteenth century crises. In fact, I have concocted two little fables (in Chapters 2 and 8) that recapitulate a lot of history within a comparatively short compass (to be comparatively short is not to be short, however).

My emphasis upon challenging example sets this work apart from most comparable literature of recent vintage, which more often traffic (if they supply "fer-instances" at all) in specimens like "dog" and "doorknob." Such choices trace to the tacit assumption that, at some fundamental level, "all concepts act alike." But this (very classical) presumption will prove much in dispute in these pages.

If I can tell this part of our tale correctly, without spoiling everything by indulging in excessive technicalities, the story of why drab terms behave badly should seem fascinating in its own right, because words will sometimes do the damnedest things.

The rest of this book pursues this basic outline in a fairly straightforward, albeit long-winded, way. As I observed in my Preface, different audiences might choose to navigate its expanses in different fashions. On the one hand, there is currently a very widespread conviction in the humanities that analytic philosophers such as myself have neglected our proper topics, which ought to focus upon grander matters than errant vocabulary. Such critics have become inclined, with increasing frequency, to "turn philosophers themselves." As I conceded earlier, many of academic philosophy's current obsessions are apt to seem strange or purposeless even to a charitable observer, but this appearance does not mean that such apparently exotic concerns do not connect quite directly with more robust stuff. Indeed, for such readers, I hope our discussion will persuade them that, like it or not, delicate undertakings within a linguistic vein are practically inevitable for us all, and that we shall do a better job within these dominions if we appreciate the necessity of keeping a foot near to the brakes of common sense before we roar ambitiously onward. In Chapter 2, I outline a cautionary calamity that has overtaken one of my favorite subjects (folklore)—a ruination that, if it is not wholly caused by impulsive philosophizing, has certainly had its axles considerably lubricated thereby. In the course of this book, we shall sometimes fuss about minutiae that may seem unworthy of the attention of analytic philosophy's less patient critics. But the proper story of how such concepts work is exactly one where little misapprehensions about descriptive practice are apt to enlarge into full scale disasters if they pass unrecognized. I hope, if nothing else, that I have written this book in a way that makes it clear that academic philosophy's attention to the details of linguistic engineering arises, in its core ambitions, from a well-motivated desire to minimize highway fatalities.

On the other hand, this book is primarily intended as a contribution to ongoing analytic philosophy, although, if that ambition were pursued too exclusively, I would surely exclude our first group of readers. Fortunately, I think that, at a slight cost in bulk, both audiences can be adequately accommodated. In the main, most of our discussion will not be concerned with philosophy in its more devotedly codified aspects, even with respect to what I have called the classical picture. The issues with which we shall generally be concerned instead take their origins within the rushing stream of everyday, practical decision making and it is largely along those familiar banks that our discussion will ramble. Accordingly, I hope that readers with a philosophical background will pardon the fact that I sometimes supply brief background details that they may consider superfluous. I feel that, since I must dutifully identify and explain sundry scientific commonplaces for the benefit of philosophers, there is no reason why the same courtesy cannot be returned and that its essential philosophical context cannot be sketched for the benefit of readers with other forms of background.

In fact, I think all of us will do well to recall the *practical motivations* that gave urgency to the philosophical study of concepts at the turn of the twentieth century, because I often feel that allied issues have been lost sight of in much recent work. Although it is usually recognized that Russell and his cohorts became exercised about concepts because they hoped to resolve substantive conflicts in other fields, it is usually presumed—quite falsely—that such troubles are long since resolved and the philosopher can instead concentrate upon a narrow spectrum of concerns (the old Don is dead, but the family business continues on). But these assumptions are plainly wrong and have sometimes led the modern work to become anemic in its motivations. The best way to document my point is simply to set forth a range of evocative examples and ask my fellow philosophers as we go along, "What do you wish to say about that?" Quite often, I think, the response will simply be, "Gee; I've not been concerned with cases like that." And if those replies are forthcoming, they mark how far we have descended from Russell's level of inquiry, for he ranged over exactly the same territory as I propose to explore. The answers I suggest will be different than his, but we look at the same landscape.

(iv)

Science should be used but not mentioned. The first precaution we should adopt in attempting to minimize conceptual misadventures is to beware of dressing every concept in common khaki. In this regard, most meditations on our subject too swiftly "overlook the impertinent individualities" of particular evaluative judgments, to paraphrase Charles Lamb's complaints about Sir Thomas Browne:

> That the author of the Religio Medici, mounted upon the airy stilts of abstraction, conversant about notional and conjectural essences; in whose categories of Being the possible took the upper hand of the actual; would have overlooked the impertinent individualities of such poor concretions as mankind, is not much to be admired.[7]

As noted above, many philosophers eagerly herd every passing appraisal of concept or attribute into immediate commonality, gathered into some great, generic corral dubbed "the domain of concepts," "the field of logical possibility, "the world of universals," "Plato's heaven" or some variant enclosure of that ilk. As indicated in our discussion of Frege's third realm, I don't consider the metaphysical connotations of phrases like these to represent matters of great consequence; I worry rather about the manner in which the critical features of specific evaluative judgments become dusted over in this indiscriminate massing of abstracta. In the ensuing bustle, we lose sight of the impertinent individualities that allow our everyday talk of "concepts" and "attributes"

---

[7] Charles Lamb, Essays of Elia, i (New York: G. P Putnam's Sons, n.d.), 122.

to serve so many useful functions in the ongoing administration of linguistic use. "I want to figure out how concepts *in general* work—how they grab onto the world—," announces the overly ambitious investigator, "for that's the only aspect of everyday conceptual evaluation that I find truly mysterious." No; the substantive information we convey when we judge that, e.g., "Archie has not fully grasped the calculus concepts" can differ subtly from occasion to occasion and we are sometimes tempted into dubious crusades simply because we have blurred together the shifting hidden complexities of these judgments. There is *less* commonality to our sundry weighings of "conceptual grasp" than meets the eye and we make a great mistake if we rush too quickly to framing general hypotheses about "how all concepts behave." Accordingly, although we must render proper tribute to the many fine services that words like "concept" and "attribute" provide, we should also recognize that these drab and unprepossessing terms occasionally act as the Uriah Heeps of language, 'umbly pretending to accommodate to our wishes whilst secretly scheming to usurp our affairs. It is probably this attention to the basic tension between the admirable and unfortunate aspects of real life conceptual appraisal that most distinguishes our discussion from that found elsewhere in the philosophical literature.

In this connection, we might observe that schematic approaches to *concept* and opinions on the nature of *philosophy* itself tend to support one another in unhappy symbiosis, particularly within the analytic tradition. Many contemporary authors regard the duty of maintaining vigilance over the "conceptual domain" as their especial charge, where the conceptual domain stands to the philosopher as does the ocean to the oceanographer. The former is simply the bloke who watches after what is logically possible rather than the Gulf Stream. Conversely, the presumption that concepts in their inherent purity require such specialized wardens greatly affects our picture of what such qualities must be like. As remarked above, this assumption seems to represent the continuing legacy of classical thinking.

But whatever its origins, I reject this tidy allocation of chores; the subjects discussed in this book seem chiefly distinguished by their *messiness*. Indeed, the natural world, it seems to me, rarely proves hospitable to disciplinary division. Even the devoted study of, e.g., the life of a sea squirt is apt to carry one eventually into chemistry, physics, mathematics and perhaps a spot of philosophy, for the backyard of every science opens out onto all the others. I agree with T. H. Huxley when he writes:

> *Science is nothing but trained and organized common sense, differing from the latter only as a veteran may differ from a raw recruit: and its methods differ from common sense only as far as the guardsman's cut and thrust differ from the manner in which a savage wields his club.*[8]

Because of their different assumptions about our subject, some readers may regard the topics treated in this book as falling outside of philosophy's proper dominion (although I doubt that they could determine exactly where our investigations should be placed). It

---

[8] T. H. Huxley, "On the Method of Zadig" in Science and Culture (New York: D. Appleton, 1882).

seems to me that such expulsion of our endeavors is predicated upon a picture of concepts and conceptual analysis that is under critical challenge here. But even if I am wrong about philosophy's proper mission, I believe this work articulates useful things with respect to its chosen topics, never mind their exact disciplinary classification.

Before we proceed further, let me introduce a somewhat awkward notation I will employ for convenience in the sequel. Quite commonly our notions of simple concepts like redness are closely associated with linguistic *predicates* such as the phrase "is red." Since we do not wish to confuse the linguistic unit "is red" with its purported conceptual underpinnings, I shall designate the concept itself in boldface rather than quotation marks. Thus I may write: **being red** (or **redness** or even simply **red**) is the concept that belongs to "is red." None of this notational barbarism is intended to convey any sort of substantive philosophical thesis. I shall sometimes distinguish real world *attributes* from the *concepts* we frame on their behalf, but I won't introduce any special notation to this effect.

I might also mention that, as the book wears on, I will largely restrict my attention to *predicative expressions* such as "is red" or "is harder than," rather than spending much times with *names* like "Vess," *descriptive phrases* like "that incredible banjoist" or *nominalizations* such as "fleet-fingeredness." This is largely because much standard philosophy of language often shifts the problems of the latter phrases onto the predicates (a paragon of this transfer can be found in Bertrand Russell's celebrated theory of descriptions) and I want to investigate the linguistic problems of concepts in their purest and least cluttered forms. If I write loosely of the term "red," I generally have in mind its predicative development as "is red."

In restricting my attention largely to predicates, I in no way share the old nominalist contention that traits represent naught but particular objects gathered under the umbrella of a common name. Quite clearly, we use "concept" in a broad manner that does not demand any alignment with linguistic items at all and there are plenty of cases where we clearly possess concepts that can be supplied no predicative expression. In stressing predicative use, I am mainly trying to bring forth the *skills we manifest* when we possess a concept, as opposed to the *contents we happen to grasp*, for one of our chief tasks is to understand better how skills and contents interrelate. In this way, my emphasis on predicate usage is really intended as emblematic of a more general range of skills. In any case, this book's ambitions scarcely stretch to the explication of every gainful employment of the term "concept," but simply hope to probe the underpinnings of a certain range of everyday forms of conceptual evaluation, and to relate this assessment to the characteristic problems of philosophical tradition.

Finally, I often write of the *directivities* and *supports* of predicates rather than employing more standard terminology such as "intensional characteristics," "normative standards" or "denotation." All of the latter come heavily burdened with classical presumptions I'd rather avoid, even at the price of sounding a bit vague. In short, I am not attempting to introduce an idiosyncratic technical vocabulary of my own in "directivities" and "supports." Rather I am trying to *evade* previously entrenched terminology of that ilk.

(v)

Ur-philosophical currents. Recent philosophical literature is commonly distinguished by the working presumption that an author ought to blast every competing vessel from the harbor before he sails his own skiff in. That is, I should first survey the very long list of the doctrines currently active on our topics of interest and then methodically dispatch them all. Such an odd methodological requirement would scarcely be tolerated in any other subject; I believe its popularity derives largely from the picture of philosophy as custodian of the conceptual (wherein any serious rival can be expected to sink under its own internal incoherence).

I shall largely decline this combat, partially because it makes for dreary reading. But there are more imperative reasons as well, which stem largely from the fact that our first obligation must be to explain why we are so interested in concepts anyway. We have already noted that other philosophers, even of the most devotedly analytic persuasion, rarely regard such studies as either deeply informative or crucial. Earlier I indicated the wide range of genuine *scientific* problems that Russell wanted to address, but almost none of the modern accounts harbor such ambitions (insofar as I can tell). Recent investigations often focus upon rather odd matters such as the question of whether a stuff much like water discovered on a distant planet properly qualifies as *being water* or not. In truth, issues of some importance do lie hidden within such queer questions, but their linkage to matters of practical concern is scarcely evident and the enveloping literature rarely makes much effort to improve the situation (I am firmly of the conviction that philosophical questions should only be pursued with one hand on the sturdy staff of cases that matter).

In this regard, I believe that Russell had exactly the right explanation for why even non-scientists will benefit from studying the potential wiles of concepts: wrongheaded thinking about these unexciting ingredients within our thinking can send any of us off on lunatic crusades. Such misfortunes do not befall only applied mathematicians who unwisely trust series expansions more than they should. That is, exactly the same factors that occasionally send the engines of scientific progress off the rails bedevil us in the pursuit of more ordinary affairs, with the consequence that, instead of having our buildings collapse or our cannon balls dropping on our own troops, we wind up ruining folklore or being unkind to elderly naturalists. Or, in the case of the explicitly philosophical, we gloomily conclude that we are permanently walled off from the external world by some intervening conceptual fog. All of these dreadful things can happen if we treat the impertinent individualities of unprepossessing words too roughly (as we shall see in the next chapter).

Indeed, although a philosophical author may fancy that the rather boring problems of concepts have been successfully delegated to the experts, it is more likely that vital issues within her favored topics tacitly rely upon subterranean assumptions about the possibilities of "clarity of thought" and the like. In this way, the most difficult problems of philosophical tradition often get quietly transported to a realm of concepts as classically conceived (the region serves as our dark side of the moon or Sargasso Sea). We should cast a more watchful eye upon the complacent attitudes typical of everyday conceptual evaluation, for that is where much of our wrongheaded thinking obtains its characteristic motifs.

Accordingly, to understand the problems of concepts adequately, we need to return to the gravels from which it all springs—to the headwaters of what might be called *ur-philosophy*: those utopian strands woven into our everyday thinking that sometimes induce us to overvalue our conceptual cards somewhat; that incline us to presume that we possess a little bit firmer hedge against future contingency than we really do. Our first order of business is to observe how *ur*-philosophy's fugitive voices can genuinely lead us astray within the idiosyncratic circuits of everyday or scientific judgment, when our patterns of thinking become diverted one way or another by their siren strains. Within the more developed and example-free presentations of philosophy, all visible surfaces have often become so highly polished that the underlying processes of *ur*-philosophical manufacture are no longer apparent and the grain that sometimes bewilders us becomes entirely hidden. There is not enough friction available to make forward traction possible.

To start our project upon grittier wheels, we must appreciate how easily humble and natural musings about concepts and attributes can insinuate themselves into our practical affairs and lead us onward to unhappy conclusion. Sometimes the process resembles a familiar species of nightmare. We have been cheerfully ambling along a pleasant country lane when we notice that our surroundings have turned grim. Now we seem trapped within some vast cemetery that sprawls endlessly over gray hills. We find nothing but huge mausoleums that honor dynasties of abstracta of which we've never heard. "Where did all these *edifices* come from?" we ask and wonder what faulty turn in the road could have led us into this disconcerting City of the Dead. It's better that we do not linger long amongst the marble but instead retrace our way back to that sunny lane.

In this conviction that the formal philosophical investigation of concepts often advances too swiftly up the garden path, I echo the allied sentiments of the philosopher J. L. Austin, who observes of a related group of evaluative words (he is discussing the sense data doctrine that each moment we are confronted with a determinate field of

directly perceived visual information):

> *My general opinion about this doctrine is that it is a typically* scholastic *view, attributable, first to an obsession with a few particular words, the uses of which are over-simplified, not really understood or carefully studied or correctly described; and second, to an obsession with a few (and nearly always the same) half-studied "facts." (I say "scholastic", but I might as well have said "philosophical"; over-simplification, and constant obsessive repetition of the same small range of jejune "examples" are not only peculiar to this case, but far too common to be dismissed as an occasional weakness of philosophers.) The fact is . . . that our ordinary words are much subtler in their uses, and mark many more distinctions, than philosophers have realized; and that the facts of perception, as discovered by psychologists but also as noted by common mortals, are much more diverse and complicated than has been allowed for. It is essential, here as elsewhere, to abandon old habits of* Gleichshaltung, *the deeply ingrained worship of tidy-looking dichotomies.*[9]

This is a beautiful encapsulation of a sentiment I deeply share, but its wisdom seems insufficiently appreciated today. For Austin and myself, the very grandeur of a sweeping philosophical thesis provides probable indication that we don't quite know what we are talking about; that the "importance" of our Grand Contention may derive from the simple fact that we have jumbled different concerns together. Presumptions that sound philosophical progress can be achieved through rarified transcendental argumentation or by thoroughly examining tabulations of "all philosophical positions possible on a topic" startle us, for such methods seem highly prone to dusting over the impertinent individualities that most likely reside at the seat of our problems. Quite the contrary, Austin and I recommend that our attention should turn as quickly as possible to the examination of concrete circumstance where our everyday forms of conceptual evaluation will display their stripes in *ways that truly matter*. Only there are we likely to find the clues to where we have wandered astray in our Great Thoughts. True; the examples we will consider in this book are quite unlike anything found in Austin's Sense and Sensibilia (for I believe we must zig-zag between technical example and ordinary life to get our job done), but we share an underlying commonality of skepticism and philosophical modesty.

(vi)

Semantic finality. However, most adherents of the so-called ordinary language movement (the school to which Austin is usually consigned) presume that we must have acquired the appropriate subtle uses of our ordinary words in the process of *becoming competent in English* (Austin's own attitudes seem weaker and more delicate[10]). Although professional philosophers frequently bungle their intricacies, it is maintained,

---

[9] J. L. Austin, Sense and Sensibilia (London: Oxford University Press, 1964), 3.
[10] J. L. Austin, "A Plea for Excuses" in Philosophical Papers (Oxford: Oxford University Press, 1961).

we nonetheless learn complex, implicit rules from our linguistic tutors that restrict "concept" and "attribute" to finer circuits of proper application. If we would only attend to these rules, it is argued, we should be able to prevent language "from going on holiday"[11] in the manner that leads to errant philosophizing.

The thesis that we learn, as part of the process of becoming competent in English, complicated layers of criteria for the application of words like "concept" or "red" has proved notoriously hard to defend. Its continuing source of attraction to certain thinkers lies in the hope that, could these evaluative epicycles be cleanly identified, many of the problematic assertions of mainstream philosophy could be cleanly dispatched. Unfortunately, there is little evidence that well-bred usage shelters such delicate and canny discriminations. Linguists, to be sure, have ably demonstrated that "proper usage" makes very fine syntactic discriminations indeed, but these most often represent the artifacts of linguistic descent rather than homegrown displays of philosophical acumen.

While I have considerable sympathies for many of the objectives that Austin and the ordinary language school set themselves, such projects rest upon an untenable view of language insofar as they demand a foundation in the notion that "our linguistic training tells us how to use notions like 'concept' properly." Certainly, the project in the present book proceeds upon the basis of diametrically opposed presumptions. In particular, the story told here maintains that many of our conceptual misadventures arise precisely because our "linguistic training" has *not* prepared us adequately for dealing with a vexatious world.

To explain what I have in mind, let us consider a more general claim that still informs many forms of philosophy of language apart from the ordinary language school. This is the tenet that I call *semantic finality*, <u>viz.</u>, the claim that, with respect to a wide range of basic vocabulary, competent speakers acquire a *complete conceptual mastery* or *grasp* of their word's semantic contents by an early age—no later than 10 or 11, say. This core content then acts as an *invariant* that underwrites many of our characteristic endeavors: "If we don't share common, fixed 'contents,'" it is asked, "how can we possibly understand what others are talking about? For that matter, how can we be sure we are addressing even the questions we pose to ourselves?" To be sure, it is conceded that, beyond their initial period of conceptual inoculation, speakers will often tinker with these early basic contents in minor ways—e.g., later we learn that the usage of "dog" can permissibly extend to cover the wider family <u>Canidae</u> and poetically stretched to embrace human feet. Nonetheless, the majority of matters we subsequently learn about dogs—that Jones' specimen down the street is an ugly brute; that they are largely color blind; that they are available in sizes smaller than squirrels, <u>etc.</u>—do not alter the stored core content of *being a dog* and can be ignored by the student of semantics proper.

It is commonly argued, furthermore, that such semantic finality by the age of linguistic majority follows as a necessary consequence of the fundamental *creativity of language*: the undeniable fact that a linguistically competent speaker can understand a

---

[11] Wittgenstein, <u>Investigations</u>, §38.

vast range of sentences she has never before encountered. Here is an explication of the latter by the linguist Ray Jackendoff:

> *The fundamental motivation behind generative syntax is of course the creativity of language—the fact that speakers of a language can understand and create an indefinitely large number of sentences they have never heard before . . . Corresponding to the indefinitely large variety of syntactic structures, then, there must be an indefinitely large variety of concepts that can be invoked in the production and comprehension of sentences. It follows that the repertoire of concepts expressed by sentences cannot be mentally coded as a list, but must be characterized in terms of a finite set of mental primitives and a finite set of principles of mental composition that collectively describe the set of possible concepts expressed by sentence . . . It is widely assumed, and I will take for granted, that the basic units out of which a sentential concept is constructed are the concepts expressed by the words in the sentence, that is, lexical concepts. It is easy to see that lexical concepts too are subject to the argument from creativity.*[12]

Indeed, Dr. Seuss relies upon this same creativity more succinctly when he explains the virtues of the letter "O":

> "O" is very useful; you use it when you say,
> "Oscar's only ostrich oiled an orange owl today."[13]

The joke, of course, is that nobody except Dr. Seuss himself (and derivative commentary such as my own) is likely to utilize the proffered "useful" sentence; nonetheless, we feel we *understand* it completely. The doctrine that the full range of possible sentential thoughts is generated by an initial stock of fully understood core concepts is sometimes called *the thesis of strong compositionality*.[14]

As such, the doctrine is very much part of what I have called the classical picture of concepts. To be sure, strong compositionality is no longer quite the overpowering dogma amongst linguists that it was some years ago—it is recognized, for example, that a wide range of linguistic irregularities are acquired by more specialized means later in learning. But, surely, there is much that is right about a basic contention of "finality"; it seems likely that there are fairly specific forms of data that a speaker must internalize in order to parse novel sentences with respect to their grammaticality and rough import.

However, for our purposes in this book, it needs to be recognized that the semantic invariants provided under such "finality" are unlikely to carry the burden that many philosophers expect them to lift. As we continue to work with our words past our hypothetical date of finalized capacity, virtually every term of macroscopic evaluation becomes subject to subsequent shaping pressures for which our training has left us unprepared. In compensation, subtle correctives and barriers creep into our language,

[12] Ray Jackendoff, "What is a Concept that a Person May Grasp It?" in Eric Margolis and Stephen Laurence, eds., Concepts (Cambridge, Mass.: MIT Press, 1999), 307.
[13] Dr. Seuss, Dr. Seuss's ABC (New York: Random House, 1963), 34.
[14] Alan Cruse, Meaning in Language (Oxford: Oxford University Press, 2000).

often quite unnoticed, with the net effect of turning our classificatory concepts in quite different directions than we originally pictured. These processes etch a finer grain into our usage that often serves as the wharfs from which *ur*-philosophical misadventures later embark.

A good deal of this book will be devoted to cases of a more substantive cast, but let us look at a familiar predicate where the effect is quite palpable. I have in mind "is a rainbow," a phrase whose revealing eccentricities will be discussed on occasion throughout this book. Here is a word that might be regarded as the ultimate linguistic survivor: like its biological equivalent, the cockroach, we can be confident that "rainbow" will remain active in English on the Day of Armageddon. Yet if ever there was a word conceived in semantic sin, it is this one, for as children we clearly assimilate its usage to that of "arch," to the extent that we liberally accept any fairy tale in which agents deal with "rainbows" as if they could be climbed, moved or located (from L. Frank Baum's Tik Tok of Oz):

> [A] *gorgeous rainbow appeared* [and the fairy]... *held out her arms. Straightway the rainbow descended until its end was at her very feet, when with a graceful leap she sprang upon it and was at once grasped in the arms of her radiant sisters, the Daughters of the Rainbow.*[15]

To parse a passage like this correctly undoubtedly requires the infusion of a fair number of "arch"-related semantic notions. Indeed, we might employ the Baum passage as a reasonable test of whether a 7-year-old child "knows the meaning of 'rainbow'" or not.

But, of course, the worldly stuff that actually props up our ongoing "rainbow" usage is nothing like an arch at all, but consists of suitably irradiated raindrops. How do we manage to keep talking profitably as adults of "rainbows" in the real world, given the

---

[15] L. Frank Baum, Tik Tok of Oz (Chicago: Reilly and Lee, 1914), 248. The illustration is by the great John R. Neill.

preposterous misunderstandings in which this term was engendered? In this regard, I recall no pedagogical sagacity on the part of my parents, estimable as they otherwise were; to the contrary, I vividly remember having the veil of "arch" lift suddenly from eyes in the course of perusing The Boy's Big Book of Science (or some tome of allied title). At approximately this same age, my mid-childhood belief in Santa Claus suffered similar ontic shock from the whisperings of an older brother, but, unlike "Santa," "rainbow" somehow regained its wobbly legs and managed to earn a very robust, apply-it-to-the-real-world continuation into my adult years. What secret flexibility allows "rainbow" to adapt so successfully? In fact, the predicate manages to soldier onward precisely because we absorb rather complicated adult restrictions with respect to the circumstances in which we can meaningfully speak of rainbow "locations" and "orientations" (we shall study the mechanics of this in 7,viii). To be sure, our original "arch"-focused naïveties linger on in fossil form, in the guise of a peculiar double standard that divides the sorts of statement we tolerate *within a fairy tale* from those that we accept *within real life, adult application*. Since these quiet restrictive controls tend to "just grow up" (like Topsy), it is quite easy to overlook their presence.

The chief mischief that an exaggerated faith in semantic finality brings to our understanding of linguistic process is the belief that all these quiet mature adjustments of context and usage don't matter to conceptual content proper; that, mutatis mutandis, the latter must remain essentially mummified from age 8 to 85. But this presumption of invariant continuation, I claim, is not correct at all and often proves the source of grievous misunderstanding. After all, when we typically wonder about the "proper content" of our concepts within the intrigues of ordinary life (or when we become scientifically confused), we are rarely interested exclusively in the invariants required to recognize grammaticality, but instead worry about matters of a larger scope. Can we trust this concept to behave acceptably when we try to bring it into an *untested domain* of application? Will we will be *led astray* if we trust old inference patterns in this new arena?

Admittedly, it is hard to fit serious issues of "behavior within untested domains of application" to our "rainbow" example, but we can feebly try. Is it ever possible for a real life rainbow to *lie on its side*, for example? Could we employ such hypothetical occurrence as a signal to alert a confederate to a secret rendezvous? The answer to both questions happens to be "yes," but little of practical consequence hinges upon the result. However, it is plainly obvious that our "untested domains of application" will matter a good deal to notions like "force" and "hardness" (to pick two terms we shall study extensively), for our buildings fall down and our knife blades dull at inopportune moments if we augur their conceptual contents wrongly. As I shall vividly detail, when we normally ask, "How should our concept of *hardness* be properly understood?," we are framing a question that reaches far beyond the range of what any 8-year-old master of the terminology knows. We portray what occurs within everyday conceptual evaluation quite wrongly if we presume it simply represents a matter of checking whether a speaker qualifies as "knowing the word's meaning."

In short, I claim that the linguistic behaviors of "hardness," "force" and "redness" display considerable affinities with "rainbow"'s manifestly weird deportment. It is merely that their finely grained oddities are less apparent to the untutored eye (but, of course, this contention remains to be proved).

With "rainbow," we also witness a basic phenomena that will occupy us in more substantive forms throughout the book: no matter how a term may begin its career, the subsequent necessity of accommodating to real world contours can cause it to migrate in unexpected directions. The term's continuing vitality may require that we absorb peculiar restrictions that arise as *natural adaptions* of misbegotten original instruction to suit the developing demands of physical circumstance. These complicating but improving coils are likely to lock in place no matter how we are have been initially instructed (our parents may have been fierce devotees of the thesis that rainbows truly are arches, but we will meekly accept the necessary adult curbs all the same). There is no reason to expect our linguistic training (which, after all, is willing to certify us as "competent masters of the concept **rainbow**" at ages—7 or so—when we still attribute material forms to rainbows) secretly *anticipates* the later adaptations in any reasonable sense. Without benefit of juvenile or parental foresight, adult "rainbow" usage regularly discards large portions of its originally allocated field of grammatical claims, leaving behind only a complexly gerrymandered residue that neatly illustrates Wittgenstein's famous remark:

> It is not every sentence-like formulation that we know how to do something with, not every technique has an application in our life; and when we are tempted in philosophy to count some quite useless thing as a proposition, that is often because we have not considered its application sufficiently.[16]

That such mature retoolings are rather commonly required merely reflects Nature's obdurate unwillingness to conform to classificatory practices that are ingenuously framed. Children, on the other hand, usually can't acquire the full complexity required unless they build upon earlier stages more naïvely pictured. The additional strictures they must eventually acquire to satisfy the world's prickly requirements represent a (fairly) predictable *adaptation* to adult circumstance, but their contours will not appear foreshadowed in what the children have actually been taught.

In my estimation, a chief service rendered by words like "concept" and "attribute" is that they provide a vocabulary that allows us to monitor and correct our usage as we slowly advance them towards increasingly demanding standards of adequate performance. To fulfill this function sensibly, our talk of "concept grasp" et al. must display considerable sensitivity to the maturational level of the speakers we attempt to evaluate. Faced with a very young child who is plainly baffled by Baum's description of the fairy on the rainbow, we might declare, "Huey probably hasn't really acquired the concept *rainbow* yet, having not reached the required Piaget level of causal understanding with respect to material objects." But an adult who fully accommodates this same demand might be reasonably viewed as conceptually incompetent: "Dewey clearly misunderstands our normal concept of *rainbow* because he absolutely insists that rainbows can't

---

[16] Wittgenstein, Investigations, 6,520.

represent banks of irradiated raindrops on the grounds that rainbows have to be things that fairies can potentially climb and no one can coherently perform that activity on smallish drops of precipitated water. Clearly Dewey mistakenly builds more into his peculiar conception of *rainbow* than should be there." Here we seem to fault Dewey for stoutly maintaining exactly the same juvenile thesis that we require as a conceptual benchmark in assessing young Huey's conceptual achievements. But we don't seem satisfied with an exclusively adult approach either: aged Louie might suffer allied conceptual criticism if, despite his stunning mastery of the optics of atmospheric display, he stares at the Baum passage in puzzled bewilderment, "I don't get it; how can anything coherently climb up a bank of irradiated rain water?" Louie may be a master of luminary science, we might sadly conclude, but he doesn't fully grasp the notion of *rainbow* as the rest of us employ it. In such subtle ways, it seems that the standards we demand of conceptual grasp adjust themselves naturally to the shifting contours carved out by "rainbow" 's quirky career.

Since such issues will concern us in the sequel, I might also remark that a concept's behaviors over long periods of historical time (the strange vicissitudes that *force* has suffered, for example) need to be approached with an allied *context-sensitivity*.

Accordingly, it simply does not appear to be true that we evaluate the contents of concepts according only to what needs to be learned by the age at which speakers are normally pronounced conceptually competent. In fact, as we witness in Dewey's case, we naturally utilize "concept" as a term to *guide* a usage along a more profitable course if it has begun to develop improperly. Dewey is grown up now; he should recognize that a proper usage of "rainbow" does not require that they must possess a frame upon which folks can clamber. So we tell him, "Dewey, you don't have this concept quite right."

To be sure, the additional restrictions we must later learn in order to continue to qualify as grasping "rainbow" 's content properly rarely affect its range of accepted *grammaticality*, in any reasonable sense of that term. As we noted, sentences forbidden in adult usage are usually accepted without cavil in fairy stories. For this reason, perhaps the devoted linguist needn't evince much interest in the phenomena of post-competence learning that I stress here. We can concede that a discrete and recognizable stage of "acquiring the basic syntactic and semantic skeleton of English" probably constitutes a seminal event within the formative etiology of a usage. If so, whatever worldly pressures further shape linguistic behavior beyond this point, however interesting they may be, needn't concern the student devoted solely to limning this hypothetical platform of early competence. But the student of philosophy—or science, music, intellectual history or any of the myriad other topics where *ur*-philosophical thinking about concepts frequently goes awry—cannot afford the luxury of such a tightly confined focus on linguistic "content." For when we typically talk about "conceptual contents" in those contexts, we rarely restrict our attention to the concerns of our narrowly focused linguist.

A chief difficulty here is that the classical picture of concepts firmly believes in semantic invariants as well—indeed, the notion is critical to its optimistic assessment of human capabilities. In turn, this conviction traces to the simple *ur*-philosophical pictures we commonly frame of our predicates, where we presume that hidden constancies underlie terms that are actually subject to considerable flux and instability. The question of why we prove so vulnerable to these *ur*-philosophical currents will serve as a recurrent theme in this book. At present, my point is simply that the linguist's competency invariants can rarely serve as the semantic contents of classical thought. After all, the latter are frequently invoked in circumstances where mentioning the linguist's competency invariants would seem like a joke. "What should we regard as the proper core of the concept *force*?" "Well, my mama taught me that a force is a kind of shove."

The root reason why we cling strongly to the invariants of the classical picture traces to a *fear of unfoundedness*: if language isn't tightly moored to constant concepts, then our projects may come unraveled. This is revealed in the nervous questions we are inclined to frame: "If we don't share common, fixed 'contents' with our fellow speakers, how can we possibly understand what others are talking about? Without continuing invariants, how can we even address the questions we pose to ourselves?" I think the only way to address these unsettling concerns is to work through an appropriate range of calming examples. But we don't develop these anxieties because we've read modern linguistics and have decided that our thoughts must be therefore restrained by the invariants it has uncovered; such worries trace to far more primal sources.

In any case, it is easy to fall into the trap of presuming that, whenever we speak of the concepts affiliated with predicates, we always consider the same underlying factors. But the rigors of matching the complexities of real life usage actually force our adult employment of "concept" to follow more complex patterns, although the various hedges and correctives that make this possible may escape our notice. In short, applicational *practice* and associated *picture* may come rather dramatically apart in our usage of "concept" (just as it does with "rainbow"), without our paying much attention to the shift.

Prima facie, it is easy to supply cases where our evaluations of what is required for the "complete conceptual mastery" of a trait shift dramatically according to context. We provided several examples involving *being a rainbow* above; here is another. A mathematics teacher might write in a letter of recommendation:

> Although it was a purely technical "cookbook" course, through her fine work Penelope has demonstrated a complete mastery of the fundamental calculus concepts and is more than adequately prepared to take courses in mathematical analysis.

Yet two hours later she might announce, in a second vignette from college life:

> Class, we must pay careful attention to these dreary δ/ε matters, because even the great Euler didn't really grasp the proper content of the calculus concepts he manipulated with such astonishing skill.

On a possession of invariants view, the discordance betwixt these two natural expressions of "conceptual evaluation" should trouble us because, by the standards we utilize in framing the second claim, Penelope "possesses" the concepts of the calculus far less ably than Euler. Not only was he more technically deft than Penelope (or anyone else now alive), he even thought correctly about "limits" to a certain extent whereas no semantic issues of this ilk may have ever crossed Penelope's mind during her immersion in cookbook rules.

In the sequel, I will often stress that real life conceptual evaluation is heavily contextual and that phrases like "mastery" and "proper content" generally focus upon the skills that are especially salient *at the stage of development* under consideration. But if we ignore this palpable sensitivity to developmental grade (which I call "seasonlity" later) and remain implacably convinced (because of semantic finality) that all key directives of predicative use lie secretly preformed within early conceptual grasp, then we will engender the somewhat mythical and elusive picture of concepts that stands at the core of the classical picture.

(vii)

Lessons of applied mathematics. Accordingly, despite my sympathies for Austin's disapproval of philosophical Gleichshaltung, the argument in this book will *not* proceed under the assumption that it seeks a *conceptual analysis* of "concept." Indeed, I think the range of words that "concept" attempts to evaluate are so varied in their impertinent behaviors that "concept" itself cannot be expected to behave in a rule-monitored way across all of its applications. Our evaluative term eventually acquires its subtle discriminations through its assigned duties; whatever initial guidance we acquired from Mom and Dad are probably simplistic in their contours.

But why do predicates sometimes behave so perversely? Here my lines of thought depart even more dramatically from Austinian emphases, for I believe the answer rests largely at the unwelcoming door of Mother Nature. The universe in which we have been deposited seems disinclined to render the practical description of the macroscopic bodies around us especially easy. Quite the opposite; applied mathematics has discovered that even physical systems of a theoretically simple composition are apt to behave in disagreeably complex ways. Insofar as we are capable of achieving *descriptive successes* within a workable language (that is, devise linguistic gambits that permit valuable inferential conclusions to be drawn or allow for prudent planning), we are frequently forced to rely upon unexpectedly roundabout strategies to achieve these objectives. It is as if the great house of science stands before us, but mathematics can't find the keys to its front door, so if we are to enter the edifice at all, we must scramble up backyard trellises, crawl through shuttered attic windows and stumble along half-lighted halls and stairwells. Add an extra term to an equation we already understand or tweak its boundary conditions slightly and we may find that we must invent entirely new fields of mathematics, with an expenditure of vast amounts of cleverness and perseverance, to extract any information at all from our slightly altered specimen. This observation—that we must continually devise unexpected stratagems to further our slow linguistic advance upon the world—represents a vital *lesson from applied mathematics* from which we can all benefit. Many working philosophers, however, greatly underestimate the inferential difficulties that frequently prevent us from reasoning readily from premises to practical conclusion. Through one swift swipe of unjustified optimism, the practical obstacles that force conceptual evaluation to turn complex in real circumstance become removed from view. If, as is the wont of many professional philosophers, one deals exclusively in schemata ("theory T," "premises P", "conclusion C," etc.), one can pass an entire career without ever experiencing the retarding obstinacies of real practicality.

The history of successful applied mathematics often provides tales of the following sort: scientists begin treating a target subject matter with terminology that they initially conceptualize according to a fairly simple picture, but they find, as its successful applications grow, that puzzling anomalies or breakdowns gradually emerge. The restrictive patterns in which their words seem wisely used do not suit their original picture of its activities at all. A painful—and often protracted—scrutiny of "how their original successes worked" may ensue, to the eventual conclusion that their under-pinnings rest upon drastically different foundations than were originally presumed; that an accurate treatment of their subject requires more delicate considerations of strategy and circumstance than were contemplated in the confident days of first beginnings. Indeed, these emergent complexities can prove so intricate that, as with "rainbow," it is virtually unimaginable that humans could have wended their way to such refinements without having first bumbled through an initial stretch of semantic naïvety. In the interim, we must sometimes bide our time patiently, while we await semantic illumination.

We should not pretend that, through armchair meditation of a sufficiently diligent sort, we might have forecast from the outset how these wavering directivities will work themselves out. Nor should we imagine that, as we evaluate such terms for "content" in the course of their developments, we can necessarily penetrate to the deepest heart of what makes them tick. Possibly in fifty or a hundred years we will better understand the sources of the pressures that mold our usage as it does, but, most likely, not now. In many ways, this plea for tempered patience represents nothing but a recasting of Quine's favorite simile (derived from the sociologist Otto Neurath, who appropriated it, in turn, from antiquity's ship of Theseus) of language requiring maintenance like a schooner at sea:

> [I]n Neurath's figure, we cannot remodel [the vessel of language] *save as we stay afloat in it . . . .The ship may owe its structure partly to blundering predecessors who missed scuttling it only by fools' luck. But we are not in a position to jettison any part of it, except as we have substitute devices ready at hand that will serve the same essential purposes.*[17]

except that I allow that the day can eventually come when our ship is completed and we recognize how all its finished parts fit together. But the utility of "concept" talk does not apply only to perfected frigates; it provides a tool we must employ in the construction work as well. And this is why our evaluations so often behave contextually; they are helping advance the carpentry at hand.

---

[17] W. V. O. Quine, Word and Object (Cambridge, Mass.: MIT Press, 1960), 124.

Accordingly, a fair amount of this book will be devoted to questions of what might be reasonably called *linguistic engineering*: given the problems that a difficult world presents, they supply viable strategies for employing language to advantageous effect in their presence. Leaning upon the hard-earned wisdom gathered within applied mathematics, I will suggest some unusual policies for resolving these difficulties, which appear to be realized, at least to first approximation, within the behaviors of certain familiar classificatory predicates. We can also benefit from the council of the engineers with respect to semantic patience: sometimes we lack the means to figure out why our linguistic mechanisms work as they do and we must wait until our understanding of supportive process improves. After all, as the great Edwardian scientist Oliver Heaviside remarked with respect to premature efforts to frame an electrical topic rigorously: "Logic is eternal, so it can wait."[18]

In fact, the lessons of applied mathematics supply several stronger morals for our project: that our optimal forms of physical description are often constructed from ill-suited materials skillfully assembled and that surface syntactic simplicity can be purchased at the cost of complex underpinnings. But we should wait until we can investigate suitable illustrations before we attempt to develop these thoughts further.

I firmly believe that, even when we retreat from the comparative rigors of applied science to the slacker demands of everyday offhand usage, the requirements of strategic complexity do not vanish, for the same physical world confronts Huxley's veteran guardsman and his raw recruit. To be sure, the sharp figures of required strategy may lie comparatively muted within the carpet of looser usage, from which adjacent patches of irrelevant assertion have been less rigorously pruned (adult "rainbow" talk is loosely segregated from "arch"-based misunderstandings only through rather gimcrack constructions). It will be my constant policy to oscillate betwixt fairly regimented examples of technical usage (to be explained, however, in accessible terms) and the looser dominions of informal physical description. It is my hope that such comparisons can best illuminate the nature of the problematic that "concept" talk generally needs to address. To be sure, the untutored novice is likely to find himself consigned to a broader range of adversarial circumstances than his superior, who can depend upon the conventions of civilized fencing to maintain a more discernible order within his own thrusts and lunges, while the recruit must thrash about in improvised response to less disciplined foes. But, again, I am not attempting full generality of description here; I cannot supply a complete inventory of every pressure that effects every bit of language. It will serve our purposes if I mange to trace out several non-classical patterns whereby language use accommodates the strategic complexities required by real world recalcitrance.

To sum up: although I agree with the ordinary language school that our *ur*-philosophical strayings are often occasioned by misunderstood words, these confusions do not stem from violations of linguistic norms laid down by polite society, but from

---

[18] This is from Heaviside but I haven't been able to retrace my source. The allusion is apparently to St Augustine: *"And yet the validity of logical sequences is not a thing devised by man, but it is observed and noted by them that they may be able to learn and teach it; for it exists eternally in the reason of things and has its origin with God."* On Christian Doctrine, bk. 2, ch. 32.

the misdiagnosis of external shaping pressures. We can't fault predicates for merely "going on holiday," for, in a language that is constantly evolving to suit novel circumstances, one word's day at the beach may prove to be another's survey of exploitable resources.

(viii)

Why study concepts? Thus, although the techniques proposed will be somewhat novel, my basic motivations for studying the problems of concepts should seem rather familiar. We must first keep in mind the fact that the classical tools that Russell and his contemporaries articulated were designed to tame the strange and unexpected behaviors of certain scientific terms. The materials they employed to this end were deftly extracted from our everyday presumptions about conceptual evaluation. The problems Russell et al. sought to remedy are quite palpable and, insofar as classical approaches have genuinely assisted in the advance of science, they allow us to witness the good offices that our words of conceptual evaluation commonly render us, even if their underpinnings have been wrongly construed. Nonetheless, when all is said and done, the classical picture of concepts is slightly too Pollyannish at its core: it is uniformly bright and cheery and fancies that, with just a little hard work and good old-fashioned soap and water, we can neatly mop up all of our messes. Looking backward to the motivating problems of Russell's era today, it now appears that the classical approach didn't manage to diagnose their underlying problems quite rightly. The characteristic failures of those misreadings suggest, moreover, that our future prospects in science are likely to be confronted with the same kinds of unexpected twists and oddities as bedeviled the nineteenth century. We must learn to live with a somewhat diminished set of expectations in comparison to those championed by the optimists of the classical era. If so, how should we look at concepts, so that our philosophical expectations on this subject can be brought in line with a less rosy appraisal of our conceptual prospects?

Indeed, a good way to understand the project of this book is to view it as simply the engine of Russell's thinking thrown in reverse (so that it becomes a kind of refrigerator). Following our strong *ur*-philosophical tendencies to regard our predicates as generally invariantly stable and otherwise amendable to "clear thinking" remedy, Russell proposes that the conceptual difficulties afflicting science should be corrected through similar expedients. One hundred years later, we now recognize that many of the central puzzles of his day cannot be wholly remedied in his optimistic manner, but trace instead to deeper and more subterranean questions of effective strategy. I maintain that the same kinds of hidden strategic factors also affect the common classificatory terminology of everyday life, albeit in less overt forms. I therefore recommend that we transfer applied mathematics' richer appreciation of the unavoidable divergences between *fond hope* and *supportive reality* back to the circles of everyday life and let this wisdom curb the strands of *ur*-philosophy that sometimes prompt us to rash enthusiasms and embellishments.

So the basic philosophical brief we set on our desks is exactly the same as Russell's: evaluate, as best we can, the prospects we confront for bringing wayward predicates and concepts under adequate management. This requires, for the reasons we have surveyed, that we study what we are about when we evaluate the contents of sundry concepts, for that is the activity of ordinary life from which this entire fabric spins.

As we have noted, linguists or psychologists frequently have quite different goals in view: determine what sorts of data need to be absorbed in order that certain basic linguistic and psychological skills be acquired. As such, these are perfectly laudable purposes and can also be fairly described as "constructing theories of human concepts." But, in accepting that description, we should not fall into the trap of presuming that such investigations are likely to prove directly pertinent to problems sketched above. That would occur only if an extremely strong version of semantic finality were to hold: that everything we normally consign to "conceptual content" is captured by the conditions of competency we acquire when we master a notion. Prima facie, that assumption should be embraced only after very cautious scrutiny.

Such animating concerns keep this book's investigations in harmony with both philosophical tradition and issues of salient practical consequence. As mentioned before, I am sometimes puzzled about the exact motivations of the contemporary philosophers who pursue the study of concepts nowadays, because their proposals have little evident bearing on the problematic I have sketched. To be sure, sometimes (as in the case of David Lewis) the point of view seems wholly classical in quality and hence can be understood as simply a fine-tuning of Russell (and I've incorporated some of Lewis' views in Chapter 3's appendix). With respect to W. V. Quine, Michael Dummett, Robert Brandom and other critics of that type, the motivating impulse is to isolate the precise manner in which the classical picture distorts a reasonable view of human capacity. I do not agree with their varying diagnoses but fully share their overriding objectives, for this book represents my own effort to carry a similar project through.

But other writings on concepts often leave me baffled. Sometimes the provocation to their production seems little more than disciplinary tropism: a new "theory of concepts" is proposed simply because "that is the kind of thing philosophers are supposed to do." There is a variant strain afoot that maintains that a "general theory of concepts" is wanted to satisfy the alleged requirements of folk psychology, cognitive science or both. I believe that serious misapprehension about the likely character of scientific theories is tangled up here, but these are issues best postponed until a suitable moment later in the book (10, iii).

However, I am reluctant to criticize such endeavors very extensively, for I am perplexed by the fact that such works rarely wander near the kinds of troublesome cases that explain, to me at least, what the primary *point* of worrying about concepts is. But I hate to frame hypotheses as to how authors might address issues they ignore, for I am not fond of putting words into other people's articles.

This discomfort with the motivational lapses of the contemporary literature explains why a fair number of pages are devoted towards placing the common focus of Russell and myself back on the table (including its original ambitions for scientific improvement). I have strived to accomplish this as far as possible with simple and homey examples, although I will also register some of the characteristic cases that have proved critical within the development of science. But if the reader finds the little parables wherein I develop this material (contained mainly in Chapters 2 and 8) boring or superfluous, they can be lightly skimmed.

(ix)

Mitigated skepticism. The exaggerations of classical thinking and its derivatives are scarcely our only concern, for there remain all those nihilistic tendencies that cluster under philosophical banners such as "holism," "post-structuralism" and "deconstruction." For better or worse, none of these can be fairly labeled as classical in intent. However, the first of these—*holism*—was engendered in the mid-nineteenth century as an attempt to counter certain forms of classical rigidity. In its original form (say, as provided in the writings of the German physicists Hermann Helmholtz and Heinrich Hertz), the doctrine was temperate in character and represented only a rather mild departure from classical orthodoxy (4, iii). But in the twentieth century, holism's more unhappy proclivities were allowed to run to wild and destructive extremes, supplying us (inter alia) with Kuhnianism and post-structuralism. Truly if these doctrines represent our only alternatives to classical thinking, we should surely cleave to the latter, following Hilaire Belloc's advice:

> *And always keep a-hold of Nurse*
> *For fear of finding something worse.*[19]

Certainly, I want my own measure of anti-classicism to be considerably more restrained than any of this. In fact, our concepts don't fail to be classical because, as holism would have it, their busy fingers weave through every doctrine we accept, but because the increasing demands of real world pressure often shift the polar compasses that guide our words silently in subtle and unrecognized ways. It is an unfortunate aspect of our culture that we are encouraged to suppose that conceptual readjustments always enter language in some sudden triumphal burst of brilliance—this prompts the exaggerated worship of "genius" to be surveyed in Chapter 8. Episodes of this ilk occur, of course, but quite often significant changes gradually sneak into a usage in small and unnoticed ways. Sometimes no assignable human agent can be credited for these little turns of screw, for it is mainly the hidden hand of Nature's obduracy that forces the directionality. Adaptively stumbling through a series of imperfect adjustments represents as significant an aspect of the natural history of words as it does with respect to the descent of biological species. Full recognition of the required subtleties of a terminology often dawns upon us slowly and it seems beyond the reach of human capacity to speed up this lengthy process of arrival significantly. Analogously to "rainbow," certain

[19] Hilaire Belloc, "Jim, Who Ran Away from his Nurse, and was Eaten by a Lion" in <u>Cautionary Verses</u> (New York: Alfred A. Knopf, 1976), 12.

developed strategies seem so inherently complex that it becomes hard to conceive how they could have been linguistically first delivered without the midwifery of misunderstanding and false optimism. For such reasons—and these considerations will be abundantly illustrated in our case histories—, *sometimes* it is wise to not inquire too deeply into the strategic workings of a successful span of usage; *sometimes* our linguistic motto should temporarily be, "If it ain't broke, don't attempt to determine exactly how it really works."

Nonetheless, such intervals of profitable neglect last only so long; eventually our semantic pigeons return to roost and we become forced to trace more accurately the true rationale whereby our usage has heretofore supplied us with proximately valuable results. And we report what we've learned in the language of "concept" and "attribute," for that is one of the chores they facilitate.

In sum, our limited capacities for far-reaching conceptual insight create a linguistic predicament that nicely illustrates what David Hume aptly describes as

> the whimsical condition of mankind, who must act and reason and believe, though they are not able, by their most diligent inquiry, to satisfy themselves concerning the foundation of these operations or to remove the objections which may be raised against them.[20]

Hume, to be sure, gloomily presumed that the semantic underpinnings of most words remain permanently sealed off from our view, whereas I maintain that we are perfectly capable of discerning their proper foundations clearly. The rub is simply that doing so can consume a lot of time and research and cannot be readily acquired through armchair musings. In the meantime, as Hume correctly notes, we must continue to "act and reason and believe." In consequence, many of the most interesting questions in philosophy of language and the methodology of science concern the issues of how we should proceed in the periods while we patiently await fuller enlightenment. But permanent pessimism aside, otherwise Hume is right: our conceptual plight *is* rather whimsical, given the pretensions to *complete understanding* we commonly entertain:

> The greater part of mankind are naturally apt to be affirmative and dogmatical in their opinions, and while they see objects only on one side and have no idea of the counterpoising argument, they throw themselves precipitately into principles to which they are inclined, nor have any indulgence for those who entertain opposite sentiments. To hesitate or balance perplexes their understanding, checks their passion and suspends their action. They are, therefore, impatient till they escape from a state which to them is uneasy, and they think they can never remove themselves far enough from it by the violence of their assertions and obstinacy of their belief.[21]

Our "affirmative and dogmatical" natures (from which none of us wholly escape) play a substantive role in complicating our understanding of conceptual evaluation—the optimism at the heart of the classical picture stems from these inclinations. As Hume's remarks indicate, we share an innate inclination to overestimate slightly whatever

---

[20] David Hume, <u>An Enquiry Concerning Human Understanding</u> (Indianapolis: Bobbs-Merrill, 1955), 169.

[21] Ibid.

security we've managed to achieve within a favored field of endeavor. A safety engineer trusts that her parameters of building tolerance are somewhat more reliable than they really are. A mathematician is convinced that his own proofs will stand forever as logically unassailable, even as he is aware that the prevailing currents of mathematical focus often swirl elsewhere in unpredictable directions. We feel instinctively convinced that we know what it's like for a stone to be *red* on the surface of Pluto, although none of us has ever visited such an inhospitable clime. Perhaps most emblematic of this basic human foible, the mere act of entering a gambling casino seems capable of reducing the most rational among us to quivering, primitive superstition, improvising implausible incantations and highlighting spurious patterns in vain attempts to convince ourselves that we can hedge, through suitable linguistic gambits, against outcomes that lie inherently beyond our control. The headwaters of classical optimism trace, I believe, to this same *ur*-philosophical spring.

As Hume observes—and the lessons of applied mathematics collaborate—, we are frequently forced to "act and reason and believe" in linguistic circumstances that lie far in advance of any satisfactory assessment of the "foundations of these operations." Given our genetic inclination to claim unmerited certainty, it is not surprising that we habitually exaggerate the strengths of the assurances we possess when we fancy we have grasped a concept adequately. Often we presume that we have gauged the long range directivities of our terms to standards higher than we should presently aspire. In truth, what we concretely know about the working bases of commonplace descriptive vocabulary is apt to prove *somewhat thinner* and to provide *somewhat weaker guarantees* with respect to future linguistic activity than we choose to believe. Nevertheless, we doggedly struggle to maintain the shifting slate of semantical considerations that might arise over the long history of a tricky word within a single and tidy folder, for that hypothesis of semantic predetermination better supports our illusions of perfect con- ceptual foresight. Rather than accepting our altering evaluations as simply the natural expression of new interests that emerge as a word ages, we fancy that its unfolding morphology must have lain preestablished, its schedule of adult organs already intact, within some originating conceptual seed. All of this latent content, it is claimed, we manage to grasp completely early in our careers and the erratic later fortunes of *derivative*, *force* and *hardness* indicate nothing beyond the pitiable fact that we sometime botch the processes of maturation. Or, when a term's patterns of unfolding prove too irregular to suit this convenient myth of preformation, we decide that its users have somehow switched, without noting the slippage, the concepts originally consigned to the predicates "derivative," "force" and "hardness" (semantic accidents that presumably occur during "moments of mental abstraction" like the one that caused the governess in The Importance of Being Earnest to mistake her infant charge for a three-volume sentimental novel). Indeed, imputations of unnoticed polysemy represent a common hallmark of classical thinking, as we shall frequently observe in the sequel. These temptations to fictive hypothesis are understandable, for if we seek to maintain the assurance that we possess the fortitude of semantic character to restrain our own usage to the conceptual straight and narrow, the lamentable straying behaviors we invariably witness in the usages of our peers can only be explained by the fact that they, due to undisciplined inattention, have permitted their words an excess of conceptual leash, leading to the shifting evaluations of "conceptual content" we have described. Whereas only experiment can decide whether a theory is *true* or not, we would very much like to believe that unadorned clear thinking can, if we are simply careful enough, inventory the contents of our various concepts completely. 'Tis odd, we wonder, that so few of our predecessors have been able to uphold this same semantic standard successfully.

Insofar as I can determine, such are the root causes of our instinctive attachment to classical "conceptual invariants." As much as anything, the long argument of this book is designed to encourage my readers to look at natural linguistic processes in terms other than these; to tell a tale of thought and language that does not recount a dirge of stalwart contents continually grasped and continually betrayed. In fact, as we'll discuss later (5,i), there is a substantial tradition of philosophical endeavor (which I will call *pre-pragmatism*) that agrees with me in these mildly deflationary ambitions. Unfortunately, most of its adherents become so carried away by anti-classical fervor that they embrace alternative visions that are "ever so much worser" in their consequences than the classical story itself (the post-structuralism of which I earlier complained is a case in point). The trick, therefore, is to weaken the classical picture of content sufficiently to bring our conceptual expectations into alignment with what is humanly feasible, without utterly shutting the door on our capacities to improve our usage in rigor and clarity.

To gain a preliminary impression of the typical manner in which we mildly exaggerate our conceptual hold over descriptive words, consider this science fiction narrative (adapted from an old paper of mine[22]). As a kid, I once saw a movie entitled <u>Untamed Women</u> in which a tribe of Druids were depicted as having emigrated long ago to an isolated South Sea island also populated, as luck would have it, by dinosaurs and ill-natured cavemen. Through their centuries of Polynesian isolation, this Druid band continued to speak a charming, although stilted, form of antique English and when the Yankee aviator heroes of our movie landed their fuelless B-29 immediately before them, all assembled Druids cried out, in a spontaneous display of collective classification, "Lo, a great silver bird falleth from the sky." To these Druids, having never heard words like "airplane" and having little contemplated the possibilities of machine flight heretofore, "bird" seemed *exactly the right word* to capture the novel object that had just settled before them. Most real life linguistic communities are rather conservative in how readily they accept new terminology, so it is not surprising that the Druids persisted in employing "bird" in the same airplane-tolerant way throughout the course of the film. And we may imagine (here I depart from the movie's scenario, which strayed in more lurid directions) that this linguistic practice perseveres even as the Druids eventually master all of modern biology and allied fields. "Yes, I recognize", an up-to-date Druid declares, "that we do not want to place great silver birds (which are mainly metallic in

---

[22] Mark Wilson, "Predicate Meets Property," <u>Philosophical Review</u> 91, 4 (1982).

composition) into the same *biological class* as other animals such as chickens. None-theless, my forebears have always employed 'bird' with a more general meaning than do the Yankees and I respect their ancestral practices. For biological purposes, the technical term '<u>aves</u>' will do nicely. But why should we follow the Yankees otherwise in their strange classifications? After all, they are also inclined to dub flightless cassowaries as 'birds,' a classification that Druids have always rejected as deviant (although we allow, of course, that these creatures belong to <u>aves</u>)."

Yet, suppose that the first Druid sighting of an airplane does not transpire in observing a vehicle aloft but instead happens when an exploration party stumbles across its downed wreckage in the jungle, its unkempt crew lounging around its hulk with their laundry draped from the ailerons. "Lo!", our alternative Druid band spontaneously decrees, "a great silver *house* lieth in the jungle." The vehicle's arboreal <u>mise en scène</u> now suggests "house" to these folks every bit as vividly as the airborne arrival had erstwhile prompted "bird". This form of usage might easily persist, leading modern Druid descendants to declare, "Of course, silver houses aren't *birds*—did you ever see windows in a bird? However, our ancestors were right to characterize these flying devices as 'houses' because they can be lived in. Our people have never intended 'house' to be employed only in the narrow, 'silver house'-rejecting mode favored by the Yankees."

We know enough, I believe, about human classificatory behavior to plausibly suggest why these alternative scenarios might arise. Specifically, in classifying novel objects we frequently search through a limited span of potential vocabulary, looking for the best possible match. "What *is* this thing?" some cranial search engine asks in the manner of the elderly critic in the Ernest Pintoff cartoon. This routine then consults some ledger prompted by the accouterments of the setting. An object that maneuvers in the sky evokes a different catalog (bird? star? UFO?) than one that sits sedately in the jungle (house? rock? tapir?) But once an identifying tag has been set, it will be held fixed in memory, even when the erstwhile airborne now rests on the ground. In this sense, the

Druids were *half*-prepared to classify aircraft, but they falsely suppose that their selection of labels was fully anticipated.

The chief point of this fable is that neither set of alternative Druids has any psychological reason to suspect that they have not followed the preestablished conceptual contents of their words "bird" and "house," although the chief factor that explains their discordant classifications actually lies with the *history* of how they happen to approach the airplane. Both groups instinctively presume that their societally established notion of *bird* has already determined within itself whether a bomber properly counts as a "bird" or not. To bolster their case, they might cite the collective unanimity of their fellow classifiers or report the degree to which everyone considered the classification psychologically routine at the time (although, admittedly, they had never seen a bird/house quite that big). In short, the Druids—in the company of the rest of us, I maintain—are inclined to presume that the *guidance* behind the classification as a "bird" or "house" lies entirely contained within their preestablished concepts of *bird* or *house*; they fail to recognize that a substantial part of the directivity actually stems from their *historical point of entry* into an enlarged classificatory domain. Here the Druidic tendency to assign excessive credit to the realm of "what we have been conceptually prepared to do" seems completely harmless, but it nicely illustrates a basic ur-philosophical mechanism that allows us to misjudge the strength of our current conceptual grasp. In the next chapter, however, we shall examine cases where allied misallocations of "preparation" encourage genuinely unfortunate forms of conduct.

As I indicated above, I am scarcely alone in claiming that the "classical picture" exaggerates, sometimes alarmingly, the "thickness" of the assurances we gather when we become competent in a word. Many of my pre-pragmatic fellow travelers have been likewise troubled by what they regard as the occult or magical characteristics embodied within concepts as classically pictured, feeling, as I do, that its doctrines disguise an uncanny overestimation of real human capacity (3,ii). Although the general tenor of such remarks is right, I don't believe that terms like "occult" or "magical" provide a sufficiently sharp diagnosis of where classical thinking goes astray. As I've emphasized, the traditional picture represents little more than the natural amplification of tendencies implicit in our everyday policies of conceptual evaluation and it is most important that we respect the fact that most of what transpires there proves on the mark and helpful.

So I think, rather than complaining vaguely of myth or magic, our little parable of the Druids supplies a better initial sense of the exaggeration that neo-pragmatists decry in classical portraits of conceptual attainment: "It is beyond human capacity to fully prepare ourselves to classify any damn thing that might come along, but we can easily *fool ourselves* into believing that we possess such secret capabilities." In our story, a small degree of uncanny ability is engendered as post-airplane Druids instinctively lump together semantic considerations that emerge as salient at *different stages* along "bird" 's career, encouraging a false picture of preformed anticipation. This common but ill-founded form of semantic blurring creates, from individually acceptable but temporally distinct, ingredients, a joinery of elements that only encourages our presumed status as masters of future contingency. The mildly "supernatural" aspects of classical concepts thus emerge when many factors, plausible and important when regarded singly, are amalgamated into unsorted unity, rather as the impossible capacities of a mythological hero might be assembled from the real virtues of scattered individuals.

As creatures of an "affirmative and dogmatical" disposition, I am often reminded of an episode from my youth. I used to stalk my neighborhood as a hooded vigilante of justice, whose trademark weapon was a foam rubber boomerang. The latter proving aerodynamically unstable, I would often strike the family automobile when I sought to dispatch a tree. But rather than entertain the unthinkable thought that the Masked Avenger's aim was other than true, I would immediately rewrite the scenario into one of surprise attack: "Ah ha, you villain," I would sneer at Dad's car, "Thought you could sneak up on me." In such a vein, perhaps, we cultivate the illusion that we maintain complete mastery over our unfolding words.

But we must acknowledge that our Druidic tale, however appealing, is make believe and that we can profitably trust our intuitions about such fictional cases only to a limited degree. Indeed, one of the worst methodological sins of analytical philosophy—and the trust that perpetuates its inherited prejudices the longest—lies in its strong inclination to treat "intuitive" but fictitious narratives as if they represented hard *evidence* for its hypotheses, when, in fact, the tales do little more than *embody* the *ur*-philosophical leanings they are meant to sustain (it is as if, like naïve Dewey above, we tried to argue that rainbows can't possibly represent illuminated banks of raindrops because in Tik Tok of Oz Polychrome the fairy manages to climb upon one). An exaggerated faith in thought experiments usually represents another facet of the persuasive influence of classical thinking.

However, we can scarcely expect to run controlled experiments featuring South Sea archipelagos colonized by Druids differently visited. Fortunately for our argumentative purposes, much real life language development displays the factors at work in our Druid story within a more sophisticated guise. The key ingredient in our fictional tale lies in its attention to the *enlargement of linguistic application*: specifically, to the latitude displayed when a usage previously confined to a limited application silently expands into some wider domain. In the manner of the mathematician, we can profitably picture these circumstances as representing a circumstance where we prolong our usage from one neighborhood of local application into another. In the Druid case, two competing *continuations* are available whereby the old usage might plausibly enlarge to take proper account of aircraft.

For several important strategic reasons that we will detail later, an evolving natural language frequently displays a strong tendency to form into parochial pockets within which old vocabulary often assumes new, localized readings. Such semantic balk-anization creates no problems as long as the *transfer of information* between pockets is carefully controlled. The general effect of this fragmentation may supply the overall employment of a descriptive term with a polycrystalline appearance (like a granite), its individual grains of distinctive application oriented at sundry angles to one another with sundry interfacial gunk lying in between. Matrix structures of this type often emerge when new patterns of usage nucleate at local sites along the boundary of some older application and subsequently enlarge to become developed crystals in their own right. Or, as an alternative to this epitaxial metaphor, we might offer Wittgenstein's:

> Our language can be seen as an ancient city: a maze of little streets and squares, of old and new houses, and of houses with additions from various periods; and this surrounded by a multitude of new boroughs with straight regular streets and uniform houses.[23]

If this is so, the general impression of conceptual underdetermination we extracted from our Druid example can be regained through studying the nucleation processes that construct these new pockets of usage, for they display a loose liberty similar to that in the story of our islanders.

Such polycrystalline cases will also exemplify, in a robust way, the shaping hands of linguistic strategy—the lessons of applied mathematics to which I have already appealed, but have only lamely explained. The Druid case is too simple to illustrate much of this, but we shall begin to explore what I have in mind with the central examples of Chapter 6 and 7.

(x)

Exaggerated worries. Despite its regrettable fictive aspects, at least the Druid case conveys some of the grit of ordinary life, rather than representing an argument that exclusively strides forward upon "airy stilts of abstraction." If we inspect linguistic behaviors from too lofty a point of view, we are unlikely to notice the delicacies of strategic adaptation I highlight here. It lies in the nature of the processes I describe that evolving concepts rarely display gross symptom when seismic shifts transpire beneath their surface equanimity; in a very real sense, our words are *too dumb* to shout alarm when they cross into essentially virgin territory (we tacitly learn to hedge and control our adult usage of "rainbow" in astonishing ways, but few of us notice these patterns as they gradually settle in). Sometimes it is easiest to appreciate the complexity of the motifs involved by looking first at explicitly scientific cases, where rather sharp demands for descriptive success have forced practitioners to pay attention to subtle detail. And, most importantly, we must never disdain the "mere example," for it is exclusively through its impertinent individualities that Nature teaches us that it will not submit to facile descriptive ploys.

Perhaps the reader will better appreciate the flavor of the investigative methodology I propose to follow, if it is contrasted with a similarly intentioned approach to our problems that I regard as less helpful. Specifically, in his celebrated commentary on Wittgenstein,[24] Saul Kripke articulates what he calls a "skeptical paradox" as to whether

---

[23] Wittgenstein, Investigations, §18.
[24] Saul Kripke, Wittgenstein on Rules and Private Language (Cambridge, Mass.: Harvard University Press, 1982).

we truly grasp a rule such as *add 2* in a fully determinant way: "How can we possibly establish," Kripke asks on Wittgenstein's behalf, "that we haven't instead grasped something that will instruct us to starting adding *four* after we exceed 2,403,756? Assuming, for sake of example, that we have never performed such a sum previously, to what factors should we appeal to indicate that our 'grasp' is certain to work in the right way with respect to these large numbers?" Or, in the terminology I have sometimes adopted here, "What non-circular reasons establish that the proper directivities of *add 2* instruct us to carry 2,403,756 forward to 2,403,758 rather than to 2,403,760?" Kripke comes up empty-handed in this regard, a result that is clearly unsatisfactory. He further suggests that we might easily worry about our grasp of a concept like *redness* in an allied way, <u>viz.</u> whether our present understanding genuinely fixes the fact that the next McIntosh apple we classify should qualify as *red*. It would appear that this skeptical exercise is designed to bring forth some regrettable occultness inherent in the classical picture of concepts, although neither Kripke nor Wittgenstein is very direct on this score.

Although this gambit probably shares the same basic purposes as our Druid example, the exact *lessons* we should extract from this self-styled *skeptical paradox* remain inscrutable (at least to me), for exaggerated doubts rarely provide a lucid road map to real life worries. Indeed, the hyperbolic quality of the skepticism expressed seems to demand that it be stamped out by some sort of sweeping philosophical decree that forever bans such worries from our consideration—a sure recipe, I think, for generating great gobs of <u>Gleichshaltung</u>. For example, certain recent philosophers (<u>e.g.</u>, Christopher Peacocke) have decided that the "paradox" can be resolved only if we demand that *being the result of adding 2 to x* possess *acceptance conditions* able to guarantee, if a speaker merely satisfies these, that she truly grasps the concept in question (related reflections motivate the sundry "criteria" favored by the ordinary language school). But plausible articulations of these alleged acceptance conditions in concrete cases do not lie ready to hand (nor are they often provided by their philosophical advocates). Insofar as I can determine, the writers in question have become convinced of the merits of their unlikely demands only because they earnestly hope to squash, once and for all, the skeptical threat raised by Kripke/Wittgenstein.

But this *can't* be the right way to treat the "paradox," if only because little effort has been made to distinguish straightforward circumstances like those of "add 2" from those that obtain in the Druid example, where the underlying directivities seem genuinely unfixed. We shouldn't—I would think—want a "solution" to the Kripke/Wittgenstein query that determines that Druid "bird" must qualify as fully *fixed* relative to airplanes as "red" does to fire trucks. Nor, for that matter, should we assimilate the command "add 2" too swiftly to "compute $e^{2\pi i}$," because the surprising story of how the proper directivities of "$e^{2\pi i}$" were uncovered involves complications of a patently different nature than obtain with the simple arithmetical order ("add 2" represents the application of an easy algorithm, whereas the extension of exponentiation to complex values involved a very delicate continuation of local neighborhoods of the type we shall investigate in 6,vi). Indeed, the tale of how we learned to compute $e^{2\pi i}$ is strange enough to have occasioned the after dinner remark of Charles Peirce's father, Benjamin:

> Gentlemen, $[e^{2\pi i} + 1 = 0]$ is surely true, it is absolutely paradoxical; we cannot understand it and we don't know what it means, but we have proved it and therefore we know it must be the truth.[25]

Indeed, although we will not study its particular case in detail here, the convoluted history of $e^{2\pi i} + 1 = 0$ nicely exemplifies the sorts of *exploratory linguistic discovery* that will greatly concern us in this book, whereas I do not think we learn much about concrete linguistic process by subjecting stalwart $2,403,756 + 2 = 2,403,758$ to artificially exaggerated doubt.

David Hume, we might remember, also contends that sweeping skeptical paradoxes can indirectly aid our attempts to frame a "durable and useful" approach to the exigencies of practical life. To be sure, Hume's extreme Pyrrhonian skeptic—someone who contends that past regularities provide no guidance whatsoever with respect to future occurrence—cannot sensibly obey his own canons:

> Nature is always too strong for principle. And though a Pyrrhonian may throw himself or others into a momentary amazement and confusion by his profound reasonings, the first and most trivial event in life will put to flight all his doubts and scruples, and leave him the same, in every point of action and speculation, with the philosophers of every other sect or with those who never concerned themselves in any philosophical researches.[26]

However, Hume claims, a more prudent soul may be inspired to frame a more reasonable *mitigated skepticism* on such a basis:

> There is, indeed, a more mitigated skepticism or academical philosophy which may be both durable and useful, and which may, in part, be the result of this Pyrrhonism or excessive skepticism when its undistinguished doubts are, in some measure, corrected by common sense and reflection.[27]

In particular, the "affirmative and dogmatical" among us can benefit from a study of Pyrrhonian meditation because:

> [C]ould such dogmatical reasoners become sensible of the strange infirmities of human understanding, even in its most perfect state and when most accurate and cautious in its determinations—such a reflection would naturally inspire them with more modesty and reserve, and diminish their fond opinion of themselves and their prejudice against antagonists.

This recommendation of "modesty and reserve" represents, in my judgment, Hume's most appealing aspect (whereas, in other arenas, he seems as prone to ill-justified certitude as the rest of us). Indeed, this milder Hume (along with the English engineer

---

[25] H. M. S. Coxeter, Introduction to Geometry (New York: Wiley, 1989), 143.   [26] Hume, Enquiry, 168.
[27] Ibid., 169.

Oliver Heaviside) might be fairly cited as a patron muse of our own investigations, which bring a tempered mistrust to bear upon the "strange infirmities of human understanding." But we shouldn't claim that we adequately understand language's problematic processes if we can't localize, to a far sharper degree than the Kripke/ Wittgenstein puzzle achieves, the sites where wary vigil needs to be exercised in the course of our real life evaluative activities. By the same token, we must robustly acknowledge the much larger set of occasions where we should not tarry in doubts, for we must never become so timidly prudential that we reject the favorable inferential opportunities, however infirmly founded, that Nature decides to cast our way. "Shall I refuse my dinner because I do not understand the processes of digestion?,"[28] Heaviside once asked rhetorically with respect to a bizarre but very successful technique he had uncovered for extracting information from differential equations (we'll survey this very interesting history in 8,viii). And he was completely right; a wise mitigated skeptic must sometimes plow ahead in lieu of adequate justification.

Despite the "momentary amazements" they afford, meditations upon sweeping forms of Pyrrhonian paradox seem too unfocused to provide concrete counsel with respect to the questions about concepts I see as crucial. Indeed, the largely lamentable career of skeptical paradoxes in philosophy has usually produced a quite opposite effect. Through their disregard for instructive example, the threats posed by the inflated puzzles often do little more than frighten their audiences into embracing noxious "remedies" that they would have never imbibed otherwise. The handiwork of such scares can be seen, I think, in the implausible "solutions" advanced in the extensive literature that has sprung up in reaction to the Kripke/Wittgenstein paradox.

My own mitigated skepticism claims that, *in patches*, real life episodes of conceptual grasp are *weaker* and *thinner* in their inherent nature than the classical picture leads us to believe. Elsewhere in language I believe the classical story proves fairly accurate to first approximation. As such, these attitudes reflect a less drastic conceptual skepticism than those advanced by my comrades in pre-pragmatism such as Dewey and Quine. But setting the boundaries of reasonable caution is not easy. After all, Hume's own recommendations for the proper scope of a mitigated skepticism would have crippled the progress of science if accepted (any study of quantum theory would have been discouraged, for example):

> *A correct judgement observes a contrary method and, avoiding all distant and high inquiries, confines itself to common life and to such subjects as fall under daily practice and experience, leaving the more sublime topics to the embellishment of poets or orators or to the arts of priests and politicians.*[29]

Indeed, when matters of methodology turn tricky and we can no longer trust the soothing reassurances promised in the classical picture of concepts, our most reliable tutor is often that of *historical example*. How have complex puzzles with respect to conceptual directivity sorted themselves out in the past? When should we be sloppy

---

[28] Heaviside, Electromagnetic, ii. 9.   [29] Hume, Enquiry, 170.

in our justifications and when should we worry about rigor? What mixture of intuitive hunch and regimented procedure should be brought to bear on a problem? We need to canvas the attitudes with respect to these questions that have earned their exponents the historical imprimatur of success. From this abundant well of example—the laboratory of real life—, we will surely extract a better appreciation of the vicissitudes of conceptual evaluation than we might ever derive from an unfocussed skeptical paradox.

Unfortunately, examples being what they are, no study of cases can offer the unswerving methodological recommendations with respect to conceptual employment that philosophizing often promises, including the optimistic classical picture. Indeed, it would be very pleasant if Nature allowed us to be more "affirmative and dogmatical" in our conceptual diagnoses. But this is what mitigated skepticism comes to: sometimes only the passage of time and punishing experience can show us the proper escape from a conceptual dilemma. In the final analysis, the most reliable advisors we have available to us are not, after a point, all that reliable.

To capture our "whimsical condition" with respect to classification and reasoning in another way, we might recall those recurrent nautical metaphors of which the nineteenth century was especially fond, e.g. Charles Peirce:

> But let a man venture into an unfamiliar field, or where his results are not continually checked by experience, and all history shows that the most [stalwart] intellect will ofttimes lose his orientation and waste his efforts in directions that will bring him no nearer his goal, or even carry him entirely astray. He is like a ship in the open sea, with no one on board who understands the rules of navigation.[30]

The basic analogy can be rendered more poignant if we remember the unfortunate sailors who had previously explored the southern oceans without the benefit of tables or a sea-going clock. Lacking the means to determine true longitude:

> Too many were the ships that dashed aimlessly and fruitlessly about, too far this way, too near that, until scurvy and thirst killed off or incapacitated so many hands that the crew could no longer man the riggings and direct the vessel; and then the ship would float helpless with its population of skeletons and ghosts; another "flying Dutchman," to ground one day on reef or sand or ice and provide the stuff of legend.[31]

All the same, such pioneering expeditions were wholly necessary; certain tasks can't be avoided simply because we haven't yet found the tools to execute them safely or efficiently. *Blundering forward* is often the mother of invention, even along the less dramatic itineraries of advancing physical description.

Accordingly, this book's basic tale is one of the "strange latitudes" in which language sometimes finds itself stalled and the means whereby its words eventually wend their ways to port.

---

[30]   Charles S. Peirce, "The Fixation of Belief" in Philosophical Writings of Peirce (New York: Dover, 1955), 8.

[31]   David S. Landes, "Finding the Point at Sea" in William J. H. Andrewes, ed., The Quest for Longitude (Cambridge, Mass.: Harvard University Press, 1996), 20.

(xi)

Our prospects. As such, our discussion may sometimes read like an improbable cross betwixt some old-fashioned meditation on Man's condition (in the mode of Hume or William Hazlitt, say) and <u>Ingenious Mechanisms for Inventors</u>, since much of our argument for a wary approach to language's complexity rests upon the subtle engineering that successful descriptive strategies mandate. Although this work is intended as a contribution to the longstanding problems of philosophy, I hope the reader may also extract some simple amusement from the curios of linguistic behavior I collect here. Any substantive book on the etymologies of language is full of the bizarre and unexpected paths that evolving words sometimes follow—how "nice" managed to mutate from a term indicating *stupidity* to one marking *pleasant aspect* [32], for example. My own cases will focus upon somewhat different arenas of adaptation than treated in such studies, but the basic factors that drive language's continuing adjustments are probably rather similar at core. To the degree we can successfully remove the blinders of <u>Gleichshaltung</u> from our eyes, the better we will appreciate the clever and unexpected ways language discovers to mold itself to a difficult world. They're not all alike; all predicates do not all work in the same way! We want to reach an outlook where we can look at a usage and exclaim, "My goodness; who could have dreamed that descriptive success could be achieved in *that* fashion?."

My <u>modus operandi</u> throughout is to focus upon important acts of *conceptual evaluation*—what *information* are we attempting to convey when we claim that Archie, Betty or Veronica relate to the calculus concepts in divers ways? In some cases, it is eventually possible to capture quite crisply exactly what is at issue, although often an explicit rendering may not be forthcoming at the moment in question (in the meantime, as we await greater clarity, our evaluations perforce assume the character of schematic guesses with respect to the supportive substratum of a usage). We really have no choice; the conceptual contents we emphasize, even with respect to the same target predicate, frequently need to differ from occasion to occasion, driven by the press of salient circumstance. This is the source of the seasonality I mentioned earlier. The classical picture attempts to tame this rowdy divergence into semantic rectitude by claiming that it merely represents different expressions of some wholly grasped but partially submerged unity, but this is a viewpoint I suggest that we resist.

Given these premises, it will come as no surprise that I do not propose to *identify* "concepts" with anything specific in this book—I have no handy package to offer the gentleman worried about the "the" 's in a box. To be sure, since the informational substance of conceptual evaluations <u>in situ</u> usually concern quite palpable issues, a would-be formalist armed with lots of n-tuples can probably construct some ramshackle gizmo from such materials that will encapsulate the most important conceptual dimensions pertinent to a selected predicate. But there is little likelihood, I think, that the next concept down the road can be built of similar bricks. This is why I think

---

[32] Robert Stockwell and Donka Minkova, <u>English Words: History and Structure</u> (Cambridge: Cambridge University Press, 2001), 157.

offhanded appeal to phrases like the "realm of concepts" can prove so pernicious—our tendency to lump dissimilar foundations together represents a much greater problem than any Fregean tendency to elevate abstracta to semi-Platonic deification.

Our common talk of "*attributes*" or "*properties*," at least as I shall employ these phrases, represents a somewhat different affair, for these terms often serve to capture the range of *objective physical traits* that determine which activities are possible in this universe of ours. These worldly features frame the backdrop against which a successful language grows and we can't understand the strategies of a usage until we map out the external behaviors to which its gambits respond. Unfortunately, the classical picture muddles these matters by generically confusing its concepts with objective attributes. But these are matters we will sort through later (5,vii); our central focus will always be on the term "concept" in its multiple roles as an *evaluator* of human capacity.

A prominent philosopher once attempted to press upon me sweeping (and rather alarming) generalizations about the "nature of science" without benefit of any illustration whatsoever. I was having trouble determining whether his claims represented vacuous truisms or patent falsehoods (stabs at grandeur frequently suffer this wobbling infirmity). Accordingly, I invited my companion to sketch how his assertions might work themselves out within the context of a concrete example. After some meandering about the bush, he eventually began discussing electromagneticism in a manner that I thought traded upon an equivocation in the term "potential." After some niggling about these issues on my part, my friend banged his hand on the table and declared, "Damn it, Wilson, *sometimes you need to look at the big picture!*"

I would expect that the discussion of the chapter now concluding qualifies as cineramic enough for anyone's tastes. Now I confront the less compliant task of persuading my readers that *sense* can be made of it! Our first order of business is to release from the shackles of Gleichshaltung some of the varieties of diverse theme that naturally emerge within the circuits of everyday conceptual evaluation and become formally codified into the classical picture. At the same time we need to gain a hearty respect for the mischievous ways in which wispy strands of *ur*-philosophy sometimes impel us upon unhappy crusades. For these twin purposes I have assembled several parables that attempt to exhibit some of the flow and eddy of everyday conceptual discussion. I suggest that we now ramble leisurely over a certain span of *ur*-philosophical terrain, upturning rocks and inspecting curiosities as we wander. As we explore my little stories, we must practice a certain measure of patience, for the territory where concepts and their kin dwell is sufficiently tortuous that the natives gleefully await the tourist who arrives with an agenda and a map.

After tracing through several examples in the next chapter of unfortunate *ur*-philosophizing, I will provide a diagnosis (borrowing standard tools from applied mathematics) of the underlying circumstances that fuel these unhappy excursions. To those with a philosophical background, Chapter 2 may seem simply like a rehearsal of the old debates about the "objectivity" of color, dressed up in greater practical salience. In truth, greater territory is covered than that, but since the chapter is rather long, some readers may prefer to skip lightly past its thickets and proceed to Chapter 3 which presents more novel material. For the interested, however, Chapter 2 supplies a fairly accurate picture of how Mighty Systems from little acorns grow and should indicate why some care in the matter of linguistic mechanics is called for before we set off to be Philosophers. Eventually, we will find, even after this point, that we have not yet drunk deeply enough of the well waters of *ur*-philosophy, so we will return in Chapter 8 for a second dose.

As indicated earlier, I have assembled as an appendix to Chapter 3 a somewhat lengthy catalog of the tenets I regard as typical of classical thinking, drawn largely from Russell's Problems of Philosophy (although supplemented with additional themes I regard as compatible with its spirit). As such, this list can be consulted now, although it makes for rather dry reading (the reader is better advised to read the original Russell, which is delightful). In the book proper, I prefer to allow the classical themes I wish to discuss to emerge naturally, in the context of the practical dilemmas that call them forth. I have appended this list mainly so that the curious won't find my continuing allusions to the "classical picture" intolerably vague.

I might indicate, by the way, that the term "classical theory of concepts," is sometimes employed in the psychological literature[33] to designate the doctrine that all of our concepts are definable in terms of restricted primitives, particularly of a sensory nature. This is a far more restrictive claim than any in my montage and is not included here.

Finally, despite the classical roster's bulk, it should, nonetheless, be considered as merely a *framework* rather than a theory worthy of the name, largely because, as it stands, it avoids making concrete pronouncements about the contents of specific concepts (as they say in Texas, it is largely "all hat and no cattle"). When the project of "filling in the contents" is attempted, the entire edifice tends to turn unstable, rather like one of those alpine resorts in the comic novels which have been fabulously turned out in the latest and most extravagant amenities, but when the first guests arrive, our hapless manager/hero finds that Princess Madeleine has been booked into a room without a working bath, which forces him to open the connecting passage to suite 137, which is unfortunately occupied by the Smiths of Omaha who need to be transferred to the fifth floor. But the Rajah keeps his harem there, and so on . . . , until the entire establishment degenerates into riotous farce. As we'll see in the next chapter, the classical realm of concepts sometimes resembles such a hotel: *redness* can't be booked in the same room with *being rectangular*, so it'll have to lodge with subjectivity, but when that happens, we lose most of the external world behind a veil. And so on to very strange conclusions.

---

[33] Gregory L. Murphy, The Big Book of Concepts (Cambridge, Mass.: MIT Press, 2002), ch. 2.

2

LOST CHORDS

Perfectly correct music cannot even be conceived, much less executed; and for this reason all possible music deviates from perfect purity.

Arthur Schopenhauer[1]

(i)

Ur-philosophy's beckoning muse. Suppose some prolonged sequence of ill fortune has reduced us to emotional rubble and we now lie collapsed upon the sofa. We put a recording of Mozart's Symphony No. 40 in G Minor on the player and, as its music sweeps over us, we are gradually warmed by the miraculous manner in which the composer registers the doleful state of the human condition yet somehow, through that very act of acknowledgment, manages to lift us from our dejection. The second movement, for example, strikes us as "divine balm applied to the wounds of the soul."[2] As we listen, we cheer ourselves with the thought, "Well, human beings often act like complete jerks, but at least a Mozart, whatever his own personal traits, can occasionally transcend our baser impulses and contribute something truly noble to posterity." We agree with Richard Wagner:

> [Mozart] *leads the irresistible stream of richest harmony into the heart of his melody, as though with anxious care he sought to give it, by way of compensation for its delivery by mere instruments, the depth of feeling and ardor which lies at the source of the human voice as the expression of the unfathomable depths of the heart.*[3]

But a loitering concern might occur to us: if Mozart's music is genuinely to qualify as a *permanent accomplishment* of the human race, mustn't this "permanence" be explained in terms of the *replication of attributes*? That is, mustn't we claim: Mozart's achievement

---

[1] Arthur Schopenhauer, The World as Will and Representation, i, E. F. J. Payne, trans. (New York: Dover, 1969), 266.
[2] A. D. Oulibicheff in Louis Biancolli, ed., The Mozart Handbook (Cleveland, Ohio: World Publishing, 1954), 367.
[3] Ibid., 368.

was to delineate for the human race a complicated but quite concrete property of *music that follows a certain score*? This trait is such that, whenever its contours become suitably realized by an orchestra, a CD player, a band of expert hummers or any of the myriad means that can provide acceptable results, the beauties of the Symphony in G Minor will reemerge within the physical universe. The reason we feel we must appeal to an *attribute* here is that the Symphony in G Minor obviously can't "preserve itself" as an ageless monument in the literal "sit there and not go away for a long time" fashion that, e.g., the Great Pyramid of Cheops achieves. The Symphony in G Minor must instead rest its special form of "permanence" upon a collection of *repeatable requirements* upon sound waves that can be realized from time to time, whenever the ambient physical conditions permit. But this seems alright—indeed, the fact that music's permanence resides in the form of a repeatable prescription makes it far easier to protect the Symphony in G Minor from the ravages of erosion than any stone edifice. The nice thing about attributes, we might decide, is that they can be forgotten about but they never really go away. Thus we find solace in the immutable existence of the attribute *adequately realizing the music of the Symphony in G Minor*.

As we begin to attend to the problems of preserving such music, we will naturally search for recipes that will instantiate the specified attributes whenever we wish. Of course, this is not easy to do—numerous examples of musical notations from ancient cultures are extant for which we have little sense of how the music they report should be properly executed or even how their intended instruments were tuned. Even with respect to conventionally notated scores from the eighteenth and nineteenth centuries major questions abound with respect to their intended execution, for our standard notehead notation misses many parameters of great musical import. One cannot trust wholeheartedly to traditions of musical tutelage because these are known to waver considerably over the years. Mechanical forms of recording seem more secure, but these are subject to the problem of preserving the correct reproduction devices—have you attempted to locate a functional wire recorder recently? And serious doubts arise whether modern miking techniques and their subsequent "corrections" conform to any defensible standard of "objective registration."

Leaving such issues aside, it might occur to us that any exclusive focus upon the mechanical registration of acoustic structures overlooks important dimensions of the preservation problem. Mustn't we attend as well to *intrinsically human problems* connected with the permanence of music? To begin with a hypothetical case, mightn't it happen that there could be people who are able to detect the physical dimensions of whatever the orchestra is setting forth well enough, but who remain stonily deaf to the properties that make the piece truly great—*viz.* to that complicated admixture of sorrow and uplift that cheered us in our despondent moments? Such unfortunate people, we might imagine, could prove superior to most of us in their abilities to diagnose the orchestra's complex aural output. They can immediately pronounce when the clarinets have added a fleeting grace note to the B♭ while we would stumble if we attempted to decompose the music's nuances so precisely. And so forth. Nonetheless, they remain incapable of understanding why we regard the music as "sad." Somehow the vital properties that truly make the Mozart great do not penetrate to these listeners at all. We might reasonably regard these people as *emotion-blind*, at least insofar as music is concerned.

Let us not confuse these "sadness"-deprived folk with the crowd who can detect the melancholy in the Symphony in G Minor ably enough but simply don't like it: "Brrr . . . I don't see why you like that gloomy stuff. Give me 'Raindrops Fallin' on my Head' any day." We may regard this second variety of musical unappreciators as philistines, but at least the central qualities of Mozart that seem so vivid to us are registered (but then aesthetically rejected) by this gang. But my emotion-blind auditors detect the presence of the "sadness" either wrongly or not at all.

If an inability to register the palpable dolor in Mozart seems improbable to some of my readers, they simply haven't traveled in the relevant circles, for it is a problem I confront, albeit within the modest orbit of my own musical interests, quite frequently. I happen to have devoted a fair amount of my spare time to recording the older fiddle tunes that were once common in the hills of eastern Kentucky. To me the *sadness* inherent in many of these tunes seems every bit as palpable as that found in classical music, but I have sometimes had the bewildering experience of presenting one of my Appalachian acquaintances to an urban audience who—gasp!—begin clapping along, as if the fiddler had just executed the "Hoedown" from Oklahoma. "Oh, that was just *wonderful*," some audience member might gush afterward. "It was so happy and lively." "Happy and lively?," I interject, "Can't you hear that what he played was the most *lonesome* thing in the world?" I will then receive a puzzled look and a stammered "Well, yeah, I can kind of hear that, maybe . . . ," as they quickly wander away. Not a very convincing response for someone like myself, who hears the melancholy quality *seared* into every note.

In truth, variations upon this same problem of deafness with respect to emotive mood occur with other forms of music as well; indeed, I selected Mozart's Symphony in G Minor (at the suggestion of Lionel Shapiro) precisely because historically it has evoked a surprisingly varied range of affective reactions—thus Volker Scherliess:

> *Each generation hears these works with different ears, and associates its own thoughts and ideas with them. Thus to Robert Schumann the G minor Symphony was a manifestation of "Grecian grace" and another writer interpreted the work entirely in the spirit of Italian opera buffa . . . ., while other listeners—and this is probably true of us today—come under the spell of this work's somber, dramatic power . . . Tragedy, grief, lamentation, suffering, despair, darkness, but also strife and demonic power—these are expressions which have been used in attempts to describe the unique character of the work.[4]*

Might it then happen that future generations will develop some universal and ireradicable variant of emotion-deafness with respect to the *sadness* in Mozart or the fiddle tunes? Certainly the cheery misinterpretations of all those present-day clappers

[4] Volker Scherliess, "Notes to Mozart, Symphonies 40 and 41, Wiener Philharmonic Conducted by Leonard Bernstein," John Coombs, trans., Deutsche Grammophon 445 548–2 (1984).

fills me with gloomy foreboding with respect to ambitions of easy timelessness on the behalf of my beloved fiddle tunes and I see no obvious reason why Mozart's music might not also fall victim to this same unhappy eventuality. If so, how do we insure the "permanence" of a music's attributes in a meaningful way? It scarcely makes sense to waste a good deal of effort and money mechanically registering melodies for the benefit of future auditors who will react to them only in incongruous ways. It is as if we laboriously compiled records of tidal highs and lows for the sake of a people who would afterward misinterpret our accumulated numbers as baseball scores.

Does this mean that the affective quality of *expressing sadness musically* merely represents a detachable, *subjective characteristic* of an auditory pattern, simply indicating a personal reaction to the music, in the manner of the impatient disinterest of the "Raindrops Fallin' on my Head" crew? Well, some philosophers maintain that the two cases are, at bottom, the same but most of us are more likely to reply, "No; a score can be played badly, in which case the sadness may drop out of it, but once the music is executed correctly, the melancholy *has to be in there*, despite the fact that some ill-starred auditors cannot respond to it. Indeed," we might continue, "the Mozart can't be what it properly *is* unless it displays the sorrow. What the sadness-deprived folk experience is merely an impoverished surrogate for the true Mozart, lacking many of its core attributes. They are like color-blind individuals who can only discriminate the shapes of things and not their hues." The proper content of the Mozart, we insist, requires a certain degree of intrinsic melancholy. We recognize, of course, that all of us are occasionally subject to musical illusions when we find ourselves in peculiar moods, for we may hear "things in the Mozart" that we later decide could not have been there: "While listening, I happened to recall a silly event and that giddiness must have led me to impose an inappropriately jaunty construal upon the music. I now realize I was hearing it all *wrong*." Spurious influences of this sort can drain the sadness from music even for the most able of listeners. But *objectively*, we are inclined to think, an extraneous attribute like *sounding jaunty to Wilson on May 1, 1977* doesn't constitute a proper part of *adequately realizing the music of the Symphony in G Minor* whereas *expresses sadness musically* seems as if it qualifies as a wholly *essential characteristic* of certain portions of the score.

Certainly, if the fuller property *adequately realizing the music of the Symphony in G Minor* could be internally divested of its sadness, the music itself would lose its capacity to cheer us on the couch. However the "true music" of the Mozart should be properly conceived, it must be thought of as something that can carry the attributes of melancholy, for such modality seems essential to the music's greatness. But now our original musings about the "permanence" of Mozart's achievement have taken an unsettling turn, for it now seems that naïvely recording the stuff *mechanically* may prove inadequate to the *point* of the preservative task, because such achievement may leave the sadness wholly behind. Does excessive attention to the accurate mechanical reproduction of straightforwardly physical attributes therefore misunderstand the true dimension of the preservational problem? Do subcharacteristics such as *expressing sadness musically* represent a vital category of trait that requires a different form of custodial attention if a satisfactory "permanence" for the Mozart is to be achieved?

Or are these neurotic worries simply misguided? What *processes* must ensue if the musical content of the Symphony in G Minor is to qualify as adequately preserved for future generations? It begins to seem as if the answer will turn upon how certain funny worries about the nature of attributes get resolved.

In these musings, we see the first stirring of *ur*-philosophical impulse. As such, they have arisen in response to prosaic worries whether certain kinds of concrete activity— here sound recording—are worthwhile or not; they were not tangibly prompted by any avid desire to wax "philosophical" about music. Like it or not, the search for a reasonable resolution of a practical problem can sometimes drag us unavoidably into a philosophical assessment of the true nature of a characteristic such as *adequately realizing the music of the Symphony in G Minor*. And shouldn't we become clearer about such *conceptual issues* before we foolishly devote long hours to an activity that may be founded in an ill-conceived picture of "musical preservation"?

When I claimed, "Like it or not, all of us must turn philosopher on certain occasions" in 1,iii, I had in mind practical dilemmas of this sort, where the basic *worthiness* of an enterprise seems as if it turns upon how the "attributes" or "concepts" critical to the proceedings should be viewed. As indicated earlier, the trick in navigating such waters is often a matter of steering successfully somewhere betwixt the Charybdis of excessive conceptual confidence and the Scylla of undue caution. In fact, it is easy to go wrong and I now wish to examine two examples, extracted from real life and pertaining to the alleged "contents" of musical attributes, where the parties in question seem to have steered their conceptual skiffs too sharply in unhappy or even disastrous directions, although in some other time and place such navigational choices might have proved fully prudent. It is only through looking at a number of humble cases of this type that we will gain a proper appreciation of the general claims I have made so far: that (i) we commonly appeal to the *contents* of sundry concepts or attributes in justifying certain choices of practical activity; (ii) that these same *directivities* can sometimes be mistakenly interpreted in an *ur*-philosophical vein. Only then will we begin to appreciate the deep tensions that are causing us trouble in our "preservative" worries.

(ii)

Objective extremism. The first set of *ur*-philosophical attitudes I wish to illustrate lie so deeply submerged that they may scarcely seem like any sort of "philosophical opinion" at all. Indeed, I will illustrate their underlying presence in the opinions of someone who, although he happened to have been a specialist in another branch of philosophy, has probably never worried about any matter readily recognizable as an issue in musical philosophy at all. His *ur*-philosophical distinctiveness lies mainly in the marked *complacency* of his aesthetic judgments. But such unfazed complacency, I will argue, is almost certainly grounded in the unwitting application of a covert picture of conceptual content to a case it does not happily suit.

The case I have in mind is this. I once heard an ethicist exhort his audience to seek the "good life" in some quasi-Aristotlean manner of "full human flourishing" (whatever that might be). In this regard, he faulted the naturalist Charles Darwin, who confessed in his old age that he could no longer bear to listen to poetry or music. "But by this stage," our speaker complained, "Darwin had already written his masterwork The Origin of Species and was now merely churning out fodder such as The Formation of Vegetable Mold through the Action of Worms. How much better it would have been had he instead devoted his declining days to the arts." The speaker's judgment was that, having allowed his "human flourishing" quotient to slip, Darwin had a lot of catching up to do.

Such condescending moralizing is indubitably obnoxious, but what does it have to do with *concepts*? As a start, we might remark—although these issues will prove of greater concern in the next chapter—upon the speaker's offhanded assumption that "big ideas" are all that really counts in science (or anything else). "For how can new vistas be conquered," our ethicist will elaborate, "except by developing novel concepts that carve up the territory in startling ways? But once these grand schemes have become articulated, we can surely leave the cleanup work to the little guys and get to work on our personal 'flourishing.' " In response, I contend that such misguided worship of the "big idea" represents one of the unhappy mythologies of our times, fostered by Romantically exaggerated forms of intellectual hagiography and chiseled into the award structures of our funding agencies and universities. A more Tolstoyan picture of intellectual advance is closer to the truth: profitable forays into new terrain often prove possible only after we have learned to classify a lot of familiar little things in subtly productive ways. More often than not, the notions that wind up transforming scientific thinking in the profoundest ways originate within the humblest little turns of the conceptual screw (sometimes virtually literally: the radical rethinkings with respect to the treatment of "geometrical objects" in applied mathematics—tensors, spacetime separation and all that—historically trace to plebeian engineering concerns with respect to the best way to calculate the final position of a machine part after it has undergone several rotations). Darwin himself was prudently aware that his "big ideas" had worth only if they could be supported by a wide range of specific studies that could supply its sweeping grandeur *with clear content*. Indeed, Darwin's little pamphlet on worms (which he knew "not whether it would interest any readers, although it has interested me") points out that the present condition of our soil, and with it all of the modern plants and animals that require its presence, would not have come into being except through the spectacular industry of countless generations of earthworms. I would think that our Darwinian critic suffers an abysmal sense of curiosity if he doesn't find *this* a startling revelation (I presume that our moralist has no true familiarity with the book's content at all). And I can only believe that, when Darwin remarks at the end of his little book on worms:

> It may be doubted whether there are many other animals which have played so important a part in the history of the world, as have these lowly organized creatures,[5]

he regards their annelid industry as an apt metaphor for the patient "little science" that Darwin himself so diligently and appropriately pursued.

Later in the book, we shall supply more theoretical reasons for expecting "little ideas" to often serve as the true agents of conceptual advance. For the time being, we should merely take the advice of Sherlock Holmes:

> It has long been an axiom of mine that the little things are infinitely the most important.[6]

However, the aspects of our critic's position that are immediately relevant to our musical worries center upon the picture of musical concepts that stands behind his unquestioned presumption that Darwin fell under some obligation to resonate more devoutly to great music. For surely the "musical content" that eluded Darwin must be *unproblematically present* if he is to be fairly chastised for having shirked it.

Indeed, Darwin himself writes as if he would accept such a reproach. Here is the relevant passage from his brief <u>Autobiography</u>:

> Up to the age of thirty, or beyond it, poetry of many kinds . . . gave me great pleasure . . . I have also said that formerly . . . music [gave me] very great delight. But now for many years I cannot endure to read a line of poetry . . . I have also almost lost any taste for pictures or music—Music generally sets me thinking too energetically on what I have been at work on, instead of giving me pleasure . . . This curious and lamentable loss of the higher aesthetic tastes is all the odder, as books on history, biographies and travels . . . interest me as much as ever they did. My mind seems to become a kind of machine for grinding general laws out of large collections of facts, but why this should have caused the atrophy of that part of the brain alone, on which the higher tastes depend, I cannot conceive. A man with a mind more highly organized or better constituted than mine, would I suppose not have thus suffered; and if I had my life to live over again I would have made it a rule to read some poetry and listen to some music at least once a week; for perhaps the parts of my brain now atrophied could thus have been kept active through use.[7]

Despite this <u>mea culpa</u> on Darwin's part, I nonetheless wonder if our moralizing moralist could have read the full <u>Autobiography</u> through. As it is *foolish* to venerate only

---

[5] Charles Darwin, <u>The Formation of Vegetable Mold through the Action of Worms</u> (New York: D. Appleton and Company, 1896), 313.

[6] A. Conan Doyle, "A Case of Identity" in <u>The Complete Sherlock Holmes</u> (Garden City, NY: Doubleday, n.d.), 194.

[7] Charles Darwin, <u>Autobiography</u> (Oxford: Oxford University Press, 1983), 83–4.

"big ideas," it requires a *heart of stone* to chide Darwin, whose entire life was a struggle against illness, for his want of artistic sensibility. Just picture the aged naturalist, squirming to enforced Tennyson or Debussy, when the poor man wanted nothing better than a few spare moments to muse about earthworms! I think, if we try to express in intuitive terms what seems so inappropriate about our moralist's censure, we should be inclined to say something like, "Oh, he's got a wrong picture of how musical sensitivity works—it is not a straightforward matter of attending to traits standing in plain view."

After all, it is a natural and somewhat unpredictable aspect of our human condition that our responsiveness to music, mathematics, comic books, sex and a thousand other topics waxes and wanes over the course of a lifetime, although we often forget how extreme the variations can be. William James was refreshingly forthright about it all:

> *Often we are ourselves struck by the strange differences in our successive views of the same thing. We wonder how we ever could have opined as we did last month about a certain matter. We have outgrown the possibility of that state of mind, we know not how. From one year to another we see things in new lights. What was unreal has grown real, and what was exciting is insipid. The friends we used to care the world for are shrunken to shadows; the women, once so divine, the stars, the woods, and the waters, how now so dull and common! the young girls that brought an aura of infinity, at present hardly distinguishable presences; the pictures so empty; and as for the books, what was there to find so mysteriously significant in Goethe, or in John Mill so full of weight? Instead of all this, more zestful than ever is the work, the work; and fuller and deeper the import of common duties and of common goods.*[8]

The root causes of these alterations of temperament undoubtedly trace to uncharted aspects of how our nervous systems age and it seems unjust to expect poor Darwin to have arrested physiological adjustments over which, in his unhappy and unhealthy circumstances, he probably had no control. Our speaker's mandated program of musical improvement should seem patent cruelty in such circumstances.

In the same tolerant spirit, it seems to me, we must pardon the shifting standards of musical appreciation that inevitably occur over a long period of societal development, even if those changes seem inimical to our own ears and tastes. It is very difficult to devise experiments that can probe the origins of emotional expressiveness in music reliably; the limited results currently available indicate that a specific manner of *expressing sadness musically* is largely culturally acquired—there seem to be no acoustical invariants that reliably evoke a *sadness* reaction, for example.[9] Because the factors that prompt sympathetic response remain hidden and mysterious, I do not understand what congeries of training and physiology allow me to hear *sadness* in those old fiddle

---

[8] William James, The Principles of Psychology (Cambridge, Mass.: Harvard University Press, 1983), 227–8. I was disappointed to discover a passage chastising Darwin for not learning suitable "habits" in his Talks to Teachers on Psychology: and to Students of Some of Life's Ideals (New York: Henry Holt, 1913), 71–3. Here moralism triumphed over James' usual capacity for human sympathy.

[9] John Sloboda, The Musical Mind (Oxford: Oxford University Press, 1985), §2.6.

tunes while others are left unmoved. These are the considerations that prompt me to wonder whether, in the not too distant future, everyone might turn permanently sadness-deaf with respect to fiddle music—that *no* musical ears will remain able to detect such emotions within my favored music. In the same vein, I imagine that were we ever to hear again the lyres of Homer's time, we might struggle mightily to discern in their cacophony the intoxicating stirrings described by the poet. The superior heft of competing paradigms for emotive expression in music can easily drive the active *possibility* of hearing fiddle tunes as sad into oblivion.

Indeed, we can easily see how such losses of apparent musical content might arise even within the narrow evolution of our own listening. For example, the probable effect of listening to an abundance of mid-twentieth century jazz and popular music is that one acquires what might be called "a hunger for major seventh chords": music begins to sound empty if the tonic is not harmonically supported by a fuller chord like C-E-G-B or one of its extended cousins. Before such expectations take hold—if we have been largely raised on a diet of folk music, for example—, tonal assemblies of this type are apt to sound rather ugly; but once we have bitten firmly on the harmonic bait, we will begin to feel fidgety if the extending tones are absent. And such an appetite for strong harmonization can, almost by itself, seriously weaken the old possibilities for expressiveness that the fiddle tunes require. Once the question "why don't we hear a Cmaj7 here?" begins to loom large, the response "how sad this sounds" may recede into unrecoverable oblivion (in fact, the affective contours of Texas fiddle music altered in much this way after World War II). There is a very real sense in which we can seem to *lose a concept* by doing nothing except *learning something else* (such "forgetfulness through learning" appears as well in the Druid case of 1,ix). This is a phenomenon that is hard to understand within a traditional approach to human understanding and it is an issue with which we will struggle throughout the book.

With respect to those tape recordings I have made on behalf of future generations who, when their time comes around, may not be able to hear it properly, I can only say: I regret such changes, if indeed they occur, but I wouldn't *fault* anyone for them.

(iii)

Tropospheric complacency. What is most striking about our Darwin critic is that he has probably never considered tempering apologetics of this ilk, for he undoubtedly suffers from that form of parochial vision that Hume satirizes:

> His own pursuits are always, in his account, the most engaging, the objects of his passion the most valuable, and the road which he pursues the only one which leads to happiness.[10]

I'm sure he presumes (without having thought much about it) that the Mozartian musical merits, melancholy and all, are clearly *objectively present* in the physical sound, although it

[10] David Hume, "The Skeptic" in Selected Essays (Oxford: Oxford University Press, 1996), 95.

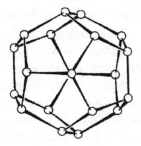

Clathrate hydrate

may require an individual of refined sensibility to perceive it properly. Of course, he grants this ability requires training; indeed, he undoubtedly prides himself in having manfully endured the mandatory drill. He will readily grant that he himself would require practice before he could spot a bird in the forest canopy as ably as Darwin. But some matters are more important than flora and fauna, he thinks, so the old naturalist can be fairly chastised for aesthetic obtuseness because the content required for proper "flourishing" is clearly out there, if only Darwin would seek the path towards it.

In my diagnosis, our moralizing critic suffers from a common form of tunnel vision in which we all, to some degree or other, participate and which needn't, in itself, bear such obnoxious fruit. The attitude in question I call *tropospheric complacency*—it represents our native inclination to picture the distribution of properties everywhere across the multifarious universe as if they represented simple transfers of what we experience while roaming the comfortable confines of a temperate and pleasantly illuminated terrestrial crust. In such a vein, we readily fancy that we already "know what it is like" to be *red* or *solid* or *icy everywhere*, even in alien circumstances subject to violent gravitational tides or unimaginable temperatures, deep within the ground under extreme pressures, or at size scales much smaller or grander than our own, and so forth. But the substantive discoveries of those who have actually probed these environments quickly reveals how shallow and hapless our complacent expectations are likely to prove.

For example, I think most of us are inclined to presume that we have a pretty good sense of what the property of *being ice* involves. Water, in fact, represents a notoriously eccentric substance, capable of forming into a wide range of peculiar structures that display admixtures of typical solid and liquid behaviors. For example,

> A chapter on crystalline water would be incomplete without some mention of a group of "ice cousins," the clathrate hydrates, also known as gas hydrates. Like the ice polymorphs, they are crystalline solids, formed by water molecules, but hydrogen-bonded in such a way that polyhedral cavities of different sizes are created that are capable of accommodating certain kinds of "guest" molecules.[11]

The author doesn't regard the clathrate structure as true ice (because it is bonded in *gauche* rather than *cis* formation), but is it clear that our everyday conception of *ice*

---

[11] Felix Franks, Water: A Matrix of Life (Cambridge: Royal Society of Chemistry, 2000), 39.

requires—as opposed to *accepts*—this distinction? (I, for one, had never thought about such matters at all). Likewise, our text indicates that in theory it should be possible to supercool liquid water until it vitrifies into a non-crystalline substance of very high viscosity structurally resembling normal glassware (in fact, many scientists regard "glasses" as different states of matter than normal crystalline solids). Should this glass-like stuff qualify as a novel form of "ice" or not? Our chemist will presumably say "no" because the stuff is not crystalline but many of us would perhaps put a higher premium on its apparent solidity. There is a popular school of contemporary philosophy (characterized by their blithe appeals to the world's alleged *natural kinds*) that severely overestimates the degree to which any of us—our societal experts or not—are presently prepared to classify the universe's abundance of strange materials adequately.

Or consider the matter of high pressure. Common materials display a remarkable ability to assume all sorts of radically different organizational structures (chemists call them *phases*) under diverse pressures (and temperatures). Indeed, gauche-bonded "ice" displays seven or eight known phases. Typically, such high pressure forms quickly revert to familiar ice when brought to atmospheric pressure. But occasionally the chemical bonds in certain high pressure phases are so strong that a material cannot easily rearrange itself back into its preferred low pressure form. A striking illustration of this type is the diamond, which truly represents an anomalous visitor to our milder dominions from the high pressure realm (the preferred, normal atmospheric pressure form of carbon is graphite; diamonds form only under extreme compression). Properly speaking, diamonds shouldn't be found near the earth's surface at all, but once volcanic forces have churned them upwards from their dens of subterranean nurture, their "unstable" bondings relax to greasy graphite so extraordinarily slowly that they qualify as "permanent" by any reasonable clock. If some analogously rugged solid form of high pressure (and room temperature) water could be formed—would it qualify as *being ice*? I do not know.

As we witnessed in the Druid case, the *manner of introduction* of a novel object can easily make it seem as if we have been fully prepared to classify it as an "X" all along—if we first learn about the clathrate hydrates from our textbook, it may never occur to us that anyone else might have reasonably considered them as "ices." It is easy to build up an exaggerated estimate of our *conceptual preparedness* from this basis alone. Few of us have probably thought much about such matters, which, as a matter of biological mercy, is fortunate because our poor cluttered brains can only bear a certain amount of information (having devoted much gray matter already to childhood memories of inconsequential television shows). What practical difference should it make to most of us that we're not presently fully prepared for a clathrate hydrate? Indeed, it is well appreciated that, in certain subjects, we do best to traffic primarily in inaccurate generalizations—"All birds fly"—and leave the penguins and kiwis to the footnotes or special occasions.

Allied to these sources of tropospheric complacency is our instinctive tendency to respond to queries about the classification of unfamiliar objects in a procrastinating vein, "Well, I can't determine from your description whether your substance is ice or not, but if you could *just show me* some of the stuff, I bet I could answer you," as if a high pressure phase of water could easily be laid out on the kitchen table. Indeed, our manifestly unwise trust that a visual presentation offers the surest key to reliable classification is rather remarkable. Consider all of those science-fiction movies—The Incredible Shrinking Man providing the great paradigm—where some human protagonist gets reduced to sub-millimeter level (and is thereby forced into battle with surly arthropods). We happily drink all this in as clearly possible, never mind the fact that human eyes shouldn't be able to focus light at that scale or that our hero can't expect to move as he does within our own gravity dominated regime. In themselves, such fantasies of "possibility" are probably harmless enough, but they can sometimes cloud our appreciation of our universe's surprising range of real variation.

Indeed, there is a passage in this vein from Nathaniel Hawthorne's "The Snow Image" that has long irked me and reminds me of the blinkered superiority of our Darwin critic:

> But, after all, there is no teaching anything to wise good men of good Mr. Lindsey's stamp. They know everything—oh, to be sure!—everything that has been, and everything that is, and everything that, by any future possibility, can be. And, should some phenomenon of nature or providence transcend their system, they will not recognize it, even if come to pass under their very noses.[12]

Although ostensibly condemning complacency of all kinds, I feel this quotation reveals a rather disagreeable vein of smugness ingrained within Hawthorne's own thinking, as he patronizes the limitations of the scientific intellect personified in the story by the clueless Mr. Lindsey. The Hawthornian "possibility" that Lindsey overlooks is that of an inanimate object—an ice statue—that becomes mysteriously invigorated by a human-like spirit. But the most striking feature of this "transcendent possibility" is its utter banality. Contrary to Hawthorne, musings of this stripe scarcely pass unrecognized—they are the very stuff of fairy tales (think of poor Sylvester the donkey encased in stone!) As such, they undoubtedly spring from conceptions of mind and soul coeval with the earliest animist religions. But excessive emphasis on these soul-like varieties of possibility runs the risk, I believe, of obscuring from our attention the genuinely surprising eventualities that often emerge in the course of clinical work with brain-damaged individuals, where our normal expectations with respect to psychology become confounded by astonishing disassociations in expected patterns of human behavior. I dare say that we are more likely to confront unexpected futures of *this* sort than any that involve supernaturally animated snow children. Such real world discoveries may leave us totally at a loss as to how our familiar psychological terminology should properly apply within their startling circumstances. If only a "soul" could jump into blocks of ice!—for in such a world the mind would indubitably possess that blessed indivisible unity upon which Descartes always insisted.

---

[12] Nathaniel Hawthorne, "The Snow Image" in Twice-Told Tales (Norwalk, Conn.: Heritage, 1966), 20.

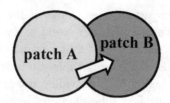

In certain modes of formal philosophy, great conclusions are sometimes reached by dwelling upon alleged "possibilities" of this kind (for example, the writings of a philosopher like David Hume are rife with what we can anachronistically dub a *cinematic conception of possibility*: if one can imagine a coherent movie of X occurring, then X must be clearly possible in some important sense). In the previous chapter, we noted the manner in which an essentially irrelevant possibility can be carried forward in the humble case of "rainbow," in the sense that the fact that fairies can climb rainbows in story books tells us little about the term's proper usage within a real life context. In fact, the *irrelevant prospect emphasized unwisely* will prove an important theme throughout this book. Through fancying themselves "masters of armchair possibility," the arrogant and cramped often convince themselves that they entertain the broadest of outlooks. In a less extreme way, the notion that philosophy's proper dominion is the "realm of conceptual possibility" is fed by these same *ur*-philosophical streams.

In general terms, we are interested in this book in what occurs when a given domain of linguistic application enlarges into neighboring territory (as occurs with Druid "bird" with respect to airplanes or "ice" with respect to the clathrate hydrates). Several natural questions arise in cases like these: To what extent are the applications in B genuinely determined by the applications already active in A? If some indeterminacy in preparation exists, what are the *leading principles* (to borrow a term from Charles Peirce) that determine how the movement from region A into B actually occurs? To what extent do the agents involved *understand* the true nature of the enlargement from A to B? In the story as I have told it, the Druid population itself views its own linguistic activities in an overly simplified manner: they simply presume, "We are merely using 'bird' in the old-fashioned way," as if the encounter with the airplane were no different in underlying character than some uncovering of a novel parrot (claims like "Oh, this simply has to be called a 'bird' " often issue from what might be called an *excess of conceptual inertia*). It is this book's contention that we frequently form pictures of linguistic development that follow this improperly simplified pattern (a disposition from which the classical theory of concepts draws much of its intuitive sustenance). In most cases, no harm is occasioned thereby, but every once in a while these proclivities represent the first steps along an *ur*-philosophical road to trouble, when our native tendencies towards tropospheric complacency load poor "attributes" or "concepts" with greater burdens of conceptual content than they can reasonably bear. As we'll eventually see (7,x), we can't properly understand what goes wrong in our musical case unless we are prepared to accept more complicated models of what can occur under linguistic enlargement.

(iv)

Tools and tasks. In the case of our critic, we witness a somewhat different species of complacency, wherein it is assumed without examination that folks of divers background will, if presented with the same schedule of training examples (region A in our diagram), naturally continue onto sector B in the same way. Indeed, our moralist has clearly presumed, "If that hard-bitten old naturalist would simply discipline himself to listen intently to Mozart and Debussy long enough, he will come to appreciate their intrinsic glories, for their manifest qualities of melancholia and clation will eventually force themselves upon him. Once these requisite models are properly grasped, their conceptual instruction will lead him to discern the same musical attributes as they appear in fresh exemplars of the aural arts." The *expressing sadness musically* aspects of the Mozart seem so *palpably present* to our critic that he can only imagine that inattentive laziness or some allied form of intellectual distraction can explain why the old man seems unable to recognize their presence in the Symphony in G Minor and elsewhere. To be sure, our moralist concedes, individuals of coarse tastes may not like the Mozart even after they discern its complete musical contours, but Darwin's problem arises from the fact he misses many of the attributes concretely present in the music, which he experiences merely as annoying noise. And such is the probable undercurrent of thinking that led us to protest in response: "But musical sensitivity is *not* a straightforward matter of attending to traits standing in plain view."

Given a certain intellectual trajectory, it is quite easy to fall into complacent, "anybody who tries hard enough can do it" presumptions like our critic's. Consider this passage, drawn almost at random from Wolfgang Hildesheimer's well-known commentary Mozart:

> No one has ever satisfactorily explained the different emotional effects of [major and minor] modes. No one will deny that, different as night and day, major and minor awaken the most opposite feelings; indeed, no other artistic discipline commands a contrast even remotely similar to this polarity, as clear-cut as turning a switch on and off.[13]

Hildesheimer is clearly oblivious to the fact that his "clear-cut polarity" arguably passes unnoticed by a sizable portion of the world's people. As the musical historian Edward Lippman comments, such tacit assumptions are typical of an older tradition of opinion in aesthetics:

> The belief in intrinsic laws of music leads . . . to a selection of a traditional repertory in which these laws prevail. The tone of [such] writings, however, is the one most typical of [older] aesthetics but increasingly out of place in a context of historical and cultural relativism, for they consider the properties they value in music to be absolute; they show little or no awareness that music exists outside their cultural horizons.[14]

---

[13] Wolfgang Hildesheimer, Mozart, Marion Farber, trans. (London: Farrar, Straus and Girous, 1982), 169.
[14] Edward Lippman, A History of Western Musical Aesthetics (Lincoln: University of Nebraska Press, 1992), 396.

"Relativism," however, is not a very useful term in this context. It is better to claim that our critic is making the mistake of treating the trait *adequately realizing the* <u>*Symphony in G Minor*</u> according to an improper *model*.

Indeed, two related possibilities suggest themselves which might prove hard to distinguish in the case of our moralizing critic. (1) He underestimates the psychological requirements for recognizing a music as "sad." (2) He treats *adequately realizing the* <u>*Symphony in G Minor*</u> according to an improperly *objectivized* picture of the attributes it represents. Since the latter doctrine is probably what Lippman has in mind under "absoluteness," let me explain it first. We cannot accomplish much, either within linguistic use or musical appreciation, unless we bring a certain range of *tools and capacities* to the table. With respect to many attributes—*being a dog* qualifies as a good example—, we can lay down a wide variety of tasks in a manner that does not require that a subject approach their completion in any particular fashion. "Pick out the biggest dog in this room," we demand and our auditors might accomplish the job in the wildest ways imaginable. With respect to most dog-centered attributes, we can be said to resemble "identical elephants," to cite W. V. Quine's appealing metaphor, as divergencies in the tools we utilize factor away:

> *Different persons growing up in the same language are like different bushes trimmed and trained to take the shape of identical elephants. The anatomical details of twigs and branches will fulfill the elephantine form differently from bush to bush, but the overall outward results are alike.*[15]

But with respect to the discernment of musical attributes, it seems harder to separate *tools* so cleanly from *task*. We know that, with respect to the parsing of the basic sounds of a language, the recognitional patterns of most speakers will become permanently fixed by an early age, making it very difficult or impossible for them to truly master the phonetic organization belonging to another tongue. Standards of "being in tune" within musical scales are likewise set by early listening experience. Sternly demanding that an auditor raised in another musical environment should learn to discern the *sadness* inherent in some favorite stretch of *our* parochial music seems tantamount to expecting that the assigned task can be divorced from all consideration of her musical toolkit. Darwin's plight, it would seem, bears much resemblance to that of someone whose ear has become previously acclimated to variant musical intervals. Those who blithely ignore these psychological divergences improperly treat *expresses sadness musically* as if it were a trait very much in the class of *being a dog*. But, surely, such assumptions operate with a wrong model of the capacities required to recognize the trait.

From a linguistic point of view, it seems natural to express the capacity-independence of the objective predicate "is a dog" in the following way. To fix the meaning of a sentence containing "is dog," we only need observe that the phrase comes regularly correlated with the objective attribute *being a dog* as its referent. Any further differences in speakers as to how they have been trained to deal with dogs or otherwise react to

---

[15]   Quine, <u>Word and Object</u>, 8.

them is utterly irrelevant to the significance of "is a dog." However, it is scarcely apparent why the doctrine deserves ridicule in this case—a simple "is a dog"/*being a dog* association does seem, at least at first appearance, to genuinely capture the true center of what is involved in canine-oriented talk.

Conceding that, it nonetheless seems rash to transfer this simple "is a dog"/*being a dog* model immediately to "adequately exemplifies the Symphony in G Minor," given that matters of recognitional capacity do not seem here as if they can be so cleanly factored away as in the case of "is a dog."

It is worth musing for a moment on circumstances where our "is a dog"/*being a dog* model would seem appropriate to "expresses sadness musically." Influenced by the Pythagorean discoveries of the correlations between the mathematical ratios of a vibrating string and pleasing harmonies, seventeenth century mystics such as Robert Fludde believed that properties such as *expressing sadness musically* represent as fundamental an ingredient in the universe's arsenal of occult forces as *being magnetic*.[16] Indeed, Fludde and his followers maintained *expressing sadness musically* could be directly attributed to sundry parts of the world order: the celestial spheres in their revolutions, for example. And *expressing sadness musically* qualifies as an objective capacity of these—after all, can't mournful music *pull the psyche* as surely as a lodestone attracts iron? This school further contends that the soul must slowly ascend through a number of stages of spiritual purification before it becomes fully open to the ambient celestial music that directly represents the universe's most vital workings—indeed, the sorrowful strains we note in the crude music of a lute or harp are regarded by Fluddeans as the feeble intimations of the true musical powers that animate the universe.

At some point we move beyond our corrupt instruments to the appreciation of something higher, albeit recondite:

> Such harmony is in immortal souls;
> But whilst this muddy vesture of decay
> Doth grossly close it in, we cannot hear it.[17]

Now if Fludde had proved correct in these suppositions, we would have good grounds for regarding physical qualities such as the Pythagorean ratios of perfect strings as the proper referential supports for our musical predicates. Courtesy of their seating in the celestial spheres, two tones can display the objective property of *being in perfect harmony* regardless of the fact that their vibrations sound irredeemably grating to any human ear. In Fludde's universe, some objective trait of *expressing sadness musically* will properly fill in the ϕ in our "expresses sadness musically"/φ scheme, although none of us are likely, in our current state of spiritual underdevelopment, to identify its instances correctly. In a milder yet similar way, and also motivated by allied Pythagorean inclinations, Newton authored a treatise on "music" that was entirely consumed

---

[16] Jamie James, The Music of the Spheres (New York: Copernicus, 1993).
[17] William Shakespeare, The Merchant of Venice in Complete Works (Roslyn, NY: Walter J. Black, 1937), 247.

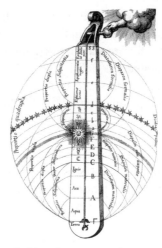

Fludde's divine monochord

by the mathematics of perfect vibratory ratios and the like.[18] Put into acoustic practice, the results would have been dreadful. Clean numbers prove Newton's harmonic guide; with respect to our merely mortal "music" there is little evidence he had much interest in the stuff.

But if Fludde had been right, actions equivalent to those recommended by our Darwinian scold would be in order: listening devoutly to horrible cacophonies of sounds becomes a true spiritual obligation.

But real music isn't like this at all. How do we correct our "expresses sadness musically" / φ tableau so that our role as variously trained auditors enters our story? The simplest counterproposal is to supply φ with *subjective values*; that is, declare that attributes like *expressing sadness musically* are properly exemplified only within a mental realm. On this picture, the sadness of a music only emerges within the conduits of our private musical experience. To be sure, we may still declare that "This phonograph record contains the saddest music," but we merely speak elliptically: we indicate that the disc stores materials likely to induce robust eruptions of the *sadness* property within the mentalities of suitable auditors. Since an attribute always needs to be instantiated within a medium and since sounds comprise the matrix that carries musical properties, sounds themselves should, under proper consideration, be regarded as psychological in their intrinsic nature (although, once again, we can extend the term to designate the air currents that serve as carriers of acoustic pattern). It is easy to find writings that happily endorse this subjectivist point of view. Thus Vasco Ronchi:

> *Sound is without doubt a subjective phenomenon. Outside the mind there are vibrations. Only when these vibrations have been received by an ear, transformed into nerve impulses, and carried to the brain and mind, only then, internally, is the sound created that*

---

[18] Penelope Gouk, <u>Music, Science and Natural Magic in Seventeenth-Century England</u> (New Haven: Yale University Press, 1999).

corresponds to the external vibrations and it is created to represent this stimulus as it reached the mind . . . Hence to identify acoustic vibrations with sound may lead uncritical young people to believe that sound is actually a physical, and not a mental, phenomenon. It might be said that the physicists did not want to prevent this misunderstanding. For, as investigators of the world without an observer, they did not like to be forced to admit that their world was without sounds, and that if they wished to study sounds, they had to return to the mental world of the auditor. The successful attainment of their purpose cannot be denied, when we ask what concept of sound is acquired by students in schools all over the earth.[19]

The reader unaccustomed to this vein of contention will surely be startled by the revelation that the objective world is without sounds. When we hear those idle jokes that revolve around "If a tree falls in the forest, will it make a sound?," we rarely anticipate that anybody, in all seriousness, will answer "No." Strangely enough, such brusque and casual banishments of the erstwhile external into the confines of pure mentality are more readily encountered within the pages of *practical handbooks* oriented to the folk who design amplification systems and who monitor the quality of printing inks than within the literature that overtly advertizes itself as "philosophical" (the latter generally attempt to mollify the radicalness of the subjectification). Indeed, our specimen quotation derives from such a source. In 7,x we shall discuss the puzzling question of why it happens that the practical folk most concerned with the physical accouterments of color and music are also the most likely parties to subscribe to quite rabid forms of subjectivism. I shall take up the issue of the philosopher's emollients in a little bit, but let us first examine the simple hypothesis that *adequately realizing the Symphony in G Minor* represents a subjective property that applies to subjective sounds.

Beginning in the late eighteenth century and in sharp reaction to views of music like Newton's, Schopenhauer and other philosophical critics supplied quite sophisticated arguments of an empirical bent that insisted that our discriminations of musical qualities must take their true seat within a subjectively centered realm. Musical objectivists have fallen prey, they claim, to the seduction of conveniently simple—but also slightly erroneous—"facts" about instrumental behavior—i.e., that the modes of a guitar string lay themselves out in Pythagorean perfection—and have falsely allowed these vibrational imposters to pass as legitimate descriptions of the true music we hear. The epigram which heads this chapter derives from such a critique.

To argue towards this end, writers of this school fastened upon the fascinating range of events that intervene in significant ways between sound waves and our musical perceptions. For example, in the mid-eighteenth century W. A. Sorge and Giuseppi Tartini both discovered the existence of *Tartini or combination tones:*[20] the fact that non-linear interactions often create harmonic vibratory components within the inner ear that are not present in the sounding instrument or the ambient air. Thus a middle C note played simultaneously with a higher G can induce spurious vibrations in the cochlea

[19] Vasco Ronchi, Optics: The Science of Vision, Edward Rosen, trans. (New York: Dover, 1991), 17.

[20] Robert T. Beyer, Sounds of Our Times: Two Hundred Years of Acoustics (New York: Springer, 1999), 20. Hermann Helmholtz, On the Sensations of Tone, Alexander Ellis, trans. (New York: Dover, 1954), ch. 7.

that will be heard as the low C note marked in bass clef, although no note in that vibratory range has actually been sounded by the instrument in question. This trick is exploited in pipe organ construction to obtain desired tones without utilizing long pipes that actually sound the note. Likewise, the perceived sound of bells is considerably complicated by this effect, among others.[21] Since these effects are unavoidable; some measure of these inner ear-induced supplements must color all of our auditory experience, motivating the composer Paul Hindemith to declare: "An interval without combination tones would be an abstract concept without being".[22] This is also the circumstance that the twentieth century musicologist Fritz Winckel has in mind when he writes in an ironic vein:

> At the root of the phenomenon of [mathematically described] harmony lies the strict periodicity of every progression. It is precisely this which must be avoided in music, as experience shows. Thus we have seen that the quite elementary entity, the sine wave, does not exist for us and that the pure intervals of the triad of simple tones do not evoke a musical experience, but on the contrary actually require a stimulating component—at least the 7th partial—in order for a vital and satisfying partial to be formed.
>
> Thus we come ever closer to the harmonic ideal, but we can never attain it since it would then elude our consciousness. . . . Experiments with synthesized sounds have established the truth of this. Periodic organization would impose a rigid law upon a work of art from the outside which would make human creative power illusory or would be prejudicial to its operation.
>
> When a musical revelation is called "divine," a very human god is meant, one who speaks to us in the idiom of fluctuating human nature, for only in the terms of these same sounds, related to us, can the soul be reached by the sense. The "harmony of infinity" will never reach our senses, and only simile can give us an idea of it.[23]

---

[21] Neville H. Fletcher and Thomas D. Rossing, The Physics of Musical Instruments (New York: Springer, 1998), ch. 21.
[22] Fritz Winckel, Music, Sound and Sensation: A Modern Exposition, Thomas Binkley, trans. (New York: Dover, 1967), 163–4.   [23] Ibid., 139.

Since these vital colorants are created within the inner ear, we can concretely witness their shaping role in the final affective contours of quantities like *sounding harmonious*. With respect to the vicissitudes of culture and development expressed in the Darwin case, we cannot directly examine the intervening factors, but their handiwork must affect the contours of a quality like *expressing sadness musically* in much the manner of the induced seventh partials of which Winckel writes. Accordingly, the proper contents of our musical traits must lie located deep within "fluctuating human nature," rather than be equated with the wholly externalized attributes provided in acoustic pattern. In short, our subjectivists argue, our naïve "is a dog"/*being a dog* model should be altered to one where the semantically supportive role of the objective attribute is replaced by a subjectively based characteristic. Indeed, the philosopher Frank Jackson has labeled theses of this ilk *location problems* because they concern the realm in which the attribute *expressing sadness musically* obtains its primary housing.[24]

Although I reject both this subjectivized replacement and its sundry semi-subjective variants, I fully agree that phenomena like the Tartini tones do demonstrate that simple objectivist models are inadequate for most musical predicates. In Chapters 6 and 7 we shall explore some methods for framing alternative models that approach our tool and task problem in a different way (however, musical language is far too complicated for this volume to describe in any completeness and so we shall largely treat simpler and better understood cases).

Earlier in this section, I suggested two related models that might lie at the root of our moralist's faulting of Darwin. The first is the objectivized picture we have just surveyed. However, our critic might very well acquiesce in the subjectively based picture but foolishly assume that being able to detect *expressing sadness musically* represents an emotional invariant available to anyone who simply puts their mind to it, no matter what their cultural and developmental background. I have no way of knowing which of these alternatives the real life critic I encountered favors, but, if he is indeed an objectivist, we see the unhappy actions—in this case, potential cruelty—to which that point of view *ur*-philosophically trends. However, we are now ready to abandon our critic and now pursue the *ur*-philosophical ills to which subjectivism leads.

(v)

Subjective extremism. One of my primary objectives in these opening chapters is to stress the ways in which our everyday thinking about concepts and attributes, as useful as it generally proves, can occasionally lead us astray. The behavior of our Darwin critic is a case in point, because his haughtiness towards Darwin represents a mixture of worship of the "big idea" and tropospheric complacency, both of which are grounded in *ur*-philosophical opinion with respect to the nature of conceptual grasp. To be sure, snobbery and patronization can find their rationales capably without the prop of

---

[24] Frank Jackson, From Metaphysics to Ethics (Oxford: Oxford University Press, 1998).

philosophical assistance, but the latter provides a dignified platform upon which such unpleasant attitudes can arrange themselves less nakedly. I began this chapter, however, with a worry about the worthiness of *musical preservation* by tape recordings and allied measures. In this respect, our Darwin critic—at least insofar as he subscribes to a "is a dog"/*being a dog* picture of musical notions—will entertain no such worries: recording captures everything objective within a music, any future misinterpreters be damned. But the subjectivist picture and its many variants do not supply such crisp affirmation of the recording enterprise.

In fact, as an amateur concerned with retaining a vein of music that will be lost unless it is now registered, I have been dismayed to discover that professional ethnomusicologists have become much less interested in recent years in old-fashioned field recording—indeed, they sometimes display a mild hostility to it—in an era where, given the accelerated rate of societal pressures, it seems most evidently required. Even more puzzling is the fact that, insofar as preservational recordings do get made, the data is often hopelessly corrupted by the musical participation of the folklorists themselves within the proceedings. What, I have wondered, has led to such counterintuitive activities? And the answer, I am distressed to report, traces to large hunks of subjectivist *ur*-philosophy about concepts and attributes. As with the Darwin critic, the blame does not lie entirely here alone, but it represents an important contributing factor. As I mentioned in the last chapter, the analytic philosophy tradition from which I derive has tended to ignore the worries that bother the folklorists and, in that respect, has not proved adequately responsive to legitimate worries about concepts and attributes that naturally emerge within the context of thinking about musical preservation—or, for that matter, elsewhere along a broad front of allied concerns that arise within the humanities. Certain folklorists have therefore elected to do "philosophy for themselves," which would represent a commendable response except that, lacking a historically inculcated sensitivity to the *brakes* that must be cautiously applied if *ur*-philosophical tendency is not to run wild, they have talked themselves into the self-destructive attitudes towards field recording that have so puzzled me. Thomas Reid, the eighteenth century advocate of "common sense," writes:

> [The exaggerating philosopher] *sees human nature in an odd, inamiable, and mortifying light. He considers himself, and the rest of his species, as born under a necessity of believing ten thousand absurdities and contradictions, and endowed with such a pittance of reason as is just sufficient to make this unhappy discovery: and this is all the fruit of his profound speculations. Such notions of human nature tend to slacken every nerve of the soul, to put every noble purpose and sentiment out of countenance, and spread a melancholy gloom over the face of things. If this is wisdom, let me be deluded with the vulgar.*[25]

Reid happens to be writing of Hume's attitudes in their most skeptical contours, but his advice applies equally well to the ill-founded pessimism that leads folklore to dismiss

---

[25] Thomas Reid, An Inquiry into the Human Mind on the Principles of Common Sense (University Park: Pennsylvania State Press, 2000), 68.

the very data it needs to cultivate. As stated earlier, the overarching imperative of philosophy should be "First, do no harm," and I am distressed that my analytical tradition has not endeavored to halt—or even retard—the wholesale destruction occurring in philosophy's name within a sister field. Worse yet, folklore's misadventures seem to possess their unhappy parallels across the modern humanities generally.

Of course, it is scarcely surprising that ethnomusicologists, who are keenly aware of the surprising variations in musical perception encountered across cultures, generally drift towards hypotheses quite different from those of our moralizing moralist. And here we witness an odd struggle that reveals a very rich vein of ur-philosophical opinion. The main text I will consider is a response to our musical preservation problem recently provided by a distinguished contemporary folklorist, Jeff Todd Titon. But Titon's position can only be understood in the context of the atmospherics of post-structuralist critique, which represents yet another influential vein of philosophical thinking that has paralyzed the humanities in recent years (it derives, however, from the headwaters of holism rather than subjectivism, as we shall soon see). To set the stage, consider the worry about the objectivity of musical fieldwork expressed by the editor (Timothy Cooley) of the very collection of essays in which Titon's response occurs:

> In the first half of the twentieth century, events conspired to undermine the confidence in Western intellectual hegemony; relativity theory and quantum mechanics undid absolute confidence in science, and the two world wars strengthened an ongoing challenge to the belief that rational thought would lead to a new and better world. The modern era was over, the science paradigm was challenged (though persistent), and in the mid-century the foundations for ethnomusicology began to shift ... [W]e have entered an experimental moment when new perspectives are needed. If the claim of an objective stance from which to analyze and compare the musics of the world's peoples can no longer be made, what can be known by the practice of ethnomusicology?[26]

To readers unfamiliar with prose of this type, the associative leaps in this passage will seem extraordinary. What conceivable relevance should the peculiarities of quantum mechanics or World War II bear to scholarly practice within folklore? Somehow the "science paradigm" is alleged to have collapsed—but what on earth is that? In fact, two interwoven considerations are raised here. (1) The worry that the conceptual categories of any purportedly "objective folklore," no matter how approached, will continue to incorporate the complacencies of mainstream Westernized music. (2) Virtually any "theoretical" classification will likewise incorporate unwittingly the prevailing large-scale prejudices of the society from which it issues and thus inherently "falsify" the data they intend to capture. Underlying both worries is a strong presumption of *semantic holism*: the notion that particular linguistic terms gain their significance only as forming part of a much larger articulated web of expressions. Defenses of milder variants on holism are common in analytical philosophy as well and we shall examine

---

[26] Timothy J. Cooley, "Casting Shadows in the Field: An Introduction" in G. F. Barz and T. J. Cooley, eds., Shadows in the Field (Oxford: Oxford University Press, 1997), 11.

several traditional exemplars in Chapter 5. But at the less disciplined hands of Titon's "post-structuralist critics," every form of social unpleasantness is apt to be holistically injected into classificatory terms of the most innocuous nature. In the folklore context, simply labeling a bit of music as a "folk song" can be readily castigated as a reprehensible political deed. After all, it is claimed, when we classify a music as "folk," we <u>ipso facto</u> demote its performer to the status of an "Other," as opposed to we imperial "I"s who appropriate their goods and exploit their resources. Consider how a well-regarded work (<u>All That is Native and Fine</u> by David Whisnant) on the past practices of folklorists begins:

> This is a book about cultural "otherness," about how people perceive each other across cultural boundaries—especially those boundaries that correlate with social class . . . In a single phrase, this book is about the politics of culture. Not politics in the formal sense of legislative act, judicial decision, or policy directive, but at the more basic level of individual values and assumptions, personal style and preference, community mores and local traditions. It is thus about the relatively intimate—but socially and politically significant—differences between the ways people talk and see, think and feel, believe and act, understand and structure their experience.[27]

It eventually wends its way to this wilting blast:

> By directing attention away from dominant structural realities, such as those associated with colonial subjugation or resource exploration or class-based inequalities, "Culture" provides a convenient mask for other agendas of change and throws a warm glow upon the cold realities of social dislocation . . . "Rescuing" or "preserving" or "reviving" a sanitized version of culture frequently makes for a rather shallow liberal commitment: it allows a prepared consensus on the "value" of preservation or revival; its affirmations lie comfortably within the bounds of conventional secular piety; it makes minimal demands upon financial (or other) resources; and it involves little risk of opposition from vested economic or political interests. It is, in a word, the cheapest and safest way to go.[28]

Notice how inoffensive words like "culture," "preserving" and "reviving" have been placed in quotation marks, which, in this context, represent the academical equivalent of the public stocks. In certain specifics, I agree with some of the criticisms Whisnant extends to the activities of certain self-styled "preservers of folk music"—indeed, I have dealt myself with the social scars left behind in some of the exact mountain communities he discusses. But I would rather credit these blunders to the obtuseness of self-promoting prigs than conclude that the entire fabric of commonsensical musical classification (constituting a "folk song" or not) is irrefragably cursed with the pernicious blinders of capitalist society. Insofar as the innocent "folk song" becomes, on occasion, incrusted with the barnacles of exploitive purpose, these extraneous deposits can be fairly easily washed away. Later (8,ix) we will discuss the many mechanisms we have available for the purpose, under the heading of *semantic detoxification*.

[27] David E. Whisnant, <u>All That is Native and Fine</u> (Chapel Hill: University of North Carolina Press, 1983), pp. xiii–iv.
[28] Ibid., 260–1.

In my opinion, indiscriminate holism of this kind represents little more than low grade philosophy of language run amuck (and rendered rather dismissively tyrannical in the bargain). Titon, unfortunately, has succumbed to the idea that most classification involves a large measure of "social construction" (a popular but rather meaningless term suggesting large scale cultural holism). Here he comments upon squabbles with respect to phrases like "folk musician" that arose in the context of a funding panel upon which he once served:

> No one, then, is free from constituting domains through interpretative acts. Instead, various interpretative communities —whether blues scholars, musicians, black historians, or folk arts programs—engage each other in a negotiation over meaning that finally is political and implicates us all.[29]

Once again, there is no doubt that certain individuals will rhetorically exploit charged vocabulary for self-serving purposes, but, as I've just stated, ordinary linguistic practice offers a variety of ways in which such gambits can be readily defused. I doubt that anyone would seriously suppose that musical classification cannot be extricated from the "political" unless they had become persuaded of the thesis through philosophical considerations. But once we bite firmly on the bait of holism, we are likely to have fallen in a ditch from which it will prove rather hard to escape.

Such, in brief, are the pathways whereby World War II and quantum mechanics become entangled with folklore in Cooley's mind. Once "everything-links-to-every-thing-else-and-the-kitchen-sink" presuppositions of this ilk are accepted, the task of being a decent musicologist becomes truly daunting, for any word uttered may unwittingly perpetuate a dastardly social order. There are many factors tangled up in Cooley's hazy melange of worries, but we will concentrate mainly on its roots in holism generically considered. We can scarcely talk coherently about a music without appealing to qualities such as *expressing sadness musically*, but in Cooley's eyes their claim to "objectivity" is very much at issue.

This is the context in which Titon offers an explicitly philosophical defense of his own practices within ethnomusicology. To catch its proper flavor and dimensions, I will quote a fairly long extract.

> Continental European philosophy since the nineteenth century regularly distinguishes between two kinds of knowledge: explanation and understanding . . . Explanation is typical in the sciences, and understanding typifies knowledge in the humanities: . . . An emphasis on understanding (rather than explaining) the lived experience of people making music (ourselves included) is paramount [to Titon's conception of a defensible ethnomusicology.] . . . In my view, music is a socially constructed, cultural phenomenon. The various cultural constructions enable people to experience it as patterned sounds, aesthetic objects, ritual substance, even as a thing-in-itself. But to say that music is a culturally constructed phenomenon does not mean that it has no existence in the world, for

---

[29] Jeff Todd Titon, "Reconstructing the Blues" in Neil V. Rosenberg, ed., Transforming Tradition (Urbana: University of Illinois Press, 1993), 238.

like everyone I know, I experience my world through my consciousness, and I experience music as part of my life world ... Playing [music with others] I hear music; I feel its presence; I am moved, internally; I move, externally. Music overcomes me with longing ... I no longer feel myself as a separate self; rather, I feel myself to be "music in the world." ... When my consciousness is filled with music I am in the world musically ... I would like to ground [this kind of] musical knowing—that is, knowledge of or about music—in musical being .... I have maintained that [in the past] we have usually sought to explain musical sounds, concepts, and behavior rather than to understand musical experience. And yet our own most satisfying knowledge is often acquired through the experience of music making and the relationships that arise during fieldwork .... If all of that is so, then an epistemology erected upon the ethnomusicological practices of music making and fieldwork as the paradigm case of our being-in-the-world, rather than upon collecting, transcribing and analysis as that paradigm case, will privilege knowledge arising through experience, ours and others'.

Post-structuralist thought denies the existence of autonomous selves. The notion of fieldwork as an encounter between self and other is thought to be a delusion, just as the notion of the autonomous self is a delusion, whereas the notion of the Other is a fictionalized objectification .... [However,] the experience of music making is, in some circumstances in various cultures throughout the world, an experience of becoming a knowing self in the presence of other becoming, knowing selves. This is a profoundly communal experience and I am willing to trust it. A representation grounded in this kind of experience would, I believe, begin to answer the post-structuralist challenge by reconfiguring the ethnomusicologists' idea of his or her own self, now emergent rather than autonomous ... Emergent selves on the other hand are connected selves, enmeshed in reciprocity.[30]

This passage assembles a heady dose of themes, some of which we will ignore or simplify at this stage in our proceedings. Specifically, there is a strong flavor of what might be called *participatory idealism* present which I'll explicate later. For the moment, let us simply interpret Titon's proposal in the simple subjectivist terms already articulated. On this reading, the fundamental hope is that, somewhere within the bloomin', buzzin' confusion of psychological happenstance, there lies a core of subjective musical experience rich enough to provide an adequate *platform* upon which the basic ambitions of ethnomusicology can be supported. The post-structuralist complaints that Titon seeks to address maintain that the basic categories of folklore falsely subject a music, even at the elementary level of its parsing as "patterned sounds," to alien standards enforced by a suspect "science model" and that even the insipid delineation of ethnomusicology as "the discipline that attempts to understand the musics of folk or other different cultures" institutes a demeaning asymmetry betwixt "I" and "Other." In response, Titon, encouraged by the directness and vividness of his musical collaborations (the forms of knowledge he considers "most satisfying"), claims that in these ranges of intense experience he becomes *directly acquainted* with the true inner nature of

[30] Jeff Todd Titon, "Knowing Fieldwork" in Barz and Cooley, eds., <u>Shadows</u>, 87–100.

the musical sample—or, at least, comes as close to direct acquaintance as is humanly possible. Furthermore, he assumes that, because of their group nature, the musical experiences of his subjects, "now reconfigured as collaborators," are likely to resemble his. Thus, if in these joint efforts he senses a music as *sad* and his chums agree in this selection of descriptive vocabulary, he can reasonably conclude by analogy that all parties will have experienced closely homologous traits within their private dimensions of subjective contour. In short, Titon feels reassured that he can point inwardly to his musical sensations and validly declare, "See! *This* experience directly manifests the true musical characteristics of this sort of piece, largely shorn of corrupting ties to hegemonic notions of 'the folk' and the like." This directly witnessed inner landscape provides an arena where "humanistic knowledge" of music can build, comparatively free of "science model" distortions that constitute the central target of post-modernist critique.

Despite Titon's gestures towards "communal reciprocity," this tale of how descriptive vocabulary might find uncorrupted inner support surely qualifies as a "private language" of the sort envisioned by Ludwig Wittgenstein. That categorization hardly establishes that Titon's proposal is wrong, for more reasonable theses have been dismissed under the "private language" heading than by any other dismissive ploy within the arsenal of analytic philosophy (claiming without further argumentation that "Your doctrine violates Wittgenstein's strictures against private language" represents the analytic philosopher's equivalent of quoting Scripture to convince pagans—and where the text cited derives from Revelations). But without engaging in such dogmatism, there is a legitimate complaint woven into these Wittgensteinian themes that seems applicable to Titon's proposal: his tale oddly *shifts* the primary support of our musical discourse into a strange inner locale which seems quite inappropriate for such a public activity. We shall return to this mislocation of support problem later.

However, I can supply a preliminary sense of what seems so disconcerting about this displacement from my own field experience. More than once I have commented "Boy, that's a sad tune" to one of my informants, only to be answered, "Yes, it's just as lonesome as hound dogs baying after the fox on an autumn night." I personally experience great difficulties in attributing profound musicalities to such events. To gain full "reciprocity" with my subjects should I spend long evenings acclimating myself to fox chases? Such a proscribed program of canine instruction seems eerily reminiscent of the diet of Tennyson and Debussy our critic would have impressed upon poor Darwin. In fact, the root sources of these two tutorial absurdities are the same: they trace to common *ur*-philosophical misapprehensions about what "understanding a trait" involves.

Stripped of its Continental finery, Titon's proposal is essentially that of a subjectivist model where the true support of the predicate "expresses sadness musically" lies situated in *inner experience*, rather supported primarily by sound waves or similar "objective" source, and where the proper basis of musical classification reflects the directly instructive character of that sensory presentation rather than involving the externally distorting constructions of a scientific scheme. If this view is correct, what consequences follow with respect to our old worries about musical preservation? From its point of view, shouldn't a scholar interested in "saving music" find ways to insure that our internalized "practices of music making" are actively replicated, rather than falling victim to false ideals of "collecting, transcribing and analysis"? Since a musical trait like *adequately realizing the music of the* <u>Symphony in G Minor</u> is manifested fully only within the realms of human appreciation, any kind of mechanical registration, whether in the guise of notation or recording machine, at best supplies a denatured prompting that, if conditions are favorable, will induce the attribute's reappearance within an auditor's subjective realm. But, as we've witnessed with poor Darwin and the folks who clap along with fiddle tunes, such prompts may fail to illicit the correct internal attributes, even though such listeners may detect everything "objective" in the recording as ably as you or I. Shouldn't it become more important for "preservationists" to learn to play the old fiddle tunes themselves and pass along its proper "reciprocity" so that the music can be readily reincarnated experientially, in the medium where its proper sadness truly lives, rather than consigning its fate, as in "objectivist" days of yore, to the fickle clutches of notation or tape recorder? Such philosophical reasoning would certainly explain the alarming alteration in the quality of field recordings I reported upon earlier.

I'm uncertain how far Titon himself would be willing to wander up this garden path (the work I know seems constrained throughout by common sense), but consider the following passage drawn from an essay that accompanies a recent issue of field recordings by prominent collectors of the 1940s (Frank and Anne Warner). Its author, Tim Erikson, has clearly bathed in philosophical waters similar to Titon's, albeit with less sophistication:

> The value in this music [recorded by the Warners], *however real it may be, can't exist outside perception and experience. It simply can't be "preserved" or materialized, though the recordings contain its echo, calling it to mind. It seems to me the only reliable way to keep something alive is to live it, thinking less about what we have and what we know and more about what we do with it . . . . In ten million years the English language is likely to have turned into something, though unfamiliar, but all the books we know, along with this CD, are likely to have gone to nothing.*[31]

Note how the phrase "thinking less about . . . what we know and more about what we do with it" echoes Titon's contrast between "explanation" and "understanding." It is not altogether surprising to discover that Erikson is a member of a little orchestra that prides itself on performing the folk songs recorded by the Warners, insuring, in Erikson's view, that songs "will stay alive" in a manner that the original performances sitting within the "dead" digital pockets of a CD cannot accomplish. This is not quite a defense for ruining fieldwork by superadded participation, but it comes close.

Such reasoning, I confess, reminds me of an apocryphal academic tale I was once told. In the dark days of the cold war, some spasm of conscience induced a governmental official to worry: "Given that our military activities may lead to thermonuclear destruction of civilization as we know it and given that we are also storing large amounts of toxic wastes with very long half-lives, how might we protect the bands of itinerants who may drift

---

[31] Tim Erikson, liner notes to <u>Her Bright Smile Haunts Me Still</u>, Appleseed APR CD 1035 (2000).

near our radioactive dumps in the post-nuclear era? Clearly we cannot presume that our doleful descendants will be able to read or even that they will continue to speak English. How can we warn them of the dangers we have left behind?" An invitation for grant proposals was sent out and the winning entry proposed that an artificial new religion should be encouraged within the region, a sect that maintains an hereditary priesthood. Such an arrangement will insure that when unwitting nomads wander near the blighted vicinity, shamans will be on hand to warn, "Mighty bad place—no go there."

As I have noted, some measure of misguided participatory urge does seem to have infected current preservative practice. But surely such interventions must prove unfortunate by any reasonable scholarly standard. After all, our original worries about musical preservation arose from the recognition that, as fresh musical paradigms crowd around us, we can easily lose the delicate ability to respond to the nuances of an older music on its own terms. By the same token, with ears educated to Mozart, Ellington and the Beatles, urban academics are unlikely to recapture the pristine rhythmic sensibilities natural to someone raised in rural Kentucky before the advent of rural electrification. If so, why should folklorists wish to burden their recordings with blundering interventions destined to obscure the crucial details that future generations will need in order to study this music properly? Indeed, although we stressed the concern that future auditors may miss musical qualities patent to us, it is also likely that some of them may discern vital differences in the music to which we are presently insensitive. Thus it is impossible to listen today to the well-intended collaborations of the 1940s between Dixieland "revivalists" and New Orleans old-timers without being painfully aware of the ruinous rhythmic and harmonic intrusions typical of swing music. However, the revivalist perpetrators were blissfully oblivious to the foreign elements they had introduced. We scarcely want *philosophy* to trump *common sense* in recommending such corruptions of the raw data vital to a subject matter, but this seems to have occurred within modern ethnomusicology to a palpable degree.

Of course, the real villain of our story is the preposterous post-modern critique that denies, upon an absurd philosophical basis, any coherent defense of reasonable scholarly activities. Titon's push into subjectivism simply represents an attempt to repel this onslaught on its own terms.

Clearly something went haywire when we offhandedly decided that the preservation of "musical content" needs to reach beyond the tape recorder. Misbegotten *ur*-philosophical impulses with respect to the basic nature of musical attributes have ratified practices that can only be regarded as wildly deleterious. We might hope that "philosophy should do no harm," but some screw has wiggled loose in this case. Indeed, folklore has generally suffered terrible drubbings at the hands of its would-be philosophers. In the 1950s the field was greatly victimized by what might be called bullies of the "theory T syndrome" (3,vii). Absurd methodological demands were placed upon folklore by know-it-alls who insisted that if "it is ever to become a discipline," ethnomusicology must turn "scientific" according to silly misapprehensions of what "science" represents (warning to the gullible: whenever a critic starts fussing unduly about "disciplines," run!) Given this deplorable prelude, it is understandable why Titon should seek an alternative to the "science model." But, in truth, the worries about objectivity trace to the straying behavior of little words like "concept" and "attribute"; no imposing edifice of counterbalancing "humanistic knowledge" needs to be erected in methodological rebuke.

Would that folklore had stayed away from the philosophizing impulse altogether. Unfortunately, the headwaters of *ur*-philosophy lie too near the centers of important things for this to prove entirely feasible.

(vi)

Amphibolic reveries. The radical subjectivization of color traits on the grounds that science has discovered that they do not happily correspond to straightforward objective qualities has, of course, proved a recurrent irritant to many reasonable thinkers. "Our color classifications have their roots in a more robust form of worldly support than that," we would rather insist. It seems an erroneous displacement of the sort just surveyed to claim that a rose is "red" courtesy of the fact that it regularly occasions outbreaks of subjective hue within human witnesses. Joseph Addison supplies a vivid rendering of the traditional subjective doctrine in one of his celebrated eighteenth century essays on the "Pleasures of the Imagination":

> Things would make but a poor appearance to the eye, if we saw them only in their proper figures and motions. And what reason can we assign for their exciting in us many of those ideas which are different from anything that exists in the objects themselves (for such are light and colors), were it not to add supernumerary ornaments to the universe, and make it more agreeable to the imagination? We are everywhere entertained with pleasing shows and apparitions, we discover imaginary glories in the heavens, and in the earth, and see some of this visionary beauty poured out over the whole creation; but what a rough and unsightly sketch of nature should we be entertained with, did all her coloring disappear, and the several distinctions of light and shade vanish? In short, our souls are at present delightfully lost and bewildered in a pleasing delusion, and we walk about like the enchanted hero of a romance, who sees beautiful castles, woods, and meadows; and at the same time hears the warbling of birds, and the purling of streams; but upon the finishing of some secret spell, the fantastic scene breaks up, and the disconsolate knight finds himself on a barren heath, or in a solitary desert.[32]

From this point of view, we make a philosophical blunder, albeit a pardonable one, if we carelessly allege a rose to be red "in the direct way"; only sensations can do that. In this regard, T. H. Huxley's later confession is rather amusing:

> I have made endless experiments on this point, and by no effort of the imagination can I persuade myself, when looking at a color, that the color is in my mind, and not at

---

[32] Joseph Addison, "Pleasures of the Imagination," no. 413 in The Works of Joseph Addison, vi (New York, G. P. Putnam, 1854), 334.

a "distance off", though of course I know perfectly well, as a matter of reason, that color is subjective.[33]

Here Addison and Huxley subscribe to the traditional sense data assumption that when a vividly colored scene is surveyed, we directly discern a visual field comprised of subjective colored patches that mentally intervenes between ourselves and the true world of uncolored objects before us. This interpolated screen of directly perceived sense data is usually called the *veil of perception*[34] by its critics and many authors, starting with Thomas Reid, have attempted, through a wide variety of philosophical stratagems, to remove its interposition within our perceptual processes. In this fashion, it is often claimed, apparently on Wittgensteinian authority, that the very idea of wholly "private objects" of sense data type represents an intrinsically incoherent conception, a theme I do not endorse myself but to which we shall return more fully later (7,x).

Although Addison and Huxley accept the revelation that no colors properly exist in nature with remarkable good cheer, it is not surprising that the Lake Poets and a wide contingent of fellow travelers from all walks of life have found such veil of perception assumptions to be utterly repugnant. How can *any* discovery of science possibly cancel the attributes that we learn of "without any other discipline than that of our daily life" in Wordsworth's famous phrase? Or, as the philosopher/mathematician A. N. Whitehead expresses the complaint:

> For us the red glow of the sunset should be as much part of nature as are the molecules and electric waves by which men of science would explain the phenomenon. It is for natural philosophy to analyze how these various elements of nature are connected.[35]

But why have so many scientist/philosophers been inclined to rob color of its status as a true attribute of the physical world we inhabit? Well, a range of considerations of variable quality can be here cited, the more subtle of which exploit the Tartini tone-like behavior of our color classifications (these are the behaviors that worry the practical books on color and will be discussed in 7,x). However, the most venerable line of thought is the simple contention that, "from science's point of view," colors seem *explanatorily inert*, in the sense that even if atoms happened to be adorned in true shades of bright red and orange, no information about these secret hues would be transmitted by light to the eye, which only carries data relevant to the manner in which the object's surface absorbs and regurgitates light waves. To explain how my lady manages to pluck the fairest flower in the garden, only the behaviors of the photons enter the story.

This is the point at which the average advocate of robust color attributes finds her opening, for she will retort: "Yes, for science's limited *predictive purposes* color attributes do not need to be mentioned, but they nonetheless comprise vital components within a complete inventory of proper external world traits. Their apparent omission within

[33] T. H. Huxley, Hume, with Helps to the Study of Berkeley (New York: D. Appleton, 1898), 271.
[34] Apparently, this popular phrase originates with Jonathan Bennett: A. D. Smith, The Problem of Perception (Cambridge, Mass.: Harvard University Press, 2002), 275.
[35] Alfred North Whitehead, The Concept of Nature (Cambridge: Cambridge University Press, 1964), 29.

science merely indicates that the latter has chosen to approach its descriptive tasks in a crabbed and circumscribed manner. To neglect the colors merely represents science's especial foible, it needn't be ours." This is the point of view from which Samuel Taylor Coleridge writes:

> In order to submit the various phenomena of moving bodies to geometrical constructions, we are under the necessity of abstracting from corporeal substance all of its positive properties, and obliged to consider bodies as differing from equal portions of space only by figure and mobility. And as a fiction of science, it would be difficult to overvalue this invention . . . But [scientists have] propounded it as truth of fact: and instead of a world created and filled with productive forces by the Almighty Fiat, left a lifeless machine whirled about by the dust of its own grinding.[36]

Unless we are driven to the instrumentalism recounted in 4,iv, a critic such as Coleridge is likely to accept that science's favored lot of attributes do appear in the external world but merely as comparatively anemic specimens within the world's full bouquet of traits. As Wordsworth expounds the thesis in "The Excursion," the purely geometrical aspects of our surroundings are "especially perceived when nature droops / And feeling is suppressed."[37] But the surer bonds of conceptualization that tie human souls to their world in robust communion lie in precisely the splendid attributes that science chooses to neglect. As L. Susan Stebbing remarks in her evocative Philosophy and the Physicists of 1937, the deniers of objective color have

> made a metaphysic out of a method . . . In so doing [the physicists] have forgotten, and philosophers do not seem to remember, that their method has been designed to facilitate investigations originating from a study of "the furniture of the earth."[38]

In the next chapter, we shall survey other forms of the widely endorsed doctrine that science, in its apparent favoring of certain descriptive concepts over old friends such as *being red*, thereby engages in some kind of odd or blinkered project cut from a different cloth than a straightforward accounting of what is to be found in the world before us (such themes ripple beneath Titon's musings on "knowledge in the sciences and the humanities" as well). I reject this "science as exceptional" thesis entirely, of course.

It is possible at this point to revert to the naïve objectivism of our Darwinian critic and proclaim that color (and musical) predicates straightforwardly report unproblematic traits of the objective world, whereas their stranger scientific brethren (e.g., "is a quark") may possibly prove justified only in an instrumental manner (this may represent Stebbings' final assessment of their circumstances, although the matter is not entirely clear). However, many thinkers have opted for a more complex response to redress our location problem that I shall dub *amphibolism*. It represents a doctrine with

---

[36] S. T. Coleridge, Aids to Reflection (London: G. Bell and Sons, 1913), 268–9. M. H. Abrams, The Correspondent Breeze (New York: W. W. Norton, 1984).

[37] William Wordsworth, "The Excursion" in The Complete Poetical Works of William Wordsworth (London: MacMillan and Co., 1930), 419.

[38] L. Susan Stebbing, Philosophy and the Physicists (New York: Dover Publications, 1958), 64.

respect to conceptual content that is admirably developed in the writings of Immanuel Kant and has become adapted to a wide variety of alternative philosophical formats, including Titon's variety of apparent Heideggerianism.

In rough terms, the general claim is that our naïve conception of "objective" concepts as *correspondent* to real world attributes is incoherent; that every viable concept must inherently involve the constructive agencies of our own minds in some irrevocable way. In its strongest form, this amphibolism embraces the full-fledged *participatory idealism* of Bernard Bosanquet:

> [T]he "world as idea" means no less than this, that the system of things and persons which surround all of us, and which each of us speaks of and refers to as the same for everyone, exists for each of us as something built up in his own mind—the mind attached to his own body—and out of the material of his own mind.[39]

This contention offers a "misery loves company" resolution to our worries about the proper location of musical attributes: *every* trait whatsoever is irrevocably laden with some degree of inherent subjectivity and, accordingly, traditional primary qualities such as *being cubic in shape* participate in the same sorts of semi-psychological hues as enfold *expressing sadness musically*. Our apparent "inner and outer worlds" should be viewed as comprised of essentially the same stuff, merely regarded from different perspectives.

The notion that we cannot coherently distinguish between the genuine aspects of the world around us and the personal constructions we happen to bring to their description is rather startling, rather as if we had been informed in a physics class that *mass* cannot be disentangled from the specific system of weights and measures (pounds versus grams) that we deck it in numerical values. Or that coordinate dependent quantities (e.g., radial distance within a scheme of polar coordinates) cannot be segregated from their more objectively seated kin (vector distance). But orthodox practice in science teaches us just the opposite: we commonly require proposed equations of state to obey sundry requirements of *frame indifference* if they expect to represent viable principles of physical behavior.[40]

Nonetheless, to many thinkers, including our contingent of Romantic poets, a mudding of the line between "objective" and "subjective" conceals a vital advantage, for they believe that our personalized grasp of amphibolic concepts allows us to participate directly, in some mystical or quasi-psychological way, in the unfolding processes of Nature herself. M. H. Abrams glosses this doctrine admirably as follows:

> Whether a man shall live his old life or a new one, in a universe of death or of life, cut off and alienated or affiliated and at home, in a state of servitude or of genuine freedom—to the Romantic poet, all depends on his mind as it engages with the world in the act of perceiving.[41]

---

[39] Bernard Bosanquet, The Essentials of Logic (London: MacMillan and Company, 1906), 6.

[40] C. Truesdell and R. A. Toupin, "The Classical Field Theories" in S. Flügge, ed., Handbuch der Physik, iii/1 (Berlin: Springer-Verlag, 1960). I do not mean to imply that frame dependent quantities are not themselves genuine quantities, but merely that we don't expect physical behavior to be sensitive to their peculiarities.

[41] M. H. Abrams, Natural Supernaturalism (New York: Norton, 1973), 375.

Or in Wordsworth's famous words:

> [M]an and nature as essentially adapted to each other, and the mind of man as naturally the mirror of the fairest and most interesting properties of nature.[42]

The neglect of this direct amphibolic bond is what Coleridge has in mind when he complains of the blinkered "scientific attitude":

> a few brilliant discoveries have been dearly purchased at the cost of all communication with life and the spirit of Nature.[43]

In a musical context, allied participatory entanglements lead to views such as those defended by Schopenhauer in The World as Will and Representation or the contemporaneous musicologist F. T Vischer:

> From the totality of these fundamental determinants we obtain the essentially amphibolic character that is peculiar to music in comparison to the other arts. Music is the ideal itself, the soul of all the arts laid bare, the mystery of all form, an intimation of the structural laws of the world and equally the fleeting, still enfolded ideal.[44]

I am unlikely to serve as the most able expositor of sentiments such as these, foreign as they are to any way that I think about the world, but the rough idea is that the deepest organizational patterns within the universe itself—given by its "structural laws"—are represented by a gradual coming into existence of ever more complex patterns, unraveling in organic growth from an "enfolded ideal." In psychologized miniature, a great piece of music will likewise blossom into parallel harmonious texture within our minds. Accordingly, as we hear a piece of stirring music, at the same time we gain a personalized intimation of the quasi-botanical pulses that drive the universe's growth. In this wise, "musical content," keeping its full quotient of inherent sadness intact, participates as both *symbol* and *exemplar* of processes that shape the external universe, while remaining directly available to each of us psychologically. "Musical content," properly speaking, represents a deeper amphibolic invariant, capable of living simultaneously in both mind and world.

Leaving aside the misty complexities of Vischer's developed opinions, I like his word "amphibolic" for the way in which the content of a descriptive concept is analogized to a variety of intellectual *salamander* capable of inhabiting the realms of subjectivity and objectivity simultaneously. As we shall see in the next chapter (3,ii), the doctrine that concepts inherently "live in two worlds" lies at the basis of what I shall call *classical gluing*. As such, related themes tacitly reappear in many classical authors who otherwise share none of Vischer's Romantic proclivities. And *amphibolic*, it seems to me, represents a useful term to designate the wide spectrum of philosophical opinion that rejects as misguided any attempt to disentangle the "objective" contents of predicates from their more subjectively informed directives, at least if "objectivity" is regarded as concerned

---

[42] William Wordsworth, Lyrical Ballads (Menston: Scolar Press, 1971).
[43] Coleridge, Aids, 289. The Philosophical Lectures (London: Routledge and Kegan Paul, 1949), lecture XII.
[44] Lippman, History, 326.

with the manner in which language finds *correlated underpinnings* within the world before us.

Indeed, softened forms of the doctrine that "attributes should not be conceived as existing independently of our structures of conceptualization" have penetrated quite deeply into the fortress of analytic philosophy in recent years. In fact, a popular epithet has been recently coined ("metaphysical realist") to stigmatize those of us resistant to the lure of tinctured insight (I shall call such doubts *anti-correlationalist* because they largely omit the "participation in the World Spirit" aspects common in the nineteenth century varieties). Gary Ebbs explicates the basic theme crisply:

> The idea behind metaphysical realism is that we can conceive of the entities and substances and species of the "external" world independently of any of the empirical beliefs and theories we hold or might hold in the future. To accept this picture, we must conceive of the relationships between our words and the "external" world from an "external" perspective. We must imagine that we can completely distinguish between what we believe and think about the things to which we refer, on the one hand, and the pure truth about these things, on the other. In this imagined "external reality," things, species, and substances are individuated by their own natures or constituting principles. This picture generates questions about what these principles of individuation are, and thus drives philosophers to theorize about the metaphysical structure of the things, species, and substances in the "external" world.[45]

Described in these sweeping terms, "metaphysical realism" certainly sounds like a foolish policy, but we should ask ourselves if we really understand what Ebbs is saying. A useful form of experimentation to employ in such cases is to lower the level of abstraction by replacing the programmatic "thing" throughout by some suitable exemplar (pick your favorite rabbit) and "species" by an appropriate choice of trait (*liking carrots*). By such substitutions we obtain:

> The idea behind metaphysical realism is that we can conceive of rabbits and their liking for carrots independently of any of the empirical beliefs and theories we hold or might hold about such mammals and their vegetative preferences in the future. To accept this picture, we must conceive of the relationships between our words and rabbits from an "external" perspective. We must imagine that we can completely distinguish between what we believe and think about rabbits and their favorite foods, on the one hand, and the pure truth about these issues, on the other. In this imagined "external reality," rabbits and their affection for carrots are individuated by their own natures or constituting principles [quite independently of our thoughts]. This picture generates questions about what these principles of individuation are, and thus drives philosophers to theorize about the rabbits and food preferences of the "external" world.

Thus particularized, I utterly fail to see what is odd about *this* position, except that the task of "theorizing" about rabbits and their favorite foods seems more the prerogative

---

[45] Gary Ebbs, Rule-Following and Realism (Cambridge, Mass.: Harvard University Press, 1997), 203.

of animal husbandry than philosophy. Our de-abstractification of Ebbs winds up expressing little beyond the banal observation that rabbits (at least in the wild) pretty much go about their own businesses, independently of how we happen to think about them. I think we should be loathe to blithely abandon our commonsensical assumption that we can sort out such issues of conceptual contribution to our "rabbit" talk quite crisply (although doing so adequately in other kinds of circumstance may require a good deal of strenuous scientific investigation).

In fact, many anti-correlationalists have recognized the justice of this complaint and have sought to establish various ersatz notions of "objectivity" consistent with their basic tenets.[46] Generally, these surrogate proposals follow Kant in claiming that a defensible notion of conceptual objectivity should turn upon our abilities to reach classificatory or truth-evaluative accord with our fellow men: proper "objectivity" in classification represents a matter of *inter-personal agreement* rather than correspondence to unsullied data. In other words, such doctrines parse the phrase "objectively based trait" as, roughly, "represents a classification agreed upon by independent agents who share identical standards of rationality," rather than resting upon any form of "directly registers facts about the target state of affairs". As witnessed in the Ebbs quotation, any unabashed appeal to direct word/world correlation is viewed with great suspicion by amphibolists.

In this regard, we must be prepared to distinguish the basic doctrine of coherent word/world correspondence from stronger claims that are commonly advanced on its behalf. In particular, straightforward classicists such as Bertrand Russell invariably assume that the nature of a given predicate's worldly correspondence is inherently *self-guaranteeing*, in the sense that once we adequately grasp a term's meaning, then we will be able to discern, after sufficient armchair analysis, the basic structure of its intended correspondence with the world. True: such correspondence may not prove successfully realized in practice; it has empirically emerged that no attribute in the universe corresponds to our old notion of *containing caloric* but at least we can recognize a priori the simple pattern of word/world ties that this concept hopes to establish. Or so Russell opines. Indeed, this presumption of a *foreseeable pattern of correlation* lies very near the core of basic classical thinking and will concern us much in the chapters to follow. In contrast, I will argue that, in many cases, the true nature of a predicate's correspondence with the circumstances it addresses may not prove obvious at all and will require dedicated research to unravel. Such alignments, furthermore, are also prone to slippage as time goes on.

But despite my reservations with respect to word/world connection as it is conceived within the classical picture, I do not think we can possibly understand the engines of common linguistic development unless we attend directly to the *patterns of genuine correlation* that gradually emerge—and sometimes fade away—during the courses of the usage's historical evolution. Few modes of linguistic behavior, even those practiced by

---

[46] Crispin Wright's project in <u>Truth and Objectivity</u> (Cambridge, Mass.: Harvard University Press, 1992) seems to be rather of this type, for example (although I find his precise motivations obscurely presented).

the most dissociated and ethereal forms of religious cult, are likely to last long if they do not embody tolerable stretches of substantive word/world coordination, if only in dedicated patches here and there. Quite commonly, these supportive correlations prove more recondite in their strategic underpinnings than we anticipate when we learn the usage and semantic mimicries are common where stretches of discourse appear to relate to the world in a much different manner than they actually do. All of these considerations represent themes that will be explored more fully later in the book— where examples will be supplied! My observation at present is simply that the indispensable idea of word/world correspondence should not be thrown out with the classical bath water in which the notion commonly swims. But that is exactly the ambition of the anti-correlationalists.

Indeed, in their eagerness to avoid an Addison-like veil of perception falling betwixt the external world and ourselves, such authors commonly succumb to an analogous doctrine on the conceptual side of things that strikes me as equally dreadful. Because they assume that idiosyncratic human construction and subjectivity represent refractory components of every form of conceptual content, they generally accept doctrines about descriptive policy that are quite unnerving in their own right. In particular, anti-correlationalists often inform us that many incompatible forms of conceptual scheme or "ways of world making" exist that can serve all of our descriptive ambitions equally well. Articulated in terms of schematic "theories," this familiar *underdetermination of theory* doctrine asserts: for any viable descriptive theory T, there will exist rivals T′, T″, etc. capable of accommodating the same set of observational consequences equally well.[47] To be sure, in the history of science, apparently competing approaches sometimes emerge that at first look quite different in their conceptual contours yet seem to accommodate the available data equally well (a locus classicus can be found in the erstwhile opposition between Heisenberg's matrix mechanics and Schrödinger's wave theory, although most real life examples are complicated by some measure of the facade problematic we shall discuss later (6,xii)). However, in most of these cases, such rivals are eventually discovered to encode the same basic physical information in mathematically different but interrelated ways (thus spectral theory reveals the bridges that carry Heisenberg's favored vocabulary over to Schrodinger's). Common sense would judge that the two sets of descriptive predicates merely talk about the same data in different ways but an anti-correlationalist approach to conceptual content cannot easily ratify this opinion. Through a strong insistence upon a neo-classical picture of semantic invariance, it is usually driven to contend that we have been supplied with two distinct "ways of world making" that describe external reality in intrinsically different terms (7,iii). To get the engines of scientific description turning, we must tacitly opt for one of these viable schemes, even if we fail to notice the conventionality of the choice we select. Or, to articulate this point of view in a different way, some choice of

---

[47] I have discussed this doctrine critically in two early papers ("The Observational Uniqueness of Some Theories," Journal of Philosophy, (May 1980) and "The Double Standard in Ontology," Philosophical Studies (March 1981)). I believe that these remarks remain essentially correct, but now consider that the problems canvassed in Chapter 4 are more central to the underlying theory T syndrome problems.

T over T' is required to *prime the pump* of science: until we have simply assumed a beginning span of T's content to be true, we lack the means to coherently test the empirical assertions that get advanced under its aegis. When common sense loosely pronounces that T and T' "talk about the same data in different ways," it merely observes, according to anti-correlationalist gloss, that schemes T and T' are equally viable descriptively. We fall into desperate muddles, they claim, if we believe that the merits of a doctrine's correlative ties to external reality can be coherently examined in its own right.

This underdetermination thesis plainly lowers an insurmountable *veil of predication* betwixt the world and ourselves, which bars us from ever determining whether the concepts we employ genuinely match the true traits of the world or not (I have just described the doctrine in its familiar theory T guise, but authors like Ebbs entertain a similar point of view without assuming so much logical empiricist apparatus). I find such uncanny doctrines with respect to descriptive capacity every bit as disconcerting as the traditional veil of perception, for we wind up walled off from the world either way (it is merely that the darkening curtain is comprised of *concepts* rather than private *objects*). I find it odd that philosophers are often cheerfully willing to accept an impediment of this ilk in their eagerness to avoid the perceptual intercessory. The Quine of Word and Object represents an excellent case in point. He is proud of the fact that he can dispense with any epistemological reliance upon "private objects" through his elaborate doctrines of ontological commitment (his opening section is entitled "Beginning with Ordinary Things"), but this apparent advantage is achieved only at the cost of a warm embrace of a quite severe form of underdetermination of theory thesis.[48] Once we have slipped down this unhappy path, we become eventual prey to the holist fables of incommensurable irreconcilability woven by Kuhn or worse. All of these opinions represent tropisms that I am eager to resist.

Such considerations are testimony to the mute manner in which the classical realm of concepts serves as a convenient Land of Nod to which overt philosophical unpleasantries can be surreptitiously dispatched. We rid ourselves of unwanted "private objects," yet we pick up uncanny "concepts" in trade. In my opinion, we have merely bartered an uncomfortable thesis with respect to sense data for an obnoxious dual with respect to concepts, whose oddities seem less evident only because we attend to their contours less. We should become more wary of these doctrinal exchanges (7,x). Certainly we should not allow scare-quoted phrases such as Ebbs' "an imagined 'external reality' " to persuade us that everyday assertions such as " 'rabbits' refer to rabbits" represent some wild-eyed form of "metaphysics" comparable to belief in astral projection. True: the standard classical picture of how we learn of these correlational relationships is distorting in its simplicity, but that error does not establish that the direct examination of a predicate's links with the world it serves does not represent a viable form of investigative enterprise.

Plainly I am no fan of amphibolism with respect to concepts. Quite the contrary, I shall develop an account of natural linguistic process that will allow us to disentangle

---

[48] W. V. Quine, "On Empirically Equivalent Theories of the World," Erkenntnis 9 (1975).

the psychological and objective strands of linguistic directivity that run together in our *ur*-philosophical thinking quite effectively, as well as giving proper recognition to a third category of strategic concern (7,ii). Accordingly, philosophical sermons to the effect that it is inherently incoherent to segregate the subjectively based aspects of linguistic shaping from their more objective counterparts do not represent music to my ears. But we will approach these matters in a different manner than suggested in this chapter (7,ii).

As we have observed, neo-Kantian lines of thought typically eschew word/world renderings of conceptual objectivity in favor of appeals to agreement within a cabal of cooperating investigators. Allied claims about the vital role of "community" in linguistic process became prominent in the latter twentieth century due, <u>inter alia</u>, to the enormous influence of Wittgenstein's <u>Philosophical Investigations</u> (a Heideggerian variation upon these strains is echoed in Titon's concern with "being enmeshed in reciprocity"). It is in this vein that Wilfrid Sellars writes:

> And there is, as we know today, a sound score to the idea that while reality is the "cause" of human conceptual thinking which represents it, this causal role cannot be equated with a conditioning of the individual by his environment in a way that could in principle occur without the mediation of the family and the community. The Robinson Crusoe conception of the world as generating conceptual thinking in the individual is too simple a model.[49]

This reads as if Robinson Crusoe could never acquire the concept *being a rabbit* if he merely dealt with rabbits and never any fellow islanders. This unlikely claim is often presumed to follow from Wittgenstein's strictures against a private language, although it is hard to find two interpreters who agree upon what those "strictures" are (Sellars' opinions, however, most likely trace to pragmatic influences such as John Dewey).

Sellars complains that it is naïve to think of "the world as generating conceptual thinking in the individual." But why? There are certain tasks that we cannot easily accomplish unless we engage in intervening runs of linguistic activity. Elementary forms

[49] Wilfred Sellars, "Philosophy and the Scientific Image of Man" in <u>Science, Perception and Reality</u> (London: Routledge and Kegan Paul, 1963), 16.

of mathematical calculation provide simple examples: it is frequently impossible to convert observations (sightings of a target object) to actions (setting a cannon to the correct firing angle) without relying upon some mediating stream of notational exuberance. For such computations to work properly, the various symbols displayed in the gunner's scribblings must display some fairly tight alignment with physical data, although these linkages may prove quite intricate in their patterns of word/world alignment (as we'll observe in concrete cases (4,x)). But surely the solitary Robinson Crusoe stranded in some bleak and otherwise unpopulated locality will retain ample reasons for devising a computational language to improve his cannon firings? If so, mightn't worldly necessity still serve as the mother of conceptual invention within our lonely outcast, Sellars' apparent asseverations to the contrary? We shall expand upon these complaints in 5,ii.

. . . .

Throughout this book, I take the facts of mathematics pretty much for granted. However, the notion that this subject must assume the role of *regulative principle* prior to any description of the world in physical terms represents a vital aspect of neo-Kantian tradition, as aptly emphasized by my friend Michael Friedman.[50] In this book I have not attempted to dabble in topics so grand as these; I have instead considered concepts entirely from a scientific realist point of view. I do believe that the easy road to neo-Kantianism has been paved, historically at least, by strong reliance upon veil of predication related claims. What its doctrines would look like without implicit classical picture premises, I cannot say.

. . . .

(vii)

Seasonality in conceptual evaluation. Let us pass in quick review over the basic themes of this chapter.

(1) We began by worrying, under the heading of tropospheric complacency, about the distortions that arise when we too quickly presume that the behaviors of the world's collection of objective attributes carry us from one setting to another in an uncomplicated manner, leading to improper expectations as to what kinds of tasks, linguistic or otherwise, can be accomplished within those extended contexts. Similar complacencies often lead to improper assumptions about the classificatory or inferential capacities of our peers.

(2) In fact, the nature of some of these expectations of carryover patently rely upon matters of human capacity or point-of-view that seem ignored in an unduly objective treatment. We employed *adequately realizes the Symphony in G Minor* as a central example.

(3) To include these missing "point of view" ingredients within an adequate model, we shifted to a picture where the support provided by the objective trait *being a dog* in

---

[50] Michael Friedman, Dynamics of Reason (Stanford, Calif.: CSLI Publications, 2001).

the semantic schema "is a dog"/*being a dog* is replaced by a subjective quantity that incorporates a measure of how the trait presents itself to us. This alteration in our scheme blocks the cavalier expectations about common capacity that troubled us in the deportment of our Darwin critic.

(4) Unfortunately, this subjectivist relocation of our predicate's directive basis seems too extreme, in that the primary thrust of its descriptive interests now seem focused upon quasi-psychological concerns far removed from the practicalities in which the predicate found its original usage (i.e., the discrimination of symphonic sounds or colored fabrics). We then explored the curious doctrines of amphibolism that attempt to mollify this uncomfortable displacement of conceptual locus.

(5) Worse yet, both objectivist and subjective approaches to conceptual content apparently force upon us, quite against the recommendations of common sense, odd policies with respect to the preservation of music and instruction in musical appreciation.

I presume the reader has found our rapid pilgrimage from wistful musing on the timelessness of Mozart into the gloom of participatory idealism rather astonishing, for we seem propelled along our journey largely by rather small worries about the true nature of "musical content." It seems as if some melodic mouse has unaccountably inflated into a philosophical elephant—indeed, a creature apt to frighten hapless critics and ethnomusicologists into improvident behaviors. Somewhere within the granary of concepts and attributes our erstwhile wee beastie has located some Wellsian food of the gods that has puffed it up into grotesque grandeur. And I have promised, in the course of this book, to develop a fuller account of why this inflation occurs.

This explication will trace the phenomenon to our deeply rooted inclination to overlook the *seasonalities* that naturally attach to our everyday tools of conceptual evaluation: viz., the factors that lead us to regard factor θ as critical to the behavior of predicate W on day 1, but later dismiss its affective importance in favor of some disharmonious consideration φ on day 2. For reasons that will emerge later, we possess a deep attachment to the notion that the contents of our concepts stay largely invariant over time. It is this strong *ur*-philosophical desire for *semantic fixity* that induces us to squash together the real but disparate directivities of θ and φ into some fictive homogenized "content" allegedly able to govern the correctness of W's employment unilaterally at all points in its career. Once the diverse liquors of linguistic change have been allowed to blend together in this ill-advised way, we will scarcely be able to discriminate the distinct manners in which they shape the behaviors of garden variety descriptive vocabulary. Once we learn to keep these reactive agencies distinct, we will be able to sort out the objective data registered in our discourses ably enough. From this point of view, exaggerated worries that classificatory terms such as "folk music" are so irremediably steeped in social prerogative that their evils can be corrected only through extreme countermeasures should seem like a scarecrow concocted from naught but the garments of philosophy of language run amuck. As we shall see later in the book, everyday conceptual evaluation regularly avails itself of specific processes of *semantic detoxification* in its efforts to keep language rolling forward along profitable rails. *Ur*-philosophical problems, such as those surveyed in this chapter, often begin in a failure to appreciate the underpinnings of these detoxification techniques properly.

This is not to claim that discerning the specific winds that effect linguistic development is an easy task. It is unlikely to represent a project that can be accomplished through armchair musing about "possible cases," in the manner that many academic philosophers still favor. More often than not, the puzzlement attaching to a particular specimen of usage stems from a mixture of physical and strategic factors that require unraveling before we can entertain any chance of understanding the unexpected directivities that influence our predicate's odd behavior. This chore generally requires a good deal of rough and tumble scientific investigation, often reaching across a very wide canvas of concerns. As we await their outcomes, we must cultivate in the meantime semantic patience as the tools required for a proper diagnosis are gradually developed. This temporary need for *forbearance* in the attribution of fixed semantic content to a predicate is responsible for the philosophical mitigated skepticism that I advocated in the previous chapter.

The next two chapters will endeavor to probe our tendencies to presume otherwise more deeply and explain more fully why we instinctively desire a greater invariance and homogeneity in "conceptual content" than our worldly circumstances allow. They also sketch how certain key schools of developed philosophical thinking have sprung up around our muddled expectations with respect to conceptual evaluation. Then, beginning in Chapter 5, I shall lay out several sample schedules of shaping influence that are apt to affect a descriptive usage and, from that vantage point, return to the basic issues of objective content that we have surveyed in the befuddled dialectics of this chapter.

3

CLASSICAL GLUE

I, whom no living beauty yet could warm,
Am now enamour'd of an empty form.
Isaac Hawkins Browne[1]

(i)

Under a predicate's sheltering wing. The fundamental source of last chapter's muddles lies in the fact that we commonly expect "concepts" to carry great evaluative burdens, yet not buckle under the freight. A frequent symptom of this overloading is that it becomes impossible to locate the trait within any satisfactory housing. The atmosphere's humble currents seem too meager a substratum to support *adequately realizing the Symphony in G Minor* in its full, melancholy glory and we begin to search for another matrix in which our property can be more suitably instantiated. We find ourselves tempted to plant our trait within subjective mentality or even ship the entire affair off to amphibolic shoals. But, in the final analysis, no proposal for attribute relocation seems wholly satisfactory and so our orphaned concept appears destined, like the boll weevil of ballad, to "keep looking for a home."

If articulated solely in these "where do these traits display themselves?" terms, our worries about "the nature of musical concepts" are apt to look rather silly, as if some peculiar game is being played with words that has nothing to do with anything important about music. Indeed, one often finds such "idle philosophizing" dismissed with scorn—even by professional philosophers.[2] But such disdain does not do justice to the deeper origins of the conceptual problems involved. Our metaphysical frivolities are symptomatic of more troublesome affairs—the surface ripple of *ur*-philosophical currents that run at greater depths. The overloading of which I've complained stems from a

---

[1] Isaac Hawkins Browne, "On Seeing a Portrait of Miss Robinson, Painted by Mr. Highmore" in Rev. Henry Phillip Dodd, ed., The Epigrammists (London: Bell and Daldy, 1870), p. 376.

[2] Aaron Ridley, "Against Musical Ontology," Journal of Philosophy (2003).

very basic inclination to overestimate our human capacities for anticipating the unexplored, especially in linguistic matters. These sanguine hopes adversely affect us all, even the most doggedly anti-philosophical amongst us. Typically, these appraisals assume the guise of presuming rashly that, because a certain group of skills have been mastered, other capabilities will follow automatically in their wake. The drab cloth in which these faulty anticipations are typically dressed is the prosaic mufti of phrases such as "has fully grasped the concept"; "completely understands the trait"; "has achieved mastery of the meaning."

To study how these mistakes arise, I will narrow much of our discussion to circumstances where some common predicate for everyday physical classification (such as "is red" or "weighs five pounds") is credited with a unitary concept as its sole reference, for in this simple alignment of *language* with *concept* we can witness a prototype for wider sorts of *ur*-philosophical error. Of course, no one presumes that predicates and "concepts" invariably align in tidy patterns: some attributes resist ready expression in language and some predicates clearly bear complicated relationships to their conceptual supports. Nonetheless, often classificatory predicates seem to capture classical "conceptual contents" at exactly the right level of grain and it is with these cases that we primarily wish to deal.

There is a second reason why we should scrutinize predicative expressions centrally in our investigations. Long ago Bishop Berkeley and allied thinkers suggested that abstract entities such as concepts and properties gain their semblance of ontological respectability through donning the reassuring garments of "general names": we mistakenly presume that a contrivance called *being a rabbit* exists simply because we know how to align the sundry individual rabbits of the world under the linguistic heading of "is a rabbit." The predicates and the rabbits exist to be sure, we are assured, but the concept *being a rabbit* itself is a fictitious go-between invented to provide a pseudo-explanation of how our practice of using predicative expressions works. "Concepts" have simply borrowed an ersatz substantiality from their more respectable linguistic cousins, the predicates. In Chapter 5 we shall examine a milder form of this anti-conceptual doctrine defended by the American philosopher W. V. Quine (who, unlike Berkeley, is not a nominalist proper because he tolerates restricted varieties of abstract object such as sets).

Unlike authors of this persuasion, I harbor no hostility to abstract objects per se. To the contrary, I will argue (5, vii) that quite extensive fields of attributes need to be accepted as robust components of the physical landscape. Unless we can appeal to these traits in a commonsensical way, we will not be able to understand how a developing language shapes itself to the contours of the world it addresses. Nonetheless, Berkeley and Quine correctly observe that a bit of repeatable syntax (such as a predicative phrase) displays an astonishing capacity to make the amorphous appear concrete. The lure of shared phoneme, after all, leads many of us to categorize crayfish with catfish as mutually "fish," despite their lack of biological or etymological affinity (the "fish" in the former represents a corruption of "crevis"). If we can understand the motives that induce us to pile up an excess of distinct capacities under the accommodating shelter of a predicative expression, we will have begun to unravel the processes behind the confusions of the previous chapter.

However, in restricting our discussion of "concepts" largely to their role in capturing the cognitive significance of various specimens of classificatory predicate, I run the risk of illustrating Joseph Addison's admonition:

> There is nothing in nature so irksome as general discourse, especially when they turn chiefly on words.[3]

But, however dry or irksome our investigations may prove, they will gain considerably in clarity and focus through this strategy. After all, even in Addison's own circumstances, many of his greatest essays partake of exactly the flavor he abjures.

In truth, I hope my readers may extract the same humble pleasures from the weird byways of linguistic process as I have myself. With respect to the book's larger ambitions, there are two varieties of human temperament that become drawn to philosophy's lair: those with a burning hunger to uncover the Secret Natures of Things and those who find such earnest yearnings puzzling in themselves and in want of some commonsensical dissolution. The best exponents of the old ordinary language school—J. L. Austin, in particular—are nicely representative of this second personality type and my own work follows in their deflationary spirit, if not their methodology. For skeptical inclinations such as ours, a warm satisfaction arises in observing the murky rendered clear, even if much of its erstwhile grandiosity gets lost in the recasting. In many ways, this clarifying impulse is akin to the delight we feel when we learn that some obnoxious social snob has secretly commenced his career in the pest extermination business. 'Tis not an entirely admirable form of enjoyment, to be sure, but essentially it is what this book has to offer.

(ii)

Classical gluing. Our first order of business is to gain a better grip on the "primitive grasp of conceptual content," as that notion appears within classical modes of thinking. In a linguistic context, the most direct and appealing articulation of the basic parameters of this viewpoint were set down by Bertrand Russell in his Problems of Philosophy of 1912. To be sure, Russell happens to be somewhat out of favor with contemporary analytic philosophers because of his breezy inattention to questions of detail. But for our purposes (which are likewise unconcerned with such specifics), Russell's presentation is perfect, for it trenchantly epitomizes the formal doctrines that blossom when the ur-tendencies of everyday thinking first become subject to the ministrations of skilled philosophical nurture. In the vivid and appealing prose of which he was a master, Problems outlines the basic set of doctrines that I call the classical picture of concepts in this book. Russell himself prefers the old-fashioned term universal as a synonym for my

---

[3] Joseph Addison, "Criticism on Paradise Lost," no. 267, Works, vi. 32.

Russell

"classical concept" and I shall sometimes follow him in this usage. As sketched in our appendix, a vast amount of supplementary philosophical foliage naturally erupts from the central stalk of classical thinking, but at present I want to concentrate upon a core process to be called *classical gluing*.

Although I believe that classical gluing (or its various doctrinal cousins) continues to sit at the center of much contemporary thinking about concepts, it has inspired a large host of critics as well. Later we shall especially consider the criticisms offered by W. V. Quine, whose complaints about the doctrine most nearly approach my own. Indeed, my own project in this book can be profitably viewed as an attempt to blend attractive elements extracted from both Quine and Russell.

The most salient feature of a classical universal is that it is conceived as *living in two realms simultaneously*. Russell maintains that a concept can both (i) report upon a specific individual's frame of mind ("Mowgli fully grasps the concept *being venomous* and finds it fearful") and (ii) register the condition of his physical surroundings ("The snake in front of Mowgli exemplifies the attribute of *being venomous*"). The twin phrases central to this "operate in two spheres simultaneously" conception of universals are *exemplify* (indicating whether the trait is manifested in the snake's physical behavior or not) and *grasp* (evaluating its status within Mowgli's psychological realm). In the circumstances where Mowgli "completely understands" a concept, Russell declares that he is *fully acquainted* with the underlying universal. Here "fully acquainted" represents one of those happy Russellian turns of phrase that aptly captures natural *ur*-philosophical opinion. Once this cognitive state is obtained, there can be no doubt as to what Mowgli is talking about or how he should reason with his concept, even though it happens that he is actually confronted with a stick or innocuous corn snake. In this assumption of *fully grasped meaning*, we see the primary roots of the doctrine of semantic finality discussed in 1,vi.

Of course, there's no suggestion here that to grasp an attribute is thereby to exemplify it: I can understand the concept of *being an ice cream cone* without turning into one. Sometimes one finds classical thinking criticized through silly observations of this ilk. Russell would appropriately respond that, in the final analysis, *grasp* and *exemplify* simply represent two distinct and primitive fashions in which a universal can act.

This "living in two worlds" behavior allows the classicist to frame a simple and appealing story of how a range of basic predicates align themselves with worldly conditions: we merely grasp the appropriate concept and conventionally associate it with suitable linguistic noises and inscriptions. To mentally associate concept and sound seems an easy task (as long as the concept itself is readily graspable); the concept can then align itself with external conditions on its own recognizance, simply by determining whether the universe's far-flung objects exemplify its requirements or not. Qua human agents, we have little to do with the latter process; our chief task is to grasp the concept squarely and maintain its correlation with suitable English. By these means, the "living in two worlds character" of our concepts provides an optimal adhesion between predicate and world, for an identity is forged along the interface between what is grasped mentally and a genuine trait of the world under discussion. If someone appeals to the alleged two world commonality of classical concepts to explain the semantical behavior of basic predicates, I say that they have subscribed to a recipe of *classical gluing*. I see this reliance as lying at the very core of traditional semantic thinking. To be sure, classical thinkers often frequently propose less direct methods for keeping terminology attached to the world (Russell's own theory of descriptions represents one of these). In such cases, we must trace through their details to determine whether they ultimately reply upon classical predicate/concept adhesions as their primary mechanism.

Let me hasten to add, however, that a view of concepts can remain essentially classical, even if the breach between a content mentally grasped and the worldly attribute signalized is somewhat widened. Many thinkers prefer to maintain that only mental *representations* are truly grasped, but allow that such representations can none-theless directly *report* upon the contents of worldly traits. As long as they presume that the report and its subject matter can manifest the same content in some primitive fashion, then I do not consider that any significant departure from basic classical gluing has been effected (such shifts merely reflect quibbles with respect to the connotations of "grasp," in my opinion). Following Frege, other philosophers have claimed that the cognitive significance of what is grasped bears some less direct sense and reference relationship to true attributive content than suggested by Russell's assumption of complete identification, but we'll postpone consideration of such variant creeds until 6,iii.

We shall survey more pointed criticisms of classical thinking later in the book, but it is important to observe that many popular attacks on its doctrines mischaracterize the manner in which classical gluing is supposed to work. For example, Quine satirizes the classical view as engaging in a "myth of the mental museum"[4] and John Dewey com-plains that we should never "assum[e] that a word has such magical power that it can point to and select the subject to which it is applicable."[5] As they stand, such remarks merely represent dignified variants upon name-calling, because epithets such as "magical" scarcely *diagnose* the distortions induced by the classical picture; they merely report the author's wish that some suitable alternative be found. More importantly, those who most loudly complain of magical powers usually muddle the discussion by

[4] W. V. Quine, "Two Dogmas of Empiricism" in From a Logical Point of View (New York: Harpers, 1961), 48.
[5] John Dewey, Logic: The Theory of Inquiry (New York: Holt, Rinehart and Winston, 1938).

confusing the processes of classical gluing with a rather different story that can be called an *intention-based picture* of predicative significance.

What I have in mind is this. There have certainly been important authors (especially in antiquity) who have maintained that the essence of assigning meanings to predicates traces to our ability to *directly will* that our otherwise "dead" words should attach to the world in a prescribed way. Here the alignment of a predicate with significance is treated on the model of naming a rabbit in the backyard hutch, except operating in multiplex: "There's the rabbit selected and I hereby wish the name 'Sniffy' to attach to it hence-forth." But with a predicate, we must implement this form of intentional act many times, even with respect to objects situated in galaxies far away in space and time: "I hereby intend my predicate to reach out to all of these things."

So conceived, a capacity to perform this prolix naming seems as if it might prove rather magical. Indeed, many writers historically attracted to this intention-centered approach to predicate significance have been positively eager to draw spiritual con-clusions from our alleged ability to summons meaningless symbols into extravagant attachment to the world. So when a theologically motivated writer such as William of Ockham claims,

> [A]n intension of the soul is something in the soul capable of signifying something else,[6]

he is on the cusp of concluding that this special activity demonstrates a spiritual capacity that arranges humans on a higher rung of the Great Chain of Being than the non-signifying monkeys. In his <u>Tractatus</u>,[7] the early Wittgenstein treats the Soul as an

---

[6] William of Ockham, <u>Ockham's Theory of Terms (Summa Logica I)</u>, Michael J. Loux, trans. (Notre Dame, ILL.: University of Notre Dame Press, 1974), 7. Also:"[T]*he spiritual element of speech, constitutes one of the greatest advantages which man has over all the other animals, and... is one of the greatest proofs of man's reason*": Claude Lanvelot and Antoine Arnauld, <u>Port Royal Grammar</u>, J. Ruieux and B. E. Rollin, trans. (The Hague: Mouton, 1975), vol. ii, ch. 1. The illustration derives from a fifteenth century printing of St Isidore's *Etymologies*.

[7] Ludwig Wittgenstein, <u>Tractatus-Logico-Philosophicus</u> (London: Routledge & Kegan Paul, 1961).

unseen presence that makes humdrum symbols "come alive" by projecting them semantically onto other things, just as a table top arrangement of kitchen utensils presently represents the Battle of Antienam because our grandfather has wished that representational relationship into being. The more hard-boiled among us are likely to dismiss such musings, in league with Quine and Dewey, as supernaturalist.

But even if views of this intention-based kind, when pressed to extremes, legitimately qualify as occult, it is scarcely fair to hang standard classical thinkers like Bertrand Russell by the same rope. Indeed, the basic genius of their portrait of universals lies precisely in the fact that a means is provided that *avoids* appeal to extraordinary mental powers of linguistic anointment. In the classical picture proper, it is not through our wills that predicates get firmly attached to far-flung corners of the universe, but simply through the inherent abilities of classical concepts to live in two different realms. It is this *commonality of manifestation* that supplies the critical glue required, not any fantastic intellectual outreach. The only chore left to humble humans is merely to correlate our predicates with universals that we cleanly grasp (apes, no doubt, grasp many concepts ably but have trouble keeping their phonemes aligned). Such simple acts of association do not demand any astonishing capacities of mental projection, but simply the intellectual equivalent of aligning one's knife with one's fork: we can "put two ideas together" easily enough. In the true classical picture, it is the concept itself, without any aid from us, that categorizes the sundry objects of the external world as lying "in" or "out" of its extension—our own feeble capacities with respect to real world naming play no role in the activity of semantic attachment at all (the classical picture does not ask us to *name* every rock that sits in a galaxy far away). We obtain a story of predicate/world connection that resembles Noah and the dove: predicate and universal get aligned here on the deck of the Ark, but the latter then flies away on its own to survey (and classify) the great, unreachable universe on our behalf.

We may grumble suspiciously about this story, but it is hard to see immediately where any magical powers come into it. I do regard classical thinking as substantially exaggerating human linguistic capacities, but complaints of occult capacities do not diagnose the nature of the misapprehensions ably. We shall return to these issues of "naming with a predicate" later.

(iii)

Conceptual directivities. Characteristically, Russell discovers his prototypical universals by locating them as the semantic supports for certain key predicates, finding them, as it were, under the leaves of linguistic cabbages. Here is a typical passage that displays the vein of thinking I have in mind:

> Suppose, for example, that I am in my room. I exist, and my room exists, but does "in" exist? Yet obviously the word "in" has a meaning; it denotes a relation which holds between me and my room . . . The relation "in" is something which we can think about and

understand, for, if we could not understand it, we could not understand the sentence "I am in my room".[8]

Clearly the predicate "is located within" possesses an unambiguous meaning in English; it does not constitute unsupported nonsense as exemplified by the pseudo-sentence "I am bib-a-lollie-boo the room." But what underlying feature here separates meaningfulness from gibberish? Russell's view (in The Problems of Philosophy) is simply that "is located within" is *directly supported* by the "universal" *being located within* whereas no comparable underpinnings prop up "am bib-a-lollie-boo."

This passage, I think, represents an important line of argument and it helps to understand key elements in the thought of anti-classical critics if we ask, "In what respects is this author willing to challenge Russell in this passage?" This is not to say that Russell has rendered his own principles entirely transparent. In many of his other writings Russell is quite happy to declare that many predicates are not backed up by universals in this simple way, but require some roundabout pattern of semantic connection. Indeed, in Our Knowledge of the External World[9] (which is roughly contemporaneous with Problems), Russell assumes a position that requires that "is in" be treated in a very circuitous manner. As we shall learn a bit later in this chapter, basic tensions lie deeply ingrained within the classical picture that render the assignment of settled content to many familiar predicates quite unstable—Russell is scarcely alone in his wobbling.

Incidently, the reason Russell selects the relational predicate "is in" rather than, e.g., "is a dog" is because he is concerned to evade the attacks of Berkeleyian nominalists who claim, "There is no need to posit a universal behind 'is a dog'; it merely means 'is biologically similar to Lassie.'" Russell's celebrated retort is: "Perhaps, but surely the universal *being biologically similar* is required to back up the latter predicate."

Once the paste pot of classical gluing has been arranged upon his workbench, Russell finds a ready tool for accomplishing an astonishing variety of intellectual chores. He seems to have located an Archimedian perch from which he can: determine how rigor and trustworthiness should be cultivated within scientific investigations; explicate the conditions required for speakers to understand one another; fix the exact role of philosophy as a form of intellectual endeavor; explain where our estimations of conceptual possibility come from, and so on, running through the lengthy list of proposals outlined in this chapter's appendix. The beauty and elegance with which all this is achieved is both astonishing and admirable. It is truly a pity that the sorry world in which we have been deposited won't permit Russell's policies to be fully realized.

Worse yet, Nature expresses her unwillingness to conform to Russell's aspirations only in a sniveling and underhanded way. Rather than straightforwardly denouncing his errors, she introduces small cracks and fissures into practical descriptive usage in manners that are hard to spot yet render Russell's claims to have established a sound House of Science and Philosophy effectively worthless. Put another way, she'll allow

---

[8] Russell, Problems, 90.   [9] Bertrand Russell, Our Knowledge of the External World (London: Routledge, 1993).

Russell to pontificate all he wishes in print or within the halls of the university, but if he should ever try to build a bridge based upon his recommendations, she'll make it fall down at an inopportune moment.

In point of fact, the Russellian doctrines listed in our appendix, lengthy as they are, do not constitute a proposal definite enough to be considered as "an account of conceptual behavior," but provide, at best, the *shell* or *scheme* for such a doctrine, with most of its crucial innards as yet unsupplied. For the theses listed provide few instructions as to how the blank slate of conceptual concept should be concretely filled in for real life predicates. In fact, historical efforts to provide the missing materials in pivotal cases have been commonly frustrated, and these dismal episodes have inspired a rich set of classical excuses to explain why the classical picture experiences so much trouble in fulfilling its promises. It is for these reasons that I usually label schematic demands like those listed in the appendix as a *picture* of concepts, preferring to reserve epithets like "theory" for less skeletal accountings ("picture," in my usage, generally suggests a *schematic sketch* of a situation, whose required concreteness has been largely omitted—the term does not necessarily express reproach, but simply a demand for something additional).

In the classical tradition, the conceptual content associated to a predicate—the same stuff that binds it to the world—is intended to serve as an *invariant core* that controls the instructive directivities that attach to the predicate. As explained before, I employ "directivity" as a non-technical means for capturing the loose bundle of considerations that we might reasonably cite, at various moments in a predicate's career, in deciding how the term should be *rightly applied.* Such directivities emerge, for example, in the replies we offer to questions such as the following.

(a) *Is this stone really red? Well, why don't you simply look at it in a good light?*

(b) *Is the pressure extremely high in this portion of the fluid or not? Why don't you measure its value with a pitot tube?*

(c) *Is the pressure extremely high in this portion of the fluid or not? Why don't you calculate its value from the boundary conditions using finite differences?*

Note that response (c) differs from (a) and (b) by citing an *inferential policy* rather than an *observational technique*; we shall worry later about the comparative importance of these two varieties of directivity.

Russell and his band of fellow classicists promise us that tidy organization can, in principle at least, be installed upon the great mass of directive ingredients that typically emerge within the chaotic courses of everyday usage: each predicate can be assigned a crisp conceptual content that will answer all of these "Am I employing X rightly?" questions briskly and steadily (since real life is untidy, classical writers invariably acknowledge a range of pragmatic reasons why a run of everyday discourse might be spared from their improving ministrations). But once a proper conceptual hygiene has been practiced, the predicates cleansed will henceforth prove admirably well behaved (unless mistreated by their human handlers). Their core conceptual contents will codify which everyday directivities stand close to the proper meanings of the phrases and which stand further afield as mere empirical associations. Michael Dummett has this assumption in mind when he writes with respect to linguistic meaning generally:

> *A conception of meaning . . . is adequate only if there exists a general method of deriving, from the meaning of a sentence as so given, every feature of its use, that is, everything that must be known by a speaker if he is to use that sentence correctly.*[10]

Although I will argue that such demands for "derivation" are quite wrongheaded (10, iv), we must concede that the ways in which we talk about "concepts" in everyday life prima facie suggest, as long as their contours are not scrutinized closely, that classicism's expectations with respect to invariant contents appear quite reasonable. After all, we commonly offer evaluative claims such as the following:

(d) *It doesn't make sense to call a ruby "red" if it doesn't look so in proper light.*

(e) *The equations upon which the finite difference calculations are based track the proper significance of "pressure" quite closely, whereas the responses of a pitot gauge are subject to many unwanted disturbances and often prove quite inaccurate in comparison.*

Indeed, it is from humble assessments such as these that the notion of a "classical conceptual content" spontaneously springs. In truth, there is a natural *seasonality* that accompanies these forms of intellectual evaluation in their everyday appearances—we answer the same question in different ways on different days—, but we are usually insensitive to its presence and instead assume that some invariant core acts to resolve our directivity questions in a steady, classical manner. And it is from here that Russell's picture obtains its considerable *ur*-philosophical credentials.

Accordingly, if we ignore the seasonalities of real life conceptual evaluation and agree that we can grasp rich bundles of guiding content and hold onto them invariantly over long stretches of linguistic time, then we will have allowed Russell all the wherewithal he requires to construct the mighty mansion characteristic of classical thought. And this is a house that promises many domestic comforts, with respect to both philosophy's prospects as a discipline and science's ability to shield itself against the shocks of unwelcome discovery. Under the first banner, we can confidently announce that philosophy's anointed task is to serve as overseer of the conceptual domain; under the second, we can promise that dedicated intellectual discipline can install a tidy order upon the otherwise messy processes of scientific investigation.

(iv)

Custodians of the conceptual realm. It is within Russell's Pollyannish assurances with respect to "clear thinking" that the classical picture's most secretly invidious elements lie. But these issues need to be addressed in a delicate manner, because hasty

---

[10] Michael Dummett, "What is a Theory of Meaning?-II" in Truth and Meaning, Gareth Evans and John McDowell, eds. (Oxford: Oxford University Press, 1978), 137.

opponents of classical thinking often talk themselves into brusque doctrines that are "ever so much worser" in their practical consequences than anything Russell suggests. Recall, from the previous chapter, Jeff Titon's account of how a funding committee squabbled over the implications of "folk artist." Normally, we should expect that the methodological injunction "Let us define our terms properly before we engage in profitless debate" might help matters, although we are all familiar with situations where, for some reason, it doesn't. But Titon, like many intellectuals today, has decided that such improving gambits merely represent rhetorical aggression, a debating society form of warfare by other means. Such opinions would be utterly destructive of fruitful discourse if practical people truly believed them. We really shouldn't attempt to dispatch the comparatively mild exaggerations of classical thinking with a sledgehammer.

But something fishy resides on Russell's side, nonetheless. Consider this specimen of the improving frame of mind, extracted from Russell's friend, the mathematician G. H. Hardy. He is writing about the nagging problem of *divergent series*: expressions that don't seem to make any obvious sense, yet have frequently allowed mathematicians and physicists to make great advances by pretending that they do ("Divergent series," Abel once wrote, "are the devil"). Hardy:

> It is plain that the first steps towards such an [improvement] must be some definition, or definitions, of the "sum" of an infinite series, more widely applicable than the classical definition of Cauchy. This remark is trivial now: it does not occur to a modern mathematician that a collection of mathematical symbols should have a "meaning" until one has been assigned to it by definition. It was not a triviality even to the greatest mathematicians of the eighteenth century. They had not the habit of definition: it was not natural to them to say, in so many words, "by X we mean Y". There are reservations to be made, to which we will return in §§1.6–7, but it is broadly true to say that mathematicians before Cauchy asked not "How shall we define $1-1+1-\ldots$?" but "What is $1-1+1-\ldots$?", and that this habit of mind led them into unnecessary perplexities and controversies which were often really verbal.[11]

On the one hand, we must surely concede that Hardy has made a substantive contribution to his subject through the new definitions he lays down, yet, at the same time, some subtle hint of unearned superiority wafts through phraseology such as "they had not the habit of definition." "But, Professor Hardy," we may retort, "although you have made great improvements, the rocks on which you stand upon are not radically superior to their's. Your discoveries are just as prone, in the fullness of time, to the winds of happenstance, for the twin afflictions of perplexity and controversy represent permanent fixtures of the human situation." It would be fair, in many respects, to regard Hardy's condescension towards his elders as weakly comparable to the smug manner in which the critic of the previous chapter chides Darwin for failing to appreciate Tennyson.

---

[11] G. H. Hardy, <u>Divergent Series</u> (Oxford: Oxford University Press, 1949), 5–6.

Although my chief concerns in this book will lie with basic predicates of macroscopic physical description, not those of pure mathematics, the basic critical question we should ask is much the same (although its detailed answer may be quite different): what limits should we realistically set upon our human capacities to settle the governing directivities of our predicates? And it seems to me that we must walk a finer line, tinged in a gentle skepticism, than Russell allows, taking care to not tumble into radical sloughs of despond either: we *can* often improve an investigation gone astray with "Let us define our terms properly before we engage in profitless debate," but we can't work miracles thereby.

In fact, Hardy is being unwittingly vague as to exactly what constitutes "setting a definition," a fact to which other writers of his time were more sensitive (these issues will come up again in the next chapter). And subtle elisions of this type mixed with misplaced confidence provides a dandy medium upon which the muddles of *ur*-philosophy happily breed, as I shall begin to document in the next chapter. Pace Russell and Hardy, we have no means at our disposal to prevent conceptual troubles from ever occurring, but we can limit their damages to a considerable degree.

Before I explain what I have in mind here, let me briefly return to another aspect of Russell's picture that was mentioned above: the notion that philosophy should serve as *steward of the conceptual realm*. Although this view (or some variation thereof) remains prominent in academic circles, I will generally confine myself to scattered comments in its regard, for my unhappiness with such opinions can easily be discerned without the reader requiring a constant rat-a-tat-tat from my little drum.

The general shape of the objections I shall offer to classical thinking and its sundry ambitions takes the following form: although we possess a variety of effective methods for tweaking language into better form when it strays off course, any attempt to settle its rails as securely as Russell wishes will generally prove downright foolish, even if the project can be accomplished. Profitable descriptive practice often demands strange strategies that we are unlikely to anticipate in advance and we often need to rely upon Nature's own Delphic but improving guidance to do better. After all, we scarcely want to forgo the road that leads to the castle and the princess in favor of the path that leads to the trolls and the bog, simply because Bertrand Russell forgot to put the former on his map of possibilities. And in the remaining chapters I will argue why this is so, both on the basis of basic considerations drawn from applied mathematics (Chapter 4) and from a direct analysis of the real life sources of *ur*-philosophical mishap (Chapters 6 and after). From these investigations we shall obtain a more guarded appraisal of what is actually possible within the dominion of linguistic improvement.

In the remainder of this chapter, I will mainly discuss classicism's problems from an internal point of view: particularly, the difficulties in fleshing out its contours beyond the bare skeleton presented in the appendix. The purpose of this internal examination is not to proselytize, for I doubt that a single classical mind will be turned thereby, but to gain a warmer impression of how its typical devices of self-protection operate. I also want to comment upon the regrettable tendency, common amongst classicism's most ferocious critics, to seek anti-classical imitations of its most pernicious features.

The main symptom of classical difficulty, from my vantage point, is revealed in its struggles with what I shall call *conceptual overloading*. I will first explain the phenomenon in metaphorical terms and then supply several substantive exemplars in the sections following. I have already conceded that Russell has built a very fine mansion, but it remains an empty shell at present, for we've not attempted to put any furniture in its rooms. When we begin this process—that is, assign concrete allotments of predicative content to specific words—, unhappy tensions begin to emerge: the grand piano in the parlor warps the floorboards, which then cracks the upstairs walls, which ruins Grandma's old settee in the bedroom and so on. Each attempt to arrange a room in shipshape order invariably creates difficulties somewhere else. This phenomenon of gradually escalating disasters (in the mode of the old vaudeville routine, "No News or What Killed the Dog?") represents the *overloading* I have in mind. Its inescapable emergence prevents the house of classical content from serving as a satisfactory domicile; our mansion appears delightful only in the palmy days when we haven't tried to live in the joint. In spite of Russell's assurances otherwise, we must accept conceptual instabilities as the unavoidable inconveniences intrinsic to linguistic life, not simply some docket of minor irritants to be eventually extirpated through a dedicated schedule of home improvements (as classical optimists valiantly assume). And our purpose in this book is to study the structural mechanics that explains why any form of linguistic domicile is apt to behave like this.

All the same, any classical critic of my stripe, who is honest with himself, should sheepishly allow that Russell's original edifice, before we moved the furniture in, represents an exceptionally alluring account of the roles that everyday forms of conceptual adjudication might perform within our intellectual lives—gee, won't it be nice to live in a fine home like that? By comparison, the alternative point of view outlined in this book will seem, to anyone who values sleekness and beauty, ramshackle and sprawling in comparison (representing, perhaps, the philosophical equivalent of the Winchester mansion). But this domestic disorder is not my fault!—it's not I who has rendered the real life behavior of language and its ongoing evaluation so convoluted and shifting.

In fact, a deep reluctance to relinquish the shapely contours of the classical account often spoils the efforts of thinkers who set their caps to dethrone Russell-like thinking: they scramble to reconstitute, by other means, the pleasing uniformity and completeness characteristic of the rejected picture. In particular, the most enticing element within the classical narrative, from which most of its other attractions derive, lies in the *controllable semantic invariant* that "core conceptual content" provides, <u>viz.</u>, the notion that predicates carry with them relatively permanent bundles of directivities which are open to our inspection and modification. Antagonists commonly reject Russell's tale of semantic adhesion as ontologically suspect, yet rarely question the methodological prerogatives that controllable concepts render feasible. For example, many writers influenced by Wittgenstein have urged that the classicism's brute primitive *grasps the concept* **being red** should be replaced by the societal surrogate *grasps the communal standard applicable to the term "red."* Such proposals are usually motivated by a desire to avoid the uncanny grasp of naked universals featured in Russell's thinking as well as alleviating veil of perception concerns. As such, these proposals plainly qualify as *anti-classical* in theme. But such authors invariably leave untouched (insofar as such issues get addressed at all) the key methodological privileges that accrue to the classical picture, for such critics presume that their communal dependency will manifest a controllable invariance comparable to that of its displaced classical rival. I believe that such approaches thereby miss the central locus of classical distortion, which lies in the unsustainable methodological optimism it encourages, rather than ontological excess <u>per se.</u>

This timid inclination to imitate the comforts of classical housing warps even the thinking of a Quine in unfortunate ways, even though he otherwise represents the author who best appreciates, in my opinion, the doctrines that must be relinquished once the assurances of classical gluing are abandoned (to be sure, he frequently runs to extremes in his critiques, but even these usually contain substantial nuggets of probity). Specifically (as we'll discuss in 5,xii), his attempts to explain everyday conceptual evaluation in terms of *mapping to a home language* represent a misguided attempt to incorporate a large degree of Russellian organization within his own schemes. But to classicism's blandishments of tidiness, we should say "no" more firmly; it is exactly our *ur*-philosophical mania for the immaculate that occasions our worse confusions.

Such factors often make the proper classification of anti-classical imitators of classical privilege, if not ontological substance, rather difficult—should they be considered members of an extended "classical tradition" or not? For clarity, I shall generally confine my use of the phrase "classical picture" to the doctrines outlined in the appendix, but I usually expect that my criticisms will reach to their anti-classical fellow travelers as well. These matters are further complicated by the fact that relatively few discussions focus upon the practical issues of rigor <u>et al.</u> central to our studies. Indeed, I consider this lack of comment upon methodological implications to be the most damning feature of the rival anti-classicisms with which I am familiar.

For allied reasons, I decry the current tendency to presume that the problems of concepts or universals can be satisfactorily discussed in terms of *generic examples*; such attitudes reveal a comparable blindness to the fundamental issues with which we should be most concerned. For example, the discussion in Jerry Fodor's <u>Concepts</u>[12] focuses exclusively upon samples such as *being a dog* and *being a doorknob*. For somewhat different reasons, neither specimen is likely to reveal the subtle strains upon "content" that will be highlighted here. To readily observe the seasonal shifts in predicative directivity central to our concerns, we must usually examine descriptive predicates that have become subject to a larger degree of heightened demand upon their performance. The evaluative phrase "is hard" supplies a good example of what I have in mind: quality manufacturing requires that industry press its discriminations of *hardness* evaluations to finer exactitude than we normally require in ordinary life. As this refinement process occurs, the fissures and fine grain symptomatic of anti-classical behavior begin to emerge clearly (this specific example will be discussed in some detail in 6, ix). If *doorknob* displays little evidence of the same textures, it is only because we've never attempted to

---

[12] Jerry Fodor, <u>Concepts: Where Cognitive Science Went Wrong</u> (Oxford: Oxford University Press, 1998).

push its discriminations to comparable standards of accuracy (the exceptional semantic stability of the species predicate "dog" traces to other sources and will be considered at a later point (5,ix)).

Indeed, such omissions in contemporary discussions of concepts explain why I prefer to employ Bertrand Russell as my chief paragon of classicism, rather than some more up-to-date candidate. In his formative era—the latter part of the nineteenth century—, both physics and mathematics had become mired in a morass of subtle but important methodological troubles. Russell serves as an admirable representative of a class of broadly educated thinkers who became drawn to philosophy of language precisely for the help it promises with respect to the authentic dilemmas that arise in these disciplines. And, in this regard, classical methodology appeared, for a considerable span of time, as if it offered a genuine escape route that could liberate Newtonian physics from its clouds of confusion (it is only now, one hundred years later, that we recognize why classical improvement policies do not prove completely satisfactory in this case). What Russell wrote was sometimes sloppy and inconsistent, but he always kept his eye upon the wider world around and, in league with the other intellectual giants of his era, he deserves much praise for his attention to the practical. In our own thinking, we would do well to imitate his example. Later philosophical generations have been inclined to luxuriate in the house that Russell built (or some facsimile thereof) while simultaneously forgetting the earthy problematic that precipitated its construction in the first place. This decoupling from motivating concern often leaves modern philosophical disputes churning in idle disengagement from any behaviors that might suggest something amiss in their appealing lines of thought.

It is important to note that the key ingredients of classical thinking are largely present in earlier writers such as John Locke, having been plucked from the same *ur*-philosophical veins as Russell later excavates. It is merely that the richer set of methodological crises that had emerged by Russell's time renders the practical advantages and disadvantages of classical thinking more readily apparent.

....

Insofar as the concerns emphasized in this book go, appeals to the grasp of *communal standards* offer no improvement over the internalization of conceptual contents favored in orthodox classicism. Indeed, I think only a loss in clarity is the likely result of such a swap. What, after all, are the "communal standards" for employing the predicate "is red" like? The only plausible response, known to all competent speakers, is "declare something to be 'red' only if your community is likely to believe that it is red," which scarcely seems any improvement over the primitive grasp of *redness* favored in orthodox classicism. To be sure, as I'll outline further in 7, x, our real life employment of "red" does demonstrate a fine-grain pattern that is critical to its successful employment. Yet, unlike the doctrines of Austin and his school, I do not believe that most competent speakers ever become aware, even implicitly, of this filagree through their linguistic training; such patterns are rather forced upon us gradually through the silent guiding hand of adaptation to practicality.

....

(v)

Wandering significance. Before I expand further upon the topic of classical overloading, it will be helpful to sketch my own point of view through a simple analogy.
When I was a very young boy, I was fascinated by a cheap early reader entitled Scuffy
the Tugboat and his Adventures down the Big River.[13] In Scuffy (which, like all evocative pieces of juvenile literature, plays deftly upon our neurotic fears of getting lost and
transmogrifying into adulthood), a little toy boat, capable only of navigating the circuit
of a bath tub with its rubber band motor, dreams of "achieving greater things": in this
case, a paddle within some quiet neighborhood brooklet. But even this mild expedition
proves beyond Scuffy's control and our protagonist soon finds himself helplessly swept
into ever mounting torrents, amid lumberjack log rafts and through floods and locks. At
the very last moment, just before he is swept forever out to ocean, his owner prov-
idently rescues Scuffy, having been miraculously able to augur the little boat's likely fate.

The largely unforeseeable directivities that shape our vocabularies to higher standards of adequacy operate much like the natural forces that drive poor Scuffy onward.
True: our rubber-band powers of semantic self-determination play their limited roles
within these histories, but far more powerful will be the interplay between water and
riverbed that pulls our language onward to improvement. It goes without saying that
the directivities that are useful within the bath and brooklet are unlikely to matter much
within the roaring rapids. Nonetheless, the shifting schedule of instructions our predicates will confront connect with one another organically: the specific directions in
which each word currently needs to lean will become apparent at each stage in its long
descent. But these diverse forms of affective influence will enjoy their own seasons and
no persistent classical core will steer our classificatory term completely to its estuary.

Conceptual overloading occurs when we attempt to retell Scuffy's story in a
homogeneous manner, where exactly the same factors are claimed to guide his motions,
whether up and down the river or at home in the bathtub. And then our narrative begins
to turn inconsistent: his rubber power powers are perfectly adequate; no, they're not; he
is carried along in a laminar flow; no, it's developed turbulence, and so on.

Quine, I might observe, favors a picture of linguistic evolution not wholly unlike this
one of mine—it is evident in both his famous discussion of the analytic/synthetic distinction and his frequent citing of Neurath's boat (nautical metaphors naturally occur to
points of view that emphasize evolutionary development). The main divergencies
between Quine and myself concern the natures of the formative currents we expect to
encounter along the rivers of unfolding usage. It is here that Quine makes the mistake of
copying the homogeneity of classical methodological thinking too closely, for he wants
our words to be driven onward largely through adherence to general improving maxims
("Set your affairs in the simplest regimented order," etc.). These policies allow Quine to

---

[13] Gertrude Crampton, Scuffy the Tugboat and his Adventures down the Big River (New York: Random House, 1946).
Illustrated by Tibor Gergely. Gergely also provided the pictures in The Boy's Big Book of Fire Engines, another key
element in the early literary shaping of my psyche.

advertize a schedule of smooth sailing comparable to Russell's, and its prospects for success are no more realistic than his.

Among all the directivities that can potentially buffet words to and fro in their courses, there are certain patterns of guidance that submit themselves more readily to our conscious control and allegiance. If we like, we can fairly easily bring our speech under the discipline of an *algorithm* or an *axiom scheme*: "Add '2' to the numeral you already have." Likewise, we can readily obey instructions of immediate impression: "Label with an 'X' any person who reminds you of Cary Grant." Determined submission to instructions of this ilk might be characterized as *personally imposed directivities*. Through a plethora of methodological strictures of this type, Quine installs a much larger schedule of self-imposed discipline in his portrait of how sound linguistic navigation proceeds than I would consider advisable. By so doing, he brings the practical ramifications of his views into close conformity with classical expectation (whereas *helplessness* stands at the center of my Scuffy metaphor).

As I suggested before, trusting excessively to such controllable fonts of guidance is not especially prudent, for such policies are likely, in the long run, to lead us astray in our dealings with the external world, rather than improving matters. We frequently do better if we entrust language's fate to semantic oracles of Nature's own devising, whose intimations we tease out by experimentally testing the waters as we go. Through obedience to these liberalized fonts of guidance, we generally frame usages of greater practicality, but the strategic underpinnings responsible for their successes will often seem opaque to us in the sequel. Our subsequent attempts to unravel these semantic puzzles typically initiate a new season in the career of a usage, leading to a number of philosophical morals that I shall collectively label as "Oliver Heaviside's lesson" (after the great electrical engineer). But it is best to wait until we turn to substantive examples before I amplify upon these themes further.

(vi)

Overloaded contents. On the story just told, directive guidance of the form "to employ 'P' correctly, consider factor X" should be expected to be seasonal in character, depending upon the place in its evolutionary development that a predicate finds itself. But the essence of the classical picture lies in the presumption that, behind this shifting array of sometimes conflicting advice, there can be isolated a *core of conceptual content* that will stand firm throughout all of the predicate's apparent fluctuations in directivity. This core supplies the essential ingredients that attach the term to the world in a semantically determinant manner and allows us to understand our fellow speakers in a common way. To be sure, in real life we are often sloppy in our conceptual attention and allow our words to drift from one bundle of significance to another, but—and this represents the critical claim of classicism—we needn't do so: by practicing appropriate conceptual firmness, we should be able to hold our predicates to a fixed semantic compass. According to my alternative viewpoint, we possess real but limited control over the wanderings of our words and should not unwisely demand more. Like Scuffy the Tugboat's powers of locomotion, our improving means are fairly meager and we typically exaggerate their real life capacities.

Let us witness the tensions that typically emerge when we attempt to assign particular predicates fixed allocations of classical content. I propose that we examine three particular specimens: "is red," "is a gear wheel" and "is hard." Great philosophical battles have been waged over the proper contents of each of these phrases in the past—disputes that I view as symptomatic of typical conceptual overloading. Later in the book, we shall return to each of these terms, after suitable diagnostic tools have been developed, and develop specific explanations for why natural *seasonalities* generate these various puzzles of overloaded content.

Let us begin with the classical concept of *being red*—viz., the bundle of content that allegedly supplies the predicate "is red" with its central significance. Intuitively, our grasp of this notion seems both immediate and not further decomposable. The nineteenth century scientist/philosopher Ernst Mach expresses this familiar opinion as follows:

> *Brightness, darkness, light and color cannot be* described. *These sensations, experienced by people with normal sight, can only be* named, *that is* designated by means of a *generally recognized arbitrary convention.*[14]

---

[14] Ernst Mach, Principles of Physical Optics (New York: Dover, n.d.), 1.

This same simplicity of grasp is on view in this celebrated passage from John Locke's Essay Concerning Human Understanding:

> But [to] all that are born into the world, being surrounded with bodies that perpetually and diversely affect them, [a] variety of ideas, whether care be taken of [them] or not, are imprinted on the[ir] minds [as] children. Light and colors are busy at hand everywhere, when the eye is but open; sounds and some tangible qualities fail not to solicit their proper senses, and force an entrance to the mind;—but yet, I think, it will be granted easily, that if a child were kept in a place where he never saw any other but black and white till he were a man, he would have no more ideas of scarlet or green, than he that from his childhood never tasted an oyster, or a pine-apple, has of those particular relishes.[15]

That is, absent the prompting of suggestive sensory experience, we will be unlikely to frame the proper contents of **redness** or **tasting like a pineapple,** but, permitted such experience, the concepts will become absorbed without remainder. And there are several aspects of "directivity" under consideration here. To *classify* something as properly "red" or not, we are directed, first of all, to consult the look of it, as long as this represents a feasible activity. And to *understand* more general statements such as "Caesar picked up the red pen," we are told to keep those same classificatory instructions before our minds' eye, even to the point of imagining Caesar as reaching for a pen that strikes us as "red."

Against this popular conception of what the proper content of **being red** is like, consider this objection from the celebrated Helen Keller, who was born both deaf and blind. She protests that she *can* grasp the concept of **redness** despite her sensory limitations and the legions of philosophers and scientists who have proclaimed otherwise. She writes in her autobiography:

> I understand how scarlet can differ from crimson because I know that the smell of an orange is not the smell of a grapefruit. I can also conceive that colors have shades and guess what shades are. In smell and taste there are varieties not broad enough to be fundamental; so I call them shades . . . The force of association drives me to say that white is exalted and pure, green is exuberant, red suggests love or shame or strength. Without the color or its equivalent, life to me would be dark, barren, a vast blackness. Thus through an inner law of completeness my thoughts are not permitted to remain colorless. It strains my mind to separate color and sound from objects. Since my education began I have always had things described to me with their colors and sounds, by one with keen senses and a fine feeling for the significant. Therefore, I habitually think of things as colored and resonant. Habit accounts for part. The soul sense accounts for another part. The brain with its five-sensed construction asserts its right and accounts for the rest. Inclusive of all, the unity of the world demands that color be kept in it whether I have cognizance of it or not. Rather than being shut out, I take my part in it by discussing it, happy in the happiness of those near to me who gaze at the lovely hues of the sunset or rainbow.[16]

---

[15] John Locke, An Essay Concerning Human Understanding, i (New York: Dover, 1959), 125–6.
[16] Helen Keller, The World I Live In (New York: The Century Company, 1908), 105.

Here Keller largely emphases what might be called the *inferential directivities* connected with "is red": she knows that *to be scarlet* precludes *being crimson*; that *being red* suggests "love or shame or strength" and so forth. The customary retort is that Keller's deductive directivities merely represent structural concomitants that are empirically associated with our central concept of *being red*. Their grasp alone is not sufficient for a *proper understanding* of the notion (shortly we shall see how Russell fleshes out this notion of "structural concomitant" in his celebrated theory of descriptions). Keller's grasp of the inferential patterns licenced by *redness* might easily exceed our own if she is better educated in the physics of colorants; nonetheless, orthodox opinion still declares her bereft of the central ingredients required in a proper grasp of *redness*.

Indeed, to even speak of the "ingredients" of *redness* seems misleading, for as Mach emphasizes, the trait seems, in some deep way, *indescribable*: we either grasp the notion in its entirety or we fail to have it all. The attributes inherent in a passage of Mozart are generally viewed as displaying an allied non-decomposability: *expressing sadness musically*, although complex in other senses, still represents a palpable gestalt without ingredients. Our intuitive conviction that many musical and color-oriented traits are *unitary* in this manner plays a central role in generating Chapter 2's various forms of attribute location problem, for *expressing sadness musically* apparently lacks any separate *layers* that can be sprinkled here and there in the world (in Chapter 7, we'll learn that this common *ur*-philosophical conviction is mistaken in important ways).

As we saw, conceiving of *being red* or *expresses sadness musically* in this naïve way is apt to lead us into extreme subjectivism and a veil of perception portrayal of how we obtain information with respect to the external world. To stem this drift, many thinkers object: "No, the core directivities of *being red* also demand that our classifications should conform, in suitable circumstances, with the classificatory opinions of our comrades in linguistic community." The hope is, by installing a dash of conformity to standards of communicative objectivity within the core content of "is red," we can keep our predicate's focus centered upon the classification of objects located in the external world, not redirected towards hypothetical private occurrences encountered only within our individual minds. Few authors of this public persuasion are willing to follow Helen Keller in her championing of our term's inferential directivities, however; she can't classify roses and fire trucks as swiftly and directly as the rest of us.

Let us now turn to the notion of *being a gear wheel*. Once again, this appears to be a notion that we grasp with a good deal of intuitive vigor. In this case, however, the core of its content seems to be wedded more firmly to its attendant *inferential directivities*, rather than to our classificatory capacities in respect to gear-like appearance. Consider the mechanical arrangement illustrated: plainly we can compute the direction in the last wheel will turn given that the driving spur turns counter-clockwise (such queries represent the stuff of which IQ tests are made). If informed that some gear-like component does not behave in the predicted way, we are likely to proclaim that the part "was not acting like a true gear," rather than overturning the usual deductive consequences of *being a gear wheel*.

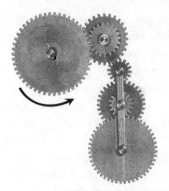

So understood, *being a gear wheel* primarily represents a geometrical classification with expectations of how two contacting bodies will displace one another. Historically, our strong conviction that we robustly grasp notions of this Euclidean class has played an important, and somewhat unfortunate, role in the early development of physics. Specifically, in the era of the mechanical philosophy, any physical classifier that could not be understood in the quasi-geometrical manner of *being a gear wheel* was commonly rejected as occult or inadequately grasped. Robert Boyle expresses this opinion as follows:

> *These principles*—matter, motion (to which rest is related), bigness, shape, posture, order, texture—*being so* simple, clear *and* comprehensive, *are applicable to all the real phenomena of nature, which seem not explicable by any other not consistent with ours. For if recourse be had to an immaterial principle or agent, it may be such a one as is not intelligible; and however it will not enable us to explain the phenomena, because its way of working upon things material would probably be more difficult to be physically made out than a Mechanical account of the phenomena.*[17]

Being a gear wheel's strong set of inferential directivities become central to this accounting of its contents, because it is primarily to our robust sense of *understanding how machinery works* that Boyle appeals. In contrast, René Descartes famously classifies *being red* as a "confused idea" precisely because the notion is inferentially non-productive: to learn that a piece of iron is red tells us far less about its potential effects upon its surroundings than to learn that it is shaped like a rigid gear wheel. True, he allows, the classificatory directivities of *being red* allow us to categorize our private sense data crisply enough, but we can infer little about the behavioral capacities of external things from its manifestations.

In fact, the desire to keep the inferential attachments of *being a gear wheel* and its purely geometrical cousin *being a triangle* integral to their intellectual content led

---

[17] Robert Boyle, "About the Excellency and Grounds of the Mechanical Hypothesis," in Selected Philosophical Papers of Robert Boyle (Indianapolis: Hackett Publishing, 1991), 153. I should mention that I portray Boyle as less tolerant of occult qualities than he was actually willing to be.

Descartes to dismiss their <u>prima facie</u> links to *classificatory directivities* as relatively unimportant. It had been fully recognized since the time of Euclid[18] that proffered proofs of geometrical propositions can be seriously compromised by the misleading appearance of a *diagram* (i.e., the famous "proof" that all triangles possess right angles). Such considerations lead Descartes to opine that our pure and proper grasp of the trait *being a triangle*, as it arises within our "faculty of understanding," is entirely non-imagistic in nature. However, because our intellects are too feeble and sluggish to pursue genuine geometrical thinking with the rapidity that life demands, God has kindly annexed a rude *displays a sensory triangle appearance* concept within a parallel "faculty of the imagination" that will assist our feeble capacities when obeying the genuine directivities of *being a triangle* proves too taxing. A longstanding tradition in geometrical thinking agrees with Descartes on this score. For example, the mid-nineteenth century mathematician Jacob Steiner includes no illustrations in his works on the grounds that:

> [S]tereometric ideas can be correctly comprehended only when they are contemplated purely by the inner power of the imagination, without any means of illustration whatever.[19]

His underlying objective is to avoid mistakes in geometrical reasoning, as well as opening a door to a projective enlargement of geometry's inferential reach, in a manner to which I'll later return (4, i). Needless to say, the predictable decline in pedagogic effectiveness occasioned by such stern policies of conceptual purity soon restored figures to the textbooks.

However, this ascendency of inferential directivities over their classificatory cousins did not remain unchallenged even in the case of *being a gear wheel*, for writers of an empiricist inclination frequently argued that their strong inferential associations are actually peripheral as conceptual ingredients. The deductive directivities should be viewed instead as extraneous associations that have become tacked onto a properly classificatory core through tacit empirical induction. Hume frequently provides arguments to this conclusion. *Being a gear wheel* cannot truly carry the rich inferential consequences that Boyle and Descartes consider as essential to its content. Why? Although we may presume that we can determine <u>a priori</u> how the interlocked wheels of our diagram will move, we are wrong in this assumption. Untutored by the forgotten teachings of previous experience, our contacting wheels might theoretically do *anything*: break into pieces, turn into butter or butterflies. But if such strange events occurred, we wouldn't necessarily withdraw our classification of our wheels as "gear teeth," but might instead report our astonishing discoveries in the form, "Gear wheels turn out to represent an unsuspected chrysalid state of butterflies." If so, then the classical core of *being a gear wheel* must consist largely in recognitional requirements, whereas its Boylean inferential accouterments get taken on board only in the courses of later empirical investigation.

---

[18] W. W. Rouse Ball, <u>Mathematical Recreations and Essays</u> (New York: MacMillan, 1962), ch. 3. Ball indicates that a missing book of Euclid presents such cases as cautionary warnings about hasty reasoning.

[19] Theodor Reye, <u>Lectures on the Geometry of Position</u>, p I, Thomas F. Holgate, trans (New York: MacMillan, 1898), p. xiii.

Considerations of this Humean ilk might be dubbed *Sherlock, Jr. arguments* (after the Buster Keaton picture), for it argues that the cinematic montage of real life experience can be conceivably edited in any wild fashion: an iron-cased state of a gear wheel can be coherently succeeded by a winged condition. Here the Humean assumes that we will still classify objects within each momentary film frame as "gear wheels" or not; hence such notions cannot carry any rich set of inferential associations as part of their invariant content.

A key motivation for denying *being a gear wheel* its usual complement of inferential associations is that, by Hume's time, it was amply recognized that opinions like Boyle's or Descartes' are inimical to progress in science. In particular, Newton's celebrated account of gravitation as an action-at-a-distance force without evident mechanical underpinnings was at first dismissed as an unacceptably "occult" explanation on Boylean grounds (on occasion[20] Newton concedes that his account is, accordingly, provisional; in other moods, he seems more inclined to defend its unsupplemented adequacy). Under Hume's radical diminishment of inferential capacity, all concepts get reduced to a priori impotence and require the supplementation of naked induction to render them deductively robust once again. From this Humean point of view, *gravitational attraction* appears scientifically on all fours with *gear wheel* (this argument should not be regarded as very persuasive, however).

This venerable dispute with respect to the core content of *being a gear wheel* may seem like a quaint antique today, but only because most of us have tacitly imbibed a late Victorian resolution of the problem in terms of *theoretical content*. A rather complex history, originally answering to serious methodological concerns, lies behind this phrase's gradual rise to prominence (we shall reopen those largely forgotten issues in the next chapter because they were resolved, from a conceptual point of view, in a rather blunt and unsatisfactory fashion). Through a subsequent process of being handed from one philosophical generation to the next, the term "theoretical content" has gradually evolved into a device for dismissing delicate issues of content allocation that an author would prefer avoiding, rather than advancing any clearly identifiable positive thesis on its own merits. As a result, billows of obscurant fog immediately envelop important conceptual topics whenever the phrase "theoretical content" is now uttered. I shall return to these matters in section (viii).

The predicate "gear wheel" undeniably displays what we might label, for want of a better phrase, an especially *warm and fuzzy content* in the sense that we intuitively feel that we *understand* the workings of devices that suit its contours vividly, in that same flavor of "Ah ha! Now I've gotten to the bottom of it all" that we express when we draw back the wizard's curtains and discover the gears, cams and rods that have produced the illusion of a great ball floating through the air. Indeed, "theoretical content" first garnered its philosophical prominence by serving as a means of expressing the thesis that science doesn't demand warm and fuzzy qualities within its explanations. In this vein,

---

[20] Isaac Newton, "Letter III for Mr. Bentley" in Isaac Newton's Papers and Letters on Natural Philosophy (Cambridge, Mass.: Harvard University Press, 1958), 302–3.

Ernst Mach and Pierre Duhem, in the company of other prominent scientists in the period 1880–1910, maintained that the progress of physics was still inhibited by allegiance to "warm and fuzzy" demands akin to Boyle's, although the constraints, by this time, had been readjusted to suit Newtonian contours more acceptably (such concerns will be surveyed more concretely later). In their critical instincts, Mach and Duhem were often right: inappropriate constraints on "acceptable conceptual content" did genuinely impede descriptive progress. However, as semantic diagnosticians, they overshot the marks required. "Theoretical content," in its unfortunate suggestiveness, represents the inheritance of these excessive opinions jumbled together with other themes that arose in that same fin de siècle scientific arena. In fact, we shall later discover that the true problems with *gear wheel*'s contents do not lie in the simple fact that conceptual "warmth and fuzziness" should not be required of a scientific trait, but that *gear wheel*'s endearing qualities are genuinely *deceptive*: they trace to directive wellsprings of a nature quite different than we anticipate and conceal, at the same time, serious lacunae in their capacities for complete descriptive coverage. Indeed, a rich set of semantic surprises can be found lurking beneath *gear wheel*'s apparently placid surface. However, these hidden motifs are somewhat subtle in character and a degree of preparation is needed to tease them out adequately. In consequence, we shall find ourselves dealing with the question "What is the conceptual personality of 'gear wheel' really like?" over much of this book's expanse (these prospects may sound dreadfully dull, but—if the assurances of an enthusiast such as I can be trusted—these shrouded surprises are genuinely surprising and will teach us much about the wayward ways of words). However, this conceptual reassessment will largely rumble on in the more technical parts of our discussion (marked with asterisks) and can be thus side-stepped by physics adverse readers.

Finally, let us briefly survey similar disputes that arose as philosophers attempted to credit the predicate "is hard" with a core budget of invariant content.

What is it for a material to be "hard"? Descartes informs us that, like *redness*, the trait merely records a disposition to occasion sensations of resistance within us: *hardness*, properly speaking, represents a quantity that directly classifies our sensations only; the notion's subsequent association with material substances such as diamonds and anvils arises only because of their tendency to arouse appropriate feelings upon contact. In contrast, not wishing to sever our grasp of classificatory notions from the physical world in this veil of perception manner, the Scottish philosopher Thomas Reid rightfully objects that, even if some uniform feeling of "hardness" exists (which is dubious), we grasp the true notion of *hardness* in a manner that is not intrinsically tied to such sensations at all:

> *The firm cohesion of the parts of a body is no more like that sensation by which I perceive it to be hard, than the vibration of a sonorous body is like the sound I hear: nor can I possibly perceive, by my reason, any connection between the one and the other . . . Hardness of bodies is a thing that we conceive as distinctly, and believe as firmly, as any thing in nature. We have no way of coming at this conception and belief, but by means of a certain*

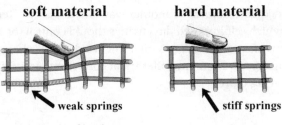

Reid's picture of hardness

sensation of touch, to which hardness hath not the least similitude; nor can we, by any rules of reasoning, infer the one from the other.[21]

Indeed, Webster's[22] informs us that a material is hard if it is "not easily penetrated" and does "not easily yield to pressure;" no propensity to cause sensations is mentioned there. But if we allow our "proper concept of *hardness*" to be purged of its extraneous sensory associations in this manner, haven't we abandoned the palpable directivities of *immediate classification* that most of us follow in learning to employ the term? Reid has placed before us an alternative directivity that we can't readily consult for the purposes of everyday classification, for it supplies a picture of activity on a molecular scale that lies beyond our immediate ken (it is also quite mistaken, but more of that later).

In essence, Reid claims that, although Cartesian, sensation-based directivities may provide the guidance that a child consults in segregating a hard rubber ducky from its softer colleagues, somewhere between six months and sixteen years, English speakers will eventually shift "is hard"'s conceptual attachments over to the externalized classifier that Reid favors. How much freedom to relieve familiar concepts from their everyday recognitional associations should we tolerate? Helen Keller has articulated a portrait of the alleged directivities of *redness* that is quite comparable to that Reid supplies for *hardness*. Can she also claim that, somewhere between 6 months and 16 years, English speakers likewise adjust the contents of "is red" to suit her base trait, although we may fail to recognize the shift? How are we supposed to adjudicate disputes of this nature?

Once again, strategic surprises lurk behind "is hard"'s exterior, but, fortunately, these are less complex than those of "gear wheel" and will be taken up as one of our first substantive examples in 6, ix. I might also mention (as Reid himself points out) that some notion of *perfect hardness* seems critical to the notion of a *rigid body*, which, in turn, serve as the basis of *gear wheel*'s special inferential capacities. There is a very interesting story tied up in these rigid body entanglements, which contributes greatly to the classical mechanics difficulties that we will survey in Chapter 4.

Following my Scuffy the Tugboat picture of language development, I see our allegedly competing directivities as emerging naturally within differing stages of a predicate's evolving career. But the story of why X emerges while Y fades needs to be told in a completely different vein than classical thinking suggests, for the contours of river and riverbed dictate the central dialectic here, not Scuffy's feeble fumblings with

---

[21] Reid, *Essays*, 57–8.   [22] Webster's College Dictionary (New York: Random House, 2001).

his rudder and inadequate motor. Put another way, Boyle, Hume, Reid, Keller and crew all squabble over which self-imposed directivities should control their predicates, when the correct answer is: *none* of them, primarily. The wind blows where it listeth and so, in the main, does language: we can only offer small corrections as we are carried along in its generally improving currents.

(vii)

Core directivities. Before proceeding further, we should take stock of standard terminology in these matters. Although I have largely cobbled along utilizing my self-invented vocabulary of "directivities" and "instructions," the factors that distinguish one trait from another are generally described in the philosophical literature as representing the concept's fund of *intensional features, cognitive characteristics* or *conceptual contents* (thus *being water* differs from *being H₂O* conceptually in that only the latter embodies the intensional feature *being chemically decomposable into hydrogen and oxygen*). I mistrust this standard vocabulary because unquestioned presumptions of semantic invariance seem etched within the very fibers of the terminology itself (especially in the connotations of "content"). Predicates display diverse personalities, to be sure, but they behave rather like human individualities: the features that seem most salient at a fixed time are apt to alter and reveal themselves in ever-changing aspects. In particular, at different stages in a predicate's career, we frequently consult substantially different guideposts as to correct usage than at earlier moments, without supposing that the term's "meaning" or "content" has thereby shifted. Since I wish to keep these facts in view, I prefer my plebeian manner of writing of the directivities pertinent to predicates, rather than trafficking extensively in classically loaded phraseology like "cognitive content" et al. (unless I happen to be characterizing an opposing point of view in the terminology it prefers). Agreed: "directivity" and "personality" sound a bit dopey, while "cognitive content" seems more up-to-date and scientific. But we should not be proud; we do not want to harden fluid aspects of language development into ersatz solidity merely for the sake of elegant phraseology.

There are many delicate issues concealed in the vagaries of "intensional characteristics" that require careful attention, although they are rather hard to explain clearly now. In this section, I will make a preliminary pass, but we'll need to return to these topics later. To begin, the associated directivities of a predicate commonly come in a wide variety of grades, some of which are quite *easy to follow* and some of which border on the totally *opaque*. I supplied a few examples of the easy-to-follow kind when I wrote of the algorithmic "Add '2' to your numeral" and its chums. A standard example of a more opaque instruction is "Add '2' to your numeral if Goldbach's conjecture is true; add '3' otherwise." Here we believe that the content of the instruction is clear enough, but we can't extract any definite *guidance* from it. On the happy day when some prodigy proves or refutes the conjecture, its hidden instructions will be liberated, as it were, but at present they remain tightly bottled up. Even more opaque are the misty intimations of "correctness" upon which we often act but can't explicate to anyone else: "I can't explain why, but I feel pretty sure that this creature should be called an 'elephant'" (directivities of this subterranean stripe will be discussed at length in Chapter 8).

We have noted, in dealing with predicates descriptive of the physical world, that classical thinkers generally wish to locate their core contents somewhere near the opaque end of the directivity spectrum. Thus Thomas Reid provides a portrait of "is hard"'s intensional core that does not provide any immediate help in allowing us to decide whether a piece of plastic properly qualifies as "hard" or not. To decide that, we will need to *scratch, tap or press* upon its exterior, operations that, as we'll discover in 6, ix, can potentially diverge in their evaluations. Reid presumably believes that the portrait of *hardness* he provides can, in a particular set of circumstances, advise us which operation proves most loyal to his central conception. Accordingly, we should be able to sort the directivities applicable to "hardness" use into a central core surrounded by the lesser, satellite considerations that we directly cite in addressing a question such as "Why did you call this block of ebony hard?"

That example highlights easy-to-follow directivities connected with *classification*. Let's now canvass a situation where *inferential directivities* prove most central. Consider a circular drumhead like a conga drum. As will be explained at greater length in 5,vii, its behavior is governed by a hierarchy of hidden traits called its *component modes of vibration*, which indicate how the membrane's complex movements decompose into a group of superimposed simpler movements that wiggle back and forth in the so-called Chladni patterns illustrated.[23] Accordingly, we can introduce a quantity expression, "height of the (0,3) drumhead mode at radial point r," that I'll abbreviate as "(0,3)(r)". But this expression will garner practical utility only if it can be supplied with numerical values through *calculation* ((0,3)(r) is not easily measured because, invariably, there will be other mode vibrations active that obscure the magnitude of our (0,3) mode's individual contribution). Often a physicist will simply obtain these values from a table or a preprogrammed calculator, but a glance in a suitable textbook shows that concrete values for (0,3)(r) are obtained through a somewhat complex layering of covering approximations (specifically, a formula for convergent series $\mathcal{S}$ supplies our numbers close to the drumhead center, but we must switch to an asymptotic formula $\boldsymbol{a}$ towards the rim, as $\mathcal{S}$ fails to provide trustworthy answers there; fairly delicate considerations determine where the crossover juncture between $\mathcal{S}$ and $\boldsymbol{a}$ must occur). In fact, we can't really employ $\mathcal{S}$ and $\boldsymbol{a}$ as they stand—they are *infinite* series, after all—and so their terms must be *truncated* at some point. But even that concession does not provide directivities that we humans can actually follow—we must round off the *real numbers* that appear in our truncations of $\mathcal{S}$ and $\boldsymbol{a}$. In short, a fair number of strata intercede between the easy-to-follow instructions of calculating with rounded off numerals and the physical quantity **(0,3)(r)** itself: the notion to which the predicate "(0,3)(r)" should properly stay loyal. **(0,3)(r)**, we would like to say, embodies the core directivities pertinent to "(0,3)(r)"'s semantic content, whereas $\mathcal{S}$ and $\boldsymbol{a}$ merely report secondary instructions.

---

[23] Fletcher and Rossing, Instruments, 73–5.

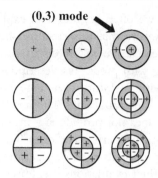

(0,3) mode

Drumhead modes

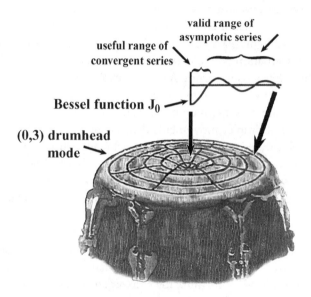

valid range of
asymptotic series

useful range of
convergent series

Bessel function J_0

(0,3) drumhead
mode

However, we should recognize that, if such interpolating directivities cannot be arranged in an intermediate place, then, as a piece of language, "(0,3)(r)" would prove of little value to us (as we'll see in V,10, physical systems possess large hordes of satellite traits, most of which are utterly unmanageable from a linguistic point of view).

Such humble considerations show why *criterial approaches* to meaning—claims that the significance of a term ought to be directly explained in terms of rules for usage—seem so implausible. We want our descriptive vocabulary to prove useful in dealing with the material goods around us, but the manipulative acts that we can readily perform as users of language (simple algorithms; looking up values in a table; classification with a measuring instrument) are unlikely to suit Nature's patterns very well in their own right (the fact that we must switch from formula *$\mathcal{S}$* to *a* provides a nice paradigm of that lack of direct fit). Accordingly, if our usage is to suit the real world's properties, our easy-to-follow directivities must be cut and pasted together according to the strategic dictates of an organizational plan derived from a less transparent directive center such as $(\mathbf{0,3})(\mathbf{r})$. It is for this reason that the classical picture typically views $(\mathbf{0,3})(\mathbf{r})$ as the core content that we *grasp* when we understand a predicate adequately, although, in terms of linguistic practicalities, we must actually *follow* the satellite directivities it spawns.

But how can we determine whether such a central core is really there or not? Perhaps we have tied a disparate bunch of easy-to-follow directivities together, but there's no higher center that genuinely binds them into coherence? We are well aware that cranks often peddle their dubious wares through exploiting the comparative opaqueness of core directivities to their own purposes. In the 1930s, feisty Alfred Lawson pioneered his own branch of physics, which he christened, unsurprisingly, Lawsonomy (at one time several colleges devoted their mission to the promulgation of this craft[24]—a large sign deriving from this era can still be seen along the highway between Milwaukee and Chicago). But in studying his proposals, the concrete directivities of use he suggests for his central conceptions (*zig, zag* and *swirl*) do not hang together by any more evident thread than "Lawson said they did." How do we determine that Lawson has not deluded himself about a conceptual center within the swarm of instructions he has issued?

In fact, cases have certainly arisen within applied mathematics that appear in their externals exactly like our drumhead case, but where the required conceptual center turned out to be non-existent. The layers of satellite directivities we arranged about the predicate "(0,3)(r)" trace to a series of formal manipulations based upon a central differential equation (i.e., assume a solution; assume separation of variables; assume a power series; assume the formula is extendible into the complex realm; assume that its main action occurs at saddle points, etc.). Applied to other differential equations that look superficially like our drumhead specification, every one of these steps is known to fail egregiously when conditions aren't right (the syntactic manipulations themselves are unlikely to complain about being applied to an unworthy equation: "If humans are stupid enough to find this 'reasoning' valuable, let 'em go ahead."). Through blind, formalistic reasoning, mathematicians have occasionally built up elaborate tissues of doctrine comparable to Lawson's corpus, entirely pieced together as a cloud of satellite directivities lacking any central sun. Indeed, there is some small danger that some of our current thinking about chaotic behavior may be based upon misleading computations in this manner, for we presently lack the theoretical assurances we would require to be certain that "there's really a there there."

An awareness that applied mathematics cannot simply provide recipes for computation without further backing but must somehow *underwrite the validity* of the procedures began to be recognized in Euler's era (1750s) and came to full flower in the mid-nineteenth century efforts of Cauchy and Weierstrass (fortunately, our computations for $(\mathbf{0,3})(\mathbf{r})$

---

[24] Henry Lyell, Zig-zag-and-swirl (Iowa City: University of Iowa Press, 1991). Martin Gardner, Fads and Fallacies in the Name of Science (New York: Dover, 1957).

can be rendered justified from this higher perspective). This recognition (which will become central in our later concerns with "pictures" and "soundness proofs") comprises a vital topic with which any adequate story of concepts needs to contend.

Fortunately, we do not need to contend with these ramifications *now*, but only bear them gently in mind as we forge ahead. However, it helps to be prepared for the following eventuality: a particular predicate "P" has adequately established its practical credentials, but our present conception of its directive core has become shaken. Somehow we must find a replacement rationale for threading its satellite standards of correctness together, a process I shall later call "putting a new picture to it." We'll find that such occasions arise fairly frequently in the career of many descriptive predicates.

In any event, tacit claims to have grasped core contents definitively commonly arise in classical thinking. Recall Helen Keller's asseverations that she understands the concept of *being red* as well as you or I (2,v). In the passage cited, she highlights her (possibly superior) command of the inferential directivities native to "is red," while simultaneously minimizing her inability to categorize colors with the naked eye in the usual manner. "Through my skills," she contends, "I approach the conceptual center of *being red* as ably as people of sight. True, I cannot detect a red apple in a sunlit room as swiftly as they, but I can reason about colors better than most sighted people. I scarcely fault their grasp of *hardness* because they cannot adjudicate its values as ably as **I** through touch." Conceptual traditionalists retort that Keller has confused her able management of satellite directivities with a grasp of its central idea: "She doesn't truly grasp the core required in *redness*'s proper apprehension, anymore than coherent concepts genuinely stand behind Lawson's 'zig', 'zag' and 'swirl.' "

From this point of view, how should Boylean complaints that Newton's action-at-a-distance *gravitational force* represents a poorly understood occult notion be addressed? "Oh, it's plain that we do understand that trait adequately," we are likely to respond. But might we demonstrate that we do? For simplicity, let's specialize our discussion to the concept **being solely under the influence of a constant gravitational force**, where we can think of a cannon ball propelled through a frictionless terrestrial atmosphere (I supply extra details in this case, because we shall revert to this example from time to time in our subsequent discussion). If we articulate the intended significance of "constant gravitational force" and "frictionless atmosphere" in mathematical form, Newtonian doctrine instructs us to write down two differential equations (within a convenient set of planar coordinates):

$$m dy^2/dt^2 = -32 \, ft/sec^2 \quad \text{(y is the ball's height above the ground)}$$

$$m dx^2/dt^2 = 0 \, ft/sec^2 \quad \text{(x is the horizontal displacement from the firing point)}$$

These differential equations resemble those implicit in our drumhead case (although their boundary conditions are of a different class) and merely embody the requirement that our cannon ball will, at each moment of its existence, decelerate downward at a $32 \, ft/sec^2$ rate (this is the gravitational aspect) but will not be impeded horizontally (because of the absence of air friction).

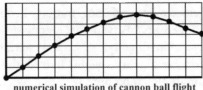

numerical simulation of cannon ball flight

Should the bare fact that we can write down these equations demonstrate that we adequately grasp the core content of *being solely under the influence of a constant gravitational force*? Not obviously, if we can do nothing more with our grasp than that, for Alfred Lawson might claim as much for his "The universe is forever in a condition of zig, zag and swirl" (he can write the claim down, but not put it to any ascertainable use). And now we confront a substantial roadblock, for the most salient and unobliging fact about differential equations is that, from an inferential point of view, they are notoriously hard nuts to crack: they do not relinquish their stored information easily, potentially rendering their practical directivities entirely *opaque*. True; it happens, in the specific case under review, that the basic techniques of freshman calculus can extract (once initial conditions are assigned) a wonderfully detailed answer, but this easy success is misleading: if we modify our equations even slightly (by including a more realistic term for the frictional resistence of the air, say), such techniques will fail us completely and we will be left staring at our modified formulae in mute impotence. As Charles Peirce once observed, differential equations "do not divulge their secrets readily and one cannot charge at them like a knight in armor."[25] Or, like Joel Chandler Harris' tar baby, we can address these refractory formulae in any manner we wish but they won't say nuttin' in return.

Mathematicians inform us that, in cases like these, we can be sure that the equation possesses a solution curve: that is, somewhere in the higher realm of inaccessible meaning the equation (plus initial conditions) inscribes a curve *e* for our projectile to follow. Unfortunately, trapped in the lowly dominion of easy-to-follow directivities, we humans don't yet have much of a clue what this *e* is like. However, there are procedures available that can *approximate e* in a fairly automatic way. In particular, there is a venerable computational technique called *Euler's method of finite differences* that will estimate our cannon ball's instantaneous -32 ft/sec^2 deceleration using an *averaged change of speed* considered over, say, 1/4 second stretches of time (the precise details will be supplied in 3,iv). This routine allows us to calculate a succession of numerical values which, if graphed and connected together by straight lines, generally provides a reasonable broken line facsimile to our cannon ball's path *e*. In this manner, we again witness a sequence of easy-to-follow directivities interposed between ourselves and the less tangible instructions conveyed within the differential equation that inscribes the proper curve *e*.

[25] Charles Saunders Peirce, <u>New Elements of Mathematics</u> (The Hague: Mouton, 1976).

But have we really provided a better demonstration of conceptual understanding in this case than Keller offers for *redness*? Haven't we merely shown that we know how to weave together a mesh of satellite directivities around *impressed gravitational force*, but without articulating the core personality sought? Indeed, this is exactly the complaint that traditionalists made about Newton's approach to gravitation: he fails to provide a truly comprehensible core concept behind "impressed gravitational force" and has only collected together a set of satellite directivities that can be followed in its absence. As we noted, Newton sometimes seems to acquiesce in the justice of this complaint, while defending the indisputable merits of the instrumental assembly he has pieced together. Even more surprisingly, Thomas Reid, who so stoutly segregates the proper content of *hardness* from its ambient indicators, allows that, in gravity's case, the needed core remains as yet unknown despite the good works provided "by the great Newton."[26]

Of course, it would be deeply injurious to scientific progress if we still believed we must continue to search for a more "understandable" core to *impressed gravitational force* in this manner, as if no tempering wisdom with respect to scientific conceptualization has been acquired in the centuries that intervene between ourselves and the Boyle who wrote "About the Excellency and Grounds of the Mechanical Hypothesis" in 1674. "Of course, we grasp Newtonian *impressed gravitational force* fully," most contemporary philosophers of language will avow. "Boyle and Reid adhere to old-fashioned notions of the ingredients required in an adequately understood concept." "But how do we distinguish *gravity*'s case from that of Helen Keller?," we ask. "Oh, that's easy," the answer returns. "*Redness*'s conceptual core involves a strong element of immediate presentation, whereas the content of *impressed gravitational force* is more abstractly *theoretical* in nature."

I find this popular response odd because it appeals to an exculpatory notion of *theoretical content* that, historically, was engendered in a confession that Boyle is essentially *correct* in his observations, but that science, for its own narrow purposes, *needn't care*. In historical fact, notions such as "theoretical content" and "understand the notion adequately through a theory" have come down to us from the late nineteenth century when various scientist/philosophers proposed that adequate "contents" for scientific predicates can be acquired entirely through *implicit definability* within a suitable body of organized doctrine (usually in the form of an axiomatic theory). The original objective of this school was precisely to prevent scientific progress from being retarded by criticism of a Boylean stripe, as well as to set practice on a firmer path of incorruptible rigor. Although appeals to "implicit definability" (which I'll explain in the next section) can be interpreted in a completely Russellian manner, the doctrine was originally intended in a quite anti-classical spirit (with respect to scientific predicates at least), maintaining that a brute capacity to string together easy-to-follow syntactic directivities is all that science truly demands of its parochial predications. Indeed, such minimalist thinking provides the critical background to Jeff Titon's contrast between the intellectual goals allegedly pertinent to "explanation" in contrast to "understanding" (as we'll see in section (x)).

---

[26] Reid, Essays, 272–3.

Accordingly, I find it peculiar that many writers today will glibly appeal to *theoretical content* as if that phrase somehow explains how we manage to grasp *impressed gravitational force* in a fully classical way.

I consider these issues important enough that the first half of the next chapter will be devoted to retracing the history and original intent of "theoretical content" in more detail. This discussion carries us further into the methodology of science than some readers may wish to venture, so let me merely reiterate that, in my opinion, fuzzy, offhanded appeals to the effect that "Oh, the content of that predicate is rather theoretical in nature" serve little evident purpose except to allow the author to evade difficult conceptual issues while fancying that some useful gloss has been offered. No: such writers need to think more carefully about what they imagine "theoretical content" signifies.

To gain a bit of historical perspective on these matters, it is worth looking at the changing fortunes of the basic notion of *energy* in the modern sense (introduced in the mid-nineteenth century as a conserved quantity involving, <u>inter alia</u>, a potential energy component). I doubt that a single prominent figure writing on concepts today would regard this notion as anything other than fully understood. But this opinion was not widely shared during the first fifty years of its usage, where it was widely regarded as paradigmatic of a characteristic known only *structurally*—that is, through its capacity to organize scientific inference in an instrumentally effective pattern (in 4,ii we'll see that one of the motives for late nineteenth century anti-classicism was precisely to argue for its conceptual acceptability). In this vein, consider William James' unshaded comment that *being an atom* or *contains energy* represent concepts that we understand only structurally and not in a more robust way:

> It is only [in terms of practical consequences] that "scientific" ideas, flying as they do beyond common sense, can be said to agree with their realities. It is, as I have already said, as if reality were made of ether, atoms or electrons, but we mustn't think so literally. The term "energy" doesn't even pretend to stand for anything "objective." It is only a way of measuring the surface of phenomena so as to string their changes on a simple formula.[27]

I find it quite striking that James presumes that such matters are known to all, as if no dispute were possible about our understanding of the trait. Why have we so much altered our evaluation of whether a core notion of *energy* is adequately grasped or not? Has some marked increase in our *knowledge* of *energy* occurred in the intervening years which might explain this reversal of opinion? No; such shifts merely indicate that acquaintance increases as the heart grows fonder, rather as Professor Higgins became accustomed to Eliza Doolittle. And such inconstancies in our standards for "grasp" and "fully understand" warn us that we shouldn't allow phrases like "theoretical content" to flit about freely in attempting to understand linguistic process, for they are apt to spread murk even as they pretend to add precision.

[27] William James, "Pragmatism's Conception of Truth" in <u>Essays in Pragmatism</u> (New York: Hafner, 1948), 167.

Closely allied with the notion of core content is another classical doctrine that I'll dub the assumption of a *canonically presented center*. Consider our everyday term "water" and its chemistry companion "H_2O." Many classicists (some alternative points of view will be surveyed in 3,viii) believe that the associated contents of "water" collect together close-to-observation directive elements that allow us to recognize the stuff in a glass; to infer that it will probably quench thirst and so forth. Nonetheless, as students of nature, we possess an abiding interest in uncovering the as yet unknown physical quality that explains why our everyday melange of directive elements holds together—to wit, the chemical quality *being H_2O*. "Here," such thinkers assert, "lie the directivities that Nature herself follows in making this stuff behave as it does. When we manage to grasp *being H_2O* in an intellectual vein, we make ourselves acquainted with these natural driving factors."

Note the swiftness of transition between instructions aimed at *language users* ("'A contains oxygen' can be inferred from 'A contains H_2O'") to *evolutionary principles* that induce *physical behavior*, i.e., causing the stuff to slosh around in a glass or to expand when frozen. If we follow the classical inclination to wed these two different flavors of "instruction" together, we might call Nature's evolutionary principles *physical directivities*. Plainly such assimilation between linguistic and physical "instructions" lies close to the heart of classical gluing, for the essence of the latter lies in the fact that Russellian universals live in two worlds simultaneously: in the realm of our psychological grasp and within the sphere of nature through its activities. By these lights, it seems natural to say that, in learning standard chemistry, we directly grasp the factors that induce glasses of water to behave as they do. If so, we might say that we have apprehended the pertinent physical directivities in a *canonically direct manner* (I will expand upon this locution in the next section). As we'll observe later in the book, doctrines of "natural kinds" generally revolve around some assumption of this general order, although it can assume a myriad of forms (7,vi).

To this day, many philosophers continue to endorse theses of this nature, despite the fact that they threaten to return us to the grip of Boyle-like strictures on understanding. "Yes, canonically direct acquaintance," it will be claimed, "represents the ultimate goal of scientific inquiry, when it can be achieved. But this desideratum is unlikely to call legitimate scientific practices into doubt, because surely we grasp the internal engine lying behind *energy*'s physical capacities in the direct way required." But why do we believe that? "We believe both that we understand the predicate adequately and that it designates a well-defined natural category." But if pressed to demonstrate our "adequate understanding," we roll forth capacities that seem suspiciously of the same character as those Helen Keller provides with respect to *redness*. Many of us believe that Keller's grasp of *redness* is not fully "adequate," but what are the telltale facts that have allowed us to improve our standing with respect to "energy" over hers in relationship to "red"?

In truth, we can turn through endless classical gyres of this type without profit unless we return the discussion of "adequate conceptual grasp" to the realm of practical methodological decision from which such locutions originally spring. And that will be our main project as we move through the book: scrutinizing our everyday evaluative words—"concept," "grasp" and "understanding"—at work within their natural settings of assessment.

(viii)

Relieving conceptual strain. Any philosophical account as resilient as the classical picture will have developed ample methods for lessening the stresses that real life places upon its favored categories. The directivities that attach de facto to the descriptive predicates are often inharmonious or out-and-out contradictory in character. I happen to believe—although substantive illustrations will be delayed until later—that this aspect of usage can often be brought under adequate control without requiring its total extirpation. Indeed, I will eventually argue, from several vantage points, that the purificational purging of affected predicates is neither possible nor desirable in common situations and we must therefore learn to live with predicates of permanently mercurial personality. Such proposals are anathema to the classical picture, of course, and Russell's approach to every problem of rigor requires that any predicate burdened with disharmony should be relieved of its excess freight forthwith. But if I am right, the situation surveyed in the previous section is irremediable—there is no viable way for the classical picture to assign stable contents to a range of familiar predicates (it is thus doomed, on my view, to remain merely a *picture* forever). No stout-hearted classicist will be deterred in her courses simply by pesky complaints such as these. Its venerable traditions have developed a wide range of excuses that explain why our everyday classifications seem laded with overloading. In this section, we shall briefly survey some of the techniques whereby this shedding of excess content is popularly administered.

Let us revisit once again Helen Keller's claim that she adequately grasps the concept of *being red*. Certainly her *understanding* of "is red" has been developed along a considerably different route of acquisition than that pursued by normally sighted folk, but she emphasizes her skills in "red"-oriented inferential manipulation. "Yes, and there's the rub," classical traditionalists will expostulate. "The trait that she truly grasps represents a classification that is centered upon a *structural role*, viz., *being a trait that differs from other qualities found within its common conceptual field in analogy to the relationships that obtain between smelling like an orange and smelling like grapefruit within their parochial field of odor.* As such, this lengthy clause presents a trait that happens to fit or describe *being red* without being identical to it, as is shown by the fact that its provisos probably accommodate *being blue* equally well. And even if Keller were to extend her account to rule out the latter quality (by including every consideration that could be cited in defense of her mastery), it seems likely that, e.g., sensory classifications available only to Martians might still satisfy Keller's conceptual demands quite as ably as *being red*.

In Bertrand Russell's evocative terminology,[28] Keller knows of the trait of *redness* only through a *description*, not by true *acquaintance*, just as we will have learned of

---

[28] William James employs very similar vocabulary: Principles, 216.

Bismarck only by reading narratives of his life in history books, not through direct personal encounter. In an elementary illustration of the same phenomenon, the concept *being of my aunt's favorite color* happens to pick out the same objects as *being red*, but its aunt-oriented conceptual contents seem palpably different from those revealed in a direct grasp of *being red*. According to Russell's famous theory of descriptions, the claim that "a̲ is my aunt's favorite color" should be symbolized logically as "$(\exists\varphi)(\varphi\text{a}\ \&\ (\forall\psi)(C\psi \leftrightarrow \psi = \varphi))$" whereas "a̲ is red" takes the simpler form "Ra." From this point of view, **being of my aunt's favorite color** and the longer characterization championed by Keller merely *describe* the conceptual contents inherent in *being red*, rather than placing these characteristics directly on display. Direct familiarity with these characteristics is possible only if, as Locke insists, they have "forced an entrance to the mind" through sensory channels.

As the roundabout description provided for a predicate becomes longer in a Kellerish manner, Russell often characterizes the trait in question as *structurally delineated*, because the target universal is picked out according to the feats it can accomplish, rather than by what it's like internally. The notion that certain bundles of conceptual delineations pick out their target concepts through structural or theoretical means represents a recurrent theme in classical unloading and a good deal of the remainder of the chapter will be spent exploring some key elaborations upon this theme.

Appeal to an acquaintance/description contrast frequently arises when some predicate needs to be relieved of an overloaded docket of divergent directivities, where the unloading often assumes the form of a distinction between *integral and supplementary characteristics*. For example, in layman's use, "force" plainly contains directive elements that run counter to one another, so part of the task of a mechanical reformer is that of sorting this mess into internally consistent bundles. Indeed, texts in classical mechanics typically lay out a sequence of notions—*force, work, momentum, kinetic energy*, etc.—that correspond roughly, in appropriate contexts, to classifications that get indifferently lumped together as "forces" in vernacular use. Remarks like the following become natural in this context: "The expenditure of effort is properly integral to the proper notion of *work*, not *force*, although sometimes the former is often improperly associated with 'force' through a process of fallible association. But when such directivities are piled together beyond the natural limits of what an integral concept can bear, we get inconsistencies." From a classical point of view, we will likely conclude that "force"'s tangled directivities result from lazy practitioners who have carelessly allowed descriptively associated traits to sneak into *force*'s proper bundle.

It will be helpful to have a slightly simpler example available that we can easily appreciate (many of us still experience trouble keeping *force* adequately distinguished from *work*, after all). Consider the phrase "weighs one pound" which, for the sake of vividness, we shall assume was coined in the merrie days of olde King Arthur. On the surface of the earth, but not elsewhere, the distinct quantities *having a mass of .45 kg* and *being under an impressed gravitational force of 4.4 nts* are pretty much coextensive. But some inattentive keeper of weights and measures back in Camelot allowed the integral directivities of these traits to commingle under the common heading of "weighs one pound." This overloaded predicative package has been passed along from speaker to speaker ever since, causing confusion along the entire twelve hundred years, although Isaac Newton eventually untangled its ill-sorted contents through his keener powers of conceptual discrimination.

In Russell's own labors, he often appeals to his distinction between knowledge by acquaintance and knowledge by description as a tool to radically shear common inferential and classificatory associations from the proper core of familiar words, frequently leaving intact only the directivities of immediate sensory classification ("looking red now"), very much in the general fashion of Hume (if not, argues Russell, "is red" will maintain an inconsistent application in both subjective and objective realms). Notoriously, such efforts at conceptual cleansing drop a formidable epistemological veil between ourselves and the world before us. Even in the writings where Russell accepts physical objects as real (rather than dismissing them as logical fictions built from sense data), he cheerfully allows our everyday classifications of *physical objects* by color to prove indirectly descriptive ("A physical object is red if and only if it possesses the unknown properties that induce red sensations within suitably situated observers"). He likewise agrees with the mechanical traditionalists who assert that we are not genuinely acquainted with the universal directly responsible for the action-at-a-distance *force* that arises between gravitating bodies—that we only possess a *structural description* of how that hidden universal happens to operate.[29] Continuing in the vein of Locke and Hume, Russell further opines that we may permanently lack the conceptual resources required to apprehend such scientific traits directly and we will may be forever sentenced to deal with them only structurally. Thus, our likely relationship to the true attribute behind "force" is confined to the same distanced estrangement that obstructs personal intimacy with Bismarck. On the other hand, Russell will probably allow (although I'm aware of no passage where the issue is discussed) that we *are* genuinely acquainted with the conceptual core of *being a gear wheel*, as is demonstrated by the warm flush of Boylean understanding that washes over us whenever we think of that idea. But here our fine understanding counts for naught, since *gear wheel*'s specifications are never truly exemplified in Nature due to her determination to be composed of fuzzy and floppy stuff instead. By sorting familiar directivites into "acquaintance" and "description" piles in this radical fashion, Russell renders his world of universals internally coherent, although the story he weaves leaves us in a chilly epistemological relationship to the universe that shelters us. However, Russell belonged to a philosophical generation that seemed rather fond of walled off isolation, for some reason or other.

But nothing in the classical picture of concepts per se forces such solipsistic assumptions upon us. Nor do its standard tools for unloading extraneous directivities (e.g., the theory of descriptions) tell us which extraneous characteristics need to be jettisoned. All of these decisions are completely up to us, insofar as the classical picture of concepts is concerned. By apportioning conceptual contents differently, we can potentially allow ourselves to be directly acquainted with a wider swatch of physical

---

[29] Bertrand Russell, "Causal Laws in Physics" in Russell on Metaphysics (London: Routledge, 2003), 189.

characteristics. Thus Thomas Reid can claim that we are acquainted with an externalized *hardness* property. Or, like many modern writers, we can insist that we directly grasp an externalized trait of *being red* (and hold that the internalized quality of philosophical tradition is instead a myth). We have observed that most authorities nowadays would presume without comment that *force* or *energy* is grasped just as firmly as *gear wheel*.

In truth, Russell is not entirely consistent on these issues. Many readers have observed an uneasy tension between the account of the realm of universals as it is sketched within The Problems of Philosophy and those that prevail in other Russellian writings of essentially the same period. In particular, a much greater stress on structurally described traits emerges in the latter, whereas the notion scarcely riffles the pages of Problems. As we noted, the physical relationship of *being in* (in the sense that I *am in* my room) is treated as directly apprehended within the pages of Problems, yet gets recast shortly thereafter as a roundabout structural notion in the Our Knowledge of the External World and The Analysis of Matter.[30] From an *ur*-philosophical point of view, we should prefer the wider democracy of universals sketched in The Problems of Philosophy, where all notions are created alike. Qua citizen of the conceptual realm, we feel that largely classificatory *being red* should be embraced as fully equal, yet not superior, to inferentially robust *being a triangle*. Likewise, *being under an impressed gravitational force* should enter our tolerant kingdom arm in arm with *containing orgone*, in spite of the fact that our stingy universe refuses to supply any instantiations of the latter. By the magnanimous lights of this conceptual tolerance, the scientific notion of *being a top quark* seems no different in kind from everyday *being a table* or *being red*, although fewer people can adequately grasp the former's contents. In our capacities as stewards of the conceptual realm, philosophers should not attempt to segregate one universal from another, in the divisive manner of a Descartes or Boyle. Instead, we should act only to repulse those hazy imposters that claim to represent clear concepts but prove themselves secretively defective in their internal constitution: *being an infinitesimal*, perhaps, or loose appeals to *represents a Principle of Democracy*.

Russell moves away from this even-handed tolerance only because he finds himself forced to do so as he struggles to assign workable conceptual contents to specific predicates in the rounds of his more detailed work on epistemology and scientific rigor. Previously undifferentiated concepts begin to fall into unwanted castes as Russell seeks responses to problems like our puzzle about "force" in the last section: what represents a reasonable demand on adequate grasp for a notion such as this? In our efforts to rid it of unwanted conceptual accretions, the tidied up product begins to look very much like a rarified quantity known only through structural description. Russell's Problems can float loftily above this unpleasantness, treating all concepts with hypocritical magnanimity, only because it confines its discussion to schematics.

Modern writers of a classical bent who write of concepts and attributes generally continue in the eleemosynary manner of Problems and will sometimes condescend, in a

---

[30] Bertrand Russell, The Analysis of Matter (London: Routledge, 2001).

chiding way, to historical efforts to segregate the realm of universals into discriminated grades of acquaintance. As in Problems, they can maintain these charitable attitudes only because they rarely attempt to install classicism's reformatory blueprint upon any muddled practical subject that has called for help in its career. But it is while in the spurs of herding real cattle that the troubles of classical overloading become apparent: only then do our chaps get torn and dirty and our canteens lost in the ravine. This is why I have stressed the fact that classical opinion, as delineated in the appendix, merely represents the *shell* of a doctrine; that we've not really provided an account of conceptual behavior until we actually fill in conceptual contents for specimen words of traditional turmoil: "red," "force" and "hardness." Though the answers that Russell, Boyle, Hume et al. provide on this score are plainly unpalatable, they should not be patronized for the demands of rank they make: confront any real life mess and see if you can do better!

In fact, we moderns do not adequately recognize the degree to which we covertly appeal to another form of conceptual unloading—or something like it—quite frequently without realizing we have done so. Specifically, we evoke that murky phrase "theoretical content" in a loose manner that leaves us with the false illusion that we still inhabit Problem's happy realm of undifferentiated universals. In hard fact, our facile appeals to "theoretical content" probably commit us tacitly to a substantially different doctrine—if we can be forced to flesh out what our exculpatory phrase actually signifies (there are several choices here, all bad insofar as the cause of conceptual democracy is concerned).

Here is an example of what I've got in mind. When we read James' comments on "energy" today, we are inclined to shrug our shoulders and declare, "Sure, the notion of *energy* contains a lot of *theoretical content*, but we've surely come to understand it *adequately through that theory*." If pressed about the import of "adequately through that theory," we may mumble about "implicit definition" or "concepts like that need to be supported by a web of theoretical doctrine." But what do we mean by those appeals?

As noted in the previous section, the notions of "theoretical content" and "understand the notion adequately through a theory" come down to us from the late nineteenth century when specific proposals were advanced to address substantive difficulties in physics and mathematics. These suggestions fell roughly in two classes: those that pursue a Russell-like acquaintance-versus-structural description program for cleaning up the overloaded contents of predicates through conceptual analysis and a superficially similar, yet motivationally quite different, policy that emphasizes the clarifying power of axiomatics instead. This second school is *formalist* in cast: it maintains that scientific predicates can gather adequate conceptual respectability through being embedded in a suitable formal system where the user only needs to understand the rules for manipulating syntax, with no higher form of conceptual grasp being required. As such, the approach rejects many of the characteristic expectations of the classical picture.

Few of us still accept the premise that axiomatics represents the universal cure-all that the formalist faction once believed it to be, but we have been unfortunately persuaded by Quine, Kuhn et al. that "theory or something like it" remains intact to sustain predicates through "implicit definability or something like it," in the diffuse form of a "folk theory," "paradigm" or "web of belief." However, genuine axiom schemes represent concrete items that can be written down on a piece of paper and their articulation, even only partial in its successes, can greatly advance the understanding of conundrums that arise in practice (thus modern work in the axiomatics of continuum mechanics has greatly enlarged our understanding of which classical physics doctrines properly link to one another). But how can we bring one of Quine's or Kuhn's hazy "theories or something like it" to the assistance of conceptual difficulties? I sometimes feel as if many philosophers have cheerfully discarded the curative tonic manufactured by the formalists as worthless, yet still wave the empty bottle around as if it represents a cure for some ill.

More generally, a fair amount of ongoing philosophy is cursed by what I like to call the *theory T syndrome*. As I'll explain more fully in the next chapter, the original intent of the formalists was quite laudable, not only because true axiomatics can help clarify a subject, but because its proponents brought an important strand of anti-classical thought into philosophy (which I'll call *distributed normativity* (4,v)). It so happens that, if the inferential structures of a domain can be organized in axiomatic fashion, then *logical connections* such as <u>modus ponens</u> and universal instantiation can seem as if they represent the *central inferential relationships* within the subject (I regard this point of view as erroneous: even in an axiomatic system, the dominant inferential structures of classical mechanics are closely tied to more specialized forms of reasoning and the particular features of differential equations). This logic-centered focus has occasioned a rather odd historical development. Many philosophers and logicians in the 1920s became convinced that quite general problems in philosophy could be profitably addressed by considering the behaviors of schematic or toy axiomatic systems (which were invariably dubbed T and T', hence my syndrome's label). Within these philosophical circles, it was generally assumed as a matter of course that classical mechanics possesses an adequate axiomatics, even if nobody could tell us either what it was or who might have accomplished the requisite deed (more accurately: a few patently inadequate proposals were sometimes mentioned as proof that the task could be done, without any attempt to weigh the merits of the proposals—see 4,iv). It was also taken for granted that the meanings of scientific words could be adequately explained by the formalist's implicit definability, although, once again, no one ever showed that this thesis was plausible for any real life predicate. This period (approximately 1920–65) represents the heyday of *logical empiricism* properly labeled (although many people still call this same group of individuals *positivists*). Eventually, its popularity faded and the philosophical presumption that the living activities of science—or anything else—could be profitably studied through formalism fell into decline (or worse: the assumption is commonly regarded with great derision today).

To my way of thinking, this history has led to several unhappy resultants. First of all, as I've already stated, axiomatic examination represents an extremely useful probative tool, even if a discipline, in the final analysis, fails to submit completely to its strictures (I consider the popular mockery of the technique misinformed). More importantly, the logic-centered drift from genuine physics over to toy schemata unfortunately directs the philosopher's attention away from mathematical structures to which we should pay the most attention (*variable reduction, equational type, asymptotic solution,* and *boundary condition*) in favor of less revealing logical structures (the logic-only portrait of theory makes philosophers fancy they understand "boundary condition" and "law," although these notions are bollixed up within most philosophical discussions). Thirdly, as the implausibilities of logical empiricist doctrine became apparent, most philosophers of science, encouraged by Quine and Kuhn, decided that language still worked "kind of like theories but not quite so formalized"—e.g., that scientific predicates are somehow buoyed semantically aloft by "paradigms," "webs of belief" or the "practices of a scientific community" (I lump such doctrines together under the heading of *hazy holisms*). This retreat from formalist precision into holist fog made it even more unlikely that the philosopher would find her way back from logic to a consideration of the more substantive inferential structures active within real life mechanical thinking. And thus we have arrived at the worst of worlds in modern philosophical thinking: rather than returning to the workshops of applied mathematics to find out how a discipline like classical mechanics is genuinely structured, we have adopted a murky picture of scientific endeavor that preserves, in a likewise murky fashion, many of the general philosophical conjectures advanced in the name of axiomatics by the logical empiricists. The lingering grip of this unproven nest of logic-centered conceptions I call *the theory T syndrome.* As often happens with diseases of this type, the folks most deeply infected with this loitering blight feel the most certain that they float free of its contagions.

While on this topic, there is a related misconception that merits deflationary comment. In rendering "theories" into schematic T's and T''s, our syndrome puffs the humble word "theory" into something quite grand, without it being exactly clear in what its grandeur consists (it reminds me of the log that was mistaken for a god in Aesop). Mild-mannered "theory," in its vernacular and scientific employments, often connotes little more than "an intriguing proposal," but it serves us well in that lowly capacity. For example, a "mean field theory" in solid state physics represents a suggestion as to how key quantities in the subject might be profitably approximated—that is, the "theory" properly qualifies as a *mathematical guess* that anticipates that the values of relevant physical variables will stay fairly closely to certain easy-to-calculate patterns. Such guesswork presently "belongs to physics" only because mathematicians haven't been able to verify, by their own stricter standards of proof, that the technique actually works (a quite large portion of so-called "physical theorizing" partakes of this "mathematical guess" status). When we prattle philosophically about "theory," however, we commonly imagine that it represents some utterly freewheeling set of doctrines dreamed up by the creativity of man and is then submitted to verification or rejection at the hands of Nature. But this picture can be quite misleading. We don't normally consider that the response "about 10,000" to the question "what is 328 times 316?" qualifies as a *theory,* but the logical status of what are frequently called "theories" in real life physics is approximately that. To be sure, the employment of mean field averaging does represent an "intriguing proposal" and *that* is why we call it a "theory."

Of course, we enjoy patting ourselves on the back by claiming, when we have an interesting suggestion to offer, that we have laid down a "theory" in some grand, if amorphous, sense, for the term carries a more impressive ring than "intriguing proposal." But we shouldn't allow this innocuous self-aggrandizement to transmogrify into the "big ideas" emphasis championed by the moralist of 2,ii. Recall how he disdained Darwin's work on earthworms as small potatoes. "But those are exactly the vineyards in which a 'theory' should labor before we should assign it much credence or cover its perpetrator in glory," we rightfully protest. Alfred Lawson, no doubt, persuaded himself that he had articulated a very fine theory; I suppose our moralist would advise him to rest upon his laurels and turn to a study of Browning. To borrow a second lesson from Aesop, it would truly be better if "theory," our originally modest gauge of accomplishment, could be restored from the pneumatic enormity it has become, after many years of energetic philosophic huffing and puffing.

Let me supply two quick illustrations of appeals to "theory" that I find counterproductive and obscurantist. Consider this episode from recent cognitive science. In learning to employ terms such as "bird" or "triangle," children pass through an initial stage where their classificatory activities seem strongly governed by statistical resemblance to some prototype set. By such standards, the child will accept some wobbly equilateral approximate as a "triangle" more enthusiastically than an extremely pointy yet correct scalene and, in the mode of the Three Men Who Went a-Hunting, unhesitatingly embrace a toad as a defeathered bird. In later developmental stages, this exclusive reliance upon prototypicality lessens and countervailing tendencies appear in the youth's behavior: "Oh, this wiggly thing looks like a triangle, but it really can't be, can it?" And such self-correction is apt to emerge spontaneously, even if the subject's prototypically founded classifications have been universally greeted with untinctured parental approbation. "Oh, the child has now begun to develop a bit of geometrical theory as counterweight," we may be inclined to say. Such descriptions are unexceptionable, I think, as long as we realize that we have merely labeled the phenomenon, rather than having supplied any account for what has transpired. But now consider the

"'theory' theory of concepts" proposed by the child psychologists Alison Gopnik and Andrew Meltzoff:

> The arguments we have advanced so far are really just plausible reasons why cognitive developments in childhood might be much like scientific theory change, in spite of the differences between children and scientists ... Within the philosophy of science, of course, there is much controversy about what theories are and how to characterize them. We have taken the modest and emollient route of focusing on those features of theories that are most generally accepted across many conceptions of science.[31]

In my opinion, the "emollient route" proposed confuses treading water with swimming: to say "the child has begun to develop a theory of birds" is simply to state that her classificatory behavior has changed. Absolutely nothing has been offered that captures how the usage has been concretely affected. And if we try to abstract an unguent "commonality" amongst all of the things properly called "theories" in science, we will come up with nothing better than "having a possibly interesting suggestion to make."

Here the psychologists have been much misled by the philosophers, who frequently chide them for merely offering intriguing observations as to how children learn *bird*, demanding instead that they produce a "general theory of concepts" (too many psychologists, I fear, have been happy to oblige). But such incitements to vacuity or blatant falsehood do not represent wise advice.

This same mythology of theorizing has tricked modern analytic philosophers into quirky methodological habits, that, from any commonsensical point of view, should seem peculiar. In the next section, we shall witness two examples: Christopher Peacocke's presumption that he can acceptably invent "terms of art" for his investigations into human conceptual behavior or Sydney Shoemaker's belief that techniques borrowed from abstract algebra represent a sensible way to approach worldly attributes. "But haven't you wished to talk about real things here?," we expostulate, "How can you simply make up your 'terms of art'?" Such attempts to brusquely barrel through delicate territory by "framing terms of art" stem, I believe, from the misconception that, within any realm, the "theoretician" is allowed to articulate any doctrine she wishes, containing any concepts, no matter how wild, she might dream up, leaving nature the subsequent task of ratifying the concoction or not, according to her caprices. Mimicking this stereotype of theorizing, philosophers, even when they address issues that they regard as entirely a priori, freely engage in methodological gambits that might be appropriate, at best, to investigations within elementary particle physics. In this ersatz vein, Peacocke and Shoemaker fancy they enjoy a liberal freedom to propose any "technical notions" they wish as long as the results "organize our intuitions about concepts" tidily.

However, in developing descriptive predicates that can deal with the *macroscopic* world with any adequacy—not only human behaviors, but simpler affairs such as bars of iron or tubes of toothpaste—, it is heartily unwise to attempt such brute force, "man proposes; Nature disposes" forays, because genuinely useful vocabulary over the macroscopic arena must usually be inched forward into better performance quite

---

[31] Alison Gopnik and Andrew Meltzoff, Words, Thoughts and Theories (Cambridge, Mass.: MIT Press, 1997), 29–33.

cautiously, taking frequent soundings from experiment as we go. This need for methodological circumspection (which should become increasingly apparent as we work through examples) lies in the fact that descriptive vocabulary with respect to complex systems usually require a rather elaborate set of *monitoring controls* to render their employment viable, at least beyond a certain level of refinement. From a strategic point of view, the results can be quite complex: many useful macroscopic classifiers succeed only by gradually erecting a rather complicated webbing of semantic support. Only infrequently can the "postulation" of some scientific genius adequately pave the way for a new macroscopic "term of art," for such pronouncements rarely provide the forms of monitored structure required for success in this realm. And if iron and toothpaste refuse to submit meekly to "theories" in the fashion imagined, how can we reasonably expect that the much greater complexities of human conceptual behavior will yield to such brute force treatments either?

(ix)

Attribute and concept. It is often forgotten that, although Russell maintains that the attribute we truly apprehend at a given historical moment may prove non-canonically descriptive in its contents, he also insists that we are usually more *interested* in the universal not yet grasped that lies behind its descriptive surface. As untutored language learners, we can grasp a conception of *water* readily only if it is framed in terms of palpable qualities of appearance and potability, but, as scientific inquirers, we are searching for the chemical trait responsible for this congeries of characteristics, even though a long quest may be required before *being H_2O's* recondite qualities became manifest (we will eventually discover that some of its instances—heavy water—do not quench thirst). To Russell, the inherent directivities of everyday *being water* and scientific *being H_2O* clearly differ, but our scientific interests will lead us to shift our attention to our chemical Johnny-come-lately as quickly as possible. By such means, Russell can explain why declarations that "*water* simply is H_2O" commonly reverberate in the classrooms of grade school science education, without that fact confounding his conviction that the two notions correspond, strictly speaking, to distinguishable universals. That is, when our pedagogues advance their casual remarks about *water* and *H_2O,* they properly claim, "The interesting chemical property that correlates in ordinary life with the superficial characteristics grouped together under the heading *being water* turns out to be *being H_2O,*" although that literal pronouncement would prove a little long-winded for the third grade mind. In short, they make an assertion that displays the same tacit logical form as "The skilled strategist of whom Kissinger has been thinking turns out to be Bismarck."

Russell's traditional point of view can be reexpressed in terms of the notion of *canonical representation.* There are a wide range of traits (*being water; being H_2O*) that agree in their classifications of everyday materials (in the usual jargon, they share the same *extension*), but are plainly distinct qua universals. Nonetheless, within this group there is often a single trait that most directly codifies the causal characteristics that makes the stuff behave as it does—scientific *being H_2O* seems clearly superior to uninformative *being water* in this respect (and a quantum analysis of the situation might provide a yet deeper level of explanation). Call this optimal choice the *canonical representative* of the entire ensemble. From this point of view, we often seek canonical representatives for salient groupings of superficial physical characteristics and this, according to Russell, is how the *water/H_2O* dialectic should be viewed (indeed, the doctrine captures Descartes' principal intent in considering *redness* a "confused idea" while avoiding the unwanted suggestion of incomplete or ambiguous grasp that "confusion" inadvertently suggests). Insofar as our earlier questions with respect to physical directivities go, Russell can reply that the canonical representative represents the central attribute around which its descriptive associates cluster as behavioral repercussions. This point of view is probably the most common in classical tradition.

Oftentimes, varied points of view with respect to conceptual content are improperly characterized as anti-classical simply because the mollifying role that Russellian appeals to "interests" play within standard classicism gets forgotten. For example, a somewhat different set of methods for relieving conceptual overloading were popularized by Saul Kripke[32] and Hilary Putnam[33] in the 1970's. But their suggestions seem to me a variation upon the classical picture, rather than providing a proper alternative to it. Or, to articulate my assessment more exactly, Kripke's specific proposals represent a mild variant whereas Putnam's opinions are mixed in their intended import. What I have in mind is the following. Both authors observe that two individuals might share all psychological directivities native to "is water" within different environments, yet the predicate itself may find itself attached to distinctly different physical attributes. They then suggest that the true semantic tie that binds predicate to property must be held in place by some form of external *causal relationship*.

Prima facie, this claim sounds like an express rejection of *classical gluing* as defined in 3,ii. However, in an effort to evade counterexamples, complaints about the vagueness of the "causal relationships" cited and an upset of conventional opinions with respect to the unwavering foundations of logic, both writers append a range of supplementary remarks with respect to a linguistic community's satellite intentions, with the net effect that their prodigal doctrine eventually returns to the fully classical fold. The sole surviving divergence in Kripke's case, insofar as I can determine, is that he now considers Russell's "trait of interest" to qualify as the proper reference of "is water," rather than embracing the layman's conception that Russell favors (which Kripke treats as merely a "mode of introduction" intermediary). To be sure, Kripke's alternative approach offers significant ramifications with respect to the analysis of modal claims (which represents his primary philosophical focus), but does not bear heavily upon the issues under

---

[32] Saul Kripke, Naming and Necessity (Cambridge, Mass.: Harvard University Press, 1972).

[33] Hilary Putnam, "The Meaning of 'Meaning'" in Philosophical Papers, ii, (Cambridge: Cambridge University Press, 1975).

discussion here. Thus viewed, Kripke-Putnam doctrine does not properly supply a rejection of classical thinking, but instead represents a reaffirmation of one of its stronger branches (Russell himself had been scared away from strong modal necessities by the criticisms of Ernst Mach and British empiricism, whereas Kripke aims to rehabilitate these discarded essentialisms).

I doubt that Kripke would quibble with this neo-classical assessment. However, a fair appraisal of Putnam's objectives within his 1974 essay "The Meaning of 'Meaning' " is more complex because he articulates his position in a manner that sounds as if he directly intends to challenge classical gluing (e.g., his blunt "Cut the pie any way you like, 'meanings' just ain't in the *head!*"[34]). But these issues matters quickly become confusing because he simultaneously advances a large number of doctrines that do not neatly cohere in any obvious fashion. Furthermore, shortly after "The Meaning of 'Meaning' " was published, Putnam's thought evolved in directions that are certainly anti-classical in character, but decidedly in the pragmatist mode we will survey in Chapter 5. But *those* opinions are incompatible with the realistically founded anti-classicism that many readers (including myself) once discerned in the pages of the 1974 essay (Putnam now rejects that reading as hopelessly steeped in an unacceptable metaphysical realism). I will return to these issues of Putnam interpretation in a moment.

Influenced by other writings[35] of the same author in his early period, many contemporary philosophers have returned to a distinction between *concepts* and *properties* (or, in Putnam's own, less fortunate terminology, "predicates" and "physical properties"). Here the general claim is that concepts represent the panoply of features that we grasp in understanding a specific predicate whereas attributes represent the physical traits that may stand behind several of these (so *being H_2O* might represent the attribute in question whereas *being water* qualifies as a mere concept correspondent to it). So expressed, the concept/attribute distinction can be interpreted as simply a variation upon the classical notion of a *canonical representative* for a family of concepts (and is so understood by writers like David Lewis[36]). As such, Putnam's distinction constitutes a familiar part of classical tradition (allied appeals appear in Locke, for example).

However, an alternative approach to the concept/attribute distinction has emerged that treats an attribute as an *abstract commonality* that lies equally behind an appropriate set of concepts, rather in the manner that the rational number "1/3" represents the commonality shared by all of its fractional representatives "1/3," "2/6," "3/9," .... Such an abstractive commonality point of view may lie latent in the opinions of those authors who believe that the notions of "concept" and "cognitive significance" represent technical notions posited by philosophers to capture an agent's mastery of language and action. Here is a specimen passage with the characteristic flavor I have in

---

[34] Hilary Putnam, "The Meaning of 'Meaning' " in Philosophical Papers, ii, (Cambridge: Cambridge University Press, 1975), 227.

[35] Hilary Putnam, "On Properties" in Philosophical Papers, i, (Cambridge: Cambridge University Press, 1975).

[36] David Lewis, "New Work for the Theory of Universals" in Papers in Metaphysics and Epistemology (Cambridge: Cambridge University Press, 1999).

mind (from Christopher Peacocke):

> [T]he term of art "concept" . . . will be used here . . . [in such a way] that if the thought
> that an object presented in a given way is φ has potentially a different cognitive significance
> from the thought that it is ψ, then φ and ψ are different concepts.[37]

Such proposals usually remark that the "identity conditions" for attributes need to be
"considered from a different point of view." The best sense I can make of these
assertions is that these writers believe that application of an appropriate *equivalence
relation* over their family of *concepts* can articulate a smaller circle of *attributes* that are
candidates to be exemplified within external reality. In other words, a common attribute
hides behind the concepts *being water* and *being H₂O*, but it isn't identical to either of
them. By approaching attributes in this abstract commonality manner, a fairer democracy of attributes emerges that avoids the scientific favoritism characteristic of Russell's
canonical representative opinions. Sydney Shoemaker, in what appears to be an
endorsement of this approach, maintains that the notion "contributes to the causal
powers of things" will carve out a suitable equivalence relationship of this ilk:

> [W]hat makes a property the property it is, what determines its identity, is its potential for
> contributing to the causal powers of the things that have it. This means, among other
> things, that if under all possible circumstances X and Y make the same contribution to the
> causal powers of the things that have them, X and Y are the same property.[38]

Such an abstractive approach plainly robs attributes of many of the thick intensional
characteristics that they display in their direct apprehension <u>qua</u> conceptual presentations, whereas the canonical representative approach leaves these grasped features fully
intact (in truth, I am uncertain whether Shoemaker truly favors this novel approach; like
many authors of an allied persuasion he is largely silent on the critical issues involved).

As we'll observe in Chapter 5, there are ample reasons why we should wish to rid
attributes of the thick layers of directivities credited to concepts in the classical picture.
However, I believe that borrowing the equivalence class technique from mathematics
represents a completely counterintuitive method for reaching this objective. As
I've already indicated, I consider all of these methodological gambits to smack of
pseudo-science.

. . . .

In mathematics, equivalence classes are often evoked to construct new structures from old, as
when Dedekind's ideals are collected together in algebra to obtain a unique factorization domain
from a ring of algebraic numbers. In this setting, the formation of classes serves to induce a
precise behavior upon the new domain based upon the facts about the old domain. To apply this
same technique to attributes merely creates an eerie sense that they comprise some ungraspable

---

[37] Christopher Peacocke, "Color Concepts and Color Experience", in Alex Byrne and David Hilbert, eds., <u>Readings on
Color</u>, i (Cambridge, Mass.: MIT Press, 1997), 51.

[38] Sydney Shoemaker, "Causality and Properties" in <u>Identity, Cause and Mind</u> (Cambridge: Cambridge University
Press, 1984), 212.

I-know-not-what hiding behind the veil of robustly understood concepts. I have similar complaints with respect to the widespread practice of imagining that a well-defined domain of entities can be circumscribed merely by introducing a suitable "criterion of identity."

. . . .

An even more radical approach, favored, <u>inter alia</u>, by Gottlob Frege, maintains that bare *extensions*—that is, the set of objects of which a predicate is true—can adequately serve as the objective *reference* that underlies a circle of allied concepts. The latter serve, in Frege's terminology, as the *senses* or *modes of presentation* that introduce the extensions to us (in his familiar analogy, the concepts *being water* and *being H₂O* resemble the two designations, "Morning Star" and "Evening Star," which both, <u>qua</u> senses, present the planet Venus to us, where that celestial body itself serves as the counterpart to the referential extension shared by *being water* and *being H₂O*). I will discuss the origins of this odd point of view in 5,vi (under the heading of "the thesis of extensionality"). Most contemporary writers (including Christopher Peacocke) modify this sense/reference doctrine so that concepts, considered as evaluators of human understanding, serve as the multiple senses that present a common *attribute* such as *being H₂O* to us as reference. This revision of Frege returns us to essentially a canonical representative point of view.

All of these proposals should be regarded as attempts to relieve the strains inherent in orthodox classical thinking with respect to conceptual contents. In particular, allegiance to an excessively thick notion of attribute makes the rationalization of standard definitional practice in science quite difficult: why should physicists be allowed to define, as they do on different occasions, "total force" as both "mass times acceleration" ($\mathbf{ma}$) and "the negative of the derivative of the applied potential" ($-\partial V/\partial x$)? Plainly, these two notions differ greatly in their cognitive significance? Or why do our grade school instructors embrace the apparent identification "*water* = H_2O"? We have already surveyed the roundabout, theory of descriptions rationalization that Russell provides for these practices, but by loading attributes themselves with less internal baggage, many philosophers have hoped that Russell's implausible stories can be evaded (some of Frege's motivation for his sense/reference distinction traces to allied worries with respect to definitional practice in mathematics). I supply a few more details on these issues in the appendix.

My own approach to these issues maintains that a reasonable notion of *attribute* (or, often preferably, *quantity*) can be defended as an appropriate sort of informational package into which the data required to characterize a physical system's potential behavior can be conveniently decomposed (I express myself rather abstractly here, because other forms of informational decomposition often prove viable and Nature seems disinclined to show any favoritism with respect to these issues of format). I'll discuss the basic issues pertinent to *attributes* more fully in 5,vi. In respect to *concepts*, on the other hand, we should resist any impulse to regard them as cognitively affective "senses," "modes of presentation" or anything else of an intervening content ilk. Indeed, the wisest policy, in my opinion, is to resist the impulse to consider "concepts" as well-defined entities at all, and instead confine our attention to the shifting manners in which our everyday standards of *conceptual evaluation* operate over the lifetime of an evolving predicate (I believe that "concept" represents a term like "Napoleon's personality"—it manifests a certain continuity over time but doesn't stay precisely fixed). We must guard against our *ur*-philosophical predilections to espy a hazy *invariance* within these evolving opinions, rather than appreciating the *natural alteration of standards* that actually emerges.

None of this denies that we must diagnose the origins of the impertinent personalities that predicates manifest over time; it merely asks that we not describe their atmospherics according to classical schemes. Instead, in trying to adjudicate the conceptual personality of a specimen predicate such as "is a gear wheel," we should draw up an inventory of the *physical information* that is captured when such vocabulary is fruitfully employed, for our first task is to map out the physical environment in which the usage achieves its practical objectives. But this is not to assume that any of the physical attributes involved in that information will map onto the term "gear wheel" in any regular or fixed way—in fact, "gear wheel" doesn't correlate neatly with any genuine physical grouping. But there are other aspects of those physical settings that explain why "gear wheel" presents the directivities it does to its employers—why it enjoys its distinctive and special personality (including Boyle's characteristic of *warm and fuzzy understandability*). In this specific case, the true source of this overall personality is rather surprising in its origins, because the component directivities we follow in using "gear wheel" correctly derive, in large part, from certain *effective algorithms for that machine design*: the reasoning rules that, in an appropriate environment, can devise an invention able to accomplish a preset task (details will be provided in 7,iv). But these formative factors behind *gear wheel*'s familiar conceptual personality scarcely present themselves to us in a classical manner: few of us *grasp* these algorithmic underpinnings in Russell's sense at all, although the manner in which we employ the predicate is tacitly shaped by these design-oriented directivities all the same. They quietly carve out the long sweep of "gear wheel" 's developmental career, rather as the great river carries Scuffy down to sea.

It is worth mentioning in this context that there is a branch of biology called *biomechanics* that pays special attention to the manners in which the physical demands of an environment interact with the abilities of the creatures who live in its midst.[39] Often the largest part of the problem in understanding an animal's behavior lies in appraising the physical constraints that present themselves to the organism, as well as gauging the strategies potentially available for accomplishing the animal's goals within these circumstances. In my view, our efforts at linguistic description confront a similarly complex arena of opportunity and effective strategy within the macroscopic realm. Often the most pungent aspects of a predicate's personality stem from the manner in which physical circumstance and linguistic opportunity have managed to reach accommodation, often making our investigations of predicate behavior rather similar in character to

---

[39] Stephen Vogel, Comparative Biomechanics (Princeton: Princeton University Press, 2003).

those familiar in biological studies of environmental opportunity. Lying along the interface between linguistic capacity and physical fact, I sometimes call these considerations *interfacial influences* in the sequel. They are not the only factors that supply a predicate with the complete personality it displays, but they are very important and have not been studied much in philosophy.

In these respects, my quasi-biomechanical recipes for unraveling the intensional characteristics of predicates are distinctly "externalist" or "naturalist" in flavor (although I do not care for either of these popular phrases much). An allied externalist orientation seems evident in the 1974 Putnam essay mentioned earlier (although this reading may not have represented his true intent). Indeed, I was a student of Putnam's in the relevant period and many of my musings can be fairly credited to (or blamed upon) the vital spark of anti-classicism that I derived from his teachings, as well as the mode of straightforward scientific realism that his essays of the same period seemed to embrace (he has subsequently denied that this realistic stance represented his fully considered point of view). Unlike the Putnam of 1974, however, I do not embrace the supplementary mechanisms of original intention (e.g., "I hereby baptize this liquid, whatever else it is, as 'water'") that Putnam includes in order to insure that predicates such as "is water" maintain invariant extensions over their extended careers (Putnam worries that, if such provisos are not guaranteed, "logic will fall apart"—see 10,v). I reject these doctrines because they seem descriptively inaccurate and inconsistent with fundamental tenets of a reasonable anti-classicism ("liquid," after all, behaves even more irregularly in its predicative fixity than "water"). In any case, the supportive fabric of facade I shall defend displays rather different characteristics than any scheme that Putnam contemplates. However, I remain deeply indebted to those early essays of his. I will return to some of these issues in 7,vi.

(x)

Explanation and understanding. Let me append a few concluding comments on issues that have been left dangling. Recall the contrast Jeff Titon draws when he compares (2,v):

> *two kinds of knowledge: explanation and understanding . . . Explanation is typical in the sciences, and understanding typifies knowledge in the humanities.*

Here is a more expansive expression of this same theme from Ernst Cassirer:

> [There is] *a type of apprehension that is contrary to theoretical, discursive thinking. For, as the latter tends towards expansion, implication and systematic connection, the former tends towards concentration, telescoping. In discursive thought, the particular phenomenon is related to the whole pattern of being and process; with ever-tightening, ever more elaborate bonds it is held to that totality. In [the other] conception, however, things are not*

taken for what they mean indirectly, but for their immediate appearance; they are taken as
pure presentations, and embodied in the imagination.[40]

In this appeal to "expansion, implication and systematic connection," Cassirer makes tacit assumptions about the holistic nature of "theoretical, discursive thinking" that are analogous to Russell's vision of a direct acquaintance/structural description divide or the views of theoretical content of which I've complained. All of these opinions are predicated upon the assumption that science is only interested in certain limited aspects of the natural world and hence frames its favored concepts in quite special ways. "This is thought's original sin, its inertia and line of least resistance," complains Ralph Barton Perry, who continues:

Just how do bodies fall and move? This is the question which for scientific purposes must be
answered; and only such answers have been incorporated into the growing body of
scientific knowledge. Who or what moves bodies, in the sense of agency or potency, is for
scientific purposes a negligible question; attempts to answer it have been, in the course
of the development of science, not disproved, but disregarded.[41]

It is Perry's belief that other forms of human conceptual endeavor are not so narrowly constrained; similar sentiments were already voiced by S. T. Coleridge. Here is a recent variation upon the same theme, a complaint by Jennifer Hornsby that philosophical reductionists falsely presume that:

any real phenomenon, however we may actually understand it, is intelligible from the
"objective, third personal perspective" that natural scientists adopt[42]

(but is *this* what a cosmologist does, we might parenthetically inquire, when she adopts a descriptive frame that moves with the observer?).

The true harms occasioned by sweeping proclamations such as these lie in their tacit encouragement of the neo-classical conceit that we can simply *peer inside* the predicates of, e.g., physics and recognize their limited contours of construction and intent. And it is precisely with respect to these self-anointed powers of a priori internal discernment that this book will be most critical. On the contrary, the words within any domain are apt to adopt impertinent individualities of largely their own choosing and behave in rambunctious ways we are unlikely to anticipate in tidy philosophical schemes.

It is common for writings of a flavor such as mine to be dismissed as "scientistic" by their "humanist" critics. I have never understood clearly in what the sin of scientism consists, unless it merely connotes an eagerness to talk about scientific fact beyond tasteful limits. But, truly, my purpose here is not to establish that "all concepts act like scientific ones"—whatever that fuzzy contention might mean—but simply to lessen the

---

[40] Ernst Cassirer, Language and Myth, Susanne K. Langer, trans. (New York: Harper and Brothers, 1946), 56. Where I have substituted "the other conception," Cassirer has "mythic conception"; he would accept a thesis of broader generalization however.

[41] Ralph Barton Perry, Present Philosophical Tendencies (New York: Longmans, Green and Co., 1921), 50, 54.

[42] Jennifer Hornsby, Simple Mindedness (Cambridge, Mass.: Harvard University Press, 2001), p. 5.

deep layers of methodological stereotype that prevent us from appreciating the varied forms of strategic engine that commonly propel *all* of our terms of macroscopic classification, whether they come extracted from science or everyday life. No "general theory of concepts" is attempted here (the only universal truth that might be fairly extracted from this book is "Words sometimes do awfully funny things"). Sweeping dichotomies of explanation/understanding contrast are more likely to hinder our abilities to appreciate the idiosyncratic patterns of predicate development than "scientism."

In any event, recall William James' claim that *contains energy* merely represents "a way of measuring the surface of phenomena so as to string their changes on a simple formula." Here he gestures towards the same divergence in *intuitive understanding* to which Russell appeals when he distinguishes between acquaintance and description: the *redness* of a sunset or the *expressing sadness musically* of an orchestral passage seems more vividly grasped than dry *contains energy*. And this conceptual aridness arises for a good reason, authors of this persuasion contend, because *contains energy* has purposefully allowed "its affective qualities to droop," to paraphrase Wordsworth, because that procedure allows science to entwine its denatured qualities in great webs of theory. Accordingly, this is why science even *likes* its central concepts to be structural in nature, for such abstractness allows dissimilar particularities to become linked together in systematic webbing ("constructing the causal nexus" is the old-fashioned term for all this organizational bustle; "building an all-inclusive physical theory" represents a more up-to-date rendering). It is these organizational ambitions that Cassirer has in mind under the heading of "expansion, implication and systematic connection." On this portrayal, it is not surprising that the warmer particularities of *being red* or *expressing sadness musically* fall by the wayside as unassimilable to architectonics. Although the vivid contents of our spurned qualities will not assist science in its contrivances, they can nonetheless supply a rich banquet of *internal relationships* upon which the artist can sup. Consider the relationships with which we must deal in graphic design: does a color seem "warm" or "cool"?; do two shades clash or complement?; does one patch induce spurious tints in another?, etc. None of these qualities or comparisons will assist the physicist much, busy as she is with the photons. But the artist or musician should care, because their mastery arises from the manner in which the internal aspects of such traits are brought together (think of the subtle forms of color harmony in which Turner trafficked).

How did such a strange story come to be so widely believed? On the one hand, its roots lie deeply posted in our *ur*-philosophical assumptions as to "what notions we understand best" and, on the other, because the scientists of the time told them so! (Perry, who was a student of William James, cites both Ernst Mach and Karl Pearson as authorities). But why would they do that? The sundry misapprehensions here entangled with "best understanding" will require the full span of the essay to address, but the odd opinions of the physicists provide the opening topic of our next chapter.

. . . .

Sometimes the poet J. W. Goethe's celebrated views on color theory and the morphological similarities of plants are cited as models of internal enterprises alternative to science's structural projects (although Goethe himself regards his endeavors as "scientific"). On this view, our deepest insights into an art form or the nature of a plant can be expressed in the form of a direct (and rather mystical) discernment of a veiled *archetype* plainly present in all of its particularized manifestations. But this knowledge should be regarded as a direct grasp of a particularized unity interior to the subject studied, not the alignment of the plant or art work under some artificially external structural quality at all. The artistic genius can discern these relationships through "concentration and telescoping" without worrying in the least as to how any of the business situates itself within the scaffolding of the causal nexus. Goethe writes:

> For though nature has the better of man, seeming to keep many secrets from him, he has an advantage of his own in that his thoughts may soar beyond nature while not fully comprehending her. We go far enough when we come to the archetypal phenomena, seeing them face to face in their unknowable glory and then turning back to the world of other phenomena. The incomprehensible, in its simplicity, manifests itself in thousands of variations, unchanged despite its inconstancy.[43]

It seems to me probable that Wittgenstein's celebrated (albeit elusive) proclamations with respect to the special mission of philosophy owe much to Goethe (whom he often cites). In his Philosophical Investigations we encounter much disdain for causal investigations "which take our inquiry on a different track" and a preference for aligning linguistic phenomena side by side in approved Goethean manner:

> [We should seek] to trace phenomena to their sources, to the point where they appear and exist, beyond which nothing further can be explained . . . Don't try to look beyond the phenomena. They are themselves the theory.[44]

I find these themes rather surprising, given his dismissal of the value of inner demonstrations in other aspects of his work.

I mention this Goethian variation upon classical grasp because, as indicated in the Preface, I am quite uncertain whether genuine affinities link my own patterns of thinking to those of Wittgenstein. It is precisely passages such as these that I find most alien.

. . . .

Appendix: Chief Theses of the Classical Framework

(1) Concepts evaluate *commonalities* in behavior that persist between objects such as the *redness* shared by a fire truck and a stop sign. *Relationships* between objects also qualify as a species of concept as well. An object is said to *exemplify* the trait if it obeys its dictates.

(2) Concepts can also capture the *mental content* of someone who entertains the appropriate ideas, as when John correctly "grasps" the concept *being red*. This claim, in conjunction with (1), indicates that concepts can plant their feet in two different

---

[43] Rudolf Magnus, Goethe as Scientist (New York: Collier, 1961), 178. J. W. Goethe, Goethe's Botanical Writings (Woodbridge, Conn.: Oxbow Press, 1989).

[44] Magnus, Goethe 168. As for Wittgenstein, I find these themes particularly pronounced in his "Remarks on Frazer's Golden Bough" in Philosophical Occasions 1912–1951 (Indianapolis: Hackett, 1993).

worlds—they simultaneously serve to *evaluate* conditions within the external world and our internal state of mental preparedness. (3,ii)

(3) Many concepts display themselves most simply by appearing *associated with linguistic predicates* as their meaning, although concepts can align themselves with other parts of language as well. Novel concepts can, furthermore, be grasped sometimes without prior linguistic handles, although opinions differ widely on the extent to which this process occurs.

(4) The conventional association of predicates with concepts provides a linkage that allows language to attach itself meaningfully to the world and to speak about objects located in faraway places and times. Conceptual intermediaries thus form the prime "glue" that ties words to the world. (3,ii)

(5) Most speakers fully obtain the concepts associated with the common predicates of their native tongue by age 10 or so—conceptual grasp becomes complete and stable after this period. They also learn rules for forming new conceptual derivatives from base concepts, e.g., *being a fake ruby* from *being a ruby*. This thesis is dubbed *semantic finality* in 1,vi.

(6) Attributions of conceptual grasp evaluate only the level of *conscious understanding* achieved by a speaker; they make no express representation as to the hidden brain mechanisms, environmental conditions or other factors that might be required before a speaker can actually manifest mastery of the concept. Accordingly, the full content of a completely understood concept displays itself in full vividness to its employers (I call this a *presentational* view of concepts in 6,iii).

(7) Concepts also codify or evaluate the key ingredients involved in *understanding* and *communication* between speakers. To comprehend one another fully, we must grasp the same concepts and bring them to mind appropriately. Attributions of common concepts also play a large role in determining whether two speakers share the same content in their beliefs.

(8) Translation between the predicates of two foreign tongues is largely a matter of locating expressions that share the same associated concepts insofar as this proves possible. Evaluation of the purpose of many endeavors, e.g., what the alchemists were trying to do with respect to the element mercury, is subject to similar provision. (10,vii)

(9) Due to the speaker independence displayed by concepts according to the above themes, they are best regarded as entities other than ourselves that we can sometimes grasp mentally. We often share concepts and these evaluations of commonality form the core of everyday "folk" or belief/desire psychology: the alleged framework of explanation that allows us to explain Alfred's plucking a peach in terms of his grasp of the notion of *eating a peach* and his desire to see that state realized.

(10) Concepts undoubtedly exist that we will never grasp, because they never occur to anyone or they exceed the capacities of the human mind to understand. Individuals of great discernment will sometimes grasp novel concepts that have heretofore eluded everyone else. (8,ii)

(11) A well-defined totality or *domain* of all possible concepts exists, even if humans have access to only a small part of it. This collection is what Frege intends by his "Third

Realm" and Russell by his "World of Universals." Their *commonality of type* allows all concepts to be treated in a homogenous fashion, giving rise to the assumption that a general logic of concept formation is possible. Accordingly, philosophical logicians can profitably investigate how logical operations and other a priori means of manufacture manage to construct new concepts through uniform rules. Such enterprises are plausible owing largely to the presentational content thesis (6), which claims that the basic ingredients of concept formation can be decoupled from whatever complications sub-conscious mechanisms supply.

(12) Concepts of attributes unrealized in our favoured physical theories also exist and should, if self-consistent, be treated on an equal footing with our own in their role as concepts. From a conceptual point of view, *being a quark* and *being phlogiston* enjoy coequal status; it is merely empirical happenstance that favors the former over the latter. (5,i)

(13) Concepts can be *manipulated* and combined into further concepts, which is the only explanation of how we manage to understand the indefinitely large collection of English predicates we can construct carrying palpably distinct meanings. This point is often described as the *creativity of language*. (1,vi)

(14) Indeed, conceptual *rules* must exist that explain how these constructions regu-larly build new notions. These rules probably can be codified into formats such as: if concepts Φ and Ψ exist, then the constructed concept $\varphi \& \Psi$ will hold of an object if and only if the component concepts Φ and Ψ both do. Such rules capture the *closure prin-ciples* integral to the realm of concepts. (10,iv)

(15) Concepts, by virtue of their internal content, stand among themselves in various relationships of *inclusion* and *exclusion*; it is this fact that allows us to grasp relations of synonymy and entailment betwixt linguistic predicates. These same contents also give rise to the many intuitions we possess about what can be appropriately attributed to a given concept or not. The primary duty of philosophy is to remain loyal to the data supplied within this fund of intuitions. (5,viii)

(16) Concepts often *emerge into consciousness* suddenly and unexpectedly. The phe-nomenology of many concepts is that they are grasped as *integral wholes*. (8,iii)

(17) Nonetheless, we retain a power to extract and adjudicate the contents of (16)'s semantic epiphanies, in the sense of being able to accurately delineate their internal relationships to other concepts we possess. For example, Einstein may suddenly discern a new, four-dimensional conception of *relativistic momentum*, but he will be also cap-able, upon sufficient reflection, of determining its sundry similarities to, and differences from, the older *Newtonian momentum*. Often this work of *conceptual analysis* proves arduous, given the many psychological obstacles that impede its progress, but, in principle, a careful thinker will always be able to discern the proper framework of conceptual connection accurately. (8,v)

(18) Concepts embody *rules* to *guide* thought, whether they represent instructions as to the proper classification and recognition of objects, salient inferential consequences or even provide the framework structure of a novel. Such guiding rails I often dub *directivities* in the text; other authors call them conceptual norms. (3,iii)

(19) The *intensional characteristics* of a concept provide the aspects of conceptual personality that distinguish one concept from another, even if the two notions hold of exactly the same real world items. Thus *being water* and *being H₂O* represent concepts true of exactly the same bundles of stuff, but the latter incorporates suppositions into its internal character that are absent in the former, e.g., that anything that is H_2O bears an integral relationship to its potential hydrogen and oxygen components. A speaker hasn't grasped the concept of *being H₂O* properly unless she recognizes this connection, whereas this demand obviously cannot be required of everyday *being water*. (3,viii)

(20) The coherence of the belief/desire psychology mentioned under (9) depends critically upon these intensional characteristics, for clearly the aspiration to own a pet unicorn is quite different from the hope of owning a pet troll, although there are no objects anywhere in our universe past, present and future that allow us to distinguish these ambitions. But clearly psychology needs concepts that can be on the lookout for such non-existent objects in different ways, a fact stressed by the psychologist/philosopher Franz Brentano. These considerations explain why the term "intensional characteristic" is often adopted as a synonym for "cognitive significance."

(21) Such characteristics fall into assignable grades: *simple* or *complex* (*being red* versus *being red or green*); *evaluative* or *norm neutral* (*being a good knife* versus *being a sharp knife*); *third person objective* versus *subjective* (*having a mass of 1000 kg* versus *regarded as heavy by Susie*); *intrinsic* versus *relational* (*having a mass of 1000 kg* versus *being hard to move*), etc. It is usually presumed that a major task of a theory of concepts is to bring some order into this melange of grades, but there is little shared agreement as to how this project should be fulfilled in detail. For this reason, no specific claims about conceptual contents appear in this outline of classicism, although the doctrine can only be regarded as a skeletal framework until such discriminations—and their rationale—are supplied.

In my estimation, these disagreements stem from the fact that specific contents cannot be inserted into the classical framework stably, a behavior I call *classical overloading*. (3,vi)

(22) It is common to distinguish between concepts that present their contents *directly* and those that merely delineate a *structural relationship* (known only by description, according to Russell). Examples are usually controversial but the apparent contrast between the direct *having a mass of 1000 kg* and the structural *representing a constant that governs a particle's response to imposed forces* illustrates the intended distinction. (3,vii) In a directly apprehended concept, the contents that capture the attribute's modus operandi lie clearly in view, whereas, in structural cases, our relationship to these same ingredients prove more distanced. It is frequently claimed (e.g., by Russell during certain phases of his career) that we never gain better than structural knowledge of many scientific traits. (3,vii)

(23) The *identity conditions* for concepts stem from their intensional characteristics: they must be the same for two concepts to be equals. (5,vii)

(24) We possess a capacity to bring concepts before our mind, to *evaluate and weigh* their applicability *critically*. In this capability, we may prove superior to animals, who

can perhaps classify the objects before them ably, but are unable to ponder whether their concepts suggest some measure of internal improvement. (8,iv)

(25) Careful attention to conceptual content is the path to achieving *clearer thinking*. We possess an ability to recognize, upon diligent reflection, whether the boundaries of a concept have been clearly delineated or not. Conceptual reflection, for example, should tell us that our usual notion of *being bald* lacks clear contours. We can either decide to plug these gaps by adding supplementary conditions or, if it seems preferable to allow the underpinnings of a word to remain partially unfixed, to reason with the term following rules that reflect those lapses. But any deficient concept can always be improved into one that is fully determinate. (8,iv)

(26) The applicability of *basic inferential principles* should stem from the internal characteristics of the concepts involved—we should expect to reason about number concepts differently than notions of color. The *soundness* of an inferential rule should be checked by insuring that in no possible relevant circumstance will the rule's application lead from a true description of a state of affairs into a false claim. In other words, the soundness of the basic rules pertinent to a predicate should be derivable from its conceptual content. Of course, once the basic parameters of how to reason with a term have been established, we can later learn many further supplements, e.g., that *being a fire engine* commonly signalizes an instance of *being red*. The validity of this last variety of inferential connection is purely empirical and is not founded in the internal characters of the two concepts involved, whereas the allied tie between *being a ruby* and *being red* probably is. (10,v)

(27) *Errors in thought* often occur when syntax is blindly manipulated by formal rules without proper regard to their support in underlying concepts. Such mistakes have often occurred in the history of mathematics, the sciences and philosophy, but they can always be avoided by insuring that the true contents of our claims are kept in view. Likewise, in language use, the meanings of various words often drift or multiply into secret polysemy without our noticing the alterations, but such meanderings could have been prevented by a more vigilant program of *conceptual hygiene*. (8,iv)

(28) Concepts are intimately associated with our notions of what is *possible*, a fact that allows us to speak meaningfully about possible but unactualized situations or "worlds." But, as noted above, the traits internal to wrong theories stand on equal feet, qua concepts, with those that happen to be displayed in our universe and so all concepts enjoy their own range of fictional worlds in which their capacities appear realized. We can unpack the intensional content of *being phlogiston* by pondering circumstances that would ensue if stuff of the required character, in fact, existed (3,iv)

(29) The purpose of *philosophical analysis* is precisely to capture the primary intentional ingredients that allow us to have such rich intuitions about conceptual possibility. Philosophers thus serve as custodians of the conceptual realm. (3,iv)

(30) The belief that a clearly delineated concept Φ applies to an object will always possess a *truth-value* (the object must either exemplify Φ or not), even if we know of no route whereby we can verify this fact.

(31) Likewise, a fully determinate concept carry will carve out a fixed *extension*—that is, the set of objects in the universe to which it applies. If Φ is a concept, then its

extension is designated by "$\{x| \; x$ has $\Phi\}$." Comparable sets will be carved out within the "possible worlds" of thesis (26) as well. (5,viii)

(32) A proper account of the *epistemology* of legitimate belief formation depends upon linkages between the internal contents of our concepts and our practical classificatory capacities. Different philosophers offer widely varying accounts of this transition.

The next range of theses concern the distinction between concepts *proper (conceived primarily as entities that,* per *thesis (2), we can concretely grasp) and* attributes *or* properties, *considered,* per *thesis (1), as the traits that become objectively manifested within a possible universe. Russell himself would have not drawn any important distinction of this type, but the presumption that single attributes may hide behind varied conceptual presentations represents a venerable philosophical theme. Here it is supposed that a common attribute of* **being H_2O** *stands behind the differing conceptual presentations* **being water** *(as grasped by ordinary folk) and* **being H_2O** *(as grasped by the chemist). (3,ix)*

To maintain such concept/attribute distinction does not necessarily represent a major departure from the classical picture, although theses (1) to (4) will require modest reformulations to accommodate for the supplementation. Classical thinkers more frequently disagree on the following doctrines than with respect to (1)–(32).

(33) *Attributes* or *properties* directly represent the traits that the objects of the natural world exemplify and which cause them to behave as they do. This is merely a restatement of thesis (1), altered to suit attributes.

(34) Attributes embody the characteristics ("physical directivities") that induce behavior irrespective of how we happen to think about them. The property of *being H_2O*, for example, could care less about the fact that chemistry is hard to learn and that many human beings deal with the traits involved in *being H_2O* through the rough and ready features of the common man's concept *being water*. (5,vi)

(35) Opinions divide as to whether everyday *being water* exists "merely as a concept" or that the trait represents an attribute of a lower grade than *being H_2O*. After all, *having a motion that heads at a 45° angle to line* $\mathcal{L}$ looks, at first glance, as if it should qualify as a rather unimportant but genuine relational attribute of a cannon ball. If we follow this line of thought, then our supposed distinction between *concept* and *attribute* tends to collapse back into unification, for a "coordinate dependent concept" now looks as if it simply represents an unimportant grade of relational attribute. Many authors who mark differences between "concepts" and "attributes" are often hazy about critical matters such as this.

(36a) How, then, do we come to know about the world's attributes? The most common (and venerable) opinion maintains that we gain this knowledge through entertaining concepts that present their contents to us in a *canonically informative* fashion. Thus when we consider the concept *being H_2O*, the physical directivities that "make it tick" appear wholly in view, for it is from the sundry characteristics of oxygen and hydrogen that we can figure out why water as a stuff behaves as it does. The plain man's *being water*, on the other hand, seems explanatorily *opaque*: it suggests no handles upon which to hang associated characteristics such as *freezing at 0° Centigrade*. Whether a concept presents its contents canonically or not can be determined on the basis of its internal content alone and, on this view, neither *being water* nor *being the favorite beverage of Carrie the teetotaler* provide such canonical representation. (3,ix)

This notion of "canonical presentation" is not the same as the "full and complete presentation" of thesis (17). *Being red* scarcely presents any underlying mechanism directly and the notion may not actually correspond to any acceptable attribute at all, but it nonetheless constitutes a paragon example of a concept we understand "fully and completely."

(36b) A less commonly adopted alternative to (36a) maintains that attributes represent the *abstractive commonalities* betwixt similarly focused concepts, i.e., that the fluid attribute we seek represents the commonality that underlies *being H₂O, being water* (in the plain man's sense), *being the favorite beverage of Carrie the teetotaler*, etc. This story has the advantage of not privileging a specific concept as canonical in virtue of its presentational contents. (5,vii)

It is often difficult to determine whether a given advocate of attributes regards them in manner (36a), (36b) or from some other point of view, despite their palpable differences.

(37) Normally, only attributes are suitable for framing induction hypotheses in science. No hypothesis should be based upon Nelson Goodman's trait of *being grue* (= being green if observed before the year 3000 or blue otherwise), for this will lead us to suppose that the claim "all emeralds are grue" is scientifically supported (with the unhappy suggestion that blue specimens exist and will be discovered in 3000). In Goodman's terminology, *being grue* does not appear to be *projectible* and this lapse disqualifies it from enjoying "attribute" status. But the notion seems conceptually coherent and should be retained within the more tolerant ranks of concepts. David Lewis expresses this doctrine by remarking that, in comparison to concepts, the distribution of true attributes in the world is "sparse."[45] (5,viii)

(38) *Being grue* fails to be an attribute contender because it represents a hodge-podge of ill-sorted characteristics: it doesn't capture a single mode of activity. Such considerations lead to the notion that *clear capacity to effect behavior* is the hallmark of a true attribute. In particular, two attributes can be regarded as identical if they accord their objects with the same range of causal powers. Such an "identity condition" is not universally accepted, however, because it appears to incorrectly identify the property of *being a (linearized) pendulum with length L* with *being a (linearized) pendulum of period* $\sqrt{(2\pi/L)}$. (5,vii)

(39) Often attributes cluster together in natural associations that merit a revival of the Aristotlean term *natural kind*. Good examples are provided in chemical substance traits like *being H₂O* or species notions such as *being a member of Canis familiaris*. Authors attracted to the natural kind notion often leave their relationships to other attributes murky. (7,vi)

---

[45] David Lewis, "New Work."

(40) Attributes grade into further categories according to the level at which they act. A quantity like *force* belongs within the range of traits *natural to physics*, which does not reach to higher level classifications such as *being a member of Canis familiaris* or *being in pain*, which are *natural to biology* and *psychology*, respectively. These subdivisions of attribute type are reflected within the vocabularies that various branches of the sciences and humanities choose as central: *physical notions* within physics; *biological notions* within biology; *aesthetic categories* within literature, etc. This doctrine is sometimes expressed in the form: different grades of attribute are reflected in the "kind terms" selected by various branches of inquiry. (5,viii)

(41) Only rarely will the attributes of one discipline prove *definable* in terms of some other branch: a famous argument about *multiple realizability* claims demonstrates that the "kinds" of psychology can't be defined in physical terms. Many authors contend that attributes form into looser hierarchies related through *supervenience*, which represents a modal condition concerning possible world manifestation that is weaker in its requirements than strict definability.

(42) Attributes are important to philosophy because the proper analysis of *key metaphysical notions* such as "law of nature," "cause," "possible world," etc. requires their invocation rather than the more inclusive concepts. The latter still prove primary in capturing the contents of a speaker's beliefs *per* thesis (7) and many of the allied tasks listed up to (30). Roughly speaking, attributes are pertinent to questions that should be addressed on a more objective basis. (5,viii)

(43) As with concepts, an attribute does not fail to qualify as a bone fide specimen simply because it fails to suit the real world appropriately. *Containing phlogiston* represents as fine a specimen of attribute as *being a quark*. Sometimes different quantities can be neatly discriminated only by considering how they behave in universes contrary to our own. It is hard to segregate *being an electric effect* cleanly from *being a magnetic effect* in our world, but it is easy to imagine possibilities where the two traits completely decouple.

(44) Certain concepts like *being red* that, at first, seem as if they present worldly traits may actually represent attributes of our mental condition first and foremost, with their physical ramifications acquired only through their dispositional behavior. That is, when we pronounce a fire engine to be red, we merely indicate that the truck possesses unknown attributes of a sort that frequently occasion our visual fields to display the attribute of *redness*. From this point of view, the differences we have drawn between concepts and (worldly) attributes are partially a distinction between *attributes manifested in the physical world* and *attributes manifested in our mental realm*. This mental location doctrine has proved very popular in traditional philosophy (Russell accepts it, for example) but is out of favor in analytic circles today. But a temptation to revert to views of this sort is very strong under classicism, so this claim has been added as an inessential inclusion to our list. (2,iv)

4*

THEORY FACADES

[M]athematics has grown like a tree, which does not start at its tiniest rootlets and grow merely upward, but rather sends its roots deeper and deeper at the same time and rate as its branches and leaves are spreading upward. Just so—if we may drop the figure of speech—mathematics began its development from a certain standpoint corresponding to normal human understanding and has progressed, from that point, according to the demands of science itself and of the then prevailing interests, now in the one direction toward new knowledge, now in the other through the study of fundamental principles.

Felix Klein[1]

(i)

Strange latitudes. In this chapter we shall first excavate the forgotten parentage that has engendered our modern conceptions of "theoretical content" and "implicit definability," which, despite many years without substantive motivational rejuvenation, wheeze onward in considerable decrepitude. If we revisit the originating concerns with the advantage of corrective hindsight, many of the theses central to this book can be briskly motivated. However, as I warned in the Preface, this particular investigative pathway may not prove to everyone's taste, for it is somewhat concentrated in scientific particulars. I will attempt to outline it all in mild and accessible terms, but the total pileup of detail may try the reader's patience. There's a delightful passage in a P. G. Wodehouse story where the disgusted uncle of one of Bertie Wooster's artistic chums threatens to sever his stipend and send the nephew off to work in the family commerce. Bertie comments upon this horrific prospect:

> *Corky's uncle, you see, . . . was always urging him to chuck Art and go into the jute business and start at the bottom and work his way up. And what Corky said was that,*

---

[1] Felix Klein, <u>Elementary Mathematics from a Higher Viewpoint</u>: <u>Arithmetic</u>, E. R. Hedrick and C. A. Noble, trans. (New York: Dover, 1939), 15.

while he didn't know what they did at the bottom of a jute business, instinct told him that it was something too beastly for words.

To which Bertie allows:

I'm a bit foggy as to what jute is, but it's apparently something the populace is pretty keen on, for Mr. Worple had made quite an indecently large stack out of it.[2]

Some of what we discuss in this chapter may strike the reader as the philosophical equivalent of working ones way up in a jute factory, although I believe these concerns provide the quickest (and, ultimately, most convincing) route to the endpoints I seek. I will keep our chapter's journey as agreeable as possible, but some readers may prefer a detour at this point, scooting onto Chapter 5 or its sequel. Every central concern I canvass here will be revisited from other vantage points later, albeit not approached in such starkly etched linguistic engineering terms (which represents the approach I personally prefer, when its details can be worked out).

At the end of the previous chapter, we were left with a puzzle. Why did so many scientists of the late Victorian era cheerfully proclaim that their descriptive purposes are limited and crabbed; that they merely hope to entrap Nature's behavior within some structural web, uninterested in deeper explanation or the robust peculiarities of color or musical experience? Why, for example, should Karl Pearson announce that physics merely dabbles in "conceptual shorthand"?

We interpret, describe, and resumé the sequences of this real world of sense-impressions by describing the relative positions, velocities, accelerations, rotations, spins, and strains of an ideal geometrical world which stands for us as a conceptual representation of the perceptual world. . . . [But] it seems to me that we are ignorant [of the nature of matter and force] and shall be ignorant just as long as we project our conceptual chart, which symbolizes but is not the world of phenomena, into that world; just as long as we try to find realities corresponding to geometrical ideals and other purely conceptual limits. So long as we do this we mistake the object of science, which is not to explain but to describe by conceptual shorthand our perceptual experience.[3]

All this appears in The Grammar of Science, published in 1892, during what is usually regarded as physics' most complacent era, before any of the oddities of quantum physics and relativity had emerged. What motives drove such extraordinary avowals?

In Pearson's case, some of the answer merely reflects personal temperament: he is apparently captivated by the exotic idealism then prominent in British intellectual circles and he is patently fond of giving his readers a stir. Nonetheless, it is easy to find expressions of essentially the same point of view within more sober sources from the same era, e.g., in Horace Lamb's straightforward primer on Dynamics.[4] Indeed, although Pearson is mainly remembered today for his work in statistics, he began his

[2] P. G. Wodehouse, "The Artistic Career of Corky" in The World of Jeeves (New York: Harper and Row, 1967), 78.
[3] Karl Pearson, The Grammar of Science (Bristol: Thoemmes, 1991), 285, 329.
[4] Horace Lamb, Dynamics (Cambridge: Cambridge University Press, 1923), 345–9.

career working in elasticity ($\simeq$ the treatment of a bar of steel as a continuously flexible substance). In fact, his pronouncement on "conceptual charts" is preceded by this passage:

> It might seem easier at first sight to explain why two adjacent ether elements "move each other" than why two distant particles of matter do. The common-sense philosopher is ready at once with an explanation: they pull or push each other. But what do we mean by these words? A tendency when the body is strained to resume its original form . . . But why does this motion follow on a particular position? . . . It will not do to attribute it to the elasticity of the medium; this is merely giving the fact a name. We do indeed try to describe the phenomenon of elasticity conceptually, but this is solely by constructing elastic bodies out of non-adjacent particles, the changes of motions of which we associate with certain relative motions. In other words, to appeal to the conception of elasticity is only to "explain" one "action at a distance" by a second "action at a distance" . . . And here no answer can be given. We cannot proceed for ever "explaining" mechanism by mechanism. Those who insist upon phenomenalizing mechanism must ultimately say: "Here we are ignorant", or what is the same thing, must take refuge in matter and force.[5]

In fact, this passage relates to physical practice in a quite definite way: it provides a *justification* for certain puzzling derivational steps that commonly appear in the routines of setting up the standard (Navier) equations for elasticity, deductions that "ask the reader," Stuart Antman comments, "to emulate the Red Queen by believing six impossible things before breakfast."[6] As such, these strange inferential procedures are symptomatic of the puzzling directivities that enfold apparently unprepossessing mechanical terms such as "force" and "rigid body." With the hindsight of a subsequent hundred years, we now recognize that Pearson-like appeals to "conceptual shorthand" simply constitute a mistake in this context—he has utilized *philosophy* to patch over reasoning gaps that should be properly filled with more sophisticated *mathematics*.

. . . .

The basic problem is that a flexible continuous body must remain flexible at all size scales and, accordingly, it becomes hard to articulate their operative principles without assuming that, at some minute level, their parts act somehow "frozen" enough to be treated as if they are rigid bodies or point particles instead.[7] Pearson is thus arguing that physicists have a conceptual right to impose the "regulative structure" of separated point particles upon our flexible stuff. But, clearly, this alleged "right" can only make sense from some idealist or neo-Kantian perspective.

I might mention that, from a physical point of view, this approach gives wrong results, for it supplies a theory of isotropic elasticity with only one material content, rather than the two obtained if the "top down" approach pioneered by Cauchy, Green and Stokes is adopted. In modern books for experts (which invariably follow the latter path), Pearson's problem is

---

[5] Pearson, Grammar, 329.

[6] Stuart S. Antman, Nonlinear Problems of Elasticity (New York: Springer-Verlag, 1995), 11–12.

[7] James Casey, "The Principle of Rigidification," Archive for the History of the Exact Sciences, 32 (1993). The principle is much employed in Lord Kelvin (William Thompson) and Peter Tait, Treatise on Natural Philosophy (Cambridge: Cambridge University Press, 1903).

addressed by formulating the governing laws on two levels of size: at the "body" level for basic balance laws and at the "point" level for constitutive behaviors.[8] But these arrangements are fairly complicated mathematically and Pearson's idealist ploy evades them by plowing past them with "philosophy." We witness further confusions of this sort within Ludwig Boltzmann's thinking in 10,viii.

. . . .

Pearson's impulse towards extravagant philosophizing represents another nice example, different in spirit from those supplied in Chapter 2, of how extraordinary conclusions can sprout from everyday practicalities that seem puzzling in some way: Pearson is certain that scientific topics A and B bear some connection to one another, but the path that connects them seems peculiar and in want of a philosophical rationale.

Pearson's specific difficulties happen to be somewhat arcane in nature, so it will be more convenient if I cite several simpler illustrations of the inferential oddities that confronted physicists and mathematicians at every turn during the Victorian era. Among these, the rise of *complex variables* (that is, the consideration of "imaginary" numbers such as $2 - 3\sqrt{-1}$) played a notable role in guiding scientific argumentation into strange, but plainly profitable, regions, leaving our Victorians frequently puzzled as to exactly what they had wrought. In a celebrated presidential address to the British Association in 1883, the mathematician (and erstwhile barrister) Arthur Cayley explicitly called for some "philosophical account" of current activities in science:

> [T]he notion which is really the fundamental one (and I cannot too strongly emphasize the assertion) underlying and pervading the whole of modern analysis and geometry [is] that of imaginary magnitude in analysis and of imaginary . . . points and figures in geometry. This [topic] has not been, so far as I am aware, a subject of philosophical discussion or inquiry . . . [E]ven [if our final] conclusion were that the notion belongs to mere technical mathematics, or has reference to nonentities in regard to which no science is possible, still it seems to me that as a subject of philosophical discussion the notion ought not to be this ignored; it should at least be shown that there is a right to ignore it.[9]

Let me supply two examples of the surprising discoveries that Cayley had in mind, drawn from geometry and engineering, respectively.

First observe that algebraic formulae supply natural *syntactic directivities* with respect to complex numbers, even if, at first, there seems no reason why anyone should wish to follow them. Consider the phrase "$2/(x^2 + 2x + 2)$," which is constructed from simple arithmetical operations. If we now plug in the value "$1 + \sqrt{-1}$" for x, we can readily compute "$(1 - \sqrt{-1})/4$" simply by following the obvious rules for complex addition and multiplication (i.e., $(a + b\sqrt{-1}) + (c + d\sqrt{-1}) = (a + c) + (b + d)\sqrt{-1}$, etc.)

Consider two circles of radius 3 centered on the x axis at respectively $(-2, 0)$ and $(+2, 0)$. To find their intersection coordinates, we simply solve their two representative

---

[8] C. Truesdell, A First Course in Rational Continuum Mechanics (New York: Academic Press, 1977).

[9] Arthur Cayley, "Presidential Address to the British Association, September 1883" in Collected Mathematical Papers (Cambridge: Cambridge University Press, n.d.), 434.

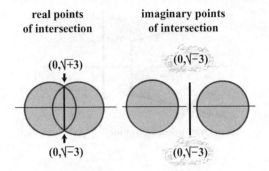

real points
of intersection

imaginary points
of intersection

$(0,\sqrt{+3})$

$(0,\sqrt{-3})$

$(0,\sqrt{-3})$

$(0,\sqrt{-3})$

equations $((x+2)^2+y^2=9$ and $(x-2)^2+y^2=9)$ by high school algebra and obtain $(0,+\sqrt{5})$ and $(0,-\sqrt{5})$. But what happens if we shrink our circles so that they no longer meet (<u>e.g.</u>, they obey the equations $(x+2)^2+y^2=1$ and $(x-2)^2+y^2=1)$? The same reasoning pattern will supply us with "intersections" $(0,+\sqrt{-3})$ and $(0,-\sqrt{-3})$. But surely it's the height of stupidity to consider points located at imaginary locations?

Well, actually, no; great advances in geometric understanding were achieved precisely through following this "stupid route," which was often viewed, in a famous phrase of Hermann Hankel's, as a "present which pure geometry received from analysis."[10] In other words, the syntactic directivities native to high school equation solving lead us into an unexpected "projective" extension of the Euclidean geometrical realm that turns out to be a rather pleasant place, actually. We shall revisit this odd episode from a different perspective in 8,iii.

Turning to engineering, a second surprising inferential extension involving complex numbers arises when we consider a circuit for controlling a telescope's orientation. By setting the left hand dial, we wish to turn the telescope to a desired position. We arrange for a current c_1 to travel from the dial setting to a motor in the telescope's base. A sensor there will return a feedback signal c_2 indicating whether the tube points in the desired direction or not. Our basic plan is to utilize the error signal e ($=$ the difference in current strength between c_1 and c_2) as our means for ordering the motor when to turn and in

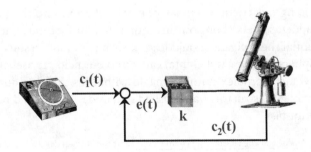

$c_1(t)$

$e(t)$

k

$c_2(t)$

---

[10] John Theodore Merz, <u>A History of European Thought in the Nineteenth Century</u>, iv (New York: Dover, 1965), 660.

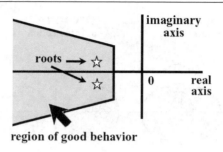

region of good behavior

what direction it should head. To do this properly we need to send the *e* current through an amplifier (marked as **k**) and then let a properly compensated result govern our motor. But the degree of amplification required will not be immediately obvious because our telescope and motor combination cannot respond instantaneously to current changes, but will instead forge ahead to a certain degree. Treated together, the amplifier plus the sluggish motor response gives rise to a total *impedance* (or transfer function) described by the formula $k/(x^2 + 2x + k)$, where k marks the strength of the amplifier.

It is at this point that complex numbers enter our story. We have noted that "$k/(x^2 + 2x + k)$" can be easily calculated for imaginary values of x, even if there is no evident reason why we should wish to do this. Graphing the new reach of our algebraic expression onto the entire complex plane, the extended results turns out to reveal, in a very piquant manner, important features about our telescope system. In particular, the complex locations of the two zeros of "$k/(x^2 + 2x + k)$" allow us to see in a single glance how long the motor will require to respond to a change in dial setting, how long it needs to stabilize upon the right location and how large will be the excessive swings it displays in the process of getting there (in the early steam engines, such runaway *overshoots* often grew dangerously large as the device's governors hunted unsuccessfully for the right stabilization). We can then design an admirable telescopic control circuit by moving these zeros around on the complex plane by choosing different values of the amplification factor k. In the case at hand, if k is set to 2, the zeros locate themselves at $-1 \pm \sqrt{-1}$, which provides a nicely cushioned telescopic control system.[11]

Plainly, extending our circuit's *impedance* $k/(x^2 + 2x + k)$ into the imaginary realm unveils many hidden secrets about our invention, but, at first blush, it is not obvious why such inferential shenanigans should lead to such admirable results. Indeed, one of the best philosophers I know (Anil Gupta) came into our field precisely through having been puzzled by such complex number magic within his undergraduate engineering courses. His impatient instructors had brushed him aside, "Oh, you'd better go see the philosophers about *that*."

---

[11] Chi-tsong Chen, <u>Analog and Digital Control System Design</u> (Fort Worth: Saunders College Publishing, 1993), 224. Philip Cha, James Rosenberg and Clive Dym, <u>Fundamentals of Modeling and Analyzing Engineering Systems</u> (Cambridge: Cambridge University Press, 2000).

. . . .

Our Victorian scientists did not confront this puzzling inferential technique in quite the format presented here, but rather in the guise of the Heaviside operational calculus. I have described the case in the present manner because it is easy to articulate briefly. It will be reencountered in its proper historical habiliment in 8,viii–ix.

. . . .

Cayley's complaint that his worries had passed neglected was not entirely true and the philosophical responses took a variety of interesting forms. First of all, there were those who, in the mode of Karl Pearson, believed that such outré inferential excursions were licenced by the human mind's need to bring the world before it under the discipline of idealized structures, even if these happen to carry us into complex realms. The sundry members of the Marburg school argued in much this way, considering themselves neo-Kantians although they happily embraced regulative ideals far more general in their scope than any that Kant had permitted (the latter belonged to a scientific generation prior to the nineteenth century blossoming of complex number guided exploration). Ernst Cassirer, in fact, framed an elaborate theory of concepts based directly upon the projective geometry paradigm:

> Here it is immediately evident that to belong to a concept does not depend upon any generic similarities of the particulars, but merely presupposes a certain principle of transformation, which is maintained as identical . . . It is this ideal force of logical connection, that secures them the full right to "being" in a logico-geometric sense. The imaginary subsists, insofar as it fulfills a logically indisputable function in the system of geometrical propositions.[12]

As the phrase "the subsistence of the imaginary" suggests, this approach presumes a rejection of straightforward realism with respect to either the physical world or mathematics, following the usual neo-Kantian inclination to treat scientific objectivity as the sharing of investigative standards between different public parties, rather than direct correspondence with empirical reality. Allied themes remain popular in philosophical circles today, although I will have no truck with them myself.

Cayley himself seems to entertain some allied regulative ideal conception himself, although he leaves his remarks too undeveloped to be certain:

> I would myself say that the purely imaginary objects are the only reality, the ὄντως ὄντως, in regard to which the corresponding physical objects are as the shadows in the cave; and it is only by means of them that we are able to deny the existence of a corresponding physical object; if there is no conception of straightness, then it is meaningless to deny the existence of a perfectly straight line.[13]

A second line of approach, more patently consistent with a classical approach to concepts, argues that our peculiar claims about imaginary points et al. represent straightforward propositions about our circuit or regular Euclidean space gussied up in

---

[12] Ernst Cassirer, Substance and Function, W. C and M. C. Swabey, trans. (New York: Dover, 1953), 82–3.
[13] Cayley, "Address," 433.

unusual form. There are geometrical claims of a familiar cast hiding behind these strange exteriors and, if we only "crack their code," we will find that assertions such as "The circles meet at the point $(0, +\sqrt{-3})$" supply sensible information about these figures, albeit expressed in an unusual way. Likewise, any talk about the imaginary behavior of our *impedance* property can also be reexpressed in perfectly ordinary terms with respect to decay of its transients, <u>etc.</u> With respect to geometry, the key historical figure behind this "unveil the true thoughts hidden beneath the formalism" policy was Karl van Staudt, who supplied elaborate and unexpected paraphrases for our claims about $(0, +\sqrt{-3})$ in the 1840s and 1850s.[14] With some doctrinal variation, both Bertrand Russell and Gottlob Frege belong to this general "true thought" tradition, as we shall discuss in fuller detail in 8,v.

At present, however, we will focus upon a third vein of doctrine that is essentially *anti-classical* in its conceptual orientation (although we will find that it experiences difficulty pressing through its opposition consistently). Such thinking eventually evolves in the general directions of the developed formalism, instrumentalism and pragmatism to be discussed later, but at the moment we want to probe the headwaters where the notion of "theoretical content" is hatched. Axiomatics, webs of belief, implicit definability and the rest of the apparatus belonging to the "theory T syndrome" grow up downstream from these spawning grounds, but let us observe such patterns of thinking in their juvenile state, so that we can appreciate how such doctrines grow from genuine dilemmas that confronted descriptive practice in Victorian times.

In 3,vi, we surveyed Boylean inclinations to regard mechanical notions such as *gear wheel* as more satisfying, from an explanatory point of view, than *gravitational force* and, presumably, either *temperature* or *chemical affinity*. By the 1880s, most practitioners would have shifted Newtonian *force* into the "satisfactorily understood" column but many still searched intently for narrowly mechanical underpinnings for *temperature* and *chemical affinity*. Indeed, today we trust that such relationships hold, albeit founded in quantum principle rather then classical mechanical doctrine. In the 1880s, great progress had been affected within both thermodynamics (that is, the theory of heat treated on a macroscopic scale) and chemistry, in patterns that entwined these two subjects with orthodox mechanics through the articulation of *chemical potential* and allied developments of that ilk. Many reasonable physicists—Ernst Mach and Pierre Duhem will be cited here—believed that the proper road to further progress lay in pressing such discoveries further. In contrast, they worried that reductive searches of a mechanist variety could retard this advance, for such efforts typically engage in crude model building with virtually no physical support and thus discourage rigorous attention to the actual ways in which materials behave. For example, it requires enormous cleverness to frame a molecular structure able to transport simple transverse linear waves, but devoted experiments can be found in utterly commonplace materials which disclose the most

[14] Ernest Nagel, "The Formation of Modern Conceptions of Formal Logic in the Development of Geometry" in Teleology Revisited (New York: Columbia University Press, 1979). Charlotte Angas Scott, "On Von Staudt's *Geometrie der Lage*," Math. Gazette 5 (1900). J. L. Coolidge, A History of the Conic Sections and Quadric Surfaces (New York: Dover, 1968).

astonishing varieties of non-linear and temperature dependent behaviors (as James Bell's excellent history shows, the experimental probing of the properties of materials truly blossomed in the nineteenth century.[15]) Duhem, in particular, realized that more precise forms of physical principle would be required if these richer realms of behavior were to be brought within the reach of applied mathematics. He was, accordingly, frustrated with the inclinations of colleagues (e.g., Ludwig Boltzmann) who tinkered with toy molecular models at the expense of laboratory realities.

Why did the molecular modelers proceed as they did? In Mach and Duhem's estimation, such tropisms represent the ill-considered heritage of old conceptual prejudices like Boyle's. To be sure, by this time no one would have listened to Boyle or Descartes in their complaints about the "intelligibility" of *gravitational force*, but a hazy descendent of those old demands must animate the sentiment that molecular explanations of temperature and chemical binding are somehow more "satisfying" than the phenomenalist level accounts developed under the sheltering umbrella of *thermomechanics*, as the richer blending of elements favored by Mach and Duhem is sometimes called. Here is how Mach saw the situation:

> *The view that makes mechanics the basis of the remaining branches of physics, and explains all physical phenomena by mechanical ideas, is in our judgment a prejudice. Knowledge which is historically first is not necessarily the foundation of all that is subsequently gained . . . We have no means of knowing, as yet, which of the physical phenomena go deepest, whether the mechanical phenomena are perhaps not the most superficial of all, or whether all do not go equally deep . . . The mechanical theory of nature is, undoubtedly, in a historical view, both intelligible and pardonable; and it may also, for a time, have been of much value. But, upon the whole, it is an artificial conception. Faithful adherence to the methods that have led the greatest investigators of nature . . . to their greatest results restricts physics to the expression of actual facts, and forbids the construction of hypotheses behind the facts, where nothing tangible and verifiable is found. If this is done, only the simple connection of the motion of masses, of changes in temperature, of changes in the value of the potential function, of chemical changes, and so forth is to be ascertained.[16]*

Such reflections led many thinkers of the period to become leery of classical pictures of conceptual context, at least within the dominions of science, because such propensities encourage ill-considered searches for warm and fuzzy I-know-not-whats, rather than focusing scientific investigations squarely on the brute facts Nature offers.

In a general way, these reasons for rethinking the basic nature of conceptual grasp are allied to those associated with the unexpected extensions of application that we witnessed in the complex number cases, because both phenomena suggest that "grasping a concept" does not represent the staid and transparent intellectual enterprise that methodologists of an earlier era had assumed. Somehow the pressures of dealing with the world around us force us to traffic in concepts that either enlarge in strange,

---

[15] James F. Bell, The Experimental Foundations of Solid Mechanics (Berlin: Springer-Verlag, 1984).

[16] Ernst Mach, The Science of Mechanics, Thomas J. McCormack, trans. (LaSalle, Iu.: Open Court, 1960), 596–7.

"organic" ways or in manners that we seem to "understand" only in an abstract and threadbare manner. "Plainly," our Victorians came to believe, "we require a philosophy of conceptual obtainment that can tolerate a freer arena for scientific creativity, no longer restrained by the shackles of Euclidean, mechanical and allied forms of inherited prejudice." To be sure, newly refurbished versions of classical doctrine such as Bertrand Russell offers can prove satisfactory in these regards as well, because he managed, through his theory of descriptions and other stratagems, to convert the traditional Realm of Universals into a more tolerant kingdom than it had previously seemed. However, let us continue to pursue formalist lines of thought for the time being.

. . . .

Let us not neglect entirely the lines of thought represented by Ernst Cassirer, because in his stress upon the *growth characteristics* manifested by predicates, he anticipates many of our Chapter 8 themes, although I regard these directivities as arising from external strategic pressures, rather than the handiwork of neo-Kantian regulative propensities.

. . . .

As is evident from the passage quoted, Mach and Duhem maintain that science should proceed at a largely phenomenological level, an implausible position for which they are best remembered today. Beneath this upper crust of somewhat crude philosophizing there lies a well-founded distrust of the specific contents credited to familiar mechanistic notions: unexpected failures of comprehensiveness in fact lurk there, as we shall see in some detail later on. Indeed, their molecular-favoring opponents were often fooled by what can now be recognized as varieties of *semantic mimicry* (e.g., the discussion of Boltzmann in 10,viii). Duhem is also aware of the fact that the circle of usual classical mechanical notions does not close in on itself in a coherent way: in dealing with the "mechanics" of any realistic material, we are quickly forced to appeal to *temperature* and *chemical potential* as unreduced auxiliary notions (I'll explain why in section (ix)). To me, this failure of closure represents an important premonition of the fact that classical mechanics secretly organizes itself as what I shall later call a *theory facade*.

I stress these specific grounds for conceptual disquiet within mechanics because they nicely illustrate how readily philosophical worries about concepts interlace intimately with practical necessities: nineteenth century physicists had arrived at a puzzling crossroads and required some methodological clue as to what developmental path to choose. As it happens, the sundry forms of philosophical response they formulated all prove exaggerated along some dimension or other, but each embodies vital considerations that we must bear in mind whenever we wonder how our descriptive vocabulary might be improved.

Let us now pursue our formalist's anti-classical leanings a bit further to see where they lead, along the path that I shall call *salvation by syntax*. For this purpose, we will begin with a pithy statement of essentials provided by the physicist Heinrich Hertz, (which should be read in conjunction with the richer views expressed by his mentor Hermann Helmholtz). Neither figure, to the best of my knowledge, shared the thermomechanical ambitions of Mach and Duhem, and were more centrally concerned to rid electrical thinking of unwanted modeling burdens. Hertz (who doesn't mark his motivations as clearly as one would like) is also properly troubled by the *lack of rigor* that infected current practice in mechanics, which represents another important contributor to the conceptual crises of the late nineteenth century.

....

I might add that the thermomechanical criticisms of traditional thinking are especially interesting for our purposes, because modern engineers continue to employ classical doctrines developed pretty much along Duhemian lines, whereas the electrical properties of materials tend to demand quantum treatments. The former situation makes it easier to recognize how trenchant many of Duhem's specific complaints about practice really were.

....

(ii)

Inferential overexuberance. In the previous section I have accentuated the positive, by emphasizing the productive territories into which predicates, freed of the burdens of traditional demands on "satisfactory understanding," can gaily lead us. At the very same time, quite the opposite can occur: well-trusted and apparently thoroughly domesticated patterns of reasoning can turn out undesirable results without warning (in some inopportune form such as a steam ship disaster). Worse yet, these failures can prove subtle in their rottenness: it can be quite awhile before we realize, "Gee, I should have never accepted *that* bill of goods." A major reason that the methodological crises of the late nineteenth century proved so difficult is that trusted tools of inferential advance were apt to turn friend or foe without warning or apparent consistency.

By Hertz' time, the corpus of classical physics had grown to large acumulation through gradual amalgamation, a process that inherently runs the risks trenchantly described by David Hilbert:

> The physicist, as his theories develop, often finds himself forced by the results of his experiments to make new hypotheses, while he depends, with respect to the compatibility of the new hypotheses with the old axioms, solely upon these experiments or upon a certain physical intuition, a practice which in the rigorously logical building up of theory is not admissible.[17]

Such developmental patterns frequently install localized sheets of doctrine that seem uneasily in tension with one another, leading Hertz to complain in his celebrated introduction to The Principles of Mechanics:

> [I]t is exceedingly difficult to expound to thoughtful hearers the very introduction to mechanics without being occasionally embarrassed, without feeling tempted now and again

[17] David Hilbert, "Mathematical Problems" in Felix Browder, ed., Mathematical Developments Arising from Hilbert Problems (Providence, RI: American Mathematical Society, 1976), 14–15. Leo Corry, "David Hilbert and the Axiomatization of Physics," Arch. Hist. Exact Sci. 51 (1997).

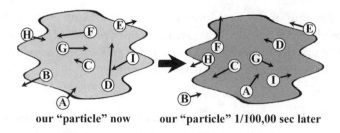

our "particle" now   our "particle" 1/100,00 sec later

> *to apologize, without wishing to get as quickly as possible over the rudiments and on to examples which speak for themselves.*[18]

Basic Newtonian notions such as *force* commonly lie at the center of such tensions. For example, in setting up the Navier-Stokes equations fundamental to the behavior of viscous fluid, many textbooks build upon the backbone[19] of the Newtonian "$F = ma$" ("the total force on a particle is equal to the product of its mass by its acceleration") and then decompose that "force" into its effective factors, including the "viscous force" $\nu\Delta u$. But it was eventually realized (first by Maxwell, I believe) that some of this applied "force" upon our "particle" could not represent the application of any true force at all (e.g., attractions and repulsions exerted by neighboring regions), but instead must express net losses or gains of momentum occasioned when more rapidly moving molecules enter and leave the appreciable volume that our alleged "particle" actually represents. As D. J. Tritton explains in his excellent textbook:

> *The same fluid particle does not consist of just the same molecules at all times. The interchange of molecules between fluid particles is taken into account in the macroscopic equations by assigning to the fluid diffusive properties such as viscosity and thermal conductivity . . . The same fluid particle may be identified at different times, once the continuum hypothesis is accepted, through the macroscopic formulation. This specifies (in principle) a trajectory for every particle and thus provides meaning to the statement that the fluid at one point at one time is the same as that at another point at another time. For example, for a fluid macroscopically at rest, it is obviously sensible to say that the same fluid particle is always in the same place—even though, because of the Brownian motion, the same molecules will not always be at that place.*[20]

In other words, the "particle" to which "$F = ma$" gets applied in fluid mechanics does *not* represent an entity that maintains a fixed mass simply by conserving its identity through time, but instead represents a more complex, ship of Theseus affair wherein a moving spatial region maintains a personality that remains trackable over time largely

---

[18] Heinrich Hertz, <u>The Principles of Mechanics</u>, D. E. Jones and J. T. Walley, trans. (New York: Dover, 1956).

[19] Newton's second law is generally read, somewhat anachronistically, as "$F = ma$," but the notion that it serves as the primary template upon which specific laws of motion are to be constructed is usually credited to Euler.

[20] D. J. Tritton, <u>Physical Fluid Dynamics</u> (Oxford: Oxford University Press, 1976), 50.

by keeping its *average enclosed mass content* constant, while meanwhile allowing its size, shape and momentum budget to vary considerably (just as the boat remained the same as its curators gradually replaced its component planks).

From the vantage point of swift pedagogy, a policy of ignoring the niggling inconvenience that some of the viscous "force" on a particle is not truly force-like in origin (or the fact that the "particles" under discussion have been tacitly allowed to behave like ships that alter their timbers) certainly makes it much easier to set the Navier-Stokes equations briskly before a classroom of largely unenthralled listeners. However, passing blithely over these shifts in the physical significance of "force" and "particle" is likely to create confusions later on, when a more advanced student is likely to have forgotten the precise details of how her acquaintanceship with the Navier-Stokes equations began. These are the very concerns that Hilbert has in mind. In the sequel, I shall call circumstances such as this, where a predicate like "force" alters its physical correlates after following the beguiling guidance of some Pied Piper analogous to "$F = ma$," *property dragging*. Such dragging will become one of our primary concerns in Chapter 6.

In Hertz' own case, his apparent concern (he is not as clear in this regard as one would like) lies with a different dragging that arises when "force" becomes cross-fertilized with "rigid body," a topic whose details will be postponed until a more opportune moment (6,xiii).

....

"Force" is remarkably prone to property dragging. For example, part of the frictional "force" that a rolling wheel encounters is due to the fact that its supporting substratum will stretch subtly under its weight, with the net effect that the wheel's journey is actually longer than it superficially appears.[21] But we typically treat the distance traveled as unaltered and correct for the extra work done by allowing "force" to shift significance slightly.

It is worth observing that, although Wittgenstein and the Vienna Circle greatly admired Hertz' preface, many of them seem to have misunderstood its physical objectives and left an unfortunate legacy of misunderstanding in their wake. Hertz is properly critical of orthodox appeals to *force* within classical mechanics because they are often *inconsistently applied,* but he nowhere criticizes the notion as metaphysically suspect, as lying too far from observation or any of the other epistemological ills that the positivists were inclined to lay at the door of *force.* Misreadings of Hertz according to these ersatz purposes are very common.[22]

....

Hertz' well-founded worry that Newtonian notions are often applied in an overly exuberant fashion represents a nice dual to Pearson's ambitions with respect to elasticity, for the latter hopes, through his appeals to "conceptual charts" and the like, to move "force" into territories that it otherwise can't reach. Specifically, Pearson needs to find some bridge between "$F = ma$" and the notion of internal stress ($\simeq$ a complicated form of directionized pressure) critical to understanding a flexible, continuous substance. For the reasons sketched in the fine print of the previous section, Pearson

[21] F. P. Bowden and D. Tabor, <u>Friction and Lubrication</u> (London: Methuen and Co., 1967).
[22] Max Jammer, <u>Concepts of Force</u> (New York: Harpers, 1962), 241–2.

believes that, in the course of this "derivation," he can permissibly replace the continuous stuff under investigation with an atomized surrogate consisting of a swarm of "molecules" that interact solely through action-at-a-distance forces. He applies "$F = ma$" to his swarm and then claims to get his original continuous substance back again by squeezing the molecular swarm together under some ill-defined "limit." Somehow this mysterious procedure magically erases the bounding surfaces of our "molecules" and replaces them all with a nice, continuously distributed gunk (this strange maneuver can still be found in many contemporary textbooks, especially those written by quantum physicists). Accordingly, when Pearson advises us that:

"Here we are ignorant", *or what is the same thing, must take refuge in matter and force,*

this passage does not merely represent airy pontification; it is intended to serve as a lubricant for an otherwise sticky transition within a nitty-gritty corner of mechanics. As remarked earlier, modern experts in continuum mechanics now believe that Pearson has employed a *philosophical maxim* to bridge over what should be properly regarded as a *mathematical gap* in his practical reasoning. They came to this conclusion after they learned that Pearson's "philosophizing" didn't help physics enter the lands of rubber or toothpaste successfully and that foundational issues in continuum mechanics needed to be addressed in a more sophisticated way, employing mathematical tools that were not available in Pearson's time.[23] Once the gap is properly filled, idealist rationales are no longer needed.

Pearson's ambitions and Hertz' anxieties nicely illustrate the kinds of methodological dilemma that often confronted reasoners in the late nineteenth century. If the problematic viaduct that carries us from "$F = ma$" to the Navier-Stokes equations is closed, will that restriction simultaneously spoil our capacity to reach the standard equations for an elastic substance? Where, along the long spectrum of derivational technique that ranges from the excessively credulous to the repressively restrictive, can the proper inferential directivities of "force" and "particle" be found?

Sober opinions (i.e., *not* Pearson's "a regulative ideal told me I can cheat") leaned towards the conservative end of this spectrum. It was optimistically hoped that the valuable parts of mechanics' accumulation could be reclaimed through hard work: if we delineate our terms precisely and stick to them, we can rid ourselves of property dragging and allied ills, as well as eventually replacing the creaky bridges to elasticity by sounder constructions. But no one presumed that the task of conceptual clarification would be easy in physics. In fact, Hilbert set this task on his famous 1899 list of problems that mathematics should address in the century to come (it forms part of his sixth problem).

But what does it *mean* to "delineate our terms precisely and stick to them"? Russellian classicism suggests the traditional answer: ponder the conceptual contents of *force* until we are certain that we have grasped an absolutely unique universal; that we have tolerated no secret wiggle room that allows the differently oriented directivities of some

---

[23] Stuart S. Antman, "Equations for Large Vibrations of Strings," American Mathematical Monthly 87, 5 (1980).

imposter concept to sneak in and drag "force" somewhere inappropriate (I call this a *"true thought" picture* of the rigorization process). But this classical recommendation means the danger of eventuating in Boylean conceptual conservatism or some allied set of stultifying requirements. A new philosophical movement became founded in this unease: perhaps "delineate our terms precisely and stick to them" ought to be addressed in an overtly syntactic manner? And thus initiates the course of conceptual salvation we now wish to trace.

(iii)

Salvation through syntax. We have posted the delicate straits through which late nineteenth century science endeavored to sail: betwixt the Charybdis of risky conceptual free creativity and the Scylla of safe but overly cautious moorings. Indeed, writ large, most of the practical concerns addressed in this book assume the form: how do we chart a reasonable course past these snares? The classical approach to concepts represents a course that passes too close to Scylla; the formalist proposals we shall now explore veer unacceptably towards the rocks (my own recommendations will combine aspects of both policies, in conjunction with a good deal of visual piloting and a frequent sounding of depths).

Put in a nutshell, the new point of view constitutes a turn of the century bargain that was struck between science and philosophy of language, an ill-starred agreement which continues to handicap our modern thinking. It runs like this. "Philosophy hereby grants science the right to practice unfettered conceptual innovation as long as it concedes that it is *up to something funny* when it describes the world in its peculiar ways: it accepts the stipulation that scientific terms do not obtain their meanings in the same classical manner as ordinary terms such as 'red' and 'doorknob.' Rather than utilizing the mechanisms of classical gluing, scientific terms promise to gain their significance entirely through indirect *syntactic ties*." If a system for employing symbols is specified in a precisely defined syntactic manner that accommodates our narrow scientific interests, then that set of terms can be regarded wholly in adequate conceptual order insofar as scientific purpose is concerned. The hope is that, with sufficient syntactic precision, the dangers of unanticipated pitfalls in our reasoning can be avoided, without any need to be constrained by the "true thought" conceptual moorings demanded by Robert Boyle and his classical chums. But what should a "system for employing the symbols of science" look like? There are several popular answers abroad here, most of which head down the unfortunate trail to holism.

In a lot of the versions to be surveyed in this chapter, strong elements of classical thinking still survive with respect to the non-scientific parts of discourse. In the next chapter, we shall review Quine's more radical proposal for painting *every* predicate with a consistently anti-classical brush.

Let us first consider an early articulation of this syntactic approach, as it emerges in Hertz, Helmholtz and other physicists of the period (who were more inclined to write of "mental symbols" rather than "predicates"). As such, the proposal will seem naïvely articulated, but we shall soon observe that a vital spark of sagacity lies concealed within these accounts, upon which we shall later capitalize.

Heinrich Hertz writes in an often cited passage from the preface to his Mechanics:

> We form for ourselves images or symbols of external objects; and the form which we give them is such that the necessary consequences of the images in thought are always the images of the necessary consequences in nature of the things pictured ... [W]e can then in a short time develop by means of them, as by means of models, the consequences which in the real world only arise in a comparatively long time, or as a result of our own imposition.[24]

Hertz is concerned to establish a right to conceptual free creativity even within the dominions of a mechanism very much akin to that Boyle favored. It happens that Hertz does not want to utilize the notion of *force* as a primary notion within his reconstruction, for he correctly realizes that its dictates inherently clash with others in the tradition of *rigid body* and mechanism (I shall explain these tensions more fully in 6,xiii). Despite popular misreadings of his objectives, Hertz does not object to *force* because it is "metaphysical" or "unobservable"—quite the contrary, his philosophy of free creativity would vigorously defend the acceptability of the *force* notion *if* its standard applications within Hertz' interests could be rendered syntactically coherent. After all, Hertz' own approach (which appeals to an abstract notion of *Gaussian work* defined over high dimensional state spaces equipped with a compass of inertia[25]) hardly traffics in "observable" notions either.

However, we can nicely illustrate the syntactic picture sketched by Hertz if we attempt a *defense* of *force* against Boylean criticism within a smaller domain where it does not suffer the debilities rightly diagnosed by Hertz on a larger scale. I have in mind the realm of *point mass physics*: the doctrine wherein the carriers of force are unextended particles that interact only across spatial distances (in the history of mechanics, this point of view is usually attributed to Boscovitch; it is these theses that rather misleadingly dominate freshman physics primers today). The rigid bodies which Hertz favors drop out of our primary picture: an iron bar will approximately keep its shape if its swarm of component point masses stay in roughly similar spatial relationships to one another, but no extended object is ever expected to act in a completely rigid manner.

Let us now conjure up some curmudgeonly opponent to complain that *force* is methodologically objectionable even within point particle mechanics because he is unable to grasp the underlying nature of its mechanical efficacy. "To claim that particle A moves particle B because a force intervenes between them supplies us with no insight into the true properties that cause these events to occur," he grumbles. Our Hertzian

---

[24] Hertz, Principles, 1.

[25] F. Gantmacher, Lectures in Analytical Mechanics (Moscow: Mir Publishers, 1970), ch. 7. Jesper Lützen, Renouncing Forces; Geometrizing Mechanics (Copenhagen: Matematisk Institut preprint, 1995).

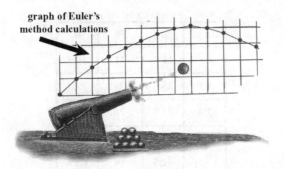

graph of Euler's
method calculations

hero responds, "In science, we do not care about 'explanation' in this fashion; we attempt to construct accurate predictions of whatever events might occur. For this task, I can lay down precise *inferential rules* that govern exactly how the predicate 'force' should be handled in the course of producing those predictions. By doing so, we will have learned how to employ the term with complete precision and that's all that matters for science's limited purposes."

Thus a sturdy redoubt against Boyle-like criticism is framed through a quick retreat up the hillside of *syntax:* "To become a competent employer of 'force,' the only notions that we need to grasp in a fully classical manner are the basic notions of *grammatical classification* and *inferential manipulation* (this word is a name; this phrase is a predicate; this sentence follows from those by <u>modus ponens</u>, and so forth). We can easily master that shallow level of 'understanding' without possessing any clue as to what deeper layers of intensional characteristics attach to 'force'. Science, for its limited predictive purposes, does not demand any deeper grasp."

In fact, within the domain of point mass physics, we can readily convert Hertz' metaphor of "forming pictures for ourselves" into concrete syntactic routine. Let our point mass be a projectile (of unit mass) shot from a cannon of rather pathetic range (I will treat a specific illustration here, but the procedure utilized will apply to any set of ordinary differential equations that can be convened under the banner of this branch of physics). Ignoring air resistance and other complicating factors, orthodox Newtonian theory instructs us that a cannon ball near the earth's surface will suffer a constant impressed gravitational deceleration of 32 ft/sec^2. From these provisos, we can immediately build suitable differential equations on the frame of "$F = ma$" (they are provided in 3,vii). Equations in hand and with a specification of initial conditions ($=$ how the ball left the cannon's mouth), we can syntactically crank out a tabulated set of numerical values that starts as follows.

Graphed on a chart as illustrated, we find that its sequential results nicely mirror the real life flight of a projectile. In fact, we have merely followed the steps prescribed in the numerical technique called *Euler's method* mentioned in passing in 3,vii. As such, the routine is immediately applicable to every point particle equation of the type contemplated.

time	elevation	y velocity	distance
0	0	50	0
1/4	12.5	42	16.7
1/2	23.0	34	33.4
3/4	31.5	26	50.1
1	38.0	18	66.8
1 1/4	42.5	10	83.5
1 1/2	45.0	2	100.2
1 3/4	45.5	-6	116.9

. . . .

Supplying the basic details, we replace "total force" F in Newton's second law of motion ma = F (mass times acceleration = total force applied) by the constant gravitational contribution $(0, -32)$ The result breaks into the component equations , $d^2y/dt^2 = -32$ and $d^2x/dt^2 = 0$. Euler's method then instructs us to construct the algebraic relationships:

$$y_{i+1} = (v_i.\Delta t) + y_i \quad x_{i+1} = (u_i.\Delta t) + x_i$$
$$v_{i+1} = -32\Delta t + v_i \quad u_{i+1} = u_i$$

which then generates our matrix if we consider a shell that is fired with an initial velocity of 83 ft/sec at an angle of 30°.

These formulae, by the way, merely codify the intuitive causal considerations that we commonly employ, in less quantitative forms, within our everyday reasoning about similar situations. Thus the left-hand equations to the left instruct us to estimate that the shell's probable vertical velocity after a small time change (say, $\Delta t = 1/4$ second) will approximately alter in such a way to produce an acceleration of 32 ft/sec² and that the shell will increase or decrease its altitude by a distance approximately equal to 1/4 of its initial and final velocities over the interval. The two equations to the left merely state that the shell moves horizontally at a constant velocity (remember that we've neglected air resistance). Such connections to everyday causal reasoning will be explored further in 9,ii.

. . . .

Plainly, the matrix of numerical data assembled by this syntactic routine provides us with an excellent stage by stage "image" of our ball's flight, in which the "necessary consequences of the images in thought" (the unfolding rows in our table or the placement of dots in our graph) correlate nicely with "the necessary consequences in nature of the things pictured" (the positions of the projectile at successive temporal moments). Our symbolic calculations "walk along" at discrete stages with our cannon ball, rather as Harpo mimicked each of Groucho's moves in Animal Crackers (indeed, Euler's procedure is commonly called a "marching method" for that very reason). But— and here is where the advantage of the pullback into syntax enters—anybody who understands simple arithmetic can fully understand our symbolic rules, even if they can't comprehend the idea of *force* to any greater depth. But this shallow "understanding" of symbolic manipulation should be all that physics requires! In a single syntactic <u>prise de fer</u>, we thereby parry the lunges of *force*'s traditionalist critics.

Here we witness the motivations that led scientists to strike their ill-starred bargain with philosophy in the nineteenth century: in conceding that scientific predicates require a thinner content than the more robust notions of everyday life, they thereby gain a permission to roam the wider boulevards of free creativity. But then critics of a Coleridgean sensibility, convinced that "science describes the world in funny ways," can cite this concession as contractual confirmation of their suspicions.

Strictly speaking, everything that Hertz desires can be achieved through Russell's classicism, for the latter allows that science will often pursue ungrasped universals under the guise of a purely structural description (3,viii). By converting Hertz' syntax instructions into a lengthy description, Russell can remain within a fully classical orbit (albeit a rather strained variety). Nonetheless, buried within Hertzian sentiment lies a somewhat inchoate criticism of classical thinking: a conviction that its picture of concepts somehow demands too much of their grasped contents, not merely within the provincial halls of predictive science, but everywhere. The classical emphasis on the richer intensional characteristics seemingly displayed by *red* or *gear wheel* constitutes some form of philosophical illusion; classical grasp does *not* represent an otherwise reasonable demand on linguistic understanding that we sometimes relax for the sake of scientific investigation (which represents Russell's official point of view in <u>The Analysis of Matter</u>). I see this vein of criticism more trenchantly suggested in the writings of Helmholtz (from whom Hertz largely borrows his philosophical doctrines):

> Natural science . . . seeks to separate off that which is definition, symbolism, representational form or hypothesis, in order to have left over unalloyed what belongs to the world of actuality whose laws it seeks. . . .
>
> The relation between the two of them is restricted to the fact that like objects exerting an influence under like circumstances evoke like signs, and that therefore unlike signs always correspond to unlike influences.
>
> To popular opinion, which accepts in good faith that the images which our senses give us of things are wholly true, this residue of similarity acknowledged by us may seem very trivial. In fact it is not trivial. For with it one can still achieve something of the greatest importance, namely forming an image of lawfulness in the processes of the actual world. Every law of nature asserts that upon preconditions alike in a certain respect, there always follow consequences that are alike in a certain other respect. Since like things are indicated in our world of sensation by like signs, an equally regular sequence will also correspond in the domain of our sensations to the sequence of like effects by law of nature upon like causes.
>
> If this series of sense impressions can be formulated completely and unambiguously, then one must in my judgement declare that thing to be intuitably representable. . . . [T]his can

only happen by way of the concept of the object or relationship to be represented. . . . [T] his is however in disagreement with the older concept of intuition, which only acknowledges something to be given through intuition if its representation enters consciousness at once with the sense impression, and without deliberation and effort . . .

I believe the resolution of the concept of intuition into the elementary processes of thought as the most essential advance in the recent period.[26]

As I read his intent, Helmholtz believes that a "residue of similarity" represents the true core content that a predicate needs to display if it is to be regarded as "intuitably representable" (= "adequately understood") and that "residue" is manifested primarily in the form of the Harpo-imitates-Groucho mirroring relationship it sets up with respect to the world. The apparent immediate understandability of *red* or *gear wheel* merely reflects the unimportant genetic fact that we are innately familiar with the inferential transitions that such predicates demand (or quickly learn them at an early age), whereas we must self-consciously force ourselves to walk painfully through the step-by-step requirements of Euler's method in order to master point mass "force" to a comparable level of skill. But that asymmetry doesn't show that *force*'s more limited set of intensional characteristics are *inferior* to those of *gear wheel* in any respect that we should care about.

This basic hunch—that classical thinking somehow demands a *thicker notion* of predicative content than is truly reasonable—reverberates through most of the anti-classical critics we shall survey in this book and lies at the heart of my own concerns as well. However, Helmholtz nowhere manages to frame a coherent anti-classical alternative that does not quickly seal us behind a quite substantive wall of predication (as his flirtations with modified Kantianism suggest).

(iv)

A home in axiomatics. It doesn't require much reflection to see that comparatively few employments of a newly minted scientific predicate can be supported in this direct, "mock the temporal evolution" of real life systems. Most forms of viable scientific reasoning assume other forms altogether. Indeed, it is far better to approach our cannon ball problem by an altogether different inferential strategy: namely, *solve* the differential equation in freshman calculus style. Here we obtain far more information about all aspects of our problem with much less fuss and without attempting to *mimic* its flight in syntax at all.

. . . .

To be perfectly explicit: (1) *Integrate* the basic equations $d^2y/dt^2 = -32$ and $d^2x/dt^2 = 0$ to obtain $y = -16t^2 + at + b$ and $x = ct + d$. (2) *Insert* the initial conditions to calculate the values

---

[26] Hermann Helmholtz, "The Facts in Perception" in Hermann Helmholtz: Epistemological Writings, Malcolm Lowe, trans. (Dordrecht: Reidel, 1977), 115–63.

for a, b, c, d to obtain $y = -16t^2 + 50t$ and $x = 66.8t$. (3) *Probe* these equations algebraically with respect to the questions we want answered. For example, if we wish to know when the ball will hit the ground, we should set $y = 0$ and solve for t.

. . . .

To be sure, few problems yield to exact solutions of this ilk, but neither was it possible in Hertz' day to utilize brute force numerical techniques like Euler's extensively (before computers, only wealthy military establishments could afford the armies of scribes required to carry out such routines to acceptable accuracy). In consequence, mathematicians devised the most astonishing bag of clever tricks to *avoid* techniques like Euler's (and, of course, our two physicists knew this well from their own work). I have utilized such calculations as an example precisely because *marching method* techniques supply a close match to Hertz' actual words: "The images in thought are always the images of the necessary consequences in nature of the things pictured." But very little reasoning in applied mathematics follows a pattern of this imitative type and a would-be formalist must develop a supportive fabric that explains "force"'s appearance within the other forms of scientific employment that do not "march along" with physical developments in any sense of the phrase.

Even with respect to Euler's method, we achieve far better numerical results if we introduce backtracking refinements (as in, e.g., a Runge-Kutta scheme) that depart from strictly imitative "marching." And, as we'll witness later (4,x), in unexpected cases, Euler's method grinds out completely erroneous answers.

With respect to his overriding objectives, quaint opinions such as Hertz' can be regarded as merely an infelicitous device for claiming that a predicate like "force" can be rendered scientifically viable through some form of syntactic support other than algorithmics. Indeed, a ready answer of this type lay close to hand in other mathematical developments of Hertz' day, specifically, within the rebirth of interest in *axiomatic organization* in the manner of Euclid's geometry: viz., certain sentences are selected as initial *axioms* from which other results follow as *theorems* by logical rules. Mightn't a webbing of axiomatics provide enough syntactic heft to keep a predicate like "force" semantically supported in *all* of its employments, rather than merely along the narrow corridors of a marching method calculation? And this syntactic answer was widely embraced, under the banners of either formalism or instrumentalism. Indeed, Hertz provided such axioms in his <u>Principles</u>, albeit not laid out with the crispness that we have come to expect since the careful labors of Hilbert and the logicians.

The doctrine that webs of axiomatics can competently support embedded predicates takes slightly different forms within mathematics and physics. In the former case, we obtain *formalism*: the doctrine that through axiomatics mathematicians set up formal enclosures in which strange congeries of predicates can comport themselves in any manner that the free creativity of the mathematician chooses, although the *interest* of this syntactic hypostasis ought to prove itself in worthy theorems. If proper axiomatic prerequisites have been set in order, rules will have been supplied that mathematicians can obey as a kind of syntactic game without otherwise knowing what their symbols talk about. In this vein, the modern writer R. E. Edwards writes:

> One may be reminded of the status of the money and property handled in a game of Monopoly: neither are real, but the rules of the game cause them to behave and to be handled in play in ways similar to real money and real property, and the players are not hindered from playing by the lack of reality.[27]

Or consider this allied observation from the early twentieth century geometer H. G. Forder:

> Our Geometry is an abstract Geometry. The reasoning could be followed by a disembodied spirit who had no idea of a physical point; just as a man blind from birth could understand the Electromagnetic Theory of Light.[28]

Here Forder contrasts our direct appreciation of *being red*'s proper conceptual content with the merely structural appreciation which "a man blind from birth" (e.g. Helen Keller) will utilize in order to mimic a more normal grasp of "is red." Within the halls of mathematics, Forder claims, we do not care about classical grasp at all and Keller can claim to understand all of mathematics without cavil.

This point of view led to great simplifications in how mathematics came to be taught. Recall that the philosophical opinions sketched in this chapter grew out of a desire to tolerate, yet safely control, the astounding enlargements that had beset traditional conceptions of what geometry or physics "should be about." In our opening section, we mentioned the peculiar complex-valued points and points at infinity that invaded Euclidean geometry in great profusion during the nineteenth century. We briefly canvassed attempts to rationalize these extensions either through hazy regulative ideals (Cassirer, but allied ideas trace back to Poncelet) or "true thought" recastings à la Karl von Staudt. The first approach was plainly too undisciplined to prevent mathematics from potentially falling into deep error, whereas the other program seemed preposterously tedious in execution and oddly irrelevant to the real mathematics at issue. So it struck David Hilbert (easily the most important figure within formalism's turn of the century triumph) and his many followers that all of these complicated "justifications" might be tidily evaded with a simple swipe of the axiomatic pen. Projective geometry, with its complex intersections, could be established with an axiomatic kingdom all its own, to which the more restrictive resources of a traditional Euclidean scheme can be profitably compared with respect to their theorems. This point of view was admirably advanced in Veblen and Young's Projective Geometry[29] of 1910 and, virtually overnight, eclipsed the boring "true thought" labors of von Staudt. Hilbert correctly believed that von Staudt had been asking too much of mathematical meaning and formalism, for the moment, seemed to supply a proper reason why. This Hilbertian

---

[27] R. E. Edwards, A Formal Background to Higher Mathematics, i (Berlin: Springer-Verlag, 1979), 14.
[28] H. M. H. Coxeter, Projective Geometry (New York: Blaisdel, 1964), 91.
[29] Oswald Veblen and John Wesley Young, Projective Geometry (Boston: Ginn and Co., 1910, 1918).

point of view is closely allied to the defense of "thinner content" with respect to physical predicates that we witnessed in Helmholtz.

It is important at this stage to distinguish between the crude formalism that Edwards apparently espouses ("working in mathematics represents a syntactically specified game analogous to Monopoly") and more sophisticated approaches such as Hilbert's own. The latter recognized that some curbs on formal procedure must be kept in place, lest the mathematician inadvertently spool out reams of worthless theorems that merely arise from some hidden internal incoherence buried within the formalism (any conclusion one likes follow by strict logic from premises that shield mild contradictions). Crude formalism of an Edwardian stripe must be supplemented with a stage of checking for *consistency or soundness* if formalist policy is to represent a viable methodological plan for mathematics, a fact to which mathematicians (who often embrace crude formalism as their preferred philosophy) are sometimes insensitive. I will return to Hilbert's legitimate concerns in section (x). Unfortunately, this adjoined necessity for checking consistency proves to be the little dangling thread that eventually unravels the comfy sweater of formalism, but we'll postpone these topics as well.

Turning from pure mathematics to the macroscopic descriptive predicates of greatest interest to us, the axiomatic recasting of Hertz' syntactic ambitions assumes the form: an adequately robust theory will set its theoretical predicates in a tight enough web of connection that such terms can be viewed as *implicitly defined* by the theory: it provides rules firm enough to govern their usage without the intercession of classical underpinnings. If we append the further thesis that the chief objective of the formalism axiomatics is to facilitate empirical prediction, we obtain orthodox *instrumentalism*.

Here we witness the philosophical center of that maddeningly persistent phrase, "implicitly defined by theory," a notion closely entwined with the "theoretical content" of which I complained in 3,vi. In fact, "implicitly defined" carries two historically established meanings and the tendency of philosophers to wobble between milder and radical pausings often generates considerable confusion. Insofar as I am aware, the phrase itself was introduced in the early nineteenth century by the geometer Joseph Gergonne, who derived it from the older idea of a quantity x that is implicitly delineated by an equation. Gergonne writes:

> If a proposition contains a single word whose meaning is unknown to us, the enunciation of the proposition is sufficient to reveal its meaning to us. If someone, for instance, who knows the words "triangle" and "quadrilateral," but who has never heard the word "diagonal," is told that each of the two diagonals of a quadrilateral divides it into two triangles, he will understand at once what a diagonal is ... Propositions of this kind, which give the meaning of one of the words contained in them in terms of others that are already known, can be called implicit definitions, in contradistinction to ordinary definitions, which can be called explicit definitions. We can also understand that ... two propositions which contain two new words, combined with known terms, can often determine their meaning.[30]

---

[30] Federigo Enriques, <u>The Historic Development of Logic</u>, Jerome Rosenthal, trans. (New York: Russell and Russell, 1968), 119–20.

Gergonnean implicit definability should be compared to the capacity to guess the meaning of a word from its context in a paragraph or being able to solve an equation explicitly for some component term (e.g., solving $x + xy = 6$ for y). As such, the original notion does not greatly differ from Russell's conception of a trait known through a descriptive route, rather than through head-on acquaintance. If I know my relative's tastes well enough, I can guess that "the vase is of my aunt's favorite color" attributes the trait of *being chartreuse* to the crockery. This mild approach to "implicit definability" does not claim that scientific language enjoys any species of non-classical semantic support; at best, the thesis reiterates the Russellian theme that the traits of deepest interest within a scientific investigation may not be truly grasped until late in the career of a theory that originally delineates them in terms of more superficial inferential characteristics.

In contrast, the radical reading of "P is implicitly defined within theory T" rests upon the instrumentalist assumption that P's syntactic webbing supplies it with an adequate "meaning." Modern writers who remain fond of phrases such as "implicitly defined" or "theoretically derived content" generally have this more radical reading in mind, albeit often left in an inchoate state.

In a physical context, axiomatic presentation alone cannot supply embedded predicates with adequate semantic content simply because the formalism isn't yet moored to physical application sufficiently. If we simply inspect our formalized principles, we are apt to not know what its subject matter is, for otherwise different areas of physics may share completely similar structures at a formal level (e.g., the well-known analogy between spring, block and dashpot mechanical systems and linear electrical circuits). In contrast, it is easy to determine what our Eulerian marching calculations concern, because palpable real world connections enter the scheme in the guise of the input and output statements that our routine grinds out (i.e., we feed the initial data "fired with an initial velocity of 83 ft/sec at an angle of 30°" into the hopper of our Eulerian meat grinder as input and it eventually grinds out the output prediction "hits the ground after 3.5 seconds 233.8 feet away"). Similar predictive inputs and outputs must be located within our axiomatics to supply its workings with a comparable instrumentalist flavor. Accordingly, many later thinkers, such as the logical empiricists mentioned in 3,viii, concluded that *bridge laws* to observation terms must be inserted as additional axioms within a physical theory, so that empirical predictions can be located as clearly defined paths of a formalism. To be sure, such bridge laws are never found within the axioms supplied in a real life physics text (such as Truesdell's First Course in Continuum Mechanics), but the logical empiricists believed that their inclusion is mandated by the need to credit physical predicates with a wholly syntactic significance. Because of this emphasis on theory facilitated prediction, the thesis of "semantic support through axiomatics" is generally called *instrumentalism*, rather than formalism, within a scientific context.

This same supplementary requirement for observation vocabulary forces most of the positivists into adopting a compromised form of *semantic dualism*: the observational predicates themselves ("is red," "is an ammeter") must garner their semantic significance the old-fashioned way: through classical gluing to genuine worldly attributes, albeit only of a macroscopic and easily observable class. Only the collection of theoretical predicates can profit from the conceptual freedom that axiomatic support offers; only these can gather their meanings in an entirely non-classical way. Some writers within this tradition struggled to evade this unattractive dichotomization, but with dubious success (Quine probably articulated the most successful attempt at a thoroughgoing anti-classicism, in a mode that we'll survey in the next chapter).

The dream of bridge principle supplementation to orthodox axiomatics proved impossible to work out. And the basic reason is rather simple: large objects like tables, ammeters and humans are complicated. In physics, we can rarely articulate a body of doctrine ably unless we deal with fairly minute objects in our fundamentals ("Physics is simpler in the small," runs the popular motto). But the objects that comprise our observations are large and bridge principle ties must reflect these quite complicated interactions. In most cases, the precise details of how commonplace measurement instruments work remain largely unknown to this day. The logical empiricist is left little choice but to allow her bridge principles to be loose and smoozy in their qualities, a trait that hardly comports comfortably with the strict axiomatics of a Hilbert. Early hopes that formalisms could be articulated that would sustain the semantic ambitions of the logical empiricists eventually evaporated away, faced with the sheer implausibility of writing down a believable bridge principle for any physical topic.

I believe that abandoning Hertz and Helmholtz' original illustrations of semantic support in terms of *algorithms* in favor of *axiomatics* was a mistake; that a vital clue to understanding how language is profitably structured has been left behind. To explain what I have in mind, it will be helpful to first extract a general notion common to a wide variety of anti-classical ways of thinking that I shall call *distributed normativity*.

(v)

Distributed normativity. Sometimes philosophical writers (e.g., middle period Wittgenstein) like to compare a smoothly running language to an effectively constructed mechanism such as a watch or efficient locomotive. Why makes this analogy so appealing? It is because mechanisms often illustrate a characteristic I shall call *distributed normativity*: some salient notion of "correctness" can be derived from the *global purpose* the device addresses. Consider, for example, the mechanical linkage illustrated, whose purpose is to mechanically calculate the natural logarithm ($\ln(x)$) of the number selected by its left hand stylus. As such, the gizmo might prove useful in equilibrating the ratio of steam to fuel flow within an engine. This global ambition of calculating $\ln(x)$ naturally induces an *internal evaluation* of the "correctness" of the device's component parts—viz., have they all been *sized properly* to allow the complete mechanism to calculate $\ln(x)$ as ably as possible? I call such standards of "correctness" *distributed* because they filter down to the components of the mechanism from its overall purpose.

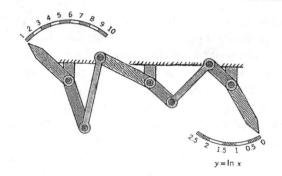

$y = \ln x$

It is a striking fact about invention within certain spheres (such as that of the planar mechanisms illustrated here—see 7,iv) that, once the basic topography of an invention has been roughed out, algorithms exist that can establish an optimal sizing of parts with respect to the purposes stated. Policies for doing so can be found in virtually every modern primer on design synthesis. A supervising engineer can therefore say to a pupil, "Oh, you've not yet gotten connecting bar 4 to the correct length yet. Fiddle with your sizings a bit more and you'll obtain a better performance." In making such claims, our tutor relies upon the distributed normativity available within this branch of engineering design. When artisans pronounce a drafting mechanism or a locomotive as "perfect," they are usually relying upon these same standards of how ably its individual parts contribute to its optimized final purpose.

After its parts have been correctly *sized*, we can likewise evaluate the "correctness" of a part's *performance*—does it move in the proper manner required to effect its global purpose.

Similar distributed normativities can be assigned to linguistic routines as well—indeed, the comparison renders the metaphor of language acting like a machine defensible. In a modern steam engine, old-fashioned valve regulation through clever mechanical linkages like the one we examined is likely to be replaced by *digital control*, where a little computer works a linguistics calculation of ln(x) from an assigned input x. Pretending for vividness that such a computer might mutter to itself as human calculators do, a linguistic calculation of ln(5) might pass through a sequence of linguistic stages such as the following:

1. *Let me guess at random that* $\ln(5) = 2$.
2. *Then* $1 + \sum 2^n/2! = 1 + 2/1 + 4/2 + 8/(3.2) + 16/(4.3.2) + 32/(5.3.2) + 64/(6.5.4.3.2) + 128/(7.6.5.4.3.2) + 256/(8.7.6.5.4.3.2) = 7.39$
3. *This guess represents a guess that is 2.39 too large.*
4. *Let me try a lower guess of* $\ln(5) = 1$.
5. *Then* $1 + \sum 1^n/1! = 2.72$
6. *This guess is 2.28 too low.*
7. *Let's try* $\ln(5) = 1.5$, *midway between the best previous high and low guesses.*

8 *Then* $1 + \sum 1.5^n \ 1/2! = 4.48$

9 *This guess is an amount .52 too low.*

10 *Let's try* $\ln(5) = 1.75$ *midway between the best previous high and low guesses.*

11 *Then* $1 + \sum 1.75^n / n! = 5.16$

. . . .

The rationale for this calculation is as follows. Begin with the equation $\varepsilon^y = x$ which codifies what $y = \ln(x)$ means. For small values of x, we can replace ε^y by the series expansion $1 + \sum y^n / n!$ which we then decide to terminate as soon as its terms become less than .01. We employ a scheme of successive approximations that frames a sequence of improving guesses as to what $\ln(x)$ might be following the flow chart supplied. That is, we systematically check our guesses at each stage by inserting them back into our $1 + \sum y^n / n! = 5$. Typically, these two sides will not match and we employ their discrepancy as a natural measure of the error in our calculation to date. We can frame a revised guess at $\ln(x)$ based upon the size of the previous error. The full procedure evinces the basic tenor of Goldilock's testing of the bears' porridges. Routines like this proceed by *successive approximation*: the pattern displays a basic computational strategy that we shall revisit from time to time in our discussion.

. . . .

A routine such as this represents an *algorithm*: a lineage of sentences (or numerical values) dictated according to precise rules, all of which serve to advance its final purpose (e.g., calculating ln (x) accurately). As such, a firm standard of "correctness" trickles down to the component sentences from that global objective, an evaluation that might potentially clash with a more classically founded notion of referential "correctness." Suppose we are trying to teach a pupil the routine and, at step 7, she writes, "Let's try $\ln(5) = 1.61$." "That's not the *correct sentence* to write now; why on earth did you write that?" we complain. "Oh, it just popped into my head," she responds, "but doesn't it qualify as the *correct answer* in any case? After all, the natural logarithm of 5 really is 1.61."

Two notions of "correct answer" are evidently in play in this dialog: a *distributed* one ("What is the correct sentence to write if the method is to achieve its final purpose?") and *directly supported* one ("Which sentences qualify as true given the normal references

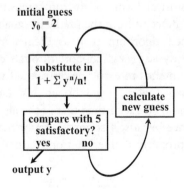

of its component words?"). I will later argue that both manners of "correctness" prove important to an adequate understanding of why language grows as it does: developmental stages where distributed normativity is dominant get seasonally supplanted by directly supported correctness and vice versa.

In the same manner as with our ln(5) calculation, the predicative purpose of our Euler's method calculations of cannon ball flight provides a distributed correctness to every sentence involving the word "force" we will be inclined to employ in such a context. For example, row five in the table of section (iii) abbreviates the claim, "After one second, the constant gravitational force acting on the projectile will have caused it to be 38 feet above the ground and traveling with an upward velocity of 18 ft/sec." In this case, it happens that this same sentence will qualify as nearby referentially correct in the circumstances posited but, as the ln(x) example illustrates, our intuitive standards of direct and distributed normativity needn't always agree. Indeed, that potential disharmony will frequently prove the origin of property dragging and the other unusual growth patterns in language we shall investigate.

Articulated in these terms, Hertz and Helmholtz hope that the distributed normativities derivable from "contributes to successful prediction" can supply a predicate like "force" with a sufficiently robust standard of "correct use" to serve science's interests; if so, the term needn't be glued to the world in any stronger fashion than that. On this picture, a predicate's usage is maintained in linguistic position through its syntactic embedding in the manner of the keystone of an arch; indeed, such metaphors are common in literature sympathetic to anti-classicism. Quine writes in Word and Object:

> In an arch, an overhead block is supported immediately by other overhead blocks, and ultimately by all the base blocks collectively and none individually; and so it is with sentences, when theoretically fitted. The contact of block to block is the association of sentence to sentence, and the base blocks are sentences conditioned . . . to non-verbal stimuli.[31]

That is, unlike the classical gluing needed to attach "is red" or "is a ball" firmly to the world, no Russellian universal must lie directly below "force" to supply it with adequate semantic heft, which it gathers instead from the syntactic instrumentalities it facilitates.

In the foregoing, I have utilized the top-down normativity native to Euler's method to illustrate the basic idea behind distributed semantic support, although, for the reasons already surveyed, most historical forms of instrumentalism claim that the applicable notion of "correctness" will descend from the predictive goals of an axiomatized theory, rather than from a localized algorithm such as Euler's method represents. Indeed, although Hertz' prose directly suggests "support through embedding within an algorithm," he almost certainly intends to extol the distributed virtues of "support through embedding within an axiomatic theory," even though this new flavor of top-down normativity proves rather different in character than that of the algorithm-derived standards.

I stress this vital difference because I seriously doubt that axiom-dependent normativity often represents a properly defined notion, partially because physical theory

---

[31] Quine, Word and Object, 11.

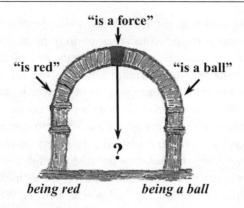

"is a force"

"is red"   "is a ball"

?

being red   *being a ball*

rarely addresses predictive goals exclusively and partially because the syntactic constraints that a set of axioms places upon usage are too weak to mark out any distributed "correctness" in themselves. Properly speaking, the "rules" codified in an axiom system represent mere *permissions*: "At this stage you may derive conclusion C if you wish." Such permissiveness allows practitioners to spend their linguistic lives endlessly extending the sequence "A", "A & A", "A&A&A", ... (where "A" is an axiom). No evident purpose is thereby achieved, but what does an axiom system <u>per se</u> care about *purpose*? In contrast, a recipe that directs a specific pattern of steps to be assembled under the umbrella of axiomatic permission is commonly called a *heuristic* nowadays. These, quite commonly, are allied to specific practical projects.

In my estimation, it is only the distributed normativities associated with focused heuristics that play a significant role in the semantic behavior of our descriptive predicates and they do so largely through a mechanism that will be called *property dragging nucleation*. To illustrate what I have in mind, consider the process, already discussed in section (ii), that pulls the predicate "frictional force" away from its original lodging over *true applied force* and deposits it upon *change in total momentum* when we shift from solid matter to the extended "particles" that arise in connection with a viscous fluid (recall that the latter gather their identities over time in ship of Theseus fashion). The mechanism that historically induced this shift is *imitative heuristics*: from roughly the time of Euler, a standardized recipe for setting up basic equations for a subject upon a "$F = ma$" framework has become canonical in physical practice. Indeed, many textbooks to this day follow virtually the same steps in setting up the Navier-Stokes equations for a *fluid* as they follow in deriving the Navier equations for an elastic *solid* (indeed, Charles Navier himself arrived at both of them in that manner, as the titles of these equations reflect). In both cases these productions are crowned by great practical success, for each supplies a model of critical importance within their respective dominions. Some species of *practical wisdom* must plainly reside in both forms of derivation: the Navier-Stokes equations couldn't have been so easily articulated if there wasn't something "right" in this borrowing of "$F = ma$" heuristics. But the "correctness" we here evoke is plainly that of a *distributed norm* descending from the utility of their common recipe: considered entirely from a *direct correctness* point of view, our derivation is "wrong" because the predicate "frictional force" does not stay in alignment with its previous signification (<u>vide</u> the discordant evaluations of "correctness" witnessed in the ln(5) example). In other words, when we blithely trust the Navier-Stokes equations on the grounds that they are "correct" for the same reasons as the elasticity equations, we unwittingly allow a heuristic strand of top-down correctness to *trump* the dictates of referent-based correctness (at least temporarily). I call this phenomenon *property dragging nucleation* because through this recipe-induced shift a new patch of classical physics comes into active development in which "force" attaches to a different attributive anchor than it had served before. We shall witness many examples of property dragging driven by allied distributed standards later in the book.

At first blush, such wandering referents seem as if they can only have deleterious consequences within a usage afflicted by them. Indeed, the classical picture takes it for granted that, had we been more vigilant in our thinking; we might have caught the drift in "force" when we moved over to liquids and therefore recommends that we strive to prove more diligent in our semantic attachments. But, oddly enough, this <u>prima facie</u> assessment isn't right: there are sound engineering reasons why distributed normativity crossovers often help a developing language remain in healthy condition. Indeed, the latter part of this chapter will articulate some of the basic reasons why this is so.

It's just as well that there's some utility in such crossovers, because, in point of fact, we lack any perfect prophylactic against their occurrence. Indeed, it is exactly here that classical thinking most plainly overextends its promises: it claims that, by simply thinking harder, we can become "more diligent in our semantic attachments." In many situations this hortatory advice will prove no more effective than the recommendation that we improve our nearsightedness by throwing away our eyeglasses. When we "grasp" a predicate according to normal standards, we engineer a thinner hold on its appropriate measures of correctness than classicism presumes and no degree of devout armchair meditation is likely to improve this situation. However, this is an unexpected moral that will require the full breadth of this book to redeem, although it represents a descendent of the same worries about "conceptual thickness" that bothered Helmholtz and Hilbert.

In any case, in the semantic tale I shall develop in this book, distributed normativity enters the story of language mainly as the driving force within the nucleation of fresh patches of usage at certain points in a predicate's career: I shelter no aspirations to employ top-down correctness as a means for supplying complete content to any predicate whatsoever. Oddly enough, if we refrain from the grander ambition of squeezing the full semantic significance of a term like "force" from exclusively distributed considerations, we will do a better job in redeeming the basic anti-classical hunches that lie latent within Helmholtz's musings (Chapter 5 will develop these propensities further under the heading of "pre-pragmatism"). And we will be able to do this in a manner that neither deposits us in holism nor leaves us stranded behind a bleak veil of predication. But before we begin to stroll along this chastened yet rewarding path, let us first ask why, in point of historical fact, thoroughgoing axiomatization did not manage to fully cure the ills to which "force" is naturally prone.

(vi)

Theory facades. The notion that certain terms might obtain their semantic significance entirely through theory-distributed means is quite pretty in conception. 'Tis a pity that the doctrine doesn't seem to be true of any real life words, which instead seem buffeted by variegated winds that blow from every corner of the compass. But, as we just observed, the narrower forms of top-down normativities associated with algorithms and heuristics can play substantive, if never completely determinative, roles within linguistic development.

In any case, the notion of implicit definability through axiomatics endured a slow and, to my eyes, rather sad, decline from the heady enthusiasm with which such proposals were greeted in the days of Hilbert. Two basic events occurred. On the scientific side, substantial attempts were made towards developing a more rigorously specified classical mechanics, largely because accurately auguring the complex behaviors of materials such as rubber, paint and toothpaste[32] required that the guidance of classical principle be considerably sharpened. The availability of computers furthermore demanded that their supportive mathematics be carefully scrutinized, because automatic computations supply absurd results when they move into regions where some derivative changes more rapidly than anticipated and other niceties of that ilk. As a result, quite sophisticated axiomatic formalisms for continuum mechanics were developed. A particularly well-known proposal of this type was advanced by Walter Noll,[33] although, for reasons I will explain in 4,viii, none of these treatments fully cover the expected domain of "classical behavior."

Most of this work was pursued within departments of engineering or applied mathematics, for physicists had meanwhile diverted their attention to quantum mechanics and relativity, which had come into prominence after Hilbert set his 1899 problem on the axiomatization of classical mechanics (indeed, their rise distracted Hilbert himself from his own efforts to resolve the problem he had articulated). Because of various mathematical analogies, the physicists gradually began to conceive of point particle physics—that is, a system of unextended masses acting upon each other over distances—as comprising the whole of classical mechanics, despite the fact that this subspeciality's inadequacies had been long appreciated. This shift occurred because the mathematics of point particles represents the chief part of classical tradition (besides electrodynamics) in which the quantum physicists took much interest. It is fairly easy to axiomatize this branch of classical thought, but many odd lacuna appear simply because the approach is too idealized to qualify as a plausible account of macroscopic matter.

---

[32] Frederick R. Eirich, Rheology, i–iv (New York: Academic Press, 1956).

[33] Walter Noll, The Foundations of Mechanics and Thermodynamics (New York: Springer, 1974). Yurie Ignatieff, The Mathematical World of Walter Noll (Berlin: Springer, 1996), ch. 9.

Meanwhile, on philosophy's side, the presumption that the essence of scientific theorizing—and the theoretical terms they implicitly support—can be captured in an axiomatic structure strongly dominates mainstream analytic thinking up to 1965 or so, as we noted in 3,viii. So firm was this faith that few thinkers paid attention to the struggles of the engineers to produce a reasonable facsimile of what was desired; the philosophers simply assumed by transcendental anticipation that axiom systems for mechanics *had* to exist.[34] When I was in high school near the end of this era, my older brother would bring philosophy of science books home from college and urge me to read them, rather than properly encouraging me in the usual frivolities of teenage life. Such works were typically filled with much abstract talk of "theory T" and its celebrated undescribed rival, "theory T'." Insofar as such schemes were ever illustrated, it was only through toy examples with axioms such as "iron rusts in water," "phosphorus smells like garlic" and the like.[35] Even as a rather unworldly youth, I knew that "phosphorus smells like garlic" could not be the stuff of which real theories are made and I expected that my first college physics course would reveal some more plausible axiomatic set. In a few paragraphs, I'll describe what I found when I got there.

Although the work of Noll and others proved very valuable within its own arena, the schematic make believe practiced by the philosophers occasioned a good deal of harm, from which our discipline has not yet recovered. In particular, the toy "phosphorus smells like garlic" examples suggested that the dominant inferential links within a "theory" could always be conceived in *logical terms*: as sundry cases of modus ponens universal instantiation et al. (any more specialized rule should be expressed as a non-logical axiom, it was argued). As I remarked before, it became common practice to conceive of "theories" in entirely generic and logic dominated terms: the "laws" of a theory represent universal generalizations of some sort; "initial and boundary conditions" supply the particularized data needed to get the laws to apply to a specific application, etc. Such terminology is borrowed from physical practice, but their significance is greatly distorted by the logic-centered focus ("boundary condition" particularly suffers this ignominy). As a result, the more substantive mathematical features of a physical treatment—the class to which its differential equations belong, for example—drop from attention as irrelevant filagree. Indeed, I have often heard academic philosophers declare that approaching the problems of science through logical scheme alone represents a great step forward, for such abstraction "allows us to determine the philosophical essence of a problem without the distracting details of substantive mathematics" (this is nearly an exact quotation from a talk I once heard, whose source I won't reveal since I regard the opinion as patently risible).

After 1965, through the criticisms of Quine, Thomas Kuhn and others, this simple faith in axiomatization and the distributed support it might supply eventually faded away, although not for altogether the best of reasons. Worse yet, a hazy sort of holism soon assumed axiomatics' former place of pride: it is still maintained that scientific words

---

[34] For a grouchy, but fair, critique of the philosophically inclined efforts towards axiomatization in this period, see C. Truesdell, "Suppesian Stews" in An Idiot's Fugitive Essays on Science (New York: Springer-Verlag, 1984).

[35] Israel Scheffler, The Anatomy of Inquiry (Cambridge, Mass.: Harvard University Press, 1963).

gather their significance through an embedding within an extended body of doctrine, but one that assumes the dimensions of a murkily delineated *paradigm, practice* or *web of belief* (Kuhn's and Quine's proposals in this fuzzy vein will be discussed in 5,xii). But we should resist drifting down these mazy trails, for they quickly lead to the dreadful post-structuralist claims of Chapter 2: e.g., the conceit that the humblest classifications of the folklorist are forever tainted by the social presumptions and privileges to which they unavoidably link, no matter what preventative precautions an agent might adopt. I regard all of these unfortunate attitudes as simply the result of having chosen the wrong fork at the crossroads of distributed normativity. All of these lingering holisms I intend to encompass under the heading of the *theory T syndrome*.

There is a standard criticism that is commonly leveled against the axiomatic picture: the so-called observation terms within a scientific practice should not be regarded as utterly free of theoretical content themselves, on the grounds that theory is required to know what an observation signifies. A little reflection shows that this is an odd way to articulate the objection. We've observed that "theoretical content" represents a philosopher's distinction originally engendered within the womb of implicitly-defined-by-theory presumption. But now that notion has somehow survived to challenge its own birthright—our modern critics conclude that *every* predicate acquires some degree of "theoretical content" from distributed sources more nebulous than axiomatics. Why not simply conclude that the original hope of semantically sustaining predicates within syntactic webbings was unrealistic? Why cling to an unmoored notion of "theoretical content" without benefit of axiomatics? The basic answer, insofar as I can discern, essentially traces to philosophy of language considerations: both Quine and Kuhn find the basic anti-classical tenor of implicit definition doctrine attractive and fear returning to the dens of out-and-out classical thinking (in 5,v we'll survey Quine's own account of semantic embedding in further details). In choosing this path, holist thinking retains many of the worse aspects of the theory T tradition, while abandoning axiomatics' admirable capacities for revealing the puzzling structure of classical mechanical thought in stable terms.

The standard criticisms just scouted apply only to the logical empiricists' determination to seek enlarged bridge principle plus physics conglomerates—they do not establish in any fashion that Hilbert's request for a philosophically unsupplemented axiomatization of classical mechanics is ill-founded. In point of fact, there are vital reasons why the real facts of usage within the classical physics realm cannot be neatly suited within the armor of an axiomatic frame, but these are completely different in character than holist critics assume. Instead, considerations of strategic complexity suggest the true reasons why practical schemes of language employment often fail to submit happily to axiomatic organization at the macroscopic level. Instead, policies of sensible *variable reduction* dictate that macroscopic doctrine is better arranged as a set of linked, but nonetheless disjoint, patches that shall be called a *facade* here. In this section, I will outline the basic phenomenology to be expected in a facade and then devote the rest of the chapter to explaining why this odd organization proves natural from a descriptive point of view (Chapters 6 and 7 will approach the same issues from another vantage point).

For orientation purposes, let me resume my tale of what occurred when I enrolled in freshman physics in search of theory T axiomatics. In the opening week, we were provided with Newton's laws, which certainly *looked* like the axioms I expected to learn (although I wondered why that "action = reaction" business was so vaguely articulated). After a few weeks, our attention shifted to beads sliding along wires and, for the life of me, I could not see how Newton's laws properly authorized the procedures we were now expected to follow (I'll detail my specific worries in 6,xiii). I asked my instructor about these, and he provided me with a very impatient "explanation" involving "internal and external forces" that didn't seem germane to my questions. The entire affair left me feeling as if I must be quite stupid. I stumbled through the course ably enough but didn't go near physics again for a long time thereafter.

Much later, when I again gathered the courage to dip my toes within mechanics' waters, I began to follow the chain of textbook footnotes that innocently begin "For more on this topic, see . . . " That policy led me into a labyrinth from which, even twenty years later, I have not yet managed to extricate myself. In particular, I quickly encountered what I like to call the *lousy encyclopedia phenomenon*, after a regrettable "reference work"[36] that my parents had been snookered into purchasing (the 1950s represented a notorious era of encyclopedia mania[37]). As a child, I would eagerly open its glossy pages to some favorite subject ("snakes," say). The information there provided invariantly proved inadequate. However, hope still remained, for at the end of the article a long list of encouraging cross-references was appended: "for more information, see **rattlesnake; viper; reptile, oviparous** . . . " etc. Tracking those down, I might glean a few pitiful scraps of information and then encounter yet another cluster of beckoning citations. Oh, the hours I wasted chasing those informational teasers, never managing to learn much about snakes at all!

In truth, this same unsatisfying process occurs in classical physics when one follows its characteristic chains of footnotes (although, unlike that boyhood encyclopedia, quite substantive amounts of useful information are gathered at each way station in the journey). Consider the popular categorization of classical physics as *billiard ball mechanics*. In point of fact, it is quite unlikely that any treatment of the *fully generic* billiard ball collision can be found anywhere in the physical literature. Instead, one is usually provided with accounts that work approximately well in a limited range of cases, coupled with a footnote of the "for more details, see . . . " type. For example most undergraduate primers in mechanics highlight a treatment that essentially derives from Newton (sometimes supplemented in the better books by allied considerations involving rigid body motions due to Euler). But such techniques can supply reasonable answers only with respect to a limited and unrealistic subset of billiard problems, as simple equation counting readily establishes (the technique does not provide enough data to resolve what happens in a triple collision, for example). Even more oddly, many of the chief events involved in a collision are not mentioned in the Newtonian treatment

---

[36] The World Book Encyclopedia (Chicago: Quarrie Co., 1953).
[37] Dwight MacDonald, Against the American Grain (New York: Random House, 1962).

at all. Real spheres distort severely under impact, as a snapshot with fast film readily demonstrates but the Newtonian scheme speaks nothing of this. In fact, we will be immediately warned of the deficiencies of the Newtonian approach if we track down a specialist text on impact by following the trail of footnotes:

> The initial approach [historically] to the laws of collisions was predicated on the behavior of objects as rigid bodies, with suitable correction factors accounting for energy losses. It is interesting to note that this concept has survived essentially unchanged to the present day and represents the only exposition of impact in most texts on dynamics.[38]

It is important to observe that the specialist texts do not simply "add more details" to Newton in any reasonable sense of that phrase, but commonly overturn the underpinnings of the older treatments altogether. In the case at hand, the entire mathematical setting is replaced: specifically, the Newtonian treatment utilizes ordinary differential equations, whereas the specialist texts employ partial differential equations of some class, which, from a mathematical point of view, represent an altogether different breed of critter. This shift allows the specialist texts to characterize the flexibilities of the balls within their treatments, although once again, several layers of coverage of increasing scope can be found along the chain of footnotes. At the next stage of detail our balls will usually be treated according to a *quasi-statical* policy pioneered by Heinrich Hertz: the collision events are broken into stages that are assumed to relax into one another in a "finds a local equilibrium" manner.[39] This method provides very nice approximations for an important range of cases, but there are plainly billiard ball events—when wave movements initiate within the balls—that fall outside its range of application. Again we can easily find treatments that take up those factors, again with mathematical and physical factors emerging into centrality that had passed unmentioned before: weak solutions and thermodynamics, in the situations when the waves form shocks.[40] High speed collisions at explosive velocities bring an entirely new range of effects in their wake.[41]

To the best I know, this lengthy chain of billiard ball declination never reaches bottom. We shall want to learn why such lack of final foundations is to be inherently expected within classical mechanics' realm.

To this end, it is useful to picture situations like this as series of descriptive patches that link to one another via "for more detail, see . . . " linkages. Patchwork arrangements of this general type, which we will frequently discover in our examples, shall be called *facades* here: they represent a basic form of the polycrystalline structuring of language mentioned in Chapter 1. Applied mathematics suggests sound strategic reasons why a practical descriptive language will sometimes assume such oddly disjointed forms. Recognizing the positive virtues of a facade is possibly the best route to understanding the general approach to natural linguistic development advocated in this book.

---

[38] Werner Goldsmith, Impact (Mineola, NY: Dover, 2001), 1.
[39] Heinrich Hertz, "On the Contact of Elastic Solids" in Miscellaneous Papers by Heinrich Hertz (London: Macmillan, 1896). K. L. Johnson, Contact Mechanics (Cambridge: Cambridge University Press, 1987).
[40] Michel Frémond, Non-smooth Thermo-mechanics (Berlin: Springer, 2002).
[41] Marc André Meyers, Dynamic Behavior of Materials (New York: John Wiley and Sons, 1994).

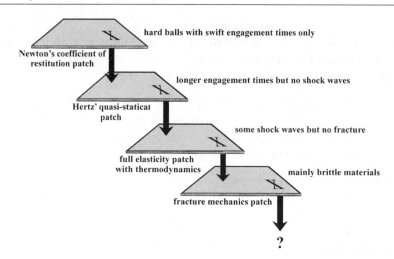

Returning to my problems with the bead sliding along a wire, it turns out that a similar chain of "follow the footnote" qualifications can be found here as well, which, in their more specialized levels, struggle mightily with the exact issues that I had raised with my instructor (details will be provided in (6,xiii)). It would have been far better for me if my instructor had notified me of this simple fact (although I doubt that he was aware of it himself), just as it is regrettable that many entry level texts improperly foster the illusion that the contents they provide handle the affairs encountered upon a pool table with perfect satisfaction (even if their footnotes renege on the promises tendered). To be sure, the capacity to steamroller over delicacies enjoys its own vital, if rude, rationale, for it allows a physicist, in Hertz' words, "to get on to examples which speak for themselves." An instructor can rapidly build up a facade that, in terms of bare-boned efficiency, may prove to be optimally effective in addressing the relevant physical events with minimal pedagogic fuss. From this point of view, it was the naïve theory T expectations I acquired from my brother's philosophy books that were at fault: I expected uniform axiomatics within a dominion that is better approached <u>via</u> patchwork facade.

But why are such methodological issues so often left enveloped in fog? Why do writers of elementary textbooks invariably adopt a tone in which "classical mechanics" is presented as a compact and neatly unified subject, well known to all, when this hyperbole merely wraps an "emperor's new clothes" obscurity over a more complexly structured situation? Such procedures scarcely invite clear thinking and needlessly scare away the many rational souls (mathematicians and lay people, as well as philosophers) who might otherwise enjoy physics.[42]

---

[42] Mark Kac comments:

In kinetic theory volumes Δv "small enough to be taken as elements of integration yet large enough to contain many particles" rendered [thermodynamics] unpalatable and even repulsive to a young mind already conditioned to look for clarity and rigor.

T. W. Körner, Fourier Analysis (Cambridge: Cambridge University Press, 1988), 176.

Part of the reason, I think, traces to the continuing hold of the classical picture of concepts—or, at least, of the *ur*-philosophical sources from which it springs—, for those ingredients lead physicists to look upon their facades in a largely classical manner. Rather than appreciating the concrete practical considerations that naturally join "force" as it is used on sheet A of the mechanical facade to its somewhat different employment over sheet B, the physicists assume that they grasp some ineffable general conception of *force* that binds these employments together in hard-to-articulate commonality: "The same notion is plainly involved in both cases—I can feel their kinship in my bones, although I can't explain its basis to the mathematicians' satisfaction." Such sentiments then engender a misty conviction that physicists enjoy special powers of intuition, while the mathematician's sharper scruples are held in undeserved contempt (Richard Feynman represented a fount of such dubious opinion). Such attitudes substitute mystic conceptual intimations for the complex, but not particularly foggy, factors that build up a facade-like structure around "force." No doubt my freshman instructor's impatience with my trifling "philosopher's questions" was grounded in some measure of this oracular arrogance as well. Like the Pearson case, such views substitute hazy *ur*-philosophy for genuine mathematical lapses.

All of this provides another illustration of the ways in which *ur*-philosophical thinking about concepts occasions unexpected harms elsewhere. At first sight, popular malarky about the physicist's "intuition" seems as if it merely represents a harmless display of self-congratulatory vanity. But such views plainly provide meat and drink for the obnoxious "big idea" prejudices of our Chapter 2 moralist and the unproductive ditherings of the teenager we shall meet in 8,i. Textbook braggadocio with respect to billiard balls can exert deleterious effects even upon faraway subjects such as the history of philosophy, where the labors of a Descartes or a Leibniz are regularly patronized for getting "Newton's rules for impact wrong" (such Newton-biased commentary is plainly insensitive to the conflicting strands that weave deeply through mechanical thought everywhere). What compulsion drives us to claim that we know more about billiard balls than we really do?—why do we regularly pretend hard things are easy?

In the days of old Hollywood, fantastic sets were constructed that resembled Babylon in all its ancient glory on screen, but, in sober reality, consisted of nothing but pasteboard cutouts arranged to appear, from the camera's chosen angle, like an integral metropolis. In the billiard ball case, we witness sheets of mechanical assertion that do not truly cohere into unified doctrine in their own rights, but merely appear as if they do, if the qualities of their adjoining edges are not scrutinized scrupulously. And this capacity for *doctrinal mimicry* is the aspect of my facades that I shall emphasize most in the sequel: they represent patchworks of incongruent claims that might very well pass for unified theories, at least, in the dark with a light behind them.

As we move forward, it is important that we look upon the virtues and vices of facades in a complimentary fashion. For engineering a descriptive language to suit complicated circumstances, a facade foundation can prove very effective. On the other hand, these structures can promote deep misunderstandings if their supportive architecture passes unrecognized.

(vii)

Variable reduction. It is common nowadays to encounter commentary such as the following: "A subject like classical physics is not really a 'theory' in the old-fashioned sense, but a *practice*, woven together by the techniques that practitioners acquire from their community." I find glosses of this type singularly unhelpful, for their import usually shifts rapidly betwixt tautology and outright falsehood. I believe many writers get drawn into the fuzzy lair of "practice" because they fancy that science's vocabulary can become entangled within a public web thereby.

Such a societal focus is apt to distract us from addressing a serious question that enjoys a quite interesting answer, at least in my opinion. As we noted, Hilbert included the axiomatization of classical physics on his famous 1899 list of problems that mathematics should address during the coming century:

> The investigations on the foundations of geometry suggest the problem: To treat in the same manner, by means of axioms, those physical sciences in which mathematics plays an important part; in the first rank are the theory of probabilities and mechanics.[43]

But if mechanics cannot be successfully regimented in this form, then it should be worth understanding why. After all, once philosophical demands that bridge principles be included in our axiomatic mix are set aside, then the call for axiomatics amounts to little more than a request for a clear articulation of doctrine. And what could be wrong in that? Surely it is better to articulate what we wish to say in crisp terms rather than reveling in the ill-defined and loose?

In a nutshell, a proper answer traces to the fact that the macroscopic objects we attempt to treat in classical mechanics are enormously complicated in both structure and behavior. Any *practical vocabulary* must be strategically framed with these limitations firmly in view. To be able to discuss such assemblies with any specificity, our stock of descriptive variables must be radically reduced, from trillions of degrees of freedom down to two or three (or smoothed out to frame simpler continua). Even systems that are quite simple in their basic composition often need to be partitioned into more manageable subsections, either spatially or temporally. For example, consider a hemispherical cup with its rim welded to a table. If we treat the cup as a continuous shell of two-dimensional metal,[44] the governing equations are simple in form, but the distribution of induced stresses will prove fairly complicated, especially near the bottom of the cup. A standard means of approaching this situation is to drop the terms from the master equations that become appreciable in magnitude only when the local bending in

---

[43] Hilbert, Problems, 15.

[44] James G. Simmonds and James E. Mann, A First Look at Perturbation Theory (Mineola, NY: Dover, 1998), 2. Their discussion implicitly begins with a one-dimensional equation where the configuration of a slice through the cup is expressed in terms of a thickness parameter h/R. But the two-dimensional shell equations from which this situation descends involve a comparable drop from three to two dimensions. This transition likewise involves a (hard to justify) boundary layer style decomposition as well. Cf. Diarmuid Ó Mathúna, Mechanics, Boundary Layers and Function Spaces (Boston: Birkhaüser, 1989).

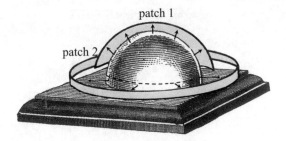

the metal is severe. If we do this, a greatly simplified formula emerges that predicts a constant stress everywhere in the smooth upper portions of the cup. However, this approximation is not reasonable near the welded rim where the material curves sharply and the induced stresses vary rapidly. So we go back to our original equation and enforce a different policy of simplification. We then obtain a so-called boundary layer equation that calculates the sharp increase in stress near the table top quite effectively. In working this out, we match the edge values of our bottom strip to those at the boundary values of our cup's top. In short, we *descriptively cover* our welded object with two *patches* of different mathematical types: the first that handles most of the interior and the second that treats the narrow band of high stress near its rim. Notice that each localized representation leaves out important aspects of the governing physics that prove important in the patch next door (through dropping relevant terms from the cup's original equations). I might also mention (we'll come back to this topic later) that the complicated join region between the two patches actually corresponds to a finite belt around our cup, even though it is represented as a simple bounding line in our reduced, two patch description.

This flavor of variable simplification is usually called *boundary layer technique* after its famous employment in the early 1900s by Ludwig Prandtl.[45] In that original context, complex equations formulated by Charles Navier and George Stokes govern the internal behavior of an incompressible fluid (such as water) that opposes shearing with a minuscule degree of resistence. So small is this friction that earlier mathematicians commonly omitted the terms that govern its influence, obtaining Euler's (frictionless) fluid laws as a result. Some simplification was required because, as a piece of mathematics, the Navier-Stokes equations represent celebrated tough customers unwilling to divulge their behavioral secrets to virtually anyone (utilizing our highest capacity computers, for example, a smooth N-S solution can be projected about 1/5 of a second into the future, after which accumulated roundoff error completely swamps the validity of our results). Unraveling the mysteries of fluid turbulence is commonly cited as one of the greatest open challenges in macroscopic physics and its issues have proved intractable mainly because of the truculent nature of the N-S formulae. Such ornery behavior encourages

---

[45] Herbert Oertel, ed., <u>Prandtl's Essentials of Fluid Mechanics</u> (New York: Springer, 2004).

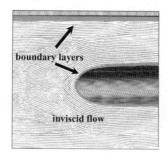

the Eulerian simplification but these frictionless simplifications display a wide variety of counterintuitive consequences—viz., airplane wings should experience neither drag nor lift (leading many nineteenth century experts to glumly predict the impossibility of heavier than air flight). However, Prandtl recognized that near pipe walls or airplane wings, the fluid must remain motionlessly attached to their surfaces, inducing a very sharp variation in the fluid velocity within a small layer along the boundary. This large change turns on the friction-related term in the N-S equations, no matter how small its coefficient of viscosity might be. Prandtl recognized that, if the fluid didn't become too turbulent, that he could reasonably join together appropriate simplifications of the N-S equations in the manner of our welded cup and thereby describe our flow in plausible approximate terms (as long as it remains laminar).

It can be helpful to picture the general problems of variable reduction in the abstract manner favored by applied mathematicians. The full behavior of a physical system can be symbolized by the motion of a point buzzing about within some high dimensional phase space, which we can portray as a complicated surface of possibly infinite dimension. As the "point" (which may represent a huge swarm of fluid molecules) moves around in the phase space, its component parts get assigned different mixtures of positions and velocities that completely fix its current state and disposition. Accordingly, a small swarm of one hundred non-rotating "molecules" will live in a phase space of six hundred dimensions! Obviously, these are too many variables to handle conveniently even on a computer. When we seek a set of *reduced variables* that can efficiently capture the *main features* of our swarm's complicated behavior, we are, in effect, looking for some simpler, lower dimension manifold to which the true

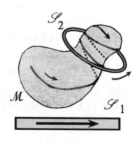

transient response:

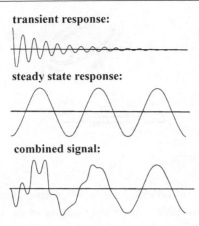

steady state response:

combined signal:

motion of our liquid will stay approximately close, at least for considerable portions of its career. In the picture, the chicken-shaped object μ is supposed to represent such a hypothetical reductive submanifold: the hope is that the interactions between fluid molecules will keep the swarm's system point buzzing fairly close to the chicken-shaped surface.

Quite often—and this situation closely resembles the problematic addressed by the boundary layer approach—, even this simplified manifold may be too hard to treat directly with computational effectiveness, so sometimes the system's behavior is further factored into different temporal epoches, matching each era to motions within even simpler submanifolds S_1 and S_2. The basic ploy is much like the decompositions of boundary layer technique, but our problem is now divided into distinct *temporal intervals* rather than the *spatial regions* of the cup case. For example, suppose we suddenly apply a steady vibration (an A 440 tone, for example) to a telegraph line. The best way to understand the circuit's reaction is to divide its behavior between the *transient response* that dominates when the early stages of our circuit's career are first applied and the *steady state response* that eventually prevails after the aftereffects of the initial disturbance have died away. Usually the transient response takes the form of a large, spiky pulse that gradually diminishes. If we pay attention to its patterns only, it can be modeled as simple decay, which we regard as occurring in the linear submanifold S_1 sitting near to our chicken planet. Eventually, our circuit will subside into a periodic forced oscillation (not necessarily at A 440) which again we treat separately within a circular submanifold S_2 whose system point travels around and around the loop. If we wish we can now erase the chicken planet as descriptively useless and regard the transient and steady state manifolds S_1 and S_2 as two large planets embedded in the larger phase space. We then picture the system point that represents our circuit as a little airplane that flies very near these celestial bodies. During the first part of its travels, our airplane hovers very near the surface of the transient planet, without landing on it, but eventually zooms off to float over the steady state planet, once again without ever completely landing. Our

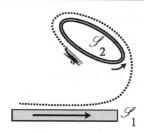

aircraft is not allowed to land on either surface because, at any point in real time, the circuit's true behavior represents a weighted mixture of the two types of response, so that some small measure of transient response always lingers in the circuit no matter how long we wait. In the usual jargon, our system point only approaches our two planets *asymptotically*. Accordingly, we calculate the smoother behaviors witnessed in manifolds S_1 and S_2 and interpolate our results boundary layer-style over the period of time in which our little airplane is busy traveling from one planet to another (so the "join region" is not treated as a singular boundary, but as a mushy segment that we characterize by simple extrapolation between S_1 and S_2). We obtain our desired reduced variables (here the degree of transient decay and the steady state oscillation, respectively) from the local geography of the planets upon which we have allowed our circuit's representative point to temporarily land.

We shall revisit this important notion of a complex behavior staying *asymptotically near* some simpler behavior from a number of points of view throughout the book; it is critical to understanding the oddities of many types of descriptive behavior.

As noted, in these circumstances we witness a *temporal form* of descriptive bifurcation, rather than the *spatial decomposition* illustrated in our two boundary layer cases, but the basic intent of the reductive strategy remains essentially the same (there are many other factorization policies possible, such as a decomposition into "fast and slow variables," but we won't pursue those here). We might also observe that in this case the "transition region" between our transient and steady state regions is treated as being of *finite duration*, rather than simply butting one asymptotic region against another, as we did in the boundary layer cases (it is sometimes useful to make the transition region larger in fluid cases as well). Usually we employ some simple interpolation scheme to patch over the transition region—we do *not* want to make any detailed attempt to describe what actually occurs in this region.

Indeed—and this is the truly striking methodological ploy illustrated in all these maneuvers—, we achieve our reduced variable simplicity precisely by sweeping most of the difficult physics into regions we do not attempt to describe accurately: I call this a policy of *physics avoidance*. And the general idea is this: if we can examine a situation from several sides and discern that some catastrophe is certain to occur, we needn't describe the complete details of that calamity in order to predict when it will occur and what its likely aftermath might be ("There's going to be a war here and the country will

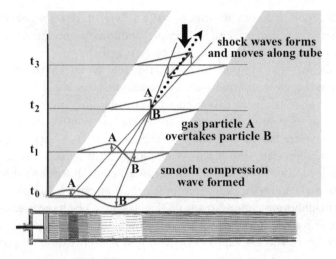

t_3

shock waves forms
and moves along tube

t_2

A

B

gas particle A
overtakes particle B

t_1

A

B

smooth compression
wave formed

t_0

A

B

Shock wave formation

be destitute thereafter"). This may sound silly, but it's exactly the policy enforced within one of the great paradigms of "physics avoidance": Riemann and P. H. Hugoniot's celebrated approach to shock waves.[46] Suppose we put our gas in a long tube and give it a violent shove on one end. There is a simple equation that describes our gas as a continuous fluid, subject to a little viscosity. Now if the initial impulse is strong enough, the faster molecules in the pulse will eventually overtake their slower moving brethren ahead and create an awful *shock wave* pileup, like the traffic snarl that would occur if our molecules had been automobiles. From the point of view of our continuous gas equation, this situation represents a descriptive inconsistency, for our equation actually predicts that our gas must display *two* distinct velocities at exactly the same spot and time (in the jargon, its characteristics cross). Prima facie, one would expect that this apparent contradiction in the mathematics will force us to abandon our smoothed out fluid description and turn to the complex details of how discrete molecules interact when forced into such close quarters. "Don't be so hasty," announce Riemann and Hugoniot. "We can accurately predict from the gas's ingoing behavior when the shock wave is going to arise and how much gas momentum will be funneled into that event. Moreover, by appealing to thermodynamics, we can also predict how the gas on the other side of the shock front will flow smoothly away from the event. And by piecing this two-sided information together, we can predict exactly how fast the shock wave will move down the tube, without needing to know the complex details that occur inside the shocked region." Thus the Riemann-Hugoniot policy sweeps what, in real life, represents a narrow but still finite region of shocked air into a two-dimensional boundary that separates regions of smoother gas. The treatment descriptively collapses a finite area of great complexity into a *singularity*: a lower dimensional boundary or point separation.

[46] James N. Johnson and Roger Chéret, Classic Papers in Shock Compression Science (New York: Springer, 1998).

Riemann and Hugoniot do not attempt to write a "law" to directly govern the shocked area's behavior; they instead employ simple "boundary condition" stipulations to dictate how the two smoother gas regions piece together.

However, the fact that a region can be descriptively avoided in this manner does not indicate that it is therefore unimportant: the condition at the shock front represents the most important physical event that occurs in our tube. It is merely that we can keep adequate track of its overall influence in a minimal descriptive shorthand, just as "a terrible war between North and South occurred in 1861–5" may supply sufficient information to appreciate the Civil War's long term effects upon our country adequately. Indeed, the whole idea of *variable reduction* or descriptive shorthand is that we are able to locate some shock-like receptacle that can absorb complexities and allow us to treat its neighboring regions in a simplified fashion. The basic Riemann-Hugoniot moral sounds like a methodological paradox when stated bluntly: a good recipe for achieving descriptive success *papers over* the physical events most responsible for the phenomena we witness! But that, in fact, is the manner in which successful variable reduction typically works.

The usual elementary physics approach to billiard balls utilizes virtually the same idea to obtain passably accurate results for simple collisions. Devised by Newton, the basic trick is to almost—but not completely—cover the history of our colliding balls with two descriptive patches, one devoted to the balls as they approach the collision and the other as they scatter away from it. But the actual events of compression and reexpansion that occur when our two balls contact one another are set within a little window that our method does not attempt to describe. Instead, we bridge over this temporal hiatus by matching our incoming and outgoing sheets according to a rule of thumb involving gross energetic qualities and a crudely empirical *coefficient of restitution* (in the simplest—and most inaccurate—treatments, one simply assumes that the balls

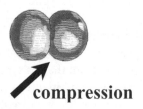

compression

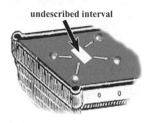

undescribed interval

are "elastic"). The rough reasonableness of such approximation can be justified by Riemann-Hugoniot style considerations, but it is plain that our method collapses the central causal events into an untreated temporal singularity. Notice how all the moments in which real spheroids display distortion have been swept into the collision singularity: Newton's treatment doesn't provide a whisper of a suggestion that billiard balls might be flexible.

But, in the long run, this approach is too crude to handle the blows encountered in, e.g., sophisticated aircraft design, where an entirely new mathematical army (partial differential equations et al.) must march on the scene like cavalry reinforcements. As we saw, in many books, the first wave of this incursion follows a strategy devised by Hertz, that breaks histories of our colliding balls into discrete stages whose compressed states are assumed to relax into one another quasi-statically. But, in typical lousy encyclopedia manner, this treatment merely represents a (very valuable) stopgap, for Hertz' recipe isn't adequate to substantive internal wave motion or truly violent impact, where shock waves will form as well.

....

We should observe that, by utilizing some important considerations of Euler's with respect to rigid body behavior, the Newtonian coefficient of restitution approach can be improved to handle oblique impacts with tolerable success and supply predictions more or less adequate to most—but not all!—standard billiard table events. But I've omitted this intermediate strategy from my story, which is complicated enough as it is.

....

And this is why we confront the complicated situation illustrated in the lousy encyclopedia diagram of the previous section. Over the real world of anticipated billiard behaviors there float several different descriptive patches representing different recipes for describing and reasoning about our real world events. The highest layer corresponds to Newtonian's coefficient of restitution strategy and covers more or less adequately an incomplete range of real world histories (hard balls with brief encounter times). When we attempt to apply this treatment to more sustained collisions, we encounter a "for more details, see . . . " link that drops us into the Hertzian plot offering a considerably different approach to similar events. But this methodology breaks down in turn for severe impacts and we are shuttled onto the considerably more complicated methods utilized in the "full elasticity" patch. And so on. Each of these local arenas share generally the same vocabulary in common ("mass," "shape," etc.), but they individually narrate rather different stories with respect to the events they cover (balls do not alter their shapes in the Newtonian accounts but they do in the other treatments; they do not transmit waves in the Hertzian picture, etc.). Typically, quite different mathematical tools supply the inferential engines that drive the reasoning within each patch.

A descriptive complex of this quilt-like pattern supplies a good example of what I intend by a *facade*: a set of patches or plateaus that are formally inconsistent with one another but are stitched together by "for more details, see . . ." linkages or other bridgework. Often the whole is fabricated in such a manner that, if we don't pay close attention to its discontinuous boundary joins and shifts in mathematical setting, we might suppose that we are looking at a theory ready to be axiomatized (recall the Hollywood sound stage analogy that motivates my choice of the "facade" label). Indeed, if those "for more details, see . . ." remarks were literally true—that is, if we truly encountered simple elaborative extensions of the treatments witnessed in higher patches—, then we might very well be looking upon a genuine theory. But in a true facade, something more radical occurs, for the patches do not cohere with one another and important physical information is secretly encoded into the discontinuous boundaries between sections, as we'll observe in the next section. To be sure, we may still feel that our local treatments "don't really clash with one another in any serious way," but this hazy impression of "family resemblance" shouldn't cause us to over-look the quite interesting forms of data registration that facade organizations permit. Unfortunately, our strong *ur*-philosophical inclinations towards a classical picture of concepts encourage us to overlook this vital informational possibility. Instead, we automatically assume, in the absence of much direct evidence, that there *must* be some lowest level treatment of Newtonian physics that embraces, in principle, the descriptive virtues of all of the higher platforms and will thereby accept a uniform axiomatization over its full basement dominion. To be sure, no one of a practical frame of mind would ever choose to toil amid the fussy mathematical complications native to this subterranean layer, but we feel certain that such foundations are down there, regardless.

Well, it is natural to make suppositions such as this, but, as we'll learn in the next section, they are probably mistaken: classical mechanics doesn't possess a lowest uni-formizing layer of the presumed type. And the chief sermon our discussion strives to preach is: that absence doesn't prevent Newtonian physics from serving as a dandy information-bearing structure in its own right. Those who posit basement chambers they've never visited should recall the gullibility of the innocent souls who observe the clever montages on the movie screen and exclaim, "My, it must have taken a lot of bricks to build a city that big."

In other words, a strong and unverified faith in classical physics' guaranteed axiomatizability generally stems from a *false picture* of how its admirable stock of predicates gather their descriptive utilities: there are important alternatives—including my facades—that have been overlooked. And that mistake, in microcosm, encapsul-ates many of the basic mechanisms responsible for the other *ur*-philosophical difficulties we explore in this book. Anytime we blithely presume that the "conceptual contents" attached to a passel of predicates behave in the simple manner sketched within the classical picture, we are in danger of building ourselves up for an awful letdown, as Fred Astaire once put it: some unfortunate *ur*-philosophical muddle may lie in the offing. That warning of optimism-induced error represents the chief message of this book, which we will examine from various vantage points throughout the book.

(viii)

A funny thing happened on the way to the formalism. Let us now explore how our two themes—facade structure and variable reduction—relate to one another. First of all, it is easy to see that any effective policy of variable reduction is apt to create a need for linked satellite treatments in the mode of the lousy encyclopedia phenomenon. These chains of connection arise because the coverage offered within a local patch can rarely reach all of the real world cases we intuitively expect to handle—if not, significant variable reduction would be likely impossible. Within the scope of any particular patch's coverage, there will generally appear black sheep that refuse to submit to the policies of physics avoidance locally practiced, simply because the physical effects we have man-aged to suppress elsewhere become quite important with respect to these prodigal cases. Their behaviors can't be profitably sectioned into simpler regions because they stay complicated everywhere. For example, suppose we have water running through a pipe. If the flow is not very intense, Prandtl's boundary layer trick allows us to factor the fluid into two regimes: near the wall and out in the free stream, where the dominant physical effects simplify in different ways (in their interfacial region, the active physics remains quite complicated but we can safely interpolate over this volume because it's fairly small). But let us now speed up our flow a little (that should be okay; the situation should belong to the same physical family as before: merely water moving down a pipe a little faster). But now our system acts as an uncooperative sibling to those considered before: the water turns turbulent and won't submit to simple boundary layer technique at all. The regions of complicated physics that we could previously confine to narrow wedges of interpolation now reign everywhere in our pipe. To describe our faster moving fluid adequately, we must regretfully leave the land of boundary layer theory and take up residence in a more complicated mathematical patch: the kingdom of the unreduced Navier-Stokes equations. Would that anyone knew exactly how we might *reason* there effectively!

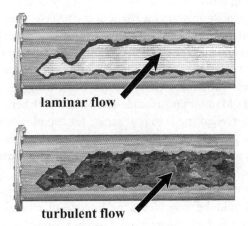

laminar flow

turbulent flow

Plainly, such black sheep cases are practically unavoidable under any policy of variable reduction: circumstances will always arise that demand that we open up internal degrees of freedom that we have elsewhere crushed into singularities or swept into approximate bounding conditions. Thus, as we drop into the lower layers of our billiard ball cascade, degrees of movement or temporal events get unfrozen within our balls that we had treated as approximately rigid or static in the platform above. Or, to vary the example, if we allow the gas in our tube to become too rarified (or if we need to examine the local shock front structure more finely), we will be forced to abandon our convenient reliance upon the smoothed out Burger's equation and must consider instead the messy statistical mechanics of a huge swarm of individualized gas molecules. Notice that this shift again completely alters both the ontology and the mathematics of the previous patch. So the customary price of practicing sound physics avoidance is that we must expect that our efforts will need to be trailed by a pack of incongruent satellite treatments, where some effort is devoted to the rebellious lambs that elude our own descriptive techniques.

....

Incidently, the physics avoidance practiced in these satellite annexes will not necessarily prove less *extreme* than those adopted within the perimeters of the Newtonian treatment; it may be simply different. Thus under Hertz' quasi-statical approach, the capacity of the balls to carry waves becomes suppressed through the background appeals to moment-by-moment equilibrium. In some circumstances, the cruder coefficient of restitution approach can supply more reliable predictions than this technique.

....

Besides the appendages motivated by black sheep cases, promising collections of physical doctrine often enlarge surreptitiously into patchwork organization through the mechanism of *property dragging nucleation* discussed earlier. It was completely natural for Charles Navier to pattern his recipe for obtaining equations for viscous fluids after his successes in setting up a model for elastic solids. But in so doing, the physical correlates of the innocent-looking term "particle" become slightly twisted, so that this classification now attaches to a more abstract invariant of conserved transport, viz. that supplied in the ship of Theseus reading sketched above. Although this subtle shift would have been impossible to recognize at the time, it becomes mandatary to pay some attention later on, as confusing ambiguities about "force" and "conservation of mass" emerge (the simplest curative is to warn researchers against borrowing results about liquids too hastily from the solids). Maintaining a facade-like bridgework between ingots of iron and tubs of water makes excellent pedagogical sense, for the ploy allows the basic map of classical success to be placed before the novice with remarkable efficiency, although a later need to compensate for the tacit property dragging through border crossing restrictions is likely to arise.

Here a toleration of property dragging should not be regarded as necessarily a mistake: a facade should be considered as an organizational structure possessing advantages all its own. Used wisely, its quilted patches can provide a platform for useful descriptive practices in remarkably effective ways, nicely engineered to evade many of the convolutions that more straightforward "is a dog"/*being a dog* arrangements would confront. Indeed, if we take our rather limited capacities for stringing bits of language together into consideration, a facade platform may sometimes provide the only descriptive scheme available to us (a theme to be developed further in Chapter 6). But the price of a facade's advantages is vigilance: we must be wary in how we shuttle information between plateaus (boundary line controls must police our inter-patch transactions). Plainly, a descriptive language built up as an incongruent patchwork cannot submit straightforwardly to axiomatization, which, by its inherent nature, provides a uniform covering of the events it seeks to describe. I submit that this consideration supplies the true reason why Hilbert's sixth problem on the foundations of classical mechanics was never fully resolvable in its originally intended terms: considered across its complete domain of intended coverage, classical doctrine can only be viewed as a remarkably efficient covering facade—its descriptive policies cannot be regularized enough to submit to proper axiomatic organization. To be sure, fairly extensive localized portions can be very usefully systematized (as in Noll's scheme for continuum mechanics), but they are neither able to claim full classical coverage nor avoid black sheep cases whose standard "classical treatment" is typically handled in other patches using different resources.

. . . .

Noll's original axiom set makes no attempt to handle fracture, extreme phase change, and many of the other situations described in the fine print of section (ix). To be sure, various tricks have been developed that bring some of these phenomena under the umbrella of continuum mechanics, but the more natural classical approach to fracture et al. appeals to discretely joined molecules. This switch in explanatory preference results in another form of foundational looping akin to those I describe in 6,xi.

. . . .

But why can't we do better? "Surely," the reader interjects, "there should be some *lowest level* of classical behavior able to cover all of our anticipated billiard events, in a manner that explains the utility of the higher patches as merely convenient approximations to its fuller story? It is only this lowest layer that we expect to axiomatize." Indeed, Hilbert (who was quite aware of asymptotic coverage) made this expectation quite clear in the comments he attached to his sixth problem (and made some prescient suggestions as to what aspects of classical doctrine might potentially serve as a lowest layer). But a surprise lies here, for such a "bottom level" lies in quantum theory, not classical mechanics at all!

If we diligently search for a lowest common layer to mechanics that speaks in a wholly classical voice, we soon encounter a puzzling *foundational looping*, where, by following the trails of "for more details, see . . . ," we often find ourselves returning to levels we'd thought we'd already left behind. I'll postpone most examples of this phenomenon until Chapter 6, but we'll observe in the next section that the shock waves that sometimes reverberate within the innards of clashing billiard balls demand that

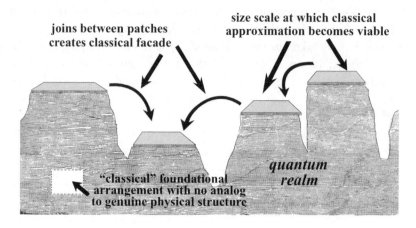

temperature and *chemical potential* be included amongst our primitive "mechanical variables," even though we might have presumed that those would have long since disappeared from our "lowest level" Newtonian physics.

Our musings on the welded cup suggest a different way of rationalizing the puzzling patterns found in classical organization: they arise as an asymptotic covering of the quantum domain, just as our two patches of simplified coverage nicely fit over our target cup. If we ask ourselves *from a quantum mechanical perspective,* "At what length scale will quantum effects supply molecules with a sufficiently robust notion of *shape* that classical modeling techniques will begin to provide useful answers?", we discover that this quantum/classical handoff occurs at many different levels depending on the particularities of the system studied. That is, molecules (or, quite often, matter collected into bundles of a higher scale of organization) must be first supplied with a trackable "shape" before any form of classical treatment is applicable. But the size scale at which these tradeoff points are permitted can vary greatly. Consider a substance such as a steel bar. Many cases of mild flexure can be modeled fairly successfully by treating the bar as a classical continuum or by appealing to sets of small classical "molecules" locked in crystal array. However, more complex phenomena in the metal require greater attention to the details of its elaborate polycrystalline matrix, where very rapid chemical changes and migrations of material occur along grain boundaries. Often these processes inherently require quantum mechanics for their understanding and these considerations force the quantum/classical crossover points to a higher length scale. Any significant involvement of electrical effects tends to do the same. Appeals to *temperature* and *entropy* are common even in the bottom level "classical" stories, because the applicability of thermodynamic principle typically reaches below the level of classical/quantum handoff with respect to *shape*. Furthermore, a survey of successful exemplars of classical "molecular modeling" shows that, for related reasons, sometimes the "molecules" selected can be modeled as point masses, sometimes as rigid bodies and sometimes as some simple flavor of flexible body (in other words, modeling practice picks no favorite among the standard competitors for serving as the "basic objects" of classical physics). Quite commonly, sundry gaps arising within the classical narratives get patched over with straightforward appeals to quantum considerations, without any attempt to construct a "classical story" for these splices (in my diagram such quantum bridge work corresponds to the gullies between the classical plateaus). The net effect of this bumpy support makes classical doctrine look like a suit of armor welded together from a diverse set of stiff plates. Considered solely on its own terms, its organizational rationale will seem elusive, but, regarded as outer fitting suitable for a quantum mechanical knight, the entire affair makes complete strategic sense as an efficient asymptotic covering. To dogmatically assume that this jumble of hinged doctrine can be regularized into an axiomatized format that employs only Newtonian terminology misdiagnoses the true nature of its descriptive successes: they are effective precisely because their sundry routines of physics avoidance neatly cover the quantum realm like an excellently tailored fabrication of buckler, breastplate and shin guard. In other words, if we purify the contents of the predicates that repose upon our facade into complete internal coherence, we will find ourselves sitting within the land of quantum mechanics, and no longer in classical mechanics at all.

But, of course, it is entirely understandable why David Hilbert and the physicists of his day would not have anticipated this assessment and would have looked to other means for resolving the surface oddities of classificatory use that puzzled the Victorians. Who might have then conceived that it is through *quantum mechanics* that classical doctrine would find its "unity"?

Occasionally, one still runs across seriously intended derivations that seek to found substantial portions of classical doctrine upon point masses or other hypothetical classical elements lying far below the length scale of true quantum/classical tradeoff. Although a justification in terms of approximation technique can sometimes be provided for these efforts in mythological grounding, quite often one has the suspicion that such endeavors are driven mainly by raw *methodological tropism*: the orbit of classical ideas must be able to close upon itself internally in complete coherence. But what little bird told our researcher that? Our experience with asymptotic coverings should persuade us that a parcel of descriptive language can prove entirely effective without such internal closure (recall how the true physics that governs our cup is not fully expressed within any of its localized patches).

In fact, the avian adviser who whispers of axiomatization (I am reminded of the trouble making parrots popular in balladry) is easy to identify: it is simply our old friend, the classical picture of concepts. The conviction that inspires our researcher is founded in the assumption that all concepts are created equal: that if coherently grasped notions can't find their application within our unobliging world, they must neatly suit hypothetical possibilities realized elsewhere (this classical democracy of concepts is enshrined within theses (12) and (28) of Chapter 3's appendix). But that faith is based entirely upon *ur*-philosophical hope, not concrete experience with wandering words.

A policy that constructs hypothetical elements to which no genuine elements of reality closely respond will be called *projection* in the sequel. As we shall observe in 5,v, critics commonly accuse the classical picture of concepts of ersatz projections akin to those of our utopian researcher. Indeed, some of Mach's and Duhem's doubts about atomism grew from the suspicion that the evidences offered in their behavior merely represented unevidenced hypostasis of this ilk. In 10,viii, we'll learn that such misgivings were frequently justified.

To properly appreciate the strategic rationale behind a facade-based usage (as opposed to merely learning how to work ably within its confines), we must recognize the manner in which its boundary arrangements (and other methods of inter-patch alignment) offer the language its peculiar effectiveness. A Niels Bohr-like complementarity between inner and exterior description comes into play, for their information bearing capacities can be traded off against one another in fascinating ways. In some situations, it is the placements of the boundaries that carry the greatest burdens in the descriptive work. The physicist Yasumasa Nishiura expresses this consideration ably:

> When we discern [a wide variety of] *shapes, we are actually observing their boundary or perimeter. The boundary is exactly the place where the state (phase) of the matter changes abruptly, or, in other words, observing the boundary enables us to grasp the shape as a whole. Information is, so to speak, concentrated on the perimeter.*[47]

Borrowing an analogy often utilized by modern workers in optics (its physical context will be explained in 6,vii), boundary region weldings often provide the vital wire frame upon which the cloth webbing of interior description gets draped.

Indeed, the lesson that we can adequately appreciate how a descriptive gambit functions only if we understand how "boundary" and "interior" work against one another has emerged vividly within many areas of modern applied mathematics. For example, modern advances in data compression and computation (I'm thinking primarily of wavelets and finite element calculations) trace to the realization that many problems can be conveniently addressed with unexpectedly simple forms of internal tools as long as they are spliced together by a suitable schedule of boundary joins. Likewise, a fruitful mode of interior description might display no easily discernible match up with physical reality, if its excesses are adequately monitored by the manner in which the problem's "boundaries" are addressed (a nice example of this behavior can be found in the Kutta-Joukowsky paradox of 6,v and 6,xiii, where the wind pressure close to an airplane wing is allowed to stretched over artificially huge distances, but the results are held in check by a subtly concealed boundary consideration).

A reconsideration of our earlier examples explains these tradeoffs: the secret to successful physics avoidance commonly confines keys aspects of the governing physics to singular surfaces and then performs the bulk of its detailed calculations only with respect to the smoother regions they hem in. We have already noted that, when we dropped terms from our original cup equation to produce the simpler equations utilized in our covering patches, we thereby left much of the operative physics behind (e.g., in

[47] Yasumasa Nishiura, Far-from-Equilibrium Dynamics, Kunimochi Sakamoto, trans. (Providence, RI: American Mathematical Society, 1999), pp. xv–vi.

the main patch we ignore the terms that dictate how the metal reacts to extreme bending). Inside each descriptive arena we concentrate upon the effects that locally dominate and ignore features that may prove vital next door. We crudely interpolate over the narrow transition band in between because all influences remain of equal salience in this region and we wouldn't be able to obtain significant variable reduction if we treated this region even handedly (this consideration, of course, is the same as gives rise to our black sheep exceptions). Nonetheless, this computationally neglected region remains quite vital to the *behavior* of the cup as a whole: indeed, its severe bending represents the chief locus where the changes wrought by welding arise.

In short, the net effect of our two-patch covering is to divide the underlying physics of the cup into factions which are allowed to rule their own duchies with their own laws. When we attempt to work backwards from these arrangements—that is, we only observe the fragments of law registered within the patches—, we will not be able to reconstruct the fuller physics that governs the cup easily, due to its reductive apportionment into fragments. Indeed, as much of the physical principle pertinent to our system is encoded in where the joins between our patches are located, rather than being directly manifested in any of the local governing equations. The moral of our reflections is then: *look to the boundaries!*

....

The manner in which locally dominating "investigative moods" greatly simplify interior logical manipulations in a Fitch-style natural deduction system illustrates a similar lesson. These matters will be discussed in 7,viii.

....

To those familiar with the manner in which "boundary conditions" et al. are routinely addressed in philosophy of science primers, it is plain that none of these vital considerations have been absorbed. For the historical reasons surveyed in section (iv), logical empiricist thinking about theoretical structure became engulfed in logic-centered concerns, allowing the richer architecture of differential equations, their required side conditions and worthy approximation techniques to wither away into invisibility. As the implausibilies of positivist doctrine gradually became apparent, many students of philosophy marched forthwith into the fogs of holism, rather than adopting the wiser course of returning to the workshops of richer mathematics. But we philosophers should place a higher valuation upon the subtle wares offered by the mathematicians, for they, tutored by demanding circumstance, have articulated a wide range of clever strategies of which the rest of us would have never dreamt, being too willing to muse in our armchairs about how the world *ought* to submit to our descriptive gambits. And it's true: if Mother Nature were truly a sporting old gal, she'd have adjusted her complex behaviors to better suit our theory T schematisms. But she isn't and she hasn't and so we must contend with her wiles in more strategically sagacious ways.

If these observations are correct, then all of those holist critics who have reacted to the failures of logical empiricism by insisting that "science represents an institutional practice, not a formalized theory" direct our attention away from the very issues to which we must attend, if we hope to understand how *ur*-philosophical puzzles arise, both in science and elsewhere. Such thinkers encourage the impression that the path to understanding mystifying policies in science is not to be reached through formal study. No advice could be further from the truth, in my estimation. The language twisting strategies I emphasize are commonly subtle and well camouflaged. Usually they can be flushed from their lairs only through fairly diligent scrutiny of a mathematical character. Indeed, much of our modern understanding of the facade structures occurrent in classical doctrine has been obtained as a side consequence of the diligent efforts of Walter Noll, Clifford Truesdell and others in their efforts to articulate a workable *axiomatization* of continuum physics able to guide current work more ably (some details of this important research will appear at scattered locations throughout this book). In particular, it is these investigations that have neatly revealed the subtle property dragging linked to rigidity and incompressibility that we shall discuss in the next section. As we'll learn over the course of this book, quite substantive confusions in traditional philosophy grow from this seemingly insignificant seed. But none of this hidden grain could have been properly recognized without the original prod of careful investigations in a strict, axiomatic vein.

....

For their own purposes, Noll et al. must cleanly segregate the role of so-called constitutive equations from the more general principles of mechanics. In standard nineteenth century practice, aspects of each were commonly blurred together through appeal to sundry geometrical hypotheses (that certain substructures behave like rigid bodies, say, or the point mass idealizations that Pearson regarded as necessary). In the short run, such tactics offer brisk derivations for the most widely favored equations utilized in traditional mechanics. At the same time, those very advantages hindered progress with respect to more rheologically complicated materials: the toothpastes and rubbers I've mentioned before. Guidance towards formulations adequate to these stuffs required a crisp recognition that traditional appeals to rigidity introduce a convenient, but intrinsically alien, element into continuum physics. To be sure, once relevant doctrine is purified in this manner, the derivation of even the simple wave equation for a vibrating string proves a rather daunting affair, ably illustrating the moral that sound descriptive practice often can't come into its own except by first passing through earlier stages contaminated in clashing directivities (a conclusion that we shall reach by many paths over the expanse of this book).

....

In much of her best work, the philosopher Nancy Cartwright[48] correctly observes the patchiness and apparent inconsistencies commonly found in textbook physics, entirely out of conformity with theory T tidiness. Laboring under the influence of the notion that "physics is a practice," she unfortunately concludes that physics merely represents a loose policy of constructing descriptive pastiches; that it fibs insofar as it pretends to supply any general or accurate account of the way things are ("lies" is her word; she also

---

[48] Nancy Cartwright, How the Laws of Physics Lie (Oxford: Oxford University Press, 1983).

invents an alternative mythology of casual narratives in the bargain). This appraisal fails to recognize the entirely coherent (and certainly not mendacious) manner in which classical physics technique ties together as an asymptotically supported facade. Indeed, Cartwright completely overlooks the labors of the large army of applied mathematicians who have unraveled the concrete rationales behind many of the techniques that puzzle her, many of which represent some variation upon asymptotic approximation. Loose appeals to "practices" rarely provide any insight into the genuine puzzles of scientific endeavor, I daresay.

By gesturing exclusively towards the amorphous expanses of webs of belief, practices, paradigms, holists encourage a naïve trust in the unfettered directivities of our everyday words when we are better advised to scrutinize what the little rascals are up to with greater diligence. I recall a drawing from an old children's book where all the King's horses and men stood proudly arrayed around a patently inadequate montage of Humpty-Dumpty, its broken pieces of shell minimally held together by rubber bands and chewing gum. Treating facades as if they were integral units displays a similar misapprehension: it doesn't matter whether we point to our gimcrack assembly and declare, "Lo! an axiomatized theory," "Lo! a set of possible worlds" or "Lo! a scientific practice." We need a few more "Lo! look what happens to 'force' when it crosses the boundary between solid and fluid."

In this regard, we should observe that the relevant mathematics inside a patch often supplies *internal warning* that it has been pushed beyond its applicable limits: when we attempt to treat the black sheep cases, we discover that we have fewer equations than variables (as occurs with triple billiard collisions) or that necessary matrices turn singular (at the "dead points" within the theory of machines) or that solutions "blow up" in finite time (as occurs in conventional point particle gravitation). In short, our inferential tools begin to squeak, "Hey, Bub, I'm experiencing a breakdown in my ability to draw reasonable consequences; you better bring some additional physics in here to correct the mathematics." When these warning balls sound, we are advised to shift to another patch for adequate coverage. As we've observed, the unhappy price of these migrations is that we are often required to redecorate our previous work in dramatically new mathematical shades—we do not simply "add a few more details" to what we've wrought; we must overturn it all in drastic revolution!

But our gas tube case also shows that sometimes these very squeaks can be cleverly exploited to temporize on a need to shift patches radically. Heeding Riemann and Hugoniot's advice, we can declare, "Let's take this mathematically impossible blowup as an *omen* that a shock wave is forming there." This ploy allows us to frame an unexpected variety of in-between mathematical patch, where so-called *weak solutions* are now tolerated along side our old formulae (the acceptance of the famous Dirac δ-function falls in place in here). Applied mathematics is full of procrastinating, halfway repairs of this ilk. Because the phraseology of the calculus can be reconfigured to encompass "weak solutions" fairly deftly, a casual observer can easily overlook their intrinsic oddities. A closer look reveals the delicate framework of controls that allow the Riemann/Hugoniot ploy to work.

It seems to me that a just consideration of the incomplete and held-together-with-paper-clip solutions we encounter in classical physical practice ought to give pause to the unbridled enthusiasm for "possible worlds" that has dominated analytic philosophy circles in recent years. If pressed, these aficionados commonly reply, "Oh, we all know what possible worlds are like: think of the billiard ball possible worlds belonging to Newtonian mechanics or the other species of physics." Presumably, the phrase "billiard ball possible worlds" is intended as a colorful way of speaking of the "models of the Newtonian laws," conceived in the fashion of the "models" studied in logic (here talk of "possible worlds" seems to serve largely as a gambit to allow the basic tenets of the theory T syndrome a longer lease on life, through camouflaging its logic-inspired structural assumptions in a fuzzy vocabulary that doesn't sound so overtly syntactic). But assuming the existence of such globally defined models flies in the face of most known facts about the solutions that the standard equations of classical physics accept (such topics enjoy comparatively few models of a global ilk and certainly not with respect to billiard balls). Furthermore, the black sheep phenomenon indicates that individual solutions rarely form into the manifolds of similar possibilities that we expect to see. For example, the Newtonian patch maintains that *two* billiard balls that clash head on will bounce away without flexure in a coefficient of restitution manner, yet, if *three* balls happen to bump, they will be treated as if they all distort internally? But how can our spread of "Newtonian possible worlds" treat these cases so differently? Likewise, standard approaches don't properly allow iron bars and buckets of water to sit together in the same patch—what sorts of "possible world" could those restrictions reflect? Plainly, the possible world enthusiast has tacitly presumed that some basement layer exists to regularize all of these treatments, but, as we've observed, that represents an entirely unproven promise. Of course, these mismatched behaviors make a good deal of sense from a facade perspective, but not from any "possible world" point of view, insofar as I can determine.

....

Considerations of modality enter physics in many interesting ways, some of which will be scouted in the next chapter. And the value of "possible world" structures in the formal study of modal behavior is undeniable. But none of this establishes that the extremely strong demands implicit in the usual notion of a "Newtonian possible world" can be rendered coherent. Too often the mere fact that physical thinking can sustain *some* modal claims is regarded as proof that the entire edifice of possible world doctrine is viable. It is as if we have agreed to do a "small favor" for a friend and it then emerges that he expects us to support all of his friends and distant cousins in opulent style.

....

I find it hard to view the cult of possible worlds as anything other than the ill-starred issue of a tacit union between the classical picture of concepts and a lingering theory T syndrome. The notion seems founded in an extremely strong form of classical gluing, stronger, in fact, than Russell himself would have endorsed, for not only is the extension of every suitable predicate held to be concretely fixed everywhere in the real universe, but in many other places as well. This second assumption seems to flow from the superjacent conviction that our "theoretical" concepts are implicitly housed within a web of theory coherent enough to accept "models." As we noted, Russell was more alive to the infirmities of articulate physical doctrine than these parties and would have refrained from such blithe assumption. Indeed, I find that many practitioners of the possible world art have almost entirely forgotten the practical motivations for investigating concepts closely that we have retraced in this book. If we ignore *these*, then virtually any contention with respect to the realm of concepts may seem possible.

Their dedicated faith in their powers of "conceptual intuition" very much reminds me of the comparable trust of physicists in their own "physical insight." In operational effect, both appeals often serve as excuses for not looking deeper into nitty-gritty mathematical details that bore them. I think such neglect typically catches up with both parties sooner or later. Our physicist might be able to hammer out a workable descriptive matrix for some revealing simple case employing elementary mathematical tools loosely, but it frequently requires a much deeper level of later critical probing at the hands of applied mathematicians before a framework is found that can extend these initial discoveries capably to complex circumstances (to be sure, certain species of physicist—Richard Feynman, say—never learn to value these labors properly, because in the meantime their interests will have shifted to some fresh topic of investigation). Perhaps our possible world philosophers will never be punished for their enthusiasms, but I doubt that their exertions will be often rewarded either, in the sense of successfully resolving the tensions that have traditionally animated philosophy. For if the observations advanced in this book are well founded, those difficulties commonly trace to the hidden turns of the screw that generate quilt-like linguistic adjustments to the descriptive problems that Nature sets upon our plates. In Chapter 1, I complained that dwelling upon storybook possibilities in Nathaniel Hawthorne's manner can impede our capacities for recognizing real world mechanisms busy right before our noses. In an allied manner, uncritical devotion to possible worlds scarcely encourages the careful scrutiny of policies for patch/boundary accommodation that I believe are helpful. However, since my objections to the milder exaggerations of Russell's classicism apply, a fortiori, to possible world aspirations, I will not beat on this particular drum excessively.

(ix)

Helpful troublemakers. Part of my mission in the previous section was to extol the virtues of facades as triumphs of efficient linguistic engineering, for fracturing a descriptive task into patches monitored along their boundaries creates a platform whereupon reduced variable strategies can exploit localized opportunities very effectively. In real life, however, facades sometimes perform these fine offices in such a discrete and imperceptible manner that, as an undesirable side effect, they create *ur*-philosophical perplexities when their structuring is misunderstood and utopian projects are plotted upon an erroneous diagnosis. As I've already noted, an unscrutinized facade can *mimic* for a true "theory" (in the sense of a body of doctrine open to axio-matization) quite capably—these matters of masquerade will prove of great importance in the sequel (such theory-imitating assemblies I call "theory facades" for that very reason). Essentially similar problems can affect the usages of everyday descriptive terms as well. Thus Chapter 7, x will argue that our troubles with "is red" and "expresses sadness musically" descend from such origins: facade-like controlling structures sur-round these predicates in a manner that is vital to their integrity but also leads their registrational capacities to follow different strategies than we anticipate.

In this section, I want to begin a short survey of the role that top-down constraints such as *rigidity* and *incompressibility* play in silently inducing property dragging and facade formation. These considerations will help us anticipate some of the puzzling phenomena we shall visit in later chapters, in which these quiet intruders happen to play a significant, if usually unnoticed, part (in Chapter 9, we shall find that *rigid object*'s oddities play a major uncredited role in generating Hume's famous perplexities about causation). By a *"top-down constraint"* I intend any requirement that requires extended matter to satisfy a prescribed condition over an extended area or extended span of time (the *holonomic constraints* of standard physics provide perfect exemplars of what I have in mind). Rigidity operates in this fashion because it requires a steel girder to hold all of its length measurements fixed, whereas incompressibility requires a flexible body to maintain its volume through any alteration in form. Any top-down constraint of this type is apt to clash in subtle ways with requirements that operate instead in bottom-up fashion (in the manner, say, of the governing equations for the iron within our truss). The cracks and joins tolerated within a facade supply enough wiggle room that these warring tensions can reach practical accommodation through their means.

Appealing to the *rigidity of parts* has comprised a vital aspect of mechanical tradition since the Greeks (consider the law of the lever et al.). Indeed, one can safely declare that, had not such invocations been regularly made, successful physics could have never gotten off the ground. And the reasons for this are quite simple: we can commonly obtain answers to physical dilemmas with remarkable simplicity if we know in advance that, e.g., the girders in a bridge will stay straight (exploitation of rigidity indubitably constitutes the most widely utilized recipe for effective variable reduction because we can usually ascertain by visual inspection that the parts in a mechanism stay approxim-ately rigid). In particular, suppose we are dealing with a truss bridge as illustrated, where the little wheels on the right signify that the unit is free to move in a horizontal direction. Utilizing nothing beyond the simple algebra of statics known to the ancients, we can readily calculate what the stresses will be at every joint of our bridge; we don't even need to know what the struts are made of—as long as they stay rigid (in the engineer's jargon, our assembly is classified as *statically determinate* for these reasons).

But let us now replace the little wheels by a clamp and the equations we have been using will suddenly lock together in *over-constraint* (adding the clamp puts an additional equation in our descriptive set and now we have too many to solve). To accommodate the new condition, we must allow previously frozen degrees of freedom to open up inside our girders: this is simply the mathematician's fancy way of saying that we must

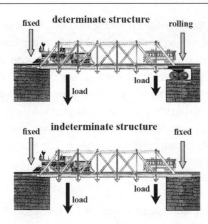

allow them to *flex*. But the rules for that require that we know how iron responds to bending, the very concern that we were able to airily dismiss in statically determinate situations. In short, add one lousy little clamp and we are forced to leave high school algebra behind and move to the land of simple calculus. This enforced emigration with respect to mathematical patch represents another illustration of our black sheep phenomenon (in this case, the troublesome flock is rather large, although human designers usually try to minimize its numbers).

Before we discuss the property dragging induced by these appeals, it might be informative if we follow our beam-related cascade a few rungs further on. In order to cobble by in our reasoning with *ordinary* differential equations alone (which is the mathematical setting in which beginning engineering primers place our clamped bridge), we must be able to collapse a three-dimensional object into a one-dimensional curve. Is that always reasonable? No: only if the beam is nicely symmetrical and its internal stresses act as if they pull along tidy fibers. Plainly, that is not always the case, and, accordingly, more complex beams will force us to collect our belonging and migrate to a patch where *partial* differential equations rule. To the non-mathematician, that little displacement sounds pretty harmless—haven't we just exchanged "partial" for "ordinary"?—but ask any expert which flavor of equation she'd rather treat! In any case, we've escalated our reasoning tools to the frame of junior year analysis class.

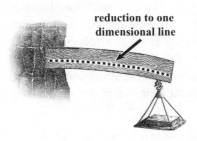

reduction to one dimensional line

However, there's worse to come. If the girders in our bridge are subject to heavy blows from passing trains, we may need to calculate the stress waves and heating that arise as a result; indeed, our old friends from the gas tube, the shock fronts, can make an unwelcome appearance. Mathematical critters such as the weak solutions we mentioned previously now roam the patch we must now call "home." And onward we go, descending to ever more elaborate basements as previously frozen degrees of freedom within our bridge open up, each further ladder conveying us downward into more fearsome regions of applied mathematics, rather like the subterranean fairylands in Hans Christian Anderson's "Tinder Box."

．．．．．．．．．．．．．．．．．．．．．．．．．

In order to carry out the Riemann-Hugoniot recipe for these shock waves, *thermodynamic considerations* must be evoked to single out the solutions we seek. That is, if we write down plausible equations for a familiar macroscopic substance like the iron in our truss, they are likely to evolve into states whose progress can be monitored only if we attend to their *temperature* and *entropy* considered as new primitive terms (in fact, allied considerations indicate that attention to chemical and electrical state is also required[49]). Standard nineteenth century mechanists, of course, tried to purge *temperature* and *entropy* from the microscopic docket of physics, but the unavoidability of shock waves often forced their readmittance into domains from which they had been previously purged (this fact provides a nice illustration of the *foundational looping* we shall discuss further in 6,xii). From Duhem's and Mach's point of view (section i), the failure of purely mechanical ideas to close into a self-consistent circle constituted strong evidence that molecular ambitions of a mechanistic cast were ill-conceived. Such behavior is not surprising from our facade perspective, because we expect the classical/quantum tradeoff to occur at varying size levels and with differing degrees of thermodynamic participation.

．．．．．．．．．．．．．．．．．．．．．．．．．

Meanwhile, the *boundary conditions* we assign our beams display an allied cascade of complexity driven by black sheep exceptions. In our indeterminate truss, we only pay attention to the averaged applied forces and turning moments. Even when we consider genuine three-dimensional beams, engineers usually describe their end conditions in very simple terms, joins between beams utilizing quite simple matching conditions. In truth, if we bind a beam end firmly with constraints as pictured, the stresses induced will be very complicated and require some of those daunting lower regions of mathematical technique to calculate. Worse yet, the relevant boundary conditions will prove very hard to ascertain: it is hard to tell what is exactly going on inside a wall or welded joint. But we can keep our mathematics at a much simpler level if we appeal to the maxim called *St. Venant's principle*, whose rationale is reminiscent of Prandtl's boundary layer technique. Near the clamping point the induced stresses in the beam will be very complex and greatly sensitive to the exact manner in which it is held fixed. But usually— but not always, by any means!—these stress complexities will die away as we move a moderate distance towards the beam's mid-section, for these internal portions react to a

---

[49] Brian Bayly, <u>Chemical Change in Deforming Materials</u> (Oxford: Oxford University Press, 1992).

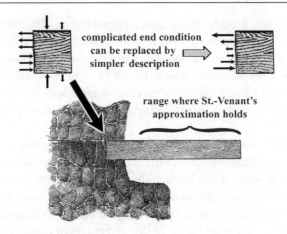

complicated end condition
can be replaced by
simpler description

range where St.-Venant's
approximation holds

wide variety of end conditions in more or less the same way (a chunk of iron in the middle of a girder is so near-sighted that it can perceive the faraway end stresses only in the crude and averaged terms we employ for a rigid indeterminate structure). Indeed, St. Venant's principle advises engineers to select these computationally simple end conditions for their problems, on the grounds that we don't really care about the finely detailed stresses near the joins but must worry greatly about the mid-section material (that's where sagging and fracture is likely to occur). But there are certainly black sheep exceptions to this recommendation.

To gain a proper appreciation for the difficulties of applied mathematics, it is worth observing that providing a precise demonstration that St. Venant's principle (or boundary layer technique, for that matter) represents a valid approximation method is apt to prove nearly impossible, simply because of those black sheep cases: the situations that required me to add a "not always, by any means!" qualification to my gloss. Mathematics, by its very nature, has a heck of a time dealing with "usually, but not always" situations, although it gamely tries, through taking thermodynamic limits, averaging, proving claims almost everywhere, etc. But these represent fairly crude expedients and the obstructions caused by rare exceptions often make the rigorous derivation of one approach to mechanics from another quite difficult (it is far easier to talk fast and bluster one's way past the hurdles, as physics instructors often do). Such derivational obstructions lie in the background of my earlier observation that much of what passes as a "theory" in physics properly possesses the status of a *mathematical guess*: considered as a "theory of beams" (of which there are many competitors), St. Venant's principle represents a stab at isolating the central effects that are expected to prove mathematically dominant in most—but not all!—situations one expects to encounter. Commonly, applied mathematicians tolerate such unproven hopes amicably, saying to the physicist or engineer: "Well, I can't quite follow how you managed to get from A to B, but I'll be happy to start over at B and investigate the applied mathematics that begins from there as a new starting point" (when one reads about a "rigorous approach to the physics of X," it usually means only a "rigorous study of some specific equation associated with X," not a study of X's wider inferential ambit). These frequent—and utterly unavoidable—interventions of derivational leaps of faith supply physical doctrine with a more loosely joined inferential character than we philosophers commonly imagine, especially if we still labor under the affliction of the theory T syndrome. But these same loosely joined aspects provide ample wiggle room in which quiet intruders like rigidity can work their property dragging wills without much fear of being apprehended.

Reflecting upon the astonishing computational advantages offered in a statical determinate bridge, we can easily appreciate the reasons why classical mechanics is fond of exploiting rigidity whenever it can manage the trick. But there's no totally free lunch: as top-down impositions, these constraints almost certainly introduce some alien element into the rest of our physical thinking, overloading its docket with more demands than it can consistently handle. To make room for the advantages of rigidity, we most likely will throw out other physical consideration we hold dear, although we may not notice the loss. This displacement phenomenon is easiest to identify in the case of the *incompressibility* constraint upon a fluid, because its acceptance forces us to suppress the *transport mechanisms* that allow the liquid to maintain approximately the same volume everywhere. Suppose we apply a squeezing pressure to a certain portion of the fluid. How are its other parts to know that they must compensate for this change in a manner that keeps the overall volume exactly constant? Well, plainly no realistic fluid can turn this trick perfectly: there must be short interludes where the overall volume is greater or less than it should be while pressure waves carry the message to outlying areas that they must adjust their positions appropriately. Placing a strict incompressibility constraint upon our fluid forces us to throw out descriptive coverage of all the physical events that occur in the intervals when the real material struggles to reconstitute its assigned volume. In effect, we must treat the temporal history of our watery stuff in a temporally gappy manner like that displayed by Hertz' quasi-statical approach to billiard ball distortion (as in that case, we remove the capacity for transmitting pressure waves). Most advanced textbooks indicate that the otherwise sound mechanical quantity of *absolute pressure* becomes undefined with respect to an incompressible fluid, which is simply the mathematician's way of acknowledging that we threw out a considerable amount of the fluid's guiding physical processes under the variable reducing heading of "incompressibility." But beginning students rather often fall into perplexities when they don't realize how much relevant physics they abandoned when they welcomed that innocent-looking phrase "let our fluid be incompressible" into their parlors. As a side effect of this concession, the predicate "pressure" gets tacitly dragged from its customary *absolute pressure* moorings.

Similar subterranean adjustments occur with respect to "force" whenever we declare that a contacting body is rigid, as when a bead is said to slide along a perfectly rigid wire. In particular, we rob the wire of any capacity to respond to the bead's incursions in proper Third Law fashion. This often unnoticed loss engenders many tensions with respect to other physical doctrines such as the conservation of energy and causes us to accept the anomalous notion of a "force of reaction" within our orbit of mechanical ideas (I won't provide details here, for we shall revisit the bead on a wire in 6,xiii). But it is worth noting, in regard to the historical events recounted earlier, that Hertz' motivations in writing The Principles of Mechanics apparently trace to a desire to resolve these conflicts between "force" and "rigid body," with Hertz favoring the latter in his own recommended architecture.

I might also mention that appeals to "rigid body" do not represent a minor occurrence within the halls of mechanics: much of the point of the celebrated approaches of Lagrange and Hamilton is precisely to provide formalisms in which the variable reducing capacities of constraints like rigidity can be exploited with maximum efficiency. But in framing these effective housings for descriptive practicality we automatically enshrine the tensions just recounted within the very timbers of our edifice. The results are not exactly facades, but they represent descriptive architecture capable of fooling their human masters quite capably.

In any case, our key observation is that quiet—and often indispensable—appeals to rigidity can easily induce property shifting nucleations of the sort we observed with respect to "fluid particle" earlier. In fact, later in the book we shall examine the generally unrecognized role that rigidity's tiny reorientations in referential compass has played in sowing significant forms of *ur*-philosophical confusion (not merely in physics, but in quite unexpected places in general philosophy as well). As remarked earlier, it is through these unnoticed nucleations that an important role for distributed normativity within linguistic development can be vividly located. To be sure, these strands of practicality typically represent a very small portion of overall usage, but their molding influence on its unfolding personality can be great nonetheless (an actress may be granted only a few lines here and there, but her little interventions may completely shape how the plot of a play unfolds). However, our chapter has already waxed fulsome (and we have a final topic to canvass), so I will postpone further pursuit of these issues until later, when we will rejoin them up under the umbrella of other considerations that affect facade formation.

To summarize a rather extended line of thought: the quilt-work assemblies I have called "facades" offer attractive platforms upon which wonderfully practical forms of predicate employment can be established. Such arrangements enjoy a substantial strategic integrity all their own: their circle of "expressive ideas" needn't close upon itself according to the "uniform platform" expectations of classical concepts. As such, we shouldn't be surprised to discover facades (or their approximates) arising fairly commonly along the streams of everyday and scientific descriptive practice. However, they can also mimic (in "theory facade" manner) for descriptive policies of a more straightforward nature and sometimes confuse their unwitting employers thereby. Classical thinking about concepts further blinds us to the significance of a facade's filagree of patches simply by insisting that we always grasp a thick wad of conceptual content whenever we adequately understand a word. This grasped content is credited with such strong adhesive powers that classical thinkers never dream that innocent intruders like an appeal to rigidity have the capacity to quietly tweak predicates from one referential attachment to another. But such sanguine anticipations are not borne out within our real life linguistic experience: such tweaking of property attachments occurs reasonably frequently and is often utterly unavoidable (it is also frequently *beneficial*, as we shall learn from the Heaviside case of Chapter 8). And it is within this specific arena that this book will attempt to assess the distortions wrought by the utopian expectations of classical thinking, without falling into excessive anti-classical gloom thereby.

....

The world of ways in which boundary joins play important roles in monitoring physical description is very wide and I regret the fact that I can only explore a few specimens in the book. But let me take quick advantage of the fine print to mention several other examples I find intriguing. Suppose we are interested in how a spray forms on the surface of a choppy ocean, modeled as a continuous fluid.

How do applied mathematicians handle such events? Starting with a smooth surface, the governing equations will gradually extend small extrusions into long spindly stalks with a ball at their end, formations that can be witnessed in stop-time photography. Unfortunately, our equations will prolong this state forever, continuing to plot an attached blob that never relinquishes its absurdly elongated umbilical tie to the mother ocean. This occurs because partial differential equations, left to their own devices, do not alter the topology of the situations they model. Plainly, some "fresh physics" needs to brought into our picture and this is commonly accomplished in a rather remarkable way. When a change in the fluid's topology looks imminent, practitioners begin investigating *two fluid* configurations that run in parallel, one containing the still attached drop and the other describing a drop of similar shape detached from its ocean. The two configurations are then tested for their respective energetic stabilities (which are determined mainly by surface tension). As soon as the two separated drop configuration reports more favorable values, we will assume that, at some point near this time, the real fluid will snap through to the two blob topology. We can picture this kind of "boundary join" as two film strips that run in overlapping parallel, where at some point in the interval ΔA, the story of our drop jumps from one strip to the other. We are practicing physics avoidance in that we do not directly describe the molecular processes that lead to drop separation, but merely cover the relevant region with an interpolating patch. Unlike our Newtonian approach to billiard collisions, this patch takes the form of a pair of transitional intervals, not an event singularity. As such, a measure of indeterminacy is introduced into our modeling because our drop will behave differently depending upon the exact moment when

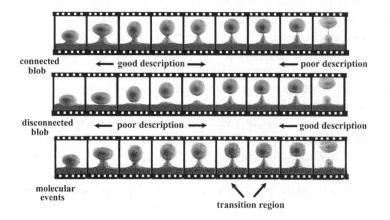

connected
blob   ← good description →   ← poor description

disconnected
blob   ← poor description →   ← good description

molecular
events   ↘ ↗
                       transition region

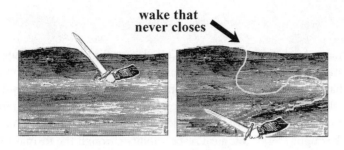

wake that never closes

the snap over occurs. A wide range of macroscopic phenomena are commonly addressed in a similar overlapping fashion, e.g., the fracture of solid materials as treated in the celebrated proposals of A. A. Griffith.

Or consider this related problem with geometrical description. Take a knife and swish it around in the bulk of our continuous fluid. In an orthodox treatment, the intruding instrument will push the free surface of the water ahead of it in its slashing, stretching the erstwhile top surface like an impenetrable but very pliant sheet of rubber.[50] Let the knife come to rest and gravity will pull the cut surface back together, leaving behind a very convoluted coil of deformed "surface," snaking through the innards of the water. This complicated story represents a proper description of our fluid's condition, because it takes a period of time before the pressures on each side of the rejoined surfaces can equalize, despite the fact that the knife, having sliced, has moved on. But, fairly soon after, our fluid will have returned to its normal, undissected condition. Unfortunately, if we believe our differential equation engines, these knife scars will never heal—a lengthy distortion of surface must remain embedded in the fluid's interior ten thousand years from now, although internal pressures will have long since equilibrated. That is, our unsupplemented reasoning tools assure us that the bosom of the ocean must retain a twisted record of every porpoise that has cleaved its crest and every victim of pirate cruelty. And this is because such equations are incapable of erasing these internal boundaries. Again the solution (which is often applied without comment) is simply to reset our modeling of ocean condition from "convoluted" to "smooth," within some nebulous interval of sufficient relaxation time. In each of these descriptive resettings, we effectively *abandon information* with respect to its previous condition, in a manner analogous to the celebrated "collapse of the wave function" in quantum mechanics.

Incidently, such considerations raise important issues to which philosophers of science have been largely insensitive. When we axiomatize a physical account, how much of its full applicational circumstances will be captured in our formalism? In particular, how are the boundary conditions and allied considerations being handled? If we formulate quantum mechanics as a theory of Hilbert spaces, the relevant boundary conditions will have been tacitly divided between the structure of the function space and additional terms in the linear operators investigated. But what *role* do these hidden elements play in maintaining the viability of our descriptive apparatus?

. . . .

[50] Richard E. Meyer, Introduction to Mathematical Fluid Dynamics (New York: Dover, 1982), 6.

(x)

The vicissitudes of rule validity. As mentioned in section (iv), Hilbert approached the issues of formalism with a good deal more subtlety than many of his followers, for he recognized that axiomatic presentation alone cannot fashion a purse from a sow's ear. After all, any doctrine whatsoever, no matter how nutty, can be laid out in impeccable Euclidean form (I once ran across a pamphlet entitled The Scientology Axioms[51] where a noted quack sought methodological respectability through the format). In particular, hidden logical inconsistencies can be milked for any conclusion we want, and many fallacious circle squarings and proofs of God's existence have rested upon these igno-minious foundations. Hilbert also realized that even great mathematicians such as Euler or Riemann sometimes went astray in assertive overconfidence; axiomatizing their assumptions would not have improved matters one whit. He therefore proposed that the syntactic consistency of an axiom scheme might be investigated through fairly elementary means—to wit, by Padoa's method, where we probe a formalism rather as we might trace the declension of dominant and recessive traits along a family tree (the idea is to grant the axioms a clean bill of health if no sentence of the form "P and not P" can possibly pop up in its chains of deduction). If a comparable syntactic completeness can also be established, then the mathematician will know that a safe syntactic play-ground has been satisfactorily established by the axiom set.

Unfortunately for this rosy picture, Kurt Gödel's celebrated incompleteness results showed that, in most cases of interest, axiomatic consistency can be established only through constructing a set-theoretic structure of comparable riskiness. This discovery thrusts the prime responsibility for delimiting the mathematician's arena of "free cre-ativity" into the arms of set theory, as expressed in the strengths of its existence pos-tulates (large cardinals and all that).[52] To be sure, many present day mathematicians dislike this dependency and in conversation frequently express philosophical opinions that seem deeply reminiscent of turn of the century faith in unchecked axiomatic support. Nonetheless, nostalgia for the good old days aside, set theory represents the final court of appeals to which all existence questions in mathematics presently get dispatched. In fact, as we'll observe in the next chapter, the existence of quantities within *physics* must ultimately address this same tribunal as well.

Even individual *reasoning rules* must be validated through set theoretic considera-tions of an allied kind and an appreciation of this dependency shall prove crucial in the pages to come. It is an unhappy, but unavoidable, fact that few rules of immediate and palpable strength can supply absolutely correct answers in all applications. Recall the technique—*Euler's method*—that we utilized in our section (iii) calculation of cannon ball flight. This represents an inferential technique of "immediate and palpable strength" in the sense that it provides easy-to-follow instructions that can be applied

---

[51] Available at www.bonafidescientology.org. Here is a sample: "Axiom 14: Survival is accomplished by alter-isness and not-isness, by which is gained the persistency called time." There seem to be no theorems, however.

[52] Penelope Maddy, "Does 'V = L'?," Journal of Symbolic Logic 58 (1993).

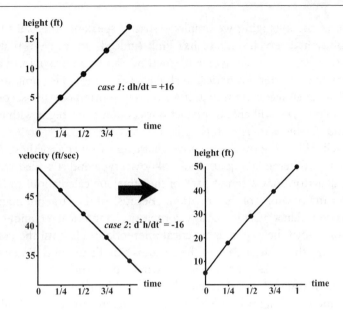

to any differential equation whatsoever and will generate bountiful results (although an enormous number of calculations may be required before any region of any appreciable size is filled in). Besides its powerful scope, the Eulerian technique is utterly intuitive in conception and, in fact, merely represents a formalization of a common variety of "cause and effect" reasoning that we employ, in limited doses, in everyday life (we shall revisit this theme in 9,ii). In fact, although nowadays we normally conceive of Euler's method as representing an *approximation technique* for differential equations, its basic steps had been in use long before the calculus was invented and provided a rough means for expressing the root conceptions behind differential equations without having such formulae available. Suppose, for example, that a rocket maintains a constant upward velocity ($= dh/dt$) of 16 ft/sec and it begins at a height of two feet ($h(0) = +2$). We immediately reason, "Every quarter second, its constant velocity will cause the projectile to climb an additional four feet." Expressed as a graph, we obtain a sequence of dots (which we connect with inter-polating straight lines) that continually increase by a factor of 4 feet, as shown. This graph simply represents a transliteration of the sequence of sentences that can be inferentially extracted by Euler's rule from the starting propositions "$dh/dt = +32$" and "$h(0) = +2$." As such, our conclusions exactly follow. But our cannon ball's circumstances are slightly different, because its *acceleration* (dh^2/dt^2) must remain a constant -32 ft/sec^2 and it starts at a height of 0 feet with an upwards velocity of 50 ft/sec. We therefore reason, "So its upward speed must change by -8 ft/sec every 1/4 second. So after the first 1/4 interval, its velocity will have fallen to 42 ft/sec. An *averaged* velocity estimate of 46 ft/sec over the quarter second interval will cause our ball to climb about twelve feet." Here we recognize that our reasoning is no longer exact because of the averaging we employ—strictly speaking, the ball's true velocity will alter slightly at every instant of its climb (indeed, we can see that, depending upon circumstances, our crude averaging method can be improved in various ways: thus are born the smarter numerical techniques that real life computer programs utilize). We may even feel certain that, if we merely shorten our 1/4 second time step to a shorter interval, we will able to predict our cannon ball's flight with any accuracy we desire. And, *for the most part*, this assumption is justified.

Let me dwell a bit more on the intuitive character of Euler's method. An engineer, confronted with a differential equation of unknown type and stranded without a programmable calculator, may attempt back of the envelope calculations of Euler type to gain a "feel for the meaning of the equation." Because of the large number of exacting calculations required in a situation of any complexity, numerical techniques of this sort didn't fully come into their own as practical inferential tools until the computer age. Nonetheless, from the earliest days of the calculus Euler's method has enjoyed a *semi-criterial* status in the sense that a teacher would presume that a student did not "understand the meaning" of a differential equation if she could not sketch an appropriate Euler's method diagram (indeed, we hold calculus novices to similar standards even today). No doubt Leibniz and Newton first assured themselves of the *coherence* of their calculus ideas by plotting broken line projectile flights as we have done here. Indeed, it is hard to see how the basic notions of the calculus could have ever been accepted had not the inferential successes of Euler's method partially paved their way beforehand. Situations like this are not rare: new terminology can only be introduced after experimentation with some inferential technique has prepared their groundwork beforehand.

Nonetheless, despite this semi-criterial centrality, in certain circumstances Euler's technique supplies egregiously unsound results, even if we make our approximations over very short intervals (its time step Δt can be set as brief as we wish).

In fact, by reformulating our cannon ball equations in what seems an entirely reasonable way, we will plot out an Euler's method chart that looks as illustrated: a levitating projectile that never falls to earth!

. . . .

Our original equations (for which Euler's technique works) were $d^2y/dt^2 = -32$ and $d^2x/dt^2 = 0$ under the assumption that 1 pound(al) shell is fired with an initial velocity of 83 ft/sec at an angle of 30°. By relying upon the conservation of energy and the initial velocity conditions, we can obtain the replacement equations $dy/dt = (2500 - 2y)^{1/2}$ and $dx/dt = 66.8$ (which are of so-called

When Eulev's rule goes wrong

first-order form whereas the originals were second order). But Euler's method graphs the latter as shown.[53]

....

Most robust reasoning techniques display unexpected bugs of this ilk on frequent occasion, as the early employers of computers discovered to their sorrow (truly dreadful consequences arose when the errors weren't so blatant and a company built an airplane relying upon the faulty calculations). To remedy this situation, applied mathematicians have learned that they must investigate, to the best of their ability, the validity of their reasoning principles from a *generic and correlational* point of view. That is, they must first model mathematically the range of physical circumstances $\mathcal{S}$ in which they expect to apply the rule and then verify whether the sentences progressively ground out by the method will unfold in proper alignment with every s in $\mathcal{S}$. Considering Euler's method from this point of view, we attempt to verify, if we can, that the technique really fulfills those "Harpo mimicking Groucho" relationships between sentence and world discussed in section (iii). In particular, we want to know: can circumstances ever arise where Harpo makes a mistake and fails to anticipate one of Groucho's moves successfully? Or, to restate these issues in less metaphorical terms, suppose we are looking at some general second order ordinary differential equation E (i.e. some equation of the same type as in our cannon ball case) with appropriate position and velocity initial values p_0, v_0. Without being provided any further details about E, we don't know what curves $\underline{f}$ will satisfy E, except that, surely, $\underline{f}$ must be a continuous curve possessing a second derivative (otherwise E won't be defined over $\underline{f}$). The set of all possible curves of this type is usually denoted C^2. So let us now consider an arbitrary curve $\underline{f}$ in C^2 and some second order differential equation E true of $\underline{f}$ (note that our specification of $\underline{f}$ and E is quite *generic*: this is all the information we are supplied about either $\underline{f}$ or E). Such a minimal specification delineates the basic *setting* of our problem. Let us now investigate how the steps directed by Euler's routine unfold relative to $\underline{f}$. A favorable situation will appear as illustrated: the Euler solution gradually wanders away from its target $\underline{f}$ as we consider increasing units of time Δt, due to the approximations Euler's method introduces (roundoff errors in our calculations will occasion even further straying but we ignore this). Nonetheless, we hope that our calculations will stay close to $\underline{f}$ (within a 2% error, say) over a decent interval and, by making the time step shorter, we can prolong the region of closeness as far out along $\underline{f}$ as we'd like. And, in the favorable cases, we can guarantee all of these things, because, by looking at the coefficients in E,

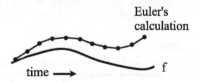

Euler's
calculation

time $\longrightarrow$

f

[53] E. Atlee Jackson, Perspectives of Nonlinear Dynamics, i (Cambridge: Cambridge University Press, 1989), §2.2.

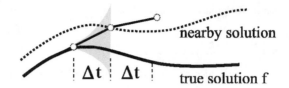

we can extract a so-called *a priori inequality* that sets up a little horn at the start of each computational step. We can then prove that the straight lines drawn by Euler's method will always stay inside these little horns over each Δt interval and thus insure, if these tolerances never open up very far, that our Eulerian broken line will stay close to f over a reasonable span, which will validate the basic reliability of this reasoning technique.

However, the degree of fluting in our little horns depends upon the equation's coefficients and a careful analysis shows that, if these coefficients fail to satisfy a certain proviso—a so-called *Lipschitz condition*—, their mouths can open up completely. If that happens, a spurious second solution to E can sneak through their opening like the proverbial snake in the grass. Euler's method, which is too stupid and automatic to distinguish good solutions from bad, may unfortunately entrain itself to this rotten intruder and produce graphs like that of the levitating cannon ball. And that is exactly what went awry in the calculation above: when we altered our original equations by what seemed like an utterly innocuous transformation (which, for other purposes, it would be), we inadvertently shifted from formulae that obey the Lipschitz condition to formulae that don't. Such lapses, whose salience was not noticed until a devoted correlative examination of the potential breakdowns in Euler's technique was performed, explains why the method sometimes fails, despite its great intuitive appeal. The upshot of our deliberations is, accordingly, this: the *proper* mathematical setting over which Euler's method supplies valid results is C^2 circumstances that *also satisfy a Lipschitz condition, not* unrestricted C^2 circumstances alone as we previously assumed (unfortunately, the Lipschitz requirement is not always easy to check). And this illustrates a developmental dialectic with which Chapter 8 will be much concerned: An unrestricted faith in rules $\mathcal{R}$ originally allows vocabulary to colonize a new patch of applications P. After detailed study of the facts encountered in P, it is decided that the validity of $\mathcal{R}$ needs to be restricted to a finer setting than originally expected in full P. Some of the *seasonality* I have mentioned in conceptual evaluation traces to the fact that the "correctness" of predicate employment must be adjudicated

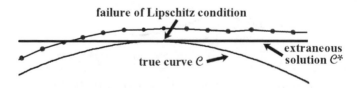

according to different standards according to which stage of the usage's development we presently occupy.

I've devoted considerable time to Euler's method, because the basic scheme of its word-against-setting investigations will prove important to us in Chapter 10, in the context of critical semantic concerns. In particular, our study of the method is *correlational* in the sense that we have investigated the Harpo-as-compared-to-Groucho manner in which the successive syntactic steps laid down by the inference technique compare with the reality that the routine attempts to approximate. The sentences churned out by Euler's rule and the temporal development displayed in $\underline{f}$ each unfold according to personalities of their own, and, as a result, the results can potentially fall out of alignment with one another: an investigation of method correctness hopes to rule out this possibility over the range of settings it examines. Our canvass is also *generic* in the sense that it depends upon very few specifics with respect to either E or $\underline{f}$. As a result, it can easily happen that applying Euler's method to a particular equation E* supplies sterling results, but mathematicians are unable to certify our conclusions because E* fails the Lipschitz requirement and they have no other means of guaranteeing that its calculations will be accurate. Some other factor allows the method to produce reliable results regardless, but we lack effective purchase on its nature as yet (sometimes roundoff "errors" allow a technique to work better than it theoretically should, through an artificial diffusion that mollifies its results with some realism). Such situations arise quite commonly in physical practice (celestial mechanics is full of them). Practitioners accept such calculations with moral certainty, yet no known proof of inferential validity certifies their results.

Much later in the book (10,iv), we shall have occasion to revisit these considerations because philosophers are familiar with studies of this general type, although only in the context of the soundness of logical rules. Unfortunately, they rarely consider how their logic-focused studies interact with the similar investigations required for reasoning techniques such as Euler's. As we'll see, the greater practical salience of the latter often effectively trumps the semantic relevance of the former in a distinctly anti-classical manner.

In the succeeding chapters, I frequently employ the term *picture* to designate a portrait of circumstances that is both generic and correlational in the manner displayed: a picture supplies a general account of how the vocabulary within a specific branch of usage matches up to worldly conditions across a range of settings (for which *mathematical models* are supplied in the manner of our C^2 functions $\underline{f}$). The illustration shows the basic elements at play in the picture $\mathscr{P}$ we have just provided for Euler's method. At the top we witness Euler's routine itself in linguistic action, grinding out specimen sentences $S_1, S_2, S_3, \ldots$ according to its mandated procedures. At the bottom we find the shifting values of the real world quantities Φ to which the predicate "P" in $S_1, S_2, \ldots$ correspond. Just above Φ I have set the class C^2 mathematical function $\underline{f}$ which serves to model Φ as its setting within the picture $\mathscr{P}$. Finally, an averaging operation (physicists call it *lumping*) converts $\underline{f}$'s and g's continuously altering values into discrete estimates pegged to each time interval Δt. If this mathematical picture

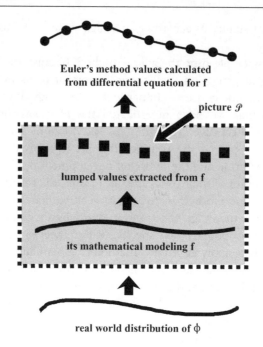

Euler's method values calculated
from differential equation for f

picture $\mathscr{P}$

lumped values extracted from f

its mathematical modeling f

real world distribution of ϕ

correctly captures the range of physical circumstances in which Euler's method is to be applied in real life application, we can then study through its evocation how closely the Eulerian derived claims S_1, S_2, S_3, ... stay true to the lumped values extracted from f̲. If the results are favorable, they supply us with a heightened confidence that our method will not play unexpected tricks upon us (such as levitating cannon balls).

The reason the purely mathematical intermediary f̲ is included in our sketch is because the picture we entertain of how a particular inferential routine works may prove *wrong*—we may fancy that an inferential routine proves successful because it relates to the world in supportive manner $\mathscr{P}$, when, in fact, its successes actually trace to the relationships mapped out in some alternative picture $\mathscr{P}^*$. Such misapprehensions will prove an important theme in the last third of this book: faulty conceptions of semantic workings represent a common facet of real life employment and we will want to learn how the ill effects of a wrong picture can be ably *detoxified*. Here we can prove mistaken with respect to either the mathematical class to which Φ correlates or the manner in which the support from f̲ travels to S_1, S_2, S_3 (both forms of mistake will be illustrated later). This is why I've drawn dashed lines in the illustration: the content of a picture $\mathscr{P}$ proper should be equated with the inner block of generic materials through which we believe the level of language connects with the physical world beneath.

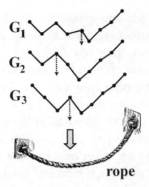

rope

In focusing upon the validity of Euler's method, I have selected an inferential routine whose unfolding syntactic steps S_1, S_2, S_3 genuinely *march along* with the shifting lumped averages of Φ (as long as the rule is utilized within its proper setting). As section (iii) observed, few successful computations relate to their subject in such a simple marching manner. As a case in point, consider the following method for computing the shape of a rope hung between two nails (its governing equation describes the influence of gravity as well as the rope's resistence to bending). Draw an arbitrary chain of broken lines between the two nails which we call G_1 (for guess #1). Compute how much energy is stored in G_1 from the governing equation. Now wiggle some little portion of G_1 a wee bit, leading to a new shape estimate G_2. Compute G_2's stored energy. If it proves less than that of G_1, then G_2 probably represents a better guess as to the cord's true shape. Otherwise, wiggle G_1 in some other way. Proceeding thus, we can grind our way through a sequence of guesses that progressively carry us, in zig-zag fashion, closer to a good approximation to the rope's hanging shape. Reasoning of this type is called a computation utilizing *successive approximations*: their routines can be compared to an archer who shoots repeatedly at a target, while an assistant retrieves her arrows and shouts back corrective hints ("A little too far to the right"; "Oops, now you're aiming too far the other way").

If our corrective instructions can be made *coercive*—that is, we force the error to become smaller on every repetition, both our archer and our broken line must zero in on a final answer (a *fixed point* in the jargon) which, if further conditions are met, will be the correct bull's-eye. The calculation of ln(5) in section (iii) represents another example of this flavor of computation.

If a real rope is draped between two nails, it will wiggle around a bit before it settles to its equilibrium rest position. So, prima facie, our computational sequence $G_1 \rightarrow G_2 \rightarrow G_3$ bears a superficial resemblance to the progressively relaxed (and lumped) states of our rope $S_1 \rightarrow S_2 \rightarrow S_3$ (indeed, our computational technique is called a *relaxation method* for this reason[54]). However, it is plainly mistaken to expect that

---

[54] F. S. Shaw, Relaxation Methods (New York: Dover, 1953).

sentences G_1, G_2, G_3 will provide *any* straightforward information about the physical states S_1, S_2, S_3 because our successive approximation calculations do not attempt to track how ropes genuinely settle into rest (indeed, the equilibrium equation in the background of our problem doesn't pretend to describe the relevant physical processes either, a point we'll examine at greater length in 9(i)). What facts about our rope do the sentences G_1, G_2, G_3 actually report? *Answer:* we progressively learn that the shape of our rope is confined to ever smaller geometrical boxes: in G_1, we effectively know nothing; in G_2, we learn that a little kink of the cord is situated lower than in our first guess; in G_3 that two little kinks are lower than our first guess, and so on. In effect, we gather data of the ilk: "Ethel must weigh between 130 and 150 pounds"; "Wait, make that 140 and 145"; "Oh, now I see that it must be very close to 143"; . . . A correct picture of our relaxation method calculations aligns each G claim with an inequality that states that our rope's position lies between limits A and B. In a proper specification of setting (which is a bit tricky to provide), we can prove that our G_1, G_2, G_3 will progressively box in the correct shape of the hanging string in all circumstances. But, clearly, this picture of how the reasoning pattern operates is quite different than the Harpo-imitates-Goucho picture suitable to Euler's method. As such, the two routines obey completely different *computational strategies*.

Oddly enough, routines of successive approximation type are sometimes wrongly pictured in a marching method manner—we shall examine several examples in Chapter 9. Very strange *ur*-philosophical opinions arise as a result.

As we saw, distributed normativity approaches to the meaning of scientific predicates are commonly called *instrumentalist* on the grounds that the theoretical frameworks in which they come embedded merely serve as "instruments for successful predication." However, I regard this terminology as misleading because successful instrumentalities, whether they be of a mechanical or a symbolic nature, always work for *reasons*, even if we often cannot correctly diagnose the nature of these operations until long after we have learned to work profitably with the instruments themselves. By a similar token, the component steps within any reasoning technique that supplies generally useful results over a varied range of settings must report genuine step-by-step information about the physical systems targeted, if only in the "successively box in the curve" manner of our hanging rope calculations. Instruments, as I have insisted, always work for reasons and worthy algorithms must keep track, somehow, of data genuinely relevant to their target systems. Indeed, the <u>modus operandi</u> of most correctness proofs validity with which

I am familiar proceed by first characterizing the (often abstract) nature of this correlated data and then showing that each step within the routine handles such information appropriately under generic conditions. This rather obvious observation will prove useful to us later.

....

A profound change in mathematicians' conception of their subject matter quietly emerged as the need for generic investigations such as these became apparent. To an early author such as René Descartes, "mathematics" (which he often calls "geometry") excludes consideration of target systems that are not amenable, in his words, to "mathematical study," where the latter phrase means something like "the manipulation of claims according to accepted procedures":

> [G]eometry should not include lines that are like strings, in that they are sometimes straight and sometimes curved, since the ratios between straight and curved lines are not known, and I believe cannot be discovered by human minds, and therefore no conclusion based upon such ratios can be accepted as rigorous and exact.[55]

In particular, Descartes insists that a genuine "mathematical curve" must obey some formula (or specified geometrical construction) that a mathematician can concretely manipulate, whereas all the other possible "curves that are like strings" belong solely to the world of physics, not to mathematics. In other words, most of the functions in the mathematical class C^2 are entirely "physical" according to Descartes, representing "curves like strings." Mathematics proper must limit itself to the opportunistic discussion of the very special physical systems of sufficiently regular description that mathematics can lay substantial inferential gloves upon them.[56] However, beginning in Euler's era, mathematicians gradually realized that they must tolerate as *part of mathematics' own dominion* arbitrary functions like the "curves like strings" that Descartes had eschewed, simply because its scope of study needed to embrace questions of the flavor, "Will this rule prove generically sound with respect to this space of functions?" This shift in approach and ontology became obligatory as it was gradually realized that commonly accepted inferential principles need to be scrutinized with considerable care given their propensities to unexpected misbehavior. In particular, Cauchy realized that questions like "Can Kepler's equation, $E = M + e \sin(E)$, be solved for E?" (i.e., expressed in the form $E = \psi(e)$) are far more delicate than heretofore presumed (earlier writers had simply assumed that such manipulations were valid, brushing away the occasional anomaly as merely an "exception that proves the rule").[57] Furthermore—and these shifts will be documented in Chapter 8—, it was eventually realized that our prima facie assumptions about proper mathematical setting for, e.g., differential equations might require readjustment: that (to cite one of our latter examples) a Sobelev class of distributions might provide a better setting for a differential equation of physics than the expected C^2. These changes in attitude arrived quite gradually, but the modern mathematician now accepts that part of her job is to establish the settings, delineated in set theoretic terms, over which given inferential principles will prove valid or not.

---

[55] René Descartes, Geometry (New York: Dover, 1954), 91.

[56] I call this the doctrine of *mathematical opportunism* in Mark Wilson, "The Mathematics of Spilt Milk," in E. Grosholtz and H. Berger, eds., The Growth of Mathematical Knowledge (Dordrecht: Kluwer, 2000) and "The Unreasonable Uncooperativeness of Mathematics in the Natural Sciences," The Monist 83, 2 (2000).

[57] Steven G. Krantz and Harold R. Parks, The Implicit Function Theorem (Boston: Birkhäuser, 2002). Ivor Grattan-Guiness, Convolutions in French Mathematics 1800–1840, ii (Basel: Birkhäuser Verlag, 1990).

It might be added that the challenges of quantum mechanics and other descriptive ills may eventually upset this portrait of mathematics' role within our thinking about physical structure, but we will work within the orthodox point of view throughout this book.

. . . .

5

THE PRACTICAL GO OF IT

For it is in mathematics just as in the real world; you must observe and experiment to find the go of it . . . All experimentation is deductive work in a sense, only it is done by trial and error, followed by new deductions and changes of direction to fit circumstances. Only afterwards, when the go of it is known, is any formal explication possible. Nothing could be more fatal to progress than to make fixed rules and conventions at the beginning, and then go on by mere deduction. You would be fettered by your own conventions, and be in the same fix as the House of Commons with respect to the dispatch of business, stopped by its own rules.

Oliver Heaviside[1]

(i)

Pre-pragmatist hunch. Although some readers will have passed its pleasures by, the previous chapter outlined the story of how philosophers of the logical empiricist school became entangled within an uncomfortable form of semantic dualism, wherein the alleged theoretical terms of science garner their linguistic significance through suspension within the webbing of theory (4,v dubbed this semantic mechanism a *distributed normativity*), whereas the regular terms of ordinary life ("is red," "is a doorknob") gain their meanings in the old-fashioned way, through direct classical gluing. Few philosophers accept this thesis in the same form today, but its atmospherics linger on, in the guise of hazy holism and what I have called the theory T syndrome (3,viii).

At the same time, many writers have challenged classical thinking quite bluntly, sensing that something exaggerated lies hidden within its ostensibly intuitive coils. Indeed, as the previous chapter also observed (4,iv), many of the nineteenth century originators of semantic dualism apparently wished to challenge classical assumption outright, but eventually capitulated halfway to its demands, in their efforts to win a greater conceptual liberty for science's conceptual endeavors. By the mid-twentieth

---

[1] Heaviside, Electromagnetic ii, 33.

century, many varieties of fully uncompromised anti-classicism had been launched and, of the many skiffs now afloat, the endeavors that tack closest to my own headings commence in what I shall call *pre-pragmatism*. With this awkward phrase,[2] I intend to rough out a loose collection of reflections upon linguistic capability that emphasizes the problematic aspects of language as it begins to shade towards impracticality. Such seat-of-the-pants hunches about language spring up coeval with the *ur*-philosophical leanings redolent of classical thought but run counter to them (our fund of pre-pragmatist percept provides the vernacular upon which the fully articulated pragmatism of a William James or John Dewey builds, as do the somewhat differently focused doctrines of a W. V. Quine).

Like classicism, pre-pragmatism has often inspired programs of philosophical thought that are extremist in their emphases, framing themselves into disagreeable holisms even less sustainable than classicism's Pollyannish optimism. However, if we concentrate upon the loose but intuitive worries that initiate these lines of thought, without hastening to covert our uneasy doubts into a grand alternative to the classical picture, we will find that commonsensical observations of great cogency lie there. And, just as Bertrand Russell served as an admirable Virgil to guide us 'round the corridors of classicism, we can invite Quine to chaperon us up the hillside of developed pre-pragmatist doctrine, for I consider his instincts as classical critic to be the equal of any. Then, when he begins his ill-advised turn towards holism, we can learn from those missteps as well and record in our notebooks, "Do *not* turn right at corner X."

Let us start with pre-pragmatist opinion in its rawest form, and then ask Quine to lead us further on. Almost invariably, musings of this type begin with the complaint that classicism's portrayal of semantic attachment is too *passive*—or, in William James' phrase, "intellectualist"—to be correct:

> [T]he great assumption of the intellectualists is that truth means essentially an inert static relation. When you've got your true idea of anything, there's an end of the matter. You're in possession; you know; you have fulfilled your thinking destiny.[3]

Although James writes here of sentential truth, his protestations apply with equal vivacity to the portrait of predicate attachment I have dubbed "classical gluing" (3,ii). Although the Russellian view renders the proper understanding of a predicate as merely a question of the grasp of the proper universal, James believes that the comprehending agent must display some fuller capacity for robust activity before the predicates she employs can acquire any tangible significance. He expects the contents of our understanding to be tied up, in his words, with "the practical difference it makes to us to have true ideas."

Quite apart from the vagaries of James' specific pragmatism, many writers have likewise urged that, in some manner or other, the classical viewpoint ignores what the

---

[2] Charles Pierce was unhappy with the supplements that William James and others had annexed to his original "pragmatism," so he invented a new term ("pragmaticism") "ugly enough not to be borrowed." In this same tradition of unattractive neologism, my coinage is designed to *remove* elements from pragmatism proper.

[3] James, "Truth," 160.

THE USEFUL SHOULD BE PREFERED TO THE AGREEABLE.

A pre-pragmatist.

physicist Oliver Heaviside called "the practical go of things."[4] Through employing language as *tools* in the accomplishment of sundry desired goals, these critics maintain, our predicates engage with worldly conditions in a more robust manner than is provided in the pallid "grasp" emphasized by classical thought. It is through the cycles of practical action that the sprockets of language become genuinely intermeshed with the gears of the world; mere armchair musing, however intense, cannot turn the trick, for it is through the achievement of concrete goals that language displays its central capacities for *performing work*. Such sentiments are often what a writer has in mind when she evokes such slogans as "concepts represent guides to action" or "meaning is use." Meditations in this vein are paradigmatic of what I consider to be pre-pragmatic thought. In 3,ii, I cited Quine's and Dewey's complaints with respect to "the myth of mental museum." Approached more sympathetically than I did there, epithets of this ilk express pre-pragmatist leanings, although they inaccurately characterize classical gluing in the bargain.

The following considerations are likely to increase our concerns with respect to classical inertness. The gurus of cults display a marked penchant for trafficking in utterly ungrounded terminology. For example, if the account offered in Martin Gardner's <u>Fads and Fallacies in the Name of Science</u>[5] is to be trusted, daft Wilhelm Reich introduced his disciples to a predicate "contains orgone" which he described, rather minimally, as "displays pure sexual energy." Over an extended period, Reich supplied his congregation with a lengthening list of natural occasions in which concentrated doses of the stuff were allegedly manifested. On objective perusal, this catalog represents a completely

---

[4] Ibid., 159. He mentions that the "go of a theory" occurs in Maxwell's writings.

[5] Gardner, <u>Fads and Fallacies</u>.

miscellaneous collection of circumstance marked by no commonality beyond Reichian whim (and a slight inclination to be bluish). Upon querying his flock, we will learn: "What causes the blue of the sky?" "Our master says 'orgone.'" "How about that blue sheen that covers a highway on a hot day?" "Dr. Reich has ascertained that it is likewise orgone." And so on. Although Reichians are capable of prattling endlessly about the "orgone containing" characteristics of everyday objects (and are even willing to sit long hours in stuffy boxes designed to concentrate the stuff upon them), the pre-pragmatist will regard their discourses as deeply defective. "This community has not taken the steps necessary to get 'contains orgone' truly *engaged* with the world. They have merely allowed the predicate to float freely above it, guided by nothing except guruish whimsy." In fact, if Gardner is right, cultists generally engineer their favored vocabulary, whether unconsciously or by design, to elude the inconvenient slings and arrows of relevant experience.

This aggregation has deluded itself into supposing that their peculiar predicates have formed a high degree of semantic attachment to the world, when no real capacity to perform linguistic work has been supplied at all. But it would seem that the classical picture cannot ratify this complaint of orgonish non-adhesion in a straightforward fashion. From its tolerant perspective, *containing orgone* should qualify as no more deficient <u>qua</u> *universal* than, say, *containing antifreeze*. After all, the learning processes whereby most of us come to grasp the latter notion do not seem dramatically different in psychological character from those that induce the average Reichian to prattle glibly of "orgone." The indulgent Russell who wrote <u>The Problems of Philosophy</u> will surely welcome *containing orgone* with open arms into his realm of universals. Of course, the classicist cheerfully allows that the empirical world nowhere instantiates this particular universal within its dominions, but this little foible of non-exemplification represents a minor detail of no particular concern to the philosopher of language.

Such an amicable toleration of rotten predicates leaves pre-pragmatists agape; surely the classical picture overlooks some essential kind of *practical grit* needed to tie words and world together in genuine alignment? They will complain, "Classical thinking makes the semantic attachment of predicates entirely a matter of diligent armchair cogitation. An orgonist can engage in such activities as well as you or I; it's what happens when we leave our plush settees that make the real difference." This, of course, is the raw objection James means to press in his complaints about "intellectualists."

But to this, the classicist will retort, "But consider the sentence 'Oscar's only ostrich owned some orgone.' That may represent a stupid thing to say, but we surely *understand* it. If so, we must have grasped concepts adequate to supporting the meaningfulness of its component predicates. But it is only this level of semantic understanding that interests the linguist and the philosopher of language. Perhaps you ought to take your complaints about the orgonists to the methodologists of science, for there is nothing to be certified as irregular in their *semantic practices*."

No true-hearted pre-pragmatist should be deflected by this familiar rebuttal. "Sure; under some construal of 'understand,' I likewise understand Lewis Carroll's 'Slithy were the barrow groves' and Little Richard's 'Wop bop a loopa; a wham bam boo,' but such toleration doesn't indicate that each isn't semantically defective in important respects. It can even happen that 'orgone' talk may remain current for a considerable expanse of time, especially if special linguistic arrangements shield it from confrontation with any practical issue, but even longevity is not proof of adequate semantic substance. Surely part of the job of the philosopher of language is to evaluate the operative directivities of sundry predicates: when they appear sufficient and, when, like 'slithy,' 'loopa' and 'orgone,' they seem inadequate."

This dispute between classicist and pre-pragmatist echoes our 1,vi discussion of the *thesis of semantic finality*, viz., that a firm grasp on many concepts critical to language is completed by the time an individual becomes judged "competent" in the employment of her tongue. Classical thinkers regard such events as important demarcation points in semantic attainment, whereas pre-pragmatists consider them mere way stations along a pilgrimage leading to more robust forms of linguistic capacity.

(ii)

Strands of practical advantage. All of these musings lean hard on the notion that linguistic activity can be said to "perform useful work," rather in the vein of a concrete mechanism such as a winch or garage door opener. Let us see if we can convert this common but entirely metaphorical comparison to a claim of any substance at all. Here are three exemplars for what such "work" might look like.

(1) An artillery officer hopes to hit a specified target based upon its geographical coordinates, velocity and wind speed. Unless proxies for the necessary computations have been built into the machinery of his cannon, our gunner must scribble a somewhat elaborate *algorithm* on a piece of paper to convert his input data into proper instructions with respect to cannon angle. Human beings simply cannot fire cannons accurately unless they engage in some span of intervening linguistic doodling, perhaps of an Euler's method type.

(2) A traveler is unlikely to navigate her way successfully to Grandma's house through a difficult and unfamiliar terrain unless she carries a written list of instructions to direct the stages of her travels. A *recipe* in linguistic form assists the performance of the task considerably.

(3) Teenage lovers will not be able to rendezvous in a fashion that eludes their families' scrutiny unless they exchange a message via faithful Nurse that allows them to *coordinate* their activities, e.g., "Meet me tonight beneath the balcony."

For want of a better term and without pretending to have identified a precise class of activities, we might loosely say that such employments display recognizable *strands of practical advantage*—viz. the achievement of certain goals requires that certain sentences fall into proper place during their execution. The "work" accomplished in each case is certified by the desired condition achieved. In the argot of the previous chapter, the sentences we string out in executing a strand of advantage each acquire a pronounced measure of top-down *distributed correctness* from their roles within the integrated routine, where a sentence may qualify as "correct" by these practicality-focused standards even if it reports a patent falsehood if evaluated by more conventional measures (vide the example of successive approximations in 4,v). Philosophical meditations with respect to linguistic work almost invariably appeal to some implicit flavor of distributed correctness.

A pre-pragmatist sympathizer will rightfully point out that few activities where "language performs real work" appear within the chatter of the orgonists. "But it is precisely within practical episodes such as these," she grouses, "that linguistic activities genuinely *entangle* themselves with the progress of worldly events. The goals desired by the speakers will not be accomplished unless the correct chain of linguistic events appears in their endeavors. If the improper linguistic act is performed, the speakers are likely to be penalized in a failure to reach their objectives. But it is precisely through the medium of these buffeting blows of reward and punishment that the physical world makes its semantic desires known to us. The idle classifications of the orgonists matter not a whit to it; they can babble like that all day and Mother Nature won't care. But if they should be so foolish as to attempt some practical purpose with the notion—e.g., build an automobile designed to run on orgone—, then they will be punished by their project's failing to budge. This is why their wily guru has encouraged his flock to employ 'orgone' largely in a manner that skates frictionlessly across the texture of the world—to affirm or deny that the distant highway is coated with orgone is unlikely to interface with any practical task the group might undertake. In this sense, their 'orgone' talk has been guruishly engineered to perform little work. But a more robust degree of pragmatic entanglement constitutes the true glue that binds more adequate vocabulary to the world, not inert armchair 'grasp' classically viewed."

Note that our pre-pragmatist can complain only that "orgone" *performs little work*. The swains and dairymaids of orgone society can arrange their secret trysts through list-based coordination: "Meet me tonight where the orgone flows abundantly." If some locale of "abundant flow" has struck Dr. Reich's fancy, our lovers can exploit that determination to mild practical advantage. One can usually eke some mild strain of practicality from the most ridiculous usage.

Intuitively, we expect that the developments of genuine recipes of practical advantage represent important *anchoring points* in the developmental history of a language: once a linguistic routine has become firmly planted in the sands of practicality, our other forms of linguistic endeavor must respect its work capacities. We will not want to abandon tools that accomplish worthy ends unless we have found superior replacements that can reach allied objectives.

Consider, in this light, J. P. Gordon's discussion of traditional practices governing the preparation of materials such as sword steel:

> *Since the subject has proved so troublesome to scientists, it was not to be expected that our ancestors would approach it in a very logical way and, in fact, no technical subject has been*

so deeply invested with superstition. A long and mostly gruesome book could, and perhaps should, be written about the superstitions associated with the making and fabrication of materials. In ancient Babylon the making of glass required the use of human embryos; Japanese swords were said to be quenched by plunging them, red-hot, into the bodies of living prisoners. Cases of burying victims in the foundations of buildings and bridges were common—in Roman times a doll was substituted. . . . [T]he science of materials, like the science of medicine, has had to make its way in the teeth of a great many traditional practices and old wives' tales.[6]

Indeed, extracting a desirable cutting tool from what was formerly hematite or native iron is no mean accomplishment, for it requires the unnatural trapping of unstable phases within the material matrix (like the diamond, much of the grain within sword steel consists largely in frozen visitors from another thermodynamic climate). All the traditional arsenal of the smithy—quenching, cold working, annealing, etc.—serves to install a very refined polycrystalline structure, delicately sensitive to impurities, within the steel, although virtually none of is mechanics was understood until well into the twentieth century. When some callous Japanese craftsman develops an effective recipe for manufacturing swords, its component stages must roughly calibrate with transmutations within the metal that are objectively required, viz. "plunge sword into belly of noble foe" reflects a need to "quickly lower outer temperature to lock in ferrite grain" (perhaps the nitrogen contents of the victim's blood aids the process in some delicate way as well). Once such a recipe is discovered, it will surely be prized until some superior replacement is found. We can only hope that, in the manner of the Roman dolls, some more humane surrogate for a "noble foe" will be quickly found (perhaps a pail of heated chicken broth). Given the centrality of the recipe—and here is where the special importance of practical advantage enters the picture—, its articulation can be expected to act as an anchor or brake on how its component vocabulary is henceforth employed. In the ameliorating circumstances described, it is even likely that the substituting bucket of brine may continue to be called a "noble foe," because of both superstitious continuity and a disinclination to be linguistically innovative. At this point a new branch of the use of "foe" commences. In my estimation, this process represents a natural way in which a usage continues from one set of circumstances into another.

To the classically minded, such episodes, although undeniable, constitute minor events within the story of language: "Oh, a simple metaphor between victims and buckets has occurred to our smithy, which inspires him to attach 'foe' to a fresh concept that is willing to accept chicken soup under its classificatory umbrella. A simple polysemy has been engendered, but it signifies little. Our smithy may not recognize his meaning change, but he would if he meditates carefully on the distinct natures of enemies and buckets of broth." Through such appeals to "changes in attached concept," classical thinkers typically avoid granting any special prominence in language to the

---

[6] J. E. Gordon, Structures (New York: Da Capo, 1981), 22.

strong classificatory directivities that often arise in connection with specific strands of practical advantage. Such techniques for *classical unloading* were surveyed in 3,vi.

But our guild of pre-pragmatists should stick to our hunches and insist that such practical directivities, even in the peculiar circumstances sketched here, represent central aspects of linguistic process and should not be dismissed as mere eccentricities.

....

Many writers maintain that *social practicalities* such as (3) are the most critical for understanding linguistic process, even to the point of denying (2,vi) the viability of "Robinson Crusoe" virtues such as (1) and (2). Although the advantages of inter-agent coordination probably lace through usage more liberally than those of an individualistic kind, they still remain fairly sparse and are apt, as we observed in the orgone tryst affair, to remain alive even with respect to highly impractical vocabulary. Although I could be happily convinced otherwise, it seems to me that any pre-pragmatic thesis that can be advanced through consideration of some (3)-type merit can be established more briskly and effectively by considering some allied (1) or (2) excellence instead, at least with respect to the range of descriptive vocabulary under consideration in this book.

After all, why should we utilize our words to please the established norms of society if applying them in some other manner suits Nature better? The guardian muse of our later chapters, Oliver Heaviside, did no such thing, for as W. E Sumpner beautifully put it:

> He was a wanderer in the wilds and loved country far beyond railhead.[7]

True, it would be hard to buy groceries if we acted like Heaviside in every quarter of our life, but, nonetheless, in descriptive work we rarely value virtues (3) so much as (1) or (2).

When I employ *strand of practicality* in the sequel, it will invariably be in the narrow sense of a *linguistic recipe* or *practical algorithm*.

....

(iii)

Linguistic engineering. In sum, pre-pragmatists accuse classical thinking of crediting inadequately attached predicates with better semantic credentials than they really merit. If so, they must also argue that we accomplish *less* in the course of commonplace regulative acts than we generally fancy. After all, there are many concrete steps we can take to redirect the currents of usage along more profitable channels: we can introduce fresh terminology, redefine old terms, set forth crisp governing axioms and so forth. From a classical point of view, the base activity involved in all of these reformatory episodes is quite simple: we align fresh concepts with our verbiage and allow their dictates to govern the correctness of every assertion uttered along our newly established branch of usage. But if completely dominating directivities can be laid down by such simple human actions, pre-pragmatism's strands of practical advantage have been

---

[7] Nahin, <u>Heaviside</u>, 219.

thereby denied any arena in which they can shape language in substantive fashion, for every critical semantic decision will have been already settled by the stipulated align-ment of predicate with concept. To get anywhere with our pre-pragmatist doubts, we must reject this contention. We can only allow that language use can be improved *to a certain extent* through such actions, for our ceremonies of reorientation rarely settle a predicate upon its future courses as firmly as the improving classicist presumes (this concern, I believe, forms the true basis of Quine's complaints about "the myth of the mental museum"). But establishing diminished expectations of this sort requires both substantive argument and striking example, for, *ur*-philosophically, we are greatly dis-posed towards inflation of our capacities in regard to linguistic management.

But how should we amplify upon these suspicions? Where do our mundane, everyday acts of corrective improvement fall short of classical expectation? Usually a strong flavor of *engineering consideration* emerges in the considerations we bring forth, at least as long as we stick to the purely intuitive level in which pre-pragmatic doubts originate.

Consider, in this vein, the problem of designing a mining vehicle for assaying the characteristics of stones encountered upon the surfaces of alien planets and shipping desirable items back to earth. If we approach this problem in a naïve way—simply dispatching machinery to Pluto that can accomplish terrestrial tasks ably—, we are likely to be disappointed in the results, for a device that employs an internal spring balance (i.e., of bathroom scale type) in its weighings will consistently supply drastically insuf-ficient "masses" to the Plutonian rocks it encounters. These errors occur because an earth-calibrated balance will measure masses accurately only if it remains in an envir-onment where the ambient gravitational acceleration remains close to tropospheric norms (and, we might add, where the local planetary surface is adequately supportive and the testing apparatus is orthogonal to its plane, etc.). *If* we happen to know the local gravitational constant for Pluto in advance, then our scale can be calibrated ahead of time so that our explorer's spring balance will produce correct values. But we may not know this "constant"—after all, buried Plutonic masses may cause it to vary significantly from one locale to another. We may need to design our mining vehicle in a more sophisticated way so that it can self-correct its classifications, perhaps by monitoring test specimens brought along for this purpose. But that skill will require a large amount of additional engineering. And there are many other potential difficulties besides erratic gravitational constants that may spoil our vehicle's registrations of *mass* as well. Whatever capacities for learning we install within our vehicle, unexpected patterns of local feedback may cause our craft to lock upon characteristics other than we desire. Each time we address any of these problems, we must burden our explorer with additional hardware and programming.

In truly alien climes problems can arise from quarters that are very hard to anticipate. Suppose we have instructed our explorer to hunt exclusively for Plutonian rubies. Pluto, however, is both a cold and ill-lit spot, well outside the range of earthly variation. The hues of beryls like rubies and sapphires depend sensitively upon scattered color center impurities in their matrix (the pure mineral is colorless). It is within the realm of possibility that the intemperate Plutonian conditions may induce a subtle shift in the crystal array, causing the local stones to unexpectedly reflect the dim sunlight strongly in the green. Likewise, beryls we would consider to be of poor quality reflect preferentially in the red in the Plutonic conditions. Even if we visit Pluto, we won't be able to see these effects, because our color vision will not be active in the low illumination; however, the altered spectral reflectances will be apparent in a time exposure photograph. Should such greenish, frozen stones qualify as rubies, for if we merely subject them to stronger light, the radiant heat will shift their delicate structure sufficiently to reflect strongly in the red as normal rubies do? Or should we say that terrestrial stones stop being rubies within Pluto's bitter climate? Indeed, we commonly allow that phase shifts induced by temperature changes alter our gemstone classifications—we comment, "These worthless beryls used to be fine rubies until Jones stupidly heated them." For that matter, should we consider our Plutonian stones in their present state to be *green* or *red* or not, in the way that we claim earthly roses remain red in the dark? I doubt that we have yet settled any of these questions, lying so far from tropospheric anticipation.

Accordingly, our mining vehicle may take great labors in exhuming "rubies" that, upon transport back to earth, appear lackluster in the tray, having spurned all of the truly desirable stones. Since *we* have never clearly pondered how color tags should be rightly assigned in such inclement conditions, it is unlikely that our extracting vehicle could have been programmed to produce classificatory results that we will invariably admire (unless a complete imitation of human aesthetic judgment has been improbably installed within its circuitry). To paraphrase the old song, our roving miner will just keep classifying right along, no matter how absurd or uncongenial we find the results.

Reflecting upon these considerations, our pre-pragmatist concludes, "Surely, we humans are not radically better prepared for universal classification than our mining vehicle. We will have endured a long schedule of training experiences at the hands of our parents that leaves us convinced that we fully grasp the concepts *being a ruby* or *being red* in every potential ramification, but, in hard fact, we will have merely assembled preparation adequate only to a narrow, local slice of the universe. Looking over the entire field of grammatical sentences that contain the predicates 'is a ruby' and 'is red,' we fancy, 'I have grasped adequate conceptual content to render every one of these claims true or false.' But this supposition is not true: the status of 'Plutonian rubies are red' remains unestablished as yet. Of course, we may encounter the sentence 'Flash Gordon picked up a Plutonic ruby' in a science fiction story and allow it to pass without cavil, but this does not show that standards adequate to the circumstances it conjures up have really been laid down. How could it be otherwise? The amount of preparatory education required to truly presage Pluto's recondite conditions would need to be fantastically detailed. In contrast, it is quite easy to raise our kids to be *complacent* and *overconfident*. Classificatory hubris established, we might not *notice* that, upon encountering the Plutonic gems, further determinations are required; like the Druids of 1,ix, we might instead 'keep classifying on' in our Plutonic mineral encounters, allowing the salience of the moment to settle the 'correctness' of our classifications (indeed, returning to our mechanical miner, it would require a huge amount of additional engineering to render it *smart* enough to announce, 'I here lay down a semantic decision') . But such 'on-the-fly' decisiveness scarcely demonstrates the prior preparation that the classical picture claims to be present. Surely the full grammatical field of English must be less tightly bonded to the world overall than that account pretends."

These pre-pragmatist considerations warn us to mistrust the intuitive thesis of *semantic finality*, as it was articulated in 1,vi: by age 12 or so, English speakers will have fully mastered enough concepts to glue an ample field of syntax, as specified in schoolbook grammar, fully to the world. "No," we should say, "we are not yet truly prepared for every recondite corner. Instead, we should cannily watch for the unmoored patches within these grammatical arrays with a vigilant eye and refuse to credit a sentence with adequate semantic credentials simply because it seems adequately 'understood' when it pops up in the confines of an adventure yarn." In other words, the classical picture claims that concepts cover every inch of advance territory in the manner of a scrupulous surveying team, whereas pre-pragmatists anticipate that our predicates often behave like the agents that the CIA frequently recruits: layabouts who fritter away their hours in neighborhood bars and then file hastily improvised "reports" when pressed by the home office.

Such concerns with respect to our genuine capacities for adequate conceptual anti-cipation should heighten our suspicion that classical thinking errs in presuming that the linguistic endeavors of orgonists do not "differ semantically" in any significant manner from our own; that our divergencies lie only in the fact that the empirical facts do not lean their way. Russell's tale of classical gluing glosses over the grit of practical entan-glement that is required to bring predicates into true engagement with external reality.

(iv)

Pre-pragmatist prospects. As the notion has been employed here, pre-pragmatism represents nothing but a vague unease with respect to the classical picture and its central notion of complete conceptual grasp. As observed in 3,vi, classicism frequently turns a bit cagey when invited to delineate the precise conceptual contents of familiar words or to specify the educational stage at which a learner comes into their possession. To be sure, classicism presumes that most speakers will have fulfilled the requirements of complete grasp for the common predicates of English by that uncertain date when they begin to be treated as linguistically competent by their peers, but it is fully prepared to wobble on these assurances as soon as the predicates at issue appear to behave in funny ways. Anti-classicists, of course, view these same "funny behaviors" as symptomatic of the errors inherent within the classical picture.

In this inventory of uneasy doubt, two basic arenas of pre-pragmatic concern have emerged. The intellectualist inertness of the classical story appears troubling, because no "capacity to perform real linguistic work" forms any part of it. And classical grasp seems to require its employers to anticipate future variation in a manner plainly beyond reasonable human capacity. Both worries, I think, are quite legitimate. Unfortunately, the activities that optimally illustrate the "work capacity" we have highlighted are quite rare in real life linguistic practice and this paucity impedes our ability to turn uneasy hunch into solid critique. At this point, most pre-pragmatists have been inclined to expand "strand of practical advantage" into some more sweeping category, such as "language game" or "useful linguistic practice," able to encompass any form of human discourse they consider legitimate. But, notoriously, such enlarged notions are hard to render clear. As F. H. Bradley rightfully complained long ago:

> But here we have once more on our hands the question of what "practice" is to mean. Any serious attempt to define "practice," would, or should, rend asunder the Pragmatist church.[8]

Indeed, if we were to draw up an impartial scorecard as to how the disagreement between classicist and pre-pragmatist presently stands, it would look like this.

On the one hand, classical gluing promises a mechanism that assigns uniform standards of correctness to every sentence that falls within a grammatically delimited field and it achieves this distribution on an easy-to-learn recursive basis. It assures us that these conceptual supports will be largely locked in position by the time a child is normally judged as competent in the language's use, although if the initial grasp is muddled, its ambiguities may need to be sorted out later. Based upon this picture of conceptual clarification, the classical scheme provides clear guidelines for how our typical problems of vagueness, ambiguity and misunderstanding should be addressed. And it achieves all of these fine things while remaining loyal to the *ur*-philosophical leanings that all of us manifest within our everyday evaluations of human conceptual behavior.

In contrast, our budding pre-pragmatism has only offered a notion of linguistic work applicable to very restricted stretches of real life discourse and whose relevance to resolving the conceptual problems of ordinary life seems quite murky. It has provided no story as to how a speaker learns its favored strands of practical advantage, of whose semantic salience most speakers seem utterly unaware. The most natural account, of course, is to claim that such routines simply get learned as humdrum facts later on, long after speakers have learned to understand their working vocabulary through completely classical pathways. But to concede this is to give up on pre-pragmatism altogether.

Frankly, the prospects for developing pre-pragmatism beyond raw hunch do not look auspicious at this stage. Clearly, a range of pressing questions needs to be addressed: (1) How can the iron grip of classical gluing be relaxed enough to allow our strands of practical advantage some arena in which they can contribute to the story of language in a significant way? As matters now stand, classicism's thoroughly effective adhesive tacks down utilitarian and frivolous patches of language with equal uniformity and regards the divide between the practical and the useless as a matter of concern only

---

[8] F. H. Bradley, Essays on Truth and Reality (Oxford: Oxford University Press, 1914), 70.

to the engineer and the homemaker, not the student of language. (2) Since the immense swatches of usage that perform no apparent work still seem patently meaningful, what attitudes should the devoted pre-pragmatist adopt with respect to this vast ocean of unexceptionable usage? (3) If pre-pragmatists elect to fiddle with classicism's approach to semantic ambiguity, how must our views of sound methodology alter, when we confront the common problems of linguistic management that the classical picture organizes under the headings of "vagueness," "ambiguity" and "misunderstanding"?

The suggestion that comes immediately to mind is that pre-pragmatists must devise some alternative mucilage of wide semantic reach and comparable uniformity, comprised of an epoxy significantly laced with stout fibers of practical advantage. Contrary to first appearance, most ordinary discourse (including, e.g., every morsel of back fence chitchat) performs useful work by the tolerant standards of this new glue, albeit of a more rarified nature than is manifested in our specimen recipes (1) to (3). Indeed, scholars who pursue "meaning is use" programs of this kind generally find that, in the final analysis, language's most egregious lapses from acceptable labor standards occur mainly in the writings of their philosophical opponents. And this quest for a better glue represents the policy that most pre-pragmatist sympathizers elect to follow—it constitutes the fatal decision that converts the pre-pragmatist into a full fledged pragmatist, a Quine, a Kuhn or Wittgensteinian enamored of "language games."

But galloping away upon such ambitious campaigns is both ill-advised and unnecessary, I think. As indicated previously, the head waters of classicism flow from the many legitimate springs that feed our everyday interests in evaluating the verbal behavior of ourselves and our fellows. On a given day, we may properly applaud young Johnny for calling the astronauts in a space station "weightless"; five years later, we may chastise him for his "error" (I'll treat this case in more detail in 6,viii). Classicism's unfortunate foible is that it assumes that none of these evaluative fountains ever need to be turned off, whereas, in real life, our talk of "conceptual grasp" et al. cycles through natural seasonalities that reflect the developmental condition of the relevant usage. Rather than rushing to find an alternative epoxy, we should instead ask ourselves critically, in reassessing the everyday semantical judgments which the classical picture treats as definitive and timeless, "Aren't there tacit issues buried here that will need to be reopened at some later time, even if they cannot be profitably addressed today?" A commonsensical look at the evolutionary history of key descriptive predicates will reveal plenty of these concerns-to-be-delayed, as well as strong motivation to approach the meandering currents of linguistic development with greater humility than classicism encourages. By examining salient examples in a suitably hardheaded manner, we can lessen the uniform flood waters of classicism enough to find the structural pilings of practical advantage once again emerging, sometimes in the mode of the *facade frameworks* introduced in 4,vi. This is not a tale of alternative adhesive, but simply a more detailed accounting of the machinery of cooperation (and lack of it) between Nature and man that often leads

Quine

descriptive language along the improving, but often mysterious, developmental paths we frequently witness.

(v)

Quine's rejection of classical gluing. Let us now invite W. V. Quine onboard to serve as foil and counselor to our endeavors. In his <u>Word and Object</u> and elsewhere, he offers a trenchant critique of classicism, yet, at the same time, invites us to accept a semantic alternative of considerable quack pretensions. Let me first delineate the basic ingredients found in Quine's alternative fixative briskly, and then turn to his attack upon classicism.

At root, Quine adopts to his own purposes the basic mechanism of predicates being supported semantically within a webbing of theory, as was described in 4,iv. The base idea is that, if we know how to manipulate syntax in response to natural conditions in a sufficiently rich way, then we qualify as understanding that vocabulary fully—no supportive Russellian universal is needed to supply further "meaning" to our term. The old logical empiricist school hoped that a governing framework of initial axioms could entwine its component predicates in enough regimented webbing that the terms will appear as if they possess classical "fully determined meaning" when looked at from afar. We rehearsed some of the familiar objections that brought these ambitions to grief, not the least of which was that the positivists discovered that they needed to appeal to classical grasp to supply their "observational subvocabulary" with adequate semantic significance, thereby initiating a torrent of journal criticism to the effect, "Well, if you can employ classical methods for 'red,' why not for 'electron'?"[9] Quine proposes a rather clever way round these difficulties, while remaining loyal to the radically anti-classical thesis that *every* predicate gathers its semantic individuality through distributed normativity alone—that is, through being held up by the threads we weave within an ongoing web of belief. He achieves this as follows. A smallish group of "observational

---

[9] Grover Maxwell, "The Ontological Status of Theoretical Entities" in Martin Curd and J. A. Cover, eds., <u>Philosophy of Science</u> (New York: W. W. Norton, 1998).

sentences" get initially attached to the world via the strands of *classificatory advantage* they offer. But this attachment occurs only at a fused sentential level, and no word/ world correlations like those assumed by classicists are put in place at the predicative level at all. General methodological principles and grand architectural desires led us to weave these observation sentences together through intermediary sentences containing other predicates, eventuating finally in a thoroughly entangled "web of belief." It is from their position within this gigantic snarl that specific predicates obtain their individualized personalities. This proposal, although it rescues Quine's endeavors from the logical empiricists' implausible reliance upon tidy axiomatics, converts his approach into a *hazy holism* of a type I particularly adjure (I'll return to these concerns later in section (xi)). For now, we will merely observe that a predicate's position within its supportive web of doctrine is regarded by Quine as providing an enlarged generalization of pre-pragmatist "work capacity" able to serve as a universal replacement for the semantical relationships favored in classical thinking. In one fell swoop, he pries every stretch of our usage from classical gluing's tight grip, simply through supplying a web-based adhesive of his own.

I'll fill in further details of Quine's scheme as we go forward, but let us now turn to his criticisms of classicism, which are best presented in a dialectic with Russell's position, as sketched in Chapter 3. At each stage, we'll see that Quine's complaints can generally be sustained in weaker measure, without succumbing to the implausible doctrines of his developed views.

To begin, let us revisit a revealing passage from Russell cited in 3,ii.

> *Suppose, for example, that I am in my room. I exist, and my room exists, but does "in" exist? Yet obviously the word "in" has a meaning; it denotes a relation which holds between me and my room . . . The relation "in" is something which we can think about and understand, for, if we could not understand it, we could not understand the sentence "I am in my room".*[10]

This simply represents an affirmation of the basic mechanism of classical gluing. Quine believes Russell's fabrication of universals must be arrested at this early stage, for once classical binding takes hold, no slip will be left in language that requires any work-based mucilage. Accordingly, Quine objects to the swift transition between the *meaningfulness* of a predicate and the *postulation* of a "universal" as its semantic support. In Quine's diagnosis, Russell's universals represent nothing more than the misguided *projection* of features belonging to the syntactic manipulation of language use onto the screen of a falsely externalized ontology. Consider the purported difference between the concepts *being water* and *being H$_2$O*. True, we do not manipulate the *predicates* "is water" and "is H$_2$O" interchangeably (until we learn certain identity statements), but this behavioral distinction can be easily explained by the normal process of differential predicate learning. It serves no useful purpose to set up mythological effigies of these lexical differences within Russell's realm of universals, where citizens *being water* and *being*

---

[10]   Russell, Problems, 90.

H_2O are claimed to dwell. Such universals comprise the linguistic equivalent of Coleridge's naïve woodsman who

> Sees full before him, gliding without a tread,
> An image with a glory round its head;
> The enamored rustic worships its fair hues,
> Nor knows he makes the shadow he pursues![11]

Uncritical acceptance of classical projection thereby lulls our thinking about language into unearned complacency—"universals" conceived in Russell's manner enjoy a dangerous "power to cloud men's minds":

> The evil of the idea idea [= the concept of a universal] is that its use, like the appeal in Molière to a <u>virtus dormitivia</u>, engenders an illusion of having explained something. And the illusion is increased by the fact that things wind up in a vague enough state to insure a certain stability, or freedom from further progress.[12]

But what maintains the predicate "is in" as meaningful if no substantive classical concept is available to prop it up? Like any admirer of distributed normativity (4,v), Quine claims that its employments are supported laterally in his web of belief like the capstone of an arch. Indeed, if all of this interlocking machinery can be regarded as properly installed, then Quine has found a sweeping reply to Russell: the true reason why a predicate like "is in" qualifies as "meaningful" derives entirely from the manner in which "is in" comes embedded within Quine's syntactic web; there is no need to plant a hypostasized universal beneath the phrase for its direct support. Russell's tale of supportive universals gets the true story of predicates almost exactly backwards, Quine thinks: because they are rendered meaningful by their place in the scheme of linguistic endeavor, we needn't saddle reality with a fictive projection of bracing universals.

> What on the part of true sentences is meant to correspond to what on the part of reality? If we seek correspondence word by word, we find ourselves eking reality out with a complement of abstract objects fabricated for the correspondence.[13]

Yes, but what about that Achilles' heel of the logical empiricists, where observational predicates seem as if they need to be classically attached to the world by classical means and then woven into the fabric of theory with unnaturally crisp bridging principles? The tidiness issue Quine disposes of through his account of the dynamics of scientific methodology, an account I find unsatisfactory but needn't concern us here. He proceeds to remove *all* predicative classical gluing from his scheme by claiming that only full-bore "observation sentences" ("Lo! a rabbit" is his favorite example) receive any worldly direct attachment and only then through a process he vaguely calls "conditioning to stimuli" (intended to be anti-classical in its causally installed character). The purpose of

---

[11] Samuel Taylor Coleridge, "Constancy to an Ideal Object" in <u>Samuel Taylor Coleridge</u> (Oxford: Oxford University Press, 1985), 122.   [12] W. V. Quine, "Meaning in Linguistics" in <u>Point of View</u>, 48.
[13] W. V. Quine, <u>Quiddities</u> (Cambridge, Mass.: Harvard University Press, 1987), 213.

this maneuver is to free the component predicates within these observation sentences from any attachments of their own to attributes or other forms of abstract object. Here is how Quine himself puts the proposal, which sets the distributed normativity at the heart of his thinking in clear relief:

> Structure is what matters to a theory, and not the choice of objects. F. P. Ramsey urged this point fifty years ago, arguing along other lines, and in a vague way it had been a persistent theme in Russell's Analysis of Matter. But Ramsey and Russell were talking only of what they called theoretical objects, as opposed to observational objects. I extend this doctrine to objects generally, for I see all objects as theoretical. This is a consequence of taking seriously the insight I traced from [Jeremy] Bentham—namely, the semantic primacy of sentences. It is occasion sentences, not terms, that are to be seen as conditioned to stimulations . . . . Whether we encounter the same apple the next time around or only another like it, is settled if at all by inference from a network of hypotheses that we have internalized little by little in the course of acquiring the non-observational superstructure of our language.[14]

As this quotation suggests, even proper names such as "Willard" or "Sniffy" fall victim to the same lack of direct connection to the world as predicates suffer under Quine's scheme. "But this is ridiculous," we complain, "if my child has decided to call the rabbit in our backyard hutch 'Sniffy,' Quine informs me that I should not assume that the truth of the claim 'Sniffy is munching lettuce' is rendered true or otherwise directly supported by the activities of said rabbit? In other words, if Russell has blundered in trusting that attributes are required to prop up the significance of 'is a rabbit,' shouldn't we equally conclude that we err in presuming that some substantive rabbit in the backyard supports the meaningfulness of the name 'Sniffy'? But, surely, such doubts are daft."

Quine's reply is that the apparent asymmetries between "Sniffy" and "is a rabbit" can be explained by paying careful attention to the restricted patterns in which we employ quantifier phrases like "there is" and identities like "is the same object as." Or, to put his point more carefully (because street corner chatter will not bear out his contentions), we will find these restricted patterns displayed when we clean up loose everyday talk following the "regimentation" dictated by proper Scientific methodology. Although this reply, in its full, gory details is quite roundabout and certainly not very "intuitive," it does produce the result that, yes, rabbits can be legitimately "posited" and, moreover, representatives of this class do correspond to the embarrassing names that our children apply to their bunny victims. But the indirect logical arrangements that render coherent this matching of *names* with correspondent rabbits breaks down in a subtle way, Quine claims, when our attention turns to *predicates*. I won't try to detail Quine's elaborate tactics here, but his distinction between the two cases rests upon his celebrated criterion of ontological commitment, whereby we should determine the "ontology" of a person's beliefs, not by looking for the direct correlates of any form of linguistic expression (even

---

[14] W. V. Quine, "Things and Their Place in Theories" in Theories and Things (Cambridge, Mass.: Harvard University Press, 1981), 20.

when such correlations are meaningful), but through inspecting the quantificational structure of the agent's beliefs (that is, we examine the sentences that the speaker advances in the idiom of "all," "some" and "identical").

I find all of these claims utterly implausible, but they prove critical to much of Quine's mature thought and the many famous theses he has championed, few of which appeal to me either. I consider these doctrines as symptoms of the fact that Quine has attempted to evade the grip of classical gluing through excessively radical tactics.

(vi)

The flight from intension. So what should we properly do? Let me observe that, although I find his web of belief story entirely implausible, nonetheless Quine's instinct that sometimes Russell needs to be answered with a spot of distributed normativity seems entirely correct (although the chore should be executed with greater delicacy than he suggests). Unfortunately, in his eagerness to prevent the ground beneath a meaningful predicate from becoming engulfed in classical kudzu, Quine's contrary policy leaves the plot entirely defoliated, with the consequence that predicates enjoy no external supportive elements beyond their ties to their syntactic neighbors. This strikes me as ontological overkill, because a moderate pre-pragmatist can allow all sorts of abstract objects to huddle in support of a predicate, just as long as they do not contribute in sum to the anticipated strength of classical gluing. Quine believes that the Russellian universals are born entirely of an illicit projection from syntax, whereas I believe that classical concepts represent a careless amalgamation of shaping elements that are generally non-linguistic in nature. We can temper Quine's anti-classical extremism considerably by simply allowing some of the "abstract objects" he bans back into our picture of linguistic process. Indeed, why, exactly, is Quine so dismissive of the basic notion of an *attribute* itself, considered solely as a parameter relevant to the behaviors of physical objects (*being a pendulum*, say), where no capacity to prop up predicates seems particularly germane to its constitution?

A full answer to these questions is rather complicated but it involves two central components. First, he worries that, were attributes allowed back in our ontological house, the noxious activities of classical gluing could soon recommence. This is a reasonable worry that we shall discuss in 5,ix. Secondly, he believes that the methodological demands of science itself have already rejected attributes et al. as ontologically odious. This assumption (for which Quine is not to blame; he has inherited the faulty conceit from philosophical tradition) stems from both a misreading of mathematical fact and history and a certain degree of simple punning. However, buried in the proper mathematical background lie considerations that raise serious difficulties for orthodox classical thinking, but they are considerably more subtle in their nature than Quine anticipates. Let us survey this second set of issues first.

On Quine's way of telling the story, "Science" has somehow decided that sets represent a better posit than properties, on account of their clearer "criteria of individuation." Although in the earliest days of the subject, logicians were apt to speak freely of properties, subsequent reflection has shown that the employment of sets only is preferable. Here a "set" is simply a bare collection of objects, with no manner of aggregation implied in their assembly. To illustrate these distinctions with a famous (albeit outmoded) example from antiquity, let us assume that creatures with hearts can stay alive only if they also possess kidneys and vice versa. If so, the two sets {x| x is a living creature with a heart} (= the collection of all living creatures possessing hearts) and {x| x is a living creature with a kidney} (= the collection of all living creatures possessing kidneys) will prove identical, because the assemblies share the same real world membership (in the jargon, they are *extensionally equal*). The fact that we can easily *imagine* a hearted creature lacking kidneys matters not; only real life specimens can render the sets distinct. According to Quine, Science sees no need at all for phony universals such as *being a creature with a heart* or *being a creature with a kidney*; indeed, our standards for distinguishing them are apt to seem rather murky. Here's how Quine tells the story in his own words:

> *Perhaps the first abstract objects to be assumed were properties, thanks again to a seren-dipitous confusion: a conflation again of essential pronouns with pronouns of laziness . . . Here is the scenario. A zoologist describes some peculiarity in the life-style of a strange invertebrate, and then adds, "It is true as well of the horseshoe crab." His "it" is a pronoun of laziness, saving him the trouble of repeating himself. But let him and others conflate it with an essential pronoun, and we have them dreaming up a second-order predicate such as "property" or "attribute" to denote objects of a new kind, abstract ones, quantified over as values of variables.*
>
> *Again a happy confusion, if confusion it was. Science would be hopelessly crippled without abstract objects . . . Even so, the pioneer abstract objects, which I take properties to be, are* entia non grata *in my book. There is no entity without identity, and the identity of properties is ill defined.* [Properties] *are sometimes distinguished even though they are properties of entirely the same things; and there are no clear standards for so doing. However, the utility that made properties such a boon can be retained by deciding to equate properties that are true of all the same things, and to continue to exploit them under another name: classes.*[15]

This withdrawal on Science's part from its former willingness to embrace traits to an enterprise that now grimly purges them in favor of sets Quine calls *the flight from intension* (I am reminded of the story of Falstaff and Prince Hal). In this context, an *intension* (see 3,iii) is any characteristic that distinguishes property-like gizmos according to any standard other than the fraternity of objects of which they happen to hold, whereas an *extension* is simply any naked set considered without regard to such sup-plementary features. In this venerable terminology, any conceptual feature to which we might intuitively point in attempting to distinguish *being a creature with a heart* from *being a creature with a kidney* qualifies as an "intensional characteristic." Into this

---

[15] W. V. Quine, From Stimulus to Science (Cambridge, Mass.: Harvard University Press, 1995), 30–40.

category fall all the directivities mentioned above as possible "conceptual contents": classificatory guidelines such as "To sort under this heading, see if the creature has a heart, rather than worrying about its kidneys" and inferential associations such as "Conclude that it probably has an artery and vein system attached." If someone were so foolish as to claim that the characteristic *containing twenty-four letters* further distinguishes the *heart* trait from the *kidney* trait, then she would be claiming that lexicographic numbering qualifies as an intensional feature as well. Of course, few classical thinkers make such a claim, although occasionally one encounters writers who fancy that the allied concepts *being both red and square* and *being both square and red* differ slightly in content (obtuse Archie might fail to infer one from the other). Quine capitalizes upon these confusions and claims that *all* intensionalities are truly of projected syntactic origin, even those of a "See if the creature has a heart" category. Our 3,vi difficulties in assigning determinative contents to classical universals represent a puzzlement with respect to the exact range of intensional features that should be regarded as intrinsic to these contrivances. Quine proposes that we simply reject as "unscientific" all questions of this ilk (such highhanded legislation contributes, of course, to the absurd portrayal of personified "Science" as a dour and unyielding scold that infects all of Quine's writings on the topic).

(vii)

Honorable intensions. This propensity to shed conceptual intensionalities is motivated by Science's methodological thirst for simplicity and clarity, Quine claims. Some molting of traditional conceptual features does undoubtedly occur at the hands of scientific practice, but Quine has thoroughly misunderstood its scope and motivating nature. However, he is scarcely alone in his confusions, because there are a range of significant facts about how properties need to be addressed in physics—or, for that matter, anywhere else—that are almost never discussed in their original and proper contours (or, at least, I have never run across a self-styled philosophical specialist in "properties" who does this). This is surprising, because many of the key observations have been fully recognized since the work of Fourier and his school in the early nineteenth century. Somewhere along the line of philosophical transmission a hazy folklore of *scientific trend* has become substituted for concrete fact and then transferred from philosopher to philosopher in analogy to the old game of "telephone," each handoff garbling the original message one stage further. In my estimation, Quine's flight from intension represents a philosophical distortion of this ilk: not a rumor that Quine himself concocted, but gossip that he has most vigorously passed along. Like many writers, Quine has a regrettable propensity to personify "Science" as a creature of Trends and Demands, a policy of which Chapter 1 complained under the heading "Science should be used, and not mentioned." But it isn't *methodology* that forces us to be cautious in how we think about the world's bouquet of properties, but simply refractory *facts* with respect to, <u>e.g.</u>, the organized manner in which garbage can lids vibrate (for such is the content of the Fourier-derived work I mentioned). But these tintinnabulations have come down to Quine muddled together with unrelated logical considerations that I shall mention later. Located downwind of Quine in this game of doctrinal telephone, the modern analytic philosopher is apt to dismiss the complaints he hears about attributes out of hand, because of the trappings of implausible trends in which the message comes couched. But this utter rejection is a great pity, for, within Quine's muddled communiqué, the unsettling clamor of our garbage lids can be faintly discerned, whereas his analytic successors hear them not and entertain extravagant fantasies of what the realm of attributes must be like. Indeed, if I were to select the single error most responsible for the oddities of current speculation in analytic metaphysics, it traces to this source: a detached unwillingness to inspect the basic victuals, within a physical property line, that Mother Nature has decided to heap upon our unsuspecting plates. Let us begin with the errors in Quine's thinking and then move on to the funny properties that hide within circular plates.

Quine's claim that physics eschews talk of—or, in his preferred jargon, "commitment to"—attributes is simply false, even by his own standards. If we look in a physics text, we will not only find particular traits discussed as such, we will encounter general definitions of what constitutes an attribute (or quantity) and quantificational appeals to great ranges of them within the basic laws of the discipline. But these are exactly the hallmarks Quine himself demands in his famous criteria for ontological commitment.[16]

. . . .

For example, the most basic laws of mechanics traffic in quantities treated only in general terms. A common manner of articulating the basic dynamic law of classical mechanics is: "For any system and any set of independent quantities x sufficient to fix its configuration, there exists a complementary set of conjugate qualities y in terms of which its time evolution can be supplied by a Hamiltonian function H and the equations $dx/dy = \partial H/\partial y$ and $dy/dt = -\partial H/\partial x$ for each x in the vector x." In the presence of so-called constraints, the generality in this claim cannot be avoided, for the usual quantities of *position* and *momentum* may not be independent for the system at hand and unfamiliar quantities may be required to fix its state. We'll see below what some of those textbook definitions of attributes look like.

. . . .

Beyond any fussing about formalities, there are many circumstances in physics where our grip on the notion of "same property" seems as fully stout (and sometimes firmer) than our handle upon "same object."

. . . .

Even in classical physics the clarity of "basic objects" with which we deal often seems subservient to our sense of how traits become instantiated over time. Thus in dealing with a fluid as a continuum, we must track the continuous flow of its "material particles" but it is generally

---

[16] W. V. Quine, "On What There Is" in Point of View.

accepted that the notion loses its utility for rarified gases when the distribution of *mass* and *velocity* over palpable volumes fluctuates too irregularly.

The most dramatic illustration of these issues can be found in the "identical particle" phenomena of quantum mechanics, where we need to evaluate portions of, e.g., low temperature helium both with respect to the number of component particles and the number of states (= complete arrays of traits) open to them. Oddly enough, the two numbers behave differently than we might expect and the particle notion cannot be accorded the higher priority.

....

It would certainly be absurd to claim that the trait *being a creature with a heart* differs from *being a creature with a kidney* on the grounds that the latter has an additional letter in its title, for it has acquired that characteristic only because it has accidently fallen within naming distance of a human being. To consider "containing twenty-five letters in its title" as a required characteristic of a trait is surely to indulge in the mistake that Quine calls *projection*: regarding an extraneous linguistic association as intrinsic to the attribute itself. We are often inclined to make similar mistakes, however, through regarding associated *computational aspects* as comprising important ingredients of functions or attributes themselves. For example, in mathematics it seems prima facie natural to distinguish the "function" $x\cdot(y+z)$ from $x\cdot y+x\cdot z$, even though they compute exactly the same values over familiar numerical ranges. Indeed, there is a sensible notion of a "structured function" available in certain domains, but mathematicians have decided that the basic term "function" should not be restricted to such a narrow class of entities (they introduce "structured functions" especially for the topics—e.g., the study of computation—where they're needed and natural). One of the prime motivations for this terminological decision is that a much richer world of unstructured functions is required to make coherent sense of the mathematics that arises in conjunction with the basic equations of physics.

Why is this? Because of the early nineteenth century work I mentioned, applied mathematicians recognized that the circle of traits vital in physics does not close under conventional grammatical strictures. In the century previous, it had been recognized (first by Daniel Bernoulli, apparently[17]) that the motion of a guitar string can be decomposed into a number of different vibrational modes that are active simultaneously and whose independent qualities determine the tonal characteristics of the string (i.e., its overtone structure). But if we inspect the natural (linearized) equation for such a string $(\partial^2 y/\partial t^2 = k\ \partial^2 y/\partial x^2)$, such a mode-based decomposition will not be evident at all, although the hidden quantities here happen to have familiar mathematical expressions from trigonometry as natural designations (e.g., the modes of our string can be expressed as "sin nx" for integer n and move as (sin nx)·(sin t)). Such traits should be regarded as *abstractly collective* in their character: they indicate that the component molecules in our wire have locked together into an archipelago of staggered modes of

---

[17] C. Truesdell, The Rational Mechanics of Flexible or Elastic Bodies: 1638–1788 in Leonhardi Euleri Opera Omnia XI (2nd series) (Turice: Orell Füssli, 1960), pt. III. J. T. Cannon and S. Dostrovsky, The Evolution of Dynamics (New York: Springer-Verlag, 1981).

movement that can each retain fixed quantities of energy within their ambits. As such, the traits must be considered as *macroscopic traits* pertinent only to the string as a whole; it makes no sense to attribute mode characteristics to a short stretch of string. Such wide scale lockings together are quite common in materials and often our capacity to understand a material rests upon our being able to tease out these global organizational patterns, which are frequently very recondite in their contours. But very few physical systems embody precisely the same sin wave modes as found in our string. Our garbage can lid conceals allied locked together qualities in its wobblings, which are likely to appear utterly random to the untutored eye, but they are not the same modes as prove important within a string or a square plate. But many systems do not possess hidden characteristics of this general type at all: a poorly manufactured violin string may contain enough non-linearities to ruin the physical salience of any decompositional modes.

....

Each mode-stored energy corresponds to the total kinetic and potential energy of a string in a sine wave configuration, *except* that we are not claiming the string actually moves in this manner, because many vibrational modes are likely active at once. One only sees a pure "motion" like (sin nx).(sin t) under improbable counterfactual conditions, although careful patterns of string damping can drain the energetic contributions of many of its neighboring modes.

Mode quantities such as these represent special cases of what are generally called *constants of the motion*: physical qualities that would normally shift value as a system evolves in time but which manage to retain constant values within the specimen under investigation. In our string,

Chladni

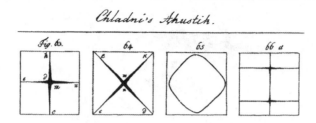

Chladni's Akustik.

the energetic value of each mode-based quantity constitutes such a constant, while its corresponding phase will alter periodically. If a complete set of constants of the motion and their corresponding phases can be found that can fix the complete state of the system, then the mathematical problem of understanding its motions can be regarded as satisfactorily solved (Hamilton-Jacobi theory operates on this basis). Unfortunately, such phase/angle quantities are often extremely hard to uncover even when they exist.

. . . .

In the late 1700s, the French experimentalist E. F. F. Chladni found that, by carefully sprinkling sand on their surface and stroking their sides with a bow, a wide range of objects such as metallic plates display a series of striking, albeit peculiar, modal patterns.[18] It was eventually realized that these sand figures represent symptoms of energetic factors secreted in the plate analogous to those found in a string, although locked together in somewhat different fashion (which is why dropped garbage can lids do not sound very harmonious). These quantities are always active in the plate; Chladni's procedures merely provide evidence of their presence (just as partially stopping a guitar string at the fifth, seventh and twelfth frets brings forth the harmonics that supply direct indication of Bernoulli's hidden quantities). Shortly thereafter, Joseph Fourier and his school, employing the technique of separation of variables, enjoyed great success in teasing forth mathematical expressions for some of Chladni's revealed qualities from the natural equations for plates and such, subject to the proviso that the objects possess a convenient geometry (squares or circles, say). Generally, these expressions took the form of series expansions, a point to which I'll return.

As we move from string to square plate to garbage can lid, etc., the functional expressions produced often turn out to be novel in the sense of not being definable in terms of previously familiar functions (the series expressions themselves don't qualify as such "definitions," for reasons I'll soon explain). Much effort in nineteenth century applied mathematics was devoted to painfully understanding these so-called *special functions* as they sequentially emerged from the basic equations of physics (hefty tomes have been written on the respective behaviors of Bessel functions, Mathieu functions, etc.). Once these functions have been mathematically located, however, we can move back to physics and predict that experiments of Chladni type will reveal their hidden

---

[18] Beyer, <u>Sounds</u>, ch.1.

presence in the systems studied. And, lo!, these predications generally hold up and many of the greatest physical successes of the nineteenth century turned, in one way or another, upon these techniques (which is why applied mathematicians often declare that, of all the discoveries in mathematics, the ones they most prize are due to Fourier[19]).

Here it is important to realize that the basic situation with the special functions of mathematical physics is much like that with rabbits: as soon as we believe that we have assigned them all suitable names, they proceed forthwith to engender a new generation that requires further labels as well. But this basic fact—which was suspected by Euler, but concretely proved by figures like Liouville—passes virtually unnoticed within philosophy today, despite the passage of approximately one hundred and sixty years.[20] Thus contemporary philosophers often write breezily of the "kind terms of physics," which they fancy will be supplied by the range of predicates grammatically definable in the "basic vocabulary" of physics. But if we understand the condition of *definability* with any strictness at all, then most qualities of a Chladni class will exceed those limits because their "definitions" must be framed through reference to special functions that provably fall outside the orbit of the strictly definable according to any reasonable choice of starting vocabulary.

Some writers seem to be confused about these basic facts through failing to distinguish adequately between what can be called *self-guaranteeing* and *non-self-guaranteeing* names or predicates. Over the real numbers, any compound of the form "n + m" is certain to possess a value once "n" and "m" have been supplied firm denotations; accordingly, "$\pi + 6.7$" qualifies as a *self-guaranteeing* expression. But this happy confidence fails even for "n/m" if "m" happens to denote 0. And much richer possibilities for referential failure emerge when we move to the typical expressions of the calculus, such as series summations (Σx^n) or integrals ($\int x dx$), whose existence is *never* self-guaranteed but always needs to be established by *proof*. One doesn't need to peruse many pages of a classic like Watson's Bessel Functions[21] to realize the great delicacy with which greatly varied scraps of non-self guaranteeing expressions must be painfully patched together before we can figure out how functions of this type behave (they include, inter alia, the modes of our garbage can lid). Mathematicians have learned, through bitter experience, to become careful about distinguishing *hope* from *proof* in the matter of physical quantities. Suppose, for example, that we have written down some differential equation motivated by physical concerns. We can hope this equation has a solution (usually, there will be many of them). If it does, that solution will carve out a large range of dependent quantities in its wigglings and we might even decide to give some of these special names, if they seem particularly important in a constant-of-the-motion kind of way. But such talk is based upon provisional faith: at unexpected moments, rather innocuous looking differential equations can fail to have solutions at all (Paul Lévy

[19] Corelius Lanczos: "If we were asked to abandon all mathematical discoveries save one, we could hardly fail to vote for the Fourier series as the candidate for survival." Elena Prestini, The Evolution of Applied Harmonic Analysis (Boston: Birkhäuser, 2004).   [20] J. F. Ritt, Integration in Finite Terms (New York: Columbia University Press, 1948).
[21] G. N. Watson, A Treatise on the Theory of Bessel Functions (Cambridge: Cambridge University Press, 1966).

found a famous example in the 1950s) and, accordingly, all our predictions of hidden quantities in our physical system will have rested upon a mathematical pipe dream. For such reasons, applied mathematicians must keep our distinction between self- and non-self-guaranteeing terms plainly in view (as mentioned before, Russell's theory of descriptions supplies a methodology for doing so). We can't create Santa Claus by writing down the expression "a fat man from the North Pole who gives toys to children" and we can't create a Bessel function through merely writing down an infinite series expression.

....

Beyond the technicalities, philosophical "kind term" enthusiasts simply get the *spirit* of how physics deals with its quantities wrong. Understanding the behavior of a physical system often requires locating independent quantities in which its behavior can be conveniently decomposed. Let me supply some details as to how this task is conceived (afterward, I'll explain the rather pungent philosophical relevance of these considerations). Suppose we have a so-called *phase space* portrait of the way in which our system evolves, where each point in the space represents our system's complete condition (or *state*) at a possible moment and where the curve that this point travels symbolizes how the system's state changes over time. Draw an arbitrary surface S across the flow of these paths, which we might think of as a bunch of sample systems laid out upon a curvy plain where each system starts with slightly different positions and velocities assigned to their component parts. Now score the surface S with an arbitrary ruling of lines A_0, A_1, A_2, etc. This ruled surface can then be regarded as the starting gate of a race we will run with our flock of slightly different systems. If we pull these scored lines up through the rest of the space following the flow, we will slice (or *foliate*) the whole phase space into thin layers rather like a piece of baklava (the surface $\mathcal{A}$ illustrates the layer cut out by pulling the line A_1, along with the flow). We have now "defined" (in terms of the geometry of the phase space; there may be no formula available!) one good constant of the motion quantity for tracking our system, namely, on which line of the starting surface did our system originally fall? In the figure, our target system starts on line A_2 with the consequence that it will forever stay on the sheet marked $\mathcal{A}$ which corresponds to a fixed value for the constant of the motion quantity just created (in the jargon, the foliation of slicings corresponds to the *level sets* of our "constant" quantity). Of course, it is a

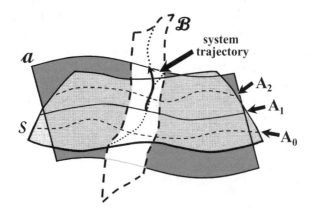

complete triviality that our system will never lose this $\mathcal{A}$ sheet value no matter how far it wanders. Nonetheless, the notion still counts as a "good quantity" in the mathematician's book (for sound reasons, as we shall soon see). We can now automatically obtain a second good descriptive quantity by simply clocking how long our system has been traveling on its sheet since leaving the starting post. However, most physical systems possess more degrees of freedom than two, so we require more independent quantities of the same ilk if we hope to pin down their full state adequately (unfortunately, I can't draw a phase space of the proper dimensions). Well, can we perhaps inscribe a second set of transversal lines like B_1 on the surface S and see if these lines can also be carried forward by the flow in such a way that they continue to cut through the $\mathcal{A}$ slicings transversally? If this is possible, then we will have found a new constant of the motion corresponding to the foliation $\mathcal{B}$. In these four-dimensional circumstances we will have then completely captured our system's unique path as lying along the intersection between sheets $\mathcal{A}$ and $\mathcal{B}$ (we merely need to indicate how long it has been traveling along each sheet to pin its present condition down completely). Now it is once again trivial that we can *start* to carve out the $\mathcal{B}$ sheets in this manner but it can easily happen that, as we follow the flow forward, $\mathcal{B}$ will begin to twist in such a way that it no longer cuts cleanly through $\mathcal{A}$, in which its ability to serve as a second constant of the motion becomes lost. Unfortunately for mechanics, this second situation occurs far more commonly than the fully foliated first possibility (which is usually described as representing an "integrable" system, although we may not be able to write down any such integrals!). Such distinctions are vitally important, because if the system's flow can be fully foliated, then it will also not behave erratically in a chaotic manner—viz., systems that start in closely similar conditions will not deviate in their subsequent behaviors too rapidly. But even if a system does not act chaotically, the secret locking together that creates its fully foliating constants may remain quite elusive. Some of the most surprising recent work on these topics has lain in the region of finding previously unsuspected "constants" hidden in long familiar equation sets. Note that all our talk of "quantities" here is determined entirely by the *geometry* of the system's behavioral flow; absolutely no heed is paid to the question of whether these quantities are denoted by familiar predicates or not.

I have gone into this detail because it sharply illustrates how far off the beam the discussions one often encounters in the philosophical literature have wandered—all the business about physics' alleged "kind terms" and so forth. In particular, David Lewis' greatly influential article, "New Work for the Theory of Universals"[22] presumes that physics would never be so foolish as to countenance quantities defined solely through having departed from some starting configuration. With this obvious fact about "kinds" in hand, he then proceeds to address all sorts of pressing chores in analytical metaphysics (that's what the "new work" in his title concerns). But he just made this prohibition up! As we've just seen, physics is quite *eager* to consider quantities defined by departing from an arbitrary line inscribed across a starting configuration.

This example is not anomalous to Lewis; virtually every piece of recent philosophical writing on "attributes" with which I'm familiar makes similar assumptions, invariably based upon features that the author believes must be displayed in the predicates that canonically represent (in 3,ix's fashion) such "kind" attributes. All this, in spite of the fact that no such predicates typically exist for most quantities of interest nor would physics have any particular interest in them in any case. Such disregard of scientific practice suggests a rigged game to me. The writings of this

---

[22] Lewis, "New Work".

school commonly appeal to "Science" to provide a "list of fundamental quantities," which are then evoked as a basis to resolve standard philosophical worries about materialism, inductive practice and so forth (it should be observed that the term "fundamental parameter," in normal parlance, *excludes* all of system's dynamical attributes). But the same authors are completely unmoved by the fact that no physics book displays any evidence of supplying the catalog they require, because they are already convinced on a priori grounds that such an accounting must exist, even if physicists are lax in bringing it forward! The culprit that engenders these mytho-logies is the theory T syndrome: under its dreamy influence, philosophers become absolutely convinced that they know what "the general shape of any physical theory" must be like, without ever inspecting an actual specimen. This scholastic hubris creates a climate where such writers issue multitudinous proclamations to the effect that "Science requires φ," where φ, upon closer inspection, looks suspiciously like an item that *philosophers* would dearly like to have.

....

Fortunately, matters are not always so bleak as this. The situation with respect to a certain class of physical systems (describable, after separation of variables, by a certain type of ordinary differential equation) took an astonishing turn in the 1830s due to the development of a beautiful theory developed by C. Sturm and J. Liouville.[23] Sturm and Liouville were able to demonstrate, through quite abstract considerations, that quantities of a modal type had to exist for a very wide class of situations of essentially the type Fourier and his school were investigating. That is, Sturm and Liouville were able to prove an *abstract existence claim* of the form: For any physical system S whose math-ematical description satisfies Sturm-Liouville conditions, there will be a family of quantities φ_0, φ_1, φ_2, ... that display the pleasant characteristics of a constant of the motion family (note: our φ are the eigenfunctions of the Sturm-Liouville operator in S's governing equation). But how did our heroes establish this general result? Given the generality of their claim, they cannot produce the desired functions by simply proffering self-guaranteeing expressions, in the mode "the function we seek is $\frac{2}{3}\sin(x+\pi)$." Instead, they squeeze in on the functions sought through a sequence of successive approximations such as those witnessed in the calculation of $\ln(5)$ in 4,x. They then prove that the function which emerges in the infinite limit of this squeezing process displays the characteristics wanted in a desirable family of mode quantities. Proofs of this nature are called *pure existence proofs* because they show that certain functions must exist without providing many specifics about what they are concretely like (e.g., where they dwell or what their name is if they have one). The G. N. Watsons of the world still have plenty of work ahead of them before they can glean how the Sturm-Liouville established quantities behave in their numerical peculiarities—that is, before they can predict what the Chladni sand patterns on a drumhead will actually look like.

....

An explicit acknowledgment of the required linkage between physical quantity existence and mathematical considerations can usually be found in any adequate formal treatment of these

---

[23] Cornelius Lanczos, Linear Differential Operators (London: Van Nostrand, 1961), ch. 7. Jesper Lützen, Joseph Liouville 1809–1882 (Berlin: Springer-Verlag, 1990).

concerns. Here are two typical specimens appropriate to classical mechanics, drawn from well-known texts by Walter Thirring and I. Khinchin respectively:

> In order to interpret the formalism it must first be agreed what the observable quantities are. The observables generally correspond to the coordinates and momenta of the particles. There is of course no reason that the coordinate system should necessarily be Cartesian; for example, in astronomy it is usually angles that are directly measured. We should therefore allow arbitrary functions of coordinates and momenta as observables, subject only to boundness and, for mathematical convenience, differentiability.[24]

Or

> In what follows, we shall often call the Hamiltonian variables $q_1, \ldots, p_s$ of a given system G the dynamic coordinates of its image point in the [phase] space Γ, and any functions of these variables [a] phase function of the given system . . . -When convenient we shall denote the set of the dynamic coordinates of the given mechanical system by a single letter P, and, correspondingly, an arbitrary phase function by f(P).[25]

Here both authors are concerned with the phase space of possible states (= *phases*) that a given physically system might potentially occupy (we already saw such spaces in our discussion of constants of the motion). For a solar system with nine (point-like) planets and a sun, this "space" will be of 60 dimensions where each point in the system represents an assignment of x, y, z locations and momenta to each component particle (six numbers required for each). However, the "coordinates" that define these states needn't be familiar position and momentum and, if constraints apply, often can't be (*position* and *momentum* will usually still have representative functions within the space, but these functions will have lost the qualities of independence desired in a coordinate choice). The linkage between *quantity existence* (which Thirring calls *observables* and Khinchin, *phase functions*) and *mathematical principle* is captured by the remark that to any well-defined mathematical function $\underline{f}$, some physical quantity φ will correspond. Accordingly, if in the context of a particular physical system S, Sturm and Liouville can establish that certain constant of the motion mathematical quantities can foliate this phase space, then physics can conclude that the physical system under examination possesses substantive hidden qualities that fix its behaviors in a Bernoulli-like way. Since mathematics, in turn, rests its own existence assumptions upon the comprehension axioms of set theory, physics settles the question of the existence of its own traits through the manner in which physical postulates interact with set theoretic principles <u>via</u> links of a Thirring/Khinchin sort. When we move to other forms of physics (quantum theory, for example), suitable adjustments must be made ("Hilbert spaces" for "phase spaces") which complicate the picture, but the general approach remains much the same.

Thirring and Khinchin concentrate upon the *dynamical quantities* of a system—those characteristics that can vary within the range of possibilities captured within a standard phase space. However, physics is also interested in a more general range of quantities than this, e.g., the area that a connected *swarm* of phase points projects onto some hyper-surface or other (the important integral invariants discovered by Poincaré are of this nature). Such a quantity is not a phase function in Khinchin's sense of a single point, but instead represents an integral over nearby configurations (such considerations often motivate additional smoothness requirements on the relevant $\underline{f}$s, such as Thirring's differentiability). Furthermore, we often wish to consider

---

[24] Walter Thirring, A Course in Mathematical Physics I: Classical Dynamic Systems, Evans Harrell, trans. (New York: Springer-Verlag, 1978), 5.

[25] A. I. Khinchin, Mathematical Foundations of Statistical Mechanics, G. Gamow, trans. (New York: Dover, 1996), 15.

larger expanses of physical possibility, as when we move to a so-called *control space* where the originally fixed parameters of our system are now allowed to vary (their masses and effective forces applied, say, or even particle number). Once again the existence of relevant quantities will be established in exactly the same "function over the control space" kind of way.

. . . .

There is a critical feature of functions obtained through limit taking of which we should be aware: the process is so brutal that the gizmo that emerges from the limit taking meat grinder can easily lose many of the key characteristics that distinguished the functions originally put into its hopper. This destruction of input characteristics is an astonishing fact that came to the fore in the course of Fourier's work. He noted, for example, that the partial sums of the series $\sin x + \frac{1}{2}\sin 2x + \frac{1}{3}\sin 3x + \ldots$, all continuous functions, lead to a broken saw-tooth function in their infinite limit, despite the fact that even great mathematicians like Euler had presumed otherwise. We cannot blithely assume, without proof, that even so basic a functional characteristic as *connectedness* will survive the operations of limit taking. Indeed, many mathematical "verities" accepted before Fourier rested tacitly upon erroneous assumptions that qualities like *continuity* would automatically persist in the functions that are established as the end products of limits. The serious foundational crises that ensued demonstrated that intuitive expectations cannot be trusted with respect to quantities constructed as limits. In the fullness of time, mathematicians were led to the conclusion (which remains standard operating practice today) that such existence and persistence decisions must, in the final analysis, be placed in the hands of set theoretical principle. These same considerations push the central notion of "mathematical function" itself into the acceptance of any arbitrary many-one alignment of a domain with a co-domain as a function, whether or not this correlation happens to be continuous, integrable or possesses any prior name as a formula. When we take a limit over a passel of continuous functions, we must first establish that its output represents a function in the modern sense and, if it happens to remain continuous et al., those further qualities must be established through proof, and not mere intuitive expectation. In other words, the fact that some individual happens to be born to an unbroken dynasty of great artists does not insure that she will become a great artist herself; she must earn that characteristic through her own deeds. And so it is for functions that comprise the scions of limit taking; we can attribute an individuating characteristic to them only if they have earned that title through their own behavior.

Of course, none of this entails that smaller classes of function-like gizmos can't be defined to which, e.g., an intrinsic notion of "rule" properly applies. We observed that computer science wants a notion of "structured function" that can distinguish the expression "x.y + x.z" from "x.(y + z)" through the former's natural association with the algorithmic ordering *multiplying x by first y and then z and then summing*. But such structured entities can be defined only in contexts where we can meaningfully talk of *pieces* with which we, or our machines, can directly compute. Rule-like characteristics of a structured function ilk are thus appropriate for simple arithmetical functions built up from finite applications of addition and multiplication but cannot be sensibly extended to cover the vastly larger universe of functions investigated in a Sturm-Liouville context. "In what order do the multiplications and additions occur in a function known to exist only through limit taking?"—the question doesn't make sense as it stands.

Furthermore, the notion of "structured function" is irrelevant—even misleading—if we wish to, e.g., count the number of independent solutions an equation accepts (say, in the course of figuring out how many distinct physical modes it will display). Our interest in algorithmic ordering only arises when we wonder how a human or computer might arrive at concrete numerical values for the quantity. In other words, "algorithmic structure" represents a characteristic that we associate with a function f largely because of the manner in which it relates to *outside systems* S: humans or their calculating machines. We can reasonably think of "structured functions" in a $<f,S>$, $<f,S'>$ kind of way, but not if we ignore the contributions supplied by S and S'. Likewise, we cannot meaningfully claim that a cannon ball displays the trait of *having its height above the ground fall into the triple digits* in isolation from an external setting S, for we must know the coordinate frame F in which the elevation is gauged. Here the fuller amplification *having its height above the ground fall into the triple digits in the frame F* reveals the tacit dependence and constitutes a fully acceptable physical relationship (in contrast, the quantity *having a particle number that falls into the triple digits* is acceptable, because its measure is independent of frame or scale). The improper allocation of conceptual characteristics that Quine calls "projection" is best viewed as a process where regular physical qualities φ acquire eerie trappings through a process of ignoring the S arising within some manner of language user pairing $\langle \varphi, S \rangle$ and further assuming that such ersatz "internal qualities" continue to attach to φ's that fail to enter into the requisite forms of $\langle \varphi, S \rangle$ pairing. Such S-dropping "projections" arise whenever a thinker blithely assumes that every physical trait manifests some form of rule-like intensionality without attending to the forms of *computation* that supply sense to such discriminations over the rather limited range of quantities for which we can *actually calculate numbers* (such näivety is akin to a child assuming that *having a cute name* constitutes an intensional characteristic of *being a rabbit*—she plainly neglects all the nameless bunnies that roam the woods). As a blatant example of this error[26] (more popular varieties will be supplied in the next section), certain authors airily announce that "The quality of being a prune or a cantaloupe is clearly *disjunctive* in its internal nature" (that is, decomposes into "A or B" pieces), without sensing any obligation that they must explain how such discriminations are to be prosecuted with respect to traits that will pass forever unnamed

[26] David Armstrong, Universals and Scientific Realism (Cambridge: Cambridge University Press, 1978).

(0,3) drumhead
mode

height at
(0,3)(r)

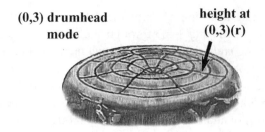

within any reasonable language or which possess a variety of equally natural predicative expressions bearing different logical structures (physics assures us that candidates for both categories abound abundantly). In the historical aftermath of Fourier's surprising discoveries (which helped reshape the entire face of applied mathematics), I find their cheerful willingness to follow whimsical "intuition" so uncautiously quite astonishing.

Returning to the themes of easy-to-follow and harder-to-follow directivities examined in 3,viii, the point I am making can be articulated as follows. Most philosophers who accept attributes at all believe that they are to be located beneath the cabbage leaves of language, as 3,iii expressed the assumption of close connection between predicates and Russellian universals. But in Sturm-Liouville cases this point of view is quite wrong. Consider our friend from 3,vii, the height of the (0,3) vibrational mode at the radial point r (which I have abbreviated as "(0,3)(r)"). This, in fact, represents the radial portion of the fifth Sturm-Liouville mode of an idealized conga drum, where the vibration occurs in three concentric rings as shown. Two basic facts should be noted about this situation. We (or, more plausibly, the practitioners of acoustics who talk about this sort of thing) would not find it useful to introduce a special predicate for "the (0,3) mode" unless we can interlace a hierarchy of accessible skills between us and the rather remote Bessel function that the trait **(0,3)(r)** delineates. That is, it is informative to learn that some peculiarly shaped drumhead conceals a set of hidden Sturm-Liouville modes, but, unless we can find some kind of computational route to its approximate values, we are likely to leave those traits unnamed by any special denominations, just as we allow the rabbits in the forest to roam generally undesignated as well. These are the considerations that lead Richard Feynman to remark:

> The whole purpose of physics is to find a number, with decimal points, etc.! Otherwise you haven't done anything.[27]

On the other hand, the distinguished Russian author (Yu Manin) from whom I extracted this exchange correctly retorts:

> This is an overstatement. The principal aim of physical theories is understanding.

Now I doubt that physics displays any fixed "principal aim," but sometimes the theoretical existence of an uncomputable Sturm-Liouville mode proves critical in itself (it

---

[27] Yu I. Manin, Mathematics and Physics (Boston: Birkhäuser, 1981), 35.

insures non-chaotic behavior) but sometimes not, as in the circumstances Feynman has in mind. There's rarely any reason to introduce "special function" titles for traits unless a tower of linguistically directive levels can be established between them and ourselves. Within the hierarchy of directivities delineated in 3,vii, only the top layer—where the truncated series expressions are located—can provide us with the blunt syntactic instructions we require for normal linguistic practicality: <u>viz.</u>, we require instructions pitched at that lowly level of vulgarity if we are to calculate the numbers Feynman seeks.

On the other hand, we can rarely start at the *linguistic side* of this hierarchy—that is, amongst the "blunt syntactic instructions"—and articulate any kind of useful predicate descriptively suited to our drum unless we are assured *by other means* that its "vulgar recipe" pieces fit together under the umbrella of some governing quantity such as $(0,3)(r)$. In the case at hand, Sturm-Liouville theory (in conjunction with basic analytic facts about series behavior) provides the "other means" that insures that "$(0,3)(r)$" constitutes a worthwhile predicate: to employ jargon I will highlight later in the book, the mathematician's existence proof supplies us with a *picture* of how our complicated practical manipulations with "$(0,3)(r)$" conform to the physical reality it seeks to match. And, in most cases, when we think about the physics of a drumhead, we concentrate upon $(0,3)(r)$'s behavior, not the odd little scraps of computation we follow in piecing together how it behaves. As George Stokes justly observed, writing in the early Victorian period when such abstract considerations were still novel to applied mathematicians (4,x):

> *Indeed, it seems to me to be of the utmost importance, in considering the application of partial differential equations to physical, and even to geometrical problems, to contemplate functions apart from all idea of algebraical expression.*[28]

In sum, the road to reliable quantities such as $(0,3)(r)$ within any context of moderate sophistication must travel through existence proofs of Sturm-Liouville type, *not* via the mere presence of linguistic predicates (which, if preexistent, often turn out to have been cobbled together somewhat wrongly, when evaluated from the higher perspective of a proper existence proof picture).

With this historical perspective in mind, let us now return to Quine's alleged flight from intensionality, which he portrays as driven by methodological cravings for ontological simplicity and so forth. We can now see that this story is entirely bogus: it is the demands of *physical experiments*, not dubious methodology, that force mathematics to reshape its primary notion of "function" along set-theoretic and rule-independent directions. It is a brute empirical fact that a large class of physical systems shelter secretive mode qualities whose presence can be verified through Chladni-type probing. Clearly, physics must diagnose, if it can, the circumstances in which such hidden traits can be expected to appear, in the form of non-trivial existence claims of the sort Sturm and Liouville provide. But these capabilities can be reached only if physics allows its

---

[28]   George Gabriel Stokes, <u>Mathematical and Physical Papers</u>, i (Cambridge: Cambridge University Press, 1883), 54.

treatment of *physical quantities* to piggy-back upon mathematics' set theoretical treatment of *function* (or something similar, such as Laurent Schwartz's distributions). In turn, the mathematician's notion of function must be enlarged in scope and stripped of improper intensionalities if it is to serve its reciprocal role as handmaiden of physical quantities adequately. This constitutes a rather tough-minded demand, for functions and physical quantities must be treated in a manner that allows them to emerge at the termini of limit taking processes, for the existence of many descriptively important properties (such as constants of the motion) can be established only by squeezing in on them through a sequence of less important quantities that are more easily shown well defined. As Fourier's surprising examples show, common personality traits often do not survive the brutal processes of limit taking and any reasonable doctrine of attributes must remain warily cognizant of this fact. For all these reasons, characteristics that belong to the quantity itself must be clearly distinguished from features that properly belong to the routines we adopt in calculating their values, as in the contrasts we have drawn between the computational layers that hover above $(0,3)(r)$ and the drumhead trait itself (which cares not whether humans can readily calculate its values or not). We begin to see strange "ghost quantities" if we do not manage to keep this cloud of ingredients well separated.

It is plainly false that the treatment of quantities outlined is *extensional* in Quine's sense: the trait **being an isolated two-particle system** has not been replaced with the *set* of systems that instantiate it, for there aren't any (gravitation weakly couples all real world systems into larger units, so its extension is the empty set). Nor do physicists equate this trait with that of **being an isolated three-particle system,** as Quine's extensionality would dictate (otherwise, the celebrated three-body problem would be very easily solved). In fact, studying the policies detailed in the fine print, we discover that the "phase or control spaces" there utilized inherently encode a good deal of modal (in the sense of "possible variation") information with respect to our systems, including the fact that two- and three-body universes behave quite differently. Quine's flight-from-intension story to the contrary, the physicist's normal approach to these issues does not represent a bizarre or denatured treatment of *property* or *quantity*. However, we must avoid painting these traits in features ("dishonorable intensions," I have called them) that properly belong to the system considered as embedded within some form of outside descriptive arrangement.

. . . .

Identifying physical quantities with functions over phase or control spaces can look unnatural at first, for it may seem as if the policy omits critical behavioral information pertinent to quantities at hand. However, this worry rests upon a misunderstanding of how ably very rich information gets encoded within the structures of our phase or control spaces, because its quantities can correspond to any (univalent) packet of information that can be possibly assembled with respect to a target system's behavior. Identifying a physical quantity with a function over such an abstract space simply indicates that the quantity φ serves as the carrier of a packet of behavioral information with respect to our target system without any further restrictions on its nature. When we wish to consider issues that hinge on how such quantities relate to our capacities to compute them, we then consider quantities that appear within a joint space that embraces our original system in interaction with ourselves (or our calculating machines, our measuring instruments, etc.).

I might mention that, even within the bounds of classical continuum physics, more delicate pressures tend to pull *quantity* and *function* out of the simple alignment discussed here, sometimes leading instead to correlations with *distributions* in the sense of Laurent Schwartz. None of this alters the basic conclusions reached here, but they weaken Quine's fictitious extensionality even further (to the point of not even making sense: distributions do not take values on point-like regions).

....

Quine, and the many others who have fled from intensionality for similar reasons, have been flummoxed by a simple historical pun. In logical tradition, it is customary to distinguish between "intension" and "extension"-based logical systems, which differ in how sentences like "All seven feet tall cowboys live in Kansas" are approached. This claim might signify that, whatever seven foot cowboys there happen to be (let's suppose there are only two or three of them), they happen to live in Kansas de facto. But it can also express the stronger contention that some mysterious factor drives any conceivable elongated cowhand to immediately take up residence in the Sunflower State. It is harder to make general sense of modal claims like the latter and the first reading represents the most natural parsing of our contention in ordinary English in any case. Accordingly, logical studies have generally emphasized the first, *extensional* reading (this venerable distinction dates back to Mediaeval monks).[29] But it is patent that *this* "flight from intension" (if it is properly so-called) has virtually nothing to do with the considerations that have led physics to interlace its treatment of quantities with set theoretic considerations.

Another contributing factor to misapprehension stems from the efforts of instrumentalists like Ernst Mach (4,i) to free the predicates of physics from unwanted intensional demands (he calls them "animistic assumptions"), leading him to deny, for example, that physics traffics in any notion of causation beyond bare descriptive formula (9,i) and other radical theses of that ilk. Although his purposes are laudable, his cure is far too radical. More reasonable attitudes towards "predicate detoxification" (for that's what Mach seeks) will be discussed in Chapter 8.

A catchy jingle like "Science seeks extensionality" is certainly easier to remember than the litany of impertinent particulars recounted here in regard to trash can behavior et al., but, unfortunately, such a slogan doesn't recapitulate the proper physical considerations with a requisite degree of accuracy.

All the same, a vital residue of correct observation resides in Quine's contention that physical quantities should not be saddled with extraneous characteristics arising from our human capacities to handle the traits in useful language. As I've stressed, our concrete linguistic activities cannot be directly instructed by $(0,3)(r)$ itself, but require monitoring through a swarm of intervening considerations that supply a map of how

---

[29] C. I. Lewis, A Survey of Symbolic Logic (New York: Dover, 1960).

our concrete computational capacities interface with the physical behaviors displayed by **(0,3)(r)**'s instances. Nothing but confusion results if characteristics pertinent only to these intervening layers get deposited upon **(0,3)(r)** itself, thereby gilding the attribute with layers of intensional paint alien to its qualities. But such misplaced "projection" is precisely what occurs frequently in classical thinking: *ghost attributes* are created in which a true physical property is cloaked in an interfacial mantle it doesn't enjoy by itself (sometimes, as we'll see, no true attribute lies at the center of the cloud at all). It is through such projection that the fundamental concept/attribute unity of the classical picture is born: features of personality properly belonging to extraneous layers get deposited upon an innocent attribute in a manner that makes it seem more concept-like than it truly is.

In these respects, Quine is right to worry about ersatz predicative projection as a prime source of *ur*-philosophical confusion. He properly observes that, merely because a predicate displays a robust meaning (recall Russell's "is in the room" argument), we can't be certain that a true attribute sits within the cloud of directive factors that allow us to employ the predicate in useful ways (we will be able to sharpen this moral considerably in the next two chapters). But after this promising beginning, Quine's narrative turns peculiar, for it informs us that authoritarian Science demands that we must behave in odd ways for the Methodology's sake: "Anytime you feel tempted to murmur of a fire truck's qualities, speak instead of the *sets* to which it belongs." Such queer instruction can only invite justified puzzlement: "Why should I do that? Being a member of a set doesn't make our vehicle do anything, whereas it's *being bright red* allows it to stand forth like a sore thumb." In truth, Quine's flight from intensionality mythology should be viewed as simply a fumbled attempt to recount the genuinely important considerations about physical quantities that Fourier and his school uncovered (somewhere along the telephone line of unreliable declination "mathematics" was misheard as "methodology").

(viii)

Ill-founded philosophical projects. It is a pity that Quine has miscast the foregoing considerations as a murky flight, for Methodological Trends are easier to discount than inconvenient facts. Indeed, modern metaphysicians (who talk freely of "natural kinds" and "possible worlds") regularly dismiss Quine's complaints as irrelevant to their philosophical concerns: "Well, maybe Science, for its own peculiar reasons, wishes to replace *being red* with a bare set, but there's no reason that any of the rest of us should imitate this bizarre policy." If Quine's story were entirely right, that retort would be reasonable. But the hard truth is Nature displays a huge inventory of vital quantities in her workings and getting descriptive vocabulary to work ably in their midst often represents a far more complicated affair than the classical picture presumes. And so, even after Quine's sundry misapprehensions have been cleared away, it remains true that the classical picture of concepts rests upon the misallocation of predicative personality in ghost attribute fashion. As such, Quine's reasons for distrusting many popular forms of philosophical reasoning remain founded in trenchant observation. Consider the following arguments, familiar to most academic philosophers.[30] (i) *Being in pain* cannot represent a physical trait because it falls outside the orbit of "kind terms" definable using the basic vocabulary of physics; instead, it classifies objects by principles anomalous to physics' favored classes.[31] (ii) At the very best, argues a famous argument of Hilary Putnam's, the *pain* trait can be characterized only at a higher logical level involving functional quantifiers (i.e., of the form "the unique φ such that...φ..." where the dotted context involves purely physical vocabulary). But that form (and the huge range of instantiations it will accept) indicates that *being in pain* merely "supervenes" over the class of physical traits but is not among their number.[32] (iii) *Being in pain* can't represent a neurophysiological characteristic because there are possible circumstances where we would judge someone to be in pain but not in that neurophysiological condition.[33] (iv) Nelson Goodman's famous oddball trait *being grue* (defined as "being green and observed before the year 2050 *or* blue otherwise") can't represent an attribute proper because it contains an intrinsic disjunctive character (as revealed by the italicized "or").[34] (v) *Being red* can't be identified with any attribute of wavelength reflection because Helen Keller will "learn something new" when she discovers "what *being red* looks like."[35] (vi) *Being in pain* can't be a scientific trait because they are all "objective," whereas the experiential characteristic is "witnessed from a point of view."[36] And so forth.

It is apparent that the arguments sampled rely upon individuating properties according to characteristics that cannot properly belong to them alone, for the same reasons that we cannot paint rule-based characteristics on general functions. Most physical attributes do not admit of any definition in finite terms, so the grammatically induced intensionalities presumed in arguments (i), (ii) and (iv) are moot; likewise, the related appeals to some hypothetical "knowability" quotient of a trait harnessed in (iii) or (v). True, if attributes happen to lie close enough to the capacities of a human being or mechanical calculator (say, the latter has a routine for calculating the values of *impressed gravitational force* under certain conditions), then natural aspects of such external treatment can be transferred onto the attributes and allow special subclasses of "structured attributes" to be articulated. It seems rather pointless to do so, but we can sensibly distinguish *impressed gravitational force as calculated by Euler's method* from *impressed gravitational force as calculated by freshman calculus*. However, it is important, at the same time, to recognize that parallel forms of intensional coating cannot be assigned to garden variety physical attributes and every one of the arguments I have listed falls woefully short in that category, in my estimation at least.

---

[30] Alex Oliver, "The Metaphysics of Properties," Mind 105, 417 (1996).

[31] Donald Davidson, "Mental Events" in Essays on Actions and Events (Oxford: Oxford University Press, 1970).

[32] Hilary Putnam, "Minds and Machines" in Philosophical Papers, ii. Jerry Fodor, "Special Sciences" in RePresentations (Cambridge, Mass.: MIT Press, 1981).   [33] Kripke, Naming and Necessity.

[34] Armstrong, Universals.   [35] Frank Jackson, "What Mary Didn't Know," Journal of Philosophy 83 (1986).

[36] Thomas Nagel, The View from Nowhere (Oxford: Oxford University Press, 1989).

I often get the impression—although I cannot prove this hunch—that many writers tacitly confuse some list of what are sometimes called "fundamental quantities"—viz., the conserved material parameters of fundamental particles: *rest-mass*, *charge*, *spin*, *color*, etc.—with the traits that might be considered as "basic" to physics. This is simply a mistake: the first list does not include any of the *dynamical qualities* functionally dependent upon *position* and *momentum*, although it is in the latter category that the great explosion in viable quantities we have discussed occurs: all of the salient ways in which a swarm of particles might lock together to induce important macroscopic characteristics within their collective behavior. And it is with respect to these that we must practice the policies of *variable reduction* highlighted in Chapter 4, which then cause the *predicates* of descriptive physics to mutate into thousands of varieties of strategic adaptation, some of which we'll visit in the chapters ahead.

Within the set of philosophical expectations popular today, there implicitly lurks a presumption to the nebulous effect that the contents of a physical theory can be articulated in essentially one way and that its terminology arranges itself into grammatical categories that reflect internal characteristics of the traits themselves (allowing us to claim that Goodman's *grue* represents an "intrinsically disjunctive quantity"). But this improbable dogma runs plainly against the fact that there are a large variety of formats in which basic physical principles can be usefully formulated, each offering their own advantages but preferring different choices of fundamental quantities (Newtonian, Lagrangian, Hamiltonian, Routhian formulations and all that). There is no indication that Mother Nature loves any of these generating choices better than the others. As already emphasized, the quantities that best capture a specific system's evolving behavior may carve up its phase space in a manner quite askew to the base quantities selected in *any* of these formulations.

It is common for philosophers to dismiss quantities they don't like (*grue* constitutes a favorite target) on the grounds that such attribute imposters are "merely mathematical" in nature and are not "truly physical" at all (writers like David Armstrong call these alleged pretenders "Cambridge quantities," for reasons I'll not attempt to explain). But it displays a great misunderstanding of physical fact to fancy that the decompositional quantities predicted by Sturm-Liouville lines of thought are likewise "merely mathematical" in character. No one believes that the overtone breakdown of a violin string does not report genuine "physical characteristics." Should the mere fact that they *sound dreadful* deprive the parallel characteristics found in a wobbling garbage can lid of similar "physical" status? I have never seen any defense of these physical/"merely mathematical" distinctions that remotely begins to struggle with these issues, although they directly reflect some of the basic empirical discoveries that have dramatically reshaped the face of applied mathematics over the past two hundred years (every physicist understands the importance of locked together macroscopic quantities).[37] Once again a fair measure of "we philosophers would very much enjoy an X, so scientists are obliged to supply us with one" thinking seems exemplified within this popular vein of unfounded speculation.

---

[37] Carver A. Meade, <u>Collective Electrodynamics</u> (Cambridge, Mass.: MIT Press, 2000).

....

No notion of "causal characteristics" in Sydney Shoemaker's sense[38] is likely to survive the manipulations of limit taking. A direct inspection of the Thirring/Khinchin treatments shows that the operative notion of quantity simply represents behavioral information about our system, in the form of numerical restrictions upon its potential motions, without any particular concern for causal etiology. The same hold for "properties," on the natural assumption that they correspond to sets over the phase space. In truth, the phrase "causal characteristic" seems to me very vague and I have no idea whether a notion like *having a center of mass motion of 6 million kg-m/sec* in application to, say, a far away planetary system, "contributes to the causal powers" of that ensemble or not (to employ Shoemaker's phrase), since it is likely that nothing is physically located at said center. However, this same *center of mass motion* certainly represents one of the key traits that we can *measure* in the system. These remarks, I might add, do not reflect a skepticism with respect to the notion of "cause" (which shall be discussed in Chapter 9), but the vaguer "causal characteristic" as philosophers employ the term.

....

Quine's characteristic manner for attacking arguments that appeal to dubious notions of "intension" generally proceeds by claiming that they presume a dubious "analytic-synthetic" distinction.[39] He articulates his (generally reasonable) complaints in this manner because he believes that all misallocated characteristics derive from syntactic shaping processes alone (that is, "is a creature with a heart" has a different personality than "is a creature with a kidney" solely because we operate upon the two predicates differently within our webs of belief). The classical thesis that universals carry an invariant content he parses as the parallel methodological claim that predicates gain their meanings through implicit definition (4,iv) from a set of fixed, axiom-like postulates. He then refutes this assumption by arguing that such assumptions shift over time and hence their implicit definability reach becomes muddied. But we should stoutly resist this implausible "all intensional features derive from projected syntactic characteristics" point of view (partially because it is closely allied with obnoxious "veil of predication" themes). It is easy to see from our drumhead example that this point of view is unnecessarily extreme: the distinctive personality that "(0,3)(r)" displays is compounded from the rich cloud of directive elements that surround the trait $(0,3)(r)$, many of which are properly extraneous to the trait itself. But most of this cloud is composed of ingredients that lie just as far from linguistic practicality as $(0,3)(r)$: e.g., the *infinite series* that asymptotically approximates $(0,3)(r)$'s values away from the center of the drum. Certainly, very little of "(0,3)(r)"'s ambient cloud has anything to do with the general methodologies of language building that Quine emphasizes, but instead buzzes about the humble practical difficulties of calculating a drumhead mode. This observation is important, because it shows that a predicate's rich personality often stems from factors that lie close to it locally (one of my projects in this book is to avoid the unfortunate holism into which most pre-pragmatists tumble).

---

[38] Sydney Shoemaker, "Causality and Properties" in Identity, Cause and Mind (Cambridge: Cambridge University Press, 1984).   [39] W. V. Quine, "Two Dogmas of Empiricism" in Point of View.

Nonetheless, Quine and I agree that the theoretical underpinnings of much modern Anglo-American philosophy rely upon the projection of ersatz intensional character-istics into locales where they don't belong. The doctrine of classical gluing encourages this illicit transfer through its basic "living in two worlds" character: features properly pertaining to syntax, approximation or mental attitude easily leach over to the world's attributes across the shared interface assumed in classical grasp. Quine and I conjointly warn that philosophers should not expect to gain much from dedicated armchair musing on our "intuitions" with respect to the nature of universals, for those hunches arrive deeply compromised in loose projections.

In 3,iv I commented upon the degree to which the edifice of contemporary philo-sophical mission and method is settled upon the unsteady sands of classical concepts. I will not pursue these themes extensively in this book (which is long enough as it is), for I am largely interested in understanding the *ur*-philosophical patterns of thought that deposit classical universals upon our doorstep in the first place.[40]

. . . .

Before leaving these topics, I should mention that quantities, here defined as real-valued func-tions over the space of phase possibilities open to the system, do not represent the only con-tainers in which information about a system's behavior can be usefully packaged, for the same basic data can be compiled into other, possibly more convenient, bundles, such as a field of vectors (= directed geometrical arrows), tensors or more exotic assemblies such as quaternion dual numbers (these provide an elegant parcel for quantifying the motions of a robotic arm effectively). The same data can be compiled in any of these ways, whose virtues vary depending upon circumstances. To be sure, employing a vector-valued measure will better reflect the symmetries natural to the situation if the behavior of the system under investigation doesn't favor some particular set of quantities as well-adapted coordinates. But once a vector-valued description is well defined, then a rather wide array of quantity descriptions will become fully defined as well. It is hard to argue that Nature herself displays any particular preference for any of these descriptive schemes, especially when we are concerned with the substantially reduced variables required for a macroscopic system in the first place.

. . . .

(ix)

Fear of attribute naming. After this lengthy, yet necessary, detour, let us return to an important question set earlier. Why, beyond simple mistaken historical assumption, are pre-pragmatists often eager to deny credence to most varieties of "abstract object," even when well-behaved specimens such as the attributes of physics are at issue? Their fundamental concern, I believe, is this. "If such gizmos are allowed back on our

---

[40] Mark Wilson, "Honorable Intensions," in S. Wagner and R. Warner, eds., Naturalism (South Bend, In.: Notre Dame Press, 1993). Mark Wilson, "What is This Thing Called 'Pain'?", Pacific Philosophical Quarterly (1985).

ontological stage, it should prove easy enough to name them and thereby refurbish the impossibly inert gluing promised by classical gluing in all its mythicalness." This worry might be called *fear of attribute naming* and I believe it drives Quine to the strange contortions typical of his mature thought.

Recall, from our discussion of how Quine avoids placing rabbits themselves in the same banned category as *being a rabbit*, that he relies upon the apparatus of phrases like "there is" and "identical" to delineate a syntactic asymmetry between names and predicates. Such lines of thought lead him to the extraordinary conclusion that we are unable to point out rabbits without presuming an embedding in those kinds of pre-existent linguistic machinery:

> As [a term of divided reference "rabbit"] *cannot be mastered without mastering its principle of individuation: where one rabbit leaves off and another begins. And this cannot be mastered by pure ostension, however persistent . . . Our individualizing of terms of divided reference, in English, is bound up with a cluster of interrelated particles and constructions: plural endings, pronouns, numerals, the "is" of identity, and its adaptations "same" and "other."*[41]

Sentiments of this sort are not uncommon in philosophy, but they should be viewed as symptoms that we have become overzealous in our pre-pragmatism.

These are the basic considerations that eventually lead Quine to his doctrines of "the indeterminancy of translation" and "the inscrutability of reference" (it's often the argumentation offered on their behalf that I find the most impenetrable). In terms of fundamental motivation, however, his general purpose seems to be one of alerting his audience to the uncemented patches of looseness that pre-pragmatists expect to find scattered through our usage. But from this point of view, Quine's diagnostic effort is not a success, because he focuses precisely upon circumstances (predicates for biological species) where there are good reasons to anticipate that the predicate/attribute binding is often fairly tight.

The motivation for this faulty tactic lies precisely in the fear of attribute naming: he believes Russell must be contested over the very ground where the case for classical gluing looks the strongest ("Give me your most favorable cases and I'll argue, even there, that the adhesive you peddle will not work properly"). But to presume this is to misconceive the true difficulty with the classical account: it isn't the connection of predicates with attributes per se that creates the distortion, but our inclination to anoint the latter with extra coats of stickiness that makes linguistic success in a predicate/attribute mode appear easier to obtain than it really is.

I'll come back to what I have in mind in a moment, but we should first observe that, very commonly, full-fledged pragmatism (i.e., Peirce, James, Dewey, Rorty) generally leans towards the thesis that thinking of conceptual evaluation as an activity that compares words with any form of external condition represents a great mistake. Thus

---

[41] W. V. Quine, "Ontological Relativity" in Ontological Relativity and other Writings (New York: Columbia University Press, 1969), 31–2.

Charles Peirce:

> *The meaning of a representation can be nothing but a representation. In fact, it is nothing but the representation itself conceived as stripped of irrelevant clothing. But this clothing can never be completely stripped off; it is only changed for something more diaphanous.*[42]

Here is a recent expression of what appears to be a similar sentiment (from the contemporary philosopher Mark Johnston):

> *Let us say that metaphysics in the pejorative sense is a confused conception of what legitimates our practices; confused because metaphysics in this sense is a series of pictures of the world as containing various independent demands for our practices, when the only real legitimation of these practices consists in showing their worthiness to survive on the testing ground of everyday life.*[43]

If I understand Johnston correctly, the noxious assumption that "The world contains various independent demands for our practices" encompasses claims as mild as "The predicate 'is a rabbit' is supposed to apply to items that possess the property *being a rabbit*" or even that " 'Rabbit' refers to rabbits" (understood in a "metaphysical way," whatever that is). But "demand" seems a strange term for describing what merely represents an innocent form of appeal to the direct normativity (4,v) that any attribute automatically authorizes: "The attribute of *being a rabbit* is a useful quantity to study, so if we want our employment of the predicate 'is a rabbit' to profit from that utility, the correctness of its applications should be judged according to their alignment with the trait." In fact, this last statement seems to me wholly true of "is a rabbit" and I believe that if we find ourselves telling a story of linguistic process that doesn't ratify such claims as correct, we should rethink our premises (even if we are heckled as "metaphysicians" by radicals as we do so).

It is easy to see that Johnston, like Quine, has wandered into exaggeration: we merely need to substitute "Euler's method" for "practices" into Johnson's "the only real legitimation of these practices consists in showing their worthiness to survive on the testing ground of everyday life" to generate a palpable falsehood. Indeed, the *better* form of "legitimation" we desire for Euler's method is a proof of its correctness (as in 4,x: a result squarely based upon the correlational studies Johnston abjures). We shall return to these issues in greater depth in 10,vi.

I believe Johnston's intent is to sever the excessive bonds of classical gluing, but it again occurs at the cost of quite implausible pronouncements about linguistic purpose. Rejecting the full classical picture does not require us to promptly embrace some monotheism of faith in some alternative adhesive (such as "surviving on the testing ground of everyday life"). It is better if we can see words and world as held together by familiar—but quite variegated—pressures, rather like the furniture that requires neither

[42] Charles S. Peirce, "Representation and Generality" in The Collected Papers of Charles Saunders Peirce, i (Cambridge, Mass: Harvard University Press, 1931), 339.

[43] Mark Johnston, "Objectivity Refigured: Pragmatism without Verificationism" in John Haldane and Crispin Wright, eds., Reality, Representation and Projection (Oxford: Oxford University Press, 1993), 85.

glue nor nails. True: such binding proves neither as tight or thorough as classicists and pragmatists promise, but the usage can muddle along well enough anyhow. The proper remedy for classical exaggeration is not to chase away every linguistic activity that looks something like classical gluing, but to mildly and patiently determine the correlational states of affairs that have actually been installed through such transactions.

In Quine's specific case, a basic tension runs through his thinking that troubles most of his readers. He begins by warmly embracing the world of science, yet he later writes as if all talk of the covariation of predicate use with attributes (or even, in his terms, sets) is scientifically untenable (most pragmatists, in contrast, never flirt with scientific realism at all). This is very odd, because science should surely be allowed to ponder the correlation of classifications and calculations with the affairs they address. Consider a sorting machine that distinguishes cans of peaches from cans of pears. Insofar as I can determine, Quine's somewhat hazy methodological strictures require us to say that "there is no fact of the matter" (a favorite phrase of his) whether our device sorts the cans by weight rather than through the patterns on their labels. But such doubts are plainly excessive—weight and label sorters operate with dramatically different mechanisms and it won't require lengthy investigation to determine what we have before us. And the evaluative locution, "This machine sorts the cans out by label," provides an excellent vehicle to report what we have discovered. But Quine, stricken with fear of attribute naming, argues, in his famous indeterminacy of translation argument,[44] that we can't determine to which features of a rabbit speakers attend as it scurries by. But why accept a philosophical position that apparently informs us that we can talk about classificatory correlations more readily in the case of tin can sorters than human beings?

Quine's thoughts drift to such extremes through a confusion of motives. On the one hand, his fear of attribute naming improperly persuades him that he must battle away the slightest hint of a "correlation," no matter how innocent in scope. On the other, he wants to render justice to our pre-pragmatist impression that the talk of the orgonists is largely unbonded to any form of concrete happenstance (although this cult may *fancy* that "contains orgone" correlates with something objective, they are simply wrong). But then, if we are fair, mustn't we concede by the same standards that most of what we chatter about lacks direct correlational credentials as well? To express the worry in Quine's preferred jargon, much of our speech activity consists in uttering "standing sentences" (= claims that qualify as correct or not independently of the context of utterance). Such assertions—some contention about quarks uttered at a dinner party, say—display no evident correlation with worldly events, no matter how well informed the commentary proves. And its predicative parts do not reveal any evident covariation with physical traits either. Such musings lead Quine to the conclusion that the only correlations displayed in usage occur at the "observational periphery of occasion sentences," in the form of sentences like "Lo! a rabbit" being murmured when the appropriate critters scamper past (and, even here, most rabbits pass our way without eliciting a single "Lo!").

---

[44] Quine, Word and Object, ch. 2.

(x)

Naming attributes ain't easy. A number of significant misapprehensions have gotten tangled together in these Quinean reflections.

First of all, he has not correctly identified where the most prominent strands of practicality in language lie, at least as suggested by the samples of 4,ii. *Ready classificatory capacity*—that is, an untutored ability to sit on street corners and pick out rabbits as they pass by—is not evidence of great practical purpose and does not facilitate the accomplishment of any otherwise unattainable human goal, which were the hallmarks of the practical advantages we cited. Consider, in contrast, our capacity to read a map or follow verbal instructions directive to the same purposes. Without the intervention of a certain interval of fussing with symbols, whether iconic or verbal in basis, we are likely to get lost in the dark woods and never make it to Grandma's house. This skill requires that the symbols in our recipe correlate with genuine worldly data in some systematic, although possibly complicated, manner. Language here serves us as a vital *instrument*, comparable to a sextant or computing machine, but, as stressed before, instruments don't work repeatedly except for reasons: they must interface with the world in some form of correlative pattern. True, we may be quite ignorant of the underlying manner in which, e.g., a Mercator map encodes geographical data (most of us are), but some mechanism of data registration must be engaged all the same if we are to derive any profitable use from the scribbles on the chart.

We also observed that, with respect to the training of comparatively permanent aspects of usage, strands of practicality often serve as the *islands of usage* around which other employments swirl, whereas mere standing-on-the-street-corner classification will not, in itself, demonstrate comparable fixity. As our cruel smithy case indicated, we do not abandon productive recipes easily, although we may improve and substitute components as we forge ahead, whereas a cult's "orgone" classifications may drift wildly with the whims of a guru, even tho' the babes in that society acquire the mastery of "orgone" completely on a par with "rabbit" or "doggie." This is plain from evolutionary considerations as well: pragmatic Mother Nature will directly reward us if we bring a better sword to battle, not for classifying passing rabbits with great finesse.

Plainly Quine has identified "practical purpose" tacitly with *prediction*, much in the manner of Hertz or the old logical empiricists. It is odd to claim that I consult a map to Grandma's house in order to *predict* whether I will arrive there or not, but Quine and his fellow predictionists attempt to reduce all practical behavior to that ambition *fixeé*. But this is unwise. The mathematics pertinent to invention or route planning often follows completely different contours than the mathematics of prediction per se, and science engages in less of the latter than we first imagine (the Euler's method calculations of 4,iii represent a sterling exemplar of a "predictive calculation," but we usually try to avoid addressing our operational questions in that manner if we can). Applied mathematicians have gradually learned to appreciate that descriptive endeavor is riddled with a great host of essentially different strategies, adopted to diverse forms of final purpose, and that close attention needs to be paid to the mathematical class to which our formulae belong.

Within philosophy, we should become more sensitive to such strategic issues, because *ur*-philosophical confusion often begins when a linguistic routine that actually pursues strategy A mimics the execution of an irrelevant strategy B (Chapter 9 will be devoted to such linguistic chameleons). Quine's vision of language as organized into a holistic web of belief presumes great methodological uniformity in our linguistic endeavors, but Chapter 4 has already illustrated some of the advantages of strategic epitaxy (for those who skipped that chapter, further illustrations lie ahead). Indeed, the best way to develop pre-pragmatist hunch is to watch for fragmentary patterns within our usage, for their filagree of boundaries and splices best reveal the degree to which efforts at classical gluing do not always succeed as expected.

Accordingly, the practical strands highlighted in our earlier musings do not confine themselves to some hypothetical observational periphery, as Quine's "immediate classification" picture would have it. The wires of certifiable practicality run liberally throughout Quine's web of belief and provide it with a more centrally supported framing than he imagines. All the same, their distribution amongst our speed acts remains quite sparse overall, just as Quine claims. One might transcribe huge gobs of daily conversation and not find a single item of authentic practicality in any of it.

We shall return to the proper treatment of this sparsity soon, but let's now address last section's issues with respect to attribute naming. The basic worry is that the classicist, sitting steadfast in her comfy armchair, might attach all of English's predicates so tightly to the world through mental effort alone that no subsequent strand of practicality might improve her accomplishments one whit (except to persuade her to switch allegiance to other attributes on occasion). The morals of the interplanetary miner of 5,iii suggest that such classical claims rest upon an exaggeration of real capacity: our classical designator might handle items like those in her living room ably enough, but she'll need to get out her chair if she plans to deal adequately with the kitchen. It is a brute fact that physical properties, considered apart from a confusion with "concepts," are not especially easy gizmos to grasp or name.

Quine's favorite example, *being a rabbit*, is atypical in these regards, because we happen to be supplied with excellent prospects for keeping a predicate in approximate alignment with its dictates everywhere. But with garden variety attributes, this is not true at all. Some seem incapable of accepting any sort of linguistic handle and, for many others, we may possess a reasonably firm grip upon their ramifications within restricted settings, but we are apt to lose them completely when they stray into other contexts. The truncated series expressions we employ for guidance with respect to $(0,3)(r)$ show the basic nature of the problem we confront: the directivities we must follow when $\underline{r}$ is small (a truncated power series) utterly fail us as $\underline{r}$ becomes bigger (because we must sum an impossibly huge number of terms to obtain useful values). We are left casting about for some new way to discover how $(0,3)(r)$ behaves for bigger $\underline{r}$. In this case, we fortunately stumble across a quite different form of expression (a divergent trigonometric formula) that allows us to follow $(0,3)(r)$ across a greater span of territory. But this is a pure stroke of fortune: divergent series are quite strange creatures and we're lucky that one of them is available to us here.

. . . .

Our divergent series supply us with excellent values for $(0,3)(r)$ if we add up only a few factors, but then feed us rotten values if we go on to more terms in the expansion (like a cagey poker player who allows us to win initially until we've become hooked on the game and then takes us to the cleaners). Such expressions gain their computational advantages in last chapter's manner: we delay consideration of our Bessel function's finer-grained complexities by shuffling them all into the many-term hinterlands and falsely promising "I'll deal with you later." Their fully convergent cousins render equal justice to all scales of functional behavior but this even handed diligence forces them to converge far too slowly for computationally limited mortals such as ourselves.[45]

. . . .

Thus, in extending the use of a predicate into new territory, a *problem of prolongation* often arises: old practical directivities fail us and we need fresh guidance to carry on. It is exactly issues of this sort that confound the classicist in her attribute naming ambitions. Rendering the observation in homelier terms, Br'er Bear discovered long ago that Br'er Rabbit is a lot easier to follow on the roadway than in the briar patch.

A little thought indicates that our average rabbit is none too easy to name either. True, we can easily denominate the bunnies we imprison in backyard hutches and other hares of special prominence as well. But what about the others? Suppose we have a solitary rabbit in the cage but two young children who have suggested competing names. "Sniffy" wins the competition whereas "Foo Foo" loses. "Sniffy" is promptly attached to our lapin prisoner. But what about "Foo Foo"? To placate its distraught champion, I might announce with baptismal pomp and circumstance, "There is a rabbit dwelling deep within the interior of Tibet that is hereby designated 'Foo Foo'." Surely, we are unable to name faraway objects in this facile manner successfully. The rabbits in Tibet are simply too distant from us to permit their designation except in coarse quantity. But our distraught daughter may supply us with motive to engage in a continuing charade of successful naming, e.g., to remark from time to time: "My goodness, Foo Foo must be growing very large; I wonder if she's getting enough lettuce," etc. But such linguistic displays do not help us in the least to connect "Foo Foo" with a genuine referent.

Even if, with some effort, we had formulated a descriptive phrase that can theoretically anoint a unique subject ("Let 'Foo Foo' designate whatever rabbit happens to squat closest to the compass point 32N, 85E at high noon local time on July 25, 2003"), it would be nearly impossible to remain loyal to such denotative dictates. However, someone might mistakenly *fancy* that she has done so. Suppose our disappointed child broods upon "Foo Foo"'s whereabouts for years and, after she reaches her majority, she sets off on a mission to locate the now antiquated creature. I hazard the opinion that anyone of such an intensely sentimental frame of mind will be disposed to settle upon a "Foo Foo" surrogate with less than perfect rigor. Rather than struggling to locate the unchartable rabbit of my original geographical specification, she will likely plump for some animal she likes. "Oh, that's the one," our seeker confidently declares; "it's got

---

[45] R. B. Dingle, <u>Asymptotic Expansions</u> (London: Academic Press, 1973).

that adorable mask around its eyes that I've always imagined Foo Foo to have." And from that moment hence "Foo Foo" will attach to this suddenly pampered animal, coupled with the firm conviction that it had been dubbed "Foo Foo" by Dad long before. Like the Druids of 1,ix, our deluded daughter remains quite oblivious to the degree of <u>post facto</u> adjustment involved in her linguistic behavior.

As is often emphasized (sometimes to exaggeration), localized biological groups are sufficiently distinguished by anatomical features and behavioral patterns that native communities around the world commonly carve up animals along species lines more or less as we do (this intercommunal commonality is much less pronounced for *family* terms such as "rabbit," which, even in English, fights a fluctuating contest with "hare" as the designation of the wider group).[46] For the sake of streamlined example, let us pretend that Quine had instead selected the species focused sentence "Lo! An Old World rabbit" rather than plain "Lo! A rabbit" as his chief illustrative example. Granted this narrowed-to-a-certifiable-species proviso, a few gestures at relevant specimens are likely, <u>pace</u> Quine, to lead to an employment that is properly described in terms of a genuine correlation between predicates and physical traits: "In this usage the attribute *being an Old World rabbit* correlates nicely with the predicate 'is an Old World rabbit,' " evaluations of behavior that should be regarded as no more problematic in nature than "In this sorting machine, the stamp 'accepted' correlates with *full can of peaches*" or "In these calculations, the output state correlates with the quality *the amount of oil optimally desired*" (I refer to the digital control example of 4,v). In Chapter 2, I described such associations as ones of simple *predicate/attribute alignment* (the pairing "is a dog"/*being a dog* was the example selected there). We should not allow philosophical crusades like Quine's to persuade us that human behavior can't be profitably discussed in such terms, because we regularly do (albeit usually in less stiffly articulated language: "In English 'dog' refers to dogs.").

....

Nor should we persuade ourselves that such activities "are possible only against a prior practice of naming," as Wittgenstein would have it.[47] Our abilities to anoint a determinate rabbit with "Foo Foo" or not seem entirely an issue of rabbit tractability, not some special degree of training on our parts.

....

However, species traits are generally unusual with respect to their *global salience*: all expected manifestations of the quality are comparatively homogeneous in their basic display (we do not expect to find instances of <u>Oryctolagus cuniculus</u> anywhere but on earth, for example). In fact, there are examples of "rabbit"-like designations that display "Foo Foo"-like qualities in their behavior. Although the fact is easy to forget, the cute,

---

[46] Jared Diamond and K. David Bishop, "Ethno-ornithology of the Ketengban People, Indonesian New Guinea" in Douglas Medlin and Scott Atran, <u>Folkbiology</u> (Cambridge, Mass.: MIT Press, 1999). Several factors make the true history of "rabbit"-related vocabulary rather complicated, but I will pretend, for sake of contrast, that it has been simple.

[47] Wittgenstein, <u>Investigations</u>,§31.

European robin

American robin

rounded bird that the English call a "robin" is not closely related to the hulking fowl that Americans so designate. Our homesick pilgrim settlers espied our native thrush and, noting its red—not even a proper orange—breast, called it a "robin," no doubt because they understandably hungered, given their crummy conditions, for a local emblem of domestic cheer. "Okay," the colonists announce to themselves, "this critter's got some color on its chest; it'll do." "Is a robin" is one of those predicates that, were linguistic process entirely orderly, would operate as a simple species designator just like "Old World rabbit" and, no doubt, if Erithacus rubecula and Turdus migratorius had freely intermingled in range, the phrase "robin" would have been forced to attach firmly to one or the other of these branches. However, the wall of the Atlantic Ocean keeps the two local employments of "robin" fairly (although not completely) compartmentalized and so the pressure to hew to a single species greatly diminishes, allowing our wayward predicate the luxury of spreading itself over two unrelated breeds, all the while presenting the appearance of an 'umble designator of a single strain. In short, when the employment of "robin" was prolonged to active use upon North American shores, a crossover in its patterns of attribute attachment occurred.

I have no idea whether our forebears realized they were employing "robin" in a markedly novel manner or, like Foo Foo's seeker (and the Druids of Chapter 1), they plowed through the critical crossover events firmly trusting they were "using 'robin' with its good old-fashioned English meaning." We shall discuss more serious cases of linguistic prolongation in later chapters where utter obliviousness to any issue of attribute shift is undeniably involved. But if these crossover episodes are not noticed, or

if they are later forgotten, their legacy can come back to haunt later generations. As a youth I remember being much puzzled with respect to a British cartoon rendering of the eponymous victim in "Who Killed Cock Robin?" The sparrows, the cranes and all the other fowl who confessed to their crimes seemed like excellent facsimiles of their prototypes, but that chubby *robin* ...? I then wondered, albeit in less sophisticated terms, whether the *robin* property was subject to some radical form of biological dimorphism.

Verbal behavior of this type illustrates a basic phenomenon that is central in our investigations: alterations in attributive correlation that arise when one patch of established usage feeds into another through some species of prolongation (I dubbed such shifts in the correlated attributes *property dragging* in the previous chapter). To be sure, our "robin" case represents an especially ephemeral and easily correctable exemplar of the process, but far more serious examples will be discussed in other chapters (indeed, the confines of classical mechanics already supplied a goodly assortment in Chapter 4). As these crossover events occur, a usage splits or otherwise forms into a polycrystalline state: a sequence of connected patches whose boundary joins need to be policed with special precautions. Among the possibilities here are the *facades* introduced in the last chapter (and destined to be reintroduced from a different vantage point in the next). Oftentimes adjacent patches that look superficially similar can operate according to markedly different underlying strategies.

In any case, the resulting usage will not display a simple "is a dog"/*being a dog* alignment, because distinct traits dominate their local ranges in a more complicated pattern (as *being a member of Erithacus rubecula* and *being a member of Turdus migratorius* do in "robin" circumstances). Why do our semantic circumstances play out so differently for "robin" and "Old World rabbit"? It seems plainly a matter of the directivities that prove most salient when the term is imported to America. Over English soil, the primary shaping factors of visible shape, mating habits and so forth mold "robin"'s employment into local correlation with *being a member of Erithacus rubecula*. But, after the phrase's voyage to America, these same directivities are no longer pertinent, because there are no fowl of that exact physiognomy or behavior in evidence, allowing a gap where the prolonging impulse, "Boy, it'd be nice to see a robin around here," becomes momentarily dominant. This *leading principle* (to borrow a term from Charles Peirce) inspires a few tentative applications of "robin" to a fresh variety of fowl with a red breast. Once this toehold has become established, the normal focusing directivities of shape and mating habits now develop a North American patch of application locally centered upon *being a member of Turdus migratorius*.

Whether property dragging actually arises in this case or not obviously depends upon rather whimsical factors: i.e., the classificatory impulses that occur to Puritan bigwigs (although it is striking how many unrelated "robins" have popped up around the world in the wake of British colonialism). In the sequel, our focus will shift to cases where the bridges between patches are comprised of more substantial stuff and address far more substantive purposes. In fact, a nice illustration of greater seriousness was provided in the last chapter. The recipe "Build up your differential equations based upon the backbone of $\mathbf{F} = \mathrm{ma}$" forms a bridge that links the branch of mechanics concerned with elastic solids to that dealing with viscous fluids. But as this crossover occurs, the term "frictional force" becomes dragged to a new correlation with a more complicated physical attribute in the bargain. We documented the considerable *ur*-philosophical confusion that was engendered by this rather silent crossover.

In 3,iii, I distinguished between liberal and tight flavors of directivity: whether the answers supplied to "Should this bird be classified as a robin?" or "Given that this bird is a robin, what conclusions follow?" are easy to implement or not. Here the instruction "Judge the bird as morphologically similar to the backyard bird back home" is easy to follow; "Check for overlap in DNA content" is not (in our Chapter 4 illustration, "Follow Euler's method" is easy to follow; "Solve this differential equation" is not). Unfortunately, as we also observed, the easy-to-follow forms of directive instruction don't lead to useful classificatory predicates in themselves, for nature rarely hews to easily specifiable pathways. At best, we can patch together *a schedule of tight directivities* that can supply more or less adequate answers through *cutting and pasting*: "Follow rule A over domain $\mathbf{D}_1$ but switch to rule B when we move into $\mathbf{D}_2$," (our (0,3)(r) directivities illustrate such cutting and pasting admirably). Such considerations show why so-called *criterial explications of meaning* (which we'll encounter at various points in the sequel) generally fail: the only standards that can be plausibly associated with "robin" or "(0,3)(r)" *as meanings* are just as hard to follow as the concepts for which they represent the "criteria."

This is not to say that the more distanced and liberal directivities do not supply useful evaluations of verbal behavior: an agent can have her attention focused just as intently by "Solve this equation" as "Follow this algorithm." It is also clear that the injunction "Consider birds <u>a</u> and <u>b</u> to be both robins if they represent the same kind of bird" will exert rather different directivities when offered in 1620 than in 2005. In the earlier time, their shared ruddy breast will immediately rouse the crossover suggestion that they might represent the "same kind of bird," whereas we grant superficial similarity much less directive credit in biological applications today. In contrast, "Old World rabbit" did not widen in range during its overseas displacement to America because there were no animals hopping around in North American arbors that excite any 1620 directivities attached to "Old World rabbit."

It is through such considerations that we should explain why our efforts to name rabbits in the backyard usually succeed, but fail for those in Tibet; why our ancestors managed to set up "Old World rabbit" in simple predicate/attribute alignment, but fail to do so for "robin." But in writing of "failure" here, it is vital to realize that a vocabulary supported in patchwork fashion oftentimes represents a *healthy state of language*, rather than constituting a mere pathology (as our robin case may wrongly suggest). Indeed, the usage that too vigorously attempts to stay in simple predicate/attribute alignment often fails disastrously as a practical syntax, having become hamstrung through its caution, whereas a patchwork vernacular may lead us onto admirable things. In the last chapter, we extracted this moral from basic considerations familiar within applied mathematics, but we will revisit the same lesson from a less technical vantage point in the chapters ahead.

....

The reader may wonder why the "robin" case has been here described as one where the predicate *changes* its worldly attachment from *being a member of Erithacus rubecula* to *being a member of Turdus migratorius*, rather than simply adhering to the disjunctive *being a member of Erithacus rubecula or Turdus migratorius*. Looking ahead to the "facades" of the next chapter, the choice will largely depend on how sentences whose contents *span* the domains of the two patches need to be addressed (in a facade, such statements correspond to the bridges of prolongation that connect the patches). In actual fact, settled linguistic behavior with respect to "robin" proves a bit complicated in this regard because we seem willing to evaluate truth-values according to rationales that are discordant from a simple facade point of view:

(1) *Some different bird has driven all the robins from my backyard* (true even in circumstances where the invaders are European robins).
(2) *The robin is a harbinger of spring* (false because the European varieties do not migrate).

As such, "robin"'s cross-patch behavior resembles the more narrowly contextual behavior exemplified in a term like "rainbow," rather than obeying true facade expectations (7,i).

In this regard, it should be borne in mind that the contours of a facade <u>per se</u> are commonly more tidy than the results of naturally evolved linguistic development accommodate. I particularly emphasize facade structures in our discussion as a simple means for illustrating how a linguistic use can be *rationally constructed* according to a different strategic drummer than a conventional "flat structure." But the vicissitudes of natural evolutionary process are likely to deposit real life vocabulary on less perfectly engineered piers than these.

In many cases, the data export restrictions between patches will prove so strongly implemented that the community may embrace no patch-spanning claims of (1) and (2) ilk. In such circumstances, the disjunctive "robin" assignment should be regarded as *descriptionally equivalent* in the manner I discuss in my "Predicate Meets Property."

....

(xi)

Ghost properties. From this perspective, we needn't rid the universe of every trace of attribute simply to prevent Bertrand Russell from nailing language too firmly to the world, despite Quine's belief that a Sherman-like campaign of eradication is required. Likewise, we needn't war against innocuous everyday claims with respect to how our words correlate with reality or repudiate every human capacity to shape the future contours of our usage in significant ways. Often we *can* name attributes ably, just as we easily denominate individual rabbits; likewise, the way we wield a predicate often *correlates* quite nicely with some objective physical quality (as a predicate, "is a dog" matches tightly to *belonging to Canis familiaris*). It is merely that our powers in these respects are not as great as we imagine, a fact that can be established through the consideration of cases like those examined in this book. As a result, we often sally into fresh patches of employment full of an unsubstantiated confidence that we are merely following the univocal instructions laid down by our *robin* concept (I called such attitudes *tropospheric complacency* in 2,iii). In many cases, this unearned self-assurance does no harm—indeed, hubris is often a required ingredient in bold advance—, but it encourages us to overlook subtle junctures and possible warnings as we plow ahead. The hypothetical Druids discussed in 1,ix do not recognize that they settle semantic issues afresh when they confront the bomber; they imagine they are simply using "bird" in the old-fashioned way. The excessive claims of classical gluing that pre-pragmatists oppose represent nothing but this *ur*-philosophical impudence writ large. Its exaggerations should be opposed with a simple challenge, "Can we really do *that* always?," rather than wholesale ontological destruction in the manner of a Quine.

The classical tradition regards its beckoning concepts as homogeneous in their contents; if they seem conflicted, it is merely because the linguistic agent has aligned multiple universals sloppily under a common predicate. But the concrete directive considerations that push "robin"'s odd career forward arise as the resultant of *conflicting pressures* that plainly trace to quite distinct origins: the behaviors native to biological classification pitched at the species level and a psychologically driven desire to have a cheery emblem of Olde England around. How the predicate "robin"'s usage stabilizes upon transport to America depends, in such circumstances, upon the delicate accident of which of these colliding considerations happens to loom largest in our colonialists' psyches. So we mustn't always presume that some single attribute sits as the central sun within some predicate's churning swarm of active directivities. Indeed, the facade structures and quasi-quantities to be discussed in the chapters ahead display simple patterns of alternative informational organization that behave like "robin" and gain great benefits therefrom.

When I became puzzled about that plump cartoon robin as a youth, it didn't occur to me that "is a robin" corresponded to anything other than a single trait: *being a robin*, I thought, just as "is a dog" signalizes *being a dog*. So I wondered about the peculiar variation witnessed in my avian trait's instantiations. What difference in diet or climate might produce the British version's extraordinarily different appearance? Or had I only happened to witness an endless stream of female robins within our yard and this specimen, finally, represents the male dimorph I'd never encountered? In these misapprehensions, I was struggling with a brand of *ghost attribute*: I believed that I grasped a cloud of predicative directivities that emanate from a single attached attribute, when I was actually viewing a combination that derives from disparate sources, having become entangled long ago through the whimsies of homesick colonialists. As such, the example is a bit different from the examples of single attributes encased in extraneous intensional features considered earlier, but the basic mechanism of mistaken projection is much the same.

It is in this vein that the exaggerations of classical thinking are most profitably diagnosed, rather than in engaging in excessive Quinean attacks with respect to the very coherence of word/world comparisons. The fact that ghost attributes often can't be distinguished from genuine attributes shows that the vaunted claims of classical gluing cannot deliver all they promise. Even with predicates (like "is a dog") that display the nicest imaginable correspondence to macroscopic worldly attributes, their funds of easy-to-follow directivities can still seem a bit unruly, for there are inevitably the odd breeds, jingoes and wolf crosses that tempt immediate classification in divers directions. At any given moment in a predicate's developmental history, the phenomenology of following the directivities of a cloud that surrounds a true attribute core can look exactly like those within a ghost attribute swarm, where no single attribute lurks within its center at all. But if we can't determine the difference through classical "true thought" analysis or other classical policies of that sort, we lose the strong assurances of steadfast behavior that form the crux of classical gluing's most alluring promises. Instead, predicate and world can easily arrange, over time, some clever strategic accommodation *between themselves* with respect to their correlative concordance, but fail to inform *us*—the self-styled masters of our words!—of the bargain they have struck. Of course, long after the deed is done, a truer picture of the arrangements they have reached may finally dawn upon us, but only at a moment when it is too late to change the deal. But, in most cases (not "robin," but the ones we shall study soon), we should sheepishly recognize that 'twas better that we were excluded from the critical plotting, because our ham-handed input would have only bungled the scheme.

In attempting to flesh out pre-pragmatist hunch in this manner, we are not seeking a semantic adhesive alternative to classical gluing, but instead attempting to understand the hard information we convey when we advance evaluative claims such as "In English, the predicate 'is a dog' picks out the physical attribute *being a dog*." As I've tried to stress in this chapter, such unshaded locutions aptly encapsulate the key facts pertinent to the employment of particularly fortunate vocabulary. But other predicates—"is a robin"—can appear altogether similar in their phenomenology, yet rest upon supportive conditions of a more complex character. When those underlying facts are recognized, we usually take account of them in our everyday descriptions by qualifying our conceptual attributions in some vein or other. Thus we might report: "In English, the predicate 'is a robin' picks out *some kind of blurring* of the attributes *Turdis migratorius* and *Erithacus rubecula*." Such descriptive adjustments do not frame a sharp distinction between the linguistic circumstances of "dog" and "robin" and preserve a fuzzy, *ur*-classical picture of how contents are grasped. But such a portrait of linguistic circumstance is misleading: warring directive factors are generally at work upon all our words and can easily reach jury-rigged accommodation in some manner other than simple "P"/φ alignment. Through painting all our predicates in a common semantic shade—as our evaluative locutions of "concept" and "attribute" superficially encourage us to do—, we readily find ourselves approaching our variegated adventures with developing predicates in the naïve mode of the Foo Foo fancier, where the "invariant qualities" of our guiding traits seem rarified and resistant to articulation: "I possess a concept of *robin* that dictates that all of these different looking birds should qualify as such, but I lack the verbal capacity to explicate its directives in any other terms than 'robin.' " We mulishly imagine that some ghost attribute hides behind our predicate's iridescent facade, giving rise to the impression that some oracle lies within our concept and whispers constant and consistent instructions to us (although we sometimes have trouble making out exactly what its Delphic intimations actually recommend).

In evaluating predicates for possible attribute alignment in real life, we commonly distinguish between terms like "dog," "robin," "hardness" and "red," that play an active role within a large number of strands of practical advantage, and those such as "orgone" or "zig, zag and swirl" (3,vii) that do not. Even with respect to the first group, we have noted that such pragmatically valuable routines thread through real life usage only sparsely. But we have also observed that, if an instrumentality works ably, whether it be of a mechanical or linguistic constitution, there will be *correlational reasons* that explain the routines' successes with respect to informational registration. Such stories do not require that the encoding assume a simple "P"/φ form. However they unfold, we should be able to ascertain what sorts of information are being handled at each stage in the discourse, although doing so may require that we first appreciate the advantages of some clever and unexpected strategic gambit. Once the coding method has been cracked, it can usually be employed as a platform for conveying information in a more general vein, even if such chatter addresses no practical purpose whatsoever.

I won't attempt to elaborate on these remarks extensively at this stage (we will return to these issues in 5,vii), but here is a prototype for what I have in mind. When a usage in applied mathematics advances into new territory, the extended applications cling at first fairly closely to the practical routines which make the virtues of their extended use evident (at first our employments only dance with the fellers what brung 'em). As confidence in its underlying viability grows—through inductive probing or by actually figuring out their informational underpinnings—, the usage will move away from the strict contours of tested algorithmic performance, usually with tolerable assurance that the extension has been safely made. Or, to recast this developmental progression in terminology of 4,v, the distributed normativity of a valuable recipe provides an entering wedge that extends old terminology into new territory. With suitable caution, a freshly established direct form of informational correlation can nucleate around this opening sliver and gradually enlarge. If so, we can then evaluate freely asserted claims over the new domain as true or not with respect to the informational correlations that make the practical recipes work.

Shall we find these truth-value appraisals valuable or not? In many instances, this question requires a complex answer. For example, if we inspect the fifteenth century writings of a Nicholas Cardano where expressions for complex numbers first appear, will evaluating Cardano's sundry claims by the lights of a modern understanding of complex numbers seem worthwhile? Yes, certainly, if we consider the computations he provides for solving cubic equations (these represent the chief strands of practicality that first brought expressions for imaginary numbers to notice). In this context we will happily pronounce, "Here Cardano has it right; over there, he has made a mistake." But this same evaluative policy may seem utterly pointless when we turn to the more speculative remarks he offers about "$\sqrt{-5}$," because they are based upon a picture of how "$\sqrt{-5}$" works that is utterly off base (we may have plenty to say about those free standing assertions, but not in the present correlational vein). But we shall return to the issues of how we wish to evaluate assertion in circumstances such as these more fully in 10,vii.

In these respects, it seems to me that wholly unnecessary dichotomizations have distorted most contemporary discussion of issues such as ours. Classical thinking promises us, through its invocation of concepts unrealistically conceived, that it is fairly easy to get our predicates attached cleanly to worldly attributes, a claim made to seem plausible largely through painting the world's true attributes in projected layers of ersatz adhesive and passing off ghostly imposters as "attributes." From this vantage point, classicists promise us that a tidy *reference relationship* exists that can tie a predicate "P" firmly to an attribute φ as long as the employers grasp φ firmly and steadfastly maintain the tie. As anti-classical critics, many of us declare this picture to be simplistic. But to do so, we needn't insist that speaking of "reference" in the course of everyday linguistic evaluation is wrong or mythological: " 'Dog' refers to *being a dog* in English" should be accepted as an innocuous expression of genuine correlative fact. On the other hand, we should also note that, if someone off-handedly asserts that " 'robin' refers to *being a robin* in English" or that " 'rouge-gorge' refers to *being a robin* in French," we will balk and ask, "Wait a minute; do you realize that a little hitch arises here with respect to *Turdis migratorius* and *Erithacus rubecula*?"

Many classical critics have felt compelled to make very radical declarations on these issues: that the "reference relationship" is a mythological notion; that it can be "naturalized" in terms of causation or allied mechanisms; that it can be understood in deflationary terms only (a position to be explained in 10,vi). Why advance such extreme and implausible manifestoes? *Answer*: "Well, as naturalists, we simply can't allow the classicist's occult notion of 'reference' to stand amongst the world's ontology and thus we need to explain away its appearance." But if *this* is the mission, we should likewise declare that *hardness, redness, gear wheelness* et al. need to be dispensed with, naturalized or deflationized away, for they display exactly the same basic behaviors as "concept" or "refers." That is, *all* of the predicates on this list—"is hard"; "is red"; "is a gear wheel"; "is a concept"; "refers to"—represent terms of macroscopic evaluation and, as such, are successfully employed only by adopting more complex and shifting strategies than simple "P"/φ alignment. In fact, it is wisest if we first figure out how "hard" and its evaluative colleagues operate, and afterward look at "concept" and, eventually, "reference" in light of what we learn (we'll discover that the oddities of the semantic evaluators merely echo the peculiar particularities of the target predicate words they treat).

Why have most anti-classical critics adopted such extreme positions? Much of what has gone amiss is surely traceable to the lingering hand of theory T syndrome, as kept alive by figures such as Quine himself. Under its influence, analytic philosophers have become thoroughly convinced that, at any moment in time, we advance grand "theories" of the world based upon some favored ideology of predicates to which we are "committed." As philosophers, our annointed task is that of walking critically through this list—e.g., "is hard"; "is red"; "is a gear wheel"; "is a concept"; "refers to"—and striking out, reidentifying or deflating any ingredient we can't wholeheartedly endorse. And this project is presented as one that only a laggard or slacker would refuse: "Step up to the plate; are you for this predicate or against?" In truth, terms of macroscopic evaluation simply can't be manhandled in this way; it is only a demented picture of "theory" run amuck that makes us assume otherwise. No, virtually every term of macroscopic evaluation has its own complex story to tell and, without much subtler clarity of purpose, it is absurd to embark upon a project of trying to decide whether *being a gear wheel* is "required in our ontology" or not. Each of our listed predicates performs valuable linguistic work most of the time, but on occasion each also misleads, simply because it functions in more complex ways than we appreciate. In the sequel, I will not supply any grand "big picture" that explains how all predicates behave—my story would be inconsistent if I believed that possible—, but I can provide simple models that demonstrate how a range of typical *ur*-philosophical misapprehensions can arise from the unexpected behaviors of particular predicates.

From this moderated point of view, we should not accuse the classical picture of mysticism, supernaturalism or "metaphysics (in the pejorative sense)," for such epithets do not reach to the true center of what is at issue. The basic ingredients encountered in classical gluing are merely the real capacities of everyday linguistic life writ large, blended into a soothing elixir that promises more than it can deliver. True, with this nostrum in hand, we fancy we can accomplish reformatory feats in language that lie beyond our capacities, but, all the same, there isn't a single ingredient in the brew that can't, if applied in a favorable setting, genuinely reorient our language in improved or altered directions. We *can* assign our rabbits silly names if we choose; we *can* look up an unfamiliar word in a dictionary and use it more appropriately thereafter; we *can* correctly guess the gist of a term by overhearing a conversation; we *can* coin a new phrase in a psychology article; we *can* taste a pineapple and devise a marker for its special qualities, even as devotedly private diarists; we *can* reorient the employment of an old predicate significantly after experiencing a "Eureka!" of sudden understanding; we *can* invent a novel measuring instrument and bend the old word "temperature" to fit its guidance. Quite dramatic improvements in usage have been achieved through each of these familiar activities. As such, they represent the everyday episodes that classicists highlight in defense of their portraiture of conceptual grasp: "You see, these all represent occasions in which we link up predicative expressions with concepts that we have just come to grasp." As critics of classical exaggeration, we should never deny that such episodes frequently occur exactly as described, but instead mildly demur, "Yes, but we can easily find ourselves in linguistic circumstances that superficially resemble yours, but where the outcomes you promise mysteriously fail to materialize. And those surprises arise because the advancement of usage is also driven by many factors that run counter to the capacities you emphasize."

Through considerations such as these, our initial pre-pragmatist suspicions with respect to classicism can be validated without needing to concoct an implausible adhesive to replace that promised within the classical picture. Nor do we need to abandon the external world behind a dim veil of predication of the sort that pragmatists or Quineans frequently erect.

....

In many ways, the policies of anti-classical criticism I recommend greatly resemble Wittgenstein's frequent injunctions to grant opponents the core validity of the capacities they highlight, while restricting their range. However, Wittgenstein seems to also believe that we can successfully ascertain those ranges a priori, by reflecting intently upon the nature of the "language game" as we have learned it. But this last thesis is completely contrary to my own opinion, which doubts that our everyday usage comes enfolded in a restrictive structure as elaborate as that of a language game and believes that, insofar as such strictures arise, we do not acquire them fully formed from our linguistic peers. Similar issues will be discussed in connection with J. L. Austin in 7,xi.

In the 1970s, a number of prominent philosophers—Hilary Putnam, Richard Boyd, Michael Friedman—claimed that science's successes would be "miraculous" if its key terms lacked reference. Such thinking, it seems to me, shares in the general tenor of the "methods which lead to true results must have their logic" point of view that we shall develop more extensively in Chapter 8, although the first position expects simple "P"/φ arrangements while the second anticipates that more complicated and localized supports may be in the offing. My own thinking began under the influence of the first school and, after reading Heaviside and others, evolved towards the second (which is standard in applied mathematics).

....

(xii)

Hazy holism. As emphasized previously, a chief attraction of the classical picture lies in the fact that its invariant conceptual contents provide a sunny vision of everyday linguistic evaluation and improvement that is elegant and encouraging in its contours, whereas the story I tell is ugly, fractured and tinctured with a disagreeable pessimism. A similar taste for tidiness leads Quine and many other pre-pragmatists to make a truly unfortunate decision at this point: they attempt to imitate the superficial sleekness of the classical evaluative story by recasting its semantic uniformity in descriptive terms they find more acceptable. In almost every case, this policy soon leads to an exaggerated reliance upon distributed normativity and hazy holism.

Camelopardal

Consider the following situation. Suppose we have been inspecting the Renaissance bestiary compiled by Edward Topsell and come across the description of the "camelopardal":

> This beast is engendered of a camel and a female libbard. . . . The head of the camelopardal is like a camel's, his neck is like a horse, and his body is like a hart's; and his cloven hooves are the same as a camel's.[48]

We may ask ourselves, as scholars frequently do, "Does the term 'camelopardal' refer to anything real?" And sometimes the intuitive answer is, "Yes, it talks about giraffes," but sometimes we decide, "No, the creature is entirely mythological, insofar as we can discern." If, like Quine, we posit that genuine correlative comparisons are illegitimate, lest we acquiesce in wholesale attribute naming, then we are obliged to construe these natural predicate/world evaluations in some other manner. Quine's solution is to claim that such appraisals properly represent commentary as to how Topsell's 1607 usage should be *mapped* into our modern tongue: "What term in English will best translate 'camelopardal' in its original contexts?" (traditional pragmatists often side with Quine in this leaning—<u>vide</u> the quote from Peirce above). Upon this basis, Quine constructs an elaborate vision of semantic evaluation that trades, in one way or another, upon this "mapping into a home language" idea. To make such appraisals justly, Quine thinks, we must thereby compare huge hunks of Topsell's language to our own, for the links that support "camelopardal" in his Elizabethan web of belief must be mapped somehow to our own, presumably with considerable allowance for intervening changes in attitude.

Views of this type have proved enormously influential in contemporary philosophy (Donald Davidson's approach to every philosophical issue seems premised on this presumption as axiom). In Quine's own hands, such opinions quickly lead to that sequence of euphonious hypotheses for which he is greatly celebrated (the indeterminacy of translation, the inscrutability of reference, the underdetermination of theory,

---

[48] Edward Topsell, <u>Histories of Beasts</u> (Chicago: Nelson-Hall, 1981), 32.

and so on). I believe that each of these theses is deeply disloyal to the pre-pragmatic instincts with which we began, which fault classical thinking precisely for the ersatz uniformity in which its strong gluing cloaks our linguistic activities. But Quine's alternative "mapping into a home language" story seems like an attempt to *imitate* classicism's univocalism within another framework.

Such accounts invariably encourage a holism with respect to conceptual evaluation, for we become obliged to take account of the vast and amorphous webbing that allegedly supports our predicates. Such doctrines are apt to prove corrosive in their intellectual consequences, as we witnessed with respect to the post-structuralists that contend that every application of "folklore" is irrevocably stained with the presumptions of Western elitism (indeed, such ferocious critics frequently claim Quine and Thomas Kuhn as avatars, albeit priests frequently faulted for their timid dispositions[49]). Quine and Kuhn gravitate to holism because, in different ways, they have adapted the old "predicates as sustained by a web of axiomatics" picture (described in 4,iv) to looser circumstances. According to the older account, a set of axioms serves as the implicit definability webbing from which science's theoretical predicates obtain their semantic support. If two scientists come to loggerheads about whether the term "force" is rightly applied or not, they can consult their axiomatic handbooks and determine whether they are using language in a common way or not. Quine's opinions about language begin in the correct observation that real life linguistic development cannot advance along such neatly charted paths. He then decides, "Still, some webbing of supra-sentential links is required to hold our predicates aloft, but that fabric can be largely woven together by the bonds of early learning, supplemented by the subsequent modifications and improvements this netting receives under the regulative shaping provided by explicit scientific methodology." Under the heading of "methodological shaping," Quine has a long list of syntactic imperatives in mind: "Assume no entity without necessity"; "Regiment your assertions into first order logical form"; "Find the simplest and broadest generalizations under which satellite claims can fall"; and so forth. As this process of doctrinal distillation continues—as science gradually organizes its sundry claims into ever broader coverage and uniformity—, our resulting web of belief will, in some idealized final state, assume the organization of an axiomatized theory where logical principle reigns supreme over all. By rephrasing the old empiricist picture in terms of this story of language growth, Quine evades the old implausibilities with respect to "bridge principles" et al., yet seems to provide every predicate with an adequate webbing of distributed support.

Unfortunately, by casting the net of implicit definability wide in this looser manner, a very large swatch of usage must be considered if we hope to gauge the "meaning" that a given predicate carries for its employers. This is how we reach the improbable conclusion that we shouldn't attempt to translate "camelopardal" without first scrutinizing great gobs of Topsell's prose. Quine's celebrated indeterminacy of translation thesis traces to his assumption that such large scale alignments between Topsell's belief set and our own will be perforce imperfect and resolvable in incompatible ways.

---

[49] Steve Fuller, Philosophy, Rhetoric and the End of Knowledge (New York: Laurence Erlbaum, 1992).

In Quine's vision, it is the driving impulse of regulative principle that relentlessly urges us to conglomerate everything we have to say into one great glob, maintained in orderly array by the far reaching and homogenizing ties of logical principle (e.g., if we accept two sentences A and B, no matter how unrelated their contents, then we must willingly embrace their conjunction "A and B" as well). But who conjured up Quine's Demiurge of Methodology? As we observed in the previous chapter, commonsense thinking in applied mathematics suggests a moral quite the opposite: sometimes our predicate employments are best partitioned into patchwork sectors for greater descriptive efficiency.

Quine's "mapping into a home language" story makes the adjudication of disputes between scientists potentially equivocal if global accord on translation schemes can't be reached. The scientific historian Thomas Kuhn arrives at an even deeper pessimism on this same score through a somewhat similar route. He begins by noting, much as we have done here, that a scientist's prevailing attitudes will be shaped by loose congeries of directive factors: the successful techniques that have proved their worth in prior puzzles; the descriptive parameters that look as if they can be capably extended, adapted or improved within fresh regimes; the set of problems that seem most central to her subject; the recent availability of analytic tools or instrumentation; a topic's perceived similarity to some field presently further advanced, and so forth. Two scientists might experience setbacks in reaching agreement on the proper application of a predicate if their backgrounds with respect to any of these directive centers prove significantly different. Kuhn correctly recognizes that these various flavors of predicative influence do not fit neatly into either the classical or formalist conception of "theory." Quite the contrary; it is common for workers to subscribe to the exactly same official set of doctrines (the "laws of Newtonian physics," say), yet nonetheless become entrapped in bitter wrangles about specific cases simply because they are differentially influenced by the "point of view" factors enumerated.

So far, so good. But Kuhn then decides, first, that his melange of factors ought to be collected together under the alternative heading of *paradigm* and this nebulous assemblage should serve as the semantic fabric from which a given scientist extracts her applicable standards of correctness for a term. Quite famously, Kuhn compares the activity of a paradigm to some encompassing gestalt that irrevocably tinctures how its victims view the world. Plainly, the impulse to gather scattered directivities into a Kuhnian bundle traces to a desire to provide a mistier imitator of classical invariant content.

Unfortunately, this story makes it quite unlikely that two scientists operating within different paradigms will truly "understand" one another, a dismal conclusion that Kuhn, famously, embraces and uses to explain the refractory deadlocks to which competing investigators often descend. This conclusion represents a depressing retreat from the goals to which 4,iii's inventors of "theoretical meaning" had originally aspired, because they had hoped that explicit axiomatics would *facilitate* resolvable scientific discussion, not decrease its likelihood. But that optimism is possible only if the governing axiomatics can be kept firmly in public view. Once we exchange "axiomatic structure" for

Quine's loose "web of belief" or Kuhn's psychologized "paradigm," a bleaker account of communicative capacity emerges simply because the supportive webbing for our predicates now resides largely hidden from scrutiny, beyond the ready reach of mutual discourse. In Kuhn's familiar phrase, the languages of two scientists loyal to distinct paradigms are then apt to prove *incommensurable*:

> These examples point to [a] most fundamental aspect of the incommensurability of competing paradigms. In a sense that I am unable to explicate further, the proponents of competing paradigms practice their trades in different worlds . . . That is why a law that cannot even be demonstrated to one group of scientists may seem intuitively obvious to another. Equally, it is why, before they can hope to communicate fully, one group or the other must experience the conversion that we have been calling a paradigm shift.[50]

This inability to "communicate" suggests that the act of convincing a fellow scientist must represent an exercise more of raw power than rational discussion, a suggestion that post-structuralists have pounced upon with loony enthusiasm (recall from 2,v that even Jeff Titon has become wrongly persuaded that an innocuous squabble about musical terminology represents a "political act"). Kuhn himself never wished his doctrines to be carried to such extremes, but he never successfully tempered the psychologized holism that brings him near such disasters either.

In fact, here is a typical specimen of holism gone wild (from Terry Eagleton's Literary Theory):

> There is no question of returning to the sorry state in which we viewed signs in terms of concepts, rather than talking about particular ways of handling signs . . . When I read a sentence, the meaning of it is somehow always suspended, something deferred or still to come: one signifier relays me to another, and that to another, earlier meanings are modified by later ones, and although that sentence may come to an end, the process of language itself does not. There is always more meaning where that came from.[51]

As is often the case in such contexts, reasonable worries about the difficulties in approaching historical texts get thoroughly jumbled up with a coarse philosophical approach to the notions of *concept* and *meaning* (vide Eagleton's opening sentence). But its "House that Jack Built" description of how words get their "meanings" nicely emphasizes the web of *horizontal ties* that hazy holism emphasizes: we can't adequately equilibrate the utterance of two speakers unless we look far into the nether reaches of what they believe, often in utterly unconscious ways. It isn't any wonder that the term "folk music" soon gets linked to "World War II" by such "six degrees of separation" standards (2,v).

But if we inspect conceptual disputes in real life, they rarely range to such cosmic dimensions, but generally focus upon fairly specific strands of practicality. This is certainly true of many of the scientific battles that Kuhn invariably describes in

---

[50] Thomas Kuhn, The Structure of Scientific Revolutions (Chicago: University of Chicago Press, 1996), 150.
[51] Terry Eagleton, Literary Theory (Minneapolis: University of Minnesota Press, 1983), 116.

paradigm-laced language (we shall revisit a celebrated case of scientific impasse betwixt the French chemist Pierre Duhem and his English rivals in 6,xiii and 10,viii). Typically, such disputes involve questions of *semantic depth*, rather than *holist horizontality*. As we observed in 4,viii, the term "force" occasioned much turmoil in late nineteenth century physics, not because the relevant parties were impeded by blinkering gestalts, but because it was then impossible to recognize the facade-like underpinnings upon which "force" gathers its semantic support. That is, both "force" and derivations from Newton's "F = ma" were central within many of the era's most sterling displays of descriptive achievement, but, en masse, these techniques were not fully harmonious with one another, creating the problems of 4,ii. Different scientists came to sharp dis-agreements about procedure, largely through favoring certain cases as more revealing of the true platform on which they believed "force" would be eventually found to rest.

. . . .

For example, party A expects that true forces will always be derivable from a potential, because conservation of energy can then be easily established, whereas party B expects that the reaction forces of rigid body thinking require a central place within mechanics' halls.

. . . .

In an argument about such matters, two opponents will critically reexamine the situations favored by their rivals: "You have interpreted the physical support for this technique in manner X, but, observe, its basic workings can be approached in my alternative manner Y." Unfortunately, in our nineteenth century physics context, no one then alive possessed the requisite physical or mathematical knowledge required to bring their disputes about "force" to reasonable resolution: beyond a point, every party to the dispute was obliged to rely upon seat-of-the-pants hunches that simply couldn't be further adjudicated at that point, although we can now diagnose the facades and semantic mimicry that introduced the confusion into their disagreements. In forming their hunches, our warring scientists are influenced by the cases they know best, which serve as the paradigms (in the old-fashioned sense of the word) upon which they draw. But this biasing phenomenon doesn't differ greatly from the fact that fans who root for different baseball teams generally have different opinions about who is likely to win the World Series. A debate about "force" can be easily hamstrung by the fact that neither party actually understands the *strategic policies* underlying its successful uses well enough to clinch their dispute. These problems generally represent localized semantic diffi-culties; the other physical doctrines they happen to entertain play comparatively little role in generating their impasse. This is why I remarked that the proper source of their disagreements lies in misunderstandings of localized semantic support, not in "force"'s horizontal ties to other words or doctrines.

Again I believe that ill-founded tropisms towards holism generally trace to a desire to imitate classical pattern within an anti-classical frame. From a classical point of view, our disputants must each grasp some concept under the heading of "force" and, if they prove stalemated with respect to the same factual situation, they are probably thinking of slightly different traits, a matter that they should be able to remedy through careful introspective analysis. Mistrusting the "true thought" aspects of this classical story, holists maintain that their semantic differences must trace to distinct embeddings within widely diffused, and essentially uncomparable, webs of supportive belief. But this is not the right way to view matters, in my opinion. Our disagreeing scientists can probably come to reasonable agreement with respect to the somewhat discordant bundle of strands of practical advantage that buzz around the problematic "force," but neither disputant has yet found a satisfactory underlying pattern that can bring this jumble into fully controlled harmony. They have their hunches and opinions on this topic, but much further development will be required before they can be properly considered redeemed. It is not that our antagonists fail to understand one another; they simply disagree on the right way out of their semantic quandary.

Kuhn's discussion does raise the important question, "How should we discuss conceptual disagreement rationally with a party whose unconscious mental processes plainly function according to pathways plainly different than our own?" In Chapter 8, we will find that reaching reasonable accord rarely requires that we must pass through these hidden and inaccessible causeways.

The rise of hazy holism in the aftermath of axiomatics' fall from philosophical grace reminds me of another youthful experience. There was a brief period when it was assumed in my boyhood circles that an optimal birthday celebration should be a triple feature horror movie weekend at the Bagdad Theater. To an impressionable youth of a logical bent, these occasions invariably constituted trauma, for the photoplays of such productions were rarely tightly scripted. I recall one film in which it was firmly established that, were fresh air ever administered to a fungus that skulked within a South American cavern, the nasty stuff would quickly grow and engulf the world. Some scientist, apparently believing that no hypothesis should evade direct empirical confirmation, did precisely that and, true to form, the gunk (which, if memory serves, looked remarkably like laundry suds) commenced its career of engulfing. The movie's hero and heroine were trapped in this same cave and, after many narrow escapes, escaped to a romantic beach and kissed passionately. "The End," the credits rolled. I sat there in the dark, stunned. "It's all well and good that they evaded that mold temporarily," I worried, "but what about the rest of us?" Although in some sense I realized that it was "only a movie," I nonetheless scanned the newspaper for weeks thereafter, on the lookout for reports of an unpleasant life form working its way through the Isthmus of Panama.

It strikes me that our current thinking about concepts in science much resembles the character of that film. The late nineteenth century faced real life difficulties with respect to method that left them perplexed as to how the correct directivities of specific notions such as *force* should be ascertained and controlled. For a time, appeal to axiomatics and implicit definability promised a brisk and simple resolution of these problems, but this proposal eventually proved unsatisfactory. Let's adopt hazy holism instead. The End. Wait a minute—you've still left that horrible fungus growing. What can we now say about the *original concerns* that prompted the worry about "force" 's odd behavior in the first place? What steps can we realistically take to stave off the unhappy troubles to which unchecked behaviors of this kind are otherwise inclined? It's hard to extract any advice of profit from the hazy holists (unless the reader regards the "advice" that ruins folklore in Chapter 2 as profitable).

The answer I will provide, while not so upbeat as that advocated by either classicists or formalists, suggests that we should learn to scrutinize the fine-grained structure of our assertions closely, watching for the subtle crossover boundaries and other structures characteristic of a facade. As I explained in Chapter 4, this task requires that we approach the issues of linguistic structuring in terms of variable reduction, asymptotic approximation and the rest of the rich array of tools that have been developed within applied mathematics, and not continue to cobble along appealing only to logical flavors of organization (or, in Quine and Kuhn's cases, some hazier form of the same). Pace Quine's assumptions otherwise, the natural progression of our evolving descriptive endeavors often leads to a division of labor within localized platelets, rather than meekly submitting to sweeping organizational imperatives of a global character. In the previous chapter, I argued for the viability of such fractured schemes through basic considerations of effective linguistic engineering; in the pages now before us, I will suggest that such patchwork structures represent patterns commonly encountered within everyday descriptive use.

6

THE VIRTUES OF CRACKED REASONING

I am not yet so lost in lexicography as to forget that words are the daughters of earth, and that things are the sons of heaven.

Samuel Johnson[1]

(i)

Interfacial accommodation. The biologist Marston Bates would begin his lectures on biomechanics with the remark, "I think I'll start with a rabbit beneath a raspberry bush and gradually get into the physics of the thing."[2] In this chapter, we shall begin a new stage of "getting into the engineering" of linguistic affairs, for we will develop a richer appreciation of the variant strategies that a system of linguistic description can adopt in representing the world about us in a fruitful manner. At several earlier points (1, ix; 4, vi), I have discussed how the employment of a group of predicates sometimes divides into localized *patches* connected by bridges of natural connection. I call such epitaxial patterns *facades* and have emphasized the manner in which their component patches are organized into a polycrystalline or quilt-like manner. In Chapter 4, I supplied an argument in a linguistic engineering vein that explains why such organizational structures often emerge as the natural prerequisites for describing complex systems with a manageable number of descriptive terms, following some successful policy of variable reduction. Without presupposing that discussion, I will now approach our facades from an evolutionary perspective that emphasizes the reasons why the characteristic etching of a facade often emerges within a usage after it has been subjected to an increasing degree of polished refinement. This point of view is entirely complementary to the variable reduction emphasis of Chapter 4, but it involves fewer technicalities.

---

[1] Samuel Johnson, "Preface," A Dictionary of the English Language in E. L. McAdam and George Milne, eds., A Johnson Reader (New York: Random House, 1964), 122. According to the editors, this is paraphrased from Samuel Madden.　　　[2] Stephen Vogel, Life in Moving Fluids (Princeton: Princeton University Press, 1994), p. iv.

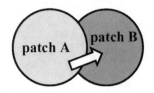

I have complained (5, ii) that the classical picture of concepts does not allot any substantive shaping role to what I have called *strands of practical advantage*: linguistic routines or recipes that facilitate the completion of some substantive goal-oriented task, to which consequences in the form of palpable rewards and punishments attach. In the next two chapters, we shall see how such pragmatic considerations can significantly color a usage, even if the affected predicates are largely employed in circumstances far removed from those practicalities (I emphasized earlier that these strands of practicality distribute themselves quite sparsely throughout a general usage). We shall also observe how these task-oriented aspects of a predicate's personality frequently force a poly-crystalline structure upon the employment as a whole. I call the bundle of factors under consideration *interfacial* because they reflect the manner in which the representational tools we have available to us (the symbols we can recognize, remember and work computations upon; the tests and observations we can easily run or make) suit the physical circumstances in which we attempt to utilize language to our benefit. With sufficient cleverness in our strategic arrangements, we can adapt linguistic tools that, in themselves, possess less than optimal qualities to our purposes perfectly. And this is what we want to study: how symbolic capacity and physical environment come into successful practical accommodation.

The next two chapters will investigate how such interfacial characteristics tacitly supply their affected predicates with surprisingly pungent flavors: the conceptual impression left by a phrase such as "is red" or "gear wheel" partially traces to strategic considerations of which we have little awareness. As such, these qualities contribute to the overall ambience of predicate personality that classical thinkers consider to be the term's *intensional content* (I prefer the homespun "personality" to the fancy "intensional content" for the same reasons that we might resort to "bugs" if we doubt that the biological taxa of Insecta and so forth are well conceived). Such factors supply excellent representatives of a wide class of affective considerations that get omitted from the story of language as it is conventionally told. "True, such factors do influence usage at the margins," it might be conceded, "but they don't play any significant role in explaining how language obtains its meaning." To convincingly turn aside such traditional dis-missals, we confront the same difficulty that Quine faced in the last chapter: the strengths of classical grasp and gluing must be lessened enough to allow other deter-minants on usage to affect the correctness of what we say. But our approach will prove more accommodating in manner than Quine's: "You classicists properly emphasize *some* of the directive elements to which we must attend in adjudicating the correctness of our linguistic responses, but you ignore others that do not always lay so patently in view."

I make no pretense—either here or anywhere else in the book—that I have somehow divined a complete catalog of every directive factor that buffets our words about (that I have traced every current that pulls little Scuffy down the river). But I will set forward some simple models of the ways in which interfacial considerations can significantly color a usage, in a manner that shows us how the pre-pragmatist doubts outlined in the previous chapter can be plausibly prosecuted. After all, not all pre-pragmatists need to grow up to be pragmatists. Or Quineans, either.

Eventually I will argue that the true personality of predicates such as "is red" should not be regarded as the simple, indescribable invariant of classical thinking, but stems, in substantial part, from a complex mixture of strategic considerations. If I can make out my case plausibly, then the root source of Chapter 2's worries with respect to music and color revolve around the fact that we tend, in our *ur*-philosophical thinking, to compress long term interfacial aspects of "is red"'s personality into features that we allegedly appreciate from the very moment we first grasp the predicate's meaning. But to assume this is to entirely misunderstand how the directivities that guide "is red"'s employment actually work.

(ii)

Representational personality. A convenient place to begin our discussion is to quickly canvas the problems as to how geographical facts with respect to a spherical (actually, slightly ellipsoidal) earth might be usefully captured within planar maps, for such practices embody, in microcosm, many of the concerns that affect practical usage more generally. As I've already stated, most of the themes emphasized in this book have been borrowed (or outright stolen) from considerations familiar in applied mathematics. Within this realm, the historical road to increasing sophistication with respect to wise descriptive policy initiated in the study of maps. So by centering our own investigations in this same arena, we can approximately recapitulate the historical path that runs: Lambert → Gauss → Riemann → Weyl → Whitehead and Veblen, with many other important contributors along the way.

As is well known, it is impossible to map terrestrial topography onto a sheet of paper without introducing considerable distortion in the result. At best, we can select a few features that we would like to register in our maps accurately and conveniently, while abandoning other critical qualities to their representational fates. For example, the descriptive quantities maximized in the familiar Mercator projections (essentially, the maps of the whole earth most commonly seen) are the "rhumb lines". That is, the compass and sextant routes that a sailing vessel might reasonably follow appear on such maps as straight lines, making the job of the navigator much easier. This specialized objective is achieved at the price of great distortions in areal representation, especially within the higher latitudes (as manifested in Greenland's extremely deceptive size upon a Mercator map). Many alternative schemes have been invented that capture areas more accurately—such as the *Hammer projection*

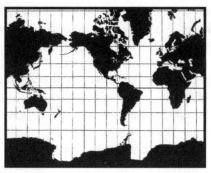

Mercator projection

Hammer projection

Goode projection

illustrated—although at the price of considerable distortions in shape (worse than on the Mercator, although its depictions of shape are not exactly terrific either). In other words, two countries that occupy equal planar area on the map on the Hammer map will possess the same square footage in real life. The equal-areal representation of shape can be improved through permitting large cutout regions within the map, as in the *Goode projection* shown, although most viewers find these interrupted lobes rather strange.

Each projection type embodies its own distinctive *personality*, which is never in complete harmony with the physical system it attempts to describe: the spherical earth. As we attempt to maximize selected representational virtues (accurate areal representation), we mislead in others (shape). And there are clever mappings—a famous early example is due to George Airy—that strike suitable compromises in how ably a range of desirable features are registered (as Airy says, they "minimize the total evil" in the

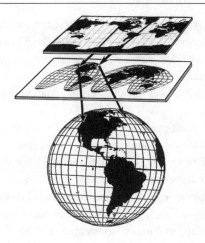

map[3]). In fact, the maps we most commonly see in everyday life represent "tempered Mercators" with their grossest distortions mollified in sundry ways (some subtle; some not—such as the common practice of omitting everything above and below the two Arctic circles).

How do we correct for these representational pitfalls in our maps? The most effective scheme is to supply a rich *atlas* of maps that cover the earth several times over, each of which is dedicated to answering questions best suited to its own personality. It is convenient to picture these complementary maps as hovering above one another, connected by fibers that link together the representations of common locales.

If asked "how does the size of Greenland compare with the United States?," we follow these fibers to *lift* our attention from the Mercator projection into an areally correct map where we can adjudicate the desired comparison by sight or measurement. But if asked, "How should I sail from Annassalik to Portland, Oregon?", we should pull back our thinking to the Mercator chart and plot our course there with a straight edge. In other words, a competent employer of an atlas will address the questions she seeks by thumbing to the right pages of the atlas, often in a rather complex fashion: a seaman plots sailing routes by combining the information supplied in several maps, often without knowing the underlying theory that explains why this bustle of procedures supplies suitable sailing instructions. In this way, a properly constructed atlas demonstrates that representational tools of a limited capacity can be cobbled together to capture terrestrial data entirely successfully, as long as we shuttle between its member representations according to a suitable *strategy of usage*.

There is a second reason why we must employ a compendium of maps: no flat map of any personality type can cover the earth without some serious irregularity or singularity arising in its alignments, such as having the North Pole stretched into a line

[3] John P. Snyder, Flattening the Earth (Chicago: University of Chicago Press, 1993), 127. Frederick Pearson, Map Projections: Theory and Applications (Boca Raton, fla.: CRC Press, 1990). J. H. Lambert's "Anmerkungen und Zusätze zur Entwerfung der Land-und-Himmelscharten" of 1772 represents the first mathematical treatment of projection.

across the top of the chart (the Mercator isn't able to capture Santa's home at all, because the increasing spacing it assigns to the upper latitudes push both poles off to infinity). To cover the whole earth without such exceptional points, at least two overlapping maps must be used and, most commonly, we see three: a modified Mercator that covers a large equatorial strip with two supplementary patches introduced for the polar regions.

Here the topological disparity between the round earth and our flat maps creates the need for a two-or three-sheeted covering, but, in fact, it is generally wiser to employ more charts of a yet smaller scale in our atlas, not only for the detail they contain but because at such scales we can better balance their representational virtues more sensibly Airy-style through attempting less ambitious coverage. Of course, for many purposes we require larger scales—if we must go around the earth in eighty days, say—, but these wider reach maps must be approached with greater caution because virtually any policy of projection goes awry in its global aspects.

Most good atlases also contain a *preface* that delineates the projections that underlie the component charts, as well as explaining the proper strategies of map employment: which map should be employed for what purpose; how longer range questions can be resolved by piecing together local map information and so forth. An able seaman can often plot his navigational routes quite capably from an atlas despite having never read the theoretical preface at its head; he understands the "practical go" of the book without the benefit of the introductory disquisition (which, after all, contains no specific geographical data). We will later find that the linguistic analogs of these prefaces play an interesting role in the story of conceptual evaluation.

In any case, the overlapping and fibered set of maps included in an atlas represent the inspirational prototype for my *facades*, for an atlas represents an evocative way to visualize the ways in which various blocks of a usage need to fit together in order to cover a subject matter effectively. It also provides a convenient picture of the strategic concerns that the concrete directivities of predicate usage need to address.

Instead of shifting to a wholly different map to resolve our questions about distance or area, it is also possible to correct for the distortions in a Mercator chart by simply supplementing the map with an adjoined recipe for calculating true lengths and areas from the quantities we can directly measure within our map. Soon after Mercator's map appeared, the English mathematician Edward Wright supplied a table of "meridional parts" designed to supply the correction factors needed to convert the distances measured on the map to proper terrestrial values.[4] This bundle of corrective factors represents the predecessor of the metric tensor later developed by Gauss and Riemann.

I find it convenient to picture the activity of these satellite recipes and reasoning algorithms as little patches that hover over our Mercator map, although they do not duplicate basic geographical data so massively as happens if we utilize a complete alternative map such as the Goode to resolve our areal questions. Perhaps an adjoined correctional routine like a table of meridional parts should be properly viewed as a

---

[4] Lloyd A. Brown, The Story of Maps (New York: Dover, 1979), 134–9; also Snyder, Flattening, 43–9.

band-aid laid over a Mercator map, rather than a true covering patch. However, I shall usually ignore these distinctions in topical administration and call them both patches.

In fact, a little reflection shows that Mercator and Goode projections that cover the same terrestrial sector are (potentially) informationally equivalent in the sense that, with a proper supplementation of band-aids, any question that can be answered in one can be resolved in the other (the qualifier "potentially" must be inserted because real maps usually differ data-wise because one will have room for symbols that can't be squeezed into its more cramped companion). Our need to shift amongst maps within an atlas, accordingly, represents a function of both *geometrical considerations* (the topological disparity between earth and plane) and our limited *computational abilities* to process the data contained within a given map effectively.

Our abilities in this regard are sometimes rather surprising in their contours, for purely psychological factors can make the construction of a good map a quite delicate affair. For example, consider the accompanying sketch of the globe and ask yourself, "How does the size of Madagascar compare with that of Spain?" The answer we provide will not directly reflect the true measured areas found on this map, but will reflect the extensive unconscious corrections we automatically make in viewing this drawing in a three-dimensional, rather than a flat, manner. Indeed, our inability to turn off this 3-D reading is so strong that most of us experience a good deal of difficulty in answering the alternative question correctly: "How does the area of Madagascar in the figure compare two-dimensionally with that of Spain?"[5] Such psychological factors often cause other maps, that are excellently designed in theory, to perform poorly in practice because we ruin their representational virtues by automatically correcting for distortions as if we had been looking at a less judicious projection such as a Mercator chart.

In any event, there are several equally acceptable ways in which we can qualify the interfacial personality that a specific map type displays: (1) in terms of the practical questions that can be easily addressed using the map, either directly or with the assistance of easy-to-implement recipes; (2) in terms of the projection scheme followed: to what qualities in the map does *length along the earth's surface* correspond? A quick look at any theoretical work on map projection will show that the recipe employed in familiar maps often follows a rather complicated encoding strategy. Our two

---

[5] In Mark Monmonier, How to Lie with Maps (Chicago: University of Chicago Press, 1996), 18, there is a striking example of how a graph can employ our automatic three-dimensional reading of the globe to surprising effect.

perspectives on map personality are complementary to one another in that the recipes for map projection are usually developed by investigating what conditions need to hold if, e.g., equal areas in the map are to correspond tidily to equal terrestrial regions. It is usually easier (because less abstract) to think of *map personality* in terms of the practical questions easily addressed by a humanly feasible routine, rather than in terms of the supporting informational coding. However, we can't have one factor without the other: unless the right coding lies in place, the easy-to-follow routines won't supply useful answers.

We should immediately observe that, although a map's personality is best conceptualized in terms of the questions it aptly addresses, it may easily happen that the map is rarely utilized for those dedicated purposes in normal practice. This observation is nicely illustrated by the Mercator projection itself, whose true personality is framed by quite arcane "how to plot a navigational route at sea if you only have a sextant and compass" considerations (it is a purely historical accident that a map designed for very specialized purposes became our canonical expression of "what the earth looks like in a map"—we will return to this intriguing topic in 7,viii). As such, such projects rarely loom large in the everyday lives of most of us (including even salty skippers who now find their aquatic ways about with the assistance of global positioning satellites). An odd background keeps modified Mercators as our central map of choice, despite its manifest non-optimality for most practical purposes. As such, the factors that keep it alive nonetheless will tell us much about how words actually survive on the bumpy currents of linguistic evolution.

Nonetheless, we should still conceptualize the Mercator's personality in task-oriented terms, because that account provides us with the best sense of the circumstances in which intemperate use of a specific map is likely to create problems. In the Mercator's case, its prominence often leads us to answer questions like "How much bigger is South America than Greenland?" quite wrongly (it is about eight times as big, but they look nearly equal on the chart). If a society retains the Mercator in active use, we should ask, "What *remedies* will these people employ to evade the poor decisions that indiscriminate employment of this map will otherwise induce?" Later we shall examine the somewhat sneaky correctives that professional cartographers have introduced to save us from gross, Mercator-guided error.

Without pursuing such complications further at the moment, we have learned enough about the quirky personalities of individual maps to appreciate why basic geographical fact about the earth is best organized as an atlas of many linked maps or, to use my alternative designation, a *facade*. Each individual map supplies its own compendium of easy-to-apply *recipes and reasoning routines*: "to compute an 'area' for Greenland, divide its representation into 1/16 inch squares, count them and divide by 256." Unfortunately, on a Mercator map, the resulting "area," tho' easy to compute, doesn't represent a particularly useful value. However, by playing the virtues of one map against another in an atlas, we can achieve an entirely admirable and undistorted impression of what the earth is really like. In my earlier phrase, we employ slightly unsuitable tools to excellent descriptive purpose.

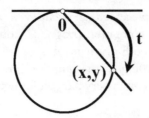

Mathematicians have gradually learned that allied notions of personality are applicable to more general forms of data registration, including systems that are overtly linguistic in character. Indeed, it is easy to shift from maps to language even in the present circumstances, simply by considering the subject of computer cartography, in which geographical facts are stored in a data base in a manner so that pertinent queries can be addressed. Here we store geographical information in the form "<F, C>" where "F" is some feature of interest (*occupied by a city*, say) and C is some variety of coordinate location. But it is usually necessary to employ several different coordinate schemes simultaneously over a given geographical region, because different forms of representational scheme offer better or worse opportunities for addressing basic tasks we might set the system (for computing areas, raster registrations are employed, which mark local squares as occupied or not, but, for route planning, vector registrations are used similar to the hub-and-spokes representations discussed below).[6] In addressing more complex questions, a computer program will shuttle rapidly between different representational registers.

Here is a simpler standard illustration of the task oriented personalities intrinsic to particular representational schemes of a linguistic type. Consider the varying merits of *regular* and *parameterized descriptions* of a figure's shape. Here a "regular description" simply assigns numerical values (x,y) to points in the manner of a Cartesian coordinate system, so that a figure such as a unit circle becomes algebraically registered by its familiar Cartesian equation $x^2 + y^2 = 1$. However, we can also put *parameterized coordinates* on the same figure. Choose a point O on the circle itself and let the parameter t mark an angular distance turned around O. From this point of view, t will generate our circle through the equation pair: $x = 1 - t^2/(1 + t^2)$; $y = 2t^2/(1 + t^2)$ as t sweeps in a circle around O. Plainly these two descriptive modes cover the same circle in different descriptive formats. Despite their informational equipollence, they present quite distinct personalities with respect to their capacities for settling vital practical questions quickly.[7] In particular, the nonparameterized equation format allows us to *test* very quickly whether a given point lies on our curve or not, which cannot be easily resolved by looking at the parameterized form alone. However, the second format allows us to draw systematically the curve's complete shape, whereas it is often hard to determine whether you have finished the graph of a nonparameterized equation (especially when its

[6] Christopher B. Jones, <u>Geographical Information Systems and Computer Cartography</u> (Harlow: Addison, Wesley and Longman, 1997).

[7] My colleague Ken Manders, in unpublished work, uses the term "representational granularity" to roughly this effect.

equations admit disconnected pieces, as can occur even with an equation of the second degree). For these reasons, computer programs commonly store equations for important configurations in both formats, despite the informational redundancy, shuttling between them according to the question presently at issue (unfortunately, finding a parameterized mate is often very difficult for figures of a higher degree, even if they indeed exist).

I believe that interfacial personality in this sense plays an important, but generally unacknowledged, role in framing the intensional characteristics of many parts of language. To this end, it is helpful to rehearse a familiar situation (to academic philosophers, at least) from this point of view, to gain a rough impression of how classical thinking typically ignores such factors or assimilates them too swiftly to ill-suited categories. In the previous chapter, we observed that Quine views the classical conception of "conceptual content" as a *fictitious externalization* of factors that properly reflect the manner in which the predicates are embedded within the web of belief that syntactically sustains them. I accept no such web nor the holism that goes with it, but I agree that the classical picture's "content" often mislocates predicate directivities that properly trace to interfacial concerns.

(iii)

Presented contents. Accordingly, let us address a much discussed linguistic circumstance highlighted by both Frege and Russell, from distinct but closely related points of view. Here the focus is usually on the behavior of *proper names* like "Gottlob" or "Ernest", rather than predicative expressions, although both authors expect their conclusions to carry over to the latter as well.

Consider this characteristically Fregean scenario (supplemented with a dash of Nathaniel Hawthorne[8]). In some New Hampshire locale "immense rocks have been thrown together in such a position as, when viewed at a proper distance, precisely to resemble the features of the human countenance." Young Ernest, growing up in the spacious valley that lies in distant view of this magnificent rock physiognomy, has, from earliest memory, called the land form in question "the Great Stone Face," which he soon contracts to "GSF." In his later rambles over rill and ridge, Ernest stumbles across "a heap of ponderous and gigantic rocks, piled in chaotic ruin upon another," which he appropriately dubs "The Big Pile of Rocks" ("BPR" hereafter). Little does Ernest suspect that GSF and BPR are one and the same. Being a lad of impeccable rectitude, Ernest records in his diary the dimensions, mineral composition, accessible trails unto, etc. of his "two mountains" in double entry for years before it eventually dawns on him that "GSF = BPR," at which point his needlessly multiple linguistic tallies can be quickly collapsed into a more compact whole. In other words, information about the same

[8] Nathaniel Hawthorne, "The Great Stone Face" in Twice-told Tales (Norwalk,Conn.: Heritage, 1966), 22. Shortly after I wrote this, the geographical original sadly collapsed, leaving me feeling guilty that I had been flippant in its description.

Modes of presentation

mountain will be registered in linguistically bifurcated form until Ernest discovers the vital "GSF = BPR" bridge that allows for a swift and substantial pruning of his alpine bookkeeping.

How should Ernest's prolix linguistic condition be rationally explained—for Ernest is nothing if not tediously rational—, given that his sentential groups concern the same subject matter? Frege makes the natural suggestion that the mountain's two available *avenues of approach* or *modes of presentation* supply the names "GSF" and "BPR" with distinct associated *senses*, viz. the traits *the geological feature that looks like a gigantic man* and *the geological feature that looks like a big pile of rocks*. It is natural to picture these senses as *arrows* that point towards the mountain in different ways. According to Frege, the fact that "GSF" and "BPR" rest upon different arrows of semantic connection makes comprehensible Ernest's disinclination to transfer information registered in GSF format into that captured by his BPR idiom. Frege further claims that their common *semantic reference* (or *denotation*) is the mountain itself, unencumbered with any consideration of how it presents itself, whereas the terms' differing senses (which he regards as a second semantic characteristic) capture the divaricate routes whereby these names reach their shared referent. When we speak of the "meaning" of a proper name in everyday talk, we may, depending upon context, fasten either upon the denotation or the sense as our primary focus of interest. In today's jargon, Frege proposes a "two factor" account of the semantic support of the name "GSF": its supportive sense (*the geological feature that looks like a gigantic man*) and the referent to which this sense points (the mountain itself).

Russell would describe these circumstances in slightly different terms, appealing to his theory of descriptions. He advises us to attend to the sentential context in which "GSF" appears, say, "the GSF is big," and reparse the whole unit utilizing a predicate that captures Frege's associated sense, arriving at "There is something which is uniquely a geological feature that looks like a gigantic man and which is also big." In so doing, Russell only associates the conceptual contents of the intervening predicate with "GSF," and does not need to bring the mountain itself into his semantic story at all (except as the object that happens to make the assertion true).

For our purposes, such differences between Frege and Russell are unimportant, for both maintain that when Ernest grasps the name "GSF," he thereby grasps in a direct way of which he is fully aware, the conceptual content conveyed by *being a geological feature that looks like a gigantic man.* With "BPR," in contrast, the associated content is instead *being a geological feature that looks like a big pile of rocks.* For both authors, these two modes of presentation represent the contents that come to mind when Ernest thinks in either "GSF" or "BPR" terms.

Either way, the presence of these contents helps explain or rationalize otherwise puzzling features of Ernest's linguistic behavior. Since he doesn't know the truth of an identity such as "GSF = BPR," it is not surprising that his diary entries will contain large swatches of needlessly duplicated "GSF" and "BPR" centered portions, despite the fact that both fragments correspond to exactly the same swatches of reality. As such, our explanation of Ernest's bifurcated linguistic behavior initially seems quite satisfying.[9]

Let me supply an important parenthetical digression at this point. Frege's account utilizes phraseology that is potentially ambiguous in its connotations. The two characteristic phrases that are commonly employed interchangeably in standard discussions—viz., "avenues of approach" and "modes of presentation"— can suggest two distinct ways of understanding what a "sense" actually represents. Must Ernest himself be *aware* of the discrepancy in sense between "GSF" and "BPR"? The phrase "mode of presentation" suggests "yes": a sense captures the manner in which the mountain *presents itself* to Ernest. However, "avenue of approach" may suggest otherwise, because Ernest might approach two objects along different routes without his being aware of any distinction. Once upon a time epistemologists were fond of devising tales where wicked people were forever fooling gullible folks like Ernest with facades that were carefully crafted to resemble true barns. Such pasteboard cutouts affect Ernest along a different avenue of approach than a true barn but deluded Ernest has no inkling of the routing whereby he is presently affected. This same ambiguity even appears within the little diagram I've sketched of our Ernest scenario: an avenue of approach is naturally symbolized by an arrow, whereas its presentational aspects correspond to the view supplied in the magnifying glass.

Classical tradition firmly insists that conceptual materials associated with two names (either via Fregean sense or Russell's theory of descriptions) should be consciously recognized as distinct by the agent in question: Ernest must realize that his two mountain presentations differ in their conceptual contents. Indeed, our defense of Ernest's rationality depends upon the fact that he is aware of both, for otherwise the fact that he loads his diary with superfluous double-entry data would be inexplicable (it would be surprising if an agent victimized by shifting barn facades would engage in parallel diary prolixity even though, unknown to himself, he actually views a multitude of objects when he believes that he has only witnessed a single barn).

Nonetheless, certain contemporary writers are inclined to understand "sense" in an avenue of approach vein, whereby the notion seeks to capture the psychological factors that explain why Ernest utilizes his two terms differently without implying that he thereby possesses any representation of their different origins (Jerry Fodor represents an example of this inclination[10]). Frequently, this school equates the arrow of sense

---

[9] Gottlob Frege, "On Sense and Reference" in <u>Collected Papers</u> (Oxford: Basil Blackwell, 1984).

[10] Jerry Fodor, <u>Concepts: Where Cognitive Science Went Wrong</u> (Oxford: Oxford University Press, 1998).

with some causal pathway that connects Ernest with his mountain in a specific way, of whose ceremonies Ernest may know very little. One finds these two understandings of what a "sense" might represent frequently dubbed as *internalist* (= presentational) and *externalist* (= viewed from an outside perspective) approaches within the recent literature.

Thinkers of a classical disposition are frequently bewildered by such externalism, maintaining that the whole point of evoking a sense is to capture the distinctive *point of view* from which Ernest regards his mountain when he speaks of "GSF." Often they throw up their hands in rhetorical despair: "If the notions of 'sense' and 'concept' are not intended to capture an individual's cognitive point of view with respect to a name or predicate, then what on earth could these notions be good for?" In this vein, the philosopher Kent Bach writes:

> As for me, I have no idea what it is to think with a concept that one incompletely understands. That is because I have no idea what it is to understand a concept over and above possessing it.[11]

At present, our interests are largely focused upon classical thought and so "sense" will always be interpreted in a firmly *presentational* mode.

(iv)

Intimations of intensionality. In the foregoing section, mode of presentation contents attached to *being a geological feature that looks like a gigantic man* and *being a geological feature that looks like a big pile of rocks* were evoked to explain classically why Ernest handles their corresponding names differently, despite the fact that only a single mountain is concerned. As such, these contents set forth *directive elements* of which Ernest is fully aware: "Why did you call that 'GSF'?"—"Well, it looks like a gigantic man, doesn't it?" However, there are other vital features of predicate personality that enter Ernest's story of which he is, at best, dimly aware, although they also direct his classificatory activities in distinctive ways. They, in fact, trace to what I have dubbed *interfacial concerns*: the arrangements required to bring representational capacity into fruitful alignment with physical fact. The strategies employed in utilizing an atlas of maps provide the basic exemplar of the concerns I have in mind and in this section I shall indicate how allied considerations play a hidden role in influencing how Ernest employs his "GSF" and "BPR." Such directivities generally display themselves only on a multi-sentential—but *not* holistic—scale, in the manner in which Ernest works with blocks of sentences containing our two names. But although his linguistic behavior is guided by such considerations, they do not represent ingredients of which he is accurately aware at all.

Turning to specifics, Ernest most likely stores his geographical data within a different kind of "map" than we have considered, which I will dub a *navigational list* (the

---

[11] Kent Bach, Thought and Reference (Oxford: Oxford University Press, 1994), 267.

store + store = store
home   barroom   home   barroom

psychologist Reginald Golledge[12] calls it *route-based knowledge*). Consider agent-centered instructions such as:

(a) *To get to the GSF, first go north from the village along Main Street.*
(b) *Look for the second trail on the right after the old manse and follow it;*
(c) *Walk about ten miles and take the middle fork of the branch under a big oak tree.*

This form of geographical registration possesses its own special advantages for achieving certain sorts of task. To plot a novel route from location A to B in list mode, we merely need to apply what is often called *toe-to-head* computation: search for some C where we know how to get to C from A and also how to get to B from C and concatenate the two subroutines (or discover some longer sequence of interpolations). True: we may not generate the most efficient routes in this manner, but we reliably get there just the same. In contrast, as anyone who lives in a city as convoluted as my own Pittsburgh knows, consulting a conventional city map can suggest as-the-crow-flies routes that appear admirably efficient on paper, but unregistered obstructions (i.e., one way streets) ruin their actual assay. In the same manner as we characterized the Mercator projection, a wide range of practical advantages and disadvantages distinguish navigational lists from conventional map registrations. In fact, computer geographical information systems generally address complicated questions through shuttling betwixt data registrations that essentially encode these two styles of map. Such rosters of computational capacity and deficiency supply a navigational list representation with an intrinsic personality as piquant as that of the Mercator projection.

Books on the psychology of wayfinding often utilize *hub-and-spoke* diagrams to symbolize such navigational list structures, for such images supply a nice picture of their representational capabilities (to be sure, we scarcely store little tree-like sketches in our head, any more than conventional maps literally lodge in our craniums). But it is easy to extend a hub-and-spoke chart by adding a fresh map of the same type to any of its nodes and such ready prolongation supplies a nice representation of the great computational advantages for easy updating that navigational list structures provide.

Depending upon education, circumstance and inclination, most of us utilize several varieties of representational map tied together in loosely coupled form. Thus we often store coarse, large scale geographical data within some semblance of conventional map format while reserving navigational list registrations for closer quarters such as a familiar neighborhood. For example, without a goodly expenditure of thought, I could not sketch any but the rudest map of the local hamlet in which I live, although relying

¹² Reginald G. Golledge, <u>Wayfinding Behavior</u> (Baltimore: Johns Hopkins Press, 1999), 9.

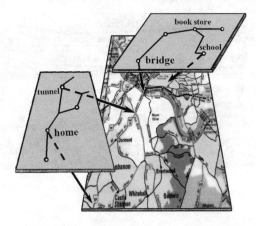

fairly exclusively upon navigational list registrations, I get around it pretty well. I do, however, retain in my head a coarsely grained conventional representation of how the sundry neighborhoods around the metropolitan Pittsburgh area distribute themselves on a conventional map and *that* I can sketch rather easily. I plan my longer journeys by first considering the large scale topographic map and then relying, where possible, upon hub-and-spoke representations for the finer details of local driving, just as a traditional mariner switches from astronomical and dead reckoning guidance while far at sea to piloting techniques when nearer to shore. Once again, a good way to picture such patterns of data storage is to install a collection of hub-and-spoke maps over a conventional map by linking fibers. To plot an expedition to a distant pizza parlor, I isolate a basic trajectory across the conventional map and then lift my thinking into the navigational list patches to obtain local driving instructions.

In Ernest's special circumstances (his provincial upbringing; the wooded setting), it is virtually certain that his local geographical knowledge will be registered in navigational list terms only—it may have never occurred to Ernest to attempt a conventional mapping of his woodland rambles and it might be difficult to construct one in any case. As he presses ever further into the fecund countryside, he readily adds on the data gleaned from his explorations as simply extension branches to established nodes (as noted, a great advantage of hub-and-spoke registrations is that they are easily prolonged, while updating and correcting a conventional map is often difficult). However, this same convenience supplies poor Ernest with no ready test for sameness of locales that lie along different branches except "Gee, this place looks kind of familiar"—a criterion that may avail little in an arboreal setting where the various pathways that converge upon a mountain share few recognizable landmarks ("Woods is woods," Ernest has sometimes been heard to complain). He might even punctiliously register angles and travel distances ("turn right 33° at the old manse and walk 5.3 miles down a straight section of trail") in his list-based diary in sufficient detail that a surveyor could compile a conventional map from its entries. In fact, theoretically, Ernest's diary and the surveyor's map might contain exactly the same amounts of concrete geographical information

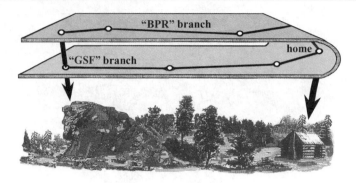

(in the sense of winnowing the set of open topographic possibilities to a smaller subset). However—and this is simply a computational failing that most of us share—, we can generally remember individual turning angles at hubs ably, but we are quite lousy at adding them up as we go (just as we cannot easily compute areas accurately from a Mercator map, although, theoretically, the requisite data is lodged there). We accordingly fail to retain a reliable impression of the total rotation we have undergone in the course of a journey of appreciable length (unless we can utilize sun-based clues unavailable to Ernest in his sylvan wanderings). Indeed, it is this same incapacity to keep track of turning angle that defeated Pooh and Piglet's pursuit of the woozle. Few of us can accomplish a surveyor's calculations in our heads and so we can easily appreciate how Ernest might volubly fill out a diary whose informational content is nearly equipollent to that provided upon a corresponding topographic map, without it occurring to him that GSF and BPR might be one and the same. Once we attempt to translate the diary data to a topographic chart, the hypothesis that GSF = BPR is likely to stand forth in glaring immediacy, for conventional map registrations are just as strong in forcing hypotheses of identity upon us as the navigational list techniques are feeble. In sum, a wanderer who utilizes only navigational list registrations is far more likely to fall into GSF/BPR mistakes[13] than the explorer who utilizes conventional map methods. This greater susceptibility does not trace to anything particularly idiosyncratic about Ernest except his environmental setting and the array of computational tools to which most human beings are limited within similar circumstances.

The mathematicians have a nice way of representing a situation like this (whose ramifications we shall explore in increasing complexity over the next two chapters). Consider Ernest at home prior to any discovery of the problematic mountain. As he ventures from his home base along path A, he builds up a patch of navigational list directives that eventually embraces the GSF; sallying forth along B, he constructs a patch covering the BPR. Since he lacks forceful criteria for identifying nodes reached along different branches, his descriptive language is inclined to develop into a two-sheeted covering of the physical topography. Accordingly, part of the characteristic personality

---

[13] Joseph Camp, Confusion (Cambridge,Mass.: Harvard University Press, 2003).

intrinsic to navigational list registrations lies in their greater tendency to develop into *multi-sheeted coverings under data prolongation* than do conventional map structures. As such, this propensity is readily detectible only in the behavior of navigational lists of wide ambit, jut as the areal peculiarities of Mercator projections are more vivid within a global map than within some small scale regional chart. Such considerations lead us to anticipate that a certain metastability might emerge within Ernest's activities that seems nicely symptomatic of the interfacial sources of his GSF/BPR confusions, but which seems inadequately anticipated within a bare mode of presentation story alone.

Here's what I have in mind. In the standard literature with respect to modes of presentation,[14] it is frequently observed that the introductory mode in which we first encounter a new object—whether it is Susie in her pillbox hat or rocks in a big pile—rarely fixes itself as the aspect under which we invariably think of it subsequently (ill-advised haberdashery, hairstyles and even geological perspectives are soon forgotten, fortunately). Frege was well aware of this drifting tendency, which he regarded as due to a (usually) pardonable shift in the name's meaning that becomes only problematic in circumstances (such as mathematics) where strict rigor requires monitoring. But enough "forgotten meaning shifts" of this type can lead to the peculiar metastability I mentioned above. It is easy to elaborate our narrative so that Ernest eventually learns that GSF has a reverse side that looks exactly like BPR and <u>vice versa</u>, without his thereby deciding that the same mountain was involved (he might mistakenly decide that, since GSF and BPR represent different land forms, some geological process must shape many New England mountains into Janus-like GSF/BPR duality). As this new information is gradually absorbed, Ernest comes to believe that GSF and BSF look exactly alike and that his original modes of presentation *looking like a gigantic man* and *looking like a big pile of rocks* can no longer be regarded as presenting either mountain uniquely. He may even forget the ontogonies of his names: "Why did I designate this rubble 'GSF'? Was it something about a great stone footwall?" But, for all their presentational equivalence, Ernest may still presume they constitute different geographical features and occupy different positions: "I agree that they look almost exactly alike, but still they're different." Through this gentle process, Ernest's bifurcated language has been advanced to a state of virtually identical presentational contents, without causing his diaries to collapse into single entry data registration or otherwise budge him from his "GSF ≠ BPR" proclivities. Of course, such informational integration may occur with calamitous rapidity on the day when it finally dawns on him, to his discomposure, that GSF is undoubtedly the same hill as BPR.

I call this hypothetical condition a *metastability* in analogy to its usual physical meaning. Recall that, with sufficient care, a glass of water can be slowly cooled to far below 0° Centigrade without its turning to ice. The water is said to then be in super-cooled or metastable condition, because, although it can retain its liquid condition indefinitely, its proper equilibrium state at that temperature is as ice. Small internal energetic barriers prevent the liquid from reaching its proper equilibrium. However, a

[14] Gareth Evans, "The Causal Theory of Names," <u>Proceedings of the Aristotlean Society</u>, suppl. vol. 47 (1973).

small infusion of external energy—a slight tap on the glass—will induce a very startling phase change, as the supercooled fluid surmounts its internal hindrances and the whole glass swiftly converts to ice. On the other hand, diamonds, among other substances, likewise qualify as technically metastable at room temperature and pressure, but much higher energetic hurdles prevent the stones from quickly collapsing into their equilibrium formation as graphite (a fact for which jewelers are deeply grateful).

My diagnosis of Ernest's situation is as follows. On the one hand, there is a class of palpable directivities such as

(1) *Classify x as "GSF" if x looks like a gigantic man*

that Ernest follows in his usage and whose nature he clearly recognizes. It is this collection of presentational directive elements that classical accounts exclusively emphasize as central within Ernest's behavior. On the other hand, there are other considerations of an interfacial character that influence Ernest's patterns of usage as robustly as the first class, but to whose underlying nature he may prove entirely oblivious. In particular, the specific navigational list capacities and limitations we have highlighted may tincture Ernest's nomenclature in layers of conceptual personality as critical to its resultant character as (1), despite the fact that Ernest himself fails to recognize this influence. In particular, the name "GSF" is also associated to the directive instructions:

(2) *Assimilate new information about the "GSF" in navigational list mode*
(3) *Plan new routes in toe-to-head manner*

The classical explanation of Ernest's geographical foibles rests entirely upon the fact that he does not associate the directivity

(1*) *Classify x as "GSF" if x looks like a big pile of rocks*

so strongly to "GSF," although, theoretically, he might. However, I believe it is equally important to attend to his directive omission of

(2*) *Assimilate new information about the "GSF" in topographical map mode.*
(3*) *Plan new routes by as-the-crow-flies computation.*

The point of my metastability fantasy is to suggest that, even if associated differences in presentational aspects like (1) have all been analogically cooled to virtually nothing at all, Ernest's GSF $\neq$ BPR troubles are likely to persist, for multi-sheeted growth under data enlargement represents the natural propensity of any policy of informational registration that restricts itself to policies like (2) and (3), without the supplement of (2*) and (3*). In fact, Ernest himself may be dimly cognizant of (2) and (3)'s contributions to "GSF"'s personality, without being able to identify their nature correctly. Suppose that, like Persephone, Ernest spends half the year in Kansas, where he works part time as an aerial surveyor. After a sufficient number of embarrassments of a "GSF $\neq$ BPR" nature, he may become positively spooked about his capacity to name objects within his New England environs. He may even attribute his propensities to misdiagnosed sources: "New Hampshire names like 'GSF' feel positively haunted in some strange way. I believe there must be some Great Wendigo in these woods that mystifies the mind, because I never make naming mistakes like these when I'm in Kansas" (perhaps Ernest will someday erect a tourist attraction on the site, comparable to the beloved Oregon Vortex of my youth). But "GSF"'s spooky personality traces to nothing more occult than the fact that Ernest enjoys ready access to (2*) and (3*) style directivities while in Kansas, but not in New Hampshire.

In such cases, we can fairly say that Ernest entertains an *intimation of intensionality* with respect to "GSF": he recognizes that some distinctive core of determinativeness flavors his term with a characteristic personality, but he is presently unable to identify its underlying nature correctly (he thus resembles the intuitive cartographers of Mercator's time, who realized that different forms of map were useful in one manner or another, without possessing any crisp understanding of why this is so). One of the chief differences between the story I tell here and classical thinking traces to the fact that our everyday evaluative talk of "concepts" et al. often revolves around such undiagnosed directive elements: as Ernest's plight makes clear, such aspects of usage often demand active management and corrective improvement and "concept" and its kinfolk provide the descriptive tools we usually bring to this task. Unfortunately, in both our *ur*-philosophical thinking and within developed classicism proper, we are inclined to assimilate my interfacial factors improperly to presentational content or deny that they play any role in the "story of meaning" at all: "Yes, your map making factors help explain why Ernest often gets confused, but they have nothing to do with what he *means* by 'GSF' ". But I urge that we consider them as important elements in the full story of language that are sui generis in their qualities.

Indeed, we should generally expect that any hypothetical segregation of "the factors that properly belong to the story of language and those that do not" will prove both arbitrary and steeped in classical picture prejudice. However, I do not wish to argue my case through situations as patently contrived as those of Ernest: we will soon move onto cases of greater robustness and practical urgency. My present purpose is simply to illustrate that the notions emphasized in the coming pages possess prima facie application even within the stock examples currently popular in philosophy of language. However, Ernest's case displays a special feature that obscures many of the issues of wider importance that we wish to investigate. It lies simply in the fact that Ernest has a *simple cure* available for the multi-sheetedness to which his usage is prone: after learning that "GSF = BPR," his diary can be readily pared back to single entry format (with "GSF" and "BPR" appearing randomly as mere stylistic variants of one another). Using the mathematician's jargon, Ernest's branched covering of his New Hampshire homeland can be easily *regularized* to a single-sheeted replacement. But the cases that interest me most (several of these were already examined in Chapter 4) are the situations where allied regularization would represent a foolish or unworkable policy, and that the repairs required to keep the potential multi-valuedness under control require a more complex format than the simple acceptance of an identity such as "GSF = BPR". As we'll see in the next section, situations of this ilk are common within applied mathematics and we'll eventually learn that similar patterns of linguistic monitoring are employed within many everyday contexts as well, although, like Ernest, we rarely recognize their presence explicitly.

With respect to the murky internalist-versus-externalist dichotomies mentioned in the previous section, my interfacial factors will be confined quite narrowly to the strategic circumstances whereby available linguistic capacities (e.g., computing a route in toe-to-head fashion) get adopted to suit physical circumstance ably. I believe that this vital range of considerations (which represents the linguistic analog of biomechanics) has been too often passed over, as philosophical authors leap rambunctiously between internalisms and externalisms (in 10,iii, I explain why I consider these divisions ineptly drawn). Or, to recast my claims in materials science analogy: between the *microscopic* aspects of language (atoms and molecules) and the *macroscopic* (tables and galaxies) lies a *mesoscopic* level of dislocations and crystalline structure. The influence of my interfacial considerations can be observed most readily within what were labeled as *strands of practicality* in Chapter 5—short runs of articulated sentences (recipes, inferential patterns) that advance extra-linguistic ends. In my opinion, these middle level considerations affect predicative character in distinctive ways that we need to appreciate better. As such, our discussion will display a mesoscopic emphasis that falls between the attention to individual word meaning typical of classical tradition and the sprawling webs of belief favored by Quine and his cohorts. In my diagnosis, it is the intimations of intensionality that arise in the middle range that most commonly occasion the familiar puzzlements of *ur*-philosophy, as well as inducing the scientific impasses that Kuhn mistakenly characterizes as the clash of paradigm-addled mind sets.

My plan in the succeeding chapters is as follows. In this chapter and the next, I will describe how strategic factors sometimes induce atlas-like structures upon usage that color the personalities of their component predicates in manners that we frequently misunderstand. Through studies of this sort, I hope to persuade my readers that we should be wary of presuming that, because we seemingly grasp a predicate like "is red" stoutly, we thereby "fully understand in what the trait consists." In the presence of unrecognized mesoscopic factors, such contentions can prove utterly misleading. After that—that is, in Chapter 8 and onward—, we shall take up the question of *wise linguistic management*: given that all the proper directivities of suitable predicate use fail to lie explicitly before us as promised in the classical picture, how should we understand our capacities for controlling usage profitably? Here I shall argue that considerable capabilities are available to us—our descriptive situation is neither hopeless nor permanently compromised—, but that teasing them out often requires considerably more investigative work than we anticipate.

Although I do not plan to discuss these issues of management and improvement extensively until we first gather better data with respect to facade-like structures, it is worth observing, before we leave Ernest behind altogether, that several tutorial paths are available that can prevent him from falling into multi-valued blunders so often. The first method is simply to expand the sets of directivities he follows, by persuading him to switch to other forms of geographical representation. Thus we might ask him to draw a topographical map of the region based upon his arboreal rambles. After a suitable interval of fumbling with protractors and rulers, we expect to hear: "My goodness! It never occurred to me before, but GSF and BPR *have* to be the same mountain! Gee, maybe I should have tried to draw a map before I promised to guide those tourists to New Hampshire's two great anthropomorphic outcroppings." Such identificational epiphany will no doubt persuade him that improved control of geographical names can be obtained by shifting data from one registration scheme into another, just as a fibered map of Pittsburgh helps correct for the weaknesses inherent in monotone manners of depiction.

But this pedagogical policy merely teaches Ernest techniques for correcting the faults endemic in "GSF"'s old personality through supplementation; we have not helped him grasp their underlying origins at all. With respect to a conventional atlas, I have already commented upon the virtues of a good *preface*, for there is a clear distinction between appreciating "the practical go" of a set of maps and understanding the mathematical *theory* that lies behind their construction (conversely, someone might easily be a whiz at the latter yet completely helpless in utilizing its data in practical circumstances). Thus Ernest might report his current state of linguistic awareness thus: "I guess if I'd tried to draw a topographical map earlier, I might have more easily avoided this embarrassing mixup, but I'm not sure why." To advance him to a deeper understanding of his linguistic woes, we should take him to some woodland café and draw a lot of pictures like those supplied in this book, for such sketches constitute a homey method for coming to grips with the governing mathematics of the situation (indeed, scribbles on napkins represent the prime vehicle whereby real life mathematicians come to understand their own theories).

The critical feature of such preface-style sketches is that they force Ernest to consider how his patterns of data prolongation *correlate* with respect to the worldly data they attempt to capture. In fact, our napkin sketches put his language use and geographical fact alongside one another in a common portrait as they unfold relative to one another (indeed, we are inviting Ernest to consider his employment in the same Harpo-imitates-Groucho vein that we discussed in 4,iii except that his hub-and-spoke techniques do not follow a simple marching method strategy). If he investigates the possibilities carefully enough from this correlative point of view, he will recognize that his weak angular registrations leave open a great potential that his maps will display improper geometries on a broad scale even if their local registrations and capacities for route-planning remain quite trustworthy. If so, Ernest will have gained an improved picture of his nomenclatural practices: he finally understands the theory behind his usage, just as Lambert first diagnosed the proper basis of the Mercator projection. This improved knowledge may induce Ernest to become more careful in working with his navigational lists, even if he never employs topographical map directivities at all.

In the sequel I shall employ the term *semantical picture* for this preface-like vein of knowledge; it supplies a specific form of linguistic fact that I regard as fully comparable to the understanding we achieved through mathematical investigation with respect to Euler's method in 4,x (observe that a rude sketch can often accurately convey the essence of a formal mathematical study, which is why I utilize so many cartoons in this book). Because of our primary interest in the causes of *ur*-philosophical error, I shall often concentrate upon situations where speakers employ terminology according to properly productive strategies yet entertain incorrect pictures of their underpinnings, in the mode of Ernest and his language-clouding Wendigo. We must actively frame semantic pictures if we hope to improve our usage through other means than brute trial and error, but it is easily possible to lean upon portraits that are quite badly mistaken or shortsighted.

(v)

Unsuitable personalities. Let us now review some classic mathematical considerations that show how subtle the issues of predicate personality can be, as well as supplying some important tools for understanding their behavior. I may delve into a few more details than some readers may ideally prefer, but I believe it is helpful to understand the natural setting in which multi-valuedness arises, rather than merely presenting the situation as an unmotivated curiosity.

A so-called *analytic function* is the sort of gizmo that we obtain when we take familiar functions over the real numbers such as addition, multiplication, logarithm, etc. and extend their reach to make sense over complex numbers. By a "complex number," I intend numbers of the form $a + bi$ where i abbreviates a hypothetical square root of -1. It turns out that the operation of ordinary multiplication (i.e., $3 \times 6 = 18$) naturally extends to the complex numbers by the rule $(a + bi)(c + di) = (ac - db) + (ad + bc)i$ (little surprise there; that is obviously the way the operation should work). This means that functions that can be delineated over an interval with a *power series* (i.e., an expression of the form $a_0 + a_1x + a_2x^2 + a_3x^3 + \cdots$) automatically extend a certain distance into the complex numbers because the series is entirely composed of simple extendible operations. Most functions that we can readily think of (unless one is a mathematician) are "analytic" in this way: they make equally good sense if applied to complex values.

In other words, the movement of an analytic function from the real line (its original home) out to the complex plane is driven by the directivities natural to addition and multiplication. As I sketched in Chapter 4, nineteenth century mathematics and physics reaped enormous benefits by following the Pied Piper of " + " and "×" in this inferential outreach, leaving the practitioners somewhat mystified at their successes. In particular, important clues to the understanding of many functions are provided by the manner in which zeros and poles form on the complex plane: places where the function either becomes 0 or infinite. To cite an example already described (4,i), the behavior of a telescopic control system is beautifully revealed in how its critical points locate themselves on the complex plane.

At first glance, analytic functions look quite ordinary in personality and many mathematicians believed falsely that they could be utilized in physical work freely. For example, even Poincaré famously declared

> *The physicist may, therefore, at will suppose that the function studied is continuous, or that it is discontinuous; that it has or has not a derivative; and may do so without fear of*

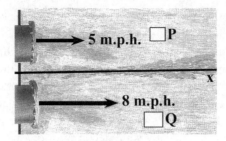

ever being contradicted, either by present experience or any future experiment. We see that with such liberty he makes sport of difficulties that stop the analyst.[15]

Here he had in mind Weierstrass' well-known approximation result: given any continuous function over an interval, there will be an analytic function that copies its behavior as closely as one likes.

However, another great French mathematician, Jacques Hadamard,[16] observed that this conclusion was not right: that analytic functions possess strong personalities that render them unfit for many types of physical application, rather as the personality of a Mercator map makes it unsuited for the accurate representation of areas. Analytic functions are *headstrong* in a manner that creates subtle tensions anytime we wish to treat the normal flow of a fluid, for example. In particular, a striking feature of any analytic function lies in its *reproducibility*. If we are told how such a gizmo behaves over some very small portion of the complex numbers, then we can completely reconstruct how it must behave everywhere else. This supplies an analytic function with a strong regenerative capacity akin to that of a flatworm—you can take a tiny slice of the critter and it will grow back all of its missing parts. But this behavior is unnatural, Hadamard reminds us, for the functions that commonly arise in physical considerations. For example, suppose we have two large hoses that dump water into a wide ocean. Let us suppose that the result is a current that moves with a velocity of 5 mph above the x axis and at 8 mph below, with a little region of turbulence in between. Now if this combined flow were describable by an analytic function (using $x + y\sqrt{-1}$ as a complex coordinate over our two dimensional plane), then we should be able to calculate the flow everywhere simply from a little piece located at p above the x axis. But this reconstructability is unreasonable, Hadamard observes, because how can our little piece at p know that the flow from the bottom hose isn't now flowing in at *10 mph*, for this change hasn't had enough time to begin affecting p as yet? Or, to put the same point another way, any analytic function requires the fluid condition at p to be fixed by its condition at q and this isn't reasonable, because it takes time for physical effects in water to propagate from one spot to another. True, the Weierstrass result says that we can approximate our physically defined function f(z) as closely as we like (within a region) by an analytic

[15] Henri Poincaré, "Analysis and Physics" in <u>The Value of Science</u>, G. B. Halsted, trans. (New York: Dover, 1958), 83.
[16] Jacques Hadamard, <u>Lectures on Cauchy's Problem in Linear Partial Differential Equations</u> (New York: Dover, 1952).

mimic g(z), but an alien *rigid personality* will have crept into the copycat g(z) that simply wasn't present in the original f(z).

On the other hand, if we have tacitly engaged in some simplification strategy like those discussed in Chapter 4, the appearance of analytic functions becomes more reasonable physically. In particular, we often make the assumption that the fluid flow is in *steady state*, where we assume that the transient patterns that arise when the water flow starts all died away and we only witness the steady state response to constant input from the hoses (transients and steady state decompositions were discussed in an electrical context in 4,vii). Strictly speaking, this steady state flow represents an idealized condition of our water, because it will take infinitely long before our transients completely die away. On this new, steady state assumption, the rigid linkage between p and q becomes physically reasonable, because we now secretly maintained our hoses at constant flows over an *infinite period of time*, allowing regions p and q ample opportunity to reach accommodation with one another (and thus allow their conditions to be deducible from one another in approved analytic function fashion).

Now there are plenty of equations that pop up commonly within physical applications that accept only analytic functions as solutions. The consequence we can extract from Hadamard's overview is that some reductive policy akin to our "assumption of steady state response" has been tacitly evoked, allowing analytic functions to sneak into the picture with their unnaturally rigid personalities. In common physical practice, silent appeals of "steady state" type walk in the door quite freely and the average practitioner often does not observe their entrance with any care (see 9,i for more on this). But, from a mathematical point of view, such considerations usually carry us from one mathematical arena to another (in our two pipe case, from equations of (possibly) *hyperbolic* type to *elliptic* sorts—distinctions to which Hadamard drew special attention). Sometimes this lack of strategic notice catches up with the student of physics or engineering later on.

Here is a classic example. Airplane wings fly in a gas of very low frictional resistence, so it seemed reasonable in the nineteenth century to ignore the frictional terms in the basic fluid equations (the Navier-Stokes equations), which are very hard to solve in any case (our Chapter 4 discussion of Prandtl's work indicated why this seemingly natural assumption was not, in fact, reasonable). Unfortunately, the simplified equations predicted that an airplane wing should experience neither "drag" (= retarding force) nor "lift" (= buoyancy upward), leading to understandably pessimistic appraisals of the prospects for heavier than air flight (despite the example of birds and butterflies). Shortly after the Wright Brothers' initial flights, however, the applied mathematicians Wilhelm Kutta and Nikolai Joukowsky developed a novel method for calculating reasonably plausible values for lift (although not drag) utilizing functions of a complex variable.[17] The resulting "circulation theory" is still commonly taught to students (although computers have rendered Prandt-like methods of calculation more practical). Their

---

[17] John D. Anderson, A History of Aerodynamics (Cambridge: Cambridge University Press, 1998), ch. 6. K. Pohlhausen, "Two-dimensional Fields of Flow" in R. Rothe, F. Ollendorff and K. Pohlhausen, eds., Theory of Functions as Applied to Engineering Problems, Alfred Herzenberg, trans. (New York: Dover, 1933).

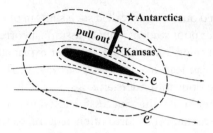

method sums (by relying upon a so-called "complex velocity potential") the shifting pressures we encounter as we encircle the wing along a nearby contour 𝓒. This sum-mation proves to have a net contribution upward from which the lift is easily calculated using a formula of Bernoulli's. So far, so good, but aeronautical students often become puzzled by the following observation. There is nothing in Kutta and Joukowsky's procedure that requires that the encircling contour where we compute our sum need lie near the wing; instead, we can pull the contour as far away from the wing as the atmosphere allows (say, to 𝓒′ as pictured). Even along 𝓒′ we will calculate exactly the same pressure summation as around the nearby encirclement (in fact, apprentices are taught to exploit this very trick, typical of so-called "contour integration," to solve the problem). "You mean," a puzzled pupil might ask her instructor, "that I can theoret-ically walk in a great circle that cuts across Asia and the Antarctic and still detect the air disturbance occasioned by a tiny plane flying over Kansas?" "Yes, of course" will be the reply, possibly accompanied by some unhelpful mumbling about Cauchy's residue theorem. It is experiences like this that prompted John von Neumann's remark: "One never really understands mathematics; one simply grows used to it."[18]

In fact, an unstated appeal to "steady state" response has been made here, allowing the rigid personality of an analytic surrogate for the real life velocity potential to enter the picture, allowing the contour to be pulled away from the wing in "state at p fixes the state at q fashion" (additional hidden subtleties lie behind the success of this peculiar inferential procedure but I'll postpone their diagnosis until later). But the Kutta and Joukowsky procedure had been long in use before its underlying support was eventually teased out by applied mathematicians.

The puzzlement of our aerodynamics student represents a nice exemplar of the pro-cesses often responsible for *ur*-philosophical confusion, as surveyed in Chapter 2. Some collection of seemingly innocuous descriptive terms—in this case, "wind velocity" and "lift"—appear in some reasoning context that is tacitly controlled by some unnoticed set of subtle strategic policies. That embedding context allows new directivities to attach to "wind velocity" and "lift" that eventuate in genuinely useful final results (e.g., reasonably good wing designs), but some of the steps in the reasoning seem mysterious to our pupil and in want of an explanation. In fact, the net effect of the incursion of analytic personality has secretly added directivities that pull the predicate "wind velocity" away from its accustomed physical significance and cause it to serve as a carrier of information of a

---

[18] David Wells, <u>Curious and Interesting Mathematics</u> (Harmondsworth: Penguin Books, 1997), 259.

more abstract and smeared out nature (I called such shifts in physical significance *property dragging* in Chapter 4). Our poor student is apt to assume incorrectly that "wind velocity" has remained fixed in meaning and will look to other explanations of her peculiar procedures, some of which can lead her very badly astray. In fact, Chapter 9 will supply several real life cases of serious misunderstandings of exactly this type.

The net moral I am after here is this. Successful descriptive predicates that show up in effective recipes and inferential procedures often acquire, as the price of their efficacy, unexpected coatings of supplementary directivities. The personalities that result can prove somewhat headstrong in character and require a compensating system of controls to prevent such words from wandering too far astray in their long range exuberance. The strategic reasons why such complications are needed often require a rather deep appreciation of how wise strategy affects descriptive practice. As such, this conclusion is exactly the same as we extracted from our discussion of maps, but transferred to more abstract linguistic circumstances.

Let us now see why the boundary line fencing provided in a facade often supplies the controls required to keep our predicate personalities operating in a generally useful fashion.

(vi)

Analytic prolongation. The headstrong personalities of analytic functions display another important feature that is intimately tied to the metastable behaviors we witnessed in Ernest's names. From what source does that rigid "patch p determining patch q" character of an analytic function spring? *Answer:* from the way that such quantities grow to cover their full domains through a step-by-step process of *analytic continuation.*

To explain what I have in mind, it is convenient to examine one of those paradoxes involving complex numbers that commonly appear in the puzzle books. What goes astray in this reasoning to "prove" that $+2 = -2$?:

$$2 = \sqrt{4} = \sqrt{(-2 \cdot -2)} = \sqrt{-2}\sqrt{-2} = \sqrt{-1}\sqrt{2}\sqrt{-1}\sqrt{2} = (\sqrt{-1})^2(\sqrt{2})^2$$
$$= -1 \cdot 2 = -2$$

A proper reply will bring out the "headstrong character" of the concept $\sqrt{z}$ (which qualifies as analytic).

I will indulge the reader's patience by first supplying some background to calculations like this. Why were mathematicians of the eighteenth and nineteenth centuries so *eager* to insure that familiar functions like square root and exponentiation (i.e., x^y) make sense with respect to complex values? On the face of it, it is scarcely apparent that a term like "$(1 - 2i)^{3i}$" should mean anything. After all, no one considers it their parallel duty to discover a meaning for the "exponentiation" of Cary Grant by Archie Leach: "Cary Grant$^{\text{Archie Leach}}$." It happens that, once the crazy foray into complex territory has been initiated, wonderful formulae like $e^{i\theta} = \sin\theta + i\cos\theta$ are discovered that have

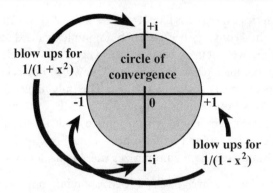

thoroughly rewritten the face of modern mathematics. But what motivated such odd sallying forth in the first place?

The answer begins in the increased understanding the extended functions provide with respect to the queer and seemingly whimsical behaviors that ordinary real-valued functions (such as employed in physics) commonly display. Specifically, many central techniques of applied mathematics rely heavily upon the expansion of key formulae in *power series*: infinitely long expressions that comprise sums of terms in powers of x (e.g., "$1 - x^2 + x^4 - x^6 + \cdots$"). Unfortunately, such summations display a perverse tendency to stop supplying meaningful values for real number inputs without apparent warning (their partial sums may diverge or, even if they do eventually converge, they do so at such a languid pace as to prove utterly useless in practice). This unreliability causes applied mathematicians a good deal of trouble, for in reasoning to other conclusions, they must avoid presuming that some function's power series converges in a region where it doesn't: carelessness in this regard can quickly generate horrible fallacies of "6/0" type. In a famous instance, Laplace supplied a "proof" that the solar system is permanently stable but its validity hinges critically on whether a certain series converges.

To display the strange behavior I have in mind, consider the simple functions:

(a) $1/(1 - x^2)$
(b) $1/(1 + x^2)$

Through formal long division, we can calculate appropriate power series for each:

(a') $1 + x^2 + x^4 + x^6 + \cdots$
(b') $1 - x^2 + x^4 - x^6 + \cdots$

Both series converge only within the narrow interval $-1 < x < 1$. But *why* do (a') and (b') *fail* outside of this span? In the case of (a), an answer is immediate on the face of it: the original function $1/1 - x^2$ can't be well defined *at* $x = \pm 1$ because it "blows up" (=becomes infinite) there. But $1/1 + x^2$ suffers no manifest impediment of this type; (b) is perfectly well defined at $x = \pm 1$. So why does *its* power series also break down beyond these limits?

As previously noted, our usual rules for adding and multiplying regular numbers extend automatically to the complex realm. This extension in turn supplies a ready meaning to power series expressions like "$1 - x^2 + x^4 - x^6 + \cdots$" (there is no difficulty in explaining what the "convergence" of such a complex-valued series should mean). But when we do this, we obtain a beautiful answer to our puzzle about $1/(1 + x^2)$: viewed over the full complex plane, it confronts "blow up" obstacles at $\pm$ i exactly like those that stymie its cousin $1/1 - x^2$ at x $= \pm 1$. The only difference between the two expressions is that (b)'s impediment is located at x $= \pm\sqrt{-1}$ rather than along the real axis. But a singularity anywhere is sufficient to limit the reliable convergence of a power series to a circular region that falls short of the blowup. In an excellent primer on these topics, Tristam Needham summarizes these considerations as follows:

> But how is the radius of convergence of a [corresponding power series] determined by f(x)? It turns out that this question has a beautifully simple answer, but only if we investigate it in the complex plane. If we instead restrict ourselves to the real line—as mathematicians were forced to do in the era in which such series were first employed—then the relationship between [f(x) and the radius of convergence for one of its power series] is utterly mysterious. Historically it was precisely this mystery that led Cauchy to several of his breakthroughs in complex analysis (he was investigating the convergence of series solutions to Kepler's equation, which describe where a planet is in its orbit at any given time).[19]

The clarity and understanding that this program of expansion to the complex plane brings to many types of puzzling behavior in analysis is truly remarkable and hence it is not surprising that mathematicians quickly became interested in figuring out how a wide range of erstwhile real-valued functions (such as exponentiation) behave when their application is pushed outward into the complex numbers. As Hadamard once commented,

> The shortest path between two truths in the real domain often runs through the complex numbers.

A value where a function or quantity becomes meaningless (as $1/(1 - x^2)$ becomes undefined at x $= \pm 1$) is called a *singularity*. The phenomenon we have just surveyed shows that, in several basic ways, such functions are sometimes "controlled" by the places where they no longer make sense! I mention this, because we'll later see that the boundaries lying between sheets of usage often act in analogous ways.

However, the circumstance that is most analogous to the Ernest case lies in the fact that, in the vast majority of cases, familiar functions are extended to complex values through a process of *prolongation*. Unlike power series expressions, a run-of-the-mill functional expression such as "2^z" or "$\sqrt{z}$" (here we intend the positive root) do not immediately inform us on their faces how they should be applied to complex inputs. Here our obliging friends, the power series, come to our assistance. It is easy to find power series expansions that match, *within certain intervals*, the real number values of $\sqrt{z}$ e.g., we can use $\sqrt{(1 + x)} = 1 + x/2 - x^2/2\cdot4 + 3x^3/(2\cdot4\cdot6) - \cdots$ for $-1<x<1$). Why not utilize this same series (which automatically makes sense over the complex

¹⁹ Tristan Needham, Visual Complex Analysis (Oxford: Oxford University Press, 1997), 64.

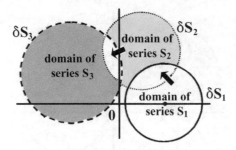

numbers) to tell us how $\sqrt{z}$ should behave on nearby complex values of z (e.g., $1 + 1/6i$)? It was through an extensive program of quasi-empirical experimentation in prolongation through sundry series of this ilk that eighteenth century mathematicians (particularly Euler) determined how many familiar real-valued functions ought to behave over complex values.

But power series are usually only *locally defined*—that is, they break down outside of limited circular domains. How do we reach complex numbers that lie beyond the dominion of our first exploratory series? One of the pleasant features about power series calculations is that they can be recentered upon different values. Suppose we moved out into the complex plane following an initial series S_1, which breaks down once we reach a boundary circle δS_1. Let c be some complex value just inside δS_1. Why not center a new series s_2 upon c and see where its new boundary δS_2 falls? (series S_2 will usually look quite different than S_1 from a syntactic point of view). If we properly skirt blow ups and so forth, we will be able to build up a pattern of overlapping circular domains in our sallying forth that will extend our original functional expression to make sense over almost all complex values (on occasion, larger natural obstacles block entry to certain regions of the full plane). This step-by-step process for pushing functional meaning from one local domain into another through appeal to overlapping series is called *analytic continuation*. Of course, we have been looking at similar pictures of *prolongation from domain* $\mathcal{D}_1$ to *domain* $\mathcal{D}_2$ for some time—they were all introduced with malice aforethought to prepare the reader for an analogy with the present mathematical circumstances.

Note that, as we scuttle outward onto the complex plane in crab-like prolongation, we are following *pathways of natural computational extension*: the guidance suggested by our familiar algorithms for addition and multiplication as displayed in the format of power series expansions. As it were, these series would really like us to move onto the complex plane in the manner they prescribe (we might borrow a phrase from the redoubtable Oliver Onions and consider these algorithmic directivities *beckoning fair ones*: temptations that pull us forward into untested terrain). In the case of complex numbers and power series, the inferential expeditions encouraged by these alluring algorithms are soon rewarded by the delightful treasures we discover in the lands beyond (including that miraculous mathematical pearl, $e^{i\theta} = \sin\theta + i\cos\theta$). Sometimes, regrettably, succumbing to syntactic enticements does not lead to such happy eventualities, but we'll not dwell on such gloomy thoughts for the moment.

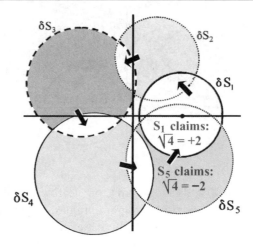

But—and this is the chief observation I am after—as we pursue our program of analytic continuation, a remarkable side effect can occur. Following a sequence of appropriate series, we can continue values for "$\sqrt{z}$" completely around the origin, starting from a region over z = +4. When our power series discs once again cover the real value z = +4 after circling the origin, the replacement series we utilize now blithely informs us that, no, the proper value of "$\sqrt{4}$" is *not* + 2, as we originally thought; it is actually − 2! If we cycle a second time around the origin (which is called a branch point) using the same kind of continuation, "$\sqrt{4}$" recalculates more happily as + 2 once again. If we are unfamiliar with this phenomenon, we will be surprised by this functional inconstancy because we might presume, from the fact that each individual power series supplies uniquely determined values to a functional expression *locally*, that the full assembly generated by the pattern of "analytic continuation" will also display unique values *globally*. But this tacit expectation often proves mistaken.

At first glance, these troubles merely suggest that we've trusted a lousy sequence of series, but further investigation reveals that the tendency to develop doubled values for $\sqrt{z}$ is quite generic, even if we utilize non-power series considerations for our prolongations. Indeed, there are many natural ways to push "$\sqrt{z}$" into complex values and every one of them displays exactly the same multi-valuedness. Furthermore, $\sqrt{z}$ is not anomalous in this strange behavior; many other familiar expressions (e.g., ln(z)) curl up into multi-valuedness as well. Some intrinsic stiffness buried deep in our fundamental rules for addition and multiplication force these instabilities in functional values as we cycle the branch points. Like it or not, if we wish to deal with such extended "functions" at all, we must learn to live with this peculiar behavior. As the mathematician J. F. Ritt amusingly writes:

> There are, however, certain questions connected with the many valued character of the
> elementary functions which [once] could be pressed back behind the symbols . . . but which
> have learned to assert their rights . . . It might be great fun to talk just as if the elementary

functions were one-valued. I might even sound convincing to some readers; I certainly could not fool the functions.[20]

Here Ritt is referring to the fact that the early mathematical pioneers often dismissed these aberrant behaviors as inferential oddities for which no disciplined overview was needed, but, in truth, the phenomena involved can't be coherently understood unless we accept multi-valuedness as natural to the internal character of the functions themselves.

In fact, such considerations show that "$\sqrt{z}$" and "ln(z)" shouldn't be regarded as true *functions* at all, if we restrict "function" to its usual meaning as a many-one mapping between domain and range. True, their standard mathematical title is "analytic function" or "function of a complex variable," but mildly inept nomenclature doesn't render them "functions," anymore than a starfish qualifies as a true fish. Many analytic "functions" manifest a twisted personality that refuses to spread out uniformly across the complex plane—in Ritt's amusing analogy, they've "got their rights" and they'll be damned if they'll lie flat for anyone.

The structural analogies to Ernest's troubles with "GSF" and "BPR" should seem quite palpable: whenever a body of data enlarges by step-by-step prolongation, there is a chance that the extensions will begin to contradict values earlier laid down. Unlike the Ernest case, there is no simple "GSF = BPR" remedy available for $\sqrt{z}$; there's no way to "uniformize" its behavior to a single-valued covering of the complex plane that doesn't include artificial rips and tears. Hidden within the personality of the manner in which we calculate roots over the real numbers lies a *torsion* that manifests itself as an inherent multi-valuedness when those rules are prolonged across the complex domain, even if we heartily wish that $\sqrt{z}$ wouldn't behave like that. Complain as we might, we cannot evade the fact that the natural behavior of $\sqrt{z}$ contains an unavoidable twist in its unfolding. Here the Muse of Mathematics offers us a tough bargain: "I'll happily supply you mortals with a gizmo that extends real-valued $\sqrt{x}$ wonderfully, but its price is that it will be intrinsically multi-valued." We cannot "fool the functions" into acting any other way. We thereby witness an Ernest-like lift in $\sqrt{z}$ that can't be cured by any simple "GSF = BPR" corrective.

Due to Riemann is an evocative picture of the torsion that $\sqrt{z}$ evinces: imagine a ramped parking lot with two floors in which we can drive around forever without running into anything (the topology of such a *Riemann surface* cannot be realized as an ordinary spatial shape within three dimensions). While we are driving on level one, the correct value of $\sqrt{4}$ looks as if it should be clearly $+2$ but, as we motor onto level two, the value -2 begins to seem preferable. Since we subsequently return to floor one after transversing tier two, mathematicians call $\sqrt{z}$ a "function of two sheets." The Riemann surface for ln(z) is even more disheartening: it is a "function of infinitely many sheets"($=$ a parking lot with a Borges-like hierarchy of levels). Of course, such Riemann surfaces represent the prototype of the branched pictures we drew for what transpires within Ernest's geographical practices.

---

[20] Ritt, <u>Finite Terms</u>, pp. v–vi.

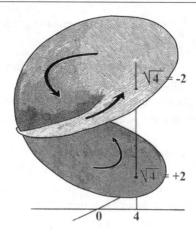

Because of the appearance of multi-valuedness, inferential principles which make good sense locally often lose meaning on a global scale. Consider the distributive property that $\sqrt{(ab)} = \sqrt{a}\sqrt{b}$. As long as we don't move too far away on the Riemann surface, distributivity does not cause problems (the calculation $\sqrt{(4i)} = \sqrt{4}.\sqrt{i} = \sqrt{(2.2)}\cdot\sqrt{i} = \sqrt{2}\cdot\sqrt{2}\cdot(1/\sqrt{2} + i/\sqrt{2}) = (\sqrt{2}\cdot\sqrt{2})/\sqrt{2} + (\sqrt{2}\cdot\sqrt{2})i/\sqrt{2}) = \sqrt{2} + \sqrt{2i}$ is unproblematic). But the identity "$\sqrt{(ab)} = \sqrt{a}\sqrt{b}$" loses clear sense if we don't confine our operations to a local region of our Riemann surface. But this limitation is violated in the fourth stage of our $2 = -2$ paradox:

$$2 = \sqrt{4} = \sqrt{(-2\cdot-2)} = \sqrt{-2}\sqrt{-2} = \sqrt{-1}\sqrt{2}\sqrt{-1}\sqrt{2} = (\sqrt{-1})^2(\sqrt{2})^2$$
$$= -2.$$

Here the operation of squaring $\sqrt{-2}$ has rotated us to onto the upper floor where the "wrong root" of 4 sits. More generally, inferential operations that are vital locally can become problematic on a more extended scale if the basic usage has been built up through a sequence of continuations from one domain to another. This is another illustration of the general moral that what holds true locally may fail globally.

From a philosophy of language point of view, the lesson of our "$2 = -2$" paradox is *not* that expressions like "$\sqrt{z}$" "can't be assigned a meaning at all" (as Frege might have claimed) but simply that their proper handling requires attention to local/global discriminations that we may not have anticipated when we first pushed "$\sqrt{x}$" out to complex values. Although mathematicians usually avoid the expressions "$\sqrt{z}$" and "$\sqrt{-1}$" (in favor of "$z^{1/2}$" and "i"), they have gotten quite used to handing the analogous multi-valuedness encountered with "$\ln(z)$." We can work with such expressions very profitably but we must take care in their proper inferential management.

Mathematicians like to anthropomorphize their subject matter and in this fashion maintain that expressions like "$\sqrt{z}$" *like living* on a Riemann surface better than on the flattened complex plane. In Hermann Weyl's famous comment:

> Riemann surfaces are not merely a device for visualizing the many-valuedness of analytic
> functions, ... but their native land, the only soil in which the functions grow and thrive.[21]

Truly, the contrast between the two-sheeted surface and the flat plane below it provides a vivid picture of the special personality that the inferential principles natural to the expression "$\sqrt{z}$" display. Such pictures of *inferential personality* will prove quite valuable to us in the sequel.

(vii)

The Stokes phenomenon. Thus far, we have considered analytic functions only in their own terms, as purely mathematical quantities. However, to reason effectively in physical circumstances, we often follow deductive patterns that look exactly like computations of an analytic function. But the latter incorporate headstrong personalities somewhat unsuited to the physical quantities we wish to discuss. How should we correct for the errors into which this mismatch would otherwise lead?

There are actually a variety of solutions to this problem, the most obvious of which will be discussed in this section, although our focus will largely shift to the other forms of solution later in the chapter.

Let's set the scene with a specific illustration. Suppose that short wavelength light from a distant light bulb strikes a completely reflective razor blade and we want to calculate how the light will reflect from its surface. Since the situation is two-dimensional, complex numbers can be employed as useful coordinates. In these circumstances it is natural to shift to a steady state treatment, because we aren't really interested in tracing the whole elaborate story of the transients that arise when the light is first turned on and then encounters the blade (this would involve very elaborate calculations greatly prone to error and the main conclusions we seek will be swamped in irrelevant filagree). In making this adjustment, we will have switched to governing equations that allow analytic functions in the door. This shift makes it difficult to express the fact that the light arrives at the blade from the upper right hand corner because, on any bounding line that can be set down, some light reflected back from the blade will mix with the incoming flux. From a technical point of view, our incoming light requirement does not represent a conventional boundary value problem, a point to which I'll return.

Arnold Sommerfeld, in famous investigations of 1894,[22] found several exact expressions for the kind of analytic function that solves this problem, including a series in Bessel functions. However, these representations prove quite impractical because computing acceptable values from them requires an enormous number of operations. As H. Moysés Nussenzveig comments with respect to the related problem of computing

---

[21] Hermann Weyl, The Idea of a Riemann Surface, Gerald MacLane, trans. (Reading: Addison-Wesley, 1955), p. vii.
[22] Arnold Sommerfeld, Mathematical Theory of Diffraction, Raymond Nagem, Mario Zampolli, Guido Sandri, trans. (Boston: Birkhäuser, 2004).

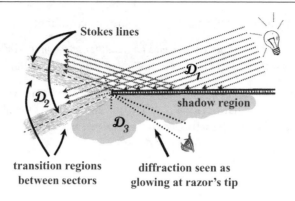

Stokes lines

$\mathcal{D}_1$

$\mathcal{D}_2$

$\mathcal{D}_3$

shadow region

**transition regions
between sectors**

**diffraction seen as
glowing at razor's tip**

diffraction effects inside the raindrops that create rainbows:

> *Computers have been applied to the task, but the results are rapidly varying functions of the size parameter and the scattering angle, so that the labor and cost quickly become prohibitive. Besides, a computer can only calculate numerical solutions; it offers no insight into the physics of the rainbow. We are thus in the tantalizing situation of knowing a form of the exact solution and yet being unable to extract from it an understanding of the phenomenon it describes.*[23]

But Sommerfeld found that, by dividing the plane around the razor into three sectors $\mathcal{D}_1$, $\mathcal{D}_2$, $\mathcal{D}_3$ and ignoring two extremely thin sectors of complicated behavior along their boundaries, he could replace his slow-to-converge Bessel function series with a much snappier series utilizing terms such as "$e^{-ikr\cos(\theta-\alpha)}$", "$e^{-ikr\cos(\theta+\alpha)}$" and "$e^{ikr}/\sqrt{kr}$". And this replacement not only permits an astonishing reduction in computational complexity (Nussenzveig estimates an advantage of approximately 15,000 to 1 in his circumstances), the replacement terms are much easier to interpret: they represent incoming and outgoing plane waves, plus a diffracted wave front that radiates circularly from our razor's edge. Indeed, in this new representational guise, we can discern that we have actually solved the problem we sought: how incoming light scatters from a razor blade (discerning facts like this represents the kind of "understanding" that Nussenzveig claims is absent in the more exact representations).

But our new representational format displays an odd behavior called the *Stokes phenomenon* (after its discoverer, George Stokes): the same calculation rules do *not* work properly all the way around the razor blade, but must be readjusted every time we cross the boundary of one of our $\mathcal{D}$ regions (which are called *Stokes lines*). That is, to compute proper values of light intensity around the blade, we must follow a sectorized policy: in region $\mathcal{D}_1$, trust formula $\mathcal{F}_1$, but once the Stokes line boundary into $\mathcal{D}_2$ is crossed, allegiance should be shifted to formula $\mathcal{F}_2$ which is obtained from $\mathcal{F}_1$ by altering its coefficients and ditto when we move into sector $\mathcal{D}_3$ (see fine print for details). But why does our inferential recipe alter in such an abrupt way—after all, the slowly convergent

---

[23] H. Moysés Nussenzveig, "The Theory of the Rainbow" in <u>Atmospheric Phenomena</u> (San Francisco, Calif.: W. H. Freeman, 1980), 69.

Bessel function series it supplants does not act in this inconstant manner? Indeed, this trisected behavior both puzzled and intrigued Stokes[24] greatly.

Our replacement series obtains its advantages through practicing *physics avoidance* (4,vii) and ignoring the complicated light behaviors within the little slices near the Stokes line boundaries. This policy lets us employ exponential terms to characterize the dominant behaviors inside the $\mathcal{D}$ patches in very simple terms. But this changeover in representational language, from Bessel term factors to exponentials, produces a change of *inferential personality*, for square root coefficients appear that alter the longer range behavior of our replacement representations. That is, to compute the distribution of *light intensity* within sector $\mathcal{D}_1$, *we directly consult* the guidance of an exponential term based formula $\mathcal{F}_1$ describing an analytic function f_1 whose natural disposition is to curl up into unsuitable multi-valuedness. We value $\mathcal{F}_1$ as a linguistic expression because the true values of *light intensity* supplied in Sommerfeld's more exact formula speak to us in phraseology we cannot easily understand, whereas $\mathcal{F}_1$ translates these oblique inferential instructions into a tongue we can better grasp (in Wittgenstein's famous analogy,[25] the former seems like an expression made for a god, not a human being). But $\mathcal{F}_1$ is willing to serve this interlocutory role for only a short span; beyond the Stokes line boundary, the analytic function f_1 begins, in Ritt's phrase, "to assert its rights" outside of $\mathcal{D}_1$ and eventually climbs away from any tracking of *light intensity*. We can picture this situation as one where f_1's twisted Riemann surface $\mathcal{R}$ lies interposed between us and the physical plane upon which the true *light intensity* function lives. Over sector $\mathcal{D}_1$, f_1 copies *light intensity* closely but lifts away after that. To curb this curling, we etch a line across the pavement of our $\mathcal{R}$-surface parking lot and announce, "Halt, Hitherto Successful Pattern of Reasoning! I will follow your dictates no more." When this $\mathcal{R}$-based line of deductive demarcation is beamed down to the plane of the razor, its projection shows up as a Stokes line. Moving into sector $\mathcal{D}_2$, we need to consult a fresh formula $\mathcal{F}_2$ for computational guidance, which happens to be composed of the same terms as $\mathcal{F}_1$ but with different coefficients. And we must repeat this retooling process once again when we proceed into $\mathcal{D}_3$.

In other words, we have managed to cover our *light intensity* function with three segments of analytic functions each living on a copy of $\mathcal{R}$. Let us cut out the "good portions" of each sheet and glue them together along the Stokes lines. The result is a single-valued covering of the razor blade, composed of three inferential patches with abrupt transitional links between them. Patchwork coverings of this type are what I call *facades* or *atlases*, in homage to the multiple globe coverings of section (iii) (a number of allied structures facades were canvassed in 4,vi–ix; I will articulate the general notion involved more formally in 7,i).

In our new *light intensity* covering facade, we harness the unsuitably rigid formulae $\mathcal{F}_1$, $\mathcal{F}_2$ and $\mathcal{F}_3$ to our descriptive purposes by first terminating their inferential directivities

---

[24] George Gabriel Stokes, "On the Discontinuity of Arbitrary Constants which Appear in Divergent Developments," in <u>Mathematical and Physical Papers</u> (New York: Johnson Reprint Co., 1966). Robert W. Batterman, " 'Into a Mist:' Asymptotic Theories on a Caustic," <u>Stud. Hist. Mod. Physics</u> 28, 3 (1997).

[25] Wittgenstein, <u>Investigations</u>, §426.

along well-chosen boundaries and then welding the results together by brute force juxtaposition. The resulting montage of functional fragments supplies a reasonable facsimile of the way light actually scatters around a razor blade (rather as a graceful statue might be soldered together from shards of an uncooperative metal: from a tolerable distance away, an assembly of flattened beer cans might look rather like Elvis Presley). If we do this properly, we obtain a theory-like linguistic instrument that represents a nice compromise between physical fact and the tools we have available to us. The moral to be drawn is that a healthy descriptive language sometimes requires a few *strategic cracks* here and there to prove seaworthy (if its hulk is welded too tightly together, the craft can't accommodate the strains that Nature sends its way, as occurred with the unlucky Liberty ships of yore).

In short, we gain great advantages by employing slightly unsuitable reasoning tools in a strategically judicious manner. Carl Bender and Steven Orszag express this theme, central to our investigations, as follows (writing of the Airy function Ai that appears in the theory of rainbows):

> [T]he reason for using an asymptotic approximation is to replace a complicated transcendental function like Ai(z) by simpler expressions involving elementary functions like exponentials and powers of z. From a practical point of view, much is gained by such approximations. However, one pays for these advantages by having to deal with the complexities of the Stokes phenomenon. The Stokes phenomenon is not an intrinsic property of a function like Ai(z), but rather is a property of the functions that are used to approximate it. The Stokes phenomenon reflects the presence of exponential functions in the asymptotic approximation.[26]

R. Meyer reiterates the same observation as follows:

> The pervasiveness of the phenomenon . . . indicates that it should be generic and possess fundamental roots . . . [It is the] natural consequence of a decision to characterize functions by the help of approximating functions whose multivalueness differs from that of the functions to be characterized . . . . The wave character of [physical distributions like f(z)] is their most important property, by far, in th[is] context and in many instances it is the only scientifically relevant property. The representation by multivalued functions is the only way in which the wave character can be displayed with great clarity . . . . The Stokes phenomenon is a necessary, and rather economical, price for the representations we need most of all.[27]

This captures the key idea I'm after in our discussion: a few Stokes cracks represents a "rather economical price" to pay for reasoning that can reach practical results in relatively short order.

. . . .

[26] Carl M. Bender and Steven A. Orszag, Advanced Mathematical Methods for Scientists and Engineers (New York: McGraw-Hill, 1978), 117.

[27] R. E. Meyer, "A Simple Explanation of the Stokes Phenomenon," Siam Review 31, 3 (1989), 435.

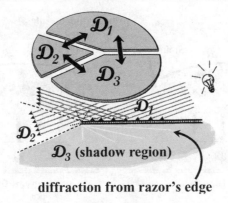

diffraction from razor's edge

Curiously, considered as an analytic function, the true values of *light intensity* E(z) themselves sit on a two-sheeted Riemann surface whose lift gets cut off by a branch cut along the razor blade (to be more accurate, *light intensity* is supplied by the square of E(z)). Sommerfeld is able to construct his exact solution by cleverly manipulating materials on the unphysical side of this surface (the technique is usually known as the method of images). Here we see the useful aspects of the artificial rigidity of analytic functions: they may induce singularities within unphysical regions that can be nicely employed to *understand* them better (this same moral is displayed in the "unphysical" aspects of *impedance* mentioned in 4,i). Sommerfeld finds the exponential series by exploiting the wonderful trick of asymptotic expansions discussed in 4,vii. Taking advantage of E(z)'s analytic personality, the integral contour $\mathcal{C}'$ can be exchanged for another $\mathcal{C}''$ lying further away from z (such a trick was employed in the Joukowsky-Kutta calculation of section v). Our revised $\mathcal{C}''$ can be made to snake through the saddle points of E(z) where most of the important activity with respect to our sum occurs.[28] We then arrive at a new expansion for E(z) that possesses many delightful properties, chief among them that we need very few terms to obtain astonishing accurate answers. In fact, the first term alone is fully adequate for most practical purposes and we thereby obtain the three sectorial formulae[29] mentioned in our discussion:

For $\mathcal{D}_1$: $e^{-ikr\cos(\theta-\alpha)} - e^{-ikr\cos(\theta+\alpha)}$

$$+ \sqrt{(2/\pi)}e^{1/4i\pi}(\sin 1/2\alpha \cdot \sin 1/2\theta/(\cos\theta + \cos\alpha))(e^{ikr}/\sqrt{kr})$$

For $\mathcal{D}_2$: $e^{-ikr\cos(\theta-\alpha)-\sqrt{(2/\pi)}e^{1/4i\pi}}(\sin 1/2\alpha \cdot \sin 1/2\theta/(\cos\theta + \cos\alpha))(e^{ikr}/\sqrt{kr})$

For $\mathcal{D}_3$: $\sqrt{(2/\pi)}e^{1/4i\pi}(\sin 1/2\alpha \cdot \sin 1/2\theta/(\cos\theta + \cos\alpha))(e^{ikr}/\sqrt{kr})$

....

In a facade of this type, the "E(z)" values we calculate with three formulae $\mathcal{F}_1$, $\mathcal{F}_2$ and $\mathcal{F}_3$ correlate with the same physical quantity (*light intensity*) on each of their three patches, although the individual inferential directivities of our formulae differ greatly from one another. In our Chapter 4 examples of "theory facades" (and in most of the cases we will discuss later), predicates generally do *not* connect with the same physical

[28] K. G. Budden, Radio Waves in the Ionosphere (Cambridge: Cambridge University Press, 1966).
[29] Max Born and Emil Wolf, Principles of Optics (Oxford: Pergamon Press, 1980), 565–75.

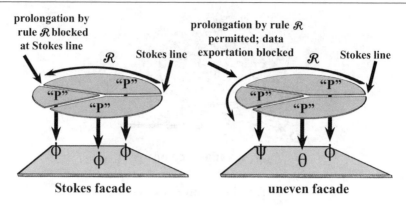

Stokes facade   **uneven facade**

quantity in each patch, but instead shift their allegiances (thus, in 4,ii "frictional force" alters its physical correlate as we move from the patch of solids into the patch of fluids). In both cases, an aspect of *alien personality* creeps into the personality of our descriptive vocabulary: it picks up the flavor of our helpful—but slightly unsuitable—assistants $\mathcal{F}_1$, $\mathcal{F}_2$ and $\mathcal{F}_3$. Faced with an inferential technique $\mathcal{F}$ of globally unsuitable personality, we have two basic choices: (1) terminate one's allegiance to $\mathcal{F}$ at some Stokes line boundary and follow other directive instructions beyond that or (2) stay loyal to $\mathcal{F}$ across that boundary but allow predicates employed in $\mathcal{F}$ to alter their attribute correlations (care in the combination of data over large scales is then required, in the same mode as we manage the inferential behavior of "$\sqrt{z}$"). An atlas of patches sewed together in the first manner I shall call a *Stokes facade*; the second, an *uneven facade* (of course, combinations of the two strategies are quite viable as well).

I will work primarily with uneven facade behaviors in this book, simply because it is easier to show that unusual forms of linguistic structure are at issue. In truth, I believe that the Stokes cases often provide the most interesting situations, for, once diagnosed, they vividly demonstrate the subtle ways in which the personalities of descriptive predicates can be strongly flavored by practical factors of which we have little or no awareness. In particular, Stokes-like situations often provide the nicest illustrations of *semantic mimicry*: language employments that look as if they operate according to simple supportive models, but actually require a fair amount of strategic machinery (such as Stokes line barriers) to remain viable. It is in this context that the linguistic pretenders I called *theory facades* in 4,vi arise: ways of talking that might very well pass for straightforward theorizing in the dark with a light behind them (to fondly paraphrase Trial by Jury once again). Indeed, the razor blade situation constitutes my original prototype for this idea, for the striking feature of Sommerfeld's approximation formulae is that they precisely delineate the traditional world of *geometrical optics*—the venerable assumption that light travels in rays—with a splendid bonus: it provides a ray-like approach to the diffraction pattern witnessed around the blade's edge (specifically, if the razor is viewed from the shadow region $\mathcal{D}_3$ where the incoming light is blocked, it will appear to glow as if a fluorescent light has been placed there). And it has become increasingly appreciated—the physicist J. B. Keller was a pioneer in this work—that, through considering further terms within Sommerfeld-like expansions, other ray-like structures can be added to our picture, such as the strange creeping rays that spin off glass globes as if they were luminous lawn sprinklers. We thus create a very useful structure of *improved ray optics*[30] situated above the wave account of light as a Stokes patchwork. But if we mistakenly approach this account as simply a "physical theory" in its own right, we will likely be puzzled by its contents, for we will soon discover that its practitioners merrily talk not only about diffraction and creeping rays, but also about sundry forms of invisible "imaginary rays" that allegedly flit about raindrops and the like. Viewed properly as a facade covering of wave optics, these strange locutions make good sense—it is merely that some of these live on the "unphysical" half of the Riemann surface employed in Sommerfeld's original treatment where they control the weakening of the beams we see (see the fine print above).

Virtually everyone familiar with these circumstances has been struck by this astonishing mimicry of ray optics; indeed, Sommerfeld himself comments in his original paper:

> In the language of geometrical optics one would interpret the formula [$\mathcal{F}_3$ for region $\mathcal{D}_3$] in the following way: rays propagate from the winding point [=our razor's edge] in the direction of the radius vector throughout the entire interior of the region, exactly as if the winding point were a luminous point . . . [O]ur eye sees it in exactly the same way, as if our visual nerves were precisely tuned to the concepts of geometrical optics. Our eye actually attributes the rays to a light source that is supposed to be at the winding point; that is, the edge of the screen appears, as seen from the region of the geometrical shadow, as a thin luminous line. This is obviously an optical illusion. . . . The error occurs because the eye forms the analytic continuation of the approximation formula [$\mathcal{F}_3$] across the transition parabola, which is not allowed.[31]

By "the transition parabola," Sommerfeld intends the thin and complicated region that is ignored in our approximation formulae. He observes that we paper over this little region (= his "form an analytic continuation") by thinking that we see a fictitious light source in its center, although no such "luminous line" really exists. That is, the light patterns inside the excluded parabolas are quite complicated, but our brains place a simple virtual image at the tip by falsely continuing the ray patterns we see in region $\mathcal{D}_3$ up to the razor's edge. However, there is no more a true light source located there than replicas of ourselves stand four feet behind the mirror. This unwarranted extrapolation provides a literal example of the "filling in" we provide when we mistake a facade for a "theory."

It is also commonly appreciated—and much important scientific work in recent years has followed this model—that working in an enriched ray optics environment provides the best setting for purposes such as telescope design. O. N. Stavroudis writes:

> So naïve a concept [as enriched ray optics] seems out of place in this age of quantum electronics. Nevertheless, whether we approve of it or not, geometrical optics maintains a

[30] D. A. McNamara, C. W. I. Pistorius and J . A. G. Malherbe, Introduction to the Uniform Geometrical Theory of Diffraction (Boston: Artech House, 1990).   [31] Sommerfeld, Diffraction, 61–2.

unique position in modern technology. It remains the only convenient means by which the gross properties of an optical system can be described in terms of its design para-meters . . . The process of designing an optical system depends almost exclusively on ray tracing . . . So we are faced with a paradox. On the one hand the procedures, the formulas used so successfully in optical design, are tailored out of the fabric of geometrical optics. On the other hand geometrical optics is, from the most generous point of view, an absurdly naïve statement of the physics of the propagation of light.[32]

When Stavroudis observes that ray ideas provide "the only convenient means by which the gross properties of an optical system can be described in terms of its design," he has in mind considerations such as the boundary condition difficulty, mentioned above, of expressing notions like "light comes in from the upper right hand corner" (we saw that this fact emerges clearly only upon our approximation facade). And there are two important considerations involved here. The first is that we can't really understand the operational core of most natural optical circumstances unless we are able to locate simpler structures like those of geometrical optics within their midst. Consider our *light intensity* predicate "E(z)." Without $\mathcal{F}_1$, $\mathcal{F}_2$ and $\mathcal{F}_3$'s helpful intervention, "E(z)" comes close to representing one of those expressions that is "fit only for a god"; it sits there quite idle until we can gather some handle upon its informational secrets. As Y. A. Kravtsov and Y. I. Orlov observe:

> [The] *available exact solutions* [of wave optics]—*valid as a rule under rather special conditions—can satisfy only a small proportion of applicational demands. This is why we turn to the . . . asymptotic techniques developed in the recent decades . . . .In this text we would like to reflect the advantages of* [this] *new direction in wave theory, which is underlaid by the principle of "spanning a wave-field cloth on a ray framework."*[33]

Indeed, this metaphor of cloth over a frame (which is very popular in optical circles) is important and the patchwork joins of my facades are intended as the basic supportive scaffolding needed to render the finer fabric of localized description coherent. The philosopher Robert Batterman has expounded on these important themes in his excellent The Devil in the Details, to which I am deeply indebted.[34]

The second key aspect is that the crux of this ray-facilitated understanding relates as much to the problems of optical *design* as to *prediction* per se. Later in the book (7,iv), we shall see that the reasoning requirements natural to design tasks are often quite different than those pertinent to prediction et al. and greatly influence the descriptive vocabulary we find suitable. As Stavroudis correctly emphasizes, ray-type ideas are better adjusted to these invention-related ends than our ordinary wave-related considerations and these inferential advantages explain the continuing centrality of ray-related notions in the thinking of anyone involved in practical optics. However, due to theory facade mimicry, the design-centered origins of this "understandability"

---

[32] O. N. Stavroudis, The Optics of Rays, Wavefronts and Caustics (New York: Academic Press, 1972), p. xii.

[33] Y. A. Kravtsov and Y. I. Orlov, Caustics, Catastrophes and Wave Fields, M. G. Edelev, trans. (Heidelberg: Springer, 1993), 5–7.   [34] Robert Batterman, The Devil in the Details (Oxford: Oxford University Press, 2002).

can easily pass unrecognized and it may falsely seem that our continuing preference for the venerable notions of geometrical optics trace to nothing more than the observation that "*light ray* represents a simpler concept than *light wave*." But this is wrong: we're simply missing the strategic rationale that supplies "light ray" with its special capacities.

In fact—and it was quite a surprise to me when I first learned this,—the intuitive understandability of *gear wheel* and other mechanical notions of that Boylean ilk trace to quite similar sources, as we shall discover in 7,iv. And these observations bring me to the nub of 6,iii's objections to the presentational view of predicate grasp that emerged in our discussion of Ernest. A predicate is often surrounded by a characteristic cloud of directivities that trace to underlying strategic purposes such as *serves the ends of invention*. We respond to these indicators of "correct usage" in our employments, yet do not always recognize their source or their underlying rationale. Put another way, "light ray" 's conceptual personality is deeply steeped in emphases that favor effective design, but we rarely appreciate the origins of that personality, anymore than we may realize that Ernest's "GSF $\neq$ BPR" problems trace to his hub-and-spoke map registrations. In that sense, we do not always understand what we grasp; we do not understand the strategic or interfacial underpinnings that supply our words with the conceptual personalities they display. The classical picture of concepts—and many of its rival doctrines as well—either pretends that the strategic underpinnings I have cited constitute part of a predicate's consciously grasped content or dismisses such considerations as pragmatic issues irrelevant to word semantics per se. Through cooption or exclusion, classical thinking misses many of the vital mechanisms involved in effective linguistic control, for the patchwork qualities of a facade often provide the framework upon which the finer fabric of localized description gets draped. Or so I will argue over the next two chapters.

However, as I noted, we will largely pursue this line of thought through the consideration of predicates whose usage unfolds over uneven facades. In the next section, I'll begin with a simple case of this sort.

. . . .

Before leaving the topic of our razor blade, there's a curious observation of some salience that we shall pursue, in another guise, in 8,x. Sommerfeld's approximation techniques supply divergent series and, as such, are not intended to provide information about the complex light patterns that appear within the narrow, parabola-shaped zones excluded above. Nonetheless, astonishing recent discoveries by Michael Berry[35] and others have shown that, by attacking these same linguistic expressions with unexpected tools, a lot of information about the omitted regions can be teased out. And this shows that our initial impressions of a phrase's conceptual content can prove mistaken in unexpected ways, for descriptive language can secretly code informational opportunities of which we are presently ignorant, but which carry those terms in utterly new dimensions once a key to their secrets is discovered. Indeed, the emergence of such hidden directivities is, essentially, the chief topic of Chapter 8.

. . . .

[35] Mark J. Ablowitz and Athanassios Fokas, Complex Variables (Cambridge: Cambridge University Press, 1997), 494–8.

(viii)

Weight. In the razor blade case, we divided our domain into sectors to prevent the strong inferential directivities attached to our series expansions $\mathcal{F}$ from dragging the predicate "light intensity" away from a steady attachment to *light intensity*, thereby creating a *Stokes facade*. But might it sometimes be better to follow the guidance of these exuberant directivities, at the price of adjusting the physical quantity to which our predicate is linked? If we police the boundaries between sectors in the right way, the resulting pattern can represent a viable form of linguistic engineering: the *uneven facade*.

Indeed, predicate usage in natural language often enlarges through *prolongation* or *continuation* in much the same manner as an analytic function obtains its full domain of definition through gradually tiling the plane in overlapping power series circles. As this occurs, uneven facades often form, frequently displaying collateral oddities such as multi-valuedness. Sometimes, it is not natural—or even possible—to cure this tendency towards multi-valuedness. Instead, it is wiser to accept that feature as part of the intrinsic nature of certain predicates, just as we tolerate two-valuedness as an unavoidable aspect of "$\sqrt{z}$"'s character. Of course, this amiable toleration of odd personality has a price; it forces us to pay attention to the sector of the facade in which we are working, lest paradoxes of a "$+2 = -2$" ilk be generated. But such controls can be instated, allowing our uneven facade predicates to serve as useful linguistic servants, even though they retain their stubbornly multi-valued personalities.

Here is a simple example of uneven facade growth that I've always found striking (the case is similar in structure to the "robin" example of 5,ix). We are warned sternly at school to not confuse the properties of *mass* and *weight*—we are instructed that the predicate "weighs 180 pounds" should be employed so that it always denotes the Newtonian vector quantity *is under an impressed gravitational force of 794 newtons* and never the mass measure *has a mass of 81 kilograms* (these quite distinct traits are commonly run together in untutored speech because they are coextensive on the earth's surface). In everyday life, we assign approximate "weight" values to objects simply by roughly estimating the difficulty of performing sundry tasks that relate to its locomotion: How easily can the object be lifted or thrown? How much it will hurt when it falls on one's head? And so forth. In circumstances demanding greater precision, we measure the object upon a spring balance. Our confusing system of "English units," unfortunately, does not sharply distinguish *force* from *mass*, but no one using the metric system confuses a *force* measured in *newtons* from a *mass* measured in *kilograms* (one of NASA's interplanetary exploratory vehicles went astray, I've been told, because several engineering firms failed to agree on whether a "pound" represents a measure of mass or weight). Our high school physics teachers often labor to liberate our thinking from any remaining *mass/gravitational force* confusions. To dramatize the instruction I have in mind, let us assume that we have taken a solemn adolescent pledge to cleave "weight" firmly to *impressed gravitational force*.

Let us now survey some of the assignments of "weight" we commonly accept outside of familiar terrestrial contexts. In school, we learn that a normally 180 pound astronaut—let's call him "John"—will only weigh about 33 pounds ($= 15$ newtons) in the moon's gravitational field, measured in terms of the local impressed gravitational force. This readjustment seems just—it is much easier to lift and toss John about within lunar circumstances and his "weight" will register "thirty-three pounds" upon a conventional spring balance. We are further told that everything is weightless inside an orbiting space capsule, including John himself. If we are devoted watchers of astronauts at work upon television, the gambols we witness in orbital contexts fully accord with our "folk physics" expectations of what ought to happen when the "weights" of familiar objects shrink to zero—erstwhile "heavy" things become very easy to lift; John registers nothing on his bathroom scale and so forth.

Suppose that somebody now invites us—perhaps in the course of a school exercise designed to illustrate the diminishing influence of terrestrial gravitation—to register upon a *single chart* stretching from the earth to the moon the "weight" that John will assume in each locale. This represents a choice of *map scale* that we rarely consider, but, at first glance, the task seems quite straightforward. Surely we know roughly what the map should look like in qualitative terms: John's "weight" should gradually diminish as we approach the moon, eventually becoming zero along some surface in space fairly near the moon. Once this surface is transversed, his "weight" vector will turn to point towards the moon, growing in magnitude until it reaches its expected value of thirty-three pounds upon the lunar surface. But now a hidden anomaly within our everyday employment of "weight" strikes us, an observation that has perhaps escaped our attention previously. Heretofore we have steadfastly accepted the contention that John is "weightless" in a space capsule located in a standard orbit approximately 250 miles from earth, a distance far shy of the roughly 200,000 miles out where his *impressed gravitational force* vector will actually sum to zero. At John's 250 mile orbital location, the impressed gravitational force will have diminished only by a few pounds from its terrestrial 180 pound value. A tame multi-valuedness has emerged: we are inclined to assign John simultaneous "weights" of both zero and 176 pounds.

What has gone wrong here? What directivities led us to suppose that John is "weightless" in orbit, despite having pledged "weight"'s troth to *impressed gravitational force*? Undoubtedly, the answer traces to circumstances allied to our earlier considerations with respect to maps. When we picture a situation involving the "weights" of familiar-sized objects to ourselves, we instinctively align our data within mental maps fixed at some middling *scale of size*: our living room, the backyard of our house, the interior of a space capsule, the floor of a lunar crater (rather as our knowledge of a city might be largely carved up into neighborhood-sized navigational lists). As we apply the predicate "weighs 180 pounds" within each of these limited domains, we expect the phrase to attach to some local facsimile of the folk physics evaluative standards that operate in the context of our living rooms and back yards. These reasoning rules provide natural prolongations of our terrestrial standards into novel settings. The resulting inclination to consider "thirty three pounds" an apt evaluation of our

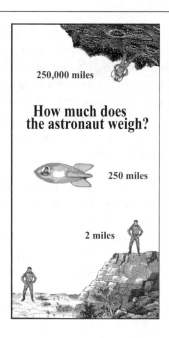

astronaut's "weight" on the lunar surface stays loyal to our high school vows. However, the same folk physics prolongation to the space station shifts "weight" to another property, roughly: *work required to move x relative to local frame*. But we do not notice this shift because John's free floating behavior, as well as that of the unattached paraphernalia around him, meet our expectations of what we should see if all *impressed gravitational force* were reduced to zero (as approximately occurs in interstellar space). And no other natural indicator of the substantial gravitational forces actually impressed upon John emerges to warn us that we have made a classificatory shift. In the face of so much apparent classificatory reward and so little discouragement, we remain well satisfied with our "weighs zero pounds" judgments. Thus the powerful inferential and classificatory directivities of our folk physics reasoning standards silently shape our space station employment of "weight" in a parochially adjusted manner, despite the fact that the predicate is thereby *dragged away* (4, ii) from its erstwhile attachment to *impressed gravitational force*. Such folk physics directivities provide the bridge or nucleation site from which a new patch of usage containing space stations grows. And, of course, even proverbial "rocket scientists" cheerfully succumb to this tempting prolongation.

Casting these circumstances as an uneven facade, our terrestrial evaluations should be regarded as defining a natural domain choice (E) which becomes attached to sheets of other scale sizes through bridges of folk physics prolongation. That is, we consider that we still "use 'weight' in the same way" when we discuss objects that fit within an elevator, an expanse on the moon (M), the interior of an orbital vehicle (O), the full span of space from the earth to the moon (S), etc. The bridges between most of these patches represent well-traveled pathways, but that between O and S is not.

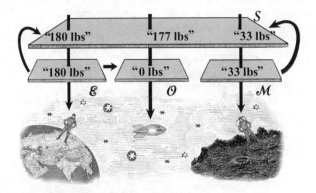

This lack of direct connectivity (as well as the fact that we rarely bother to worry about puny space stations when we think in S-scale terms) shields the palpable conflict between the S sheet evaluation and that of the O sheet from our attention. However, if a school exercise encourages us to compare the two classifications, the multi-valuedness hidden in our atlas becomes revealed, although its benign presence rarely creates trouble otherwise. Such latent-but-not-devastating circumstances should be compared to the fact that most of us believe that we know how to calculate with complex numbers ably, when, in fact, we are not adequately prepared for the shifts in context presented in 6,v's "$+2 = -2$" calculation. Working mathematicians, after running up against many painful surprises of this ilk, have learned to be careful with respect to the branch of the Riemann surface upon which their calculations unfold—data cannot be freely exported from one patch to another. Most of us have never been taught such precautions, however, and are only preserved from frequent error by the fact that the computations we attempt with complex roots generally prove quite local in character. Our shifting "weight" talk is quarantined through the same loose control.

Such roving assignments ratify our close-to-discordant inclinations to declare both that "Astronauts get some of their weight back within a rotating space station" and "Passengers merely experience a *simulation* of weightlessness within a falling elevator." When we think of subjects in a hoist, their normal terrestrial "weights" (i.e., the amount of *impressed gravitational force* they experience) remain quite palpable to us, whereas we have very little motive in the orbital context to discuss applicable values of this same quantity (they are somewhat hard to compute in any case). Our earth-to-moon diagram surprises us precisely because it forces us to consider how space station data exports to other localized applications.

I find it fascinating—linguistically, quite revealing—that our high school vows to cleave "weight" firmly to *impressed gravitational force* utterly fail to prick our semantic consciences when we follow the folk directivities that carry us from E into O. Why don't we notice the switch? Simplifying what is undoubtedly a complex situation, we possess two key criteria for attributing a "weight" to something: (i) as a folk physics parameter for gauging ease of lifting effort, etc.; (ii) through Newton's gravitational law. The great ease of applying the former (and the lack of much everyday demand to employ the

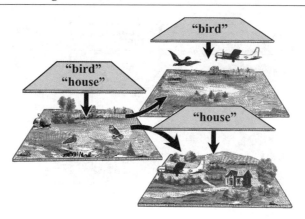

latter) encourages an initial colonization of our space station patch in a manner that favors the "zero weight" assignments. These determinations get remembered as "facts about astronauts" and such dignity as scientific verities hardens our unusual orbital prolongation into permanence. Allied processes solidify the Druids of 1,ix in their first impression classifications of airplanes. In these circumstances, the original sheet of "bird" usage as employed in innocent, bomber-free days possesses two potential continuations onto divergent sheets, neither of which the society has yet explored. But once a pathway has been actively opened by the first airplane spotters, the bridge into the alternative prolongation gets thereby closed. Accordingly, the usage never displays multi-valuedness, but the same processes of patch-to-patch prolongation are clearly at work.

These observations are clearly pertinent to the pre-pragmatist doubts canvassed in the previous chapter (especially 5,iii musings with respect to the large degree of prescient engineering required to prevent a classifying robot from suffering problems of multiple continuation). At any particular point in time, many of our predicates are firmly correlated with a smaller portion of the full universe than we *ur*-philosophically anticipate, but our inclinations towards *tropospheric complacency* (2,iii) mask the fact. We don't notice the facade unevenness of our usage because of both Druid-like satisfaction with the contextually decided continuations that we later adopt and by the unnoticed contextual controls that prevent the patches in an uneven facade from conflicting with one another. Or, to put matters another way, classical thinking tells us that when "weight" gets *property dragged* from *impressed gravitational force* to *work required to move x relative to local frame*, we must have failed to attend to the little alarms and whistles that surely get tripped within our semantic awareness when that bridge of prolongation is crossed. But this is entirely wrong: we merrily walk into the new patch in the brightest sunlight, with nary a discouraging sound or signal to dissuade us.

The classical picture errs in its exaggerations—to hold a predicate fixed to an attribute, it is not enough to meditate deeply upon meanings or vow sincerely to never shift referents. Such activities, although they often alter subsequent usage in positive ways, do not always serve as adequate prophylactic against natural shifts of "weightless astronaut" type. To be sure, the prolongation to "weightlessness" induced by our folk physics rules might have been prevented, but only at the cost of much stricter vows back in high school, e.g., we would have needed to pledge additionally, " . . . and, furthermore, I will never grant an object in a novel frame a 'weight' unless I have studiously calculated its value by Newton's gravitational law and verified the constancy of that result in every permissible change of frame" (a weekly postcard reminding us of our pledge might be helpful as well). But the directivities we consciously assume in language are rarely that ample or authoritarian, leaving ample room for property dragging temptations to silently transport our employments into shifted patches.

I believe that Quine, in his earliest inklings of the errors in classical thinking, sensed that language's attachment to the world must surely be *looser and more uneven* than tradition leads us to expect (5,i). However, rather than making his case by simply exhibiting salient examples in a manner akin to my own, Quine's thinking became sidetracked by dubious holism, with the end result that a justified caution with respect to loosened attachment blossomed into an impossible paradox of little instructive value. Rather than embracing his extreme indeterminancy of translation et al., Quine might have stayed truer to original pre-pragmatist hunch if he had simply noted, "Remember high school. We may earnestly vow to walk the semantic straight-and-narrow, but the headstrong personality of 'weighs five pounds' is not restrained by the classical gluing we fancy we have set in place."

Consider how we generally describe the situation in ordinary life when "weight"'s oddities are brought forward. We respond, "Gee, how did that happen? I knew that the concept of *having weight* mandates attention to *gravitational force*, so my attention must have slipped somewhere along the way, allowing folk physics reasoning to saddle 'weight' with a second, unnoticed meaning." Here's how an intermediate grade textbook (Glencoe Physical Science) handles the subject. Since they have earlier insisted that weight is "the measure of the force of gravity on an object," the authors claim that an astronaut's condition is merely that of "apparent weightlessness," so that when we speak of astronaut "weightlessness," we must really mean something different:

> The space shuttle and everything in it are in free-fall around the Earth, thus producing apparent weightlessness . . . But to be truly weightless, the astronauts would have to be free from the effects of gravity . . . So what does it really mean to say that something is weightless in orbit? . . . [If a scale were falling along with the object to be measured], *the scale couldn't push back on the object, so its dial would read zero. The object would seem to be weightless.*[36]

Having once explained the bridge that leads to this "second meaning," they feel free to use the "new meaning" without apology. A paragraph later they write:

> What happens to the physical condition of astronauts who experience the sensation of

[36] Charles McLaughlin, Marilyn Thompson et al., Glencoe Physical Science (New York: McGraw-Hill, 1997), 103–4 (sentence order switched). Insistence that weight = impressed gravitational force appears on p. 84.

weightlessness for extended periods of time? . . . Health tests performed on American astronauts have shown that some bone and muscle deterioration occurs during long periods of weightlessness.

Notice the segue back to using "weightless" without any qualification as merely "apparent." Having once discharged the duty of recognizing this "second meaning," our authors feel justified in employing this "second meaning" freely within orbital contexts.

As such, there is nothing greatly wrong with this mode of description; the authors are plainly telling a story of the origins of "weightlessness" comparable to my own, yet couched in our everyday terms of conceptual evaluation: "concept," "understand" and "meaning." However, that choice of phraseology readily encourages the picture that, in crossing over to "weightless astronauts," we have done something mildly untoward linguistically, as if we have confused "council" with "counsel." Built into our ordinary talk of "changing the significance of 'weight'" is the specter of an unnoticed slip into polysemy; that some semantic ball has been dropped that we had formerly grasped firmly. By tacitly installing these nonexistent events within our mental histories, we begin to wonder why we haven't better remembered these episodes of semantic adjustment. In fact, we commonly fancy that we do: "Now that you bring it up, I recall being puzzled when I first heard about astronauts being weightless." Whether such recovered memories should be regarded as valid or not, we can easily persuade ourselves that our former usage should be described as one where "we grasped two concepts of **weight** but failed to keep them distinct."

It is in these innocent ways that *ur*-philosophical errors begin. In speaking of "two concepts of *weight*," a faint whiff of absolutism and assumed control have crept into an otherwise correct account of "weight"'s developmental circumstances. We thus reassure ourselves: "If I had just thought about the matter a bit harder, I could have pulled those two notions apart." In fact, this claim is true, but it encourages the further belief that, if we had simply concentrated harder upon our high school vows, our use of "weighs five pounds" would have never become bifurcated as it has. But this is likely *not* true.

It is precisely within these humble, but slightly askew, evaluations that the grander methodological ambitions of a Russell take root: "If we simply concentrate harder upon our basic semantic vows, the confusions into which mathematics, physics and philosophy commonly fall might be eradicated forever." Such is the *ur*-philosophical stuff of which dreams are made. We shall return to these issues more fully in Chapter 8 when we consider the status of programs for rigorization in greater depth.

Before we move onto richer examples, consider the following remark, which I encountered in the psychological literature:

> [I]*f the question is why astronauts sometimes do and sometimes do not float around inside a spacecraft, the presence of gravity on earth is an important causal hypothesis that differentiates the occurrence of the outcome from its absence.*[37]

---

[37] M. W. Schustack, "Thinking about Causality," in R. J. Sternberg and Edward Smith, eds., The Psychology of Human Thought (Cambridge: Cambridge University Press, 1988), 95.

In terms of her daily usage, this author probably doesn't *apply* "weight" in a manner differently from the rest of us. But she plainly conceptualizes the notion wrongly; how should we characterize the character of her thinking? And here I suggest: she embraces an incorrect *semantic picture* of how the term "weight" relates to the world; she has no inkling that a facade-based switch occurs within orbital contexts. Unlike the majority of us, who might fairly protest that "We've never really thought about what 'weightless astronaut' means," this author clearly has—and has patently come up with the wrong answer.

In the sequel, we shall have much occasion to ponder these correct usage + wrong picture combinations more fully.

(ix)

Hardness. What is it for a material to be "hard"? Recalling the discussion of 3,vi, Descartes claims that it merely represents a disposition to occasion a subjective sensation of "resistance" within us:

> For as far as hardness is concerned, our senses tell us nothing about it except that the parts of hard bodies resist the movements of our hands when we encounter them. Besides, if whenever our hands moved in a certain direction, the bodies situated there were to move back at the speed at which our hands approach; we would never feel any hardness.[38]

In particular, there is a sensation-type called "resistance" and *hardness* simply represents the disposition to engender such feelings in the agent under appropriate conditions of testing (which will not be met if the table, say, moves away from our probing fingers). In 2,vi, we examined proposals that treat physical color classification as "dispositional" in this same spirit.

However, the eighteenth century Scots philosopher Thomas Reid rightfully objects that, even if some uniform sensation of "hardness" exists (which he considers dubious), the notion of *hardness* itself is of a distinct nature not intrinsically tied to such sensations at all:

> Pressing my hand with force against the table, I feel pain and I feel the table to be hard. The pain is a sensation (experience) of the mind and there is nothing resembling it in the table . . . I touch the table gently with my hand and I feel it to be smooth, hard and cold. These are qualities of the table perceived by touch; but I perceive them by means of a sensation which indicates them. The sensation not being painful, I commonly give no attention to it. It carries my thought immediately to the thing signified by it, and is itself forgot as if it had never been. But by repeating it and turning my attention to it, and abstracting my thought from the thing signified by it, I find it to be merely a sensation, and that it has no similitude to the hardness, smoothness, or coldness of the table signified by it.[39]

[38] René Descartes, Principles of Philosophy, Valentine and Reese Miller, trans. (Dordrecht: D. Reidel, 1983), 40–1.
[39] Thomas Reid, Essays on the Intellectual Powers of Man (Cambridge, Mass.: MIT Press, 1969), 244–5.

punching   Brinell indentor

Furthermore:

> Further I observe that hardness is a quality, of which we have as clear and distinct a
> conception as of anything whatsoever. The cohesion of the parts of a body with more or less
> force, is perfectly understood, though its cause is not: we know what it is, as well as how it
> affects the touch.[40]

Indeed, Webster's informs us that a material is hard if it is "not easily penetrated" and
does "not easily yield to pressure." Reid's point is that this requirement is straightforwardly physical in nature (he classifies *hardness* as a "primary quality," borrowing
Locke's terminology); any sensations of "resistance" are merely part of the evidential
path that helps establish a material's hardness or softness.

So who is right? In fact, neither, because our usage of the predicate "is hard" displays a
fine-grained structure that we are unlikely to have noticed, for our everyday usage is
built from local patches of evaluation subtly strung together by natural links of prolongation. More specifically, in everyday contexts we adjudicate the "hardnesses" of
various materials, both comparatively and absolutely, through a wide variety of comparatively easy to apply tests—we might *squeeze* the material or *indent* it with a hammer;
attempt to *scratch* it or *rap* upon it; and so on. In most cases, we will be scarcely aware of
the exact technique we will have employed for this appraisal: "Did I rap, squeeze or
scratch that piece of wood? I can't really remember." In fact, our choice of tests is likely
to have been suggested by the material in question: we instinctively appraise a wood
by rapping upon it, a rubber by squeezing, a metal by attempting to make a small
imprint; a glass or ceramic by rapping lightly or scratching (not by trying to make a small
imprint!). In fact, we are normally interested in comparing hardnesses mainly within
natural groupings of stuffs of generally allied characteristics, generally metals with
metals, ceramics with ceramics and so forth, although interesting crossover cases also
arise. When we are invited, as sometimes happens, to evaluate widely dissimilar
things with respect to "hardness"—a clay flower pot, say, with a Bakelite version of the
same—, we might be initially nonplused by the query and cast about for a reasonable
standard of equitable comparison. But we scarcely pay attention to this network of
evaluative parochialisms: the tests we select seem to be naturally ingrained within the

---

[40] Reid, Inquiry, 61.

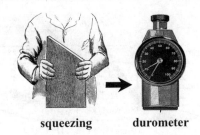

squeezing   durometer

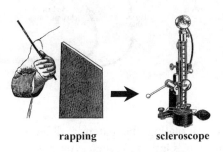

rapping   scleroscope

limited "search spaces" in which we look for a material that's "a little bit harder than specimen X."

These hidden forms of preferential technique for assessing "hardness" become quite salient as everyday informal methods become improved into the carefully calibrated forms of testing apparatus that go by such titles as Brinell or Vickers indenters (vigorous squeezing and then releasing); superficial Rockwell testing (mild squeezing and partial releasing), durometer[41] (squeezing without releasing), sclerometer[42] (scratching), scleroscope (a different instrument that raps its specimen), the Charpy impact test (hitting with a hammer) and so forth. Different manufacturing industries rely upon these different varieties of test procedure according to the materials with which they customarily work. According to The Metals Handbook of the American Society for Metals:

> The definition of hardness varies depending upon the experience or background of the person conducting the test or interpreting the data. To the metallurgist, hardness is the resistance to indentation; to the design engineer, a measure of flow stress; to the lubrication engineer, the resistence to wear; to the mineralogist, the resistance to scratching; and to the machinist, the resistance to cutting.[43]

[41] Samuel R. Williams, Hardness and Hardness Measurements (Cleveland: American Society for Metals, 1942), 327–33.   [42] D. Tabor, The Hardness of Metals (Oxford: Oxford University Press, 1951).
[43] Andrew Fee, Robert Segabache and Edward Tobolski, "Hardness Testing" in John Newby, ed., Metals Handbook Ninth Edition, viii (Metals Park: American Society for Metals, 1985), 71.

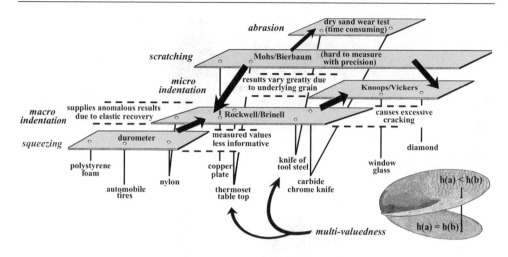

Indeed, it is only by extrapolating the patently different <u>modi operandi</u> of these professionalized tests backwards into our casual everyday patterns of hardness appraisal that we learn that we instinctively appraise different materials according to standards provincial to the stuff at hand. When we reflect on our activities, the feelings induced when we investigate how easily a plastic scratches are not much like the sensations of rapping on a steel plate, yet we may come away from both probes with the common assessment, "that stuff's pretty hard."

The end result is that our employment of "hardness" silently distributes itself into a patchwork of sheets, locally distinguished by a certain vein of probing (scratching, tapping, etc.), that sit over various varieties of material stuffs and continue smoothly into one another.[44] The basic causes for this tacit multiplicity seem similar to those observed for "weight," although ramified in complexity. As an inferential tool, most of us employ the term "hardness" informally as a generic, single-scale "folk physics" parameter that we consult in selecting a material for, e.g., manufacturing purposes. Left at this general level, the term requires further specialization before it can carry much data usefully. Accordingly, over each localized domain of related stuffs, the "hardness" parameter is likely to silently specialize on the forms of probing that prove most informative with respect to the materials and purposes locally at issue. It is very easy to see this "connected sheet structure" when we survey the developed "hardness testing" techniques of industry, because quite different forms of apparatus (of the types listed) are popular with respect to differing classes of material. The exact nature of these refinements in technique typically depend on not only the *type* of material under consideration but the *circumstances* in which an evaluation of "hardness" is likely to be

---

[44] Kenneth and Michael Budinski, <u>Engineering Materials: Properties and Selection</u> (Upper Saddle River, NJ: Prentice-Hall, 2005), 245, supply a valuable chart showing five overlapping patches of *hardness* coverage over a span that encompasses rubbers, plastics and metals.

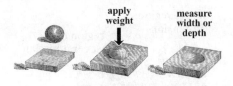

required. In particular, it is often imperative that our evaluations of "hardness" be both convenient and non-destructive.

....

For example, in manufacturing either a steel or hard candy (where delicate heat treatment is required), we need to ascertain rather quickly whether a sample from the batch has achieved the requisite "hardness," which we might test (in the case of the steel) by measuring the width of impression left by a punch of assigned load or (in the case of the candy) by determining the load at which the test piece crushes. But N. J. Mills explains why the so-called Brinell-type tests, widely employed in the manufacture of metallic items, are rarely applied to common plastics:

> Hardness tests [of indenter type] *are widely used as a non-destructive method of estimating the yield stress of metals, and hence checking whether heat treatments, or surface treatments like carburization, have been carried out correctly. Although plastics have much lower hardnesses than metals, the test is not widely used. This is partially because there are no heat treatment methods for modifying the yield stress, and partially because viscoelastic effects make the size of the indentation decrease with time. It is often important to determine the resistance of plastics to surface damage, but the tests used are more closely related to the specific applications.*[45]

That is, the manufacture of plastics rarely require the same kind of "check for a specified hardness" tests that recipes for metals and candies require. Furthermore, the tests that prove most reliable for judging the condition of a steel require that we sink a steel ball or other indenting probe into the test object for thirty seconds or so, then measure the diameter of the impression left behind. But a polymer displays a "viscoelastic" tendency to slowly fill in this cavity and, in an extreme case like a rubber, regains its original state rather quickly. If we followed the usual standards for the hardness of a steel, ordinary tire rubber would prove to be rather "harder" than cold-worked steel. There are procedures (Shore Durometer) that adjudicate "hardness" according to how far the indenter sinks into a specimen under a given load, which may be appropriate to a normal rubber, but which supply less adequate information as to a metal's ability to accommodate stresses through "plastic" deformation—that is, the manner in which it responds to dents.

In fact, pressing a ball into a substance represents a rather complex process and, depending upon the material effected, completely different effects may dominate the process. For example, some type of slightly raised lip usually forms around the edge of the ball when it presses into the material. In this small region, the material is pulled in weak tension (as opposed to the much larger compressive state that obtains elsewhere). For most metals, this wee bit of extrusion isn't important (except insofar as it makes the diameter of the crater harder to measure), but for a brittle substance such as a plastic or glass, it signals disaster: the material is likely to crack when pulled upon (as opposed to pushed). So how do we measure the "hardness" of a plastic? By taking extra

[45] N. J. Mills, <u>Plastics: Microstructure, Properties and Applications</u> (London: Edward Arnold, 1986), 163.

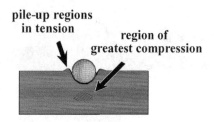

precautions (say, using a different type of indenting procedure) to suppress the tensile state or to consider the pressure at which this fracturing occurs to be a good measure of its proper kind of "hardness," despite the fact that metals don't display allied response. When Reid maintains that *hardness* is "the cohesion of the parts of a body with more or less force," he is obviously not considering whether the "force" comes in the form of a push or pull.

Nor do most of us, not forewarned, ponder the issue as we go about the business of classifying the objects around us into "hard" and "soft." Nonetheless, *likeliness of indentation* is rarely a standard we observe in evaluating a plastic whereas it is often critical in how we think about a metal, so, in this subliminal way, the differences in response type between the two materials tacitly affect our classificatory behaviors. Indeed, for a piece of sugar candy—which qualifies as a form of glass—, we usually evaluate "hardness" according to when rupture initiates.

For a soft plastic, however, it is more common to estimate its "hardness" or "softness" according to its resistance to *abrasion*. The capacity to avoid scratches is, of course, the trait that the celebrated (but rather ill-defined) Moh's test employed by mineralogists seeks to quantify (in the familiar "anything that can scratch quartz has a Moh's number greater than seven" vein). However, the tribological processes excited in any form of frictional rubbing are even more complicated and varied than occur in the quasi-static impression of an indenting object and so the prospects for articulating a successful approach to generic "hardness" in abrasive mode are dimmer (indeed, it is very hard to achieve consistent values of any precision using abrasion based tests). The situation is even more unpromising for the sorts of "hardnesses" we evaluate in *impactive circumstances*: "Boy, this glass is hard—look what it did to that bullet" or even "I didn't realize how hard water becomes when you dive into it from one hundred yards above." Of course, specialists in bullets and so forth regularly speak of "hard" and "soft" behaviors, but these employments are quite localized to impulsive time scales and are not easily exported to the more leisurely walks of life.

We might also mention that our dictionary's claim that a hard substance "does not easily yield to pressure" is either wrong or quite misleading. We normally expect that a hard object will not readily fall apart under slight pressures but a substance that coheres too strongly at the molecular level (specifically, it does not allow for sufficient "dislocation" movement) will not be able to distribute applied stresses throughout its bulk, with the consequence that even small pressures may concentrate at critical junctures and destroy the integrity of the material. To be sure, metallurgists sometimes bite this particular bullet through exclusively following the directivities of indenter focused criteria:

> *Many intermediate phases are extremely hard and brittle so that a small ingot of such a substance may often be crushed to a powder by gentle pressure in the jaws of a vice.*[46]

---

[46] R. A. Higgins, Engineering Metallurgy (London: Edward Arnold, 1983), 175.

But most of us will be inclined to react to such a discovery, "Boy, this bar really isn't very hard." Certainly an unexpected degree of microscopic mobility represents a crucial ingredient in the correlated toughness we expect in a paradigmatically "hard" material. Authors like Reid who fancy that the basic requirements for *hardness* are "perfectly understood, though its cause is not" naïvely trust to a natural *picture* of the trait wherein "perfect hardness" entails complete rigidity, whereas "hardness" in real life circumstance secretly demands some measure of dislocation mobility or other method of stress dispersion. But Reid's naïve "picture" has an unusual role to play in "hardness"'s linguistic career, which we will discuss further at opportune moments.

....

Accordingly, within a given localized specialization of "hardness," a natural tendency arises for the predicate to hunt for some more fundamental evaluative trait upon which it might locally settle, where the physical attribute in question proves especially important for the class of materials under consideration.

....

Specifically, within the most common range of steels and related metals, "indenter" type tests are commonly employed to examine the capacity to accept permanent indentation at substantial but not extreme pressures. In these central cases, the important physical variable that controls the process is the metal's so-called *yield strength*: the tension at which a plug of the stuff needs to be pulled in order to introduce a permanent set in the material (in the jargon, when "plastic" response initiates). In these contexts, *yield strength* becomes the attribute to which "hardness" locally gravitates (although such identification is completely unnatural for a glass). Indeed, it is quite common for basic texts in metallurgy to regard "hardness" as simply a "measure of yield strength" calibrated along an unusual, machine friendly scale (rather as the Fahrenheit temperature scale relates to the less convenient but theoretically better founded Kelvin scale). Accordingly, such works commonly provide tables that translate Brinell hardness values into a proper yield strength (the rule is that approximately $H = 500Y$). But within the clique of those who converse most actively about rubbers, "hardness" tends to specialize instead in the direction of *Young's modulus of elasticity*, which represents a quite different physical quantity (*yield strength* is either undefined or of peripheral importance to a rubber).

....

However, we should observe that these tendencies towards local specialization are always constrained by the restraints of practicality, for manufacturing often requires traits that can be quickly evaluated "on the fly." A good measurement of *yield strength* requires complexities that simply cannot be carried out in a factory setting, where the supervisor needs to test a sample of the metal to adjudicate whether the complete batch has completed the current stage of its manufacturing routine satisfactorily. In this regard, a Brinell hardness test can be executed many times more quickly and conveniently than standard direct measurements of *yield strength*. But convenience has its deficiencies as well. Sidney Avner comments in his <u>Introduction to Physical Metallurgy</u>:

> *The property of "hardness" is difficult to define except in relation to the particular test used to determine its value. It should be observed that a hardness number or value cannot be*

utilized directly in design, as can [yield value], since hardness numbers have no intrinsic significance. Hardness is not a fundamental property of a material but is related to the elastic and plastic properties. The hardness value obtained in a particular test serves only as a comparison between materials or treatments. The test procedure and sample preparation are simple and the results may be used in estimating other mechanical properties. Hardness testing is widely used for inspection and control [such as] heat treatment or [mechanical] working.[47]

When Avner stresses that hardness values "cannot be utilized directly in design," he indicates the fact that they do not supply the optimal recipe values we should juggle when we try to figure how a hunk of worked steel might be manufactured to meet prescribed product specifications. Avner's point is that the quick classification merits of "hardness" far outweighs its potential design recipe value.

In a related vein (although with quite different practicalities in mind), the nineteenth century geologist Alfred Werner correctly stresses the importance of "in the field" mineral classification in the course of his important early attempt to develop a "scratch test for "hardness":

It has been considered that a mineral was sufficiently distinguished when its uses were described or when its components were given, or when it was classified here or there according to certain reputed principal characters. This resulted in the neglect of complete and correct descriptions of minerals according to external characters. This has been carried so far that scarcely a single mineral is to be found described in a manner in any oryctognostical textbook that it can be immediately recognized and clearly distinguished from others resembling it. Yet this is the most essential part of an oryctognostical system— I would rather have a mineral ill classified and well described, than well classified and ill described.[48]

Such divergent demands of practical advantage help capture our predicate within the orbital basins of competing testing techniques, but, once a particular attractive center has become established, local affinities between instrumentally determined "hardness" and salient focal quantities like *yield strength* will issue precisified directivities that typically pull "is hard" into specialized alignment with the latter. It is not surprising that we accept these local departures from a strict alignment of "hardnesses" with the values of instrument readings, for we generally regard our measuring tools as imperfectly assigning numbers to preexistent quantities anyway. Indeed, any real life test procedure supplies fluctuating values, so in considering how the quality of our tests might be improved, we naturally begin to ask, "what intervening factors spoil perfect correlation between instrument reading and the trait we're really after?" Thus in David Tabor's well-known primer on The Hardness of Metals (where "indentation hardness" is mainly

---

[47] Sidney Avner, Introduction to Physical Metallurgy (New York, McGraw-Hill, 1974), 24. Hans-Georg Elias, An Introduction to Plastics (Weinheim: Wiley, 2003), 214.

[48] A.G. Werner, On the External Characters of Minerals, Albert Carozzi, trans. (Urbana: University of Illinois Press, 1962), 99.

in view), the comment appears:

> It has long been recognized that the formation of the indentation itself leads to an increase in the effective hardness of the metal so that the hardness test itself changes the hardness of the metal under examination.[49]

Indeed, our village smithy—if we still have one—can tell us that "work-hardening" (banging on it with hammers and rollers) "toughens" or "hardens" a metal; that it thereafter requires a more vigorous tug to make it stretch permanently. Tabor is merely expressing the fact that the usual processes of indentation testing must increase (or, occasionally, soften) *yield strength* through work-hardening and that it would be desirable to minimize this factor within a "good hardness test." From a linguistic point of view, we here witness a funny, but quite common, chain of shifting dependencies: the salience of indenting tests for metals is needed to pull "hardness" into approximate alignment with *yield strength* (a correlation not valid for other materials), but, once this local tie has become established, the "hardness"/*yield strength* link can then haughtily criticize as "imperfect" the criteria that originally engineered its alignment. Nonetheless, we can't understand these local precisifications properly unless we see them as arising within a larger field that gets divided into patches largely on the basis of salient test procedures.

Surveying our results, we see that two sorts of mild multi-valuedness have crept into our "hardness" classifications. The first occur when two distant patches on the *hardness* manifold cover common materials in divergent manners, as arises amongst certain tabletop coverings and high grade knife blades. And then there are also the evaluative shifts that can occur under a shift in the *scale of application*, similar to that which induces the property dragging witnessed in the astronaut example. Thus Tabor informs us, on the one hand, that "hardness" needs to be regarded as a measure of how the *surface* of a metal responds to indentation yet observes a few pages later that such interventions supply a somewhat misleading estimate of the "true hardness" of the bulk metal *below* that surface (the remark quoted is of this type). If pressed about such apparent discrepancies, such an author will likely retort with some annoyance: "Oh, you know perfectly well what I mean: the claims I've made are entirely consistent with one another." And this is a perfectly fair response, for we can easily follow the adjustments on the manifold of *hardness* that he expects of us. Such shifts in scale-contextual content commonly occur when classificatory terms that we acquire in conjunction with some simplistic, Reid-like pictures of their supportive underpinnings get adjusted to differently sized arenas where variant schedules of physical events dominate. Indeed, we shall survey another philosophically celebrated example of the same phenomena ("solidity") in the next section.

This is perhaps a convenient time to recall a substantive divergency that separates the present point of view from that favored by the ordinary language movement (1,vi). We agree that a term like "hardness" is governed by a complex schedule of localized

---

[49] Tabor, Hardness, 15–6.

controls in a manner that allows Tabor to advance apparently contradictory claims in a coherent and informative way. But the ordinary language school also presumes that such communicative capacities can only be acquired as "rules" imbibed in the course of becoming competent in English. I, on the other hand, view the facade of *hardness* as gradually framing itself as an initially unspecialized term adapts itself, in a fairly predictable manner, to the individualized aspects of the materials around us. True: such developmental processes occur naturally only because we share a general psychological proclivity to approach favored topics in small and restricted domains and because we are able to juggle rapid shifts in "investigative mood" while scarcely noticing the lifts and pullbacks. But these capacities merely provide the general mechanics that allows a specific facade structure to form with respect to a term like "hardness": the exact patches that form within its atlas will be determined quite directly by the diverse manners in which bulk matter responds to our practical manipulations. Whatever "rules" our linguistic peers might have originally taught us in regard to "hardness" prove quite beside the point, soon overcome by the impertinent particularities of metal, ceramic and plastic.

....

Departing the forge momentarily for the circles of academic philosophy, it is common nowadays to claim that a trait like *redness* represents a so-called *response dependent quantity*, as if this updating of Locke's "secondary quality" represents a clear category of metaphysical gizmo. It is true (as observers like Michael Dummett[50] have usefully noted) that terms like "redness" (and, by extension, our "hardness") must remain tethered to various forms of "on the fly" classificatory judgment lest their basic utility be utterly destroyed. But this basic fealty to swift practicality in no way prevents local specializations of the Tabor sort from naturally arising, since instruments never cover their fields of practicality perfectly. Closer attention to example will usually disclose a complicated webbing of localized refinements upon the employment of typical "on the fly" classificatory words and that "response dependent quantity" is probably not a well-defined category into which either attributes or concepts can be sorted.

Our experience with "hardness" also demonstrates that it is generally very difficult to pin "response dependence" to a precise standard: no matter how carefully we think we have specified the testing conditions demanded, unexpected factors are likely to render the "response" engendered ambiguous in import. My favorite example lies with the newly discovered *shape memory* alloys.[51] Indent these metals and they will likely fill back in most of the impression left by the punch. However, change their temperature slightly and they will "remember" the deeper punched impression and quickly revert to it. So what is the proper way to assess their Brinell "hardness," because who would have dreamed that such dramatic thermal effects might intervene in the "response"?

We might also observe that, within modern engineering manuals, the availability of competing terminology such as "toughness," "yield strength" and "Young's modulus" tends to drive

[50] Michael Dummett, "Common Sense and Physics" in The Seas of Language (Oxford: Oxford University Press, 1993).

[51] A. V. Srinivasan and Michael McFarland, Smart Structures (Cambridge: Cambridge University Press, 2001). Vladimir Boyko, Ruvin Garber and Arnold Kossevich, Reversible Crystal Plasticity (New York: American Institute of Physics Press, 1994).

"hardness" from its erstwhile centrality of usage according to the same basic processes of banishment that are familiar from historical linguistics:

> *An interesting case concerns* **autumn** *and* **harvest**. **Harvest** *is the native Germanic word, cognate with German* **Herbst** *"autumn." However, after the Norman conquest, the upper classes adopted a great many French words, including* **autumn**. *This borrowing promoted a semantic shift:* **autumn** *became the normal word for the season, while* **harvest** *was reserved for the agricultural labor the peasantry would have been performing at that time.*[52]

As this supplantation takes effect, the venerable cries of "this kind of steel is pretty hard" mainly reverberate in the work places of practical manufacture nowadays, and less commonly trip across the tongues of the engineers who design the materials that are manufactured there. In the halls of planning, "yield point" and "Young's modulus" instead reign supreme.

....

(x)

Linguistic management. The domain localization just discussed is driven by factors similar to those that affect "weight": originally unspecialized standards for evaluating and reasoning about "hardness" gradually adjust themselves to the impertinent particularities of metals, plastics, minerals, cutting tools, foams, etc. through locally advantageous accommodations. But the various classes of materials do shade into one another and, accordingly, questions of how to appraise the comparative "hardnesses" of a metal and a ceramic, say, frequently arise. Generally this occurs with respect to borderline stuffs, such as the new ductile ceramics utilized in space vehicles, where the materials have been especially contrived to incorporate metal's more desirable features. Here conventional indenter-type tests seem immediately appropriate for their assessment and provide a smooth overlap that ties the domains together (in the manner of a power series continuation).

But if we attempt to discuss the comparative hardness of materials that lie in widely separated portions of facade, we can become startled by the disparity in standards locally applied. Indeed, sometimes practitioners accustomed to the stabilization of "hardness" natural to, e.g., a rubber become nonplused or overtly critical of the procedures commonly applied to a steel when such matters are brought to their attention. For example, J. R. Partington reports a complaint that Brinell hardness really measures plasticity rather than hardness[53] whereas D. Tabor, situated quite differently, remarks that the duroscopes of the rubber industry merely capture the elastic response of a material, rather than its proper fully plastic response (as measured nicely by a Brinell tester). Yet another range of central interests prompts the remark that scratchability represents a "truer," but harder to calibrate, standard of *hardness* than the more

[52] April McMahon, <u>Understanding Language Change</u> (Cambridge: Cambridge University Press, 1994), 180.
[53] J. R. Partington, <u>An Advanced Treatise on Physical Chemistry</u>, iii (London: Longmans, Green and Company, 1952), 234. This represents a 1909 opinion of Kurnakow and Schemtschuschy.

popular indenter-based tests:

> *Obviously Moh's Scale would be inadequate in the accurate determination of the hardness of such materials as metallic alloys, and rather different types of hardness test have been developed for such substances. Such instruments as the Turner Sclerometer (which attempted to measure surface "scratchability") were soon abandoned in favor of machines which measure the resistance of the surface layers of a material to penetration by some form of indenter rather than the surface hardness defined in terms of abrasion resistance. Whilst these current methods are less representative of true surface hardness they offer means by which surface properties can be compared much more accurately.*[54]

Again arguments on the other side of this question can be easily found, <u>viz.</u>, that the basic process of abrasive scratching brings so many physical traits "unrelated to hardness" (such as a capacity to shear into whiskers) into play that no suitable criterion for *hardness* can lie in the direction of such tests. Of course, all of these locally biased favoritisms have "some truth to them," in the same vein as dedicated partisans of "$\sqrt{4}$ is really $+2$" or "$\sqrt{4}$ is really -2" can each produce legitimate local directivities properly belonging to the overall concept of $\sqrt{z}$. Arguing too vigorously over these issues shows that the global behavior of the terms in question has not been adequately understood. In the final analysis, *square root* and *hardness* both live on twisted Riemann-like surfaces assembled from the manner in which their local evaluative sheets patch together.

In such cases, somewhat disconcerting displays of multi-valuedness of "$\sqrt{4} = +2$ and -2" type can be produced if diverging sheets can be located that happen to cover situations in common. For example,

> *John has become completely weightless in orbit and, if he doesn't eat his rations, he's likely to lose five pounds soon.*

Patently, we are not encouraging the inference that John will soon weigh -5 pounds. Or (elaborating upon considerations mentioned above):

> *A material can theoretically be made so extremely hard that an ingot will be unable to redistribute even the mildest applied stress and will thus lose all of its hardness and will crumble into dust under the slightest probing of any kind.*

It is interesting that, despite their surface appearance of paradox, that both of these claims can be interpreted sensibly, simply by keeping mental track of the patch of the covering surface we are operating upon. Our ability to take account of these rapid adjustments in proper context is the central topic of 7,viii.

The inclination to tolerate mild forms of multi-valued description is symptomatic of a predicative use that has been framed by allowing patches of local utility to *continue* into one another through following some salient directivity (here: the common reach of some instrumentally facilitated practical test). As the great mathematical utility of $\sqrt{z}$ abundantly establishes, multi-valuedness needn't impede the utility of a usage at all; it

---

[54] R. A. Higgins, <u>Properties of Engineering Materials</u> (New York: Industrial Press, Inc., 1994), 79.

merely demands some flavor of borderline control, whether delicate or coarse, restrict how local appraised data is to be carried from one patch to another. Often, the natural limitations of our "search spaces" to narrowed classes of material provide sufficient control to ward off the incoherencies that would arise if we began to evaluate metals, plastic sheeting and swimming pools encountered at high velocity with respect to their comparative hardness, because we employ vastly different evaluative measures in all of these cases. But it rarely occurs to us to make such comparisons (look about your surroundings—can you align everything you see upon some uniform scale of increasing hardness?) In other words, our abilities to shift contexts quickly along with the rarity with which we compare discordant patches provides an adequate data control to keep our usage of "is hard" from collapsing into deductive incoherence, just as similar loose preventatives usually keep us from falling into "$\sqrt{z}$"-induced error (to avoid mistake altogether, we would need to understand "branch points" and "cross cuts" and all the other apparatus that the expert employs to keep her "root of z" usage on track). As the "weight" example also shows, the isolation between distinct patch specializations needn't be stringent or perfect to serve adequately in practice.

....

Ray Jackendoff, in his Foundations of Language, employs examples of roughly this sort as an excuse for "pushing 'the world' down in the mind of the language user,"[55] whereby he means that a descriptive linguist ought to regard a predicate like "is hard" as unproblematically single-valued because we allegedly think of it that way. I doubt that this is true, but surely a story needs to be told as to how we manage potentially paradoxical language so deftly. And I do not see why this task should not fall to "semantics."

More generally, there is a puzzling disposition amongst many philosophers to presume that concepts like *weighs five pounds* possess some pristine content that can be completely detached from everything we know about the moon, astronauts and such. Such ploys are often used to create an imaginary philosophical conserve free from the annoying gnats of factual counter-example. Thus: "It's an a priori fact that objects retain their weights as they move about the universe." "But that's not true." "You deny my claim only because you confuse the scientific conception of *weight* with that the ordinary man employs; my a priorism holds only of the latter." But I don't think there are many "ordinary men" around who don't take the moon into better account than our a priorizing philosophers allow.

....

If a term's usage strongly rests upon instrument-linked considerations, it is not surprising that it will often display a multi-valuedness induced by patch-to-patch con-tinuation. Should we conclude that "hardness" behaves in this manner precisely because the tie to instrumentalities lies intrinsically etched in its core semantic contents? The thesis that *hardness* essentially represents a *dispositional trait* represents a common classical diagnosis of the behaviors witnessed here. But consider the complementary career of "temperature," a term that was originally transferred in the early seventeenth century from everyday evaluations of balminess (as in "the temper of the air") to cover

---

[55] Ray Jackendoff, Foundations of Language (Oxford: Oxford University Press, 2002), 303.

the readings of sundry measuring instruments (early thermometers or thermoscopes that, in fact, half acted like barometers). As such, it supplies a rare case of a common evaluative term whose entire developmental career can be neatly surveyed[56] (that is, the concept of *temperature* proper had little currency beforehand—in contrast to the quite different notion of *heat content* which surely did). Depending upon the working fluid employed, thermometers in themselves supply inherently incongruent readings in a manner like that of our sundry hardness testers. As is well known, the nineteenth century flourishing of thermodynamic thinking allowed Lord Kelvin to articulate an "absolute" approach to *temperature* that freed it from the shackles of instrumentation. Could this eventual emancipation from dispositionality have been foretold from the intrinsic character of *temperature* in 1620? I don't think so, although the classical point of view typically imposes such a requirement.

. . . .

To make matters worse, Kelvin's absolute treatment only applies to conditions of thermal equilibrium, whereas we commonly utilize the notion across situations of a considerably more general character. But great controversy continues to this day as to the proper significance of the notion in such unsteady circumstances.

. . . .

Wouldn't it be preferable if *hardness* could be brought to a condition of similar improvement? We would then achieve a one-sheeted covering of all materials and all inconvenient multi-valuedness would vanish. In this regard, in a library sale I once ran across a discarded volume from 1942 called Hardness and Hardness Measurements by Samuel Williams that casts this semantic proposition in rather poignant human terms. Williams, who was a physics professor at Amherst College, had been drawn into the investigation of hardness through being assigned to some metals research committee after the end of World War I. One of the motives that led him to devote his ensuing labors to his chosen topic traced precisely to the hope that some "absolute scale of hardness" could be uncovered, untethered to the restricted whimseys of specific types of apparatus:

> There are all sorts of conceptions as to what constitutes hardness if one may judge by the methods employed to measure it. There is hardness as measured by penetration, by scratching, by resilience, by machinability, by yield point, by electrical and magnetic properties and by many other related physical properties. Even if the definition of "absolute hardness" cannot be realized, it will be of some value to make clear, for instance, what sort of hardness is needed in a good cutting tool for use on a lathe, a milling machine or a shaper. The machinist knows he has a good cutter. Can the physicist tell him exactly why? Such a possibility would be a real step forward in our understanding of what hardness means. The importance of understanding the significance of hardness tests cannot be overestimated. It is a conservative statement to make, that no other tests on materials

---

[56] W. E. Knowles Middleton, A History of the Thermometer (Baltimore: Johns Hopkins University Press, 1966), 17.

approach in numbers the tests made under the name of "hardness tests" as the criteria whereby materials may be classified or selected for special purposes.... In spite of the various factors which affect the resistance to motion arising from friction, we have worked out a fairly satisfactory measure of friction which we express as a coefficient of friction. Can one work out a coefficient of hardness?[57]

And here is his dream:

The crying need is that out of the many ideas now involved in the term "hardness" there should eventually emerge an absolute method which will enable hardness to be included among the specifications of the physical properties called for in the finished product . . . This seems to the writer a goal which we should strive in the measurement of hardness, the ability to express hardness in terms of an absolute scale rather than in relative terms.[58]

As "temperature"'s ascendent career demonstrates, this is not an unreasonable ambition. Indeed, authors of Williams' era commonly write of "hardness" in a way that presumes that such a uniformizing partition must exists, even if we can't presently specify how it is constituted. Thus Frank Sisco's 1941 Modern Metallurgy for Engineers remarks:

All indentation hardness methods fail to measure the hardness of the undeformed metal. An exact mathematical correlation of the various hardness numbers is therefore impossible as the amount of strain hardening varies from test [method] to test [method] and, furthermore, varies from metal to metal.[59]

Sisco simply takes it for granted that some "absolute hardness" obtains that is imperfectly registered in the known test techniques. Optimistic passages of this ilk have largely vanished from more modern books, because the huge variety of unrelated microscopic mechanisms that can potentially intervene in an act of probing such an indentation or scratching has become manifest. Samuel Williams, unfortunately, labored in an era just before the importance of dislocations, phase changes, thin layer lubrication and the army of the other critical notions of modern materials science became recognized, crushing in their wake any hope that an absolute "hardness" of the sort Williams sought might be produced. Here is a typical modern assessment:

In spite of the theoretical work done on hardness, hardness cannot be considered a fundamental property of a metal. Rather it represents a quantity measured on a arbitrary scale. Hardness measurements should not be taken to mean more than what they are: as empirical, comparative tests of the resistance of the metal to plastic deformation. Any correlation with a more fundamental parameter, such as the yield stress, is valid only in the range experimentally determined. Similarly, comparisons between different hardness scales are meaningful only through experimental verification.[60]

---

[57] Williams, Hardness, 5, 10.
[58] Ibid., 8.
[59] Frank T. Sisco, Modern Metallurgy for Engineers (New York: Pitman, 1941), 133.
[60] A. A. Meyers and K. K. Chawla, Mechanical Behavior of Materials (Upper Saddle River, NJ: Prentice-Hall, 1999), 160.

Williams is right to emphasize "the importance of understanding the significance of hardness tests," but appropriate answers can only be supplied on a relatively localized basis. Williams, in fact, had an uneasy premonition that his labors might eventuate in anticlimax, for, as he modestly states in his preface:

> This treatise . . . is a pioneer attempt on the part of a physicist to clarify concepts of hardness, . . . if such a property exists.

(The emphasis is Williams' own). He makes the further observation that, as language, "hardness" is destined to prove a great linguistic survivor; that its associations with swift practicality guarantee that the term will never vanish utterly from the colloquial vocabulary of anyone who works with materials. He writes (quoting from <u>The Republic</u> at the end):

> One other question might be raised at this point: Would it clarify our thinking if we eliminated the word "hardness" from our scientific vocabulary? The author's point of view is that the term is so firmly entrenched in our everyday vocabulary, that we would not go far before we began to use it again. For instance, one could talk about measurements of penetration, or measurements of machinability, scratchability and other terms already used. Whatever we call it, back of each measurement is a something which we might just as well call hardness from the start and keep on calling it that, until we know something about it.
>
> Socrates: But why should we dispute about names when we have realities of such importance to consider?
> Glaucon: Why, indeed, when any name will do which expresses the thought of the mind with clearness?

But, as he fully recognizes, the "something in the back of every measurement which we might just as well call 'hardness'" can (and does!) turn out to rest upon a more complicated framework than a simple physical property. A set of evaluative patches linked together in a Riemann surface-like way through instrumentally inspired continuations supplies a simple model of the semantic platform upon which a predicate like "is hard" is deposited. We thereby obtain the foundations for a workable, practical usage, but with no single physical property supporting the predicate everywhere, hence no "absolute hardness."

Williams' desire for a notion of *hardness* that is single-valued everywhere is the analog of the "BSF = GSF" cure that resolved Ernest's problems with his multiply registered diary (in the next chapter I say that Williams seeks a *flat covering* of "hardness"'s dominion). But we can't always uniformize a multi-sheeted usage and leave it intact. That is, we can easily slice away the upper floor of the Riemann surface for $\sqrt{z}$, but we are left with a function with a big jump in it, which precisely ruins the analytic personality upon which all the characteristic virtues of $\sqrt{z}$ all depend. Likewise, if we attempt to align "hardness" with some uniform microscopic characteristic (such as the *slip initiation* involved in *yield point*), we will have selected a trait that is unimportant and hard to recognize in the case of a rubber or plastic. Permitting our term a facade-based behavior preserves its practical utility over a much wider range of applications. Indeed, these considerations illustrate my 5,ix observation that true physical attributes are often hard to track: over the steels, *yield point* sits plainly on the roadway, but, with respect to the rubbers, it has jumped back into the briar patch. In the absence of a traceable critter, we must instead employ terminology that hews more closely to the results of attainable measurement and develop a facade-founded usage that, to rework Williams' thought, "we might just as well call 'hardness' from the start and keep on calling it that."

Contrasting the unequal fortunes of *hardness* and *temperature*, we find that we possess no reliable clues, after we have first gotten a descriptive predicate up and running, as to how its ensuing usage will finally stabilize. Facade or flat structure—that is largely an arrangement that word and world must work out between themselves, and they are likely to apprize us of their decisions only long afterward. Fervent desires for uniformity or single-valuedness on the part of the phrase's human initiators count for comparatively little in this regard. For it is largely on the forges of practicality that a useful fit between our available linguistic tools and the physical facts gets gradually hammered out and our feeble original intentions or the semantic vows we frame in intervening years are unlikely to deter this pattern of final accommodation substantially from its appointed courses.

To return briefly to the terminology of 5, ii, I consider claims like "material $\underline{a}$ is harder than material $\underline{b}$" as direct grist for the mills of *design recipes* that contain clauses like "To build a better widget, select the hardest material possible compatible with other design desirata." It is through these means that 5, ii's strands of practicality play an important role in erecting the bridges that carry us from one local patch of grass into the neighboring green pasture.

Returning to our opening discussion of Descartes and Reid, *hardness* proves to be neither a simple physical quantity nor a constant sensation, but an informational package with characteristics <u>sui generis</u> of its own. In the next chapter, we shall ponder how such novelties (which arise commonly in descriptive life) should be considered in respect to their "objectivity" or "subjectivity." It will not surprise the reader to learn that our Chapter 2 troubles trace to this source.

Certainly, with respect to Descartes' assumptions, it is a mistake to presume that any invariant "feeling of hardness" exists. Descartes has clearly projected a hypostasized mental condition backwards from our linguistic behavior into our brains: "I call all of these things 'hard,' although they lack physical commonality; <u>ergo</u>, they must induce similar sensations within me." Reid is right to insist that this assumption is plainly false. Indeed, it is unlikely that many instances of our gathering data with respect to hardness involve any appreciable measure of "sensation" properly considered, for the brain continually monitors our physical surroundings in ways that make no impression upon consciousness. If my familiar oak desk suddenly turns flabby, I will notice it, but no allied recognition is required if the brain feels assured that there is no need to revise its standing estimate that "the desk is hard."

In the quotation above, Thomas Reid expresses an <u>a priori</u> faith that *hardness* must represent a single underlying characteristic of matter (he lists it as one of the primary

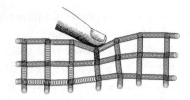

qualities that materials must display), an apriorist position considerably more radical than Williams' mild (and, in his time, reasonable) hope. Almost certainly, Reid has some kind of quasi-mathematical sketch in mind, like that illustrated. And he likely murmurs to himself as he contemplates this rendering: " 'Tis obvious that some single number must measure how deep that depression will form under an applied loading and, surely, that is what *hardness* must represent. And this is something we perfectly understand." But we now recognize that these musing are utterly naïve: what conditions of slip will apply between indenter and surface? will there be plastic flow in the material underneath the depression? will fracture initiate beforehand? etc. Reid has plainly conjured up a *mythological picture* of *hardness*'s circumstances. But such associated pictures should not be dismissed as simple mistakes, for they comprise an important part of the full atmospherics of intensional personality that surrounds "is hard." Oddly enough, these "wrong picture" associations typically persist, even when we understand the relevant physics better. In this regard, we should bethink ourselves of our old friend "rainbow," that carries an unyielding association to concrete arches, no matter how well versed we become in the verities of illuminated rain drops. As such, this wrong picture provides a kind of "latent DNA" to a predicate, upon which fresh sprouts can readily grow if properly nourished (as when we write stories of rainbows wherein fairies clamber over their surfaces). Indeed, faulty pictures often propose continuations into neighboring patches that practical employment then firms up as useful (we shall consider an example in the next section).

In the case of *hardness*, Reid-like pictures of simple bounding surfaces entangle its fortunes intimately with those of *being a rigid body*, a great conceptual rascal whose misadventures concern us greatly at other points in the book (4,ix and 7,iv). In particular, an accompanying notion of *absolute hardness* springs into being, which comprises the hypothetical trait that allows such substances to stay rigid, come what may. That such a notion hovers on the edge of paradox is well known from old puzzles about "What happens when an impenetrable body is struck with an irresistible force?", as immortalized in Johnny Mercer's "Something's Gotta Give" (less remembered is the fact that the St. Petersburg Academy set the topic as a prize in 1720 or that Newton trusted so much in the absolute hardness of molecules that he thought angels were required to juice up the universe when hard body collisions ran it down[61]). Later in the book, we devote more attention to these *mythological picture* associations, because they comprise an important part of the story of linguistic life.

[61] Wilson Scott, The Conflict between Atomism and Conservation Theory (London: MacDonald, 1970).

(xi)

Foundational looping. "Hardness" 's multi-valuedness may remind astute readers of a venerable philosophical controversy. In her well-known <u>Philosophy and the Physicists</u> of 1937, L. Susan Stebbing chastised the astrophysicist and science popularizer Arthur Eddington for claiming (rather innocuously in my opinion) that a floor board really "has no solidity of substance" because it is composed of elementary particles that lie comparatively far apart. Stebbing maintains that Eddington is illegitimately using "solid" in a manner untrue to its proper linguistic significance:

> What are we to understand by "solidity"? Unless we do understand it we cannot understand what the denial of solidity to the plank amounts to. But we can understand "solidity" only if we can truly say that the plank is solid. For "solid" just is the word we use to describe a certain respect in which a plank of wood resembles a block of marble, a piece of paper, and a cricket ball, and in which each of these differ from a sponge, from the interior of a soap bubble. and from the holes in a net. . . . But there could not be a misuse, nor a figurative use, unless there were some correct and literal usages. The point is that the common use of language enables us to attribute a meaning to the phrase "a solid plank"; but there is no common usage of language that provides a meaning for the word "solid" that would make sense to say that the plank on which I stand is not solid.[62]

To be sure, it can be legitimately questioned how ably the passage criticized functions as deft popular science, given Eddington's propensity to mix metaphors drawn from different scale sizes (which is part of her complaint). However, the Stebbing excerpt (which is well known in philosophical circles) suggests some stronger thesis to the effect that any predicate must possess some fixed core meaning from which all of its sundry subsidiary uses must descend. But it should not be allowed, she imagines, that such satellite uses can completely overturn the parent discriminations from which they derive: usages rarely bite the hands that feed them, at least in cases like this. And some of Austin's remarks on the legitimate use of "red" have a similar flavor.

Stebbing's overriding purpose is to argue that the deliverances of science must err (or be regarded purely instrumentally) whenever they reverse the opinions of everyday common sense, a thesis we already visited (2,vi) in connection with the Lake Poets and the philosopher A. N. Whitehead (from whom she draws explicit inspiration). Indeed, the problem of reconciling the apparently divergent "worlds" of science and common sense is usually dubbed "Eddington's two-tables problem" in contemporary analytic philosophy, tracing to another passage that Stebbing singles out for allied criticism (of course, the problem is much older and much of Eddington seems directly derived from Karl Pearson's <u>The Grammar of Science</u>).

In my diagnosis, these puzzlements trace to a natural scale dependent lift in the employment of "is solid," closely related to those displayed by its cousin "is hard." In

---

[62] Stebbing, <u>Physicists</u>, 51–2. She is criticizing Sir Arthur Eddington, <u>The Nature of the Physical World</u> (Ann Arbor: University of Michigan Press, 1958).

fact, this lift (it is actually a "drop" because we shift from tables to a molecular scale) is facilitated by the fact that we associate "solidity" with a mythological picture ill-suited to the real factors that render common materials "solid." In his Essay, John Locke offers Reid-like remarks on "solidity" that evoke the basic contours of this faulty portraiture nicely. On the one hand, he writes:

> If any one asks me, What this solidity is, I send him to his senses to inform him. Let him put a flint or football between his hands, and then endeavor to join them, and he will know . . . The simplest ideas we have, are such as experience teaches them us; but, if beyond that, we endeavor by words to make them clearer to the mind, we shall succeed no better than if we went about to clear up the darkness of a blind man's mind by talking; and to discourse into him the ideas of light and colors.[63]

In the process, we come to understand fully the necessary effects of *solidity*, without being informed as to its causes (which, Locke gloomily speculates, may lie beyond human comprehension):

> [S]olidity consists in repletion, and so an utter exclusion of other bodies out of the space it possesses . . . Upon [this] depend their mutual impulse, resistance, and protrusion.[64]

In more mathematical terms, I would express this claim as: material objects are solid if and only if they preserve their volumes under all interactions, as long as they remain integral. As such, this is a sense of "solid" which allows an incompressible liquid to prove a solid, which, I believe, is how Locke intends it, although the example readily shows that "experience" cannot supply us with any constant *sensation* of such "solidity."

With respect to real world behaviors, Locke's repletion picture is completely erroneous, because an everyday solid steel block utilizes a very complicated set of molecular mechanisms to supply its macroscopic self with its familiar obdurate demeanor. Busy dislocation movements, which are dramatically compressive, are required to diffuse applied stresses quickly so that they never become localized upon critical bonds (recall the unhappy dislocation-less ingot that crumbles to dust under mild pressures). Indeed, in a piece of steel a fair amount of outright recrystallization also takes place under external invasion, involving the swift transport of component materials from one site to another.

Nonetheless, Locke's picture of "solidity" directly embodies an important *mathematical classification* that can be applied at all size scales: that of a space filling, volume conserving flow of finite measure. This is plainly the "directivity" that Eddington follows in lowering "is solid" down to a microscopic scale. As such, it represents a completely natural continuation of the predicate's use. In the process, "solid" thereby acquires the harmless multi-valuedness to which Stebbing objects. For example, I might readily reexpress my observations about the importance of molecular diffusion in a steel by the claim:

> A piece of steel remains a solid partially by not acting like a solid on the molecular level: its matrix must be swiftly permeable by a variety of molecular species.

---

[63] Locke, Human Understanding, i., 156–7.
[64] Ibid., vol. i., 154.

Or:

> A solid can support loads only by not acting like a solid along its surface, but by allowing the asperities of actual contact along that surface to contract and flow at their tips.

Of course, a rather dramatic degree of "property dragging" is involved here; "is solid" has redeposited itself upon physical characteristics that have virtually nothing in common with the qualities that allow a macroscopic block of steel to be a solid. Sometimes words are willing to follow almost any excuse to conquer new territories of application and such wantonness seems evident here. But, however distressing we find these expansionist tendencies, they represent common patterns of linguistic extension and we are stuck with them.

When Stebbing claims, "There could not be a misuse, nor a figurative use, unless there were some correct and literal usages," how are we supposed to evaluate the circumstances just surveyed? Why shouldn't the false Lockean picture behind Eddington's continuation qualify as the term's "correct and literal use"? It seems odd to dub such standards as a "figurative use" when they represent a far more venerable mode of evaluation than anything I can extract from the true story of the qualities that lie at the base of our successful macroscopic employments of "is solid."

Stebbing's problem traces to the semantic assumption, utterly emblematic of classical thinking, that some single, unified package must represent the base conceptual content belonging to *being solid*, from which any askew employments must spring as either derivatives or mistakes. No: a complicated bundle of incomplete and conflicting directivities comes integrally associated with "is solid," some of which spring from associated mythological pictures and some of which trace to the practicalities of useful classification. *All* of these ingredients (and others) enter into *solid*'s conceptual personality, even if they individually prove faulty in some respect or other, for the ongoing usage relies upon them all at various points in its developmental pilgrimage. Sometimes this mass of directive elements reaches mutual accord along a twisted surface of application, with the result that applications of "solid" on different size scales may prove utterly discordant when compared directly to one another. Stebbing's accusation that Eddington must have made some *mistake* in his usage thus traces to the classical assumption that correctly applied "original meaning" always carves out a single-valued employment everywhere. But it is the stark lesson of "$\sqrt{z}$" that this ain't true, however much we might wish it otherwise ("we can fool ourselves, but we can't fool the functions").

(xii)*

Mechanical torsions. In the foregoing, I have generally written as if a multi-valued usage commences with a single sheet of usage settled in place, to which further patches later attach through prolongation. As such, the description probably suits the etiology of "weightless in orbit" fairly well. More commonly, there is no clear historical priority in the order in which the local sheets form and the patchwork structure of the facade is better viewed as having gradually emerged as initially rough inferential recipes and measurement techniques become refined under the increasing demands of practical achievement, rather as a polycrystalline granite slowly solidifies out of a homogeneous melt. As this stabilization process progresses, a fine filagree of protective cracks begins to divide the usage into interconnected platelets, allowing a multi-valued platform to emerge, but with no particular patch serving as parent to the rest. This formation through localized specialization undoubtedly provides a better account of how *hardness*'s ontogeny has unfolded.

In Chapter 4, we observed that great philosophical disputes about the nature of concepts arose in the context of late nineteenth century mechanics, driven by the many practical concerns that troubled physical practice in that era. In 4,ii, we briefly surveyed some of the oddities that baffled our Victorians. In many of these circumstances, the problems arose from subtle forms of the *foundational looping* we have just surveyed. Although they knew it not, the classical mechanical framework in which they toiled comprises an excellent illustration of an *uneven facade* that cannot be regularized without ruining its integral descriptive utility. As various illustrations scattered throughout the book will show, doctrines of quite different supportive natures can mimic for one another quite nicely within classical physics, giving rise to a collected bundle of useful assertions that can seem—if we don't look too closely—as if they constitute a unified theory. In truth, however, we confront what was dubbed a *theory facade* in 4,vi: an uneven pile of pasteboard cutouts that ably masquerade, from selected angles, for an integral metropolis. As such, its atlas structuring is secretly subject to substantial degrees of property dragging, but these semantic displacements occur in quiet ways, allowing unsuspecting observers to assume that mechanics' predicates have remained steadfast in their attributive attachments. Because of their faith that predicates will hold in place as long as they are "clearly thought through," none of our nineteenth century philosopher/scientists imagined that their puzzles might represent the basic symptomatology of a language that irresistibly twists under prolongation in a manner akin to $\sqrt{z}$'s incurable behavior: that the torsions in their practices stem from the fact that a framework of predicates can conform, at a macroscopic and asymptotic distance, to the facts of an underlying quantum world only by arranging themselves upon an uneven facade (or some more complex form of interfacial support). Accordingly, our Victorians often devised grander (and, from a commonsensical point of view, more upsetting) explanations for the loops they confronted, such as instrumentalism or the embrace of obligatory descriptive idealization in a neo-Kantian spirit. But, as often proves the case afterward in cases like this, the proper succor for methodological discomfort is not to be obtained from radical panaceas, such as those offered by Professors Pearson, Mach or Duhem, but simply through a patient unraveling of the language's supportive environment to greater depths of grubby detail. In classical mechanics' case, these "grubby details" are of a startling nature themselves: the strange ways in which quantum mechanics supplies "volumes" to molecules and surprising facts about how partial differential equations behave. It is therefore entirely understandable (and I hope this sympathy comes through in my prose) why the Victorians should have been totally

Duhem

unprepared for the backdoor manners in which a welter of surface anomalies in methodology (the "fine grain" of which I often write) gradually emerged within their most successful practices. *We*, however, should be better forewarned against such eventualities in *our* own futures: the misadventures of the late Victorian era should warn us that premature philosophizing can unhelpfully lull us into assuming that we understand methodological puzzles whose true sources trace to deeper roots.

In this chapter's concluding sections, I will sketch several of mechanics' characteristic loops, as well as diagnosing their origins in facade-related terms. These studies can help us resist the allure of the holistic proposals of Chapter 5, in that plain mystification about the semantic underpinnings of descriptive predicates is more often the true source of seemingly intractable scientific disagreements than Kuhnian "paradigms" or Quinean "intractable bodies of implicitly defined concepts." The reader who has little taste for these topics may prefer to advance to the next chapter, where our notion of facade is formulated in schematic terms and its attendant features are brought to bear upon philosophical difficulties like those surveyed in Chapter 2.

Let us look at one of those turn of the century debates which is commonly glossed in "clash of paradigms" terms (inter alia, by Duhem himself). In his The Aim and Structure of Physical Theory, Duhem complains of the practices of English practitioners such as Lord Kelvin and Clerk Maxwell, who drift inconsistently from one physical *model* to another, without attempting to set any clear set of underlying principles behind these constructions:

> No doubt what is exact and truly fertile in the work of Maxwell will one day take its place in one of those systems in which thoughts are conducted in order, in the image of Euclid's Elements, or of those majestic theories unfolded by the creators of mathematical physics. But Maxwell most assuredly was not seeking that . . . [T]he French mind, thirsting for simplicity and unity, is stupefied [by Maxwell's] careless logic. In all manner of things, the French demand a system.[65]

---

[65] Pierre Duhem, "The English School and Physical Theories" in Essays in the History and Philosophy of Science, Roger Ariew and Peter Barken, trans. (Indianapolis: Hackett, 1996), p. 64. Also: Duhem, The Aim and Structure of Physical Theory, Philip Wiener, trans. (Princeton: Princeton University Press, 1954).

point mass

rigid body

flexible body

By "system," Duhem is simply insisting upon the same intelligible order that Hilbert requested in his sixth problem on mechanics as discussed in 4,iv.

But what on earth does having a "French mind" have to do with axiom systems? Notoriously, Duhem accuses the English of possessing "ample but weak" minds that derive from their regrettable propensity to seek mental pictures for the systems they treat.[66] Through their dependence upon imagery, Duhem expostulates, the British evade the obligation to articulate their working principles as concrete and coherent doctrines (Duhem thinks that their "ample minds" instead consult a large library of familiar mechanical situations, which they employ as templates for other applications). Frankly, much of this "visualizing minds" rhetoric should be dismissed as an unpleasant fossil of the national chauvinism typical of Duhem's time (needless to say, he determined that the minds of his countrymen behaved more capably than their British and German competitors). To be sure, a certain amount of cognitive science literature has been devoted to probing the visualizing tendencies of sundry physicists.[67] This material is interesting, but its merits run orthogonal, in my opinion, to the discrepancies that Duhem correctly notices: <u>viz.</u>, the puzzling cheery British acceptance of *foundationally circular explanations*. They instead presume, according to Duhem, that

> a physicist will logically have the right first to regard matter as continuous and then to regard it as formed of separate atoms, to explain capillary phenomena by forces of attraction acting upon stationary particles, and then to endow these same particles with rapid motion in order to explain heat phenomena.[68]

Indeed, foundational looping is quite apparent in the great British classics such as William Thomson's (= Lord Kelvin) and Peter Tait's <u>A Treatise on Natural Philosophy</u>. I'll supply several examples in a moment.

Let me first supply a little background for orientation. In mechanical tradition, there are three basic classes of objects which can lay some claim to mechanical

---

[66] Some of this derives from Pascal's differently centered discussion in the <u>Pensées</u>.
[67] Arthur I. Miller, <u>Imagery in Scientific Thought</u> (Cambridge, Mass.: MIT, 1986).
[68] Duhem, <u>Aim</u>, 101.

centrality: (a) the *point mass*, a singularity bearing mass and charge, but with no extended geometry whatsoever; (b) the *rigid body*, a space-filling volume likewise carrying mass and charge but permitting no distortion in its internal geometry; (c) the *flexible body* or continuum, like (b) except capable of flexure at every size scale whatsoever. Here we might think respectively of an isolated massive point (a), a stiff iron bar (b) and a flexible plate or a liquid (c). In the quotation, Duhem correctly observes that the British are cavalier as to which of these should be regarded as foundational to the rest, seeming to switch bases according to pure convenience.

Let me insert a clarifying reminder (the issue was already mentioned in 4,iv). For extraneous reasons mainly connected with the rise of quantum theory, an approach to "classical physics" involving *point masses* as its sole basic objects has become canonical in most modern introductory textbooks, but, historically, this was not so. More commonly, some mixture of (b) and (c) was favored, for the reasons such as those discussed in the fine print. When Kelvin writes,

> We have long passed away from the stage in which Father Boscovich is accepted as being the originator of a correct representation of the ultimate nature of matter and force,[69]

he is rejecting point masses as foundational, for that is precisely the view that Boscovich[70] first advanced (Newton's own formulations are ambiguous on these foundational issues). However, these issues are considerably muddled by the fact that confusions about infinitesimals commonly lead our nineteenth century authors to write about "point masses" where they really mean "an internal point within a continuous body."

. . . .

Specifically, if we model an isotropic piece of steel as a collection of point atoms in Charles Navier's original manner, a single elastic constant emerges from the computations, whereas adopting A. Cauchy's top-down, continuum based derivation allows two, in greater conformity to experiment (although this took a long time to establish definitively). Cauchy himself pursued both approaches, but the continuum-founded approach is now canonical in all modern texts. In fact, George Green and George Stokes should be credited with championing the superiority of Cauchy's top-down approach, an opinion that Kelvin here echoes (so much for the greater consistency of the "French mind"). It should be mentioned that, by employing molecular ellipsoids, Poisson duplicated the two-constant account, so Kelvin does not intend to rule out that foundational possibility.[71] In his deeply instructive <u>Encyclopedia Britannica</u> article "Atom," J. C. Maxwell pointed out that point-atom swarms were unlikely to remain stable and would not display fixed spectra, observations that would be correct if quantum mechanics hadn't decided to intervene.[72]

However, these reasons for rejecting point masses get muddled by the confusions about "freezing" and infinitesimals to be discussed below. In a modern context, we usually allow *fields*

---

[69] Lord Kelvin in Robert Kargon and Peter Achinstein, ed., <u>Kelvin's Baltimore Lectures and Modern Theoretical Physics</u> (Cambridge, Mass.: MIT Press, 1987), 107.

[70] Roger Joseph Boscovich, <u>A Theory of Natural Philosophy</u> (Cambridge, Mass.: MIT Press, 1966).

[71] C. Truesdell, <u>Essays in the History of Mechanics</u> (Berlin: Springer-Verlag, 1968). A. E. H. Love, "Historical Introduction" in <u>A Treatise on the Mathematical Theory of Elasticity</u> (New York: Dover, 1944).

[72] J. C. Maxwell, "Atom" in W. D. Niven (ed.), <u>The Scientific Papers of James Clerk Maxwell</u>, ii (New York: Dover, 1965).

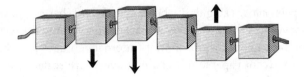

like that of electromagnetism to comprise a fourth, <u>sui generis</u> category but such opinions were not common in the era we discuss.

As noted in 4,i, the vital background of Duhem's concerns (and those of Mach as well) has much to do with the methodological limitations that an artificial allegiance to mechanism places upon electrical research and thermomechanics, a topic to which I'll return in a moment. The malarky about "kinds of minds" obscures the basic soundness of Duhem's position on *these* issues.

....

Let us now examine the type of puzzling cycle that Duhem has in mind by considering a standard derivation of the equations for a vibrating bell. The simplest form of bell is a flat plate in the form of a gong (like the giant cymbal that opened movie swashbucklers produced by the Arthur Rank Corporation) and so we will deal with that. A recent passage by Phillippe Destuynder and Michel Salaun summarizes the standard Victorian approach to a plate pungently:

> *Kirchhoff and Love have suggested that we assimilate a plate to a collection of small pieces, each one being articulated with respect to the other and having a rigid-body behavior. It looks like those articulated wooden snakes that children have as toys. Hence the transverse shear strain remains zero, while the planar deformation is due to the articulation between the blocks.*[73]

In other words, we have elastic strings running through holes in the middle of our little "snake" pieces that explain why our plate tends to straighten itself out when distorted. These strings don't like being stretched but the rigid pieces of our "snake" don't mind slipping by one another. Sometimes one finds that these hypothetical blocks are called "molecules" by Victorian writers while the "strings" get relabeled as "intermolecular forces" (although to a surprising extent writers like Kelvin are content to call them "strings"). As such, our assembly of molecular blocks provides us with an excellent understanding of basic plate behavior in a manner of which Robert Boyle would have wholly approved. However, this model requires us to keep track of far too many blocks to be practical in any fashion, so our Victorian primers usually take an ill-defined limit that squashes blocks, strings and interfaces together in a manner such that a fourth-order differential equation miraculously emerges (the *biharmonic equation*) that mathematically describes a plate that is continuous and elastically flexible everywhere. That is, an ensemble of jointed rigid bodies (2) somehow transubstantiates into a flexible

---

[73] Phillippe Destuynder and Michel Salaun, <u>Mathematical Analysis of Thin Plate Models</u> (Paris: Springer, 1996), 10.

Kelvin

continuum (3). Achieving radical ontological metamorphosis through "mathematical derivation" should seem like a peculiar process (it is). However, if we regard our final differential equation as merely reporting an *averaged smoothing* of the block-based model (in some hazy sense of "average"), perhaps we can still regard our rigid blocks as providing "the real story of what's happening in the plate." Final differential equation in hand, we can uncover in Sturm-Liouville fashion (5,vi) the characteristic vibrations of our gong, allowing us to predict how "musical" it will sound, and so forth.

So far so good; perhaps *rigid bodies* can be adopted as the base physical elements that underlie all of classical physics. After all, we have just successfully accounted for the flexibility of a steel gong by assigning it a molecular decomposition into rigid blocks. In Kelvin's time, it was difficult to advance much direct evidence of molecular structure within a bell or steel plate <u>per se</u>, but the known behavior of heated gases <u>et al.</u> provided strong argument in favor of particulate structure.[74] Consider the findings of spectroscopy. If we hit molecules with radiation, they respond to the interaction with characteristic resonances such as witnessed when a wobbly jelly or a metal shell is struck. To manifest the limited variety of spectra witnessed in our spectroscopes, such vibrating molecules must embody special eigenvalue modes derivable from their special geometries. There are quite a few possibilities of materials that might instantiate the spectral behaviors witnessed: jellies, elastic bodies and fluid vortices. The English explored all these possibilities for molecular media with great vigor. But whichever of these routes is pursued, we have fallen into an odd *explanatory circle*: at the outset we posited rigid, block-like "molecules" to explain the *flexibility* of a macroscopic bell, but we've just decided that real life molecules are not so rigid after all, but behave more like . . . bells!

It is really the fact that English scientists cheerfully embrace such blatant varieties of *ungrounded foundational looping* that constitutes the striking feature of their procedures,

---

[74] Maxwell, "Atom." In fact, the situation is even more puzzling, because nature seems to provide us only with a very limited variety of molecular bells, seventy or so in number. We simply do not encounter the variety of spectra in nature that we otherwise would expect if *any* old bell shape could exist at the molecular level. Maxwell cited this restriction upon the natural manufacture of atoms as cosmological evidence for a creator:

> *a number of exactly similar things cannot be each of them eternal and self-existent and must therefore have been made and the phrase "manufactured article" . . . suggest[s] the idea of their being made in great numbers* (p. 484).

rather than any hypothetical tendency to visualize (every smidgen of reference to "blocks" et al. can be removed from our story and replaced by dry talk of the Kirchhoff conditions[75] $\sigma_{xz} = \sigma_{yz} = \sigma_{zz}$ without altering the looping one whit). Duhem aside, I do not see mental picturing and unfounded looping as intimately related. But let's look at another example before we pass judgment on the British toleration of foundational circles.

. . . .

Practices become even odder when we generalize our plate model plate to curved *shells* such as the Liberty Bell, for our erstwhile *rigid* blocks need to be credited with an inherent elastic resistence to altering their curvatures: that is, we treat them as flexible plates. It is worth observing that real pieces of steel are markedly polycrystalline in nature, suggesting that approximately rigid elements might become behaviorally dominant at a higher scale of grain, as indeed happens to a certain extent. But the blocks in our plate model bear no descriptive relationship to the real grain structure of a metal.

By the way, although in books written for quantum physicists one still encounters derivations of equations for continua that proceed by squeezing together discrete elements, the practice is almost universally rejected by experts in classical topics as incoherent.

. . . .

In 4,vi, I mentioned that the standard approach to billiard ball behavior found in undergraduate textbooks tacitly presumes that our balls stay entirely rigid throughout their collisions. But this claim is patently erroneous and this simplified description served as one of our first illustrations of *physics avoidance*: the policy for reducing descriptive variables through compressing behavioral complexities into boundaries and singularities (the rigid body approach treats the actual moments of contact in this manner). As the customary cost of this policy, we witness the *lousy encyclopedia phenomenon* (4,ix): the covering of a range of basic physical behaviors by a number of incompatible patches of physical description, tied together by lifts that we follow when a given patch proves unable to handle a specific case adequately. As we gradually explore the patches of wider coverage, we find that our balls eventually begin to be modeled as flexible elastic bodies, that convert the kinetic energy of collision into internal stresses that both cause the balls to compress, and send waves crisscrossing their interiors. Such problems cannot be resolved until plausible mathematical conditions are set along their moving boundary interface (do they stick or slide and by what rule?). Qua mathematics, these requirements pose truly horrific problems and applied mathematicians generally try to avoid their complexities by tarrying in some simpler lower sheet of modeling (through crude estimates such as: if the balls don't stay in contact long enough for more than a few internal waves to cross the balls, then a modified Eulerian approach is acceptable[76]).

---

[75] Arthur Boresi and Omar Sidebottom, Advanced Mechanics of Materials (New York: John Wiley and Sons, 1985), 452.

[76] W. J. Stronge, Impact Mechanics (Cambridge: Cambridge University Press, 2000) covers a wide range of phenomenon using largely close-to-rigid-body treatments.

Unfortunately, even this fuller treatment (presuming that we can resolve our moving boundary line problem adequately) cannot cover our expected range of behaviors completely, because in extreme cases our descriptive mathematics begins to demand that more physics be inserted in the problem. In particular, in vigorous collisions *shock waves* akin to those of 4,vii form inside our balls and our governing equations break down for those cases. However, the standard way of "adding more physics" is to evoke *entropy production,* which means that vocabulary such as "temperature" and "heat" needs to be added to our covering descriptive vocabulary. With high heats, chemical phase changes often initiate inside the material, radically affecting how they respond to strain (we tend to be unaware of the fact, but it is estimated that innocuous acts of mild rubbing or bending can locally heat the surface of steel to a temperature of more than 4000°C, at which point the material considerably alters its structure[77]). So the notion of "chemical potential"—the propensity to alter behavioral phase—seems as if it needs to be included in our descriptive arsenal as well. Nor have we yet handled the somewhat different hysteresis effects (= change in behavioral response after a history of bending) that arise due to dislocation pileup in our rapidly flexing balls (some modern continuum approaches use fancy ideas from differential geometry to model these factors). Finally, we've not yet installed any mechanism that allows our balls to shatter, although this, of course, occurs in real life with unlucky balls.

In other words, if we start with virtually any family $\mathcal{F}$ of similar macroscopic situations (balls colliding on a billiard table), we will find members f* of that family that cannot be adequately handled by the prevailing formalism T that works reasonably well for much of $\mathcal{F}$. Instead, we find that the f* require attention to physical traits (*temperature, chemical condition*) that T pushes off to the sidelines when it considers those members of $\mathcal{F}$ that it can adequately treat. In other words, T contains descriptive holes that prevent it from handling all of $\mathcal{F}$ in a satisfactory way, asking us to consult a quite different account T' for other $\mathcal{F}$-like situations. As we observed in 4,i, it was Duhem and Mach's policy to recommend that physics should keep looking for a formalism that embraces as much of $\mathcal{F}$ as possible.

But this policy quickly conveys us into very deep and difficult mathematical waters, lying far beyond what was available to the Victorians. As they stand, the "purity" of Mach and Duhem's insistence that physics treat every descriptive variable affecting the family $\mathcal{F}$ on equal terms becomes excessive: "Everybody knows that temperature and heat are occasioned by the vibrations of molecular units, so why must I include 'temperature' et al. as primitives in my physical treatment? Likewise, we know that plasticity and hysteresis tie to dislocational defects within the underlying crystal structure, so why should I accept a weird torsional term[78] as basic in my formalism? Such a policy of evenhandedness pushes us to ridiculous complications very quickly." Foundationally,

---

[77] F. P. Bowden and D. Tabor, Friction and Lubrication (London: Methuen and Co., 1967).

[78] Here I refer anachronistically to modern "continuum theories of dislocations" as a means of indicating the extremities to which pursuit of a Duhem/Mach policy carries one. Such formalisms are very useful, but they are, conceptually, quite strange. Analogous work on liquid crystals might be mentioned in this context as well—a theory that Duhem did know about.

we will hope that *temperature* and all the other non-mechanical notions that cause such complexities on the macroscopic level might drop away in favor of a purer mechanics at the molecular level.

In fact, it is reasonable to make this assumption, but the rub is: those molecules must manifest *intrinsically quantum behaviors*; they cannot truly act like thorough-going classical systems. Why? If we try to employ *classical molecules* to found the visible behaviors of our macroscopic billiard balls, our supportive molecules must sometimes act like the fashion of flexible geometrical blobs able to attract and repeal one another. That is, our molecules must act as some cousin to billiard balls themselves. But then we become obliged to characterize the *classical material* of which these molecular units are composed. But any plausible choice will exhibit shock waves and, as we saw, those eventually demand appeal to thermodynamic notions to settle their behavior. Accordingly, another painful foundational loop has emerged (one quite familiar to Duhem who worked in these areas). Quantum mechanics supplies its molecules with size and flexibility through altogether different policies and thus can better serve as a statistical foundation for macroscopic behaviors. In short, the realm of classical mechanical concepts does not *close upon itself internally*: its local methodologies are given global coherence only through support by an intrinsically quantum substratum.

We reached this same conclusion via a different route in 4,viii: there we noted that the full facade of classical doctrine makes good asymptotic sense as a valuable set of reduced descriptive variables suspended over quantum reality, but not one that can stand alone as a coherent "theory" in the logician's sense (that is, the contours of the classical atlas can't be regularized into a flat covering without ruining many of its practical utilities). Classical molecular modeling cannot gain a handhold on Nature until quantum mechanics stabilizes its fuzzy electron clouds sufficiently that they can be tracked as smoothly shifting classical shapes (where and how this tradeoff occurs represents a subtle and ill-understood matter to this day[79]). Worse yet, we cannot restrain the quantum/classical tradeoff to a fixed size scale: materials chemically interact with one another at many different scales in manners that require a consultation with quantum principle to predict the outcome of their skirmishes.

But what options lay open to Duhem and Kelvin, our two representative nineteenth century physicists who had not an inkling of the facade-like justification of their discipline? As we saw, Duhem maintained that the only sensible way to interrupt our foundational looping is to choose the *sheet of the broadest coverage* achievable, which represents a thermomechanics that treats deformable bodies (not swarms of point particles or rigid bodies) as the proper raw materials of mechanics, along with an acceptance of many non-mechanical quantities as primitives on an equal par with "stress" and "strain," despite the false pretensions of the kinetic theory of heat to eliminate thermal notions from physics.

In fact, Duhem's continuum-focused recommendations—with some important caveats—are exactly those adopted by most practical engineers who work with

---

[79] Richard F. Bader, <u>Atoms in Molecules</u> (Oxford: Oxford University Press, 1994).

macroscopic materials today, generally operating within a framework nicely articulated by Walter Noll in celebrated work of the 1950s (in fact, Duhem himself favored a variational approach to foundational issues that belongs to the same general family as Noll's). In this approach, we do not attempt to explain behaviors through dropping to lower levels of modeling and averaging; instead, we try to find *constitution equations* that articulate how, e.g., a piece of iron will behave under all conceivable regimes of local stress, strain, temperature and chemical environment. We gain considerable engineering *reliability* thereby simply because the *constitution equations* are usually based upon fairly direct schedules of laboratory testing, rather than speculative molecular models.

A central reason for adopting this non-reductive approach lies in its comparative safety: traditional attempts to construct molecular models for real life substances often supply seriously inaccurate results, simply because of the many subtleties that intervene at every conceivable size scale between the molecules and us (i.e., dislocations, crystal grain; interfacial extrusions, etc.). The advent of computers has led to greatly improved successes in the molecular modeling vein,[80] although many barriers stand between molecular modelings and believable computations.

In truth, any philosophical commitment to never lower our sights below the size scale at which we began looking at matter is plainly excessive (even though Mach often comes close to holding such a position). If I understand him rightly, Duhem never adopted an exclusively hard line with respect to any modeling at subscales (he knew full well, for example, that steel displays a rich microstructure visible under a microscope and that understanding the phase changes of its components helps enormously in auguring the overall behavior of the macroscopic material). I believe his considered opinion probably resembled that advocated by Karl Pearson (4,i) except in tolerating a wider range of foundational predicates: before the project of physical description can get underway at all, the physicist must *artificially intervene* with some basic conceptual structure of her own choosing to impose upon amorphous reality, even if she recognizes that her choice possesses features discordant with what we actually witness. I call this a philosophy of *essential idealization*: the scientist must select some artificially crisp set of conceptual units to prime the pump of physical description, upon whose basis she can then frame empirical descriptions of laboratory events. As we saw, many English physicists such as Pearson adopted this "essential idealization" position to argue that *point masses* can be adopted as the basic elements of classical mechanics, despite the fact that we don't really believe that anything is actually composed of them (the diligent reader may recall Arthur Cayley's comments about "shadows in Plato's cave" from 4,ii). Here's how Karl Pearson expressed the thesis:

> *I feel quite sure that to assert the real existence in the world of phenomena of all the concepts by aid of which we describe phenomena—molecule, atom, prime-atom—even if* [they be admitted] *ad infinitum, will not save us from having to consider the moving thing* [we utilize in our mathematical treatments] *to be a geometrical ideal, from having*

---

[80] Rob Phillips, Crystals, Defects and Microstructures (Cambridge: Cambridge University Press, 2001).

to postulate [a fictitious entity which] *is contrary to our perceptual experience* [of a continuous world].[81]

Here is a less flamboyant expression of the same doctrine from the great elastician A. E. H. Love (of plate modeling fame):

> *The necessity for a simplification arises from the fact that, in general, all parts of a body have not the same motion, and the simplification we make is to consider the motion of so small a portion of a body that the differences between the motions of its parts are unimportant. How small the portion must be in order that this may be the case we cannot say beforehand, but we avoid the difficulty thus arising by regarding it as a geometrical point. We think then in the first case of the motion of a point.*[82]

In this regard, Kelvin himself generally favors the *rigid body* as base element, which he defends through a so-called *principle of rigidification* (which partakes of some of the flavor of "essential idealization").[83] Duhem, of course, believes the proper choice of base elements are found in the richer pastures of thermomechanical concepts. He eventually decides that only "metaphysical opinion" can decide these issues, although he plainly regards the thermomechanical approach as preferable.[84] He believes that the tendency of the English to visualize situations in terms of machinery or points accounts for their metaphysical preference for points and rigid bodies, not realizing thereby the unscientific prejudices that prompt these choices.

. . . .

The key contributing factor behind this essential idealization philosophy traces to Victorian confusions over infinitesimal elements in mechanics. As explained in the fine print to 4,i, formulating proper laws for the continuum situation in a mathematically coherent manner is rather tricky and everyone in the nineteenth century was forced to cobble by with inadequate resources, for which "essential idealization" served as a convenient—but totally wrongheaded—excuse.

. . . .

[81] Pearson, Grammar, 298. Pearson's thesis, although it is hazily expressed, is that one must always expect to begin with mass points in a mathematical treatment of nature, although, as more precise experimental information is gathered, the size scale at which these "ideal elements" will be introduced will require readjustment to ever lower levels (from *atom* to "*prime atom*" and beyond). A similar opinion of Boltzmann's will be examined in 10,viii. In fact, one does not get the number of elastic constants right if only mass points are utilized in the modeling, although Pearson, judging by his comments in A History of the Theory of Elasticity and the Strength of Materials (with Isaac Todhunter (New York: Dover, 1960)), never accepted the empirical disconfirmation of Navier's "rari-constant" approach to the subject. Stephen P. Timoshenko, History of the Strength of Materials (New York: Dover, 1983), ch. 5.

[82] A. E. H. Love, Theoretical Mechanics (Cambridge: Cambridge University Press, 1906), 2. Love represents a good example of an author who apparently espouses a Boscovitchian point of view, yet allows his "particles" to lie in contact with "reactive stresses" between them, doctrines that are not consistent with a strict mass point of view (cf. pp. 347–52). Love's later A Treatise on the Mathematical Theory of Elasticity (New York: Dover, 1944) is much clearer on fundamentals and he admits, "The hypothesis of material points and central points does not now hold the field" (p. 14). His "Note B" (pp. 616–27), however, suggests a lingering personal nostalgia for mass points. His "Historical Introduction" provides an admirable précis of nineteenth century investigations.

[83] Thomson and Tait, Treatise. James Casey, "The Principle of Rigidification," Archive Hist. Exact Sci., 32 (1993).

[84] In future work, I will explain why Duhem worries that the scheme of mechanics proposed by Hertz might prove empirically indistinguishable from the thermomechanics he favors.

Although I have stressed the *reliability* considerations that lie on thermomechanics' side of the register, from another point of view, that school is too eager to embrace more abstract forms of physical doctrine that might tolerate excessive possibilities of an entirely spurious nature. For example, thermomechanics' desire for descriptive completeness assures us that we can freely appeal to *temperature* even in fully non-equilibrium situations, a claim that many knowledgeable observers distrust. Likewise, Duhem believed that mechanics' formalism should accept rotational strains in addition to the usual sorts, but is it fruitful to open up fresh fields of mechanism in this direction? I see Kelvin's preference for models as arising from a reasonable *agnosticism*: beyond certain limits, we lack firm mechanical principles we can fully trust, so it is safer to directly verify that the behaviors we assume can be genuinely exemplified within a physical system at some macroscopic level. In fact, some measure of Kelvin's caution remains the prevailing norm in thermodynamic work today. Although formalizations of thermomechanics in Duhem's vein are rightly regarded as valuable, only the zealots maintain that it satisfactorily hems in classical behavior at a macroscopic level.[85]

In short, although Duhem describes matters in such a vein, the impasse between "French and English ways of thinking" does not seem primarily one of a clash of visions driven by incompatible paradigms or metaphysical allegiances, but traces instead to simple *semantic hunches* about reliability in mechanics that simply could not be settled with the facts then available: Duhem and Kelvin each responded to *symptoms* of the foundational instabilities inherent in classical thinking but neither enjoyed any awareness of their quantum causes. The *hunches* that they each cultivated with respect to how to proceed with safety—along a thermomechanical path in Duhem's case; with tempered agnosticism in Kelvin's—have been accorded a split verdict in history's evaluation that neither party could have possibly anticipated.

And this is why it is important to revisit their struggles in a vein that is suitably sympathetic to everyone caught up in the confusion: different forms of methodological wisdom are embodied in their diverse ways of thinking that do not represent "ways of viewing the world" so much as seat-of-the-pants premonitions with respect to marching into virgin territories as wisely as possible. In such cases, it is sometimes impossible to tell where the path of prudence lies (even though the classical picture would like to assure us otherwise). Indeed, in one of his less dogmatic moments, Duhem nicely describes the basic dilemma that confronted nineteenth century physics in his <u>Evolution of Mechanics</u>:

> *What route would* [the articulation of mechanical ideas] *take? Several paths lay in sight; the entrance to each was wide open and quite smooth; but hardly had one gone along a path than one saw the causeway shrink, the track of the route become unclear; soon one would see no more than a narrow path half hidden by thorns, cut across by bogs, bounded by abysses . . . Where is he who would be carried through to the end desired, who, one day, would come upon the royal way? . . . He who sows therefore cannot judge the value of the*

[85] I. Müller, "Entropy in Nonequilibrium" in A. Greven, G. Keller and G. Warnecke, eds., <u>Entropy</u> (Princeton: Princeton University Press, 2003). R. S. Rivlin, "Red Herrings and Sundry Unidentified Fish in Nonlinear Continuum Mechanics" in G. I. Barenblatt and D. D. Joseph, eds., <u>Collected Papers of R. S. Rivlin</u> (New York: Springer-Verlag, 1997), 2765–82.

grain; but he must have faith in the fertility of the seed, in order that, without faltering, he may follow the furrow he has chosen, throwing ideas to the four winds of heaven.[86]

The only error in this description is that Duhem tacitly assumes that "the royal way" will assume the form of axiomatizable doctrine, when, in fact, a bumpy facade stands at road's end. As I stated, that unanticipated eventuality produces a mixed verdict on their methodological disputes, for Kelvin's unwillingness to stay loyal to a single brand of seed looks rather prescient as well (in 10,viii we shall consider some other important aspects of their debate, again with sympathy for the hidden semantic tangles that lay beyond anyone's capacity to unravel at that time).

I have found that modern historical evaluations of the debate over molecules and atoms in the late nineteenth century are often insensitive to underlying issues of the sort surveyed here. Far too often, the era's confrontations are adjudicated in entirely the "big ideas" mode of our Darwin critic: Who was right about molecules? Whose dogmatic philosophizing retarded scientific progress? Generally, *those* questions can't be answered with any reasonableness, but we can recognize the *local pulls* that made Kelvin or Duhem advance in the directions they did. And these differed between the two parties, but not in a manner that the opposing sides failed to understand or whose rationale couldn't be communicated. True, some fancied that high flown philosophies might settle the case on one side or another, but that assumption was wrong and merely retarded attention to the issues that genuinely matter. In truth, everyone needed to heed Heaviside's reminder: logic is eternal, so it can wait for awhile.

Indeed, we should really be sympathetic to their philosophizing propensities as well, even though it veers to dogmatic extremes, for such musings were prompted by genuine practical difficulties in locating where the trustworthy core of the notions of classical physics lie. And who could have reasonably anticipated the strange manner in which those tensions eventually resolved themselves? However, as modern academic philosophers and historians we can properly fault *ourselves* for having not profited better from classical physics' travails by continuing to see such disputes as battles of paradigms and webs of belief, rather than the puzzled reactions that arise when unsuspecting practitioners confront the delicate filagree of patchwork arrangement typical of successful classes of reduced variables.

....

In thinking about issues such as this, we should beware of succumbing uncritically to a philosophically driven tropism for *unsupported filling in*. As we noted, classical molecules enter the scene only after quantum electronic clouds manifest enough trackability to be approximated by spreadout classical continua. Sometimes one encounters articles that, by some hook or crook, attempt to derive the laws for these classical continua by filling their insides with point particles or rigid bodies, despite the fact that, in reality, no correlative structures of that ilk exist at all within the original electron cloud.[87] But what is the purpose of this reconstructive make-believe,

[86] Pierre-Marie-Maurice Duhem, The Evolution of Mechanics, Michael Cole, trans. (Alphen aan den Rijn: Sijthoff and Noordhoff, 1980), p. xl.
[87] A.I. Murdoch, "A Corpuscular Approach to Continuum Mechanics: Basic Considerations" Arch. Rational Mech. (1984).

even if the mathematics can be made to work? Well, certain responses might be reasonable (e.g., the filled in points might supply a nice basis for an approximation technique), but, in many cases, the researcher seems motivated by naught but a desire to build complete universes in a classical concept mode: the field of classical physics ideas must be made to display internal closure, even at the cost of relying upon fill-in entities of no real life descriptive merit. I think that such ill-motivated stories improperly divert our attention away from the real world causes of the odd prolongations that build up mechanics' uneven facade in the first place.

As we'll discover in 10,viii, Kelvin and his friends were sometimes fooled by allied forms of unsupported filling in: they explained phenomena through molecular hypotheses where the "molecules" in question represent "filled in" projections from entirely macroscopic behaviors. This is an accusation that Duhem frequently makes against their procedures and later advances in applied mathematics have largely proved Duhem right.

....

(xiii)*

Beads on a wire. In this concluding section, I will collect together several stray strands that have been left dangling. In particular, let us return briefly to that bead sliding along a wire that occasioned so much freshman anxiety in my axiomatically-oriented psyche. To recall the situation described in 4,vi, I had been baffled by some apparent incongruities in how my instructor expected me to employ *force* in such circumstances, but was brusquely dismissed when I inquired about them. In our earlier discussion, I alluded to my eventual surmise that the problems stem from "force"'s uneven performance upon the bumpy atlas of classical mechanics, but did not provide an analysis of my specific bead-on-a-wire difficulties. Let me now do so, as they provide a nice illustration of the themes developed in this chapter.

Let's begin, as we did in that unnerving class of long ago, with a point-mass approach to mechanics. From this point of view, the basic objects of the mechanics are spatially separated punctual particles that are never allowed to touch, presumably due to some strong (but never specified) repulsive force that prevents them from colliding. All forces that act between our mass points must perforce be of "action at a distance" type—whether attractive or repulsive, they will reach from one particle to another across a span of empty space in the manner of Newtonian gravitation. Indeed, our textbook made the quite specific assumption that every force arises as particles act upon one another in pairs, that these forces are always directed along the line connecting the two

masses and that their strengths only depend upon the separation of the points and are not sensitive to, e.g., their relative velocities. With considerable fudging, such requirements can be read into Newton's own fuzzy third law of motion, albeit with a considerable degree of anachronism involved, and that is the position my freshman text, in fact, adopted. If these restrictions are placed upon the permissible forces, then the conservation of energy can be derived as a theorem within this brand of mechanics. And, once again, that is the policy my instructor glibly followed.

. . . .

More specifically, the laws of motion for a particular collection of particles are constructed upon an appeal to "Newton's second law" $\Sigma F = ma$, where ΣF represents the (vector) sum of all the individual forces that act upon a selected particle p, m is its mass and a supplies its current acceleration (i.e., d^2q/dt^2 where q codifies p's position). By forming corresponding $\Sigma F = ma$ formulae for every particle in our bunch, we get a set of equations that (we hope!) will move our point mass swarm forward in time in a reasonable manner. In the form discussed here, the Third Law requires that all forces F_i be derivable from a potential V_i. If so, the system's complete energy budget can be represented as a conserved sum: Σ kinetic energy $+ \Sigma V_i$. Newton himself did not believe in energy conservation of this type.

. . . .

Let us now invite our bead and wire system on stage, just as my instructor did a week or two after presenting "Newton's Three Laws of Motion" in the above guise. For vividness, let's assume that both bead and wire are made of the same metal. Given that we have adopted a point mass framework, we should expect that our sliding assembly will be modeled by a collection of mass points as pictured (the squiggly lines symbolize the strong connective forces). As the ring slides onward, it is deflected from its headlong course to follow the cable's contours, alterations that must be achieved through some schedule of attractive and repulsive forces. At any moment in time, it is easy to calculate the exact quantity of *total force* required, which is traditionally called the *force of reaction* that the wire exerts against the bead. Assume that we've calculated the total force f_0 required to pull the bead into its proper positioning A if it arrives with a certain velocity v_0. Suppose we now send our bead scooting along its course at a faster clip v_1. As the ring now approaches A, what force of reaction f_1 must it feel now? Obviously, f_1 must be of greater vectorial strength than f_0 because a stronger force is now required to decelerate it to being situated at A with a speed of v_1 along the tangent. This requires that the masses in our wire must be able to calibrate the collective force they exert upon the bead in a manner that is *sensitive to its velocity*. But, wait a minute, just two weeks ago didn't we require that no forces may act like that, lest our derivation of the

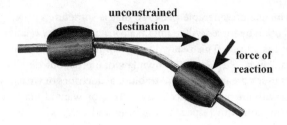

conservation of energy be ruined? Because his prose is often murky, I cannot be certain, but I believe that allied worries lie behind Hertz' concerns about *force* in his <u>Principles of Mechanics</u>.

In fact, we here witness a perfect illustration of how property dragging gets induced through prolongation. Specifically, although we pretended that we have modeled our wire as a set of point masses, in fact we treat it as an integral *rigid body*—as what is traditionally called a *constraint* in mechanical tradition. But as we welcome **rigid body** into our parlors in this fashion, the personality of **force** is secretly made to shift considerably, for the appeal to constraints acts as a mathematical axle around which *force* computations rotate from their original point particle status as important determinants of physical behavior to that of mere after-the-fact calculations in the manner of *determinate structures* in traditional statics (<u>vide</u> 7,iv). That is, if we had remained in the arena of unvarnished point mass arrangements, as my physics instructor sketched them, we would always need to provide substantive principles to govern every force discussed, in the manner of Newton's law of gravitation (that is, a governing rule must be applied for each component force F_i in $\Sigma F_i = ma$). Our governing recipe operates in the functional direction "calculate force → find the motion." But, in accepting the wire as a constraint in our sliding bead case, we are led to assume that the resulting movement is totally known and now our calculations become one of "force of reaction" type: "find the motion → calculate force." As noted, this backwards flavor of computation enjoyed a long history in mechanical tradition, for **rigid body** notions are central in Greek thinking about statics. But this reversal in the calculational employment of F = ma acts as the bridge that drags "force" away from its original point mass significance (with the attendant restrictions that guarantee conservation of energy) into some adjacent patch where "total force" garners a somewhat different meaning (from a point mass point of view "total force" must now connote some averaged quantity of the ilk *total exerted force as averaged over bead motions lying in a velocity range between v_0 and v_1*). Matters are made even more confusing by the fact that conservation of energy can be reinstalled as an *independent principle* within our new patch of point masses-plus-constraints. My instructor was trying to nudge me into this new patch of physics, but he needn't have fudged his derivations to get me there. It is fair to say that most primers in classical mechanics introduce constraints to their readers in some form of underhanded way.

....

There is a useful principle of venerable pedigree that is sometimes called *generalized inertia*: a rigid body not acted upon by forces will maintain a constant velocity relative to any geometrical constraint to which it is affixed. This really doesn't come very close to being true, but it's still quite useful. It looks "kinda like" Newton's own law of inertia, but can't be derived coherently on that basis (it bears much greater affinity to traditional doctrines of virtual work which likewise sneak in constraints without much notice). My instructor wanted me to employ generalized inertia as an approximation with respect to the bead and wire, which is a wise policy, but he needn't have told such a phoney story about its status.

....

Note how the utility of "F = ma" as a useful rule serves as the bridge connecting our mildly discordant patches. This behavior casts an interesting light on Henri Poincaré's celebrated claims that some subset of physics' claims are *true by convention*: that some stock of principles must be assumed as correct before the work of empirical description can get underway (the conventions "prime the pump" of mechanics in my earlier metaphor). Poincaré is best remembered for arguing that Euclidean geometry displays this conventional status, but he also argued that Newton's second law F = ma displays the same qualities:

> It is by definition that force is equal to the product of the mass and the acceleration; this is a principle which is henceforth beyond the reach of any future experiment.[88]

Here he is thinking of "F = ma" in an entirely "find the motion → calculate force" mode, which is only appropriate over the patch of constraints, and then transferring the conclusion illicitly back to the original sheet of point particles without constraints. But, quite generally, the property dragging capacities of innocent looking appeals to "rigid bodies" quite flummoxed our nineteenth century authorities.

Logical positivists such as Rudolf Carnap (4,iv) were quite taken with Poincaré's conventionalism and believed that major hunks of their implicitly defining axiomatic schemes enjoy that status. Because Quine essentially smudges the positivist's tidy "theories" into messier "webs of belief" (5,v), he famously remarks that our opinions represent a "gray lore, black with fact and white with convention."[89] He believes that the two shades can't be sorted out due to the rough hewn and holistic manner in which the web enlarges (his famous critique of the analytic/synthetic distinction rests upon this basis[90]). However, we now see that what goes wrong with conventionalism is often more localized than this: as a single statement, "F = ma" can enter into sundry computational recipes in different ways and this fact alone induces it to span patches where its former property correlations have become grayed through attribute dragging. In my opinion, the basic validity of Quine's rejection of analytic/synthetic clarities can be sustained without appealing to holism of any kind, but simply in terms of the non-classical looseness of predicate/world ties that tolerates diverse forms of patch-to-patch prolongation.

[88] Henri Poincaré, Science and Hypothesis (New York: Dover, 1952), 104.
[89] W. V. Quine, "Carnap and Logical Truth" in The Ways of Paradox (Cambridge, Mass: Harvard University Press, 1976), 132.
[90] W. V. Quine, "Two Dogmas of Empiricism" in From a Logical Point of View (New York: Harper and Row, 1963).

I might observe that predicates such as "momentum" are prone to similar dragging, especially as innocent-looking words like "virtual" become attached to them. Edoardo Benvenuto comments in his excellent Introduction to the History of Structural Mechanics:

> The word "momentum" goes through several stages of meanings and interpretations and becomes in itself a puzzle to be solved and a source of connections and analogies . . . [I]n the case of "momentum," the word persists, a fixed term around which different concepts revolve. This lexical stability deeply influences thought because it makes the term a part of history, welds it to tradition, and at the same time stimulates debate to clarify the definition . . . . The word itself plays a hidden role because of its sheer persistence. It is the custodian of manifest intentions, an object of hermeneutic research, and a spur to historiographic reflection about the nexus rationum inter se.[91]

There is a second aspect to the bead on a wire case that illustrates another general feature we have discussed, especially in 4,vii. Appeals to "rigid body" represent the imposition of top-down schemes for reducing descriptive variables (that is, we utilize the fact that certain aspects of a complicated system's macroscopic behavior are already known to us to simplify how we reason about the aspects we don't yet know). To this, some degree of physics avoidance needs to be practiced: complicated behaviors need to be swept into minimally described singularities and boundary line regions. A quite blatant form of this reducing avoidance is central in the bead and wire case, but it is rare that it excites any comment.

Let me return to my puzzled freshman condition of decades ago to illustrate what I mean: "Gee, if the bead and wire are made of the same stuff, isn't the bead likely to begin sticking to the wire on close approach, rather than sliding along without restraint as we're supposed to assume?" And it's true: if we put two clean pieces of completely smooth metal into contact, they do bond quite tightly. The main reason we (fortunately) don't witness more of this welding in ordinary life is that surfaces are quite uneven, their bounding layers display a very complex chemistry and their surfaces are dirty due to the greasy atmospheric crud that cloaks virtually everything around us. Kenneth Ludema comments:

> One of the mysterious aspects of research reports and published papers is that sliding surfaces are discussed as if they have no contaminant or other substances on them, whereas virtually everyone else, educated and uneducated, child and adult, knows that such substances are ubiquitous.[92]

In fact, these unnoticed surface complexities greatly decrease the ferocious adhesion that metals would display if allowed to approach in the naïve manner of our point particle model (much of the true story of what happens depends greatly upon intrinsically quantum principle per the observations of the previous section).

---

[91] Edoardo Benvenuto, Introduction to the History of Structural Mechanics, i (New York: Springer-Verlag, 1991), 18.

[92] Kenneth C. Ludema, "Friction" in Bharat Bhushan, Modern Tribology Handbook, i (Boca Raton: CRC Press, 2001), 218. Duncan Dowson, History of Tribology (London: Professional Engineering Publishing, 1998), ch. 11.3.

But all of these difficulties have been simply swept away by fiat in the bead case. As a vital policy of useful variable reduction, appeal to frictionless constraints is a good idea, but we must expect some hidden dragging in predicate attachment as the price to pay the Pied Piper of useful descriptive practice.

. . . .

To anticipate some themes emphasized more explicitly in 7,iv, some of the special salience of constraints within mechanics traces as much to considerations of *design and planning*, rather than *predictive accuracy* per se. The distinction nicely emerges when non-holonomic constraints are considered, such as a skater sliding down a frictionless plane under gravity. Under these conditions, standard mechanics predicts that our skater will never reach the ground, but instead pirouette in cycloids forever. And it is true: to the degree that we can implement simulacra of frictionless sliding, we witness approximations to these unexpected behaviors. But in most cases where we worry about skates and the like, our true interests center upon *efficient guidance*, where the attack angle of the skate is presumed under the skater's control. And it has been recently realized that questions of this ilk demand a subtly different mathematical formalism than our prediction problems, an observation that has inspired much deep research.[93] In this shifted context, the friction that attends the movements of any real life skate is not being discarded for the sake of predicting strange behaviors that we're never likely to witness, but as a mathematical annoyance that impedes our computation of optimal steering path without altering the final results much. In other words, "frictionless skate" does not acquire its characteristic mechanical personality from predictive contexts, but from those of design and planning.

. . . .

Finally, let us briefly return to the Kutta/Joukowsky paradox of 6,v. Here we witness another stunning example of property dragging, but of a somewhat more complicated ilk. Roughly speaking, what occurs is this. In air flow around an object, viscous effects will cause its boundary layer to separate from the body, creating an ample wake. However, the Laplace equations with which Jutta and Joukowsky work descriptively lift away from this base situation in three critical ways: (1) steady state flow is assumed, in which all true time development is suppressed in their equations; (2) all account of friction is neglected; (3) they attempt to deal only with streamlined bodies such as airfoils possessing a sharp trailing edge. But then our authors allow

---

[93] A. M. Bloch et al., <u>Nonholonomic Mechanics and Control</u> (New York: Springer, 2003), 23–5. Donald T. Greenwood, <u>Advanced Dynamics</u> (Cambridge: Cambridge University Press, 2003), 316–23.

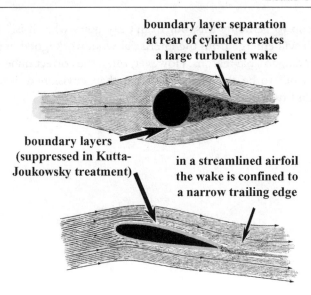

boundary layer separation
at rear of cylinder creates
a large turbulent wake

boundary layers
(suppressed in Kutta-
Joukowsky treatment)

in a streamlined airfoil
the wake is confined to
a narrow trailing edge

(in the form of a *Kutta condition*) a jump along just the trailing edge line where the wake of an airfoil can be expected to lie. This allows them to find a solution to the simplified Laplace equation alone that is fairly close to the nearby free stream pattern that Prandtl's boundary layer theory would calculate around the airfoil outside of the thin boundary layer itself. To get the lift on the wing, we should compute the total upwards force within this nearby free stream layer. But because we have exchanged the true **upwards force** quantity for an analytic function replacement, we can take advantage of the latter's rigid personality to set our circle of integration conveniently far away from the wing—over Antarctica in my 6,v description. In real life, of course, the viscosity of air and the fact that our wing does not truly fly in steady state circumstances would have utterly wiped away such uncorrupted South Pole notification of our plane. In other words, our Joukowsky-Kutta description lift tacitly drags the predicate "upward circulation force" away from its original locus and replaces it by some subtly rigidified and smeared-out-between-Kansas-and-Antarctica surrogate. But it isn't any wonder that beginning engineering students frequently become mystified when they are instructed ("ordered" is more like it) to follow Joukowsky-Kutta directivities, without being offering a clue as to what happens to their descriptive predicates in the process.

Our sundry classical mechanics examples nicely display the manner in which the *personality of a predicate*—its overall conceptual feel—can be significantly flavored by background factors (such as hidden lifts to analytic functions or design agendas) of which the linguistic agent affected possesses no real inkling, although the concrete directivities she follows in utilizing her predicates in practical circumstances trace to these undiagnosed sources. Her current awareness of these strategic factors partakes, at best, of only those *intimations of intensionality* that I discussed in Ernest's case: "There's some good reason to be doing this, but I can't say quite what it is." Eventually, the underlying truth will emerge from the interfacial woodwork; good linguistic schemes succeed only for diagnosable reasons. However, often the correct underpinnings of our intimations turn out to be completely different than envisioned, just as Pip's great expectations didn't turn out as anticipated either.

7

LINGUISTIC WAYFARING

[T]he human mind, when it sails by dead reckoning, without the possibility of a fresh observation, perhaps without the instruments necessary to take one, will sometimes bring up in very strange latitudes.

J. R. Lowell[1]

(i)

Atlases and facades. With apologies for a short stretch of dry schematization, let us resumé some of the behaviors examined in the last chapter within a simple model. This framework is intended to be rather rough and ready—any real life case is likely to exhibit special features beyond its reach. Indeed, the scheme will not even cover all of the examples within the present book (e.g., circumstances where the semantic reading of a predicate shifts dramatically within a single run of reasoning). Nonetheless, having at hand a formal picture of an organizational structure that sometimes arises in language will be useful in our attempts to understand the exact manner in which classical thinking distorts our understanding of natural linguistic process. Virtually everything I express here can be easily extrapolated from our prior discussion.

What I have in mind is simply a collection of the ideas that I have already been using under the heading of a *facade* or *atlas* (I use the two terms interchangeably). We begin with some basic domain $\mathcal{D}$ of physical fact that we wish to cover in a linguistically profitable fashion.

....

I chiefly have in mind what the physicist calls a *phase space* (= the set of possible behaviors open to a given type or physical space) or a *control space* (possible systems organized by material parameters such as the range of solids we might wish to evaluate for their "hardness"). Some further details on these ways of thinking were provided in 5,vi.[2]

---

[1] James Russell Lowell, "Witchcraft" in Among My Books (Boston: Houghton, Mifflin, 1881), 128–9.
[2] Anil Gupta has a similar idea in mind in his "frames"—cf., his "Meaning and Misconceptions," in Ray Jackendoff, Paul Bloom and Karen Wynn, eds., Language, Logic, and Concepts: Essays in Honor of John Macnamara (Cambridge, Mass.: MIT Press, 1999), 15–41.

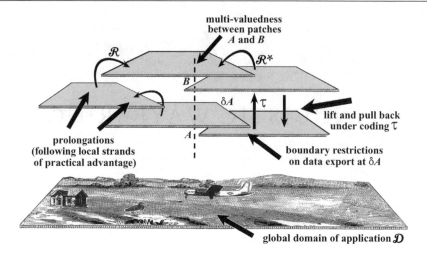

multi-valuedness
between patches
A and *B*

$\mathcal{R}$

$\mathcal{R}^*$

B

δA τ

A

lift and pull back
under coding τ

prolongations
(following local strands
of practical advantage)

boundary restrictions
on data export at δA

global domain of application $\mathcal{D}$

....

Over $\mathcal{D}$ we erect basic *patches* (or *sheets*) A_i, corresponding to localized flat maps, on which some basic physical vocabulary of predicates, relation symbols, names and limited mathematical and logical resources will be made available (e.g., truth-functions and quantifiers over the subdomain covered in the patch). Over each A, each predicate "P" will correspond to one or more attributes in $\mathcal{D}$ below, under the condition that if attributes φ and ψ are both assigned to "P" on A, they will act coextensively within the region of $\mathcal{D}$ it covers (the intention is to allow "weighs five pounds" to align with both *mass* and *impressed gravitational force* over the earth's surface). Allied requirements hold for the relation symbols. Names like "a" are required to pick out unique and fixed referents, however, no matter in which patch they occur. There will also be a set of local recipes and reasoning tools $\mathcal{R}$ attached to each patch such as modus ponens, Euler's method or policies for selecting a "hard" material in manufacturing. I'll explain the structural importance of $\mathcal{R}$ in a moment.

Each patch A possesses a natural boundary δA which is marked by the fact that, once we move beyond δA into B, then some of the predicates in A will either shift to new property alignments within B or the reasoning tools $\mathcal{R}$ native to A will no longer lead to sound expectations. Thus, in a *Stokes facade* case (6,viii), a rule may continue to make sense outside of A in that it presents conclusions that we might accept, but boundary crossing restrictions warn us that such beguiling directivities should not be trusted beyond δA (the need for reasoning cutoffs of this type was dubbed the *Stokes phenomenon*). Likewise, in an *uneven facade* (6,ix), restrictions along δA may inhibit the free importation of sentential data brought in from patches outside A, as is required to keep most of the examples provided in the last chapter coherent (in a pure Stokes facade, however, such import remains legitimate). Later, in section (viii), we shall study some of the unusual ways in which these restrictions on data importation get implemented within naturally arising facade-like contexts. Of course, mixtures of both types of boundary control are easily possible.

These conditions set up a local grammar upon **A** but we will often be interested in how language behaves over the joins or continuations that connect the patches, which can be smooth, abrupt or overlapping. Quite commonly there will be a small group of recipes or rules of inference $\mathcal{R}$ that prove central to creating this bridging between **A** and **B** (the role of power series in analytic continuation represents our prototype here). Patches can also sit partially astraddle of one another through fibered connections established by common names. Connected to these will be translation principles τ that regulate how data shall *lift* from one sheet to another (as in the different covering maps of 6,ii). We demand no specific topology in how our atlas of covering patches fits together, so it may be possible to move through the patches in a multi-valued manner.

An atlas of essentially one patch, that covers its whole domain adequately will be called a *flat structure*; it is essentially the linguistic platform that the classical thinking expects to see, once a language has been cleansed of its undesirable ambiguities. We'll discuss these matters further in section (iii).

In our prior discussion (6,v), we noted the need for a *preface* or *picture* of our atlas's workings: viz., a schematic overview of how the patch-to-world relationships unfold in the facade. We observed that an agent might be able to employ an atlas quite capably from a practical point of view, yet entertain an erroneous picture of its descriptive workings. We shall be particularly interested in such situations here, for many common varieties of *ur*-philosophical error trace to this cause. In particular, we shall be much interested in *semantic mimicry* where some facade-like construction passes, amongst its employers, for a flat structure: it looks very much like the "first-order theory" of the logicians if we don't scrutinize its oddities too closely (pretenders of this ilk are called *theory facades* for this reason, even to the point of suggesting the title "facade" for any patch-like linguistic platform). In the latter part of the book, when we shift from primarily considering a language frozen in its current organizational state to asking how it is likely to alter as time goes forward, the role of these pictures will become quite critical.

A facade assembly should be regarded, in analogy to the two-sheeted Riemann surface for $\sqrt{z}$, as a *strategically informed platform* upon which a stable linguistic usage can be settled, instead of following the flat structure model of uncomplicated "is a dog" / *being a dog* alignment. As long as a speaker respects the boundary divides marked by ∂A, she can employ an unevenly founded language to freely express what she wishes locally, while exploiting the boundary restrictions between regions to create an overall employment that may prove more effective and efficient overall (demanding that the predicate "P" stay attached to attribute φ even over domains where φ can't be easily tracked deprives "P" of a utility it might otherwise gain through a bit of harmless property dragging). The basic worry that animated the worries of Chapter 5 is this: how can strands of practical advantage (represented here by our inferential rules and recipes $\mathcal{R}$) influence the semantics of a language, given that most of its employments are not "practical" in any obvious manner at all? Our facade models demonstrate simple

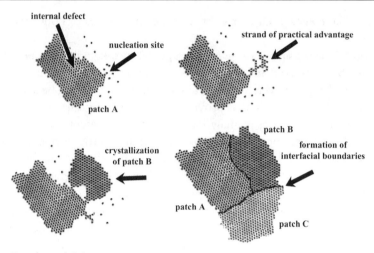

internal defect

nucleation site

strand of practical advantage

patch A

patch B

crystallization
of patch B

formation of
interfacial boundaries

patch A

patch C

Facade nucleation

ways in which such concerns can be addressed, without needing to puff up "practical advantage" into some amorphous notion of grander scope such as "language game" or "practice." Specifically, the *range of practical validity* for the $\mathscr{R}$ rules determines both where the boundaries ∂A need to be set and which boundary crossing restrictions on inferential rules and data importation should be enforced there. Here the "practical correctness" of our $\mathscr{R}$ rules must be explicated primarily in what 4,v calls a *distributed normativity* vein: the top-down, instrumental "correctness" of the rules $\mathscr{R}$ marks where the reach of the *direct normativities* internal to each patch (<u>i.e.</u>, the sentences made true under the local "P"/φ alignments) must terminate. To be sure, in an uneven facade where the rule $\mathscr{R}$ stretches validly across the boundary that divides patch **A** from patch **B**, $\mathscr{R}$ will usually qualify as "true" according to the direct standards internal to both **A** and **B**, but this shared upward correctness will not represent $\mathscr{R}$'s most significant semantic trait. Rather it is $\mathscr{R}$'s instrumental *correctness* over these domains that is responsible for connecting patches **A** and **B** together in the first place. In this fashion, a stitched together atlas represents a structural hybrid between the standards of semantic evaluation natural to a standard first-order *theory* in the logician's sense and those claimed for a Quine-like *web of belief*, where the central notion of "correct assertion" derives entirely from the distributed normativity of inclusion within a growing body of theory (in our facades, however, our appeals to *distributed correctness* derive entirely from short strands of practical advantage $\mathscr{R}$, not from sweeping banks of holistic belief). The fact that ordinary "P"/φ standards of sentential correctness reign within each patch allows a speaker ample room for free, even completely impractical, assertion, while still reserving an important shaping role for the imbedded strands $\mathscr{R}$. We will later see that it is around these $\mathscr{R}$ rules that fresh patches of usage often nucleate.

To put these matters in a nutshell: the local patch-centered freedom of expression found in the facade of *weight* allows stuffy Uncle Fred to recite permissibly, for hours and hours with uncheckable enthusiasm, endless correct terrestrial facts about "weight" without directing any of this phonetic torrent to remotely useful purpose. He can do the same with respect to the innards of space stations, but the correctness of what he says about these matters will be adjudicated according to space station-adjusted standards of what "weight" there indicates. Fred himself may lack any inkling that such evaluative shifts govern his assertions nor will he notice the special distributed "correctness" of the folk physics reasoning rules that induce these adjustments. It is also unlikely that he recognizes the hard-to-evaluate status of his wide domain assertions ("I'd rather lose weight in a space station than on the moon").

I reiterate that the profitable use of language can be structured in manners far more complicated than a facade (we shall see a hint of some of these in section (viii)). Our earlier discussion showed that much of the ongoing discussion in philosophy of language has been hampered by unnecessary shibboleths (semantic finality; the free creativity of language; the theoretic nature of descriptive contents, etc.) that hem in the possible options unreasonably and prevent our recognizing perfectly natural forms of linguistic structuring for what they are. Facades happen to provide a particularly easy means of appreciating how language can march to a different drummer, even if the complications of real life behavior must be somewhat shoehorned to fit their schematic contours tidily (it is important to cite real life cases to advance my contentions, but real life is invariably messier than any facade!).

Or, to put my claims more accurately, much of a language's potential usage is likely to be currently *formless*, in that it is not yet settled how its terminology should be employed over domains as yet rarely visited. As those extended domains become eventually colonized, Druid-style (1,ix), we can sometimes anticipate that the usage will assume the approximate form of an uneven facade if the underlying environmental advantages strongly favor that alternative.

Throughout our ruminations, it is important to acknowledge that usages can be straightforwardly situated upon a flat platform, whereupon every predicate stays in unique alignment everywhere with a single attribute below, in approved "is a dog"/ *being a dog* manner, and with no boundary line curbs on reasoning required. Indeed, such flat platform support undoubtedly represents a more common occurrence in normal usage. Nonetheless, we can't always prevent alternative structures like my facades from arising, even in circumstances where they are totally unexpected and unwanted. As I've emphasized, many of our *ur*-philosophical difficulties trace to our proclivity to presume that our predicates sit on flat platforms with respect to worldly events, when, in fact, they need to obey some alternative organizational pattern.

To claim that a usage sits upon a flat platform does not imply that its unitary predicate/attribute correlations are installed there by the mechanisms of classical gluing, however; no particular explanation of how its correlational relationships have become locked into position is implied by the flat structure attribution at all. Quite frequently, these issues have been greatly muddled in philosophical discussion. In 2,iv, I mentioned that the philosopher Gilbert Ryle satirized what he calls the "Fido"/Fido view of language, which, adjusted to suit our concentration upon predicate behavior, becomes the "is a dog"/*being a dog* view of language. Many philosophers have borrowed this dismissive characterization from Ryle. As often occurs when unflattering labels are concerned, the basis for complaint is generally left murky. When a writer complains of "Fido"/Fido thinking, what does he have in mind? Is some unacceptable mechanism like classical gluing presumed to form a tacit part of the view dismissed? Or is the author instead *demanding* that such a mechanism be supplied? (this may have been Ryle's own position). Or does our critic regard the mere possibility of <u>de facto</u> unitary correlation as incoherent, in the manner of many pragmatists and neo-Kantians? Or is it his intention to observe, as I have argued here, that sometimes predicates line up with worldly correlates in a more complex fashion than a simple, flat platform arrangement? Often I can't really determine.

Whatever the intent of other writers, my own position is quite clear: circumstances where a group of predicates sit in simple "is a dog"/*being a dog* relationships to the world are rather common, a fact that generally, but not always, represents a desirable semantic situation (almost certainly the predicate "is a dog" itself falls within this happy state). In other circumstances, there are repair steps available that can bring a previously uneven usage into flat structure accord, just as it is easy to name rabbits imprisoned in backyard hutches (our very real capacities for redefining terms and laying down axioms often address these needs). Sometimes, a formerly rocky usage will naturally evolve into a unitary pattern, without our taking any overt steps on its behalf at all (biological classifiers pitched at the species level are often self-correcting in this manner, for example). All the same, there are plenty of circumstances remaining where more complicated alignments arise, either because we don't bother to enforce the corrective steps required to keep the word's alignments unitary (the wanderings of "weighs five pounds" and "robin" might have been remedied by language police of a sufficient vigilance) or, more importantly, because such arrangements are incompatible with practicality (this is the lesson of reduced variables (4,vii) and our classical physics examples).

Since flat platform arrangements are desirable for many purposes, it is often an improvement when a usage can be *regularized*, in the sense that some nearby linguistic platform can be located that retains all the practical utilities of an uneven atlas, but restrains its predicates to constant attribute associations. In our "weight" case, finding a suitable regularization is quite easy: we simply need to stick "apparent" in front of "weightless" every time we employ the phrase within a space station context. Most classical thinkers acknowledge that real life usage often behaves unevenly, but they tender a strong methodological promise: by thinking about the meanings of our terms carefully enough, any uneven usage can be regularized. It is within this kind of assurance that the deepest exaggerations of classicism lie. Sometimes it is simply beyond human capacity to fulfill such projects, for classical thinking extrapolates our limited "name the bunny in the backyard hutch" capacities far beyond their true reach. We can *strive* for better, but lack the means to *guarantee* them.

Thus it is important to recognize that, for a variety of very basic reasons, flat platform arrangements sometimes cannot be implemented in a usage without ruining its practical capacities: like "$\sqrt{z}$," many of our most successful classificatory terms obey a growth logic of their own and they simply won't tolerate being straightened out by some goody-goody linguistic reformer (this is why 4,viii's failure to find an axiom scheme that can adequately embrace classical "force"'s twists and turns represents such an important illustration of our central concerns: an axiomatization, after all, is intrinsically a delineation of word behavior over a flat platform).

. . . .

Standard techniques for variable reduction suggest many other interesting models for predicate support. Thus, sometimes a complex behavior can be divided into a set of "fast variables" modulated by an envelope of "slow variables." Such factorization allows predicates native to a certain flavor of behavior (such as "frequency") to invade other domains where they do not directly belong. And there are many different cases of this type (an astonishing example, mentioned elsewhere in the book (9,iii), is the manner in which Norbert Weiner's "generalized harmonic analysis" explains the application of "frequency" to naturally occurring light). But none of these alternative forms of predicate "borrowing" fit my atlas structure format without undue strain.[3]

. . . .

(ii)

Quantities and quasi-quantities. When applied mathematicians consider a complicated physical system $\mathcal{S}$ that they hope to describe in a useful way, they often conceptualize the possible quantities pertinent to $\mathcal{S}$ in the manner of an architect's elevation drawings, where a prospective building is projected onto blueprint planes arranged before its front, side and top sides. However, if the target object is irregular, it may prove preferable to project its features onto more unusual planes, so that we can examine how it looks from more oblique angles. In dealing with the earth, we usually like to project

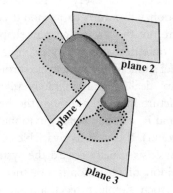

[3] Igor V. Novozhilov, <u>Fractional Analysis: Methods of Motion Decomposition</u> (Boston: Birkhäuser, 1997).

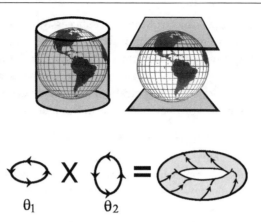

the polar regions onto maps in a different manner than the temperate belt (we employ tangent projections for the poles but a cylindrical projection for the tropics), although this choice entirely reflects human settlement patterns on the globe rather than any intrinsic geometrical feature.

More abstractly, in dealing with a physical system in a phase space, a given physical quantity (following the standard treatments surveyed in 5,vi) will slice the system's possible conditions into curvy sheets on which the quantity in question retains a constant value ("foliate" and "level surface" are the official words). In trying to describe a given system effectively, we try to choose descriptive quantities so that the system's temporal changes of state will look simple according to the slicings adopted. As discussed under the heading of Fourier decomposition in 5,vi, the wiggling motion of a guitar string can be decomposed into different vibrational modes or *eigenfunction traits* which operate independently of one another (here they look like sine waves). Rephrased in terms of the relevant physical quantities, this decomposition means that the condition of our string at any moment can be completely fixed by the quantities that supply numbers to its eigenfunction energy E_i and phase variables θ_i, because once we know how much energy has been poured into each of the string's countable patterns of vibrating (that is, we are told how *loud* its ith overtones are for every i) and the time is also known at which this ith eigenfunction vibration will return to its home configuration (this is what θ_i registers), then we can figure out the string's current shape and wiggling velocities by simple addition. From a geometrical point of view, this means that our string will remain confined to fixed level surfaces of E_i and cycle through the θ_i slicings in a simple closed pattern (since we are dealing with infinitely many degrees of freedom, the situation is impossible to picture literally, although the behavior of a string with two effective energy modes E_1 and E_2 can be *roughly* envisioned as a point that wraps itself around the contours of a donut). If descriptive variables of this admirable geometrical simplicity can be found, physicists usually regard the system's behavior as adequately understood (the chief ambition of Hamilton-Jacobi theory is try to construct finite dimensional equivalents of such simple projections—as *action-angle* variables—when they exist).

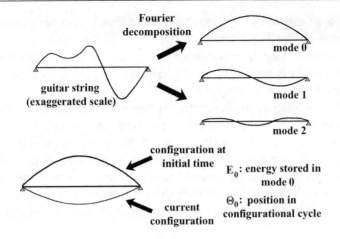

The Fourier variables E_i and θ_i can be considered as *naturally adapted* to our string because its behavior looks especially simple when mapped onto their covering surfaces. However, it is common to utilize these same E_i and θ_i descriptive terms even if they no longer correspond to the temporal behavior of our target system in any tidy way (Fourier analysis remains a common tool in signal analysis even when its sine wave patterns bear no natural relationship to the signal). Thus, because of uneven manufacture or larger vibrations, the sine wave patterns in many kinds of string will no longer conserve modal energy individually but will exchange it amongst themselves. The blueprint projections for such a non-linear string will not look simple at all: its E_i and θ_i quantities will vary quite unevenly over time. Nonetheless, we may still choose to describe our string in terms of its shifting E_i and θ_i values, because we happen to have a nice Fourier analyzer in our lab that can compute E_i and θ_i values swiftly from the string's shifting set of configurations. Indeed, registering our string's shifting array of E_i and θ_i values often proves a more economical vehicle for data storage than if we register its more multitudinous collection of **height** and **velocity** values (in more recent years, it has become common to project our irregular string's configurations into wavelet values that are even more efficient in terms of the storage demands they exact, although they don't represent especially *natural traits* of any physical system[4]).

The point I'm after is this: there are many sets of projected quantities available for the description of a physical system, some of which are particularly well adapted to its behavior and some of which are not especially so, but are convenient for other reasons. Which decomposition we choose to favor can be a function of *natural adaption* to the system (as with our original guitar string), our *own convenience* (the Fourier analyzer in

---

[4] C. Sidney Burrus, Ramesh Gopinath, Haitao Guo, <u>Introduction to Wavelets and Wavelet Transforms</u> (Upper Saddle River, NJ: Prentice-Hall, 1998).

our lab) or some mixture of the two (ignorance should also be thrown in the mix, because sometimes wonderful natural descriptive variables exist for a system, but we've not stumbled across their framework yet). The fact that a great liberty in selecting descriptive quantities is wanted in physics explains the generosity implicit in the text-book approach to *physical quantity* outlined in 5,vi.

Thus far I've written about descriptive systems that primarily aim to capture and possibly predict the *temporal evolution* of our system. But, as we'll see, a shift to other dominating concerns can radically change the character of the descriptive variables that most interest us. In particular, as we'll later see (7,iv), a focus upon *design* (= the invention of a device to meet assigned specifications) can alter the quantities we emphasize as desirable in remarkable ways (from a mathematical point of view, a design objective relocates our target system to a different mathematical arena than an interest in prediction).

In a related manner, as we observed at considerable length in 4,vii, the basic need for substantial variable reduction in macroscopic situations affects the character of our descriptive variables greatly, including all of those quantities that get swept into boundaries and other descriptive singularities. As the argument in that chapter outlined, facade-like organization represents a common outcome of these reductive necessities. A good choice of reduced descriptive vocabulary will often see its patch boundaries *naturally adapted* to its target subject matter as well (the shock wave case of 4,vii supplies an apt example of this). The personalities of the descriptive predicates left active will generally be affected in consequence.

As noted before, the range of predicates potentially useful with respect to a target system needs to be set at an extremely generous scale, stretching beyond the sets definable in any fixed terminological basis (when reduced variables are concerned, we must tolerate predicates that cover informational packets even broader than those delineated in 5,vi). Many philosophers currently believe that descriptive practice in physics is far more exclusive in its contours than this, expecting that any physical system can be adequately captured in terms of a very small set of "natural kinds." Why they believe this seems to arise from a mixture of *ignorance* (never looking in a suitable textbook), *confusion* (mixing up fundamental material parameters with dynamical attributes) and *wishful thinking* (the "natural kind" point of view facilitates many grand metaphysical projects). I have already complained about these matters in 5,vi to which I refer readers who may have skipped that section.

However, the issue of greatest salience to our immediate concerns is this: how should we think about the conceptual personalities of the predicates we employ to describe a physical system? Let's look at the simplest case first, where our predicate "P" correlates with a given attribute φ in tidy "P"/φ alignment. Suppose we look at an array of reports involving "P." Two distinct forms of question can naturally arise in this case: (1) Do our "P" reports accurately reflect the target system's φ values or not? (2) How has "P" happened to align itself with φ in the first place? In addressing this second question, we must discuss some of the factors of choice and convenience we have already canvassed, <u>e.g.</u>,

"Energy and phase are naturally adapted to our string's behavior": "I happen to have a fast Fourier analysis program on my computer," etc. Some of these considerations have plainly less to do with the specifics of the target system's own behavior than our capacities to place language upon useful quantitative planes—matters that hinge upon the measurement and computational resources we have available to us (I will often call these our *available descriptive tools*). And mixtures of the two concerns are clearly possible.

Circumstances that demand uneven facades make this divergence in concerns quite vivid, for we face circumstances where materials enjoy their own attributes without problem, but we, as potential employers of language, must scramble mightily to erect screens to which a suitable set of reduced data can be projected. Sometimes the best we can supply constitutes a bumpy and multi-valued affair. We can appreciate the gap at issue by reviewing concrete circumstances where we are forced to settle for a somewhat compromised descriptive format. In fact, this is precisely the situation that arose in 6,x in connection with Samuel Williams' quixotic quest for an "absolute" measure of *hardness*. So let us ask ourselves, in a manner comparable to our treatment of the favorable capacities of quantities like E_i and θ_i, how we should look upon the informational presentations provided in an uneven facade setting? Once again it helps to look upon the situation geometrically, in terms of elevation drawing-like projections from our target system. The only notable novelty that uneven facades introduce is the fact that we've now elected to map information about our target systems onto a *multi-sheeted covering* of their domain. But, if we utilize such multiple registrations with requisite care, they can prove quite useful. Here are two simple examples.

(1) Let us wrap a sheet of paper around the mid-section of the earth a number of times. Sometimes the resulting projection can be employed to useful purpose, e.g., to plot a narrative of how the journeys of two circumnavigating pirates compare. In this multiplex packaging, temporal information has been coded together with geographical data (which we can later disentangle because of the general eastward trend displayed by our privateers).

(2) In describing a fluid effectively, we simplify its description by blurring its molecular positions into a smooth continuum. As noted in the fine print of 4,vii, when

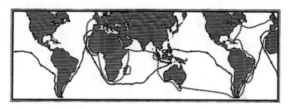

Circumnavigating pirates

the fluid surface forms droplets, we need to run two overlapping descriptive sheets over these events in which the drop-to-be is described as, respectively, attached or unattached. In this overlapping interval, our reduced descriptive vocabulary simultaneously assigns two incompatible "shapes" to a single molecular configuration. But this abrupt multi-valued covering isn't particularly surprising: it is the natural side effect of the policy of variable reduction we have chosen.

More sophisticated (and substantial) toleration of multi-valued coverings can be found in the frequent use of complex variables in physical applications, some of which we've already examined. Indeed, it was the fact that "$\sqrt{z}$" serves as such an upstanding descriptive citizen in the applied mathematical world that led us to first realize that "Gee, being multi-valued ain't so bad as long we control the data registered by the right strategies" (this is the lesson the Stokes phenomenon teaches).

However, we shouldn't consider the physical data registered on a multi-valued sheet as a true *physical quantity* in quite the same way as if the data had been registered on a flat platform: as with our pirates, the multi-valued registrations need to be decoded a bit before our privateers' actual properties at each point in time can be recognized. On the other hand, our multi-valued data registrations plainly supply physical information in a format *not all that different* from an orthodox delineation of its qualities. After all, there are a large variety of useful formats in which the requisite physical data gets compiled in alternative packages—as vector-valued or tensor-valued fields, for example—which likewise aren't "quantities" in our textbook sense either (these were discussed in 5,vii). Likewise, any toleration of *singularities* in a projection can ruin a representation's status as a veritable quantity in a harmless manner: *horizontal position* on a modified Mercator map fails to represent a proper quantity overall because the North and South Poles are each assigned infinitely

many values (i.e., they blow up into lines under the projection and do not map to a unique numerical *horizontal position* value there). Indeed, many would-be quantities of averaged origin (*stress*, for example) spoil their quantity status by blowing up to infinity at selected locales.

Accordingly, I propose that the uneven data registration supplied by *hardness* be designated as a *quasi-quantity,* to be understood as bearing the same relationship to certifiable physical quantities as $\sqrt{z}$ bears to regular single-valued mathematical functions (unfortunately, items like $\sqrt{z}$ are commonly called "functions," even though they're not single-valued; properly speaking, they should be considered "quasi-functions"). Qua data registration, the deliverances of a quasi-quantity should be viewed as *objective* (or nearly so) as those provided by a true physical quantity; we've simply introduced a few kinks in the projections whereby we gauge our target system.

In considering *hardness* in its information reporting role, we have left entirely out of consideration the factors that lead us to adopt a platform of this kind. As we saw, the basic answer traces to our inability to link practicable determinations of "hardness" together in a flat structure format. True, materials undoubtedly possess true qualities that hover somewhere in the vicinity of *hardness*'s uneven facade, but we are unable to place a useful label on any of them (like rabbits in the wild, generic quantities are not easy to name). But this failure on our part has only indirect relevance to the materials we seek to describe. The entire discussion of Chapter 2 was riddled with a false dichotomy between *objective* and *subjective* traits (a mysterious category of amphibolic qualities made its fleeting appearance as well). But, plainly, those discussions muddled the issues we have distinguished here, hampered by *ur*-philosophical inclination to compress every issue of predicative personality into a single *ghost attribute*. But Chapter 2's narrowed patterns of thought do not match the ways in which we employ "objective" in the ordinary evaluation of information presentation. Suppose somebody publishes an accurate map, along with the relatively simple formulae that define the projection scheme utilized. Is the data in the map "objectively presented"? I would think so. Will the data still prove "objectively presented" if a wrong projection formula is supplied as a preface or omitted altogether? Here we are likely to comment, "This map maker doesn't realize what he is doing," but regard the data itself as nonetheless unscathed. And what should we say if the cartographer has carefully studied the vagaries of human psycho-logy and has concocted a very irregular projection precisely designed so that untutored observers will generally draw generally correct inferences with respect to the com-parative sizes and locations of countries? In point of fact, many of the general purpose world maps we see nowadays *are* compiled in this last fashion—and their weary designers have long ago given up on the vain effort to tell us exactly what they've done by publishing their projections (which cannot be supplied by formulae). Does this blatant dependency upon a projection selected solely, because of the quirks of human psychology render the map's data less "objective"? Well, not in any clear sense, espe-cially since its employers extract more objective answers from such maps than they would from regularly plotted projections. It seems wisest to characterize the data supplied in all of these maps as equally objective—none of it has been *corrupted* by human foible—but the formats that suit us best are partially determined by subjective (= directly reflective of human psychology) considerations. Intelligent use of a strange map shouldn't be regarded as distorted and the same moral holds, I think, for information reported in quasi-quantity form.

But the ghost attribute proclivities of classical thinking discourage us from making these simple distinctions and thereby hangs the tale of Chapter 2.

(iii)

The veil of predication. Throughout our discussion I've endeavored to unpack the classical thinker's fused "intensional content" into its operational significance with respect to our everyday evaluations of predicate employment: to what considerations do we point when asked guidance questions of the sort, "How do we decide whether we've applied 'hardness' rightly in these circumstances?" (I've used the term "directivities" as an informal means of collecting these varied responses into a loose bundle). I've suggested that the answers we provide rarely stem from any single source, but instead reflect a range of influences that can potentially shape our usage. Specifically, the answers we provide with respect to "hardness" sometimes represent issues of the local correctness of particular "hard or soft" registrations and sometimes involve broader considerations with respect to the manner in which the strategic platform on which the terminology rests pieces itself together. Generally, we approach these flavors of inquiry differently, although we deal with them all in a common language of "correctness," "conception" and "understanding."

To illustrate what I have in mind, consider the query:

Has this Brinell indenter supplied a correct value for this piece of steel's hardness?

Depending upon the context, this inquiry might invite a local response such as:

No, the metal wasn't properly cleaned and the value can't be trusted.

On the other hand, the same question in a different context might prompt this generalized dialog between Samuel Williams (the optimistic reformer of the last chapter) and some more up-to-date scientist, familiar with the great complexities of material alteration under indentation and etching. Williams replies to our query:

Operationally I'd have to say "yes" at the moment, but I hope to uncover a better understanding of **hardness** *comparable to* **absolute temperature** *by whose lights Brinell measurements will require correction.*

His modern respondent disagrees:

I'm afraid that the corrections you seek are a mirage, for it's not possible to set up a practical employment of "hardness" in the general neighborhood of current usage that *doesn't rely upon specific testing instruments as primary within local patches. It would be nice if it were otherwise, but our conception of **hardness** simply can't be made into a universal material parameter comparable to **absolute temperature**. Theoretically, quantities of the desired type exist, but we have no means of attaching a language of any utility to any of them. Instead, we simply have to accept as final those **hardness** values that can be obtained through the complex blurring of qualities that occurs within the operations of a specific form of testing instrument.*

And the nature of this dispute can be characterized in my own terms as follows. Both Williams and his respondent recognize that current employment of "is hard" follows effective directivities (= standards of fairly trustworthy measurement and inference) that lay themselves over the patchwork of a theory facade. Williams hopes to discover some less test instrument-specific instructions that hover near "hard"'s current guiding directivities, yet correct them and allow us to stretch a flat structure over the entire field of bulk solids (rather as the thermometer-dependent "temperatures" of early physics became subservient to *absolute temperature* after the rise of thermodynamics). Such a nearby flat structure, if it can be found, can be regarded as *regularizing* our present day, uneven "hardness" facade. Both parties agree that the predicate "hard" would thereby gain considerably in its practical utility, but our modern physicist believes that the missing directivities Williams seeks simply do not exist—that the physical processes potentially in play when we test for empirical "hardness" are too varied to be leveled into the univalent directive instructions Williams seeks. If we ill-advisedly align "hard" to standards so rarified that they can never be applied, we will have deprived the term of its current utility and thereby moved it in the unmoored directions of "orgone" and "zig-zag-and-swirl." Or, to frame a more probable outcome, reform minded folks might give lip service to an idealized treatment of "hardness," but their actual employment will stubbornly follow the venerable furrows of testing instrument directivity, rather as "weighs five pounds" doggedly hews to its old operational ways despite our high school vows.

Both Williams and his opponent are perfectly clearheaded with respect to the issues under dispute and the sorts of data required to settle the matter. Our atlas picture merely clarifies the concerns under discussion in the same manner that allied geometrical

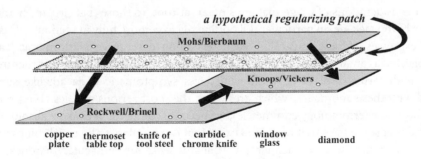

Regularizing a facade

portraits help us appreciate the tangle of considerations that relate "Bessel function" to the truncations that supply the term with enough numerical directivities to gain a practical foothold in usage. Viewed in this way, it is not surprising that localized directivities can piece together globally in unexpected ways (Bessel "functions" themselves prove irredeemably multi-valued over the complex plane).

Having not been privy to my words of advise, Williams and his imaginary respondent do not pursue their discussion in my recommended vocabulary, but harness the everyday argot of "conceptual content" to the chore. Thus, our respondent declares to Williams:

> You believe that the concept of **hardness** contains hidden resources that we've not yet been able to articulate.

Or:

> You hope to replace our current conception of **hardness** with something more absolute.

Although, by classical lights, these two assertions claim different things, within the context of our physicists' discussion, they prove entirely equivalent and articulate the same issues that I would describe in "regularizing a facade" lingo.

As explained earlier in the book, our disputants can employ "concept" profitably in this manner precisely because "concept," like "hardness" itself, represents a word of shifting focus, capable of highlighting significantly different directive factors depending upon the issues under discussion (such chameleon-like adaptability I have previously called a *seasonality*). As such, "concept" and its kindred evaluative terms prove of great assistance in our attempts to manage an ongoing language (the discussion with Williams represents a case in point). But ur-philosophical errors can initiate if we improperly regard "the concept of *hardness*" as reflecting some hard nugget of completed content with which we constantly deal from the first moments in which we grasp the word. Just as with "hardness," we tend to credit the evaluative predicate "concept" with a much greater degree of constancy than it really displays. Indeed, the two replies to Williams cited embody a slight hint of invariant suggestion, but these unhelpful connotations do not impede the substance of their conversation.

However, such proclivities towards ill-assumed invariance play a substantial role in stirring up the controversies about music and color surveyed in Chapter 2, which reflect considerations of usage that are similar, at root, to those that situate "hardness" upon a bumpy facade. In particular (I'll outline my case more fully in section(x)), "is red" and "expresses sadness musically" each rest upon descriptive platforms at least as complicated as an uneven atlas (indeed, they are surely more byzantine in structure). As such, both terms display the multi-valuedness symptomatic of facade-like support. Faced with these anomalies (which are where the most serious puzzles about musical and chromatic terminology commence), we become confused. We take it absolutely for granted that some base trait of *redness* exists that can regularize the usual things we say with "red" in precisely the manner that Williams seeks for vernacular "hardness." We believe this because we feel certain that we know exactly what the required trait is like in "is red" 's case: it is that quality that is *directly presented* to me when I grasp what "red" means ("Surely, if I fail to understand *redness* completely, then I'm unlikely to understand *anything*"). But we later discover that we are easily led to evaluate the colors of material things in different ways by obeying the salient directive indications that flow from our concept of "red". "Gee, I understand *redness* itself perfectly," we mutter, "so perhaps I have been applying it too carelessly to improper things. What sorts of objects *genuinely* meet the trait's internal requirements?" And thus we find ourselves stranded in the bewilderment of the location problems of Chapter 2, with their attendant worries whether *redness* inherently represents a physical, mental or amphibolic classifier. We've compressed the personality native to the predicate "is red" into a ghost attribute and now we can't find the attic rooms it is supposed to haunt.

But this is where we go astray, for the predicate "red" is swayed by a swarm of multiple directivities and doesn't reflect any core unity at all. As with "hardness," "red" (most of the time) conveys substantive physical information about its objects (roses, fire trucks, neon lights, etc.), but the nature of this information differs widely from target system to target system. The word's behavioral oddities stem from the same basic circumstances as engender those of "hardness": we lack the tools to settle a predicate of comparable utility on anything other than an uneven platform patched together through natural continuation. The mild inconveniences so occasioned do not greatly compromise the local objectivity of the physical information conveyed, but they do require us to take $\sqrt{z}$-like precautions in working with claims about "redness" especially over a wider scale.

As stated before, I am heartily disinclined to *identify* "the concept *redness*" with any definite collection, for it is best to regard "concept" itself as a loose repository of the seasonally shifting directivities that attach to predicates, much in the mode of a human personality. Since such shaping factors can potentially spring from the most unexpected quarters, we should beware of fancying that we've cataloged them all through tying "concept" too firmly to some subsets of influences. However, insofar as we are mainly interested in how "is red" manages to serve as a carrier of objective information with respect to physical objects, we can say that it accomplishes the task in the form of a *quasi-quantity* (or some more elaborate informational package of the same ilk).

One of the morals implicit in our tale of Ernest (6,iv) is that single-stranded practices of employment are especially apt to enlarge into multi-valued coverings, whereas a more varied diet of lifts into other varieties of descriptive covering can curb these tendencies considerably. In "red" 's case, we are firmly convinced that we can correctly tell whether something is red or not *by the look of the thing* and are loathe to relinquish this opinion to any other authority. That favoritism is all well and good—we do build up an admirable usage through this headstrong reliance on a single technique—but we shouldn't be surprised that the results become informationally multi-valued over a wider scale. So the next question we should ask is: given that motley patches have been stitched together in this monochrome fashion, how does our usage of "red" correct for the problems that this policy engenders?

As noted in 5,x, classical thinking opposes my portrait of wispy and disjointed predicate personality by compressing a term's varied sourcing into a *ghost attribute* of greater substance and homogeneity. Once such projected items are installed in *ur-philosophical* position, they become full-bodied obstructions to seeing the physical world rightly. On a classical view, to think at all, we must embrace fully formed concepts. But then the predicates whose directivities we most easily follow—e.g., "is red" as in "Turn right when you see a red house"—, seem as if they can't be supported by the physical attributes responsible for telling the apples and fire trucks of the world how to behave, whereas, possibly, their cousins, the shape predicates such as "ellipsoid" and "prismatic," can. So *redness* now seems as if it represents a trait that we grasp perfectly competently, but one that presents its supportive physical information to us in a cloudy or confused way (exactly as Descartes alleged). We then regard such concepts as providing modes of presentation that capture their target physical attributes only from a distanced viewpoint or adopt the Russellian position that we know most of the world's important universals only through indirect and ephemerally attached descriptions. We see ourselves walled up behind an insurmountable veil of predication, unless we hold out the slender hope that someday we may stumble across the happy concepts that present their worldly supports directly and canonically (in 3,viii, we noted how fuzzy invocations of "theoretical content" muddy these waters further). By compressing the buzzing pushes and pulls of predicate shaping into solidified cores in the classicist's manner, the universe becomes populated with an overabundance of ghost attributes that we must sort through patiently, searching for canonical presentations rather as Diogenes once hunted for an honest man.

So we should avoid looking upon concepts as completed conceptual bundles, through whose smoky interiors we dimly peer at the world. The mere fact that we employ hammers and saws that are not perfectly suited for the tasks in which we employ them does not mean that we can't execute exactly the carpentry we desire in our final results. Just because no flat map can capture all terrestrial geographical data perfectly does not entail that we cannot employ a multiple-paged atlas ably, with no misapprehensions arising with respect to the earth's true qualities at all. Indeed, we understand the geographical data registered in an atlas far better than we understand the maps themselves, for we are often oblivious to their most basic characteristics. We employ Mercator maps frequently, but many of us fail to notice the greatly inflated size they accord to Greenland, simply because we have learned to switch automatically to other pages in our atlas whenever we wish to compare the size of countries (this observation, unfortunately, more truly suits demographers and other professionals; most of us lay people are protected against areal errors through stranger provisions—see section, (viii) below). We needn't compensate for the Mercator map's intrinsic distortions (or even notice their presence) if we have acquired, without being sharply aware of the fact, habits of map employment that never attend to those aspects of the projection (this point is similar to the familiar observation that the brain doesn't need to invert the upside images formed on our retinas because our cognitive systems don't register the top/bottom orientational aspects of the retinal data in the first place). By shuttling between somewhat imperfect maps in a sage manner, we needn't be inconvenienced by their individual representational flaws at all and may remain unaware of their presence. So it is misleading to say we "see through" the flaws in a Mercator map; instead, we engage in a battery of routines that lead us to scarcely attend to their problematic features at all.

With respect to predicate use, we typically pursue the easy-to-follow pathways of ready classification and inference available to us, aided by domain specializations that protect us against unwise data exportation. We are generally unaware of the detailed features of these skills even though we rely upon them constantly. As Charles Peirce sagely advises us with respect to everyday reasoning:

> For the methods of thinking that are living activities in men are not objects of reflective consciousness. They baffle the student, because they are part of himself.
>
> "Of thine eye I am eye-beam"
>
> says Emerson's sphinx. The methods that thinking men consciously admire are different from, and often, in some respects, inferior to those they actually employ.[5]

In particular, like most animals, we possess remarkable capacities for establishing geometrical conclusions swiftly: we anticipate how two objects will occlude one another when shifted to new positions, etc. The exact subconscious routines and rules of thumb that allow us to reach these determinations remain largely unknown at this time; comparable feats on computers usually require elaborate algebraic routines. Nonetheless, we employ these reasoning capacities constantly, not only to keep track of the comings and goings of the palpable material objects in our neighborhoods, but, through transferred reasoning (8,iii), to uncover conclusions even in topics that have little to do with geometry per se. Every mathematician appreciates the wisdom of trying to picture her problems in geometrical terms,[6] however recondite the true subject, and the dozens of little drawings in this book are testimony to my own faith that allied transfers help us keep track of linguistic processes as well. But although we utilize these routines of geometrical reasoning constantly, as "living reasoning methods they baffle us," because nobody knows quite how we accomplish what we do (someday a clever team of mathematically sophisticated psychologists will figure this out).

Our strong reliance on these opaque reasoning routines clearly plays a vital important role in building the facades we utilize. In fact, our strange prolongation of "weight" into the space station setting is intimately entwined with these geometrical tools, as is our allied inclination to regard a small portion of a moving fluid as a "trackable object" (because its contours can be stained with ink and traced in its progress through the liquid). Both of these inferentially driven prolongations unwittingly create facades for "weight" and "frictional force" (for details on the latter, see 4,ii). We rarely suffer for this quilt work because compensatory policies quickly come into play that insure that the data registered within the facade-supported vocabulary of "weight" and

---

[5] Charles Peirce, "The Critic of Arguments" in The Collected Papers of Charles Sanders Peirce, iii (Cambridge, Mass.: Harvard University Press, 1933), 252–3.

[6] Jacques Hadamard, The Psychology of Invention in the Mathematical Field (New York: Dover, 1954).

"frictional force" get wisely utilized. Neither maps nor predicates must be considered in representational isolation; they must be evaluated in light of the management strategies that tie them to other descriptive structures or in an atlas otherwise curb their exuberant proclivities.

Ernest's case (6, iv) shows that employments that enlarge their applications through monochrome technique display a tendency towards multi-valuedness. Generalizing the moral, any predicate use strongly dominated by a single strand impulse to continue into fresh territories is apt to find itself situated upon an uneven facade later on. Such considerations are plainly applicable to "red": our *ur*-philosophical conviction that its directive essence is fully captured by the way things look is likely to leave us with an incredibly complex descriptive platform on our doorstep later, complemented by a large host of unnoticed boundary line corrections required to remedy the platform conflicts that our headstrong prolongation would otherwise leave in its wake. With respect to *redness*, conceptual simplicity is surely in the eye of the beholder, because, objectively viewed, the platform on which "red" rests itself is quite rococo in its windings. But, for all that, considered in its information convening capacities, *redness* behaves much like the large host of other averaged quasi-qualities upon which we regularly rely in characterizing the behavior of everyday objects. As such, "red" 's complexities of platform formation neither obscure nor corrupt the data it codifies with respect to physical objects, any more than the large map size of "Greenland" needs to compromise our appreciation of the island itself.

Staring at an apple and declaring, *"This* is what *redness* is really like," provides us with no sense at all of how "red" operates as a successful, facade-supported informational package, nor any appreciation for the compensatory controls required to keep that structure viable (quite the contrary, it encourages us to underestimate the importance of such issues). What we are really doing, when we gaze intently at the apple in our hand, is announcing a determination to follow certain easy-to-follow classificatory directivities wherever they may lead, forgetting about the considerate crew of linguistic engineers that need to follow after, quietly erecting firewalls and escape ladders to prevent our usage from falling into descriptive incoherence (unfortunately, their compensating architecture rarely protects us from the temptations of *ur*-philosophy).

For these reasons, I very much oppose the idea that a concept provides a *mode-of-presentation* leading to an attribute. We follow practicable directive elements in deciding when and how to employ a predicate, but we are not thereby presented usage with a *portrait* of the attribute on those occasions. Attributes do not possess "conceptual presentations," canonical or otherwise; in framing a predicate usage wisely, we hope that we have patched our linguistic capacities together in a manner that suits the attributes active in the situations before us. Sometimes, in a successful employment, a predicate "P" will line up with a true attribute in tidy "P"/ϕ fashion, but it can just as easily correlate with a quasi-attribute or some more complex form of informational package. After detailed investigation, we can eventually determine how physical information happens to be registered within a usage (whether in "P"/ϕ mode or otherwise), but we can't *see* how it operates simply by staring intently at what we fancy is a "presentation" of its supportive concept and sorting through its contents.

Perhaps the points I am stressing will become clearer if we examine a few characteristic passages from Nelson Goodman, who positively revels in the veil of predication assumptions I reject. Like many philosophers (Quine among them), Goodman likes to compare an assortment of descriptive predicates to a coordinate system for *location*:

> Consider . . . *the statements "The sun always moves" and "The sun never moves" which, although equally true, are at odds with one another . . .* [W]*e are inclined to regard the two strings of words . . . as elliptical for* [statements pegged to specific frames of reference] . . . *Frames of reference, though, seem to belong less to what is described than to systems of description: and each of the two statements relates what is described to such a system. If I ask about the world, you can offer to tell me how it is under one or more frames of reference; but if I insist that you tell me how it is apart from all frames, what can you say? We are confined to ways of describing whatever is described. Our universe, so to speak, consists of these ways rather than of a world or of worlds.*[7]

As commonly happens, Goodman misuses the term "frame of reference": in normal physicist's parlance, that represents a moving space and not a set of *coordinate axes* (which is how Goodman needs to be interpreted). Regardless, this claim is certainly true: if we want to predict the future location of a satellite, we'll need to put some coordinate numbers on its position (although, for other purposes we will wish to delay this imposition as long as possible using vector notation, for the reasons surveyed in 5, vi). Observe that Goodman implicitly assumes that this coordinate choice will have been put *fully in place* before the process of satellite description can begin. In fact, such "fully assembled" assumptions are not generally valid even in circumstances that involve spatial location (e.g., the adaptive coordinates described in the fine print below) and such presumption is plainly inappropriate when we deal with quasi-quantities that piece themselves together through patch-to-patch prolongation in the manner of *hardness*. But let us first follow Goodman a bit further in his reasoning. He is certainly right to presume that the imposition of specific axes commonly has much to do with our own idiosyncratic choice and comparatively little to do with the satellite itself (recall our earlier discussion of naturally adapted descriptive quantities; rarely do Cartesian location coordinates qualify as such). "So how," Goodman, in effect, asks, "can we remove, the artificiality of the framework we have imposed in order to talk concretely about our satellite?" One approach (suggested by the mathematician's approach to vectors and such) is to search for the invariants that support the varied coordinate assignments we utilize. But this road, if pursued generally, can quickly lead to an uncanny and pallid universe consisting of traits whose internal characteristics we know only distantly and structurally—i.e., we reach that eerie, distanced picture of external things that Russell embraces and Coleridge abjures. Accordingly, Goodman rejects such odd enterprises of conceptual distillation

---

[7] Nelson Goodman, Ways of Worldmaking (Indianapolis: Hackett, 1978), 2–3.

and cheerfully decides that we are forever confined behind the walls of our sundry conceptual frameworks, even when they blatantly conflict with one another:

> *Talk of unstructured content or an unconceptualized given or substratum without properties is self-defeating; for talk imposes structures, conceptualizes, ascribes properties. Although conception without perception is merely* empty, *perception without content is* blind *(totally inoperative). Predicates, pictures, other labels, schemata, survive want of application, but content vanishes without form. We can have words without a world but no world without words or other symbols.*[8]

But our discussion of maps reveals the faulty impression that Goodman's coordinate analogy leaves. He presumes that we must first see a "world" cloaked in the idiosyncratic features of a specific representational scheme and then attempt, if we can, to ascertain the realities that lie behind it. But we have just observed that we are only dimly and partially aware of the concrete directivities that build up the platforms upon which our descriptive predicates gradually arrange themselves. We automatically utilize the way that a soft acrylic plastic responds to a fingernail tap as our chief means for deciding whether it should be called "hard" or not (rather than testing for abrasion), but we enjoy little active awareness, at this level of adaptive specificity, of the directive routine we consult. Using Peirce's felicitous phrase once again, *rapping* is only barely an "object of reflective consciousness". As we gradually fill out our knowledge of the macroscopic world in this adaptive manner, we build up monitored strategic platforms capable of dealing with a wider range of materials while collecting data in locally advantageous ways. Usually, our recognition of facade structure lags considerably behind our knowledge of local fact, for the same reasons that we presently know a lot more about the temperatures of things than we understand the detailed workings of our mercury thermometers (the behavior of fluid metals like mercury is complex and rather mysterious). That the old church bell in the wildwood tolls an A note above middle C is known to many, but that a very strange, ear-induced prolongation of tonal classification from strings to bells is also involved probably escapes every party within hearing. That stainless steel enjoys a Vickers hardness value of about 260 is long established fact; that "hardness" must remain forever confined to an uneven facade has been generally accepted only since the 1950s (and that judgment may conceivably be reversed). The Goodman passages presume that we grasp concepts in fully framed analogy to a Cartesian coordinate choice, but this is misleading: the global facades of **hardness** and **sounds a tone of A** build themselves up in the active rounds of classifying and reasoning with their corresponding predicates.

Goodman's coordinate comparison is also inept because (leaving Ernest's degenerate example to the side), our facades usually include linked sheets that feature specialized forms of local capacity and we shuttle between theses in lifts and pullbacks exactly in the manner of a skilled employer of an adequate geographical atlas. In the end result, we needn't suffer a distorted or blinkered understanding of our intended subject matter; we have merely utilized a large collection of imperfect tools in a sagacious manner.

---

[8] Nelson Goodman, Ways of Worldmaking (Indianapolis: Hackett, 1978), 6.

Goodman is justly celebrated for his emphasis on the differing representational capacities of various species of icon, but he fails to recognize the monitoring role that compensating controls play in preventing patch-centered idiosyncracies from compromising the overall worthiness of an atlas-like descriptive system.

I am not quite sure what Goodman intends under the heading "talk imposes structures," but the pesky handiwork of the theory T syndrome seems in evidence. If its implicit definition assumptions (3, viii) with respect to our descriptive vocabularies were really true, the supportive webs of belief that such holisms require would provide a better analogy to Goodman's prefabricated coordinate axes. But it is the little practicalities ("A Shore scleroscope provides a good way to compare a steel to a hard ceramic") that primarily quilts the patchwork of *hardness* together, whereas grander theories of the nature of the quality like Reid's play little formative role (mercifully, because, in this case, such musings will lead us unprofitably astray). As Chapter 4 indicated, the macroscopic world is too darn complex to submit happily to the rigid maneuvers characteristic of blunt theory T imposition: we do better to tolerate a few cracks here and there in our descriptive platforms as we gradually cobble them together.

The struggles of Helmholtz and Hertz described in 4,iii are worth recalling in these respects. Their main objective is to argue for a relatively unconstrained freedom of *descriptive choice* in physics, in the face of constrictive demands on the understandability of the notions chosen. Such considerations lead them to declare, in the context of treating a system's temporal evolution, "All that's needed to fulfill this task adequately is that the right predictions be reached" (like many commentators, they tacitly ignore the fact that science is obliged to other forms of descriptive task besides simple prediction). If these chores can be adequately achieved through brute syntactic stipulation, either by algorithmic rules or formal axiomatics, then science will have discharged its descriptive obligations. From this point of view, the old-fashioned demands for deeper "understandability" can be rejected, for anyone can understand the syntactic procedures required to employ an axiom system. However, a vocabulary learned in such a rarified manner seems as if it must lack the more robust impressions of conceptual personality that we witness in the terminology *we really understand,* such as "red" or "gear wheel." This line of thought leads our physicists into that fateful capitulation surveyed in 3,x: "Well, I guess that Science, in its own affairs, doesn't really care about understanding in that robust sense." It leaves poor Helmholtz hoping that some faint "residue of similarity" between thought and worldly traits will have been preserved.

Regarded from a sympathetic point of view, Helmholtz and Hertz are urging that a system of descriptive predicate usage should be regarded in good order if its techniques are strategically well adapted to useful purpose. They are right to think this, but they have further assumed that the required "system" can be easily set forth in an "once and for all manner," through laying down axioms or similar inviolate rules. But the realm of the macroscopic seldom yields readily to blunderbuss treatment. Suppose we attempt to get some new descriptive predicates up and running through axiomatic technique. As we move through new territories of application, we'll soon find ourselves making sneaky compromises and tolerating a lot of hedging provisos. After enough tarnished traffic of this ilk has passed, we'll find that our erstwhile pristine predicates have become thoroughly dusted in the soot of everyday life and will have emerged with rich personalities quite comparable to those of "is hard," "is red" or any other long domesticated term. And this is why Wordsworthian claims that "theoretical terms" display pallid personalities is mistaken: after suitable buffering, they become as admiralty adapted to the delicacies of Nature as "red" or "expresses sadness musically."

....

My thinking on these matters is much influenced by the standard Veblen-Whitehead approach to differentiable manifolds,[9] wherein we install the proper attributes on our manifold by shuttling amongst an atlas of charts in a mutually correcting fashion.

It is also worth mentioning, with respect to Goodman's assumptions about coordinate frames, that it is sometimes necessary—general relativity provides many cases in point—to investigate material behavior within some space whose global geometry is not known in advance, but must be plotted out marching method style at the same time as we calculate the material events that occur in its midst. Because the complete geometry is not known on an a priori basis, we cannot know how many patches will eventually be required to cover its surface, but must instead build up systems of *adaptive coordinates* as we march forward. It is sometimes difficult to determine what sort of descriptive frame we have built when our numerical work is finished: whether the singularity we see in our charts reflects a true feature of the underlying geometry (such as a black hole) or merely represents an artifact where our coordinate building recipe has reached the limits of its usefulness (serious concerns of this ilk plagued the early days of relativistic cosmology). It is an allied necessity for assembling coordinate frame and worldly fact in delicate tandem that I stress in the paragraphs above.

....

I will add a final remark with respect to Chapter 2's false dichotomy between objective and subjective traits (the amphibolic category can be included here as well). In my story, humanly idiosyncratic factors can shape the *personality* of a predicate like "red" greatly without compromising its capacities as a carrier of objective information. Quite commonly, the most important considerations behind the formation of complex facades have little to do with issues of subjectivity (in the sense of human psychology) at all, but are determined by the measurement and inferential tools readily available at a macroscopic level. *Hardness'* atlas, for example, seems to be framed almost exclusively by considerations of this type. Accordingly, the basic shape of a usage often represents an *interfacial compromise* between the physical attributes of the systems under investigation and the measurement and inferential capacities available to us—true subjectivity may hardly enter the picture at all. Or, to express the situation in a slightly different way, target systems and human capacity join together to create an *environmental opportunity* upon which a productive usage can be founded, if we only have the wit to discover it (the psychologist James Gibson[10] coined the neologism "affordance" to roughly this

---

[9] Oswald Veblen and J. H. C. Whitehead, The Foundations of Differential Geometry (Cambridge: Cambridge University Press, 1932).

[10] James J. Gibson, The Ecological Approach to Visual Perception (Boston: Houghton Mifflin, 1979).

effect). Over time, a usage can gradually settle itself upon a specific form of strategic platform without our recognizing the underlying reasons for the adaptive shaping at all (situations of this sort will interest us much in the sequel). In such cases, the study of natural linguistic process partakes of much of the flavor of the biomechanics mentioned in 3,ix.

This is not to say that manifestly psychological factors do not sometimes affect the growth of a facade considerably: without their influence, the platform of "weight" would have surely not become pieced together as it has, and the same is plainly true for the complex platforms on which "is tuned to A" and "red" sit. Even so, corrective mechanisms such as inferential prohibitions along boundary line barriers generally compensate for the facade features of genuine psychological origin and the resulting usage will serve our practical objectives with respect to the material goods around us admirably. In philosophical discussion such as Goodman's, such questions of strategic engineering are rarely considered, thereby leading to a very exaggerated conception of the degree to which "subjective" factors distort the behaviors of our key classificatory notions (the criticism I offered of Frege-like explanations of Ernest's behavior in 6, iv is precisely that such accounts ignore the contributions to word personality that arise from deficiencies in the manner in which navigational list maps accommodate geographical facts). I'll return to the philosophical effects of neglecting interfacial considerations later, in 10,iii.

(iv)

Machinal ideas. Rather than continuing to write abstractly about how ghost attributes get built up through predicative enameling, it may prove more enlightening to plunge immediately into the details of a substantive example where a common term of familiar usage displays an extremely pungent personality that we feel we grasp quite firmly, but whose true underpinnings trace to sources completely different than we expect. However certain "jute factory" aspects of our discussion may persuade some readers to skip ahead to the next section. The specific case I wish to probe in this vein is our old friend *being a gear wheel* (actually, I will chiefly discuss several of its simpler mechanical cousins, for reasons I'll explain later). In particular, we intuitively assume that *being a gear wheel* represents a potential characteristic of material objects such that, if a group of objects really exemplifies traits of *being a gear wheel* type, we will "really understand why the collection acts as it does" (I called these "warm and fuzzy feelings of understanding" in 3,vi). As we noted, early mechanists such as René Descartes and Robert Boyle were utterly convinced of the patent applicability of machinal ideas to the world, to the degree that Descartes argued that God must have planted these basic categories in our heads as part of our intellectual birthright (shape classifications such as *gear wheel* seem prima facie of the same type as those found in Euclidean geometry). Courtesy of this divine implantation, we come fully prepared to understand the world's workings rightly, as long as we are able to puzzle out the specific blueprints for whatever piece of worldly clockwork happens to attract our attention (in this department God refrains from feeding us answers a priori).

Unfortunately, the world does not operate in *gear wheel*'s way—she instead follows the lead of less palpably understood traits such as *being acted upon by an action-at-a-distance gravitational force* in her real world operations. "But this is mere empirics," many philosophers declare. "Descartes' clockwork universes are *conceptually viable,* even if they are never actualized." Indeed, the many modern enthusiasts of "possible worlds" commonly cite "the clockwork universes of Newtonian physics" as prime illustration for the unactualized possibilities central to their thinking (thus betraying some confusion, for whatever "Newtonian physics" comes to—vide 6,xiii—, it surely does not supply us with clockwork universes). Such is the *impression* that the familiar personality of "is a gear wheel" readily leaves.

But now let us look at the underlying reality. I shall present my conclusions in capsule form first, then fill in some of the details. "Being a gear wheel" has substantial real world application, but only when confined to an extremely narrow set of configurations that I'll call its *prime patch.* This patch is hemmed in by various odd requirements that we rarely notice, such as *containing the equivalent of at least four linked parts* and *having those parts move in parallel planes* (look around the house, virtually everything you would intuitively dub as a "mechanism" obeys those provisos). The special salience of the prime patch and its attached limitations lies in the fact that especially effective recipes for mechanism invention, improvement and diagnosis are available here, all of which I'll generally lump together under the heading of *design purpose.* To render such recipes possible, we must lift away from the mathematical setting pertinent to *causal prediction* in a quite dramatic way—that is, we alter our governing terminology in a manner comparable to—but more radical than—the shift between parameterized and unparameterized descriptions of a circle that was presented in 6,iii. As noted there, this shift in setting greatly facilitates the answering of important classes of practical question and the everyday terminological prominence of "gear wheel" should be regarded as a descriptive displacement induced by questions mainly pertinent to *design* rather than *prediction.* Our intuitive impression that "we really understand *gear wheel*" derives from the fact that we tacitly utilize these design-oriented recipes quite actively in our everyday thinking about household objects, often in manners whose intervening steps we do not consciously notice. Our belief that we apprehend the "causal characteristics" demanded by the attribute *being a gear wheel* is almost completely mistaken, for most aspects of true causal process have been *descriptively purged* as we lift our thinking into the design-directed orientation of our prime patch. In the jargon of 4,vii, significant *physics avoidance* occurs within this transfer of descriptive setting and *true temporal process* represents one of the central aspects of gear teeth behavior that becomes most strenuously avoided within our prime patch. This is not to say that the errant term "cause" doesn't sneak back into usage within our prime patch, but the term severs its erstwhile connections with *causal process* in this reemergence (we'll discuss these aspects of "cause"'s wandering behavior more fully in chapter 9). Such factors entertain the illusion that "gear wheel"'s range of potential application is far wider than it really is (another factor is the degree to which we have ourselves cluttered up the surface of the earth with mechanisms, another surprising observation to which I'll return). Cartesian inclinations to see all of Nature in mechanism-related terms constitutes a paradigmatic example of *tropospheric complacency* (2,iii) at its most extreme: we see machines everywhere because we've built so many of them ourselves. In fact, no treatment relying upon gear wheel and its machinal companions can prove conceptually closed: sooner or later we will confront the *lousy encyclopedia phenomenon* (4,ix) that shuttles us into notably different descriptive settings.

If I can tell this story rightly, we should come away from our investigations with the conviction, "Once matters are laid out like that, I can see that the usage of 'gear wheel' has always been tacitly hemmed in by these prime patch provisos, although their formative role generally slips by unnoticed," as well as the cognate impression that our sense of *conceptual understanding* has thereby been turned upside down: "Gee, those design factors represent the true sources from which the characteristic intellectual feel of 'gear wheel' actually springs." The total conceptual package is intrinsically multiplex in its sources, although, from a phenomenological point of view, its composite personality typically strikes us as simple and undivided in character (just the traits of *redness* or *expressing sadness musically* strike us representing something entire). And it is this impression of *unified character* that caused the opponents of gear teeth's mechanical prominence so much philosophical trouble (I have in mind the defenders of *free creativity* within scientific endeavor such as Hertz and Helmholtz (4,iii), who wanted to carve out a permanent liberty to employ less warmly "understood" traits such as *action-at-a-distance force* if the empirical facts warrant). Unable to explain *gear wheel*'s "well understood" personality as a misdiagnosed expression of design oriented ingredients, they wrongly concluded that, to equalize the competition between "gear wheel" and other predicates, scientific terms only need to display a very thin and entirely prediction-focused conceptual personality. But this concession is wrong on two accounts. First, every predicate automatically acquires a richer conceptual ambience from human-centered factors of usage, although not in a manner that compromises their informational content. Secondly, predictive purpose is not the sine qua non of scientific intent; for most purposes, we prefer addressing chief practical questions by *avoiding* predictive tracking as best we are able.

The basic analysis of *gear wheel*'s ontogeny that I shall sketch is not original to me; it was offered by the engineering innovator Franz Reuleaux[11] in his Kinematics of Machinery of 1876. To be sure, Reuleaux couches his own discussion in the traditional language of classical concepts and attributes and this slant applies a slightly flowery cast to what, at base, represents a hardheaded analysis of conceptual development. Indeed, Reuleaux includes long passages of overt philosophizing in his text and we shall have occasion, at various points in this book, to sample portions of his very intriguing discussion. His work created a revolution in the teaching of mechanical design and

---

[11] Franz Reuleaux, Kinematics of Machinery, Alexander Kennedy, trans. (New York: Dover, 1963).

Reuleaux

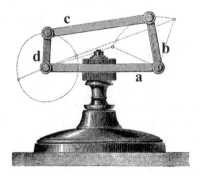

his text was widely read. I shall sometimes borrow the phrase *machinal ideas* from Reuleaux' translator to cover the family of notions under examination.

Let us now turn to some of the details. I should begin with a clarification of terminology. A *mechanism* is commonly defined as "an assembly of moving parts performing a complete functional motion."[12]

The base idea is that some attached source of movement (a human hand or a motor) will activate part of the assembly (its crank if the applied movement is circular) in a manner that generates an altered motion or transferal of applied force arising at some follower point elsewhere in the mechanism. Thus in the illustration, an input arc motion applied to the crank (d) will pull any follower point on (b) in a back and forth rocking motion (placing the follower point off the center line of the follower bar can generate quite astonishing curlicues). Normally, neither the attached motor nor the objects upon which the linkage acts are considered to comprise part of the mechanism proper. Often "machine" is employed (especially by Reuleaux himself) in a synonymous manner.

---

[12] Webster's.

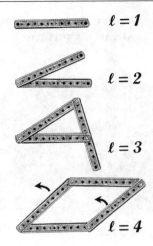

$\ell = 1$

$\ell = 2$

$\ell = 3$

$\ell = 4$

A device like that pictured (the illustration is from Reuleaux) is called a "four-bar linkage" because—well, just count its differentially moving pieces. Standard terminology classifies its sundry parts as "base link" (a), "cranks" (d), "transmission bars" (c) or "follower bars" (in the circumstances pictured, they are considered "rockers"). In the sequel, I will focus upon notions such as *being a follower link* as my chief exemplars of machinal conceptual personality, rather than *being a gear wheel*, simply because the latter introduces complications (recounted below) that are irrelevant to our concerns. Plainly we understand these simpler linkage notions in the same vivid manner as we do "gear teeth."

Let us now explain why such four-bar configurations have a special place in the field of mechanism. Suppose we have a box of little girders that we can hinge (or *pin*) together with screws, exactly as in the Erector sets[13] of my youth (such rods are usually called *links*). Utilizing just one of these links ($\ell = 1$), we have a mere rigid body; with two ($\ell = 2$), we have an uninteresting hinged contraption; with three ($\ell = 3$), either another floppy open chain or an immobile assembly (called a "structure") with special features of its own, but not the ones we seek. However, if we screw together four girders ($\ell = 4$), something remarkable happens: we have constructed a *closed kinematic chain* that can cogently deliver motion or force from one point of application to another, as well as altering the character of its application (turn the crank in a circle and the rocker bar will pound percussively). In short, we have just pieced together the simplest *mechanism* possible, according to dictionary definition quoted above. Closed kinematic chains possessing a greater number of moving parts are also possible, subject to the Gruebler requirement $3\ell = 2(p + 1)$ where p is the number of pinned joints.[14] More

---

[13] Manufactured by the A. C. Gilbert Company, who also produced the noble American Flyer toy train.

[14] This formula is usually called Gruebler's equation. Arthur Erdman and George Sandor, Mechanism Design (Englewood Cliffs, NJ: Prentice-Hall, 1984), 17. Devices with more degrees of overall freedom are usually regarded as mechanisms as well, their extra freedoms being controlled by more than one attached motor. But I've ignored these complications here.

**supply very similar readings
at earth's surface**

An ill-posed problem

general devices that employ gears, cams, conveyor belts and so forth can be usefully studied according to the instantaneous linkage mechanism they represent at each stage in their unfolding motion. This, in fact is how *being a gear wheel* finds its place within our orbit of machinal ideas, but we shall evade the complications of instantaneous mechanisms.

Before we look at the special utility of these notions, let me offer a few remarks on the mathematical contours implicit within many kinds of design problem.

If we are interested in treating the temporal evolution of a physical system, we want equations that will march the system forward from a selected initial state to its subsequent configurations (amongst the simpler partial differential equations, formulae of hyperbolic type are wanted for this task). These are commonly called *direct problems* by applied mathematicians. But in mechanical design we usually set ourselves an *inverse problem:* how must a machine be configured so that it will move through certain proscribed positions or transfer force according to a given schedule of mechanical advantage? To resolve such queries, we must be able to explore the effects of varying our system's structural parameters (e.g., we want to know what the effects of a follower bar that is three inches shorter—conditions like this usually remain fixed within a conventional direct problem). Furthermore, we must find some way of taming the *ill-posed character* of our indirect problem, a characteristic that indicates that our prescribed positions can be achieved by many mechanical systems quite unlike one another (a simple prototype for an ill-posed problem is supplied in gravitational detection of buried mineral deposits: given any simple mass distribution below, there are many doppelgänger layouts that give rise to very similar surface patterns). Unless we can eliminate the unwanted "answers" from our search space beforehand, any design recipe is likely to become muddled by the toleration of spurious complicated answers (we want the simplest answer to our design problem but what does a dumb algorithmic method understand about being "simple"?—recall

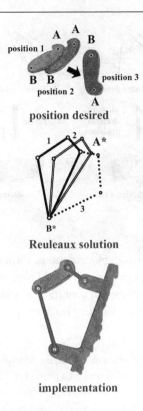

position 1

position 2

position 3

position desired

Reuleaux solution

implementation

from 4,x the problems Euler's method faces if some extraneous solution slips in at a non-Lipschitz point).

Such considerations suggest that we must discard conceivable alternatives rather dramatically to obtain a tractable design problem: indeed, it is very helpful if we require our solutions to be two-dimensional in character (a point to which I'll return). Finally, we expect *rigid bodies* to prove central in our deliberations because pieces of iron that approximate that condition are quite durable and because we can greatly decrease the operative number of degrees of freedom through such requirements.

With these preliminaries completed, how do we argue that the heart of "follower link" 's personality lies within design considerations, rather than predictive intent? Let's consider the simple problem of designing a door latch. We begin with three moving pieces and a base, and first concentrate upon the catch piece (which will become the follower link in the mechanism we produce). With a four bar linkage, we can freely designate three positions through which we'd like this part to move. It is natural to select these as follows. In position 1, the notch of the catch rests upon a restraining bar not shown. In position 2, the catch lifts itself free of that bar and finally, in configuration 3, the catch folds to an "open" position away from the catch bar. This sequence of three positions constitutes our *design objective*. To solve our problem we first draw a simple

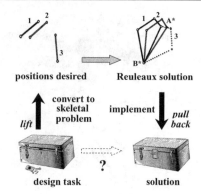

positions desired   Reuleaux solution

lift   convert to skeletal problem   implement   pull back

design task   ?   solution

line inside our catch piece whose placement is rather arbitrary. As Reuleaux first realized, the *essence* of our design problem is one of persuading this line to move through its three comparable positions. Once our ambition is restated in that simplified way, we can easily devise a perfect solution to our problem.

. . . .

First, pick a length and locate the common center B* whose distance from B in all three configurations is the same. This means that if we run a pinned rod between B* and B, its turning will carry the end point B through its required three positions. If we can also get point A to pass through its required positions, we will have solved our problem. But, plainly, we can locate an A* that, like B*, is equidistant from A in all three positions. We then set up a base housing for our mechanism between A* and B*, including a large enough cutout so that our moving parts don't bump into it, and then connect B* to B with a short bar. Finally, if we attach a handle or push button to serve as crank at A*, its turning will carry our latch through its prescribed positions perfectly. And, in fact, you'll find mechanisms of roughly this type in many door clasps. Here we have a perfect design recipe. It was traditional to find A* and B* by trial and error with a ruler, but such kinematic requirements can be easily written down as algebraic equations and solved, so a genuine linguistic algorithm is available here.

. . . .

What we have found is a general strategy for *lifting* our design problem into a simplified mathematical arena, by (1) undressing our problem so that we become concerned solely with skeletal diagrams of a rather ethereal character, (2) applying an easy algorithm to this lifted representation to solve our design problem, and (3) pulling back our results back to earth to serve as a blueprint for a real door latch, which we once again array in the mortal flesh of plates, bolts and wood (we have witnessed these lift-and-pullback schemes already, in 6,ii and will visit them again in 7,iii). It is this critical lift into a simple two-dimensional setting that makes our ill-posed original task resolvable.

Reuleaux's central insight lies in the fact that these skeletal redescriptions and their applicable reasoning rules form the key tools that lead to effective invention (or *design*

synthesis, in the preferred jargon). Indeed, every modern engineer imbibes Reuleaux's technical vocabulary virtually with their mother of invention's milk, whereas, hereto-fore, the designing machinery had been rendered much harder because this mathem-atical core had not been cleanly isolated—because, in Reuleaux's words, the "general identity is not seen through the special variety":

> In earlier times men considered every machine as a separate whole, consisting of parts peculiar to it; they missed entirely or saw but seldom the separate groups of parts which we call mechanisms. A mill was a mill, a stamp a stamp and nothing else and thus we find in the older books describing each machine separately from beginning to end . . . [But] thought upon any subject has made considerable progress when general identity is seen through the special variety;—this is the first point of divergence between popular and scientific modes of thinking.[15]

Here Reuleaux has in mind the fact that the two mechanisms pictured, although they appear quite differently clad in their metal housings, supply identical skeletal diagrams when lifted into Reuleaux representational patch. Sometimes even a mechanical wizard such as James Watt, who, in Reuleaux' assessment, possessed a genius' eye for the perfected mechanism, could not see the shared plan underneath the disguise of extraneous haberdashery:

> Watt has evidently not recognized [the common occurrence of a certain steam engine mechanism], at which one cannot wonder, given the uncouth garb of timber beams and hammered rods in which the elegant mechanism was at that time disguised.[16]

---

[15] Reuleaux, <u>Kinematics</u>, p. 9.   [16] Ibid., 5.

On the other hand, Watt often came quite close to Reuleaux's reductive insights, for the skeletal diagrams he drew in the process of perfecting his parallel motion linkage look very much like Realeaux's own. Indeed, Reuleaux maintains that Watt achieved his astonishing litany of clever inventions mainly because his lines of thought generally ran rather close to the mathematical framework required, without Watt's realizing this fact.

Once this obscuring filagree of irrelevant metal housing is cleared away through representing mechanisms as skeletal diagrams, the would-be inventor will be able to deal with the essential nature of such devices in their naked glory:

> all carry[ing] on, partially before the bodily eye of the student and partly before the eye of his imagination, the same never tiring play. In the midst of the distracting noise of their material representatives they carry on the noiseless life-work of rolling. They are, as it were, the soul of the machine, ruling its utterances—the bodily motions themselves—and giving them intelligible expression. They form the geometrical abstraction of the machine, and confer upon it, besides its outer meaning, an inner one, which gives it an intellectual interest to us far greater than any it could otherwise possess.[17]

As indicated, Reuleaux maintains that great inventors succeed precisely because their "intuitive genius" is able to distinguish the essential and inessential components within a design, even if they cannot articulate this diagnosis clearly. But the critical directivities of design improvement still guide their searches—it is easy to see how an untutored inventor might piece together our door latch by trial and error, eventually stumbling across the same part sizings as we have (the strong directivities of mechanical improvement are always active, but lacking a good representational language such as Reuleaux provides, we may need to scuffle around a bit to follow its indications). Especially skilled inventors such as Watt will be drawn to the optimal latch sizings almost unerringly because their thought processes conform to Reuleaux's recommended design principles with fair approximation already, albeit remaining dressed "in the uncouth vocabulary of 'timber beams' and 'hammered rods.' " As such, Watt will be unlikely to be able to communicate to others how he has found his inventions, lacking any public vocabulary of suitably focused inferential directivity. Reuleaux views his own task as that of the systematizer who articulates the hidden "Science" towards which his fellow inventors have been blindly groping:

> [Kinematics] in its essence, in the ideas belonging specially to it, has been left indistinct, or made clear accidently at a few points only. It is like a tree which has grown up in a dark tower, and thrown out its branches wherever it could find an outlet; these, being able to enjoy the air and light, are green and blooming, but the parent stem can only show a few stunted twigs and isolated leaf-buds.[18]

In short, fine opportunities for directed design have always lay nascent in the mathematics of closed kinematic chains, or, in Reuleaux's phrase, the essential machinal

---

[17] Reuleaux, Kinematics, 85.   [18] Ibid., 2.

ideas were destined eventually to "crystallize" into an integral and self-contained descriptive patch of their own:

> *Very gradually each invention came to be used for more purposes than those for which it was originally intended, and the standard by which its excellence and usefulness were judged was gradually raised. An external necessity thus demanded its improvement, and from this cause machinal ideas slowly crystallized themselves out, and gradually assumed forms so distinct that men could use them designedly in the solution of new problems.*[19]

In other words, Reuleaux maintains—and I believe he is correct—that a hidden hand of algorithmic opportunity silently shapes our thinking about mechanism design, whether we recognize its persistent directivities or not:

> *I believe I have shown . . . that a more or less logical process of thought is included in every invention. The less visible this is from outside, the higher stands our admiration of the inventor,—who earns also the more recognition the less the aiding and connecting links of thought have been worked out readily to his hand.*[20]

And further:

> *And today, just as formerly, the inventions must still be arrived at by a mental process; and this forms the problem which it must be the chief aim of theoretical Kinematics* [= in Reuleaux's use, the science of mechanism] *to resolve. So long as it could not reach to the elements and mechanisms of machines without the aid of invention, present or past, it could not pretend to the character of a science, it was strictly speaking mere empiricism— (sometimes of a very primitive kind),—appearing in garments borrowed from other sciences. . . .* [My] *case . . . is rather that it will become possible to introduce into machine problems those intellectual operations which science everywhere else pursues in her investigations . . . Invention, in those cases especially where it succeeds, is Thought: if we then have the means of systemizing the latter, so far as our subject goes, we shall have prepared the way for the former. Goethe,—who had so great an interest in the inner nature of everything that could enlarge the circle of our ideas,—expresses himself in the following noteworthy sentence: (Everything we call Invention, discovery in the higher sense, is the ultimate outcome of the original perception of some truth, which, long perfected in quiet, leads at length suddenly and unexpectedly to productive recognition.)*[21]

And, by cleanly isolating the core descriptive parameters required within the lifted patch of our diagram, Reuleaux has successfully isolated the mathematical sinews that lay behind much successful invention within the field of planar mechanism.

. . . .

To put the point more accurately, suitable design algorithms are available for so-called "planar mechanisms": devices in which all working parts move in parallel planes, in contrast to "spatial mechanisms" such as a robotic hand. Here the relevant mathematics is far more intractable, although much recent work has been expended in this direction, hoping to find

---

[19] Ibid., 231.   [20] Ibid., 8.   [21] Ibid., 20–1.

**local history of how the
crank is turned ("here")**

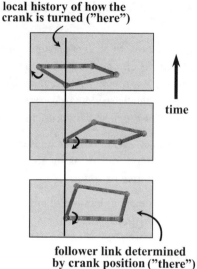

time

**follower link determined
by crank position ("there")**

a viable spatial analog to Reuleaux' planar achievements. In this vein, Jack Phillips writes (demonstrating that Reuleauxian mysticism has not yet faded from departments of mechanical engineering):

> Between machinery and ourselves there is a kind of membrane, wrapped around the machine itself... A spade, we say, is a spade. To describe spade, indeed to invent spade, one must actually approach the thing and push through with a firm geometric imagination.[22]

....

We should also observe that our machinal ideas are useful not only for *design*, but for *diagnosis* as well, as when we try to figure out what's wrong with an ailing door latch (such issues will return in a surprising context in 9,iv).

Let's now examine the sense of "deep understanding" that so captivated the early mechanists. With what intellectual capacities is it associated? Certainly a large measure must be connected with the skills of design and diagnosis, for those are the capacities that masters of machinery such as Watt most patently evidence. In fact, in dealing with real life mechanism, we are rarely concerned with predicting how the device will run if we leave it entirely alone. Instead, we usually assume that some form of motor (often called a "prime mover"!) attaches to some link pair (e.g., as a crank) which then drives the rest of the mechanism in allegiance with its dictates. This should be considered a *"here versus there" form of determinism* because we are not fixing the system's behavior from a *starting state* (as in our cannon ball calculation), but from a *local history* (how the crank turns over time). As such, the Euclidean geometry of our rigid bars fixes, on the basis of the local crank movement history, how the rest of our mechanism moves

---

[22] Jack Phillips, Freedom in Machinery: Introducing Screw Theory, i (Cambridge: Cambridge University Press, 1967), 1.

(certain breakdowns in this determinism can arise, as noted in the fine print below). In normal prediction, however, we seek a causal process whereby the past state of our mechanism fixes its future states. In this regard, our fund of machinal ideas provides no help at all. Newtonian dynamics can provide the story we seek, but generally at the cost of rejecting our base assumption that our bars remain perfectly rigid (Newtonian rods must react internally by bending to the Coriolis forces that get generated as the links accelerate while turning through their cycles). In other words, to request a *dynamic treatment* of our mechanism automatically carries us into another patch within the overall facade of classical physics where our chief machinal ideas are rendered inapplicable. To be sure, our machinal predicates can be very weakly prolonged into these patches but only as merely phenomenological classifiers—we can call a shape a "gear wheel" even if it is serving as a doorstop or flywheel, leading us to assert mildly multi-valued claims such as:

> *This gear wheel is not truly a gear wheel, but properly a flywheel.*

But in these extended applications, "gear wheel" plainly loses that diagnostic "understandability" that made it so beloved by Boyle.

. . . .

Mathematicians call a temporally specified starting state a *Cauchy condition*, whereas our local history qualifies as a degenerate form of *boundary condition*.[23]. Their different mathematical natures are displayed in the appended drawing. In Cauchy determinism, data on the starting state slice fixes what happens on all slices above, whereas in here-versus-there cases, knowledge of stretches of the local history fixes behaviors elsewhere on those slices where we don't know the local history. Exceptions to such here-versus-there determinism can be found in a device's possible *crossover* (or "dead") points, where two solutions become open to the mechanism.

. . . .

Indeed, think of the things we say to ourselves when we believe we understand a mechanism well: "Ah ha! now I see why these parts have to be here. They convert the handle's steady turning into a rapid lifting action"; "This piece isn't doing its job properly; it ought to move through a smooth arc at this point in its cycle." None of these are predictive statements; they are *diagnostic* in nature, displaying a distributed normativity derived from implicit design objectives.

A further indication that a genuine interest in causal process or prediction runs somewhat orthogonal to the core directivities of mechanical thinking is displayed in the often expressed puzzlement that machine thinking engages in "idealizations" that scarcely seem justified when confronted by the real article. We have utterly ignored friction at the joints and the Coriolis forces that trouble high speed machinery considerably; we do not worry about wear at all, beyond making our contact points minimal (*lower pair contact*, in Reuleaux's jargon). We don't even worry whether our moving parts will bump into one another and instead freely allow them to melt through

---

[23] John Earman, A Primer on Determinism (Dordrecht: Reidel, 1986).

one another in plotting the movements of the four-bar arrangement shown. Some of this is conveyed in Wittgenstein's musings on the odd notion of "possibility" that emerges when we think about mechanism:

> But when we reflect that the machine could have also moved differently it may look as if the way it moves must be contained in the machine-as-symbol far more determinately than in the actual machine. As if it were not enough for the movements in question to be empirically determined in advance, but they had to be—in a mysterious sense—already present. And it is quite true: the movement of the machine-as-symbol is predetermined in a different sense from that in which the movement of any actual machine is predetermined.
>
> . . . And what leads us into thinking [like this]? The kind of way we talk about machines. We say, for example, that a machine has (possesses) such-and-such possibilities of movement; we speak of the ideally rigid machine that can only move in such-and-such way.—What is this possibility of movement? It is not the movement, but it does not seem to be the mere physical conditions for moving either—as, that there is play between socket and pin, the pin not fitting too tight in the socket.[24]

More dramatically, we can reasonably ask, "What is the value of plotting a possible movement that is so 'idealized' it isn't even realizable?" And the proper answer is: "We make these diagrams to discover the proper sizing of parts in design and need to locate our search within as small a space as possible. Once proper sizings have been found, then we can return to the task of finding a metallic instantiation that won't be too badly hampered by joint friction, Coriolis forces et al. Given the planar nature of our design, we can usually prevent collisions by stacking our moving pieces in layers. But the short answer to the question is: our 'machinal possibilities' relate primarily to a lifted stage within an effective design process, not to genuine temporal prediction."

Reuleaux is fully aware of the special character of machinal thought and believes that everyone who properly catches the feeling of its dedicated vocabulary possesses some hazy recognition of this fact.

> [T]he sense of the reality of this separation [of the theory of mechanism from general Newton-style mechanics] has been felt not only by engineers or others actually engaged in machine design, but also by those theoretical writers who have had any practical knowledge of machinery, in spite of the increasing tendency in the treatment of mechanical science to thin away machine-problems into those of pure mechanics.[25]

In our terms, the patch of mechanism sits above the facade of regular mechanics as a specialized little domain, into which we lift our reasonings whenever we want design-related questions addressed, but which we should otherwise avoid (think of mechanism's prime patch as like an equiareal map in an atlas).

---

[24] Wittgenstein, Investigations, §194. It is known that Wittgenstein knew Reuleaux's book. Michael Nedo, Ludwig Wittgenstein: Wiener Ausgabe (Vienna: Springer-Verlag, 1993), 14. Alfred Nordmann, "Another Wittgenstein: The Scientific and Engineering Background of the Tractatus" in Perspectives on Science 10, 3 (2003).

[25] Reuleaux, Kinematics, 30.

This last observation becomes quite pungent once we reflect on the extremely narrow sliver of physical reality to which machine thinking is properly suited. Let's go back to our box of little girders and reflect: if we were to *randomly* attach these links together with pins, the result are most likely to turn out *under-or over-constrained*, in the sense that they do not comprise a closed kinematic chain, but instead represent something floppy or stressed (in algebraic terms, the associated kinematic equations only accept one solution, as opposed to many or none). If many pieces are concerned, it takes some foresight—recall Gruebler's equation—to piece together a true mechanism, a fact that Reuleaux trenchantly notes:

> [T]he machine never, or scarcely ever, comes to us as a ready-made production of nature, but as something we ourselves have made.[26]

In other words, we see a lot of mechanisms around us simply because we have made a lot of them ourselves, radically transforming the surface of the earth in much the manner of Darwin's earthworms (indeed, I found Reuleaux's observations equally startling when I first read it). Or, to put the point more exactly, we witness many mechanisms at close quarters simply because we and the processes of natural selection have together made a lot of them, evolution often blindly following the same basic design imperatives as we do. But other than through these origins, mechanisms proper are quite rare in nature (Reuleaux can only cite certain wobbly rock formations as somewhat feeble instances). True mechanisms are thus situated on a delicate and narrow saddle betwixt over- and under-constraint, where a unique opportunity for profitable design opens up that requires very simple mathematical tools (algebraic rather than differential equations).

And this is the tiny region that the prime patch of machinal ideas properly covers. Contrary to plotting the course of natural events, we should view our mechanical contraptions as running antagonistically to Nature in the sense of the Pseudo-Aristotle (who seems to be primarily thinking of architectural design):

> In many cases, in fact, Nature works against man's needs, because it always takes its own course. Thus, when it is necessary to do something that goes beyond Nature, the difficulties can be overcome with the assistance of . . . mechanics . . . ; as the poet Antiphon put it, "Art brings the victory that Nature impedes."[27]

All of this, of course, turns the traditional "universe as a warehouse of machinery" portrait utterly on its head ("the world turned upside down," as the old song goes). Authors like Boyle and Descartes were simply fooled by a mixture of *tropospheric complacency* (expecting faraway conditions to resemble local circumstances) and *semantic mimicry* (facade-situated terminology can often mimic flat structure behavior quite ably), leading them to see an entire universe of machines on the basis of a man-made local anomaly. Such copycat circumstances arise more frequently than one might expect

---

[26] Ibid., 52.   [27] Benvenuto, <u>Structural Mechanics</u>, p. xviii.

and will comprise the central topic of Chapter 9. In that chapter, I'll diagnose the borderline shifts that facilitate this mimicry in finer detail.

Before resuming our primary line of argument, I should like to comment briefly upon the manner in which Descartes' physical program is popularly discussed, as in William Whewell's dismissive comment:

> Of the mechanical truths which are easily available in the beginning of the seventeenth century Galileo took hold of as many and Descartes of as few as was well possible for a man of genius.[28]

In particular, Descartes is often criticized unfavorably in comparison to Newton for "having gotten the laws of impact wrong" and so forth. In truth, virtually everyone who labors in the bumpy vineyard of classical mechanics sooner or later runs into strange difficulties; Descartes merely had the misfortune of primarily cultivating mechanism's end of the garden. We will take up these issues from another perspective in 9,v.

A large part of the blame for these confusions should be laid at the door of *rigid body* for this extremely useful, but essentially top-down, device for variable reduction never harmonizes perfectly with the other classifications of classical physics with which it commingles (see 4,ix). Once we have called upon *rigid body*'s reductive offices, strains with other mechanical notions are bound to arise, if only in subtle ways, and will eventually require some corrective responses in the form of protective facade boundaries. Indeed, we should immediately grow suspicious whenever appeals to rigid bodies are heard within the halls of mechanics, just as we should mistrust the friendly stranger who offers us abundant "altruistic" assistance: "Uh oh, there'll be a price exacted for all this help later on." In truth, Newton and all the other great masters of classical physics succumbed to the charms of *rigid body*'s silvery tongue every bit as gullibly as Descartes; it is merely that they didn't wind up so visibly fleeced in the end.

As I have observed in a number of places (eg. 3,x and 5,vii), central aspects of contemporary philosophical debate are strongly animated by longstanding, but utterly unproven, assumptions with respect to the circle of intensional characteristics that the concepts of physics allegedly manifest: that they concern themselves only with brute causation, not internal explanation; that they invariably classify only from a "God's eye point of view"; that they never express norms, and so forth. I hope that covering the plainly teleological "norm" of design objective within *gear wheel*'s secret conceptual personality supplies some warning that this popular pasture of ideas is founded upon misconception; that such metaphysical conceits merely represent a heritage engendered by classical conceptual thinking and great gobs of faulty folklore about science.

(v)

Lifts and free assertion. At several points in our discussion, we have discussed the value of *representational lifts*: situations where data is shifted from one linguistic format

---

[28] William Whewell, History of the Inductive Sciences, i (New York: D. Appleton, 1859), 390.

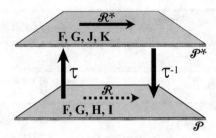

to another so that specific forms of question can be more readily addressed. A basic paradigm of this behavior is displayed in the respective virtues of parameterized and unparameterized approaches to geometrical figures (6,iii), although the allied advantages of switching between topographical and hub-and-spoke maps forms a central aspect of the tale of Ernest (4,iv). We can picture these lifts as patches $\mathcal{P}$ and $\mathcal{P}^*$ that sit above one another, distinguished by somewhat different vocabularies and different sets of valuable reasoning tools $\mathcal{R}$. When a shift is made, data registered upon $\mathcal{P}$ is translated (by rules τ) from the parochial $\mathcal{P}$-vocabulary into the format favored by $\mathcal{P}^*$ and then exported so that the $\mathcal{R}^*$-rule methods native to $\mathcal{P}^*$ can address the subject matter at hand. Plainly, this exchange is a two-way street and data may shuttle back and forth many times between $\mathcal{P}$ and $\mathcal{P}^*$ in the course of thinking about a given topic. In our everyday geometrical reasonings, such transfers are accomplished with virtually no awareness on the subject's part of their busy activities: in Peirce's phrase, the lifts "are not objects of reflective consciousness." I find it convenient to employ the term *mood* to distinguish the different personalities of investigation that transpire within $\mathcal{P}$ and $\mathcal{P}^*$: when we inquire whether a point lies precisely upon a geometrical figure specified by a set of equations, we address the topic in a different mood than if we had asked what the entire figure looks like.

I've not emphasized such patch-to-patch *lifts* much in our discussion, wishing to concentrate upon the *continuations* that stitch adjacent sheets together in a facade. However, the practical necessity for lifts in representational format forms a vital part of the structural considerations that explain why language adjusts as swiftly as it does to improving developmental pressures. In this section, I want to concentrate upon two issues: (1) how swiftly and deftly we accomplish these often unnoticed mood shifts; (2) the capacities for free assertability that exist within each local patch of mood.

I was first impressed by these twin considerations in discussing with my brother George[29] (and, later, with Jeffrey King) how anaphora (= the use of pronouns and "dummy names" in keeping a monolog brief and coherent) is handled in ordinary talk. Such devices are nicely illustrated in the ways that we commonly reason about logical matters. Assume we know the twin premises

(a) *All frogs are green.*

---

[29] George Wilson, "Pronouns and Pronomial Descriptions" Philosophical Studies 45 (1984).

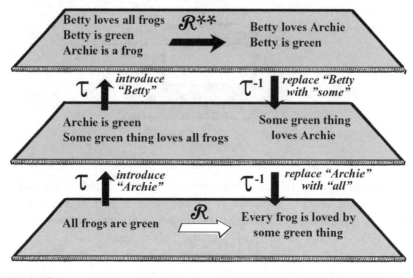

Logic lift

and

    (b) *If any green things exist, one of them will love all frogs.*

On their basis, it can be established that if any frog exists, then something green will love it. If we try to persuade an audience that such conclusion follows, we may utilize "dummy names" in our argumentation, as follows:

> *Assume that we have an arbitrary frog; for sake of argument, let's call him "Archie." Our first premise tells us that Archie has to be green. That shows that green things do exist, and so some green thing must exist that loves all frogs. So let's dub this new frog-loving green thing "Betty." Now clearly Betty must love Archie, so we certainly know that Archie is loved by something green. But Archie could have been any old frog whatsoever. Hence, we actually know that every frog is loved by some green thing.*

Such behaviors are well understood by logicians, who have worked out nice formalizations of the processes, in the form of the Fitch-style natural deduction systems discussed in the fine print. In the argumentation supplied, several lifts in mood are involved, all of which involve shifting between generalized statements like (A) and statements that contain dummy names such as "Archie is green." Thus in the second lift pictured, introducing "Betty" allows us to recast "Some green thing loves all frogs" as two simple statements: "Betty is green" and "Betty loves all frogs."

    Theoretically, given strong enough rules $\mathcal{R}$ in our lowest patch, we would be able to obtain our final conclusion ("Every frog is loved by some green thing") while remaining entirely within the bottom sheet, but, in fact, it is far more efficient to reason through mood shifts as indicated. The basic formal advantage of these shifts is that quantifier phrases such as "all frogs" and "some green thing" are hard to handle directly but, by introducing dummy names and pronouns, we shift into a simpler setting where the applicable reasoning rules $\mathcal{R}^{\star\star}$ are largely truth-functional in nature. Of course, to not incur fallacies, the shuttling principles τ and τ^{-1} must be properly specified (a not entirely easy task: a popular logic primer famously went through many editions with incorrectly formulated τ rules). For our lifts to work, we must assign the upper patch statements a somewhat misleading grammar: in its twice lifted context, the statement "Betty loves Archie" secretly carries the informational content "Some green thing loves all frogs."[30] In mathematics generally, there is a great premium in devising notations in which claims of a hidden complexity become coded in a manner that their surface reworkings obey very simple algebraic rules: we strive to dress complexities in simplicity's clothing.

So skilled are we in these rapid mood shifts that a certain *paradox of the elementary logic teacher* is engendered thereby. In logic, we employ the quantifier symbols $\forall x$ and $\exists x$ to symbolize the notions of "all x" and "some x." As such, the meanings of these symbols are utterly simple: they simply mean "all" and "some." Nonetheless, our students often fail to render a simple discourse like our frog example into symbolic notation successfully. Why? Insofar as I can determine, it traces to the fact that we perform most of our thinking about generalities in shifted mood terms, rather than relying upon "all" and "some" formulations exclusively. If confronted with a complex statement involving a lot of entangled "all"'s and "some"'s, we are likely to say to ourselves, "What the hell does this mean?" and proceed to sketch little diagrams involving a lot of little "Archie"'s and "Betty"'s. Rarely do we make mistakes in keeping track of the mood shifts required to insure that our "Archie" and "Betty" statements register the right quantities of information. In this respect, our students routinely handle data in a complex format that is far more sophisticated in its working principles than the simpler "all" and "some" packages that logical instructors vainly ask them to learn.

As such, logical arguments of this type are targeted to a practical objective: obtain the conclusion desired from the premises stated. Once such a project has been articulated, it is easy to supply heuristics that generally tell one how to set up the shifted moods to accomplish this objective. But in our conversations on these topics, my brother and I noticed that such mood shifts could also be utilized in monologs of no discernible purpose whatsoever, as long as the interpretational standards natural to each patch of mood are respected. In other words, windy Uncle Fred can coherently deliver a monolog of random opinions on frogs as follows:

> Consider any old frog; let's call him "Archie" for simplicity. I bet he's an ugly old brute; I really hate critters like Archie. But you know there's always some sucker that loves varmints like that; suppose her name is Betty. Such a fool undoubtedly will leave her lily pond as soon as Archie hops by. When she does, all of her sisters will be heartbroken . . . .

<u>Etc.</u>, <u>etc.</u> Clearly Fred is merely droning on, with no especial object in view. If we wish to challenge his reveries, we can fairly complain, "Come on, Uncle Fred, you don't really

[30] Jeffrey King, "Pronouns, Descriptions and the Semantics of Discourse" <u>Philosophical Studies</u> 51,2 (1987).

believe that every frog leaves all the sisters of some animal that loves it heartbroken?"
But this is simply the τ^{-1} translation of his last improbable pronouncement.

There is an important general moral to be extracted from this discussion. As observed
in section (i), the operative boundaries of component patches ∂A (including our lifted
varieties) are intertwined with the reach of applicable reasoning tools $\mathcal{R}$ and carry
restrictions on the transfer of data between patches. However, reasoning instruments
generally work for reasons: some strong ties between the sentences utilized in the
successful applications of $\mathcal{R}$ and genuine information about the world must lie in place,
even if we picture the nature of this relationship quite wrongly. But the same ties that
support the validity of $\mathcal{R}$ within a patch must also carve out a wider platform A over
which a considerable degree of free assertion, in Uncle Fred's fashion, can be
accomplished within the limits of ∂A and evaluated according to the same standards as
we apply to our dedicated $\mathcal{R}$-based activities. Often new patches of localized free
assertion nucleate around some initial $\mathcal{R}$-strand of practical advantage.

How wide these platforms will prove is another matter. I know of a degenerate case
or two[31] in which the shelving proves scarcely wider than the original strands of
practical advantage carved out by the $\mathcal{R}$ rules, but generally considerably more scope
for free assertion obtains than this. On the other hand, if we possess a markedly wrong
picture of the patch's underpinnings, we are likely to make assertions that do not stay
within the limits of ∂A and their assessment becomes more problematic. How we
typically approach the evaluation of usage in these circumstances represents a complex
subject to which we shall return in Chapter 10.

························

To mark out the structure of our shifting moods, Fitch-style systems typically employ vertical
scope lines with little shelves on their top that mark "all that follows to the right of the line
depends upon the investigative assumption that rests upon the upper ledge." In these terms, our
frog argument formalizes[32] as follows.

1.	$(\forall x)(Fx \supset Gx)$	Premise
2.	$(\exists x)Gx \supset (\exists x)(Gx \;\&\; (\forall y)(Fy \supset Lxy))$	Premise
3.	Fa	Premise
4.	Fa $\supset$ Ga	Universal instantiation (1)
5.	Ga	Modus ponens (3, 4)
6.	$(\exists x)Gx$	Existential generalization (5)
7.	$(\exists x)(Gx \;\&\; (\forall y)(Fy \supset Lxy))$	Modus ponens (2, 6)
8.	Gb $(\forall y)(Fy \supset Lby)$	Premise
9.	$(\forall y)(Fy \supset Lby)$	Simplification (8)
10.	Fa $\supset$ Lba	Universal instantiation (9)
11.	Lba	Modus ponens (3, 10)
12.	Gb	Simplification (8)
13.	Gb $\&$ Lba	Conjunction (11, 12)

---

[31] E.g., the Micronesian navigator example in Mark Wilson, "Wittgenstein: Physica Sunt, Non leguntur,"
Philosophical Topics (1999).

[32] Notation from Merrie Bergman, James Moor and Jack Nelson, The Logic Book (New York: McGraw-Hill, 1998).

14.	$\quad\mid\quad\mid\quad(\exists y)(Gy\ \&\ Lya)$	Existential generalization (13)
15.	$\quad\mid\ (\exists y)(Gy\ \&\ Lya)$	Existential instantiation (8–14)
16.	$Fa \supset (\exists y)(Gy\ \&\ Lya)$	Conditional proof (3–15)
17.	$(\forall x)(Fx \supset (\exists y)(Gy\ \&\ Lyx))$	Universal generalization (16)

Here the two movements into the inner scope line boxes represent our "get rid of quantifiers" lifts to higher patches, whereas the termination of these same lines represents the τ^{-1} restoration of the supplanted quantifier structure.

Semantically, we can easily supply appropriate truth-values to every line in our proof either *translationally*, in Quine's manner of step-conditionals, or by context-sensitive *interpretative rules*, as Jeffrey King provides. Both methods continue to work even if we dealing with Uncle Fred's ramblings rather than a directed proof.

Although our capacities to reason quickly gain much from such lifts into simpler syntactic settings, it becomes a drawback for certain purposes that local sentences are now longer *freely exportable*: sentence S on patch 1 cannot be automatically combined with S' from patch 2. It is sometimes possible to create new kinds of "object" that keep internal track of the patch to which they belong. In algebraic geometry, an important characteristic of a figure is the number of times a line will cut through it. Unfortunately, there are special exceptional points linked to the figure such that a line through them will intersect it in fewer spots. To avoid these exceptions, traditional geometers formulated their claims thus: "Any line through a *generic point* of the figure will . . . " Plainly, "generic point" represents a quantifier-like phrase linked to the specifics of the figure under examination. Algebraic geometry is much interested in the properties that figures assume when they combine and, plainly, the candidates for generic points will alter under this operation. For essentially free exportability reasons, modern geometers convert a figure F's old set of generic points into a single cloud-like "object": F's generic point. This strange ploy makes it easier to articulate what happens when geometrical figures combine. Kit Fine[33] has proposed a similar "make a funny object out of it" ploy to our logical dummy names "Archie" and "Betty." I don't see any particular advantage in addressing our present circumstances thus but such possibilities underscore the moral that a given patch of usage can potentially enlarge into a continuation that readjusts the original grammatical class to which the old vocabulary was suited.

We might observe that nineteenth century works on logic frequently describe the importance of mood shifts within argumentative investigation fairly ably, while they pay insufficient attention to expressing generalities in conventional single sentence, nested quantifier form. While we moderns frequently patronize the old logic primers for their expressive insufficiencies, we should recall that the informational content of our nested qualifier assertions can be recaptured as a series of simple sentences linked together through chains of lifted moods.

. . . .

(vi)

Evolutionary shaping. In Chapter 3,ix we surveyed sundry themes that entered philosophy in the 1970s under the leadership of Hilary Putnam and Saul Kripke,

[33] Kit Fine, <u>Reasoning with Arbitrary Objects</u> (Oxford: Blackwell, 1985).

generally under the heading "How natural kind predicates gain their attribute references through baptism." Such doctrines generally assume that some initial set of speakers might point to a gaggle of samples and announce, "Materials of this sort shall henceforth be called 'gold.'" It is then claimed that such ceremonies are sufficient to lock "is gold" permanently onto the "natural kind property" *being a lump of the element with atomic number 79*, even though the original speakers had no conception of that trait as such. Insofar as the intent is merely to observe that *sometimes* we can name attributes in simple "rabbit in the backyard hutch" fashion, I have no objection to these claims, although I believe such authors greatly exaggerate the range of cases in which "P"/φ alignment can be engendered through simple pronouncements.

However, the writings of this school also frequently appeal to supposed "causal connections between the speaker and the kind," whose details I won't extensively rehearse, but their general effect paints a picture where some natural kind stuff beckons to speakers after a baptism: "Come refine your usage unto me." Here the idea is that, however inept the original delineation of "P" may have been, some form of "causal interaction" will, over time, carry "P" into closer correlation with φ. It is not clear that either Putnam or Kripke themselves subscribed to a thesis, for they simultaneously ascribe intentions to language's employers much in the vein of Russell's remarks upon "the attributes in which we are mainly interested" (3,ix). However, other writers have certainly embraced the "siren call" picture quite warmly.

Here is a good example (from the American philosopher of language, Ruth Garrett Millikan):

> A substance concept causally originates from the substance that it denotes. It is a concept of A, rather than B, not because the thinker will always succeed in reidentifying A, but because A is what the thinker has been conceptually, hence physically, tracking and picking up information about, and because the concept has been tuned to its present accuracy by causal interaction with either the members of A's specific domain or with A itself, during the evolutionary history of the species or through the learning history of the individual.[34]

In other words, over time speakers will learn to track displays of a substance concept like *being water* ever more ably, adjusting their discriminations as they proceed in a manner that asymptotically stabilizes on a capacity to recognize the stuff reliably. Such directive focus towards a unique attribute will be installed, Millikan thinks, through a mixture of natural evolutionary pressures, societal training and learning from mistakes.

Like Millikan, I believe that linguistic development is profitably approached in quasi-evolutionary terms (I have reported my own affinities with biomechanics). The thesis that a usage will be shaped by various directive opportunities that gradually emerge along the interface where environmental conditions meet the linguistic tools available to speakers lies at the very core of all my thinking about natural predicative

---

[34] Ruth Garrett Millikan, "A Common Structure for Concepts of Individuals, Stuffs and Real Kinds" in Margolis and Laurence, Concepts, 541.

development. However, this shaping pressure assumes a far more localized form than the diffuse beckonings that Millikan describes, a picture that rests, in my opinion, upon a woozy conception of "causal process." Indeed, her frequent appeal to the metaphor *tuning* displays the problems I have in mind. In genuine tuning—when one electrical circuit synchronizes itself with another—, the latter must emit some discernible signal: i.e., display irregularities with which the other can become entrained through some species of non-linear interaction.[35] But it is not the least bit clear how *being water* broadcasts any kind of "signaling" of the sort postulated. Indeed—and this theme runs implicitly through the discussion of "strands of practical advantage" in Chapter 3—, we should expect that the major shaping of evolutionary pressures will be directed primarily against the *specialized forms* of linguistic routine that directly facilitate palpably useful projects. As that chapter's discussion of the callous Japanese craftsman observed, a specific manufacturing recipe will frequently adjust its recommendations, over time, to reflect fabrication policies that issue in objectively superior swords. But it is naïve to expect that the vocabulary employed in such local recipes will automatically follow suit elsewhere in the usage, because no clear pressures encourage it to do so.

The fact that facade-like structures represent entirely viable forms of linguistic engineering should make this point clear, because their cracking into patchwork allows other parts of a usage to isolate themselves from the shaping pressures that locally force adjustments within recipes for sword manufacture. Millikan presumes that her processes of amorphous tuning will generally induce a language to settle tidily upon single attributes in the same manner that classical thinking encourages, whereas, in fact, Pied Pipers of a more complex structure (such as a facade) may very well whistle its evolving patterns forward. As the evolutionary biologist Stephen J. Gould emphasizes,

> But ideal design is a lousy argument for evolution, for it mimics the postulated action of an omnipotent creator. Odd arrangements and funny solutions are the proof of evolution— paths that a sensible God would never tread but that a natural process, constrained by history, follows perforce.[36]

Atlas-organized stabilizations indubitably qualify as "odd arrangements" and I see no clear reason why Millikan's "tuning" should invariably favor simple "is a dog"/*being a dog* alignments over these. In fact, our capacities for shifting evaluative context swiftly (as highlighted in the previous section) indicate that mechanisms are available that allow a usage to readily elude evolutionary pressures to settle its predicates upon unitary attributes, except in localized patches. I recently ran across an example that demonstrates this evasion of univalent tuning in simple, but striking, terms.

Let us first fill in some background, which resembles that of "robin" in many respects. I once read in a linguistics text that "hazelnut" and "filbert" are synonymous. With respect to my own semantic grasp, this represents a claim I could neither verify or reject. Indeed, I have toiled my long span of years without meditating much upon either

[35] Arkady Pikovsky, Michael Rosenblum and Jürgen Kurths, Synchronization (Cambridge: Cambridge University Press, 2003).

[36] Stephen J. Gould, The Panda's Thumb (New York: W. W. Norton, 1980), 20–1.

filberts or hazelnuts, registering only that they represent congruent sorts of edible item, one or both commonly appearing in "mixed nut" assortments, with nary a clue as to which is which or whether they are indeed different. Here I find myself to be in good company, for informal survey establishes that most of my peers plod ahead in similar nescience.

We might expect that those who traffic more devotedly in these goods (our "societal experts," Hilary Putnam calls them) will have resolved the question crisply, for won't such agencies need to know exactly what a filbert is? However, a brief search extracts only a rather equivocal ruling from the expected authorities (in this case, from the Diamond Food Company):

> The hazelnut and the filbert are considered the same nut by some, cousins by others.[37]

Insofar as I can determine, this inconclusive remark is supported by usage elsewhere and leads one to wonder why some reformist board of agricultural commissioners would have not laid down firmer edict on this score. Further research reveals unanticipated reasons why nut growers might actively resist such reformatory pressure.

It appears that the terms "hazelnut" and "filbert" originated within different regions of Europe with respect to a different range of local species. In later years both terms were prolonged to cover the fruits of plants of the same genus (corylus) native to the North American woodlands and elsewhere, although not with equal strengths of application. Specifically, the native American varieties (C. americana) and (C. cornuta) seem to have been generally called "hazelnuts" by the English settlers, but, after American commercial production began in earnest in the Pacific Northwest during the early twentieth century, the Spanish giant filbert (C. maxima) was imported (and often graphed to C. cornuta roots to protect against disease). The nuts locally produced are generally larger (because of dedicated cultivation) than their European cousins. The American growers automatically transferred the locally preferred label "hazelnut" back to this product. This terminological specialization allows these farmers to draw a contrastive distinction with their European rivals, even though the fruit they now harvest is genetically much closer to what they will still call a "filbert" if it has originated in Eurasia than to any of the native American hazelnuts. In short, the "filbert/hazelnut" distinction, which began as an approximately biological distinction, has naturally continued into one that is largely geographically centered.

If we represent this usage as an atlas structure, we obtain a set of interconnected sheets more or less as shown. "Hazelnut" and "filbert" both begin their careers in species/geographically localized employments but are eventually continued to cover all specimens of the genus within Europe (mainly because the southern varieties were readily accepted as superior replacements for culinary purposes in England when they became available). In the meantime, "hazelnut" becomes preferred in its eventual extension to North America by our English settlers, mainly because of the resemblance of the native plants to those of England, although "filbert" is more weakly continued

---

[37] www.diamondnuts.com

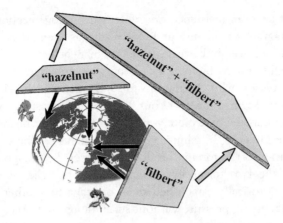

there as well (mainly by naturalists). Growers in Oregon and Washington become interested in cultivating "hazelnuts" and eventually discover that the imported C. maxima provides a better product. In the process, they continue "hazelnut" back onto C. maxima stock as long as it appears in its new habitat. Faced with competition from a smaller European form of C. maxima, this new localization allows the Oregon-centered applications of "hazelnut" to contrast with Spain-centered applications of "filbert," a term that has not experienced such a strong continuation into North America. In the manner of "harvest" and "autumn" (6,x), "filbert" is locally discouraged in its weaker North American continuation because of the local preference for "hazelnut."

This background in mind, let us examine a rather astonishing response (from a formal point of view) to the query, "Are the two nuts the same?" I obtained this specimen from the website of The Nut Factory of Greenacres, Washington:

> Hazelnuts and filberts are the same nut. Technically, the hazelnut is raised in Oregon and Washington on a bush that produces the nuts in late October. Nowhere else on earth is a commercial hazelnut crop grown. The filbert is a cousin of the hazelnut. The name filbert was used because the nut is found in Turkey, Greece, and Italy—all Mediterranean countries and all tied to Christianity. The filbert bush blooms in February on St. Filbert's day and the name "filbert" was a local term for the plant. The name was extended to the nut and over 90% of the world crop of filberts/hazelnuts is grown in these countries and exported throughout the world. The filbert is a smaller nut and many bakers in other countries grind them up to use as a powder ingredient for breads and pastries when the almond prices are high. The Oregon hazelnut is much larger and the finest variety is called a Barcelona. Nut roasters prefer the hazelnut because it looks much better in a roasted mixture. We usually only have hazelnuts, although sometimes we carry filberts. I could easily carry both.[38]

Logical pedants will complain that the opening assertion ("Hazelnuts and filberts are the same nut") insures that it is *impossible* for our grocer to not "carry both," given that he

---

[38] www.thenutfactory.com

carries either. But such an insistence on inferential homogenization is unfair to the author, for the paragraph can be interpreted as readily as a remark about the weightless astronaut who is losing weight because of dietary inadequacy.

From any straightforward biological or otherwise unbiased classificatory vantage point, there is no reason why the nuts from Oregon should be classified differently from the nuts from Spain. However, our Pacific Northwest grower patently does want them so discriminated, not merely in terms of size ("our bigger filberts"), but in terms of label (the Oregon "hazelnut" versus the Spanish "filbert"), so that his customers will not stray to the foreign product. However, he recognizes the biological pressure to force the two terms into alignment, so he relies upon a schedule of shifting from one historically established sheet of usage to another in a manner that acknowledges the unifying pressure without succumbing to it. Accordingly, he begins by acknowledging an appropriately negative answer to the biological question: "Are filberts and hazelnuts defensibly different from a natural history point of view?" But to implement his desired "the nuts from the Pacific Northwest should be preferably called 'hazelnuts'" objective, he employs the context-shifting word "technically," which does not signify anything "technical" in any proper sense, but merely indicates, "I will now shift to the preferential manner in which industry folks such as myself talk." He then encourages his readers to rehearse the developmental history of the two words in order to establish rationales for more restricted patches of employment. In essence, he invites his readers, "But now think of how you would employ 'hazelnut' or 'filbert' from this alternative perspective" (such is the localizing purpose served by his remark, "The name 'filbert' was used because the nut is found in Turkey, Greece, and Italy"). As observed in our other examples, we are quite skilled in quickly adjusting to such shifts in platform (indeed, our author evokes the natural size scales at which we apportion our geographical knowledge to encourage a sheet of use centered upon the Mediterranean region). When the danger of ambiguous localization arises as these maneuvers are encouraged, he employs makeshift locutions (e.g., "filberts/hazelnuts") to indicate the context he wants. Eventually, through installing an awareness in the reader of geographically specialized patches, he extracts a rationale for calling only the Oregon nuts "hazelnuts." From a formal point of view, these maneuvers represent the structural equivalent of marking the sheet upon which we want "$\sqrt{4}$" evaluated.

The sum effect of these contextual lifts blunts the classificatory pressure implicit in the original query, "Cut to the chase: are filberts really different from hazelnuts or not?" By encouraging several sheets of descriptive interpolation between, our Northwest grower can successfully evade any reformatory zeal to plant "filbert" firmly upon the local nuts: neither term evolves to simple "P"/φ alignment. And we have every reason to expect that such "odd arrangements and funny solutions" are rather common in language's developmental story.

There are a variety of common misunderstandings afoot that falsely encourage many philosophical writers to imagine that processes of tidy "P"/φ tuning occur more commonly than they really do. Specifically, popular slogans that can be intelligibly

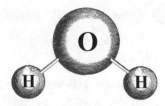

interpreted only as rough schemata often get mistaken for reports on crisp forms of predicate/attribute alignment. Good examples of such loose catchphrases are the frequently cited "property identities" beloved of contemporary analytic metaphysicians, e.g., "water $= H_2O$", "temperature $=$ mean kinetic energy", "lightening is electrical discharge". Let me mainly discuss "water $= H_2O$" (the situation with the others is even less favorable[39]).

Much of the trouble lies in the fact that it is unclear what sort of attribute the alleged *being H_2O* represents. True, we can frame a vivid picture of the familiar H_2O configuration, but its actual classificatory relationships to the substances we normally label as "water" are quite complex. Because of its great affinity for forming hydrogen bonds, the actual structures present in liquid water are enormously complex and are not perfectly understood to this day. *Being H_2O* itself is best regarded as not representing a direct classifier of much of anything, not even single molecules, but as instead providing an emblematic banner under which a family of behaviors involving some form of the familiar doglegged H-O-H structure loosely marches (put another way, *being H_2O* classifies chemical *pictures* ably, but not actual chemicals). On the other hand, the phrase "H_2O" readily transfers, through natural prolongation, to most of the complicated family of macroscopic classifications that we make under "water" 's heading. Thus we are willing to lift the seemingly molecularly focused "H_2O" up to a macroscopic scale when we say,

> *Carrie Nation's favorite beverage is H_2O.*

Likewise, we cheerfully pull our normally macroscopically centered applications of "is water" down to the molecular level:

> *A water molecule has become attached to this sugar molecule by a hydrogen bond.*

However, sometimes these shifts in scale interfere with one another. For example, for a parcel of normal water to manifest its common properties, a fair portion of it must disassociate into H + and OH − ions. We might plausibly report upon this fact as follows:

> *For water to manifest its familiar behavior, a small percentage of its H_2O content must disassociate into H + and OH − ions.*

[39] Mark Wilson, "What is This Thing Called 'Pain'?" Pacific Philosophical Quarterly (1985).

For reasons of clash of scale, it seems rather odd to say:

> *For H_2O to manifest its familiar behavior, a small percentage of its H_2O content must disassociate into $H+$ and $OH-$ ions.*

Here the first "H_2O" is clearly continued downwards to a molecular scale through "water"-linked prolongation from our macroscopic level classifications, whereas the phrase "H_2O content" stems more directly from its basic molecular level classifications, producing an apparent clash. Although the perspective shift is abrupt enough to sound peculiar, we still find ourselves able to interpret the claim in the context adjusting manner of the previous section (or, for that matter, in the Stebbing/Eddington disputes of 6,xii). If someone is genuinely puzzled by our assertion, we will explain, "Oh, that first 'H_2O' merely serves as a synonym for 'parcel of water,' but not the second."

In fact, how we classify macroscopic stuffs as "water" is itself inclined to be quite complicated, for we are willing to say both:

> *A piece of celery is not water whereas sea water is.*

and

> *By chemical composition, a piece of celery represents a purer specimen of water than sea water does.*

although we probably prefer to express the latter as:

> *By chemical composition, a piece of celery represents a purer specimen of H_2O than sea water does.*

We are even less inclined to rephrase the first as:

> *A piece of celery is not H_2O whereas sea water is,*

although I think, even here, we will let it pass.

In sum, we employ a rich variety of evaluative platforms that involve both "water" or "H_2O," often with a local preference for one or the other, but accompanied by a basic willingness to substitute one term for the other as long as contextual clashes do not arise. So neither term designates any clear or fixed form of attribute, but have instead married their linguistic fates together so that both terms shuttle around a rather complicated descriptive facade more or less interchangeably. The loose schema "water = H_2O" reports upon this terminological interchangeability, without providing evidence that *either* predicate has become definitively tuned to some fixed chemical attribute.

For such reasons, the French chemist Paul Caro is concerned that we not allow the emblematic and rather unrealizable "water molecule" to mislead us about what the phrase "pure water" actually classifies in real life practice:

> *There is an abundance of residual "non-water" that is always found in water, even the most carefully treated to please us. This repudiates the absolute notion of an ideal pure*

water in [our society's] collective imagination [that acts] haunted by it . . . However, we must reconcile ourselves to the fact that, as a physical object, pure water does not exist, especially not in Nature![40]

Indeed, as Caro also observes, official standards for "pure water" do not hinge, contrary to what we might expect, directly on the percentage of contaminants, but upon the question of whether certain delicate color and odor tests have been met (analogous to wine tasting). Such standards for "water purity" are natural because the potability of the macroscopic stuff requires some alien ionic cooperation—a hypothetical material composed of cleanly identifiable H_2O components is not remotely what we seek in the "pure water" line.

None of this indicates that "water $= H_2O$" is not a useful slogan, just as "classical physics $=$ billiard ball mechanics," understood appropriately, can impart useful information to an auditor as well. But we shouldn't allow such schemata to blind us to the frequent complexities of macroscopic classification, although, I believe, that has generally proved the trend in many veins of contemporary philosophy.

. . . .

A related phenomenon of loose categorization occurs when writers blithely assume that "the set $\{x|x \text{ is red}\}$ represents the extension of 'is red.' " We can usually parse what they mean, but only by reading " $\{x|x \text{ is red}\}$ " as representing some kind of stylistic variant upon "red."

. . . .

(vii)

Nostalgia for lost empire. Many opinions of an otherwise anti-classical cast favored in today's philosophical circles are marred by an inclination to cling to the unrealized promises of classical thinking. In particular, a presumption that phrases like "the unique reference of predicate 'P' " can justly be applied to language in most cases strongly persists, even though the writers in question will have rejected the apparatus of classical gluing that renders such arrangements credible. In cases like Millikan's, this neo-classical conservatism is expressed as an eagerness to uncover some natural process of "tuning" able to duplicate the expected reference relationships of traditional thought, without expending enough effort beforehand in establishing that such alignments genuinely exist (authors who conceive their projects in this manner often describe their endeavors as attempting to "naturalize the reference relation"). In contrast (but revealing the same tropism with respect to unique reference talk), authors of a self-styled deflationist cast maintain that claims like " 'cow' uniquely refers to cows" represent innocuous but unavoidable truisms about syntax that express nothing vital as to how bits of language correlate with creatures of a bovine inclination (recall from 5,viii Mark Johnston's

---

[40] Paul Caro, Water, Patricia Thicksum, trans. (New York: McGraw-Hill, 1993), 34.

contention that such "objectivist" interpretations represent wild-eyed metaphysics). "Instead," these authorities insist, "we merely state an inviolate rule that connects quoted bits of syntax with unquoted bits. That is, the claim supplies us with a logical license that allows us to infer 'Mickey talked of cows' from the assertion 'In English, Mickey used the predicate "is a cow." ' " Quine has called these quotation mark removing transitions *rules of semantic descent*: he likewise declares that "Clarabelle is a cow" and "Clarabelle possesses the trait designated in English by 'cow' " provide the same information, albeit lodged in different syntactic housings (rather as "Clarabelle was milked by Quine" represents a reordered variation upon "Quine milked Clarabelle").

Prima facie, such claims of guaranteed unitary reference seem quite odd from either camp, given the way that language behaves in real life. Suppose, in the midst of discussing our filbert/hazelnut puzzle, some kibitzer versed in the philosophical literatures just cited offers the following unhelpful suggestion: "This much we truly know: 'hazelnuts' refers to hazelnuts." To this we could reasonably object: "Actually, 'hazelnut' sometimes refers to Spanish filberts, as when these Oregon growers advertize their wares." On the other hand, we likely feel that we are not *obliged* to describe the situation in this manner either: we can also concur, "I suppose you're right, but the practice of using 'hazelnut' another way is pretty firmly entrenched in certain Oregon-centered contexts by now."

It seems to me that such freedom of referential attribution indicates that "refers" operates in a context sensitive manner very much like that of "concept" or, for that matter, "hardness" and "weight." In complex situations, our applications of any of these predicates generally follow whatever prime directivities seem presently salient, despite the fact that we might very well employ the same terminology in an incompatible manner on some other occasion (in short: "reference" talk often displays a *seasonal* character). Although I think the oddities of "filbert" and "hazelnut" are most clearly captured in my special vocabulary of "facades" and "patches," the same basic linguistic facts can be adequately registered within our ordinary talk of "concept" and "reference"—courtesy of the fact that these evaluative terms display facade-like behavior themselves (recall the dialog about "hardness" in section (iii) where all of the essential issues under dispute were addressed in an entirely traditional—but contextually shifting—format).

I find it odd that most anti-classicists strive mightily to insure that the phrase "the unique reference of 'P' " should apply in situations where I think we should properly comment, "Gee, the language/world relationships that here affect 'P' are rather complex and various directive factors are active that point the phrase in different directions. Any talk of 'P's reference' in such circumstances will need to manifest a considerable sensitivity to context." Why are such critics eager to install an ersatz tidiness upon word/world relationships, given that they mistrust the classical gluing that might have made those relationships more easily achievable?

Part of the answer is simply that, through not examining considerations of the sort canvassed in this book, such authors are unaware of the irregular ways in which words like "hardness" and "force" behave. But some of their penchant for reinstating classical uniqueness plainly traces to fear of a methodological bogeyman: the worry that:

> *If predicates are not assigned crisp attributes as references, standard logic will crumble and all of our standards of clear thinking will dissolve into ineffectiveness.*

Some of the pertinent issues are rather complex and much of the remainder of this book will be concerned with rejecting the essentially classical picture of "clear thinking" that is here presumed. We won't be able to discuss such concerns adequately until we investigate some salient real life situations where a developing language requires methodological management in the absence of an established picture of its strategic workings (this is the primary topic of the next chapter).

For now, let us simply ask what the worry "standard logic will crumble" might mean in the context of facade-like behavior. Consider the simple *rule of conjunction*: from A and B, established separately, infer the conjunction A & B. In patchwork circumstances, we should not accept any such principle on a global scale, for facades are designed to prevent unwanted free exportation from sheet to sheet of this kind (this is the basic lesson of how we must deal with "$\sqrt{z}$," let alone the other multi-valued terms we have examined). It will be objected, "Oh, you're not supposed to employ the rule of conjunction in those cases, because the preconditions aren't right." Exactly, but how can we be sure that such "preconditions" have been met? A classicist will respond, "Peer deeply into the hearts of your predicates and verify that they are utterly well defined" (in 8,v I call this a *true thought picture* of rigorization). But most anti-classicists doubt, just as I do myself, that such introspective endeavors always provide feasible answers, maintaining instead that such classical trust stems from an improper picture of concepts (in truth, most contemporary philosophers of language rarely worry about the problems of improving rigor at all, although they should).

But if we don't accept old-fashioned "true thought" responses here, how should we address the question, "When should we infer A&B from A and B?"? Loose appeals to "preconditions" without further explication supply an empty response tantamount to "Infer A&B from A and B when that inference is reasonable."

In this regard, we might observe that Quine supplies a crisp answer to our query, but the advice he supplies is plainly unwise. He maintains that rules like conjunction comprise the firmest of the regulative principles that we utilize in the course of attempting to whip our ragged web of belief into shipshape condition: "If A&B is missing from the web, install it there forthwith." He famously allows that the methodological demands laid down by such trusted logical principles might be conceivably relinquished, but only in circumstances of dire extremity such as quantum theory might inflict upon us.[41] Well, what about hard steels, $\sqrt{z}$ and filberts? Do their behaviors push us to those same cataclysmic extremes, in also refusing to submit to the conjunction rule's imperatives? Of course, this is a silly way to describe matters, but Quine shouldn't have attempted to defend "logic's status" through demanding global forms of linguistic

---

[41] W. V. Quine, <u>Philosophy of Logic</u> (Cambridge, Mass.: Harvard University Press, 1986).

organization that are patently unwise with respect to languages that describe macroscopic objects in practical ways.

In point of hard fact, the rules of first order logic are scarcely the most important inferential tools within our quiver: inherently unreliable principles such as Euler's method and our unconscious routines for reaching geometrical conclusions rapidly give us the most inferential bang for our buck with respect to reaching pragmatically valuable results. To optimize the inferential power of these rules, it is often required that the usage fracture into a quilt-like facade subject to boundary line restrictions on data exportation. We shouldn't spoil an admirable inferential arrangement dependent upon facade-monitored controls simply because some two-bit overreacher such as the conjunction rule wishes those bounding lines weren't there.

. . . .

Quine approaches "logic" in essentially the syntactically narrowed manner that Descartes views "mathematics" (4,x). In fact, we should adjudicate the correctness of any syntactic rule by seeing how it behaves "semantically" over generically specified mathematical settings, if we are capable of carrying out such an investigation.

. . . .

The proper response to our concerns, it seems to me, is that the justification of logical rules should be considered relative to a diagnosis of setting (in the sense of the generic pictures discussed in 4,x), in exactly the same manner as we approach Euler's method or any other inferential principle of that type. The classical tradition mistakenly accords logical reasoning—and largely it alone—an exalted status of permanent firmness superior to the other members of our everyday arsenal of reasoning tools. But our prospects for clear thinking will not crumble as a result of dethroning logic's primogeniture a little bit, for we must learn to be somewhat more patient in our semantic deliberations (citing Oliver Heaviside again, "Logic is eternal, so it can wait"). But we will gain a better feel for these issues after we observe the processes of "achieving clearer thinking" at work within salient sample cases (our "status of logic" issues will be revived in 10,v).

The thesis of the preeminence and permanence of logic represents but one of a large range of doctrines where current day philosophers seem overly eager to replicate many of classicism's alluring promises without founding them in traditional gluing. This, it seems to me, explains why so many anti-classicists are eager to reconstruct some reason why most predicates will possess unique referents—even if its price (as amongst the so-called deflationists) trivializes the content of the claim altogether. There is an admirable classical edifice that our anti-classicists wish to uphold, so they search for bricks that can neatly replace the classical building stones they have removed.

In Millikan's case, as well as the related school of causal theorists, such ambitions lead them to emphasize wispy relationships that aren't present in the real world in any palpable sense. Consider one of Uncle Fred's rambling speeches again. Attempting to trace some substantial causal pathway between what he now says about "frogs" and real life amphibians or their characteristics seems a fruitless endeavor: Fred doesn't need to be deeply tied to a topic to pontificate endlessly upon it (this is really the point Quine wishes to make when he worries about the unconditioned nature of standing sentences). By studying Fred's activities in isolation, we'll never get a clue as to what sort of information is transmitted in his claims. However, if we observe speakers whose linguistic acts engage more productively with the world (in using a strand of practical advantage to genuine purpose), we can generally determine the nature of physical data carried by the predicates within its routines, although such diagnosis is often rather difficult. If we wish, we can then *evaluate* Fred's claims for correctness under those standards without assuming that what he says is entangled with worldly events in any manner similar to that of the practical folk. As we shall see in Chapter 10, we are often interested in evaluating linguistic performance in this manner, but we usually gauge the "contents" of the usage simultaneously from a variety of other perspectives as well.

Indeed, consider the orgonists—those gullible followers of William Reich whom we met in 5,i. If asked "To what does their 'orgone' refer?", we might answer with equal plausibility depending upon circumstance:

(1) *To nothing really; it's entirely a figment of their imaginations.*
(2) *To some kind of blue liquid somehow connected with sexual activities.*
(3) *To Raleigh scattering in the sky; to surface layer mirages on highways; to woodland foxfire and other phenomenon that Dr. Reich has idiosyncratically anointed as "orgone."*

Each of these evaluations reports useful information about "orgone" and Uncle Fred's ruminations about "frogs" can be approached in similar veins as well. Classical thinking attempts to distill these multiple approaches into a single "conceptual content reference" for "orgone" (or "frog"), from which our three answers derive as specializations. Such a collapsing of evaluative concerns is *ur*-philosophically encouraged by "reference" and "concept" 's capacities to tailor their focus in an obliging and facade-like manner to varying matters of current interest. If so, it's surely a great mistake to seek "a naturalistic account of the reference relation" in situations where no unique form of evaluation is singled out. In this book we have investigated predicative behavior within a fully naturalistic spirit, but we do not expect to redeem all of classicism's alluring promises and ambitions in the bargain.

(viii)

The contextual control of data. A useful method for appreciating the "odd arrangements and funny solutions" that linguistic developments often obtain is to study predicates that have clearly begun their careers on the wrong foot and have needed to right themselves as they have successfully peregrinated onward. We shall investigate a number of cases in the chapters to follow but it will be useful for our discussion of music and color vocabulary if we quickly survey a familiar example now.

Dew bow

The case I have in mind is our old friend from Chapter 1, "is a rainbow." There we observed that this usage is born, perhaps of necessity, of misunderstanding: the predicate begins its enterprises clothed in the borrowed inferential directivities pertinent to "is an arch." Somehow these ill-suited garments must be eventually shed if "rainbow" is to remain viable. Clearly this occurs, for the predicate remains a robust participant in useful adult classification. But how does it manage to transcend its ill-starred birthright?

Approaching these issues in facade-like terms is helpful, although the adult accommodations that we reach with "rainbow" are probably so loose and fuzzy in their borderline controls that the phrase "atlas structure" implies a greater tidiness than suits the true proceedings. Let us begin by isolating some of the true things we assert about "rainbows" as adults, not subject to any childish illusions that rainbows can be climbed or that their "feet" allow pots of gold to be located there. All of the following can be regarded as wholly correct if uttered in an appropriate context:

(a) *A rainbow presently straddles those two hills.*
(b) *A bow has been present over the past hour, but its visibility from different vantage points has fluctuated greatly.*
(c) *Every primary bow has a large number of secondary companions, most of which are rarely, if ever, visible.*
(d) *If you look in the lawn sprinkler outside, you'll see an unusual bow that lies on its side rather than standing upright.*
(e) *Stan and Ollie both saw rainbows at the lake the other day but they were located in different parts of the sky.*

Each of these claims supplies real information about the world before us, but the nature of what is learned differs greatly between them. (b), for example, supplies direct data about the irradiated state of some long-lasting, distant shower independently of what is witnessed at any particular location and (c) comments on the nature of the light scattered, even when the caustics are too feeble to be seen. But (a), (e) and, even more dramatically, (d), supply salient information about the *landscape* against which the light display appears. With respect to (d), on rare occasion when we see a rainbow lying against the morning dew, within a lawn sprinkler or in clouds from an overflying airplane, the optical display unfolds against a background that recedes from the observer. Such depth clues sometimes induce the brain to interpret the image

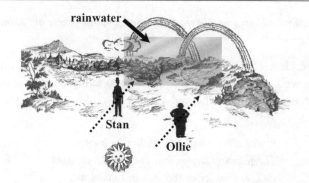

discerned as conforming with this recession with the result that the bow looks as if it lies prostrate upon the grass or clouds.[42] The exact psychological triggers that occasion these startling collapses seem to be rather delicate and ill understood; certainly no behavior centered in the water droplets themselves has anything to do with it. But this phenomenon appears so robustly to all observers that we have no inclination to describe these circumstances as ones where "The rainbow presents an *illusion* of lying on its side."

Around each of these sample sentences we can build up little platforms of allied descriptive employment, <u>viz.</u> the permissible classificatory remarks that capture similar sorts of data ("A partial arch of rainbow appears over the lake" will belong on the same sheet as (a)). The number of little arenas of fruitful application that we can consign to adult language in this way is astonishingly wide and varied. Plainly, they will not support unfettered *free exportability*: true sentences cannot be carried willy-nilly from one patch to another unless their representational structure has been modified in some fashion. Thus correct (e)-centered claims about the bows seen by Stan and Ollie should not inferentially force us to restate (a)-patch claims as "*Two* rainbows presently straddle those two hills," on the grounds that Stan and Ollie have witnessed this phenomenon and they clearly discern distinct bows (such a line of reasoning is completely kosher with respect to concrete arches). Here the problem is rather like that of "weightless": we do not want appraisals of rainbow number centered upon local observers to infect reports involving observers separated by appreciable distances. So some filagree of boundary line barriers must be installed to prevent the strong directivities natural to "arch" based reasoning from shuttling data between patches in an unhappy manner. But how, in real life, are these roadblocks implemented?

This passage comes from a popular explication by Rutherford Platt:

> *Of all the beautiful things we see in the world around us a rainbow comes closest to being purely imaginary. You cannot touch it because it has no material substance. If you walk towards it, it retreats and you cannot get closer to it. If you go away from it, it follows you*

---

[42] M. Minnaert, <u>The Nature of Light and Color in the Open Air</u>, H. M. Kremer-Priest, trans. (New York: Dover, 1954), 129.

like your shadow. No two people see the same rainbow—there are as many rainbows as there are observers.[43]

These claims are clearly intended to discourage projects involving rainbow reaching and climbing, as well as (in the last three sentences) faulty reasoning of Stan and Ollie type. Gerard Manley Hopkins devoted an entire poem on the subject of rainbow roadblocks:

> *It was a hard thing to undo this knot.*
> *The rainbow shines, but only in the thought*
> *Of him that looks. Yet not in that alone,*
> *For who makes rainbows by invention?*
> *And many standing round a waterfall*
> *See one bow each, yet not the same to all,*
> *But each a hand's breadth further than the next.*
> *The sun of falling waters writes the text*
> *Which yet is in the eye or in the thought.*
> *It was a hard thing to undo this knot.*[44]

Through first claiming that each person sees their own private bow—a contention that constantly reoccurs in explications of this sort—and, secondly, suggesting that these "multiple bows" fall along observer-linked parallels ("each a hand's breadth further than the next"), the incorrect Stan and Ollie inference is blocked, but in a manner that allows two communicating observers to calculate (usually correctly) the direction in which their companion is likely to report a bow sighting. So it isn't that we can't utilize the information reported in (a) within (e)'s, local patch, but its format must be adjusted through translation before it is exported. Such forms of "blockage and corrected reinstatement" function as rough and ready analogs for the border crossing rules we examined in the Stokes line cases of the previous chapter.

Here's another display of similar inferential tinkering, extracted from Frank Forrester's 1001 Questions Answered about the Weather:

> *How far away are rainbows? It is like asking how far away is the constellation of Orion, the Hunter. The stars forming the pattern are each separated by vast distances . . . So, too, with the rainbow in its entirety. It is as near and far away as the drops themselves, spaced dimensionally from drops nearby to those further away. A rainbow may exist in a small spray of fountain a few yards away or arch grandly against a thunderhead some two miles distant.*[45]

This claim is intended to discourage, inter alia, searches for pots of gold at rainbow-allied locations.

---

[43] Rutherford Platt, "What is a Rainbow?" in F. Lanier Graham, ed., The Rainbow Book (New York: Vintage Books, 1979), 75. Carl B. Boyer, The Rainbow (Princeton: Princeton University Press, 1987).

[44] Gerard Manley Hopkins, "It Was a Hard Thing to Undo this Knot" in Poems of Gerard Manley Hopkins (London: Oxford University Press, 1967).

[45] Frank Forrester, 1001 Questions Answered about the Weather (New York: Dover, 1981), 177.

It is curious that such inferential roadblocks are typically articulated as "facts about rainbows" (in Rudolf Carnap's jargon, they are expressed in "the material mode"), rather than as restrictions on allowable syntactic movement:

> Beware of reasoning with the phrase "rainbow location" in the manner appropriate to "arch location"

although, arguably, the latter claim might be less misleading. The rough fix quality of these repairs is also notable, for our "rainbow" explainers make no concerted attempt to keep the "facts" they cite consistent within their writings. Forrester, for example, will cheerfully accept as correct, in the appropriate context, claims like our (a) ("A rainbow presently straddles those two hills"), even if the mass of originating rain water centers in some quite different locale. Nor will Platt and Hopkins raise cavils about our Stan and Ollie claim (e).

Insofar as I can determine, "rainbow facts" such as

(f) *Bows flee or follow us as we move*
(g) *Everyone actually sees their own rainbow*
(h) *You cannot touch a rainbow because it has no material substance*

naturally occur to their authors in the context of asking themselves: "How might I explain to my readers that observers often do not agree upon rainbow locations? Well, rainbows are odd, half-mental things, so if I claim that each observer only sees a private bow, then no faulty inference shall be made in regard to locations extracted from some other person's report." Since this subjective curative too completely suppresses viable inferential connections between reports, compensating "facts" are then supplied ("but the observers will see their private bows along parallel orientations"). However, once the landscape of naïve "rainbow" talk has been sufficiently sprinkled with warning flags, our authors are content to lapse back into our customary policy of allowing two nearby observers to see a common bow (rather as their Nut Factory counterpart lapses back into his "filbert $\neq$ hazelnut" predilections once his biological obligations have been faithfully acknowledged). And these loose policies for monitoring "rainbow" talk are entirely adequate to its real-life difficulties: as long as warning "facts" (f–h) lie posted along the most popular border crossings, adults will be satisfactorily protected against the most likely forms of deductive failure. Such rough-and-ready correctives are neither elegant nor foolproof, but they seem sufficient unto the occasion (as noted below, supplying a truly accurate road map of viable "rainbow" use undoubtedly represents a very difficult task in descriptive linguistics).

In related ways, it is virtually impossible to consult any textbook written for technicians training for work in accurate color reproduction and not find passages such as the following:

> Colors, as mentioned before, are not real, and the world in front of us is not colored. Our brain creates the sensations of colors, most often from certain quantities and qualities of

electromagnetic radiation which strike our sensory organ. This is an explanation that is not easy to accept at first . . . [46]

As such, this sounds like the extreme subjectivism that exasperated our Lake Poet critics in Chapter 2. However, it is important to recognize that such claims, in context, serve a vital practical function: our color technologist must install some inferential roadblocks that will otherwise impede her abilities to go about her appointed tasks effectively. So her textbooks employ "rainbow"-like correctives to steer her rightly, although couched in *ur*-philosophical trappings that are less than ideal. I think professional philosophers often fail to recognize the practical need that such statements address and view them as merely "scientists trafficking in philosophical matters beyond their trained capacity" (which is true, in a way, as well).

Certainly, popular writers on the subject of rainbows often succumb to silly hyperbole, e.g., Fred Schaaf in his Wonders of the Sky:

> *Even though we can trace the path of light rays through their reflections and refractions and understand how the rainbow operates, the phenomenon strains the capacity of ordinary language to discuss it . . . [L]anguage is not used to dealing with a thing of the rainbow's nature—the only way the rainbow can fully enter language is in a strange blaze and twist of wordage which is a counterpart of the bow's own weird glory in the sky.* [47]

But it seems inappropriate to blame Nature for an oddity that is largely of our own making (Schaaf patently indulges in the sin of ersatz projection). After all, it is *our* psychology that induces us to first patch together an atlas of representations that links sundry facts salient to the condition of rain water to data that chiefly depends upon the distribution of depth clues within a vista. It is this human foible that forces us to install a protective filagree of roadblock "facts," so that we do not become too muddled by the gimcrack structure our brains have insisted upon assembling. The data we have registered remains objective enough, but the format in which it has been placed is entirely of our own doing. It isn't as if something truly peculiar, like a platypus, has waddled before us.

The whole story is complicated by the fact that an ongoing awareness of our original, "arch"-like picture of "rainbow" will be retained as a kind of linguistic fossil within our thinking. Indeed, without the capacity to reopen borders we must close in real life applications, we will not be able to interpret fairy tales and the like appropriately (as noted in Chapter 1, we will probably not be considered as "competent in the concept" unless we can do this). And this mummified structure often supplies a form of *latent DNA* upon which fresh forms of adult use can potentially nucleate. Ask yourself: are there circumstances in which the following claim might be true?

I just saw a rainbow endways on.

Of course, in fairy tale worlds that would be possible, but are there any real life conditions that might strongly elicit such a report and serve as a nucleating strand upon

[46] Rolf G. Kuehni, Color: Essence and Logic (New York: Van Nostrand Reinhold, 1983), 7.
[47] Fred Schaaf, Wonders of the Sky (New York: Dover, 1983), 11.

which some little patch of allied usage might grow? I think the only reasonable answer is, "It's hard to tell. It seems quite unlikely, but I wouldn't have anticipated the psychological mechanisms that provide a robust site for 'I saw a bow lying upon the ground' either." Such continuing interactions between new and old pictures of "rainbow" 's support make mapping out the true contours of the term's real world applications a quite daunting challenge. Because the biomechanics of such a task are so patently difficult, linguists have been inclined to relieve themselves of such chores by brisk appeals to "Linguistic agents think of 'is a rainbow' as designating the simple attribute *being a rainbow* and so that is where I'll leave my description of the term's 'semantics' as well" (a recommendation of this type by Ray Jackendoff was cited in 6,xi). However, I think it is plainly false that most of us "think of 'is a rainbow' " in that way: we are generally aware that the term behaves rather oddly. More importantly, we should not accept such a feeble rationale for ignoring some of the basic difficulties that language must confront in its attempts to deal with the world profitably. Or, to allow linguists a latitude to study the subjects they prefer, no one interested in the origins of *philosophical problems* should accept such a rationale, for our *ur*-philosophical adventures often begin when common words begin to wander for reasons we do not fully understand.

In his later writings, Wittgenstein offers many puzzling remarks about the "philosophical grammar" pertinent to ordinary usage, without providing clear explanation of what he means. In previous work of mine,[48] I suggested that the term *working grammar* might be appropriate to the facade-like controls that restrict how "rainbow" talk can be profitably employed within an adult setting (in contrast to the apparent or *surface grammar* pertinent to its original borrowed-from-"arch" origins). This notion of working grammar is particularly useful, I believe, when imposed controls allow terms to function in entirely different categories than they seem to fit (as when the name-like "Betty" of section (v) secretly codes for a quantificational notion such as "some green thing loved by a frog"). In this book's investigations, I have generally stayed within the orbit of tamer behaviors, so that they can be approximately fitted to simple facade models (whose local, patch-centered grammar is entirely orthodox). As observed above, "rainbow" 's behavior is too irregular to fit such structures closely and its bounding line controls are implemented in a more rough and ready manner than the term "grammar" suggests.

Perhaps in contrast to Wittgenstein's own point of view (see my remarks on this matter in the Preface), it should be abundantly clear that usage as we witness it in real life talk invariably transgresses the boundaries marked here, for no other reason than, like garrulous Uncle Fred or the orgonists, we are not always punished for ungrounded or impractical talk. We should only expect the filagree of border line divisions (or more convoluted varieties of working grammar) to appear in a language as it is developed to sufficiently demanding standards of practical performance.

---

[48] Mark Wilson, "Can We Trust Logical Form?," Journal of Philosophy 91 (1994).

loxodromic path

A beautiful illustration of some of these points can again be found within the very rich subject of world maps. In a well-known and deservedly popular exposition of cartography, David Greenhood writes:

> In a map we have not a lifelessly perfect diagram but a human confession. Man goes up against certain phases of the universe to adapt them to his needs, and must publically admit that for all his cleverness there are some things he cannot do. A map may have a grandiose caption, "The World," but if you know how to look into its depths you will find a goodly amount of human humbleness in it . . . [C]artographers have resorted to many antics, ingenuities, and devices in the age-old endeavor of making the most of the world as it is: round and immense and impossible to see all at once.[49]

In our earlier discussion (6,ii), we noted that the geometrical mismatch between a piece of paper and a slightly ellipsoid earth requires adjustments in how information about the latter is to be registered in the former. Different forms of map (which Greenhood calls "flat children of a round mother"[50]) offer specific advantages for specialized tasks such as navigation or population estimates. We observed that the realm of terrestrial geographical fact can be approached through flat maps in a completely unskewered way as long as a rich atlas of interlaced and complimentary maps adapted to these special purposes is utilized—indeed, these considerations provided our first model for how slightly unsuitable reasoning capacities can be adapted to utterly unbiased ends through an appropriate strategic utilization. The less than perfect qualities of our maps needn't deter us if our deliberations never hinge on these map-housed idiosyncracies.

However—and this is the subtle observation that Greenhood makes—, our real life employment of maps is not so perfectly adjusted as my rosy description suggests. In fact, most of us entertain a deeply ingrained preference for a quite unsuitable form of map: the familiar rectangular *Mercator projection*. In truth, this is a very odd choice of projection to have become canonical for the likes of us, because the Mercator is designed—and was explicitly so labeled by its creators—to maximize trustworthy sailing paths for the mariners of long ago. That is, a straight line on a Mercator chart

---

[49] David Greenhood, Mapping (Chicago: University of Chicago Press, 1964), 119. Text slightly rearranged.
[50] Ibid., 113.

Mercator chart (straight line = rhumb line path)

gnonomic chart (straight line = shortest distance)

represents a rhumb line or path of constant compass heading ("Keep a heading of 20° north by northeast"). If such a path is graphed upon a globe, it supplies a strange looking (loxodromic) curve curling around the North Pole and, obviously, does not represent the shortest distance between points on the earth at all. On a gnomonic chart, a straight line does mark the shortest distance of connection (an arc of great circle connection), but we "don't like the looks of that one." The only practical advantage of the Mercator lies in the fact that, at sea, it was generally wiser, before modern times, to follow a longer path of constant bearing than attempt a shorter route, except on very long journeys (it is more important to not get lost than to sail entirely efficiently). As their skills improved, navigators employed lifts between the two maps to plan better routes: the shortest distance path is first found as a straight line on a gnomonic chart and then transferred as a curve to the Mercator whose contours are then approximated by constant bearing segments, which provide the vessel with its schedule of sailing instructions. The recent availability of global positioning satellites has greatly altered modern navigational techniques, leaving the poor old Mercator without much practical use for anybody.

Nonetheless, we love it still and will scarcely exchange it for anything else. Why a map specialized in the odd manner of the Mercator has gained such a lock upon our cartographical affections is a matter of some dispute, especially since it did not occur immediately. Undoubtedly, some of our strange passion reflects a preference for tidy rectangles; some of it traces simply to the fact that our great-grandparents also cherished the map (perhaps because of the erstwhile importance of frigates to national economies). Whatever the reason, we like it and will scarcely budge in our preferences. Professional cartographers, of course, have objected vociferously to the Mercator for many years, but largely to no avail (S. Whittemore Boggs):

> [U]se of the Mercator projection for world maps should be abjured by authors and publishers for all purposes . . . [N]o man ever saw or will ever see a world that has much

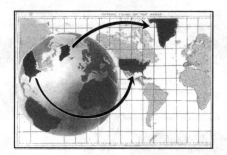

resemblance to the Mercator map and the misconceptions it has engendered have done infinite harm.[51]

No doubt "infinite harm" is professional cartographer's hyperbole, but many a schoolchild has certainly been misled about the size of Greenland ("Why shouldn't it count as a continent, Mr Chips?; it's bigger than Australia"). Nor can I gauge the misery wrought when Canada misrepresented its areal position within the British Empire in a notorious commemorative stamp of 1898.[52] In vain, do cartographers insert areally correct Goode maps in well-intentioned school atlases. Confronted with a map admirably designed for a specific purpose, we laymen fail to heap praise upon its clever deviser, but reject it with distaste in favor of the misbegotten Mercator: "The earth doesn't look funny like that". As Greenhood comments, it is only parties of practical purpose that are likely to mend their Mercator-entrenched ways:

> *Perhaps the essential lesson to learn from the "odd" projections is that it is logic which makes them weird. And the logic of each is its utility. They seem less fantastic as soon as we make appropriate use of them. What is really weird is* [our untutored] *limited conception of the world, and maybe by dint of some irony deep within the mathematical nature of forms the fantastic appearance* [that well-suited maps] *have for us at first is only a gentle ribbing of us for our still somewhat inflexible minds.*[53]

---

[51] Mark Monmonier, Drawing the Line (New York: Henry Holt and Company, 1995), 21.

[52] Doreen Massey, "Imaging the World" in John Allen and Doreen Massey, eds., Geographical Worlds (Oxford: Oxford University Press, 1995), 43.   [53] Greenhood, Mapping, 170.

What he has in mind is this. For whatever *ur*-philosophical reasons, we have decided that "the World" should appear only one way in a map and, through custom, we generally favor those that look, at least superficially, very like the old-fashioned Mercator projection. Such intransigence has created a complex situation where kindly cartographers have been forced to develop convoluted stratagems for preventing the "inflexible minds" of the rest of us from falling into desperate geographical error whilst still fancying that we have remained true, throughout it all, to our Mercator loyalties. Thus they quietly circulate policies to newspaper art departments that ban inappropriate maps within statistical displays (we usually think such graphs "look funny," but have grown accustomed to them in these contexts, and they genuinely prevent us from extracting grossly mistaken conclusions from the charts). In the same paternalist spirit, general purpose atlases for home consumption include world maps that look sufficiently like the Mercator so that our rejectionist hackles are not excited, but which secretly rely upon compromised projection recipes, so that a straight line segment does not symbolize the same geographical characteristic everywhere on the chart (in other words, the representational character of the line is stitched together in a piecemeal manner). Just the other day I visited my son's school and espied a wall map with its upper corners subtly tucked and I thought to myself, "Ah ha, the Good Samaritans of cartography are quietly watching over our welfare" (they didn't bother to correct for the southern distortions in their map, figuring, I guess, that no one is likely to reason much about Antarctica). In many sophisticated maps, extensive psychological tests are run to see how people extract information from world maps and a complex projection is then devised to optimize for our peculiarly human map reading propensities.

Although in an ideal situation, a good map should have its projection recipe published in a preface, so that the knowledgeable reader can employ the map as wisely as possible, our weird obstinacies create an inverse situation: the paternalistic cartographer must slip better maps upon us while preserving the illusion that we are looking at a "trusty old Mercator." Supplying a reliable key in the preface is unlikely to do any good and might occasion harm if its readers conclude, "Well, this certainly isn't a good atlas; it doesn't employ the Mercator projection." Certainly in the case of modern maps specifically tuned to human psychology, it is virtually impossible to publish a projection formula, as the mapping itself is not computed in a sufficiently regular manner. It is worth observing, <u>per</u> our 7,iii discussion of Nelson Goodman's views, that the *objectivity* of the data in these psychologically adjusted maps is not compromised; indeed, we *perform more ably* with these charts than with more regularly registered depictions.

All of this suggests that the true state of play in descriptive language is apt to be a mess, especially if we are not pressed to higher standards of potential performance. Linguistic analogs of Gould's "odd arrangements and funny solutions" are likely to prove the norm and, even then, the funny solutions may really constitute half-way measures: crude fixes that correct only for the most egregious problems, allowing ample opportunity for the *ur*-philosophical winds to fan the unremedied oddities into great paradoxes on occasion.

....

Interesting work has been recently pursued in linguistic circles wherein frozen grammatical units of greater length (such as "for the sake of," retained from archaic sources) comprise a library of prototypes upon which other constructions build. As the linguists Kay, Fillmore and O'Connor explain in an important paper:

> As useful and powerful as the atomistic schema [= recursion founded at a lexical level] is for the description of linguistic competence, it doesn't allow the grammarian to account for absolutely everything in its terms. . . . [A] list of exceptional phenomena contains things that are larger than words, which are like words in that they have to be learned separately as individual whole facts about pieces of the language, but which also have grammatical structure, structure of the kind we normally interpret by appealing to the operation of the general grammatical rules. This list is not merely a supplement to the lexicon: it contains information about fully productive grammatical patterns . . . One of our purposes in this paper is to suggest that this repository is very large. A second is to show that it must include descriptions of important and systematic bodies of phenomena which interact in important ways with the rest of the grammar, phenomena whose proper understanding will lead us to significant insights into the workings of language in general. A third is to make the case for a model of linguistic competence in which phenomena of the sort we have in mind are not out of place.[54]

Possibly the notion of a library of prototypes provides a good basis for understanding the behavior of "rainbow" in a manner more detailed than I have attempted here.

Considerations of space have also precluded a more active investigation of the specific capacities for linguistic management offered by the partial-coverage-via-lifts structures displayed in 6,iv's navigational lists over a topographical map arrangements (or, for that matter, in 4,ix's lousy encyclopedia coverage of billiard ball behavior). Obviously, the obligation to only address a limited range of questions within a specific patch, lifting harder queries into a linked patch of alternative representation, allows for great efficiencies in data storage and management. For example, consider 6,iv's neighborhoods-over-Pittsburgh arrangements once again. If we have this data structure available and are asked to answer some question not wholly covered within the local neighborhood maps, we usually proceed by *interpolation*. That is, if asked, "Do you reckon there's a pizza parlor between Oakland and Squirrel Hill?," we might reply, "Well, in both Oakland and Squirrel Hill pizza parlors can be found every 1/4 mile and the two neighborhoods lie a few miles apart, so I think it is reasonable to expect that you can find a pizza in there somewhere" (we witnessed allied interpolation in the boundary layer techniques of 4,vii). Plainly, many issues cannot be reasonably settled in this way and, accordingly, we must sometimes reply that we lack the data required to address the query. Notice how the lower sheet information about the distance between Oakland and Squirrel Hill is utilized in conjunction with the local neighborhood facts concerning pizza parlor density.

It seem possible that certain forms of what is loosely called "reasoning with vague predicates" operate within allied settings. Consider the predicates "is bald" and "is not bald." From psychological studies, it seems likely that we possess a neighborhood structure for various prototypes of baldness: Yul Brynner smoothness; pronounced widow's peaks; very thin in the back, patchiness in front; and so forth. Likewise, we assign a comparable topography to hirsute plenitude: full afro; duck tail; early Beatles cut, <u>etc.</u> These two neighborhoods, with their

[54] Paul Kay, Charles Fillmore and Mary Catherine O'Connor, "Regularity and Idiomaticity in Grammatical Constructions" in Paul Kay, <u>Words and the Grammar of Context</u> (Stanford, Calif.: CSLI Publications, 1997), 5.

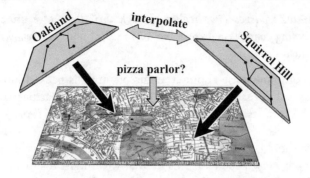

parochial "bald"/"not bald" terminologies, can be regarded as situated incompletely over a wider (but less commonly utilized) map of individuals with $\underline{n}$ hairs upon their heads (for all variable n). Certain general questions—perhaps "Does everyone admire Ronald Coleman's mustache?"—can be answered through interpolation: "Yes, all the hairy and all the bald folks do and no inclination towards dislike can be witnessed in either quarter." But sometimes it can't: "Well, all certifiably hairy and bald folks do but there also seems to be a marked shading off in admiration as hair number decreases (resp., increases), so I would be leery of drawing any interpolated general conclusion here."

....

(ix)

A paradox of classical grasp. In this section, I shall summarize some of our earlier findings and provide some preliminary indication of the topic of *linguistic management* with which our final chapters will be centrally concerned. Let me begin with a fundamental classical objection to pre-pragmatist approaches such as ours (the complaint lies latent in the doubts articulated in 5,iv).

> *I don't see how elaborate frameworks such as your facades can play much role in shaping our real life linguistic procedures, because we usually plan out our strands of practical advantage by first developing a* descriptive model. *That is, if we seek a linguistic scheme that will facilitate some practical end, we first model the physical setting before us and only afterward plan out the string of sentences that will prove strategically appropriate within the setting constructed. But this two-stage process presumes that we will have already tied our descriptive predicates "P" to traits φ exemplified within the modeling and hence their semantic behaviors will not reflect the strategizing stage of planning in any way. True, no real life traits comparable to φ may actually exist, but their posited modeling characteristics nonetheless capture the true core of "P"'s conceptual content, in a manner entirely prior to the plotting of the multi-sentential linguistic units you have emphasized in your facade story. In other words, our modeling procedures* <u>per se</u> *can usually be factored apart*

from the strategic purposes we realize through their first stage utilization. But these purpose-excluding concepts represent the unvarnished and well-behaved universals that classical thought demands.

A good deal of common sense about strategic planning is packed into this reply, but it also embodies subtle exaggerations of our linguistic capacities that provide the little chinks in the fortress of classicism that allow unruly invaders such as facades to creep in.

First of all, we should be cautious in how literally the metaphor of model building is read. In Chapter 4 we discussed Hertz' claim that we "model reality" in language and it is clear, in that historical context, that he is thinking of the process of constructing *scale models* as an analog means of obtaining predictive values for large systems as yet unbuilt (the little boat in the water trough that is scaled up to predict the ocean liner at sea; Ernst Mach often writes in a similar vein). But it is comparatively rare that we practice "model building" in any manner such as this (we can't literally build little boats in our heads or on pieces of paper). In 4,iii I discussed the lumped mimicking provided by Euler's method as the closest match I could provide for what Hertz describes, but it is clear that its procedures are too stiffly algorithmic to suit what he would like.

As we instead saw, a capacity to answer a broad range of questions about a subject generally requires that a rich variety of tools be cobbled together. As stressed in 6,ii, the reasoning and measuring techniques we employ generally embody "intrinsically unsuitable personalities" (the stiffness of an analytic function is our prototype) whose effects are mollified by skillful recombination and monitoring. More generally, any viable predicate "P" must be surrounded by a shifting cloud of easy-to-follow directivities in the sense of 3,vii, lest the term become a predicative analog to 5,v's "Foo Foo"-in-Tibet (i.e., a phrase that we fancy correlates with an attribute but whose implementations we can never recognize nor reason about with any effectiveness). Instead, "P" must situate its use in a manner that maintains genuine information bearing capacity with respect to physical systems (not necessarily on "P"/φ terms), yet retains a sufficient fund of easy-to-follow directivities from which we can extract linguistic good on occasion. Our sundry examples suggest that achieving this midrange placement is not always an easy matter, simply because the world's behaviors do not conform to human linguistic abilities in any ready way. So with any descriptive predicate, strategy potentially comes into play in settling how linguistic abilities should be tacked together to suit objective information in a viable manner.

True, we don't think of this cloud of easy-to-follow directive skills as "representing the core of the concept" or "the reference of the predicate," nor should we. As a cloud of associations, they alter over time, with additions and departures occurring in Ship of Theseus fashion. Nonetheless, a reasonably sized surrounding cloud must be present to provide the word with its robust word/world engagements, as well as contributing to its conceptual personality: the factors that make "P" seem different from "Q." Commonly, we entertain more or less adequate *pictures* of how these practical skills relate to some informational core, but sometimes we don't. For example, in the important case of the swift, intuitive calculations of future positioning typical of everyday geometrical reasoning, we natively employ a large battery of reasoning skills of whose underlying nature we have little conception or understanding (from Euclid we obtain a nice picture of the traits we want to discuss with our geometrical predicates, but we don't understand in any detail how our intuitive techniques of practical reasoning relate to this underlying portrait, although we can safely assume that they somehow do). Nonetheless, these inferential capacities are so developed (probably by our biological heritage as hunters and gatherers) that we can swiftly and effectively resolve a rich and varied array of questions about a geometrical tableau arrayed before us. Indeed, in everyday life, when we claim to consider a topic "in terms of models," we are often indicating that we ponder its parameters in lifted geometrical terms and rely thereby upon our capacities to reason amply and ably with our geometrical conclusion seekers.

As "rainbow" nicely illustrates, sometimes we *borrow* selectively from solid geometry's library of proven routines for "P" 's use, which we progressively winnow and adapt through trial-and-error experimentation until we produce a gimcrack construction adequate to the physical information at hand. As these alterations proceed, we lose the assurances of both well-tested *reliability* (we know that intuitive geometrical reasoning applied to terrestrial solid objects is generally safe) and soundly pictured *support* (as supplied for Euler's method in 4,xiii). Through our adaptive tinkering, "P" has wandered too far from these known shores of safety. As "P" 's usage builds up in these manners, through borrowing and prolongation, confusion and mistaken applications can emerge in their wake, forcing us to reconsider "P" 's semantic situation if vital matters hang in the balance. Typically we must look deeper into the cloud of directivities that surround "P," even to specialized applications that we formerly reckoned as entirely ephemeral to "P" 's "conceptual core." Sometimes we find that an unexpected but enlightening picture of "P" 's behavior can be framed, by threading through the old directivities in a new way, even if heretofore we have accepted "P" for its utilities without seriously pondering how it manages to carry effective information. Few of us have ever subjected "rainbow" to substantive semantic scrutiny, recognizing only that "it's a funny kind of word," nor are we will likely do so, as few vital issues turn upon their resolution. But in mechanical engineering, people die when bridges collapse and so we must scrutinize the fundaments of its working vocabulary more carefully. This is why the impulse towards a dedicated reappraisal of picture arises most clearly when "P" 's prospects for being pushed to a higher degree of applicational skill are under investigation (as occurs in Williams' study of "hardness").

In short, a problem in *language management* arises: old predicate "P" is plainly the instrument of genuine utilities ("is hard," "force," "gear wheel"), but unknown obstacles prevent it from improving its functionality. How do we proceed from here? We then search for directive answers *conducive to improvement*. Having now mapped out the facade-like patterns in which predicates sometimes find themselves stranded, in our final chapters we want to investigate how we self-consciously deal with usages that have wandered into complications and impediments. It is at this point that "concept" and its evaluative kinfolk (such as "understanding" and "truth-condition") enter our stage, for they serve as our prime everyday vocabulary for adjusting the courses of language along more effective rails.

To avoid confusion, let me separate two themes that run throughout this book. The first is, as just stated, that we employ "concept" as an evaluative tool for locating the special directivities salient to linguistic development. On the other hand, such management efforts should not prove excessive, for overzealous fussing can easily ruin a delicate word/world alliance rather than improving it. We must develop a *mitigated caution* with respect to our capacities for auguring language's profitable avenues of opportunity. But this second set of issues needn't concern us now (they will emerge in 10,i, inter alia), for we want to concentrate our attention upon the *locutions* in which we *register* our intentions with respect to predicative redirection, whether those policies are wise or foolish.

In these respects, we are often confused by "concept" 's activities, for the word displays what might be called (following Heaviside) a *spotting property*: we ponder the full cloud of directivities surrounding "P" and pick out the ones that seem the best indicators of the improvements we seek ("manner M is the truly proper way in which 'P' should be used"). Due to the manner in which descriptive language commonly evolves under patch-to-patch prolongation (such patterns will be called *canonical developments* later), the directivities we spot as salient may not remain constant over time and will instead display seasonalities naturally tied to developmental stage. For example, at certain stages our "conceptual spotting" isolates elements conducive to framing a picture of "P" 's usage, but at other times our semantic attention will be more operationally focused (10,vi).

As vehicles for linguistic monitoring and management, "concept" and its chums serve important functionalities. Unfortunately (for this propensity supplies the *ur-philosophical roots* for classical thinking about concepts), we rarely attend to the shifting seasonalities of our evaluations and improperly regard the factors we highlight as representing an unyielding "conceptual core" to "P," which is viewed as an invariant content around which the rest of the cloud of buzzing directivities flock as mere dependent instrumentalities. To be sure, many common predicates—"dog" and "doorknob" were cited earlier—lead such lucky or unchallenged lives that little evaluative instability is witnessed over their histories of usage, supplying nice simulacra of preserved conceptual invariance. However, for descriptive predicates of the sorts we have emphasized—wandering words that occasion serious scientific or philosophical confusion—, such stabilities are not the case and the root causes of their fluctuations in evaluation require closer scrutiny and appreciation.

Contrary to classicism, we do *not* supply reports on the full personality of the predicate when we make a "conceptual spotting" judgment: they represent reports of a search for the *indicators of fruitful improvement*, not full linguistic accounting. However, as observers of linguistic process, we can't adequately understand the evolutionary progression of our phrase unless we attend to the wider package of associated directivities, even to details that the user of the word may regard as utterly ephemeral to its "core content." As students of linguistic process, we want to understand why factor X looms large to speakers at developmental stage S: why X is currently spotted as "central to 'P.'" Such transitory judgments can be entirely appropriate developmentally, without it being the case that factors X come close to exhausting or adequately explaining the "conceptual personality" that "P" displays. Indeed, speakers often fall into *ur*-philosophical confusion precisely because, in their eyes, the X far outshine the more subdued—but real—considerations responsible for shaping their evolving use (as happens, for example, when we are bedazzled by "great ideas" myopia).

In fact, a general tension obtains between "P"'s present descriptive capacities and those we would like to see it assume in a prolongation. Properly effective strands of practical advantage must be tailored to the particularities of application, but we reasonably hope that "P" can transcend those shackles in future employments. And very often it can, but sometimes we are badly mistaken. In evaluating "P" qua concept, we typically spot those directive indicators that promise the extensions we seek as closer to the "true core" of "P". For example, Descartes and other mechanists do not *want* "gear wheel" and allied notions to be chained to human design capacity, but wish to see it freely roaming through the universe, classifying whatever it sees willy nilly without regard to lowly human design purpose. If asked about "gear wheel"'s central core, mechanists will speak only of "rigid shape," "conserved motion" and allied phraseology, even though such appeals only serve as incomplete *gestures* towards the illusive "core" they desire. If, having read Reuleaux, we ask, "Are your notions meaningfully applicable only to assemblies containing at least four rigid members subject to Gruebler's limitations?," they will indignantly reply, "No—our concepts apply to *everything*." But, in this case, such semantic ambitions merely represent wishful thinking: "gear wheel" can be successfully prolonged off the narrow sheet of designed planar mechanism only a certain distance, secretly tethered by the very computational advantages that render it so useful within its original Reuleauxian patch (this theme will be further developed in 9,v). And this represents a nice prototype of the tensions we shall address in the latter part of the book: as we attempt to improve the scope of "P"'s classificatory reach, we isolate indicators of improvement X that we regard as fundamental, whereas factors Y that are truly critical to "P"'s fullest measures of capacity languish unnoticed within the outer layers of the directive fog about "P." And these semantic "mistakes" needn't arise from carelessness or delinquency: they simply represent the less fortunate aspects of the zig-zagging processes whereby we improve language through successive approximation.

These considerations give rise to important ambiguities in "the study of concepts," as we hear that phrase employed by philosophers, psychologists and laymen. In each case, we must isolate the central concerns at issue: "Are you primarily interested in the sources that contribute to "P"'s perceived predicate personality or only the processes of *directivity spotting* that proceed under the rubric of 'concept' in everyday life? Or are you wondering how lexical data related to 'P' is stored in the subject's brain?" (and there are many other potential topics of focus available, none of which should be confused with the others). For our own endeavors—especially, our efforts to understand why *ur*-philosophical mishaps arise—, we most want to understand the reasons *why* "concept" practices its temporary favoritisms: *why* restricted ingredients get seasonally selected as central amid a much richer cloud of accompanying directivities. We need to study these processes of linguistic management at work in real life affairs and so we shall prosecute our investigation, as always, by probing revealing case studies (inter alia, the astonishing career of Oliver Heaviside's operational calculus).

Before we do so, we should briefly return to the problems with which we began in Chapter 2, even though any detailed examination of their origins lies beyond our present capacities. This will comprise the topic of the concluding sections of this chapter. In these regards, the behaviors of "hardness" and "weight" (and, as we shall soon see, "sounds an A tone" or "red") illustrate a certain paradox (or peculiarity) of classical grasp. In moments of uncertainties of usage, when we wonder in which directions "P" should be extended or corrected, we look for directive instructions that we can, in fact, follow with profit. That is, we seek answers of the form: "In situations S, the correct way to employ "P" is to follow $\mathcal{R}$," where $\mathcal{R}$ represents some classificatory or inferential instruction we are able to carry out. Sometimes we can be adequately satisfied only if this answer is provided at the lowest levels within the hierarchies of more-or-less-easy-to-follow directivities discussed in 3,vi: we need to be told that we should apply "P" if it looks red to naked eye; if it scratches feldspar; if the gauge on the duroscope reads higher than "5.4", etc. And we sometimes make allied demands upon reasoning principles: "Give me a truncated series for which I can actually obtain concrete values for 'P'." Everyone is familiar with this nitty-gritty frame of mind: it is what we adopt when someone provides us with excessively abstract travel instructions: "Your directions are too complicated—just tell me the color of the building I'm looking for."

None of this is surprising, because a language obtains practicality only if such low grade instructions can sometimes be given: they represent the baseline guidance whereby we acquire the "practical go" of a useful term. Sometimes we employ "understanding" to cite our *approval* of this lowest rung of directivity: "Ah, now *that* is finally an instruction I can truly understand" (recall how Hertz retreated in 4,iii to basement level instructions of this type to persuade skeptics that "theoretical terms" could be adequately "understood" through such syntactic instructions alone).

. . . .

Another vital way in which we utilize easy to follow directivities is evident in the following remark from Richard Feynman (he is overstating the claim that it is easy to anticipate the results of the pure mathematician, but that boastful purpose needn't concern us):

> As [the mathematicians are] *telling me the conditions of the theorem, I construct something which fits all the conditions. You know, you have a set (one ball)—disjoint (two balls). Then the balls turn colors, grow hairs, or whatever, in my head as they put more conditions on. Finally they state the theorem, which is some dumb thing about the ball which isn't true for my hairy green ball thing, so I say, "False!"*[55]

Here is a nice illustration of the way that "mental models" framed in terms of easy-to-follow classifications like *redness* can supply us with a form of "understanding" for a recondite topic.

[55] James Gleick, Surely You're Joking, Mister Feynman (Toronto: Bantam, 1986), 70.

Again this lifted device represents an important technique that we employ when we cobble together mixed techniques to obtain a more practical language.

....

For most predicates we regard these low level directive necessities as merely approximating practicalities distantly associated with "P" 's "core content." Sometimes, however, we need to scrutinize them carefully, for if we have heretofore conceived the "core contents" of "hardness" in Thomas Reid's manner, we have overshot the mark of plausible support: there is no simple attribute to which the more concrete (if often unnoticed) directivities we consult in practical usage represent "approximating practicalities." By recasting the predicate's support as that of a quasi-attribute spread across the facade described in 6,ix, we create a more sustainable picture of how physical fact and practical usage successfully intermesh. But to ascertain how this new supportive atlas should be articulated, we must first examine how various low level skills fit together: we must have first discerned, "In ascribing 'hardness' to a soft plastic, indentation with a truncated cone supplies its most suitable measure, but that test is not optimal for a metal." Speaking in the conceptual vernacular, we report, "The concept of *hardness*, properly scrutinized, must emphasize slightly different forms of response to indentation betwixt metals and soft plastics, for, to apply the term with evaluative plausibility, we must repress lip formation in the plastic as much as possible but not in the metal. A truncated cone test achieves this first objective better." In short, we "spot" far different considerations than did Thomas Reid, although he articulated an evaluation that was entirely natural to his time.

However, when we turn to the predicates that occasion confusions of the sort surveyed in Chapter 2, we are most steadfastly inclined to cite easy-to-follow directivities in speaking of the "core contents" of "red" or "expresses sadness musically." This is because it strongly strikes us, *ur*-philosophically, that our basic discriminations of hue and tonality come to us as *simple unities*, within whose innards there is scarcely room for external purpose to creep. We insist that the "core contents" of color terms (and their musical comrades) designate indivisible traits with the unanalyzable characteristics described by Moritz Schlick: that they represent properties

> which do not admit of closer definition: for we get our knowledge of them only from direct experience . . . . [I]t is impossible for me to explain to a person who has been born blind, by means of a definition in words, what I experience when I see a green surface.[56]

When Titon points inwardly to his sense of "music being in the world" in 2,v, he attempts to single out, as a means of spotting improvement, pure classificatory directivities that stand free of the unpleasant external purposes that infect our societally tinged classifications.

In many ways, such asseverations merely reflect the fact that directives couched in terms of vocabulary like "red" or "sounds an A note" represent easy-to-follow

---

[56] Moritz Schlick, Space and Time in Contemporary Physics (New York: Dover, 1963), 76.

instructions to which we commonly appeal when uncertain as to how to proceed with some problematic phrase. As such, we scarcely want to complain about how "red" or "sounds an A note" function in these important everyday capacities: we like 'em fine just as they are. However, we do not employ either predicate in that channeling manner alone, for we expect both words to also serve as good *modeling parameters*, in the sense of conveying sound information about the physical objects around us. And it is at this juncture that we confront the annoying fact of life that this book continually stresses (it was called "the lesson of applied mathematics" in Chapter 1): the more that we associate certain predicates strongly with easy-to-follow directivities, the less likely that they will be able to achieve any sort of *physical modeling* in a "P"/φ manner. On the contrary, their methods of data conveyance are apt to evolve in complex directions from a strategic point of view, in the manner of the facade arrangements we have considered or something even more baroque.

The basic reason is simple: nature's attributes are not easy to trace through their different environmental manifestations and so we commonly face a tradeoff. The more we render the question, "What is the right way to employ 'P' here?", easy to answer within a *localized* situation, the more likely it becomes that complex controls of a boundary crossing line ilk will be needed to prevent these parochial employments from interfering with one another in unhappy ways. The overall complexity of a representational task represents a *descriptive constant* that reflects the deftness with which our available linguistic resources comport with the environmental facts. We cannot evade complicated procedures if that's the only choice Nature offers us (and she commonly does at the macroscopic level). However, we can often, if we prefer, hide some of these complications in global forms of structuring that we may not *notice* so vividly. From many points of view, such an engineering tradeoff represents a sensible policy, but one that renders its beneficiaries potentially liable to the pitfalls of ur-philosophical befuddlement.

Over the hill, within the next pasture of fruitful modeling, the vegetation is different and our wandering predicative tag "P" must eventually evolve to a different species of descriptive varmint, prevented from interbreeding with the old population by the knoll between. Such was the growth-and-specialization pattern we witnessed for "hardness," but its illustrative moral applies to many predicates that prolong themselves from one patch to another with a strong insistence that basic classificatory directivities be largely preserved. Indeed, Ernest's case and $\sqrt{z}$ both taught us that usages built up largely through single-strand prolongation are apt to prove multi-valued, whereas predicates whose shaping directivities are more varied in character suffer such effects less by comparison. But, to paraphrase R. Meyers, the corrective repairs required for "hardness" or "red" seem a small, and rather economical, price for an otherwise very convenient set of data representations (their main drawback lying in the fact that they lead us too readily into philosophical confusion).

In short, a basic *paradox of classical grasp* arises: by cleaving too firmly to the local road of easy understandability, we commonly engender global usages that are quite hard to understand. When we devise a planning model, we are hoping to wend our linguistic way successfully amongst macroscopic physical objects, not seeking to follow directivities for their own sake. These considerations quickly prove *being red*'s undoing in any ambition that it can serve as a strategically simple or straightforward modeling predicate, because our biases towards accepting its easy-to-follow associations are especially strong. Nonetheless, we rightly do not abandon our attempt to convey physical content with the predicate, so we wind up with a particularly complex supportive platform in its semantic underpinnings that I'll attempt to describe, as best I'm able, in the section ahead.

The spotting property of "concept" creates great *ur*-philosophical confusion with respect to "red," because we have a strong disposition to emphasize its easy-to-follow directive value when asked. This emphasis doesn't make the rest of its controlling cloud of borderline hedges go away; we're simply not inclined to look at them. But it is precisely to this fine filagree of patchwork boundaries that we must attend if we expect to understand how "red" can function as an effective modeling parameter, albeit not one that operates in simple "P"/φ fashion.

We can therefore provide a more detailed reply to the hypothetical critic quoted at the start of this section. There are many perfectly common predicates—"weight," "hardness," "red," "force"—with which we "model the world's behaviors" *in the sense* of serving as perfectly legitimate and well-engineered carriers of physical data. These phrases, however, are *not* engineered according to the simple "P"/φ expectations that our critic tacitly assumes when she evokes "modeling." To understand how they do work, we must appreciate more warmly the nitty gritty considerations of strategy that allow us to cobble together an effective, if complex, representation of data, all of which our opponent would like to brush aside. True, like Uncle Fred, we often employ these words with no particular agenda in mind, but that airy freedom does not establish that the underlying nature of a word's characteristic conceptual personality can be rightly understood unless we dig back into the lower levels of practical directive (where surprises often lurk, as we have seen). Our interlocutor clearly imagines that more predicates achieve their "modeling" in "P"/φ fashion because she indiscriminately collapses satellite directivities onto ghost attribute cores, in the faulty manner discussed in 5,viii ("concept"'s spotting behavior greatly encourages such inclinations, of course). But such simplifications of linguistic process will only return to haunt us, in the form of *ur*-philosophical confusions that arise precisely from the strategic fine grain that we have prematurely repressed.

. . . .

A rather large contemporary literature is devoted to allegedly *response-dependent concepts*,[57] of which *redness* is one. Here the easy-to-follow directiveness of "is red" is highlighted, but without much recognition that structural compensations must be introduced to supply the predicate with effective modeling capacity as well. It should also be observed that, qua ingredient in easy-to-follow instructions, "red" et al. are tropospherically limited: we may fancy that "On Pluto, look for red stones" or "If shrunk to one millimeter size, look for the red spider" represent clear directives, but we are mistaken.

. . . .

[57] Philip Pettit, "Realism and Response-Dependence," Mind 100, 4 (1991).

Version 1:

Version 2:

(x)

Redness. After, lo!, these many pages, let us return to the tensions that occasioned the practical dilemmas of Chapter 2, which mainly concerned musical preservation, although colors were mentioned as well (with respect to our present concerns, the problems engendered are similar in underlying character). I can scarcely do justice to any of this class of very complex classificatory notions, but the simple tools we have developed (facade, prolongation, etc.) can capture some of the important underlying mechanisms responsible for the *ur*-philosophical difficulties in which we found ourselves.

Goethe writes in his Theory of Colors:

> The conclusions of men are very different according to the mode in which they approach a science or branch of knowledge; from which side, through which door they enter.[58]

Generally, these "different conclusions" won't be manifest at an early stage of the usage, when whatever tensions emerge generally get dismissed as sports or oddities not worth worrying about. It is only as the demands of improved performance press themselves upon our activities or as overtly novel phenomena are encountered, that an originally homogeneous use will begin to crystallize into clearly segregated patches of dedicated usage. In fact, if we survey the reasons offered on behalf of the thesis that musical or color properties are subjective in nature in Chapter 2, the best arguments hinge upon real divergencies in purpose that pull the relevant vocabulary towards different specializations. Recall the peculiar phenomenon of *combination* or *Tartini tones* from 2,iv, where two simultaneously sounded pitches induce, through inner ear nonlinearities, the impression of a concurrent lower note. Suppose, then, that some musical composition

---

[58] J. W. von Goethe, Theory of Colors, Chas. Eastlake, trans. (Cambridge, Mass.: MIT Press, 1970), p. lxi.

purposefully exploits the "combination tone" phenomenon, by utilizing a descending sequence of preparatory chords that strongly suggest a terminating completion on a low C note. Is it correct to describe the final sound pattern as one that "contains a low C tone" or not? Or, to convert our question to one of musical notation, which is the correct way to mark this cadence? Clearly different directivities of appropriate continuation will seem natural, depending upon the context in which the question is raised. A *performer* will likely respond,

> No, you're not supposed to play any note there; it is just a trick to create the illusion of a true tone.

because her emphasis fastens upon the question, "How should I be guided to finger this piece correctly?", whereas a *composer* might comment,

> Yes, the note in question represents a critical part of the theme, without which its coherence will be lost. It turns out, because of that odd inner ear business, that no instrumental note needs to be directly played to get the tone in question to appear.

This parting of the classificatory ways arises because the associated purposes of *reproduction* and *planning* here diverge, each pulling the predicate "sounds a low C tone" with them. As such, both responses seem natural in their context and scarcely subject to cavil. However, authors often become rather tongue-tied, in "rainbow" fashion, when they try to describe the Tartini phenomenon in language unaffected by associated purpose. Here's how Brian Cotterell and Johan Kamminga describe the matter:

> Aural illusion is probably important in our perception of the primary note of Western bells . . . What is usually identified as the primary note of a Western bell is an octave below the fifth partial, a note that cannot be picked up by resonance. The primary note of a Western bell has no physical reality; its existence is only in the listener's mind.[59]

But these authors scarcely intend to assert that a bell normally denominated as sounding a low G tone is actually ringing a C note, despite their apparent insistence that the G tone "exists only in the listener's mind." If pressed, no doubt they would reply, "Well, we merely wanted to indicate that the acoustic basis for describing a bell as 'sounding a G' is quite different than for a piano," thus retreating from a single-valued endorsement of either way of describing the bell.

An even greater curiosity affects the chimes in grandfather clocks, which, in the context of the other chimes, possess ascertainable tones, but, heard in isolation, divide listeners into two groups: those that hear it as a C and those that hear it as an F (the vibrational spectrum itself doesn't favor either interpretation clearly).[60]

Once again, we witness the secret role that overriding objectives such as *design* or *recipe* play in shaping the personalities of descriptive predicates that superficially appear entirely purpose independent.

---

[59] Brian Cotterell and Johan Kamminga, <u>Mechanics of Pre-industrial Technology</u> (Cambridge: Cambridge University Press, 1990), 273.

[60] Arthur H. Benade, <u>Fundamentals of Musical Acoustics</u> (New York: Dover, 1990), ch. 5.

Color talk is rife with blatant continuation discrepancies of this type. As Aristotle realized, the color in which a given patch of surface appears can vary dramatically according to surrounding circumstance, a behavior that is partially a function of the complicated way in which the brain assigns color tags to things. Such circumstance often leads us to evaluate the "true color" of a region according to opposing standards. Thus a weaver might respond to objections that she didn't make a patch of tapestry pure white[61]:

> *These regions in the tapestry really are white in color; it's their surroundings that make them look pink and I can't do anything about that.*

To prove her case, she might employ a "finder" or *reduction screen*: an isolating screen of gray paper that allows only the locus under contention to be seen.[62] Artists often use this device to locate the "true colors" of things so that they can be painted accurately.

On the other hand, if, employing a darkened tube, we isolate a bit of brown cloth severely from its surrounds, its complexion will take on the rather surprising aspect of a glowing orange disc (its so-called *aperture appearance*), which may induce us to comment:

> *This spot in the cloth is really brown in color; it's the lack of normal surroundings that makes it look so orange.*

Here again, which manner of evaluation seems most natural is likely to depend upon our initial aim. Individuals mainly focused upon manufacture and replication will find finder type views as most revealing of "true color," whereas those concerned with design will likely conclude that "true colors" need to be conceptualized in environmental terms—what the patches look like in the setting in which they are placed (so, in contrast to our weavers, we will declare, "surrounding a erstwhile white patch of tapestry with a strong complementary causes its color to turn pink"). Indeed, one finds occasional squabbles about whether "brown is really a dark orange" in the color literature.[63] But the fact that color talk commonly becomes multi-valued in this manner does not show that the data locally is not fully "objective," according to any reasonable construal of the term.

But multi-valuedness poses a substantial pedagogical challenge to authors of textbooks on color technology because they must train their audience to follow a narrow pathway of practical advantage devotedly, without being waylaid by divergent classifications that lie along other branches of the full platform belonging to "color." For example, in many contexts concerns of accurate color reproduction (e.g., manufacturing a dress to expectations) are paramount and improved success in these enterprises can be obtained by developing the basic directivities associated with the painter's finder in

---

[61] M. E. Chevreul's celebrated researches into color were motivated by these problems. Paul D. Sherman, Colour Vision in the Nineteenth Century (Bristol: Adam Hilger, 1981), 68.

[62] Walter Sargent, The Enjoyment and Use of Color (New York: Dover, 1964), 17.

[63] C. James Bartleson, "Brown," Color Research and Application 1,4 (1976).

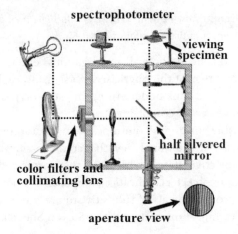

spectrophotometer

viewing specimen

half silvered mirror

color filters and collimating lens

aperature view

more sophisticated ways (through a matching of photometrically measured tri-stimulus values and all that). But to instill the right conduits of reasoning, they must notify their pupils that attempting to establish a precise "color" tag for an object in folk physics mode, without considering the illumination in which it currently appears, represents a rather vain activity—or, at least, an enterprise that does not contribute effectively to a satisfactorily printed blouse. This is why paradoxical sounding maxims commonly appear in such treatises:

> If there is one principle that needs to be kept in mind while becoming proficient in color engineering, it is that "Colorimetry does not describe what a person sees!" Colorimetry is fully enveloped by the technology of colour matching.[64]

Indeed, the author's "envelope of colorimetry" represents a fairly exact correspondent to what I call a local patch of usage. On the other hand, a manual written for the planner—a graphic designer, say—will include completely different instructions and will likely warn its audience of how misleading "darkened tube" evaluations can prove. As in our Tartini tone case, the divergent interests of *reproduction* and *design* pull color vocabulary onto different locally specialized patches.

It is, of course, possible to believe that the parties of variant purpose have become confused and declare that they are no longer pursuing the "path of true color." Oddly enough, Goethe himself falls prey to this inclination:

> The literally practical man, the manufacturer, whose attention is constantly and forcibly called to the facts which occur under his eye, who experiences benefit or detriment from the application of his convictions, to whom loss of time or money is not indifferent . . . —such a person feels the unsoundness and erroneousness of a theory much sooner than the man of letters, in whose eyes words consecrated by authority are at last equivalent to solid coin;

[64] Danny Rich, "Instruments and Methods for Colour Measurement" in Phil Green and Lindsay MacDonald, eds., Colour Engineering (Chichester: John Wiley and Sons, 2002), 20.

than the mathematician, whose formula is always infallible, even although the foundation on which it is constructed may not square with it.[65]

The claim is that Newton, in his optical investigations, enters the domain of color through the wrong door—that of the physicist who seeks to calibrate every phenomenon in numerical terms (and thus underwriting the stereotypes of "science's limited purposes" of which I have earlier complained).

But this is hardly a viable position, so our modern manuals often imitate the example of "rainbow" and endorse the traditional view that color classifications, in their primary applications, classify subjective states, which are then projected outward to sundry forms of derived physical object classification, depending upon the agent's specific concerns. This is why strong avowals of subjectivism are often offered by quite sophisticated writers in the technical manuals, such as Steven Shevell:

There is no red in 700 nm light, just as there is no pain in the hooves of a kicking horse.[66]

However, if we accept such subjectification, we are likely to wonder how our "derived color concepts" manage to get restored to their former physical habitats and can be easily led to the hypothesis of Hippolyte Taine:

All our sensations of color are thus projected out of our body, and clothe more or less distant objects, furniture, walls, houses, trees, the sky, and the rest. This is why, when we afterwards reflect on them, we cease to attribute them to ourselves; they are alienated and detached from us, so far as to appear different from us . . . In fact, as far as we are concerned, this operation is but a means: we pay no attention to it; the color and the object denoted are what alone interest us. Consequently, we forget or omit to observe the intermediate steps by which we localize our sensation; they are to us as though they did not exist; and we thereupon consider that we directly perceive the color and colored object as situated at a certain distance off.[67]

In effect, he proposes that our brains must engage in a task very much like the heinous project of "colorizing" black and white movies in garish hues, a tale of data processing that is surely erroneous.

By the mid-twentieth century it is generally realized that Taine's picture must be wrong, but most authors in the professional color field seem at a loss as to how the misapprehension should be rectified. Thus the contemporaneous color expert W. D. Wright first writes as if carrots obtain their orange coloring through "projection" from private appearance:

More remarkable still, we are not aware introspectively of the image being within us, unlike many other of our sensations and experiences; instead, by some feat of mental projection, we see the carrot out in space in its proper place on the bench.[68]

[65] Goethe, Colors, p. lxi.

[66] Steven Shevell, "Color Appearance" in Steven Shevell, ed., The Science of Color (Amsterdam: Elsevier, 2003), 150.

[67] Nicholas Pastore, Selective History of Theories of Visual Perception 1650–1950 (New York: Oxford University Press, 1971), p 182–3.   [68] W. D. Wright, The Rays are not Colored (London: Adam Hilger, 1968), 23.

But he then realizes that no special "feat of mental projection" is likely to be involved, but that whatever occurs with colors must arise in "an identity" with shape and location as well. He continues:

> Now since we use the same retinal receptors and the same visual cortex for both our space and color perception, there is clearly the opportunity for . . . an identity to be established between the carrot and its color . . . And so it is that the color of the carrot is carried out into space in intimate association with the three dimensional image of the carrot and acquires as real and objective existence as the carrot itself.

Of course, this is possibly even worse, because it reads as if the carrot itself has been projected outward by the mind as well, whatever that might mean. Wright feels sufficiently uncomfortable with his own words that he comments:

> [I once wrote that] color . . . .[exists] only when the information is finally interpreted in the consciousness of the spectator. . . . I want to look beyond this concept.[69]

I find his revised account equally confusing, but L. A Jones expresses a related view more crisply:

> Only a mechanistic view, however, would confine consciousness and color within the body of the observer. It seems more useful, and in accordance with human experience, to adopt the view that consciousness includes all the aspects of the observer's environment. Color, in particular, is most usefully considered to be a characteristic of our environment as we see it rather than an inaccessible, incommunicable sensation within us. Color is not, however, a characteristic of objects, but of the light which enters our eyes from them . . . [S]ince [our] manner of speaking conveys information about the environment, it seems well to attribute color to the agency of the environment that is most directly involved.[70]

I am not wholly sure how to interpret Jones' "consciousness includes all the aspects of the observer's environment" except in some neo-idealist manner, but let's set that odd comment aside. As I otherwise understand him, Jones claims something like this: there is a patchwork of interconnected color concepts that categorize physical objects, private sensations, light patterns, etc. Perhaps any of several notions within this circle might be designated as primary, but the generating choice that best accords with our conviction that "color is objective," but which runs the least risk of confusing the student of color reproduction, is the classifier of electromagnetic wave pattern.

Still left in place, insofar as I can determine, is the traditional assumption that, underneath it all, subjective invariants of color classification exist and support the sundry forms of physical and optical taxonomy. But this venerable thesis paints a portrait of probable brain process just as misleading (albeit more subtly) as Taine's tint displacement. Recall, from earlier discussion, that it is unlikely that any uniform sensation of *hardness* exists and that our actual applications of the "is hard" tag tacitly

[69] Ibid., 20.
[70] L. A. Jones, "The Concept of Color" in Committee on Colorimetry of the Optical Society of America, eds., The Science of Color (New York: Thomas Crowell, 1953), 45–6.

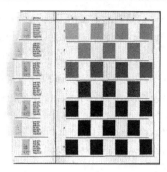

Color card

embody delicacies of handling (scratching versus squeezing, etc.) that we are unlikely to notice, if our attention has not been directed to the fact. Something very similar holds, it turns out, with respect to "sensations of determinate red." It is an evident fact of ordinary life that we can remember the colors of familiar objects in our household with remarkable accuracy, being able to make discriminations that number into the high thousands. But how do we manage to achieve these feats? Unless forewarned, we are almost certain to endorse the following picture: after sufficient direct inspection, we retain, as Humean memory "impressions," a rich pallette of private hues, which we then consult when we want to recall or even reconstruct the original object. However, the brain probably does not handle the requisite data in that way at all.

Consider this remarkable passage from the great color expert R. M. Evans:

> An observer is always aware of the illumination although seldom consciously. He has simply trained himself to take such action, automatically, as will maximize his perception of the object properties. We see this in everyday commonplace actions. If a glossy surface reflects light back into his eyes, he will move his head, change his position, or, if possible, pick up and hold the object until the reflection disappears . . . In other words, he deliberately manipulates the illumination to see the object color; these are two separate things to him but he is usually interested in only the color . . . .There is an amazing discrepancy in the ability of any person to discriminate between two juxtaposed colors and his ability to pick a color from an array that will match one he has just seen separately. Even after fairly short delays his ability is quite poor and after considerable delay it can only be described as bad unless he is trained . . . Even for fairly good observers, positive identification of a color without a comparison is of the order of thirty colors, and for the naive it may be more nearly eight to ten . . . . In everyday life the colors of objects are not stable and there is no point in trying to assign an exact color to an object; accordingly people do not attempt to train themselves in this respect . . . This does not change either his belief, or, in fact, his direct perception that the color he sees is a property of the object.[71]

---

[71] R. M. Evans, The Perception of Color (New York: John Wiley and Sons, 1974), 199.

In other words, under fixed light in side-by-side comparison, we can sort samples into a very large number of equivalence classes or, to put the matter another way, we are able to look at a fat book of colored chips such as artists employ and competently distinguish all of them. We should regard this capacity as a notable skill we possess with respect to a quite particularized species of task.

On the other hand, our ability to remember the shades we have seen, in absolute terms, proves to be astonishingly poor and hence cannot be the path whereby we recall household colors. How do we do it then? As Evans indicates, usually we subject our specimens to sundry patterns of manipulation, generally parochial to the stuff in question. With a metallic vase, say, we typically rotate the object to study its spectral appearance at different angles (or, to employ the technical term, survey its gonio-chromatic aspects). We usually fancy that we are merely "attempting to see its true color through the sheen," but, in fact, it is the schedule of appearance change under manipulation that we actually store away. This is why we usually think of "gold" as "not exactly a yellow," even if, from a particular angle a brass vase might cast exactly the same momentary appearance as a ceramic specimen we will call "yellow." Such "embedded in manipulation" techniques can allow us to reconstruct some cherished room from childhood with excellent chromatic accuracy, even though the absolute color tags we utilize in the process are quite small in number.

In sum, "object color" (that is, the redness of a ceramic vase) denominates a rather abstract constellation of capacities that involve a range of specific and comparative "color appearances" (localized judgments of the sort "after a rotation from upright to 45°, the lip of the vase looks yellower"), usually mixed with some measure of non-colorific consideration such as the crispness of images reflected from its surface (such non-spectral attributes loom so large in the "equivalence class" classifications of many cultures that their languages contain few words for incontrovertible hue classification at all). Let us say that objects which behave alike under this set of tests belong to the same *color class*. Our usual evaluations of an object's "color" reflect these determinations of color class fairly closely. As such, it is not easy to assign these behavioral classes a systematic set of finely discriminated identificatory predicates and we don't really try except at a fairly coarse level. To be sure, we will consider a painted wall as exactly "lime green" if it matches the card so marked from the paint store under a range of common illumination conditions, forgetting that a metallic vase might fully satisfy that same matching condition yet be considered of a manifestly distinct color (because its overall color class is so different). But as long as we stay within the ambit of household walls, we may be able to sort them out using a very refined system of color predicates based upon a chart from the paint store because such walls mainly stay fixed, undergo compara-tively modest variation in their ambient illumination and display less chromatic vari-ation upon manipulation than a metallic vase. But if we move to a larger selection of objects—taking all domestic items in sum, say—, then there is no uniform schedule available for aligning their intuitive color classes in one-to-one correspondence with chips from the paint store (or any richer color atlas of the same ilk). This, I believe, is what Evans intends when he writes, "In everyday life the colors of objects are not stable and there is no point in trying to assign an exact color to an object." It is also evident that our color technologists-to-be will be greatly hampered in their professional enterprises if they continue to think of "colors" exclusively in color class terms, so it is not surprising that they must be notified that the "color" terminologies employed in manufacturing context do not anoint objects with satisfactory color class tags (vide the quotations from technical manuals already cited).

The philosopher C. L. Hardin argues that Evans-like considerations indicate that physical objects, properly speaking, do not possess "colors" worthy of the name:

> Common ascriptions of colors to objects are, for common purposes, correct enough, but there are, in the nature of the case, no such ascriptions that are both precise and correct. We are not, in the final analysis, entitled to say that, as a matter of fact, a determinately green physical object is determinately—for instance, uniquely—green. . . . It is not as if there were no plausible alternative to all these Ptolemaic epicycles [that crediting physical objects with determinate colors entails]. There is, and it is simply this: render unto matter what is matter's. Physical objects seem colored, but they need not be colored.[72]

However, this is not the right way to look at these matters. It is fairly easy to assign quite systematic color tags to colored paint chips if they are compared against one another within fixed and narrow conditions of illumination (a light box, say). Because we generally picture how color vocabulary functions in a naïve way, we automatically assume that the labels established within this specialized environment can continue to serve as suitable designates for the color classes of these paint chips wherever else they wander in life. But that assumption is wrong and we will soon become bewildered by the fact that chips indistinguishable in the original setting behave quite differently in their "out of the box" behaviors (i.e., if they are so-called metamers). Nor will we have properly anticipated the difficulties we will confront in bringing our metallic vase into happy congruence with any of our chips. But these should be regarded as problems equipollent to the extendability of a local coordinate system: the fact that longitudes and latitudes that work well near the equator break down when we near the poles. We don't conclude that the earth doesn't exist simply because we can't prolong coordinates over its entire surface in the simple manner we anticipate! Ur-philosophical expectation robustly anticipates that the color classes of objects can be easily named but this presumption stems from mistaken assumptions about the nature of the kind of information conveyed by a color class label: the data provided is quite different in informational character than that gleaned from isolated light box viewings. Setting an object within a specific color class embodies a much richer set of comparisons which do not submit to simple, systematic labels except in fairly coarse terms. To be sure, we have limited capacities to improve our color class registrations by several means. First, as in the case of our painted walls, we can sometimes project our delicately varied color chip tags usefully onto the color classes of a small range of well-behaved materials that do not shift dramatically in illumination conditions. Secondly, by attending explicitly

---

[72] C. L. Hardin, Color for Philosophers (Indianapolis: Hackett, 1988), 90–1.

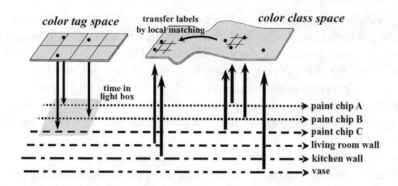

to a range of the material's goniochromatic behaviors, useful systems of denomination for "color class" can be laid down for localized classes of material (this is apparently common practice for automobile enamels[73]) by utilizing lengthy n-tuples of color tag comparisons (e.g., <lime green from 0° forest green from 45° . . . >). But labels of the latter type are considerably more complex in their degrees of freedom than we anticipate, because in undemanding ordinary life we are content to describe our cars as "sort of lime green."

The accompanying sketch illustrates the facade-like structure I've just described (although the diagram may be harder to follow than the prose). On our base domain lie the life histories of various materials we hope to classify with color tags of a "lime green" type. Here I've marked as sample objects several paint chips, a vase and a few household walls. Every object history will be mapped into the color class patch and assigned some coarse color classification: "basically red," "sort of lime green," say (for convenience, I tacitly assume that our sample objects do not change their color class behaviors over the interval we investigate—viz., none of our paints fade). I've drawn the color class space as bumpy, to symbolize that it does not take coordinate systems (= systematic assignments of refined predicate tags) easily, although coarse neighborhoods of its surface are designated as "sort of lime green," etc. During a period of their career, our paint chips hang out within a light box and, while there, a color technician correlates their localized behaviors with systematic predicate tags following standard procedures for assigned colorific labels (there are many possible techniques here; I'm imagining a match to the samples supplied in a Munsell color atlas where we obtain hue-value-chroma labels such as "5R 5/10"). Unlike our coarse color class patch tags (which classify *objects*), this second patch categorizes *behaviors* and only with respect to paint chips inside light boxes. However, the detailed predicative designations set up in the color tag sheet can be transferred over to a smallish region of the color class patch in two steps. (1) Push the tag coordinates over to the paint chips considered as enduring objects, as symbolized by the grid in the middle of color class' warped surface. (2) Transfer this paint chip grid over to the region where the walls are registered by comparing chips and walls under designated

[73] Roy S. Berns, <u>Billmeyer and Saltzman's Principles of Color Technology</u> (New York: John Wiley and Sons, 2000), 13.

lighting conditions (e.g. "Do chip and wall match at noon?"). The results of this second coordinate transfer are marked as the little grid on the left side of color class patch, a feat that allows us to assign our walls with a more detailed set of color predicates. But these prolonged predicate labels fail to cover our vase nor can they be prolonged to do so. By varying the comparisons utilized under (2), our paint chip coordinates can be intelligibly transferred to other small groupings of objects under the color class sheet (printed cloths, say), but these labelings are likely to prove incongruent with those laid down for walls.

If this schematic story is roughly correct, then we have explicated some of the mechanism that leads to misunderstandings of the type surveyed in Chapter 2. In the everyday vernacular, we classify physical objects with tolerable effectiveness under coarse, wide extent "neighborhood reach" predicates. However, we convince ourselves *ur*-philosophically—say, by holding a tomato before us like poor Yorick's skull—that we should be able to set up a more detailed collection of labels that can more nearly capture the exact simple and unified "color" we believe we grasp in the vegetable. We decide that it can only be the coarseness of everyday language that prevents us from refining our neighborhood groupings into finer discriminations; we believe that, in theory, the ambit of a "color predicate" ought to be shrinkable in a manner that covers exact chromatic points. We then find that, within light box settings, precise tags of the sort expected can be locally assigned and we immediately assume that these refined color labels should be able to classify all materials happily. But we quickly generate severe forms of multi-valuedness because our transferred label applications do not cohere in the unified fashion that we a priori anticipate. And so we become puzzled. As an easy escape from our difficulties, the subjectivist recommends that it is primarily a mediating sensory appearance that is properly classified as color bearing, not physical objects (although we believed that we were interested in the traits of vases and walls when we began our musings). Accordingly, motivated by little more than errant predicative behavior, the familiar veil of perception drops over our proceedings, wherein we fancy that we peer at wall and vase through a screen of intervening "presentations."

But this isn't the right diagnosis. It is rather that the packets of information conveyed in a color class predication are both richer and of a different order than those supplied in localized behavioral comparisons (of a light box ilk), but we are predisposed to regard the two forms of informational registration as "being of the same type." Normal English usage adds to the confusion because, in our everyday labelings of color behavior, we employ specializations of the same vocabulary that we also utilize for color class purposes. That is, in ordinary language, the Munsell "5R 5/10" will be described as "a medium brick red." And so we falsely anticipate that it should be possible to label physical objects—and not specialized behaviors— as a "medium brick red" or not. If matters weren't already confusing enough, the fact that the more refined behavioral tags sometimes *do* accept limited prolongations onto enduring objects (as occurs with respect to living room walls) makes our puzzlement even deeper. In truth, such weirdness is simply the natural consequence of the odd ways in which language naturally develops: manifest practical advantage leads refined labels such as "medium brick red" to crystallize by localized specialization from what was formerly a small field of coarse labels mainly used to report upon color class discriminations. As these usages further develop, their incongruities become manifest and need to be resolved through some form of boundary line quarantine. Lacking our alternative vocabulary of "patch" and "facade," the color books tend to address these confusions through implausible favoritism: "Color predicates really label sensations," etc. But this coarse resolution has problems of its own, for it suggests misapprehensions of the *segmentation type* I shall discuss in a moment.

From my alternative point of view, both subclasses of everyday color classification embody a good deal of indubitably objective information about their target objects and behaviors, but of quite different types depending upon the patch locally covered. The sundry location worries surveyed in Chapter 2 all start with assumptions of the type: "Our familiar concept of *redness* represents a quantity of simple and constant content. But what kind of object can directly instantiate the qualities it demands?" In so thinking, we have projected a complicated form of predicative behavior onto a ghost attribute and now wonder what halls it haunts. If we favor the subjectivist answer, our perceptions become clouded with a constant veil of intervening presentations, for our phantom attributes require some diaphanous gauze to serve as a movie screen upon which they can be projected.

To be sure, there is nothing wrong in using "red" to express "I am seeing red flashes right now," for that represents a prolongation of usage equally natural to the two we have considered. On the other hand, we shouldn't automatically presume that we know whether the more specialized "I am seeing medium brick red flashes right now" represents a viable usage or not: the situation seems to me much like that pertaining to "I see a rainbow edgewise on."

In other words, seemingly innocuous classical assumptions to the effect that "*redness* represents a simple and indivisible classificatory concept" can easily embody quite misleading implications with respect to the way that our brains process information about the world before us. In particular, the constant content picture suggests that the brain first classifies the objects we see under very detailed chromatic discriminations—in a more refined way than even *being medium brick red*—but then stupidly forgets these results shortly thereafter, remembering only the more expansive category *bright red*. But this story is misleading; the ways in which we actually remember objects is far more entangled with other measures of behavioral performance than our simple picture of chromatic amnesia suggests.

Here's an allied (and quite well-known) example of how our *ur*-philosophical inclinations trick us into presuming that our brains handle data in a manner different than they really do. We view a mosaic of colored dots that form into elephantine configuration. We naturally assume that the brain first establishes the hues found in each local dot and then proceeds to identify, upon that basis, the whole that they compose. But no: our visual system generally resolves overall illumination questions first, before it definitively interprets any local patch as determinately "pink" or "white." This unexpectedly backward computational ordering can be witnessed in vivid operation anytime we recognize a region in a movie as brown. Two regions on the screen may readily reflect light of identical composition, but one appears a reasonably bright orange while its companion looks brown. Why? Because the brain first frames a hypothesis with respect to the prevailing illumination available in the scene depicted before it tags any local patch as a "brown" or "orange."

Engineers who have attempted to construct mechanical visual systems able to recognize familiar objects have likewise learned that this unexpected "broad scale recognition first, before details are definitively tagged" strategy works best for many forms of classificatory task. Consider this warning, drawn from a book on machine recognition. By "segmentation," the author (E. R. Davies) means the problem of discriminating data belonging to an object in the image from that of adjacent objects or its background. He comments that, intuitively, we expect this to be easy:

> When objects are large and do not possess very much surface detail, segmentation is often imagined as splitting the image into a number of regions each having a high level of uniformity in some parameter such as brightness, hue or texture. Hence it should be straightforward to separate objects from one another and from their background, and also to discern the different facets of solid objects such as cubes. Unfortunately, th[is] concept of segmentation . . . is an idealization which is sometimes reasonably accurate but more often in the real world it is an invention of the human mind, generalized inaccurately from certain simple cases. The problem arises because of the ability of the eye to understand real scenes at a glance, and hence to segment and perceive objects within images in the form they are known to have. Introspection is not a good way of devising vision algorithms, and it must not be overlooked that segmentation is actually one of the central and most difficult practical problems of machine vision.[74]

That is, we imagine that the coffee cup can be readily segregated from its surroundings because the cup is white and the background dark, when, in truth, the distributions of light on both cup and background are apt to display a wide range of gray tones of very similar intensity. Through very complicated calculations the eye and brain first identify portions of the display as "cup" and "background," and, on this basis, subsequently assign the cup the local color tag "white" and the background "dark." Our intuitive expectations get the order of informational dependencies that our brain follows in inversion.

With respect to our color class case, it is precisely the *ur*-philosophical inclinations that stand at the core of classical thinking that occasion these misconceptions: the assumption that I can be totally sure of the contents I grasp when I fully understand a predicate like "is red," for this innocent-looking thesis falsely assures us that our color vocabulary can be refined in impossible directions (classical thinking forever offers guarantees it shouldn't).

Faulty semantic pictures generally encourage misguided projects and, in this light, consider the pedagogical reforms suggested by A. H. Munsell, the early twentieth

[74] E. R. Davies, Machine Vision: Theory, Algorithms, Practicalities (San Diego, Calif.: Academic Press, 1997), 79.

century deviser of a well-known color tag space still in common use today (it was cited above). Patently guided by the *ur*-philosophical picture of color language improvement sketched above, Munsell proposed a grueling program whereby children would learn an "orderly and scientific" color vocabulary to describe the physical objects around them through diligent study of his atlas:

> *Clear mental images make clear speech. Vague thoughts find vague utterance. The child gathers flowers, hoards colored beads, chases butterflies, and begs for the gaudiest painted toys . . . [At a certain stage in his development] if he wants to describe a particular red,— such as that of his faded cap,—he is not content to merely call it red, since he is aware of other red objects which are very unlike it. So he gropes for means to define this particular red; and having no standard of comparison,—no scale by which to estimate,—he hesitatingly says it is a "sort of dull red." Thus early is he cramped by the poverty of color language. He has never been given an appropriate word for this color quality, and has to borrow one . . . which belongs to edge tools rather than to colors.*[75]

Pity the wretched child expected to recall absolute tints from the Munsell charts and continually failing! The affinities to the enforced discipline that our moralist would inflict upon poor old Darwin are quite tangible: "You must be hearing every individual note of the Debussy properly, but your soul has so far failed to react to them properly. So let's listen to the piece one more time." But this expectation (and chastisement) is undoubtedly based upon a false picture of how the brain recognizes musical intervals, comparable to Munsell's mistake or that of the fellow who is certain that we recognize the pink spot before the elephant.

"So does *being red* represent an objective property or not?" The first observation we should make in this regard is that the predicate "is red" spreads itself over a rather complicated atlas of naturally connected sheets and locally corresponds to quite different forms of evaluations, to the degree that its target objects are not even of the same type (behaviors under illumination lie below the color tag sheet; material objects under the color class patch). Considered as a whole—and since its usages continue into one another naturally, it should—*being red* can't qualify as a true attribute at all, but more nearly corresponds to an informational package of quasi-attribute type. But, from this point of view, it manages to encode physical information quite nicely, albeit in a shifty and multi-valued way. True, the ways in which its parcels of usage piece together very much have the signature of human capacity written all over them, but that fact alone doesn't mean that the data entered upon those sheets has become thereby corrupted.

On the other hand, framing an accurate picture of the strategies whereby such a language operates is quite difficult (the story told here for color words is undoubtedly grossly oversimplified), and, as long as such issues remain wrongly conceived, we have not fully understood what our language is telling us. But these are not veils we cannot penetrate; the enterprise merely demands hard diagnostic work.

---

[75] A. H. Munsell, A Color Notation (Boston: Munsell Color Company, 1919), 11.

(xi)

Naturally evolved linguistic systems. In our opening chapter, J. L. Austin's warnings about the dangers of *Gleichshaltung* within our *ur*-philosophical musings were cited with approval (1,v). The target of his original remarks is the old-fashioned doctrine of *sense data*: in its most traditional form, the thesis that what we concretely perceive at any given moment represents a two + dimensional mosaic of colored patches, through which all of our visual information about the physical world filters. When we see a hippopotamus, we properly detect an array of hues (the sense data) that suggest *hippopotamus* to us. This contention, despite the fact that for centuries it was regarded as nearly an epistemological platitude, is accepted by few academic philosophers today. After our discussion of color vocabulary, it should be evident that many parallels exist between this creed and the classical picture of concepts and that my own account of what goes wrong with the latter has important affinities with Austin's complaints with respect to sense data.

In fact, the two doctrines represent *natural duals* of one another, in the sense that both assure us that the contents of our thinking manifest themselves to us through narrow and controllable avenues of *mental presentation*. Whatever evidence I have for the existence of that particular hippopotamus must come to me through the medium of all those colored blotches (and allied smells and feels); likewise, whatever conceptual contents control my "is a hippopotamus" inferences and categorizations must present their directivities to me in the form of instructive packages to which I can rationally react. In short, sense data represent the intermediate tableaux that mediates our perceptual recognitions, while concepts provide the intervening presentations that monitor our inferences and provide the templates for our classifications. When, in either form of activity, we go astray and reach conclusions that require reappraisal, traditional philosophic thought assures us that we can retreat, at least in theory, to our baseline presentations to locate the lacunae that have occasioned our mistakes. Such a picture of epistemological happenstance and subsequent repair suggests that the philosopher should serve as the community expert especially skilled in tracing conceptual mishap— the fix-it man who, with trained eagle eye, can spot the critical spots where our attention flags and we lose our ways conceptually. In short order, we conjure up a mission for philosophy exactly like that Russell favors in all his work.

Accordingly, if something is wrong with sense data thinking, something is surely amiss with the classical picture of concepts and any successful critique of the first must somehow deal with the latter as well. But rarely do we find the two doctrines philosophically joined in this manner (I see no recognition of the affinity in Austin, for example). This absence of linkage is surprising to me because the first refuge of the advocate of sense data, when challenged, is to flee to the sheltering arms of classical concepts (that is, abandon any claim to perceive articulated perceptual objects in favor of holistic predicative replacements: "It seems to me as if something with hippopotamus-like traits stands before me"). Any assault upon sense data thinking is likely to remain unconvincing if we refuse to hunt our rabbit after it jumps into the briar patch of the realm of universals. But *surely* we know what's wrong with old-fashioned sense data doctrine, for virtually no one (except freshman philosophy students) believes in it anymore? Can't we learn from that established wisdom what is wrong with the classical concept parallels? Unfortunately, the exact reasons why sense data are so widely dismissed today remain murky to me, for the explanations often advanced seem more dubious on their face than does sense data doctrine itself. For example, many writers embrace the neo-Wittgensteinian thesis that psychological attributions stand in need of public criteria, but this criticism is hard to embrace if one doubts that firm bonds can be forged that link "criteria" to *any* known linguistic practice. Other schools contend that the assumption of objects privately conceived can be ruled out through consideration of the social character of language or by pondering the skeptical paradoxes that their acceptance engenders, but, as I have several times complained, I find none of these veins of complaint at all compelling. In my opinion, the most telling objections to traditional sense data theory (in its mosaic of colored patches guise) lie in the recognition that its doctrines encourage certifiable misconceptions with respect to *human capacity and routine*, in the manner of our discussion of color vision (any <u>a priori</u> insistence upon an intermediate stage of mosaic tableau forces "pink spot adjudicated before elephant" misapprehensions upon our understanding of the sequences of perceptual recognition). From this point of view, talk of sense data <u>per se</u> isn't inherently objectionable—to jocularly declare that "I am seeing pink elephant shaped sense data" seems innocent enough—, but traditional doctrine should be faulted largely in its *operational aspects*: it leads us to incorrect appraisals of the perceptual tasks that can, and cannot, be accomplished.

Here is a somewhat ephemeral task that proves plainly unworkable, although sense data thinking assures us otherwise. Consider the familiar notion of a *visual field*: the immediate array of shape and shade that we believe we experience at any given moment. Except in dim light, it is our general impression that our visual fields are "everywhere colored" and most of us assume, unless we have been forewarned, that we should be able to determine what their individual shades might be (indeed, such assumptions frame the *ur*-philosophical basis for developed sense data doctrine). In fact, if an object of unknown color is introduced into the periphery of our span of visual awareness, we often cannot determine its hue (or much of its shape, for that matter). This incapacity seems the result of the fact that most of the cones responsible for color vision lie concentrated near the focal center of the retina. As Richard Feynman explains the matter:

> *Another interesting effect of the fact that the number of cones decreases as we go farther to the side of the field of view is that even in a bright light color disappears as the object goes far to one side. The way to test that is to look in some particular fixed direction, let a friend walk in from one side with colored cards, and try to decide what color they are before they're right in front of you. One finds that he can see the cards are there long before he can determine the color.*[76]

[76] R. Feynman, R. B. Leighton, Matthew Sands, <u>The Feynman Lectures on Physics</u>, i (Reading: Addison-Wesley, 1963), chap. 35–3. Apparently, Feynman somewhat exaggerates, for some degree of chromatic peripheral vision is available, but

Because we strongly feel, as virtually a phenomenological given, that our visual field is "everywhere colored," we naturally presume that localized color patches must also be present in the outlying portions of our vision as well. Indeed, this conclusion seems to follow by the same logic that guarantees that if a board is colored purple everywhere, it will be purple in every extremal inch of its bounding edges. Clearly, one of the problems with traditional sense data thinking, as we find it in Locke or Hume, is that it encourages such "visual field like a colored board" beliefs.

But what is the potential practical harm in that? In an adventure story we might read:

> Although in the corner of his field of vision Zack espied the familiar yellow of Speedy flying to his rescue in their old Velociprator, he moved neither his eyes nor his head lest his captors look in that direction as well.

Plainly, its author labors under the misconception that Zack's visual field entertains properties similar to those of a painted placard or vantage point (in the sense of those "photographic opportunities" sometimes encountered in tourist regions). Indeed, however sophisticated in the contrivances of visual perception we are, most of us are amicably disposed to read such a passage through without wondering about its realizability (any more than our appreciation of The Incredible Shrinking Man is dampened by brooding as to how the hero manages to *see* those spiders at that size scale). In fact, we might be fairly accused of not understanding the concept of a "field of vision" if we couldn't accept such storybook passages without cavil. Such is the latent DNA (7,viii) connected with "visual field" that we allow to pass without cavil. Nonetheless, we will find ourselves in serious trouble if we venture amongst cannibals with an intent to copy the clever plan obtained from the Zack and Speedy stories: no rescue operation ought to be predicated upon a capacity to spot colors out of the corner of our eyes. Of course, this proposed eventuality is quite unlikely, for such escape schemes are likely to remain the stuff of fiction merely. However, it is plain that our Zack and Speedy naïveties are part and parcel of the previous section's segmentation misapprehensions: the faulty *ur*-philosophical assumptions about processing order that create great hurdles for the engineer attempting to duplicate human recognitional capacities. Here it is evident that sense data prejudices occasion real life difficulties. Likewise, faulty (and possibly cruel) educational efforts such as Munsell's also stem from variants upon the colored mosaic picture. It is well known that allied expectations lead to erroneous expectations with respect to artistic design: traditional doctrine suggests that the visual field through one eye should be easily and convincingly sketched. As the attached etching from Ernst Mach's The Analysis of Sensation[77] attests, rather queer products emerge upon their assay (it is striking that Mach himself seemed untroubled by the palpable oddity of what he presents as accurate depiction). In cases like Georges Seurat's allied experiments in design realization, the results can prove quite glorious, but it can hardly be alleged that their outputs conform to the plotting that engenders them.

not enough to allow Zack to carry out his scheme (his recognition of pattern will be quite degraded as well). Thanks to Bob Schwartz and David Hilbert for help on this point.

[77] Ernst Mach, The Analysis of Sensation (New York: Dover, n.d.).

It was fully realized by the late nineteenth century that simple sense data atomism of a Humean or Cartesian ilk is not viable for essentially reasons such as these, and many philosophers and psychologists subsequently embraced forms of phenomenalism that refrain from advancing virtually any anticipatory claims at all. For example, the self-styled American percept theorists of the 1930s[78] regarded as a *percept* the entire impression of scene glimpsed in the specious present that they would describe in terms such as "X enjoys a percept of seeming to have an hippopotamus before him." Here the wholesale transfer of doctrine into the dualized realm of universals becomes quite plain, for the percept that X experiences is simply that of recognizing that the universal *being a hippopotamus in front of me* is supplied by the current scene as a *mode of presentation* in the sense of 4,iii.[79] In my view, such doctrines become too cautious: they no longer advance any opinions with respect to the concrete tasks that should prove viable upon the basis of limited perceptions of scene. Thus our percept theorists become leery of addressing the most coarse-grained query, "If I see a large banana suspended immediately before me in the manner of MacBeth's dagger *and* my visual field also seems colored everywhere, will I be able to assign a hue to my banana?" "We'll have to leave that question to the empirical psychologists," our percept theorists reply, "because nothing much follows from the logic of 'it seems to me' statements." Such retreats often appear in philosophy: a patch of doctrinal homeland is valiantly defended through shrinking its protected boundaries to infinitesimal compass.

---

[78] Roderick Firth, "The Percept Theory" in R. J. Swartz, <u>Perceiving, Sensing and Knowing</u> (New York: Doubleday, 1965).

[79] Current advocates of crisply defined senses, it seems to me, often teeter on the cusp of accepting virtually everything that an old-fashioned percept theorist would have wanted.

In my opinion, the worse feature of these ghosts of departed doctrine is not that they are *wrong*—how could they be, really?—, but that they are *distracting*. Old-fashioned sense data theory proposes ascertainably wrong answers as to, <u>e.g.</u>, how a colored fabric might be profitably designed based upon component threads of known shades, but at least it provides a response to important and difficult questions, with which, <u>inter alia</u>, modern technologists in color engineering continue to struggle.

When we turn to the conceptual side of the "presented content" ledger, our *ur-philosophical* misapprehensions are not wholly of a mistaken-picture-of-brain-processing ilk, but trace in equal or greater measure to an undernourished appreciation of the *varieties of descriptive strategy*: the forms of linguistic support that can be used with profit in the face of an environment of complicated, macroscopic stuff. Indeed, it can be foolish to plunge prematurely into the nitty gritty details of how the brain implements a given computational policy, if some completely different pattern of informational processing is, in fact, at issue. Indeed, our main focus in this book is precisely one of expanding our set of strategic options—this has been my intention in highlighting the importance of the predicate/world interface.

The fact that "concept" and allied words serve as our chief everyday vehicles for discussing emerging problems with respect to linguistic management also makes narrow questions of psychological process somewhat subservient, in an overall balancing of relevant concerns, to the strategic factors I emphasize. I will return to this theme again in 10,iii after we gather more data on the problems commonly confronted while reshaping an existing language to achieve better task performance.

In dealing with the classical picture of concepts, my approach is exactly the same as that recommended for the sense data picture: trace the faulty operational assumptions to which the traditional portrait gives credence. In many ways, our critical task is easier here than for sense data, because misapprehensions about conceptual contents adversely affect our plotting in far more departments than do errors in perceptual doctrine and the classical picture doesn't enjoy a further dualized realm in which it can retreat. Instead, it can protectively rely only upon the policies of conceptual unloading surveyed in 3,viii, and it is hard to provide stable answers in this department.

Returning to our comparison with J. L. Austin, he is sometimes interpreted as defending a so-called *direct realism* which, in this context, represents the opinion that "is an elephant," in all its employments, constantly designates a property pertaining to physical objects. When I make a claim that appears otherwise—"I am seeing pink elephants"—, direct realism informs me that I am still asserting something of relevance to pachyderms, albeit of a shaded variety: "Elephants are such that, if any of them were pink, they would occasion reports such as I am now inclined to supply." Although philosophers have seriously entertained this implausible position, I do not read Austin this way (although he is not optimally clear on this score). I instead see his argumentation as tending towards the claim, "Our use of 'elephant' fits into a *natural linguistic system* where, from the mere fact that only a sensory appearance is at issue in the context of a correct 'pink elephant' report, we shouldn't conclude that similar intervening appearances remain in play as invariants behind our more canonical 'elephant' reports."

So articulated, the position resembles my own insistence that atlas structures can support predicative usage in a manner that does not rest everywhere upon constant supports and that any assumption that some common "attribute of hardness" or "attribute of pinkness" underpins them all constitutes a mistaken invocation of a mythical ghost quantity. Most of the time our two predicates are supported by locally salient physical information about material objects, although they each possesses natural prolongations that convey quite different sorts of data (including that of a largely psychological significance in the case of "I see pink dots before my eyes").

On this reading, the main divergence between Austin's thinking and my own traces to the question: how does the "natural linguistic system" get established? His answer (with which he sometimes expresses discomfort) is: it comprises the barrage of worthy rules that we learn from our elders in the course of becoming competent in the usage of ordinary English words like "elephant" and "red." As such, this thesis represents the Achilles heel that occasioned the downfall of the popularity of Austin's program, for it is virtually impossible to find proof of such "system" in the workings of ordinary grammatical consideration. Quite the contrary, language's syntactic discriminations positively encourage metaphysical excess, if anything. Worse yet, if we consult a real life "plain man" (whom Austin generally evokes as the hypothetical repository of wisdom about proper "system" usage) on our questions, we are likely to encounter a party eager to embrace the most fantastic opinions warmly: "Yeah, there really aren't any elephants; everything is just an idea in the mind" (every instructor in "Introduction to Philosophy" is all too familiar with these forms of "plain man" response). Austinian reliance upon the wisdom inherent in language use must be in error, because that is precisely where the streams of *ur*-philosophical thought initiate.

This is why *natural linguistic system* is better developed along the engineering lines I have suggested: as a coherent strategy for organizing data that can be expected to emerge in a usage as demands upon performance increase. The fact that atlas structures (or similar strategic organization) need only be crudely and partially implemented in undemanding contexts (in the manner of "rainbow") helps explain why their required contours needn't appear firmly etched in conventional grammatical discrimination and why they often become evident only in comparatively technical employments. From this point of view, we can't expect to provide a plausible natural linguistic system story for sense-data discourse unless we simultaneously attend to the behavior of "concept" talk as well.

Nor should we wish to imitate Austin in his reliance upon linguistic intuition: the idea that the structure of our concepts will reveal themselves to us if we only ponder what we would likely say in various projected circumstances. Quite the contrary, such a policy is patently unwise, for it immediately tosses our tennis ball into the court of *ur*-philosophical thinking, where the classical picture of concepts is certain to win. Instead, it is fundamental to this book's point of view that we commonly find ourselves in circumstances where our usage gradually molds itself to improved contours under the guidance of strategic considerations of which we have little conscious inkling and where we lack suitable appreciation of the tacit directivities that nudge us forward to these improvements. In such cases, our language has gotten ahead of our understanding and a very careful study of how it operates is required before we can catch up to its adjusting contours. In the meantime, we should scarcely trust our intuitions about the claims we might offer in hypothetical circumstances, for those premonitions will merely repeat our old misapprehensions back to us (in the manner of Zack and Speedy stories or the fairies that caper upon rainbows). Unfortunately, although many of the better parts of Austin's thinking have vanished from the contemporary scene, the thesis that philosophy should remain in the business of setting our scattered intuitions in good order continues very much with us, a malingering residue of the classicist vision of our intellectual prospects.

There is a last affinity between sense data doctrine and classicism that merits mention. Because both theses insist that our rational processing of information must pass through the needle's eye of presented contents, they must concoct stories that explain why our actual procedures and capacities seem to follow contrary patterns. Both doctrines wind up claiming that intervening mental episodes fleetingly appear that satisfy their demands upon "presentation" yet are quickly forgotten or otherwise mishandled by their alleged witnesses. Wittgenstein remarks upon such issues:

> When we do philosophy, we should like to hypostatize feelings where there are none. They serve to explain our thoughts to us. "Here explanation of our thinking demands a feeling!"—it is as if our conviction were simply consequent upon this requirement.[80]

Titon's emphasis on the intensity of his musical feelings in 2,v is surely of this order: he wants the objectivity of his musical categorizations to be directly legitimated within the *psychological states* he shares with his fellow musicians rather than obtained from the accuracy of the physical data conveyed (in this case, the operational irrelevance of his feelings to the categorizations at issue is poignantly revealed in the sheer unlikeliness that his mates will share the psychological states he regards as probative[81]). As to outright psychological hypostasis, this characteristic extract from William James' <u>Principles of Psychology</u> comes very near:

> Suppose we try to recall a forgotten name. The state of our consciousness is peculiar. There is a gap therein; but no mere gap. It is a gap that is intensely active. A sort of wraith of the name is in it, beckoning us in a given direction, making us at moments tingle with the sense of our closeness, and then letting us sink back without the longed for term . . . And the gap of one word does not feel like the gap of another, all empty of content as both might seem when described as gaps. When I vainly try to recall the name of Spaulding, my consciousness is far removed from what it is when I vainly try to recall the name of Bowles.[82]

Obviously, we are aware that we are hunting for a name, but the presentational demands of traditional agency require that the missing word appear as "a sort of wraith" before the mind's eye.

---

[80] Wittgenstein, <u>Investigations</u>, §598.
[81] Poor Rickie in E. M. Forrester's <u>The Longest Journey</u> was apparently privy to one of these conversations, although he could not keep his mind on it at all.   [82] James, <u>Psychology</u>, 243.

Within the context of conceptual appeal, the analogous interpolated states assume the forms of unnoticed episodes of meaning alteration ("Some fog must have clouded the physicists' minds when they allowed 'force' to fall off its semantic rails in application to rigid bodies") or what I earlier termed intimations of intensionality: directive impulses that drive our behavior but whose nature we can't quite diagnose ("Classifying this creature as a *bird* strikes me as evidently correct, although I can't really explain why"). Indeed, in classical thought "concepts" frequently appear as *wraiths of attributes* (this is why they were dubbed *ghost traits* in 5,x). Through such predicative metempsychosis, classicism weaves a Jamesian tale of a stream of conceptual life in which we fleetingly grasp abundant contents, but forever forget what they are, distractedly failing to attend to their cultivation and misunderstanding their messages: in general terms, bungling the management of the riches we have been too swiftly supplied. In my contrary estimation, we typically have less firm content in hand than we imagine and must be on the perpetual lookout for fresh supplies, just as our interplanetary explorer must learn more as it trundles across unaccustomed landscapes.

How we should rationally go about the business of *concept management* with fewer corrective resources at our disposal than are affianced under classicism will constitute our chief topic in the remainder of this book. I will open our discussion with a lengthy parable similar to the classification focused narrative supplied in Chapter 2, but now constructed to evince basic data relevant to the problems of rigorous reasoning and hazy intensionalities. I shall first allow its natural dialectic to unfold in *ur*-philosophical context without much kibitzing on my part and then, beginning in 8,vii, present my own diagnosis of how we have been led astray.

8

SONG OF THE MASTER IDEA

Up! To your feet! Leap! Listen!
Listen! Awake! Break your chains! Be! . . .
Be off! Run after the one who quickens you!
You are going to mistake me for yourself, you will think I am yourself . . .
Your eyes will see what I want to see.
Your ordinary intelligence will amaze itself;
It will discover such ways that to yourself seem mad.
You will say what surprises you.
You will find yourself, having done your utmost,
You will not understand your own perspicacity.

Paul Valéry[1]

(i)

The beckoning concept. Suppose we are confronted with an indolent teenager, who has decided he is destined for intellectual greatness yet refuses to do his homework sets. A page of routine calculus exercises lies before him but our ward sees no reason for working on any of them. "What might I possibly gain by doing so?," he complains. "There are already a lot of people around who can do that sort of thing—indeed, it is easy to buy a computer program that can do it all better than I ever will. What purpose will be served if I waste my precious time on such drudgery?" After all, our youth believes a special conceptual destiny awaits him and so he idles about his bedroom, expecting inspiration to strike.

To be sure, he allows that when Newton and Leibniz first worked out calculations of these types, they displayed penetrating insight into important matters. "But," he

---

[1] Paul Valéry, "Song of the Master Idea" in <u>Selected Writings of Paul Valéry</u>, Louise Varèse, trans. (New York: New Directions, 1950), 83.

demands, echoing the "big idea" orientation of our Darwin critic (2,ii), "why should I study anything except the basic definitions, because that's where all their insight is contained?" And so he glances at such passages for a few minutes. "I've got it," he confidently announces, with a faraway look in his eye. "Limits."

We protest in response that he has an entirely wrong idea of what conceptual attainment amounts to, but he reaches for one of the many science-fiction collections (his primary fount of literature) scattered across the floor and locates a salient passage in Clifford Simak's "Huddling Place." In that tale an ailing Martian genius is alleged to have devised a new

> concept . . . that we cannot do without. That will remake the solar system, that will put mankind ahead a hundred thousand years in the space of two generations. A new direction of purpose that will aim toward a goal we heretofore had not suspected, had not even known existed. A brand new truth, you see. One that never before had occurred to anyone.[2]

The Martian devised this grand notion through pure cognition—"philosophical thinking," in fact—and in the story he informs a friend:

> I have worked on [my idea] for years, starting with certain mental concepts that were first suggested to me with the arrival of the Earthmen.

"See! That's what I'm after," says our teenager, "the Great Idea of which nobody but me has any inkling."

"But thinking is not such a passive activity as that story makes out," we protest. "Novel ideas are born of hard work—one cannot simply 'open one's mind' and great ideas will automatically seep in. And the labor required is exactly of the calculus problem type that you spurn." We assure our malingering subject that the philosopher

---

[2] Clifford Simak, "Huddling Place" in Robert Silverberg, ed., Science Fiction Hall of Fame (New York: Avon, 1970), 227.

John Locke was right when he wrote:

> Would you have a man reason well, you must use him to it betimes; exercise his mind in observing the connection between ideas, and following them in their train.[3]

We quote the great electrical engineer Oliver Heaviside: "I have developed my ideas by getting the go of them and have so learned them that way."[4] We appeal to the authority of every mathematics teacher we have ever met. In our eagerness to motivate, we even stoop so low as to cite one of those obnoxious mottos beloved of American captains of industry: "Genius is one per cent inspiration; ninety-nine per cent perspiration."

But our sedentary adolescent is not swayed. Surely, as practical parents, we are right to oppose the influence of such juvenilia upon our teenagers—it can only encourage their lazy lounging about, awaiting a visitation of conceptual brilliance in the manner of the hapless protagonist of Henry James' "The Beast in the Jungle." But when we stop to ask, "What is it about concepts that makes this so? ", we hesitate, for there is no ready body of plausible *philosophy* available to back up our entreaties.

To be sure, we could readily quote for our pedagogical purposes some pragmatically oriented writer from older times such as John Dewey or Ernst Mach. But who could possibly accept their accounts of conceptual attainment today? Mach, for example, often treated conceptual learning crudely as a mental freight elevator that lowers formerly conscious routines into unconsciousness:

> A concept cannot be passively assimilated; it can be acquired only by doing, only by concrete experience in the domain to which it belongs. One does not become a piano player, a mathematician, or a chemist, by looking on; one becomes such only after constant practice of the operations involved. When practice has been acquired, however, the word which stands for the concept has a different sound for us. The impulses to activity, which are latent in it, even when they do not come to expression or do not appear in consciousness, still play the part of secret advisors who induce the right associations and assure the correct use of the word.[5]

Or to explain why we expect the right answer to an addition problem:

> Let us consider the simple concept of the sum $a + b$, where a and b may first be assumed to be whole numbers. This concept contains the impulse to count onward from b numbers from a in the natural series' when the last number is $a + b$. This act of counting forward may be regarded as a muscular activity which is always the same in all cases, however different, and the beginning of which is determined by a and the end by b. Through variation of the values of a and b, an infinite number of cognate conceptions is created.[6]

Of course, such an explanation is ridiculous (and the suggested scheme is absurdly rigid and inefficient). Our teenager is surely sophisticated enough to react with guffaws to

---

[3] John Locke, Of the Conduct of the Understanding (New York: Columbia Classics in Education 31, nd), §6.
[4] Heaviside: see motto to Ch 5. The second quote is from Thomas Edison.
[5] Ernst Mach, Principles of the Theory of Heat, Thomas J. McCormack, trans. (Dordrecht: D. Reidel, 1986), 382.
[6] Mach, Heat, 382.

such implausible pronouncements: "Oh, come on, Dad—you and Professor Mach are trying to tell me that I learned the concept *being a lion* by acquiring the habits of behavior that might induce the 'latent impulses to activity' appropriate to such creatures? That's ridiculous—all you and Mom did was to show me a bunch of picture books containing cute drawings of the beasts, none of which involved any reaction pertinent to confronting a real life lion."

Given that our teenager allows, at least on a "Great Ideas of Civilization" level, that he ought to possess some understanding of the calculus concepts, we might try to argue that appropriate homework skills represents a sine qua non of concept possession in this instance. From time to time, allied insistence upon *criteria for conceptual competence* becomes popular in philosophical circles. For example, the contemporary writer Christopher Peacocke in his recent A Study of Concepts maintains that for a speaker to genuinely possess the concept *conjunction* (that is, the notion of "and" in logic), she must satisfy the following "possession condition":[7]

> Conjunction is that concept φ to possess which a thinker must find transitions that are instances of the following forms primitively compelling, and must do so because they are of these forms:
>
> $$\begin{array}{ccc} p & p\varphi q & q\varphi p \\ \underline{q} & \underline{} & \underline{} \\ p\varphi q & p & q \end{array}$$

In passing, we might observe that certain writers find criterial rules of this ilk especially distinguished because they allegedly supply instructions for "language entry and exit," in the sense that leftmost rule tells a speaker upon what basis she may *start* using sentences employing "and"s if she hadn't been using them previously, whereas the other two show her how to *stop* using the connective once it is already active in her discourse (there's a large literature in philosophy of logic that debates whether these "entry and exit" features are important or not).

If such claims can be generalized to other words, perhaps we will find a philosophical lever to pry our slothful teenager from his couch. Perhaps we might demand that the "acceptance conditions" for the calculus concept of a *total derivative* analogously require:

> Derivation (that is, d/dx) is that concept to possess which a thinker must find transitions that are instances of the following forms primitively compelling, and must do so because they are of these forms:
>
> $$\begin{array}{cc} dx^{n+1}/dx = c & \int^x f(y)dy = F(x) \\ \underline{(n+1)x^n = 0} & \underline{f(x) = dF(x)/dx} \end{array}$$

[7] Christopher Peacocke, A Study of Concepts (Cambridge, Mass: MIT Press, 1992), 6 (prose slightly modified for ease of reading). He suggests that we can also delineate weaker "acceptance conditions" that might "accept" me as a "competent employer" of the calculus concepts, although I fail to qualify as a true "possessor" (his motivation seems to be Hilary Putnam's elm example).

If these Peacockean criteria are reasonable, our teenager's inability to recognize these inferential forms as "primitively compelling" would brand him as an inadequate possessor of the term "derivative"'s conceptual core. Since applications of these patterns constitute the meat and drink of his spurned homeworks, following these, on a Peacockean view, represents the surest test of conceptual competence.

Unfortunately, our teenager, through diligent historical study, has provisioned himself ably against criterial onslaught. "Ah ha!" he retorts, "the fact that you consider those crude *rule-centered* forms of activity as central to conceptual grasp clearly demonstrates the degree to which you entertain a shallow and unreasonable vision of intellectual reach, as opposed to the deeper insight so ably captured in 'Huddling Place.' In fact, your second 'conceptual requirement'—the 'anti-derivative' condition ($\int^x f(y)dy = F(x):.f(x) = dF(x)/dx$—has suffered numerous bumpy shifts in evaluation over the history of the calculus. As a general principle it is clearly false, but the exact degree to which it proves *false* varies according to the exact manner in which the fundamental concepts *derivative* and *integral* are understood. Isn't it clear, in fact, that a claim of this sort needs to be *semantically justified*—its correctness must be checked against appropriate definitions of 'derivative' and 'integral' in terms of limits, infinitesimals or the like? True, in the dawning days of the subject, the anti-derivative condition may have been vaguely regarded as 'definitional' of 'd/dx,' but this posture was temporarily acceptable only because practitioners of the time possessed a very hazy notion of what 'd/dx' should signify. As soon as minimal conceptual precision appeared on the scene, the reign of the 'anti-derivative' condition was over. Indeed, the long line of thinkers stretching from Cauchy to Lesbeque and beyond have progressively probed ever deeper into these matters and it is yet uncertain that mathematicians have found the very best way of unpacking the basic calculus concepts. Hence even studying, as I dutifully have, the standard freshman textbook definitions may not do proper justice to the traits truly at the center of this subject, but at least meditating upon those explications better approximates the efforts that would be needed, on Simak's model, to penetrate to their conceptual core. But no such benefit can possibly accrue to an illegitimate insistence upon the superficial and faulty 'cookbook' criterial rules you highlight."

This, we must acknowledge, represents an effective retort on our ward's part (would that some of the effort devoted to historical research had been expended upon getting the damned problems done). His insistence that *semantic insight*—that is, correct identification of the core meanings of the central vocabulary—should trump any *criterion based upon application* seems right in this case. Indeed, it is theoretically conceivable that Laurent Schwartz, Jan Mikusiński or any of the other modern architects of "generalized notions of derivative" were no more adept at working routine beginner's exercises than our goldbricking teenager; they might have simply inspected the old-fashioned "limit" definition and observed how it might be improved. In my own case, I must sheepishly confess that, although I read a lot of mathematics books for pleasure, my cookbook calculus skills have become so atrophied that I might possibly flunk any "possession conditions" of Peacockean stripe.

Note the characteristic phrase in the quote from Peacocke above: "a speaker must find transitions of the following kinds *primitively compelling*." Our teenager correctly objects that virtually no inferential transition can be granted this "primitive" character (as opposed to "being semantically justified"). The superficial retort—"at some stage justifications must come to an end"—does not entail that they must terminate in coarse inferential criteria. Indeed, in 10,v, we shall examine perfectly natural circumstances where a fully competent "possessor" of the *conjunction* concept might rationally assert, "Look, I'm not sure why one shouldn't detach 'A' from 'A and B' in these circumstances, but I know we shouldn't."

Like much modern philosophical writing on concepts, Peacocke's book is distinguished by its paucity of examples. There is a simple reason for this, I think: plausible "possession conditions" can be articulated for few everyday notions.

(ii)

Semantic epiphany. The sad truth is that, however much we might hope that philosophizing of a criterial flavor might rouse our teenager from the couch, it seems to be the concepts themselves who resist any facile demand that they immediately go to work. It is an undeniable fact that episodes of what might be called *semantic epiphany* play an important (albeit occasional) role in our intellectual life and our teenager has seized upon these as his model for *concept possession*. Often, just as our teenager avers, we *do* seem to grasp fresh concepts of considerable novelty and richness out of the blue or suddenly apprize old ideas in deeply enhanced hues, without any clear grounding in prior intellectual labor. Sometimes conceptual grasp seems to arrive as a *voluntary visitor*, to adopt Thomas Paine's metaphor. He writes of "thoughts," but his words apply equally well to concepts:

> Any person who has made observations on the state and progress of the human mind, by observing his own, cannot but have observed that there are two distinct classes of what are called thoughts: those that we produce in ourselves by reflection and the act of thinking, and those that bolt into the mind of their own accord. I have always made it a rule to treat those voluntary visitors with civility, taking care to examine, as well as I was able, if they were worth entertaining; and it is from them I have acquired almost all the knowledge I have.[8]

Or consider this passage from the poet Paul Valéry:

> In this process, there are two stages. There is that one where the man whose business is writing experiences a kind of flash—for this intellectual life, anything but passive, is really made of fragments; it is in a way composed of elements very brief, yet felt to be very rich in possibilities, which do not illuminate the whole mind, but which indicate to the mind, rather, that there are forms completely new which it is sure to be able to possess after a

---

[8] Thomas Paine, The Age of Reason (Secaucus, NJ: Citadel, 1973).

certain amount of work. Sometimes I have observed this moment when a sensation arrives at the mind; it is as a gleam of light, not so much illuminating as dazzling. This arrival calls attention, points, rather than illuminates, and in fine, is itself an enigma which carries with it the assurance that it can be postponed. You say, "I see, and then tomorrow I shall see more". There is an activity, a special sensitization; soon you will go into the dark-room and the picture will be seen to emerge.[9]

We can find similar remarks in every field of creativity. It is precisely in Valéry's "snapshot now; print later" vein that the mathematician Henri Poincaré described, in a famous passage, how a key notion in his theory of the automorphic (= his "Fuchsian") functions came to him:

At this moment I left Caen, where I was then living, to take part in a geological conference arranged by the School of Mines. The incidents of my journey made me forget my math-ematical work. When we arrived at Coutances, we got into a break to go for a drive and, just as I put my foot on the step, the idea came to me, though nothing in my former thoughts seeming to have prepared me for it, that the transformations I had used to define Fuchsian functions were identical with those of the non-Euclidean geometry. I made no verification, and had no time to do so, since I took up the conversation again as soon as I sat down in the break, but I felt absolute certainty at once. When I got back to Caen, I verified the result at my leisure to satisfy my conscience.[10]

Likewise, Henry James described the origins of "The Spoils of Poynton" thus:

[M]ost of the stories straining to shape under my hand have sprung from a single small seed, a seed as minute and wind blown as that casual hint for "The Spoils of Poynton". . . . . I instantly became aware, with my "sense of the subject," of the prick of inoculation; the whole of the virus, as I have called it, being infused by that single touch . . . There had been but ten words, yet I recognized in them, as in a flash, all the possibilities of the little drama of my "spoils."[11]

In Poincaré's case, we can even suggest an exact hypothesis as to what might have transpired. In Hermann Schwarz' well-known book[12] on conformal mapping (which utilizes the "linear-fractional" transformations of which Poincaré writes), there appears the accompanying diagram of how the squares on a plane will map into the interior of a circle (such "tesselations" are familiar to most of us because of their centrality in the art of M. Escher). It was also well known (due to discoveries of Arthur Cayley and Felix Klein) that the characteristic features of the classical variant geometries depend upon how their straight lines intersect the "line at infinity." Poincaré merely needed to

[9] Jacques Hadamard, The Psychology of Invention in the Mathematical Field (New York: Dover, 1954), 17.

[10] Henri Poincaré, "Mathematical Discovery" in Science and Method, F. Maitland, trans. (New York: Dover, nd), 53–4.

[11] Henry James, "Preface to 'The Spoils of Poynton'" in The Art of the Novel (New York: Charles Scribner's Sons, 1934), 119, 121.

[12] H. A. Schwarz "Uber die conform Abbildung meh-fach zusammerhängender ebener Flächen," J. F. reine u. angeur Math., 83 (1869). See also John Stillwell, Sources of Hyperbolic Geometry (Providence, RI: American Mathematical Society, 1996), 13. Also: "Introduction" to Henri Poincaré,' Papers on Fuchsian Functions (New York: Springer-Verlag, 1985).

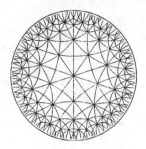

observe: "Gee, if I were to stretch Schwarz's bounding circle out to infinity, I should get the situation expected in hyperbolic geometry." Such a thought can easily come to one in a sudden kinesthetic impression of Schwarz' tessellation being pulled outward towards the far horizon.

In such a case, Mach's "secret advisors" below stage no longer seem to be reading from a previously written score; they are improvising riffs of their own. Indeed, not only do such moments of epiphany introduce us to new concepts, sometimes they reorient well-entrenched vocabulary along some "new direction of purpose." Rather than deciding that our words have changed their meaning thereby, we often feel that we have managed to *peer more deeply* into the heart of a notion that we had previously scrutinized in a shallower way. In this vein, we might claim, "Einstein realized that the fundamental notion of **momentum** could be understood in such a way that it did not need be shackled definitionally to **mass times velocity**." Or: "In considering the concept of *angle* from the point of view of cross-ratio, the French mathematician Laguerre found the manner in which this previously recalcitrant geometrical notion makes sense over the complex realm."

As remarked before (4,i), in the mid-nineteenth century a number of methodological crises arose as old notions in physics and mathematics suddenly erupted into remarkable forms of unexpected extension. Often the phenomenologies that prompted these prolongations fit our teenager's conceptual percepts admirably. Quite astonishing blossomings appeared, for example, within the strange evolution that carried conventional Euclidean geometry along the "higher geometry" route that eventually led to modern algebraic geometry (the history I have in mind is *not* that of the gradual acceptance of non-Euclidean geometries, but rather the manner in which familiar Euclidean doctrine became transmogrified into an extended "projective" realm containing bizarre points and lines at imaginary locations). The recognitional episodes that prompted these prolongations arose in several interrelated forms: <u>inter alia</u>, that of overt *algorithmic opportunity* and a more nebulous, yet striking, form of *gestalt suggestion*. Both provide material that apparently supports our teenager's conceptual picture in a very vivid manner, so it is useful to have a few examples before us. I'll begin with the *gestalt suggestions* first.

Suppose we have two triangles whose vertices are connected by lines meeting a single point p as illustrated. What will happen to the *legs* of these triangles if they are

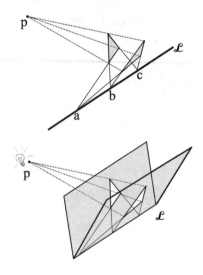

extended sufficiently far? Unless the reader has some familiarity with such problems, no immediate answer is likely forthcoming. But now look at our two-dimensional diagram in a *three-dimensional mode*, so that point p becomes a light source that sends rays from the corners of one triangle sitting on one plane to the other, sitting on some other plane, as shown in the second picture. Regarded from this point of view, the second triangle becomes the projected shadow of the first. We immediately see that if the legs of the nearer triangle are prolonged, they must meet their shadows along the line where the two planes cross. If we simply collapse this three-dimensional situation back to the original two-dimensional plane, we discover the answer to our question: *the points of intersection of the corresponding extended feet must meet along a common line* $\mathcal{L}$, a non-trivial geometric result called Desargues' theorem.

Here we do not witness the emergence of a totally new concept in Simak's sense, but plainly a "new direction of purpose" has been added to our stock of Euclidean reasoning principles. The mathematician George Pólya, discussing the gestalt shift aspects of a related geometrical problem, comments upon the "secret advisor" feel to realizations of this sort:

> The spontaneity is a very characteristic feature but rather hard to describe. If it happened to the reader that, from the entanglement of lines and letters in the figure . . . , the image of the parallelepiped "jumped" out at him unexpectedly, he will understand better what is meant by inspiration, how it is possible to interpret the sudden appearance of an impressive idea as the whispering of an inner voice, or a warning given by a supernatural being.[13]

In an allied vein, the celebrated Italian geometer Federigo Enriques writes:

> Nothing is indeed more fruitful than the increase of our intuitive powers made possible by this principle [the translation of different forms of intuition into one another

---

[13] George Pólya, <u>Mathematical Discovery</u>, (New York: John Wiley, 1965), 59.

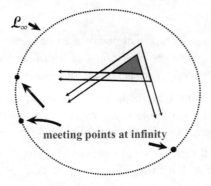

through some abstract mapping like the collapse in dimensionality we utilized].
*It seems as if to the mortal eyes with which we examine a figure under a certain aspect
there were added a thousand spiritual eyes enabling us to contemplate so many different
transformations.*[14]

Although I shall not detail the rest of the story here, considerations of an allied ilk
persuaded nineteenth century geometers that such redirected "spiritual eyes" could
discern aspects of geometrical figures—invisible imaginary points and lines— that had
lain utterly undetected heretofore. As the intersection of our two planes moves away
from the light source, the projected triangle begins to look simply like an enlarged
image of its source, becoming exactly so as the two planes become parallel. Collapsing
down to two dimensions, our construction suggests that the feet of two inscribed
triangles ought to meet at suitable points lying on a common line $\mathcal{L}_\infty$ at faraway
infinity (here the reader should imagine the production of these points of infinity as
arising in a movie-like fashion, where our two planes gradually become parallel and
collapse onto one another). By similar lines of quasi-cinematic reasoning, we can
persuade ourselves that non-overlapping planar circles must continue to meet in two
points of *imaginary location* with coordinates such as $(1, \pm\sqrt{-3})$ (actually, our circles
intersect in four imaginary points, because two more lie on the line at infinity).
Through following directivities of this ilk, the early founders of projective geometry
(J.-V. Poncelet, Jacob Steiner) took the initial steps that eventually blossomed into a
massive enlargement of the world of classical geometry. In this greatly enriched
picture every spatial figure not only contains large numbers of imaginary points and
lines in addition to its usual Euclidean parts, but also enjoys utterly unseen components living in completely invisible dual realms. To bring the unseen into view, we
employ movie-like transformations that rotate the unseen parts into view, pull the
portions at infinity into finite range or "blow up" the complications hidden in the
infinitely small. As Stephen Kleinman characterizes their picture of this "higher

---

[14] Enriques, Development of Logic, 124–5.

geometry":

> Here and elsewhere one senses that many classical geometers had a platonic view of figures like conics. There are ideal conics of which we see only shadows or aspects like their point sets and their envelopes of tangent lines.[15]

Although they are often more circumspect in their metaphors, contemporary geometers pretty much visualize their subject matter in much the same way, although they study topics of greater generality. Thus Keith Kendig writes in his excellent primer:

> Hence it turns out that what we see in [real space] is just the tip of an iceberg—a rather unrepresentative slice of the variety [=figure] at that—whose "true" life, from the algebraic geometer's viewpoint, is lived in [a much richer projective realm].[16]

It is hard, without a longer survey of basic geometric consideration, to convey adequately the astonishing sense of *deeper insight* into basic Euclidean fact that such patterns of reasoning bring to traditional geometry. Once these improvements are adequately appreciated, however, it is hard to resist the *ur*-philosophical impression that our newly discovered quasi-cinematic lines of thought prove *deeply loyal* in some murky way to the original Euclidean concepts of *point* and *line* from which these unexpected directivities depart. This impression of conceptual fidelity seemed so vivid to the early discoverers of projective geometry that they commonly spoke of the "organic unfolding" of previously dormant portions of concepts, as if these new directions of reasoning had lain all stored up in *point* and *line* all along, preformed in seedling guise. In exactly the vein that Stephen Kleinman describes, Jacob Steiner claims that the new methods of projective reasoning reveal a generating organism lying hidden behind the facade of orthodox Euclidean fact:

> In this manner one arrives, as it were, at the elements, which nature herself employs in order to endow figures with numberless properties with the utmost economy and simplicity; the organism, by means of which the most heterogenous phenomena in the world of space are united one with another.[17]

The thesis that certain individuals possess a capacity to discern the hidden morphologies of tightly enfolded concepts represents a characteristic theme within English Romanticism and German Idealism. Such metaphors may seem quaint today, but the root phenomenon of suddenly grasping a "new direction of purpose" through probing meditation upon predicative content is quite real and represents an issue that must be addressed in any adequate appraisal of conceptual evaluation. Modern biographers commonly praise scientists for their "deep physical intuition" or novelists for their "penetrating insight into moral ideals" and, in my opinion, such claims commonly partake of assumptions similar to the appeals to "organic unfolding" often heard in the 1800s, albeit now expressed slightly differently (we shall return to these issues in 8,xiii).

---

[15] Steven Kleinman, "Chasles' Enumerative Theory of Conics" in A. Seidenberg, ed., Studies in Algebraic Geometry (Providence, RI: Mathematical Association of America, 1980), 133.

[16] Keith Kendig, Elementary Algebraic Geometry (New York, Springer-Verlag, 1977), 5.

[17] Robert Edouard Moritz, On Mathematics and Mathematicians (New York: Dover, 1958), 315.

Certainly, such botanical metaphors prove febrile music to the convictions of our teenager. "See," he announces, "Simak is right about deep conceptual insight—it comes from looking into the heart of concepts. Extraneous doctrinal attachments, such as those dreary calculus problems you would inflict upon me, are of no moment. After all, Steiner's 'organic completion of geometry' corrected the strict letter of almost every important preexisting Euclidean doctrine. What's important is the *living spirit* of the concepts, not dry fact."

If we are fair, we must allow that a serious challenge to staid pictures of conceptual evaluation is contained in these considerations. Although contemporary mathematicians rarely appeal to mystic "organic growth" nowadays, they commonly announce "the proper setting for the concept X is . . . " and then brutally castigate the dunces who leave the problem in the awkward form in which it was originally posed. But what directive considerations *ratify* the "properness" of the "proper settings" they favor? In hard fact, the considerations cited usually resemble factors like those that concretely inspired Steiner to write of "Nature's organic organization." Accordingly, our teenager is right: concepts sometimes seem to naturally point beyond—and sometimes throw over—the contours in which they were initially nurtured. At unpredictable moments, old predicates cry out to continue into new domains of application in natural ways that may nonetheless surprise us, slumbering as we often do in tropospheric complacency.

. . . .

Many writers in the late nineteenth century were aware of the uncanny experience of witnessing an established pattern of conceptual thought unexpectedly unfold in the natural but hard to explain manner of projective geometry. The notion that concepts sometimes embody richly formed, yet partially disclosed, personalities underwrites the metaphorical impression that concepts represent visitors from an autonomous third realm, to employ Frege's phrase. More recent philosophers are less inclined to write in such extravagant terms, partially because many of them seem less aware of the basic phenomenology of unexpected conceptual guidance and partially because the popularity of formalism and instrumentalism have conspired to dull the impression that something remarkable occurs within these episodes.

. . . .

Returning to our teenager, his emphasis upon semantic epiphany has the salutatory effect of tempering the popular exaggerations of the *semantic finality* which we discussed in 1,vi: i.e., the thesis that speakers normally acquire a complete grasp of key classificatory concepts by an early age—no later than 10 or 11—, after which they may frame a variety of derivative conceptions, without altering the conceptual core of what they learned early on. "On what other basis can we possibly understand the very large range of sentences that we do?," defenders of semantic finality often ask. "But," our youth correctly observes, "think of all the moments when some vital spark of semantic insight reveals that a whole society has heretofore gotten the conceptual essence of a popular predicate quite wrong. They may have even regarded certain claims as criterial for the application of 'P,' but the epiphany reveals that those assumptions were not true to P's proper significance. Sometimes we can get hold of a concept in a crude enough fashion to hold its terminology fixed, but its richest conceptual dimensions still lie hidden from view—and that is what the probing intellect can recognize. We hold riches in our hand, but have failed to perceive their proper value." Certainly, our teenager is right that, in such circumstances, we invariably follow the newly emergent strands in addressing the directive question, "What is the correct way to employ the concept X here?" We show little disposition to remain loyal to old reasoning patterns simply because we were originally judged "competent in P" upon their basis.

Unfortunately, insofar as our efforts to wean our teenager from his indolent schoolwork practices go, we have not achieved much progress. Indeed, the phenomenon of semantic epiphanies suggests that important progress often does emerge in sudden blasts of conceptual gestalt: new ways of looking at old matters stemming entirely from a "novel way of looking at things," in which patient drudgery at calculus homeworks plays no apparent role. But this concession to epiphany abundantly ratifies the dreamy worship of "big ideas" shared by both Darwin critic and teenager, where the dedicated toil we deem vital has become removed to the negligible sidelines.

(iii)

Intimations of intensionality. I mentioned that the projective revolution in geometrical thinking was inspired by certain *syntactic sources* as well. If we consider some of these, a possible manner of answering our teenager is suggested. Many important discoveries in mathematics have originated when idle doodling with symbols has turned up intriguing results. As Ernst Mach writes, alluding to a famous remark of Euler's:

> The student of mathematics often finds it hard to throw off the uncomfortable feeling that his science, in the person of his pencil, surpasses him in intelligence,—an impression which the great Euler confessed he often could not get rid of.[18]

For example, although I indicated that a transformational rationale similar to that for points at infinity can be supplied for imaginary points (such as $(0, \sqrt{-3})$), such extension elements were not, in fact, originally suggested to geometers in that manner.

Instead, the germ that prompted their emergence lay along the path of *blind algebraic manipulation*. As Descartes showed, most common geometrical constructions can be easily registered in algebraic form. Thus, if we want to locate the chord where two circles meet, we solve their respective algebraic equations for common values and then write a formula for the line segment $\mathcal{L}$ between them. But even if our two circles don't overlap, blind algebraic manipulation will supply a line $\mathcal{L}$ *between* the circles that allegedly runs through the *imaginary* "meeting points" $(0, +\sqrt{-3})$ and $(0, -\sqrt{-3})$. This "chord" retains many of the same properties as a normal example (e.g., any third circle centered on $\mathcal{L}$ will cut $\boxed{e_1}$ orthogonally). Somehow our

---

[18] Ernst Mach, "The Economical Nature of Physical Inquiry" in Popular Scientific Lectures (LaSalle, Ill.: Open Court, 1986), 196.

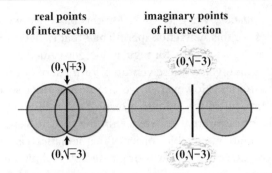

"pencil"—that is, our inclination to carry onward with purely syntactic reasoning patterns—has led us into fertile pastures, despite passing through various deductive hiccups where we don't recognize what some of our expressions mean (e.g., "the point $(0, -\sqrt{-3})$"). Historically, such entirely algebraic considerations suggested to Poncelet and his followers that the Euclidean plane contains imaginary points like $(0, -\sqrt{-3})$ (to be sure, Poncelet himself, as a self-styled "synthetic geometer," later obscured those origins, but we'll ignore this historical complication). Once this computational gambit was accepted, it became possible to devise other rationales for imaginary points and lines, but, in origin, the impetus for their acceptance came as "a gift from algebra."

We have mentioned this case briefly already, in 4,i. Here we witness a curious prolongation of word meaning that is driven largely by "suggestions arising from the symbols themselves" in the words of the geometer H. Baker:

> But now the way seems open to us, still further to generalize the Abstract Geometry, with the help of suggestions arising from the symbols themselves, using the words point, line, etc., in a proper sense consistent therewith ... [T]wo questions naturally arise: (1), Is there any geometrical utility in this extension? (2), Is it legitimate to use the postulated properties of the abstract points, lines, etc., in order to prove relations existing among the real points, lines., etc., that is, relations which can be stated without any reference to the abstract elements? ... And, as [happens] in the case [of complex numbers], it may be said, briefly, that experience has amply shewn that the gain in the generality of the statements of geometrical fact, and the increased power of recognizing the properties of a geometrical figure, enormously outweigh the initial feeling of artificiality and abstractness ... [T]he introduction of [extra] elements may well have assisted the constructive faculty [of ingenuity]; that this may happen is, indeed, one of the discoveries of the history of reasoning.[19]

Mathematics, in fact, is filled with unexpected extensions of this general flavor. For example, we might fool around with Leibniz's notation for multiple derivatives—$d^3f(t)/dt^3$, say—and, in a spirit of intellectual whimsy, ask ourselves: "What would happen if I

---

[19] H. F. Baker, Principles of Geometry, (Cambridge: Cambridge University Press, 1922), 143–4.

changed that '3' to a *fraction,* viz. $d^{\frac{1}{2}}f(t)/dt^{\frac{1}{2}}$?" On this basis, we would explore various "formal consequences" of this syntactic experiment—that is, determine where selective shards of the rules that work properly for regular, integer-valued derivatives will lead us. Much trial and error is involved in these ventures, for most of our pencil-led scribblings will dead end in absurdities. However, if we choose the right schedule of provident inferences, we may find ourselves led into a territory that displays an interesting, and perhaps valuable, flavor of internal coherence. In the case at hand, it turns out that a relatively obscure but quite intelligible corner of mathematics[20] is devoted to fractional derivatives of the type described and was originated by Leibniz himself in exactly the notation-inspired manner I have sketched (it is often remarked that the advantage of a "good notation"— Leibniz's $d^3f(t)/dt^3$ versus Newton's f''',—lies in the fact that it inspires fruitful forays of this sort). It happens that fractional derivatives <u>per se</u> have not proved of the greatest mathematical moment (although close analogs in functional analysis are important and Heaviside employs them as passing ingredients within the operational calculus). However, Euler's explorations within the complex realm partake of the same doodling character and those discoveries have completely reshaped mathematics.

From a philosophical point of view, such pencil-led prolongations seem quite mysterious and at odds with traditional opinion in regard to mathematics' allegedly <u>a priori</u> status. Attempts to supply a reasonable rationale for these curious endeavors formed a vital part of the nineteenth century family of worries about concepts and scientific methodology that were highlighted in Chapter 4. The main focus of our own discussion centers upon predicates of macroscopic physical description, not those of pure mathematics, but later in the chapter we shall find that syntactically inspired prolongation often plays an important role in the developmental history of our favored predicates as well.

Although "our pencils sometimes surpass us in intelligence," they are not, for all that, necessarily all that intelligent, for such reasonings proceed by trial and error exploration utilizing various flavors of *quasi-algorithmic routine* (in the case at hand, plowing ahead with the tedium of high school algebra). And this suggests a possible mode of convincing our teenager of the conceptual merits of his calculus homework. First, we observe that many of his episodes of epiphany represent the results of *transferring reasoning algorithms* profitable in a familiar domain into an unexpected application (in the Desargues' theorem case, lifting a two-dimensional problem into a three-dimensional setting). From an operational point of view, we can diagram the basic structure of our illustrative epiphanies in the following way. We begin in a problem domain $\mathcal{D}$ with which a natural group of inferential principles $\mathcal{J}$ are associated (say, $\mathcal{J}$ = intuitive reasoning about two-dimensional objects). Our epiphany shows us that under the mapping τ many $\mathcal{D}$ problems can be lifted into a new arena $\mathcal{D}^*$ enjoying its own set of parochial inferential tools $\mathcal{J}^*$ (<u>e.g.</u>, our tools for *three-dimensional* reasoning). These $\mathcal{J}^*$-based patterns then carry us to interesting conclusions which, if we're lucky, can be pulled back into $\mathcal{D}$ to

---

[20] K. B. Oldham and J. Spanier, <u>Fractional Calculus</u> (New York: Academic Press, 1974).

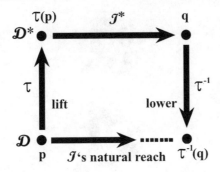

supply results that we cannot easily reach by direct methods. Indeed, we have already commented upon the virtues of such "lift and pullback" techniques on many occasions (e.g., the various map transformations provided in 6,ii). Our imaginary point case can be fitted to this model as well, where we lift from ordinary geometry into a domain of purely syntactic manipulation, which we can then pull back to produce novel geometrical predications.

Indeed, the main difference between our Desarguean epiphany and our pencil-guided prolongations lies mainly in the degree to which we are *aware* of the component stages within a transferred reasoning process. In the latter case, every step we take is painfully laid down on the paper before us, whereas the epiphany strikes us in a sudden gasp: "Wow! Look at that." But, perhaps, the two processes are not all that different underneath. We have already observed that we scarcely understand the mental processes whereby we rapidly reach our everyday life geometrical conclusions: somehow our animal heritage allows us to perform inferential feats that presently elude the capacities of known computer routine (to be sure, the conjectures we form in this manner are not always accurate). Plainly, these powerful "conclusion reachers" can be usefully adopted to unexpected subject matters—in the case at hand, other areas of geometry, but also to topics that have no intrinsic connection with lines and figure at all (as remarked before, mathematicians and physicists commonly force their topics into artificial geometrical guise simply so they can reason about them more effectively).

An interesting case of a *halfway epiphany* comes from statistics. With respect to counterfeited money, the hallmarks of fraudulent manufacture do not usually lie in the locations of individual details—a poor rendering of Andrew Jackson's nose, say—, for these are apt to vary comparably amongst legitimate tender as well, but in the ways that such variations *statistically correlate* with one another. Printing presses behave in such a way that, when Jackson's nose shifts to the left, the inscription "twenty dollars" tends to move as well, in a manner that supplies a probabilistic fingerprint of its origin. Unfortunately, it is not easy to train the human eye to directly recognize these correlations in a dollar bill and comparatively little further assistance is provided if the carefully measured congruities are listed in long tables in standard, multi-variant statistics fashion. A would-be redresser of monetary malfeasance—a Treasury agent,

say—will simply stare dumbly at this opaque blur of numbers and conclude nothing about the measured money's origins. To improve inferential performance in this arena, enterprising statisticians asked themselves: is there any form of data with respect to which humans are particularly skilled in recognizing multi-variant pattern? *Answer*: the way we recognize human faces through family resemblances, manifestly one of our most delicate and highly evolved perceptual skills (Pascal notes our skills in multi-variant similarity long ago: *Two faces are alike; neither is funny by itself, but side by side their likeness makes us laugh*).[21]

What will happen, our statisticians asked, if we *code* the unrevealing monetary deviations as facial features, such as those illustrated[22]? Under this transmogrification, salient hypotheses positively leap to the attention of our Treasury agent: "Just look at these creepy visages; they can only be the handiwork of old Flattop and his gang." Here we witness a halfway epiphany: we cannot effect the transfer from counterfeit bill data to face plot except through painful computation (by hand or machine), but once the lift is made, our native facial recognition tools suggest fruitful classificatory hypotheses to us with the full immediacy of gestalt epiphany.

---

[21] Blaise Pascal, Pensées, A. J. Krailsheimer, trans. (London: Penguin Books, 1966), 34. James Russell and J. M. Ferandez-Dols, eds., The Psychology of Facial Expression (Cambridge: Cambridge University Press, 1997).

[22] Bernhard Flury and Hans Riedwyl, Multivariate Statistics: A Practical Approach (London: Chapman and Hall, 1988), 173.

We readily appreciate that large amounts of *unconscious calculation* of geometrical insight type must lie below many cherished episodes of conceptual insight. Many of these must involve a fair degree of *algorithmically directed search* as well. We recall the celebrated folks who can, <u>e.g.</u>, correctly determine on which day of the week February 19, 1920 fell, without any evident awareness of how they managed to discover this fact. They may report that little more than a spontaneous "Monday!" comes to mind, although their brains must have engaged perforce in a huge amount of raw calculation to find this result, possibly utilizing interpolation from a lengthy list of remembered temporal landmarks. In this vein we recall the eighteenth century calculator Jedediah Buxton[23] who was "drunk with reckoning" for a full month while he unrelentingly worked out a lengthy problem on volumes (perhaps "reasoning intoxication" credibly applies to the syntactically driven forays of Euler or our projective geometers as well).

Although some of these spontaneous reasoners are of otherwise low intelligence ("idiot savants"), many are able mathematicians like A. C. Aitken:

> I have noticed also at times that the mind has anticipated the will; I have had an answer
> before I even wished to do the calculation; I have checked it, and am always surprised it is
> correct. This, I suppose (but the terminology might not be right), is the subconscious in
> action. I think it can be action at several levels; and I believe that each of these levels has its
> own velocity, different from that of our ordinary waking time, in which our processes of
> thought are rather tardy.[24]

Indeed, Aitken left behind an interesting account of how he cultivated his computational prowess, practicing clever routines until they gradually assumed the phenomenology of a Delphic oracle that ran swiftly ahead of his conscious ploddings.

Perhaps these observations suggest an adequate riposte to our teenager's idling, for we can now argue that his favored displays of "deep conceptual insight" must piggyback upon a richer arsenal of less clever, but inferentially resourceful, routines occurring on a largely unconscious level. Thus in the case of Poincaré's epiphany, some form of semi-automatic verification of analogies between the behavior of the automorphic maps and non-Euclidean congruences must have attended his sudden "Ah ha! That's it" conviction: "Gee, if we stretch out Schwarz's diagram, notable correlations with non-Euclidean geometry begin to emerge" (Poincaré, it should be noted, was very adept at calculation in an unconscious vein). It is also reasonable to presume, in many a literary effort or painting, that some allied measure of unconscious routine must survey and align its embryonic parts so that the mutual coordination described by John Ruskin is likely to emerge:

> A powerfully imaginative mind seizes and combines at the same instant, not only two,
> but all the important ideas of its poem or picture; and while it works with any one of
> them, it is at the same instant working with and modifying all in their relationships with
> it, never losing sight of their bearings on each other; as the motion of a snake's body goes

---

[23] Steven B. Smith, <u>The Great Mental Calculators</u> (New York: Columbia University Press, 1983), 171.
[24] Ibid., 27.

through all its parts at once, and its volition acts at the same instant in coils that go contrary ways.[25]

Such organized beauty rarely arises by accident; some reasonably systematic routine of *searching and checking* must have verified and installed the symmetries in the final product, although we may be consciously aware of little more than a "Everything checks out alright down here, Capt't" signal from below. In this vein Henry James writes:

> *I was charmed with my idea, which would take, however, much working out; and because it had so much to give, I think, must I have dropped it for the time into the deep well of unconscious cerebration; not without the hope, doubtless, that it might eventually emerge from that reservoir, as one had already known the buried treasure to come to light, with a firm iridescent surface and a notable increase in weight.*[26]

Plainly, much of this unconscious "assembling and checking" must involve the temporary weighing of rough sketches of hypotheses or unworkable alternatives: the "wreck and rubbish" that Thomas Carlyle's "mute workmen" of the unconscious will sweep away, if we only allow them their privacy:

> *Nay, in thy own mean perplexities, do thou thyself but hold thy tongue for one day: on the morrow, how much clearer are thy purposes and duties; what wreck and rubbish have those mute workmen within thee swept away, when intrusive voices were shut out!*[27]

Such "clearer purpose and duties" can be greatly aided by the raw combinatorial experimentation that the Victorian biostatistician Francis Galton describes as:

> *certain steps of thought, certain short cuts, and certain far-fetched associations, that do not commend themselves to the minds of other persons, nor indeed to [ones] own at other times.*[28]

Unanticipated structural commonalities between two apparently unrelated disciplines can come to our attention in this fashion: our "mute workmen" fan out searching quite randomly for coarse similarities and then report upon the salient correlations they come across. Given the happy but unpredictable results that can be achieved by unusual forms of transferred reasoning, it is no wonder that our "mute workmen" go in for a fair amount of "far-fetched" cut-and-paste work. Galton again:

> *When I am engaged in trying to think anything out, the process of doing so appears to me to be this: The ideas that lie at any moment within my full consciousness seem to attract of their own accord the most appropriate out of a number of other ideas that are lying close to hand, but imperfectly with the range of my consciousness. There seems to be a*

---

[25] John Ruskin, The Literary Criticism of John Ruskin (New York: Da Capo, 1965), 11.

[26] Henry James, "Preface to 'The American'" in Art, 22–3.

[27] Thomas Carlyle, Sartor Resartus (New York: Charles Scribner's Sons, 1921), 194. It seemed to me that Palmer Cox's Brownies supply an excellent picture of the workmen at their labors.

[28] Francis Galton, Inquiries into Human Faculty and its Development (London: J. M. Dent, n.d.), 147.

presence-chamber in my mind where full consciousness holds court, and where two or three ideas are at the same time in audience, and an antechamber full of more or less allied ideas, which is situated just beyond the full ken of consciousness . . . The thronging of the ante-chamber is, I am convinced, altogether beyond my control; if the ideas do not appear, I cannot create them, or impel them to come . . . I gather, after some inquiry, that the usual method among persons who have the gift of fluency is to think cursorily on topics connected with it until what I have called the antechamber is well filled with cognate ideas. Then, to allow the ideas to link themselves together in their own way, breaking the linkage con-tinually and recommencing afresh until some line of thought has suggested itself that appears from a rapid and light glance to thread the chief topics together.[29]

Indeed, as I was just about to fall asleep after struggling with a section of this book, I heard a wee voice from the antechamber timidly suggest, "What about model rail-roading? Will that help?" (No, unfortunately).

All in all, our discussion suggests three ways (there are, of course, many others) in which the "mute workmen of the antechamber" might perform useful work, most of it driven by algorithm or other quasi-automatic routine: (1) search for structural simil-arities betwixt unrelated fields; (2) attempt inferential lifts based upon these common-alities; (3) check the success of the experiment by searching for inferentially uncovered similarities. Most of this work doesn't require great intelligence on the part of our workmen, but the fields of similarities in which they search must be rich.

It often happens that our conscious awareness of these endeavors may take the form of little more than a relatively unspecific *hunch*: "Some deep affinity links subjects X and Y, although I can't yet say in detail what it is." Indeed, a large amount of mathematical work since the time of Dedekind and Kronecker has been driven by the impression of a deep affinity between the behaviors of algebraic numbers and fields, although a sharp

---

[29] Ibid., 147–8.

diagnosis of their perceived commonality resisted precise articulation for a hundred years. On a far more modest level, much of my own thinking about language has been guided by an equally amorphous impression of commonality between successful paradigms of scientific calculation and the more general patterns of ordinary usage.

In 6,iv, I dubbed our comparably hazy impressions of a predicate's governing directivities as *intimations of intensionality*: we feel quite certain that predicate "P" needs to be applied in manner X, without our being able to specify exactly how or why. As a humble illustration, a few years ago my (real) son was in the habit of carting home from school quite esoteric spelling words: those words of marginal employment where we feel that we "kinda knows" their meaning, but where we cannot readily specify what it is. So we might look up the term in a dictionary and the following process of internal monologue often transpires: "Can *that* be right? Let me see: in what contexts have I seen this word used? Dum de dum (= *thinking sounds*) . . . Can those situations be made to fit this definition? . . . Dum de dum (= *more thinking sounds*) . . . .Ah ha, *now* I see how it works; *now* I've got it precisely". But I don't really know what my brain is up to while this "dum de dum" process takes place; it seems to be engaged in some incredibly rapid process of dredging up dimly remembered employments of the word and reexamining their content to find a linkage to the dictionary offering. Insofar as I entertain any awareness of these verifications, it merely arrives as a curt reply from the boiler room, "It's all right, cap't, everything checks out down here." Nonetheless I usually walk away from these common yet uncanny experiences with a justifiable confidence that I have obtained a clearer handle upon a formerly wobbly word (almost certainly my mute mental assistants rearranged their workshops in a more elaborate manner than simply installing the dictionary entry into memory).

Plainly those conceptual episodes emphasized by our teenager, where such subterranean calculation suddenly credits an old term with a "new direction of purpose," can emerge in this mute workmen manner as well. Considered phenomenologically, it will seem as if an old concept has suddenly unfolded new appendages, in exactly the "organic" vein that Steiner and others highlight. The fact that we can become legitimately convinced of the "correctness" of a predicate's novel application, while remaining unable to articulately specify why, represents an important topic too often neglected in the philosophical study of everyday conceptual evaluation.

We can now argue in favor of calculus homework drudgery as follows. "Deeply reorientational insight of exactly the type you prize commonly feeds, even in quite unexpected topics, upon the transferred grasp of rich structural pattern. But the latter can be gained only at the price of having once worked through a large number of routine exercises. The minions who whisper novel insights to us from antechamberal doorways require large pools of syntactic pattern through which they can sort and compare. By refusing to complete your homeworks, you starve your subconscious search engines of precisely the material they require for unexpected conceptual success." In short, we now offer our teenager an improved version of the "conceptual elevator" justification for brute algorithmic practice that we surveyed in connection with Ernst Mach in our opening salvo.

(iv)

Our spying attention. Although we have plainly scored some consequential points, our ward remains ready to parry our thrusts, for, without further supplement, the story we have told supplies an implausibly automatic portrait of intellectual endeavor. It is simply not true that (in Galton's words again):

> [The role of consciousness] *appears to be that of a helpless spectator of but a minute fraction of automatic brain work.*

We scarcely want to encourage in our teenager a view of powerless self as creepy as that famously articulated by Arthur Rimbaud:

> *It is false to say: I think. One ought to say: I am thought . . . For I is an other. If the brass is roused to a bugle call, that is none of its doing . . . I attend to the blossoming of my thought: I look at it:* I listen.[30]

Indeed, as Dutiful Parents, we should blanche to discover that we have adopted Rimbaud as a role model for youth, however much we may admire the poetry.

"But," our teenager now interjects, "we must plainly *judge* the 'blossoming of our thought' by standards higher than are likely embodied in the epiphany itself. You have not factored the importance of *discriminating discernment* into your portrait of conceptual grasp." It is upon these powers of "higher comprehension" that our youth plots his own program of conceptual improvement—a schedule that, needless to say, eschews much tedious calculus practice. Indeed, after an account of inspiration on a railway carriage much like Poincaré's, Henry James writes:

> [M]y *account of the origin of "The Pupil" . . . will commend itself, I feel, to all imaginative and projective persons who have had—and what imaginative and projective person hasn't?—any like experience of the suddenly-determined absolute of perception. The whole cluster of items forming the image is on these occasions born at once; the parts are not pieced together, they conspire and interdepend; but what it really comes to, no doubt, is that at a simple touch an old latent and dormant impression, a buried germ, implanted by experience and then forgotten, flashed to the surface as a fish, with a single "squirm," rises to the baited hook, and there meets instantly the vivifying ray. I remember at all events having no doubt of anything or anyone here; the vision kept to the end its ease and its charm; it worked itself out with confidence.*[31]

Note the metaphor of *the spying intelligence* where a subconscious proposal "meets instantly the vivifying ray" in James' imagery of angling. Perhaps our mental activities can be divided according to the manner in which a discriminating higher self sits *in appraisal* with respect to the inventive but slightly crude suggestions of the

---

[30] Arthur Rimbaud, Collected Poems (Harmondsworth: Penguin, 1962).

[31] Henry James, "Introduction to What Maisie Knew" in The Art of Criticism (Chicago: University of Chicago Press, 1986), 324.

subpersonal workmen in the outer rooms. Paul Valéry endorses this hypothesis quite straightforwardly:

> It takes two persons to invent: one forms the combinations, the other chooses and recognizes what is desired or relevant among the set of products of the first.[32]

And Leibniz provides another variation on the theme:

> Mariotte says that the human mind is like a bag: when you are thinking you are shaking the bag until something falls out of it. Hence there is no doubt that the result of thinking depends to some extent on chance. I would add that the human mind is more like a sieve: when you are thinking you are shaking the sieve until some minute things pass it. When they pass, the spying attention catches whatever seems relevant.[33]

Given the low intelligence required for their search and compare activities, our mute workmen can be expected to produce many valueless suggestions as well (there is a popular theory of dreams[34] that maintains they represent our brain's continuing attempts to supply some localized and pieced together "coherence" to random brain excitations whose imports would otherwise get suppressed, during our hours awake, by censors that better attend to the global plausibilities of the scenarios we devise). From large quantities of possibly stupid suggestion our higher intelligence eventually winnows the meritorious. "And so," our teenager concludes, "I plan to focus my intellectual training upon learning to sort conceptual wheat from the chaff, for only my spying attention possesses sufficient intelligence to discern whether the raw products of the unconscious or the pencil have any 'true thought' backing behind them."

In other words, our teenager plans to wait until his conceptual courtiers press an entry from antechamber to awareness, whereupon he will regally adjudicate what they have to offer. He may daintity weigh the intellectual goods presented before him according to explicit standards of coherence, agreement with preestablished principle, and other norms of rational classification. If they pass muster, they may be subjected to further intellectual operations and improvements, such as abstraction or logical combination. "It is here that a mind shows its conceptual discernment and greatness," our youth proclaims, "for we should not take credit for that which is subconsciously assembled." Indeed, it is precisely our teenager's operations of inspecting, weighing and constructing that comprise the basic ingredients in the "variable and incalculable processes" described in this passage from the nineteenth century psychologist G. Romanes:

> Objectively considered, the only distinction between adaptive movements due to reflex action and adaptive movements due to mental perception, consists in the former depending on inherited mechanisms within the nervous system being so constructed as to effect particular adaptive movements in response to particular stimulations . . . Reflex actions under the influence of their appropriate stimuli may be compared to the actions of a

---

[32] Stanislas Debaene, The Number Sense (Oxford: Oxford University Press, 1997), 171.
[33] Pólya, Discovery, ii., 62.   [34] J. Allan Hobson, The Dreaming Brain (New York: Basic Books, 1988).

machine under the manipulations of an operator; when certain springs of action are touched by certain stimuli, the whole machine is thrown into appropriate movement; there is no room for choice, there is no room for uncertainty . . . But the case of conscious mental adjustment is quite different. For, without at present going into the question concerning the relation of body and mind, or waiting to ask whether cases of mental adjustment are not really quite as mechanical in the sense of being the necessary result or correlative of a chain of physical sequences due to a physical simulation, it is enough to point to the variable and incalculable character of mental adjustments as distinguished from the constant and foreseeable character of reflex adjustments. All, in fact, that in an objective sense we can mean by a mental adjustment is an adjustment of a kind that has not been definitely fixed by heredity as the only adjustment possible in the given circumstances of stimulation.[35]

Romanes does not claim that our "variable and incalculable" flavors of thought spring to life truly free of mechanical predetermination, but insists that their conscious manifestations, considered solely as they emerge within our unfolding streams of consciousness, do not conform to any manifest patterns of conditioning or "superinduced act." William Hazlitt draws a similar contrast in his Table Talk:

Those persons who are much accustomed to abstract contemplation are generally unfitted for active pursuits, and vice versa . . . Some men are mere machines. They are put in a go-cart of business, and are harnessed to a profession—yoked to fortune's wheels. They plod on, and succeed . . . A man may carry on the business of farming on the same spot and principle that his ancestors have done for many generations without any extraordinary share of capacity . . . If he has a grain more wit or penetration than [his neighbors], if his vanity gets the start of his avarice only half a neck, if he has ever thought or read any thing upon the subject, it will most probably be the ruin of him. He will turn theoretical or experimental farmer, and no more need be said . . . A plotting head frequently overreaches itself: a mind confident of its resources and calculating powers enters on critical speculations, which, in a game depending so much on chance and unforeseen events, and not entirely upon intellectual skill, turn the odds greatly against one in the long run.[36]

(This patently represents an unjust slur on the intellectual requirements of agrarian life, but we will let that objection pass, for a valid point about algorithms is contained therein, to which I will soon return).

In Chapter 5, we observed how Quine freely appeals to inter-sentential "conditioning" in situations where there is absolutely no evidence that usage displays any connection of that ilk at all. Through their eagerness to combat the classical picture, pre-pragmatists often install veins of distributed normativity (4,v) across discourses that manifest none of the same. Here's a curious example, offered in a 1930s psychology primer written by one of John Dewey's disciples (Frank Lorimer). The following represents a verbatim transcription of a 2-year-old girl musing in her crib, presumably sometime in the

---

[35] G. J. Romanes, Animal Intelligence (London: Kegan Paul, 1904).
[36] William Hazlitt, "On Thought and Action" in Table Talk (London: J. M. Dent, 1908), 102–3.

Christmas season:

> Pretty—the moon shines—pretty, cold moon—yes—watch the moon— mama come too—
> go in car, travel Breslau, yes?—first the aunt comes in, child comes in, too—gives hand,
> great-aunt, yes, yes, Hilde comes too—morning first go stores—buy things, buy butter—
> Hilde runs all alone, bring pretty cloth home, so—three pretty soldiers, rumtumtumtum—
> the red soldiers have the rututu—then we go away, aunts come again—doll carriage buy,
> will buy a pretty doll carriage.[37]

To Lorimer, these babblings present a clear exemplar of "conditioned response," whereas it seems patent to me that the child is simply *practicing* phraseology that she believes would be appropriate with respect to a drifting range of circumstances as they happen to come to her mind (my mother tells me that I privately rehearsed usage at a comparable age in just such a way). The applicable descriptive phrase here is *appropriate*, not *conditioned* in any reasonable sense of the term. Rather than evidencing any automatic response to associated ideas, the child is rehearsing, in the form of play, for social eventualities that she hopes to face. Indeed, with respect to the portion containing "Hilde runs all alone," I would conjecture that the child is imagining herself running before the adults and hearing admiring comments on her ambulatory skills from her witnesses, rather than indicating any "conditioned response" that *she* would make in the circumstances. That is, I believe I *understand* the child's linguistic activities in the sense that I recognize the circumstances she is attempting to model, but I have no hypotheses to offer as to what has *caused* her to consider such matters now. Whatever the proper springs of Hilde's babbling, they surely do not conform to any "conditioned" pattern.

Crudely deterministic strains of this ilk have largely evaporated from linguistic thinking since Noam Chomsky wrote his famous critique of B. F. Skinner in 1957, but allied proclamations of "a new Science of ideas" seem to spring eternal. And each time that a new shoot of thesis emerges from the fertile ground of algorithmic inspiration, sensible commentators point out their obliviousness to the "variable and incalculable" facts of human conceptual thought. Thus S. T. Coleridge complains of David Hartley's associationist psychology (which resembles Hume's):

> I almost think, that Ideas never recall Ideas, as far as they are Ideas—any more than Leaves
> in a forest create each other's motion—The Breeze it is that runs through them; it is the
> Soul, the state of Feeling—[38]

To be sure, such critiques often run to unjustified extremes of mystical claims with respect to unmoored "human spontaneity." For example, Sir William Hamilton (the Scots philosopher, not the Irish mathematician) contends:

> An exclusive devotion to physical pursuits exerts an evil influence in two ways. In the first
> place, it diverts attention from all notice of the phenomena of moral liberty, which are
> revealed to us in the recesses of the human mind alone; and it disqualifies from appreciating

[37] Frank Lorimer, The Growth of Reason (New York: Harcourt, Brace and Co., 1929), 80.
[38] Basil Willey, Samuel Taylor Coleridge (New York: Norton, 1972), p. 96.

the import of these phenomena, even if presented, by leaving uncultivated the finer power of psychological reflection, in the exclusive exercise of the faculties employed in the easier and more amusing observation of the external world. In the second place, by exhibiting merely the phenomena of matter and extension, it habituates us only to an order in which everything is determined by the laws of a blind or mechanical necessity. Now, what is the inevitable tendency of this one-sided and exclusive study? That the student becomes a materialist, if he speculate at all.[39]

Here Sir William has thoroughly muddled *algorithm, theory, predictability* and *determinism* together, but the notions are quite different. A classical physicist will commonly employ algorithms in her work, as indeed any of us will, but most of the time her responses will prove as "spontaneously creative" as those found in any other form of human activity. As we noted, even a fully deterministic mechanics will not disclose its secrets according to any easy, preordained pattern. Alter a differential equation modeling but slightly and our poor physicist may need to improvise an entirely new branch of mathematics before she can extract useful information from its recalcitrant confines. Nor is it true that our scientist is only interested in "ordering events within the causal nexus"—in 9,i we shall observe that a common strategy for assaulting the fortress of physical theory is to abandon our ambitions with respect to temporal tracking and focus upon equilibrium or steady state conditions instead. Doctrines that science only concerns itself with causal explanation or prediction trace to misapprehensions about physical methodology comparable to Hamilton's (3,ix).

Nonetheless, leaving these misconceptions aside, we can agree that it is unwise to presume that human activity generally follows rigidly prescribed behaviors when no particular evidence to this effect is apparent. Voltaire comments upon the rarity of rigidly regulated behavior as follows:

I know of only two kinds of immutable beings on the earth, mathematicians and animals; they are led by two invariable rules, demonstration and instinct; and even the mathematicians have had some disputes, but the animals have never varied . . . Assemble all the rabbits of the universe, there will not be two different opinions among them.[40]

As Voltaire well knew, even this limited sample is exaggerated (unlike Sir William, he is being humorous). When a mathematician constructs a proof, she is no more driven by obvious algorithm or animal instinct than Voltaire himself was when he composed Candide. Most demonstration is not rule-*driven* at all, although the results will need to *conform* to standards of proof.

On the other hand, we should not overlook the obvious fact that sometimes we *want* our own thinking to follow algorithms. When we balance our checkbooks, we do not thereby relinquish our moral liberty, but we surely wish our unfolding stages of calculation to conform to strict rules. Whimsical episodes of deviation from additive and subtractive routine should not intercede within our financial calculations, no

---

[39] William Hamilton, The Metaphysics of Sir William Hamilton, Francis Bowen, ed. (Cambridge: Sever and Francis, 1863), 22.   [40] J. F. M Voltaire, Philosophical Dictionary, H. I. Woolf, trans. (New York: Alfred A. Knopf, 1928), 83.

matter how deeply creative those interventions might otherwise prove. While we may otherwise act as airily unregulated in our linguistic activities as any Isadore Duncan, our patterns ought to stay as sternly regimented as a cash register while we're about the business of carrying an algorithm through, for such routines must be allowed to walk through their full pattern or else we will be unable to trust their answers.

I stress these facts because it is mainly within these narrowed domains of freely elected algorithmics that I discern the hand of distributed normativity within language. The facade-like models of the previous chapters demonstrate that these narrow veins of appearance can play important roles in the overall shaping of wider forms of linguistic behavior.

All the same, we can fairly conclude that linguistic activity submissive to obvious algorithm is usually, at the same time, freely elected and the moments where we consign our sentences to its governing authority are few. And it is here that our teenager finds his opening, for clearly some form of *conscious activity* is required to adjudicate the value of the raw product of algorithm or crude comparative search, whether it emerges from our pencil's doodlings or subconscious antechamber. "But to adjudicate whether these products conform to C's proper content," our teenager objects, "we must consciously already understand the concept C in some deeper and more definitive way." If our youth had indulged himself in the philosophical literature of the past ten years, he might speak of the "conceptual norms" we grasp as masters of the concept C: the alleged standards whereby the raw products of subconscious fancy are rationally adjudicated (I will return to this vein of thought in 10,iii). Once again, he sees no reason to suppose that this higher conceptual calling requires distasteful homework tedium. So it is solely in the lofty activities of "inspecting, weighing and constructing" that our ward proposes to devote his intellectual career. His purpose is to cultivate discernment in "the spying attention that catches whatever seems relevant," not to toil in grubby algorithmics.

(v)

True thought rigorization. A little reflection on reasoning principles can further strengthen our teenager's case. As we noted, easy-to-follow algorithms or recipes rarely prove wholly reliable, for most truly effective reasoning procedures sometimes go astray even if they perform well on most problems (many dutiful algorithms must be allowed their occasional Mardi Gras to compensate for their periods of Lent). These foibles of failure create a need for scrutinizing their product according to a higher grade of semantic discernment. As 4,x observed, this heightened canvass commonly assumes the form of a *correlational investigation* of potential failures in generic models framed according to a *picture* of how predicate "P" obtains its content. Although I've mentioned the importance of such "pictures" here and there in our discussion, we've not yet considered their positioning within our patterns of conceptual evaluation adequately. From our teenager's point of view, these episodes of reevaluating the imperfect products of a productive reasoning pattern display the deeper, albeit rarified, forms of *conceptual grasp* of which he proposes to become a master. As such, his perspective arises from an essentially *classical* assessment of the value of "semantic pictures." Our chief task in the later sections of this chapter is to scale back these traditional assumptions without, at the same time, underestimating the reconstructive value that readjusted correlational pictures genuinely offer.

Indeed, it was only in the nineteenth century that mathematicians began to adequately realize that the patently unacceptable products of otherwise reliable reasoning tools could not be airily dismissed as mere "sports" or "exceptions that prove the rule," although practitioners of earlier generations had tended to ignore anomalies in that manner. Some better story of how we might determine whether concepts have been "correctly applied" or not was plainly required. In fact, attempts to struggle with this problem in mathematics led to two basic proposals: a classical approach akin to our teenager's (which I'll dub *true thought rigorization*) and an opposed trend (*formalism*) that is engendered in the conviction that the classical school puts undue burdens on conceptual development. The tensions between these approaches parallel in many respects the oppositions we have already investigated between *traditional grasp* and *syntactic instrumentalism* (in Chapter 4) and between *classical gluing* and *pre-pragmaticism* (in Chapter 5). In my estimation, forming a proper appreciation of the value of "pictures" is critical to discovering a middle path of philosophical moderation that runs between these two extremes. Since this is the policy I intend to pursue in the second half of this chapter, it will be helpful if we temporarily abandon adolescent parable in favor of the history of science, to gain a warmer appreciation of the contrast I have in mind.

Because of the fact that a good notational choice can render previously elusive inferential processes virtually routine, it was common to write of the desirability of turning some "art"—activities where important discoveries are uncovered only through the efforts of intuitive and uncharted genius—into a "science," where the elusive guiding principles of intuitive discovery are codified into some strictly regulated calculus guaranteed to guide employers of no especial insight (or computers) to desired ends (for more on this art/science contrast, see section (xiii)). The mathematician Gauss explains:

> In general the position as regards all such new calculi is this—that one cannot attain by them anything that could not be done without them: the advantage, however, is, that if such a calculus corresponds to the innermost nature of frequent wants, everyone who assimilates it throughly is able—without the conscious inspiration of genius which no one can command—to solve the respective problems, yes, even to solve them mechanically in complicated cases where genius itself becomes impotent . . . . Through such conceptions countless problems which would remain isolated and require every time (larger or smaller) efforts of inventive genius, are, as it were, united into an organic whole.[41]

But it is unlikely that the "sciences" ($\approx$ algorithmally developed techniques) we develop in this manner are likely to prove *completely reliable*, at least in the guise in which they are first uncovered. We must remain warily cognizant of the possibility that a

---

[41] Merz, Thought, 724.

routine's suggestions may require *corrective check* of some sort or another. In this vein, T. J. McCormack in his introduction to <u>Lagrange's Lectures</u> employs the charming phrase "short-mind symbols" to designate the vocabulary employed in such a mechanical calculus and comments, once again citing Euler and his pencil:

> *For the development of science all such short-mind symbols are of paramount importance, and seem to carry within themselves the germ of a perpetual mental motion which needs no outward power for its unfoldment. Euler's well-known saying that his pencil seemed to surpass him in intelligence finds its explanation here, and will be understood by all who have experienced the uncanny feeling attending the rapid development of algebraic formulae, where the urned thought of centuries, so to speak, rolls from one's finger's ends.*
>
> *But it should never be forgotten that the mighty stenophrenic engine of which we here speak, like all machinery, affords us rather mastery over nature than an insight into it; and for some, unfortunately, the higher symbols of mathematics are merely brambles that hide the living springs of reality.*[42]

Note McCormack's phrase: "no outward power for its unfoldment"—he alludes to the observation, central to Hertz' thinking in 4,iii, that we can learn the inferential rules to direct the correct unfolding of "short-mind symbols" *purely syntactically*, without any need to consider what they mean. His point is simply that these syntax-based directivities can potentially conflict with the proper conclusions we would reach if we diligently attended to the predicate's underlying significance. And so a necessity for tradeoff compromise is often required. If we complicate the operations of our calculus so that it never leads to unhappy results, its refined workings may become so intricate that its mechanics will no longer prove of any utility to us. *Example*: our sorting of dollar bills by face plots patently represents a rough and ready routine and will sometimes cast suspicion upon completely blameless currencies. But if we complicate our methods of data translation so that family resemblances between our new "faces" *never* tempt us to assemble monies into spurious families, we will probably lose all advantages of the face plot technique in the process. "Faces" that never mislead are likely to prove as opaque in their saliences as the original money. Mathematicians commonly declare that the value of a mechanical calculus of this type is purely *heuristic*: its routine merely suggests interesting conclusions to the observer who then falls under an obligation to establish them by more defensible means. We shall return to this "heuristic" versus "proof" distinction in a moment.

---

[42] Thomas J. McCormack, "Introduction" in J. L. Lagrange, <u>Lectures on Elementary Mathematics</u> (Chicago: Open Court, 1901), p. v.

The observation that reasoners can easily be led in unhappy directions by trusting excessively the rushing flow of syntactically inspired directivities is very old—<u>vide</u> Descartes' lamentation in his "Rules for the Direction of the Mind":

> [The dialecticians] *prescribe certain formulae of argument, which lead to a conclusion with such necessity that, if the reason commits itself to their trust, even though it slackens the interest and no longer pays a heedful and close attention to the proposition inferred, it can nonetheless come to a sure conclusion by virtue of the form of the argument alone. Exactly so; the fact is that frequently we notice that often the truth escapes away from these imprisoning bonds, while the people who have used them in order to capture it remain entangled in them . . . Wherefore as we wish here to be particularly careful lest our reason should go on holiday while we are examining the truth of any matter.*[43]

Of course, syntactic exuberance is often pardonable:

> *The great creators of the infinitesimal calculus—Leibniz and Newton—and the no less famous men who developed it, of whom Euler is the chief, were too intoxicated by the mighty stream of learning springing from the newly-discovered sources to feel obligated to criticize fundamentals.*[44]

But the mathematician who blusters forward exclusively in this "drunk with reckoning" manner is likely to leave a lot of muddles for his descendants to sort out (as the subsequent history of the calculus amply reveals).

The classical picture of concepts advises that a just reasoner should be always prepared to retreat to what McCormack calls the "living springs" of thought—the realm of *conceptually supported* cogitation we employ when we don't indulge in automatism and truly think through the steps involved in our calculating. That is, if we have lifted an intellectual problem into the transferred dominion of some syntactic routine, we must be prepared to pull our thoughts back to their original contours. In this vein Gottlob Frege writes:

> *When we examine what actually goes on in our minds when we are doing intellectual work, we find that it is by no means always the case that a thought is present to our consciousness which is clear in all its parts. For example, when we use the word "integral", are we always conscious of everything appertaining to its sense? I believe that this is only seldom the case. Usually just the word is present to our consciousness, allied no doubt with a more or less dim awareness that this word is a sign which has a sense, and that we can, if we wish, call this sense to mind. But we are usually content with the knowledge that we can do this. If we tried to call to mind everything appertaining to the sense of this word, we should make no headway. Our minds are simply not comprehensive enough. We often need to use a sign with which we associate a very complex sense. Such a sign seems, so to speak,*

---

[43] Descartes, "Rules for the Direction of the Mind" in <u>Philosophical Works of Descartes</u>, Elizabeth Haldane and G. R. T Ross, trans. (Cambridge: Cambridge University Press, 1968), 32.

[44] Konrad Knopp, <u>Theory and Application of Infinite Series</u>, R. C. H. Young, trans. (London: Blackie and Sons, 1928), 1.

a receptacle for the sense, so that we can carry it with us, while being aware that we can open this receptacle should we have a need of what it contains.[45]

He allows that the discovery of important conclusions while motoring about solely by "heuristics" represents an essential part of practical technique:

[S]cience would come to a standstill if the mechanism of formulas were to become so rampant as to stifle all thought. Yet I would not want to regard such a mechanism as completely useless or harmful . . . On the contrary, I believe that it is necessary. The natural course of events seems to be as follows: what was originally saturated with thought hardens in time into a mechanism which partially relieves the scientist from having to think. Similarly, in playing music, a series of processes which were originally conscious must have become unconscious and mechanical so that the artist, unburdened of these things, can put his heart into the playing.[46]

But such a mechanization of inference should never be allowed free reign; its products must be eventually compared against "clear thoughts present to our consciousness in all their parts." This view captures what I call *true thought rigorization*: it supplies the basic recipe whereby the classical picture of concepts seeks to cure scientific ills.

In non-mathematical contexts as well, as when William James asks how Hegel managed, without discomfort, to pen some of his less intelligible passages or when Anthony Trollope marvels on how the tautologous bores of the world cheerfully make long speeches that convey no discernible information, they both hypothesize that the perpetrators of these pointless speech acts must be running on syntactical autopilot: buoyed by the warm sensations of flowing oratory, such individuals do not bother to establish a conceptual grounding for their words. Real thought has been abandoned, they suppose, leaving only a crust of "mechanical speech" on top. As M. V. Eggler claims:

The feeling of the words makes ten or twenty times more noise in our consciousness than the sense of the phrase, which for consciousness is a very slight matter.[47]

The key to reinstalling rigor, then, is to, in Frege's phrase, "call the submerged sense to mind." In the classical tradition, it is usually presumed that this can be done if we simply think about the subject hard enough, just as we found the error lurking in Euler's routine by careful inspection back in 4,x. Naïve accounts such as John Locke's maintain that "thinking harder" merely represents a matter of unwrapping our syntactic packages of underlying thought carefully, whereas neo-Kantians regard the process as one of subjecting our untutored thinking to the discipline of stricter intellectual norms or "regulative principles." When Frege himself writes of "signs serving as receptacles for meanings," a Lockean picture is suggested, although, in fact, he elsewhere seems to hold that the rules of proper logical inference serve as regulative principles to force would-be

---

[45]   Gottlob Frege, <u>Posthumous Writings</u> (Chicago: University of Chicago Press, 1980), 209.
[46]   Gottlob Frege, <u>Philosophical and Mathematical Correspondence</u> (Chicago: University of Chicago Press, 1980), 33.
[47]   William James, <u>Psychology</u>, 270. He references <u>La Parole intérieure</u> (Paris: 1881), 301.

thoughts to assume more respectable "contents."[48] But it is not uncommon to find that advocates of "clear thinking" are themselves not very clear about what they believe is required for the desired state of "true thinking" to manifest itself.

However, these seemingly innocuous classical expectations can quickly lead to strange and uncomfortable demands upon mathematical practice. For this reason, a contravening approach to rigor arose that opposed the true thought picture in a manner closely akin to the rejection of Boylean standards for "understanding" in science surveyed in 4,iv (if the reader bothered with that chapter). This new point of view emphasizes formalist considerations and contends that the predicates of mathematics, at least for a wide range of vital notions, can be supplied with an adequately precise "content" that is moored in axiomatics rather than "true thoughts." Through this stratagem, the often excessive requirements of orthodox classical thinking could be eluded.

To see what I have in mind, let's return to the antics of the geometers who pushed stodgy Euclidean geometry into the inflated projective dominions sketched earlier. As we noted, these unexpected extensions were originally justified by their early advocates as the "organic unfolding" of Euclid's ideas in a quasi-botanical mode. In the 1840s, however, the German geometer K. von Staudt supplied a different justificatory account whereby the assertions of the extended geometry are treated as disguised forms of regular Euclidean assertion according to a long and rather strange schedule of definitions ("points at infinity" are covert ways for talking about sets of parallel lines, etc.). I mentioned this program previously (in 4,iii), which proposes, in brief, that the claims of the extended geometers, when decoded into suitable "true thought" components, represent nothing but old-fashioned schoolbook geometry dressed in peculiar trappings. There is considerable evidence that Frege was inspired in his own work on mathematical foundations by this von Staudt precedent,[49] which fully adheres to a classical understanding of "rigor" as *follows an unblemished progression of impeccable thought*. Indeed, demands from traditionalist philosophers that geometry's strange practices require true thought underpinnings linger into the early twentieth century (Hastings Berkeley's Mysticism in Modern Mathematics of 1910 provides a good example of this species of literature[50]).

If the truth be told, much of this definitional bustle can seem rather *pointless*, besides proving tedious beyond endurance to read. After all, with a suitable schedule of definitions, we can likely embed the extended geometry within the theories of capital expenditure or model railroading with equal success, for it is hard to find reasons why one vein of interpretative "true thought" ought to be favored over another. Gestalt epiphany and syntactic excursion inspired the original emergence of the new geometry, after all, but those episodes provide little clue as to what the true thought interpretation

---

[48] Thomas Ricketts, "Concepts, Objects and the Context Principle" in Thomas Ricketts, ed., The Cambridge Companion to Frege (Cambridge: Cambridge University Press, forthcoming).

[49] Mark Wilson, "Frege: The Royal Road from Geometry" in William Demopoulos, ed., Frege's Philosophy of Mathematics (Cambridge, Mass.: Harvard University Press, 1995).

[50] Hastings Berkeley, Mysticism in Modern Mathematics (London: Oxford University Press, 1910). Julian Coolidge's position in Geometry of the Complex Domain lies neatly straddled between von Staudian and Hilbertian points of view (Oxford: Oxford University Press, 1924), ch. 8.

of a projective claim should look like (an "organic growth" story such as Ernst Cassirer provides in Substance and Function better accords with the originating phenomenology of the situation, although such an approach offers little handle upon rigorous procedure). It is not surprising that commonsensical practitioners such as Hilbert eventually asked themselves, "Do we really need to engage in all this von Staudt folderol except for occasional technical purposes? Isn't it clear *on the face of things* that the new projective notions are fully worthy of intensive study? Can't we tell a better story of rigor's requirements that doesn't force us to engage in this roundabout rigamarole?"

As we noted in 4,iii, the contrary proposal of Hilbert and his school is formalist in spirit. Complexified projective geometry can be set up as a worthy mathematical dominion in its own right through axiomatization, to which the neighboring system of regular Euclidean geometry stands in various interesting relationships. Each form of geometry implicitly defines (3,vii) its own set of concepts that can stand on their own parochial merits; the projective realm needn't be transliterated into Euclidean terms to earn mathematical respectability (although such coding has mild technical interest). Hilbert's impatience with the idea that the foundations of geometry require "true thought" moorings is revealed in a note he sent to Frege, where he complains:

> If one is looking for other definitions of a "point", _e.g._, through paraphrase in terms of extensionless, _etc._, then I must indeed oppose such attempts in the most decisive way; one is looking for something one can never find because there is nothing there; and everything gets lost and becomes vague and tangled and degenerates into a game of hide-and-seek.[51]

Indeed, loyalty to von Staudt's approach (which seems to have been reluctantly accepted as the right resolution of the status of projective geometry beforehand) seemed to have vanished virtually overnight after the publication of Hilbert's Foundations of Geometry in 1900 and was quickly replaced by a mocking disdain for von Staudt's complexities.[52]

. . . .

Even Felix Klein, inclined to be sympathetic with von Staudt, writes:

> If this entire construction is prolix and bothersome, in comparison with the ordinary real geometry, it can . . . supply incomparably more . . . . In most cases, to be sure, the application of this geometric interpretation notwithstanding its theoretical advantages, might create such complications that we should be satisfied with its theoretical possibilities and return actually to a more naïve standpoint: a complex point is the aggregate of complex coordinate values with which, to a certain extent, one can operate as with real points.[53]

. . . .

[51] Frege, Correspondence, 39.

[52] "In proving that geometry could, conceivably, get along without analysis, von Staudt simultaneously demonstrated the utter futility of such a parenogenetic mode of propagation, should all geometers ever be singular enough to insist upon an exclusive indulgence in unnatural practices"—E. T. Bell, The Development of Mathematics (New York: McGraw-Hill, 1945), 348–9.

[53] Felix Klein, Elementary Mathematics from an Advanced Standpoint: Geometry, E. R. Hedrick and C. Noble, trans. (New York: Dover, n.d.), 129.

In our earlier discussion of this episode (4,iii), we noted that appeal to formal axiomatics alone eventually needed to be supplemented by a program of definability within set theory, due to the fact that strong formalisms could not be proved self-consistent otherwise. In doing so, the technique superficially appears as if it represents a return to true thought demands (and is often so understood by contemporary philosophers who accept without cavil that "The only concepts mathematics treats are set-theoretic in content"). But a comparison with the von Staudt alternative shows that the "contents" of geometrical predicates have become considerably thinned out in comparison to earlier views.

Although our primary concern is not with the predicates of pure mathematics in this book, I believe that Hilbert's instincts in rejecting excessive true thought demands are correct, but it is a mistake to suppose that, to evade these demands, the overall personality of our predicates must become as attenuated as the formalist story suggests. Instead, a different story of what occurs in semantic detoxification is wanted, a matter to which we shall revert later in the chapter.

(vi)

Teenage victory. Returning to our main themes, in our teenager's thinking we plainly witness the reentry of the "great ideas" worship discussed in connection with the Darwin critic of 2,ii: it is only the pure grasp of important concepts that matters, not the fussy details of how they are applied in concrete circumstances. If we are honest with ourselves, we should recognize the insidious magnetism of this conceit. Most of us would like to know everything there is to know, but, as this state is plainly unattainable, we cultivate rationales for skimming off the essential cream of a subject, leaving the more objectionable aspects of toil to the little fellow. In this fashion, we fall into a form of intellectual self-regard that generally carries foolishness and error in its wake: the attitudes of those who know little of substance but fancy themselves to be great judges of intellectual capital when it is presented to them ("I can recognize a good idea when I see it" is the characteristic mantra of the self-important). Scions of wealth who ascend to great offices through family ties commonly develop arrogancies of this ilk: they persuade themselves that they enjoy some distinguished "intuitive insight" that guides their decisions faultlessly onward, much to the chagrin of the better informed underlings forced to endure such whimsies. Through his pulp fiction philosophizing, our ward has come to cherish unmoored conceptual attainment of exactly this undesirable type.

In a similar way, many physicists (cheered on, no doubt, by the hagiography of popular journalism) fancy that they possess miraculous powers of "physical intuition" that allow them to peer deeply into the inferential future without needing to bother with the annoying gnats of rigorous mathematical foundation or other pesky curbs upon their farseeing Genius. The often admired braggadocio of Richard Feynman strikes me as of this category, having fallen into the trap of

"considering Invention either avowedly or tacitly as a kind of revelation" (to preview Franz Reuleaux' opinion from section (xiii)). Any considered inspection of the past history of physics reveals that, although the intuitive Genius often propels his or her subject into brave new territory, fairly soon the path ahead becomes nettled and indistinct and distress calls for a competent applied mathematician are soon heard from the underbrush. In fact, the Heaviside case we shall shortly rehearse is exactly of this type.

Similar mystic exaggerations of intellectual capacity are to be found everywhere. For example, consider Ruskin on J. W. Turner's ability to render coastlines plausibly:

> ... but every intelligent spectator will feel the difference between a rightly-drawn bend of shore or shingle, and a false one. Absolutely right, in difficult river perspectives seen from heights, I believe no one but Turner ever has been yet; and observe, there is NO rule for them. To develop the curve [of a river] mathematically would require a knowledge of the exact quantity of water in the river, the shape of its bed ... and even with these data, the problem would be one which no mathematician could solve but approximately. The instinct of the eye can do it; nothing else.[54]

In my opinion, we should be wary of misty appeals to "instincts" and "intuitions" in this vein, for they convert neutral placeholders for complex processes we do not yet understand into positive capacities of occult contours. The false confidence in conceptional supremacy that such conceits encourage often blinds their victims to the seasonal alteration of chores that we shall outline as necessary in the next section. It is better, when we don't yet understand precisely *why* we have selected an alterative, to confess our ignorance and not attempt to fill the gap with idle "intuition." To rework Thumper's maxim from the movie <u>Bambi</u>:

> If you can't say something substantive, don't say nuthin' at all.

---

[54] John Ruskin, <u>The Elements of Drawing</u> (London: The Herbert Press, 1991), 93.

....

We might observe, without wishing to diminish Turner's magnificence in the least, that Ruskin's claim is surely exaggerated. In fact, to recognize lapping water as such, it seems fairly likely that the "instinct of the eye" normally accomplishes the pattern recognition by some sort of data analysis akin to the Fast Fourier transform. Apparently, viewers presented with a cartoon "water" simulated in such a mode are sometimes more likely to identify it as "real water" than some photographed, authentically aqueous comparison.[55]

....

In any event, the fact that we must often scrutinize the proffered fruits of our sub-conscious deliberations according to higher, "true thought" standards plainly encourages our teenager's conception of his intellectual destiny: in evading his home-work, he can more effectively devote his time to developing the "characteristically human" aspects of his conceptual discernment to their highest capacities. Consider this charming passage from Noah Porter, formerly president of Yale and the author of The Elements of Intellectual Science of 1885. He invites us to consider an isolated islander who has never seen a goat before. Porter claims that, in such contexts, human beings, but not animals, display an ability to ponder issues such as: "I am instinctively tempted to apply the concept *horned hog* here, but am I right in doing so? Should some fresh notion need to be derived to fit a weird animal like this?"

> *Herein lies the difference between the act of a brute and the act of a man in perceiving objects that are alike. In one sense, the brute may perceive what is similar as readily as a man; in some cases, even more quickly, for his senses may be more keen . . . But the brute does not attend and analyze as does a man. Hence he cannot discriminate so as to abstract; or, at best, the degree and range of such efforts must be very limited . . . [In contrast to animals, however,] we do not judge by a mechanical and superinduced act of the intellect, which, finding two names of notions, proceeds to fasten them together . . . . [Instead, w]herever there is a [genuine, humanly created] notion, there is an implied act of judgement . . . It is an organic thing, expressing the very essence of the act that gave it being, and capable of being developed into similar though more complex products. It is like a seed, which is a miniature plant, having come from a plant and ready to spring into a plant; or it is like the cell which is the organized and organizing element of development in vegetable or animal life.[56]*

Porter evidently believes that, buried deep within the individual concepts *hornedness* and *hogness,* lie conceptual seeds able to generate (under the nurture of human scrutiny) an apt classifier of the genus Capra, because their "organic outgrowth" will produce a richer product than the simple conjunctive quantity *being both horned and a hog.* Fully approving of Porter's recasting of the Great Chain of Being, our teenager regards algorithmic exercise sets as "mechanical and superinduced acts," whereas he plans to

[55] T. J. Kung and W. A. Richards, "Inferring 'Water' from Images" in Whitman Richards, ed., Natural Computation (Cambridge, Mass.: MIT. Press, 1988).

[56] Noah Porter, Elements of Intellectual Science (New York: Charles Scribner's and Sons, 1885), 331–59.

focus solely upon "attending and analyzing" as only a human being can. To our youth, the final discernment of conceptual merit represents the chief and highest task reserved to our human selves, positioning "consciousness as a fighter for ends or a director of strategies" (to cite Gerald Myers' pithy summation of William James' view).[57] The mute workmen within our subconscious selves send up coarse packets of somewhat defective judgment, but we, in our conscious wisdom, schedule these modules so that they better serve the ends of refined conceptual understanding. Plainly important insights into conceptual behavior genuinely lie entangled in this "director of strategies" conception— as Chapter 6 observed, success in linguistic endeavor is often a process of adapting mildly unsuitable tools to excellent descriptive purpose. Yet something has gone awry as well, because such musings encourage our teenager into adopting an unsuitable "skim off the cream" picture of conceptual obtainment apt to lead to a quite ineffectual adulthood. The hidden hand behind this *ur*-philosophical villainy, I shall argue in the next section, is our old nemesis, the classical picture of concepts.

But let us momentarily capitulate to our teenager's view of conceptual attainment. What exactly does he propose to do when he "inspects, weighs and constructs" proffered packages of doctrine according to his "true thought" standards? In asking this question, we immediately confront the hollow core of classical doctrine as described in 3,vi: we are not told, with any concreteness, what the conceptual contents of specific predicates are like. And this uncertainty induces a wide spectrum of possible opinion as to how the engines of conscious conceptual discernment operate. Beginning with a conservative, traditional treatment, John Locke optimistically regards the affair as largely one of unpacking and rearranging our mental suitcases:

> Another faculty we may take notice of in our minds is that of discerning and distinguishing between the several ideas it has. It is not enough to have a confused perception of something in general. Unless the mind had a distinct perception of different objects and their quantities, it would be capable of very little knowledge . . . How much the imperfection of accurately discriminating ideas one from another lies, either in the dullness or faults of the organs of sense; or want of acuteness, exercise, or attention in the understanding; or hastiness and precipitancy, natural to some tempers, I will not here examine: it suffices to take notice, that this is one of the operations that the mind may reflect upon and observe in itself.[58]

Indeed, this is the picture of conceptual decomposition into rock bottom, pre-articulated components conjured up within the "true thought" narratives sketched by nineteenth century rigorists such as Thomas McCormack. As such, this Lockean conception of "clear thinking" doesn't render adequate justice to the "deeper understandings" highlighted by our teenager, where dowdy Euclidian notions like *point* and *line* suddenly begin to call for radical application to complexified entities or a novel *goat* organically emerges from the classificatory miasma of Porter's islander.

[57] Gerald Myers, William James: His Life and Thought (New Haven: Yale University Press, 1986), 64.
[58] Locke, Essay, i. 202.

Switching to William James' Principles of Psychology, we encounter a dramatically different orientation, although, oddly, James writes as if he is merely reiterating Locke (he enthusiastically endorses the relevant portion of An Essay on Human Understanding with the injunction "Read the whole section!"). James writes:

> The function by which we thus identify a numerically distinct and permanent subject of discourse is called CONCEPTION; and the thoughts which are its vehicles are called concepts . . . Each conception thus eternally remains what it is, and never can become another. The mind may change its states, and its meanings, at different times; may drop one conception and take up another: but the dropped conception can in no intelligible sense be said to change into its successor. The paper, a moment ago white, I may now see to have been scorched black. But my conception "white" does not change into my conception "black." On the contrary, it stays alongside of the objective blackness, as a different meaning in my mind and by so doing lets me judge the blackness as the paper's change. Unless it stayed, I should simply say "blackness" and know no more. Thus, amid the flux of opinions and physical things, the world of conceptions, or things intended to be thought about, stands stiff and immutable, like Plato's Realm of ideas.[59]

As I understand him, he is scarcely claiming that we perform our conceptual discriminations in Locke's tidy matter, where we simply segregate the ingredients that lie before our mind's eye crisply. Instead, James believes, much like his brother Henry, that an adequate account of *why* a given person categorizes presented experience exactly as he does can probably be charted, if at all, only by throughly tracing the clandestine cookery that stews within "the deep well of unconscious cerebration." Our conscious classifications occur only as these hidden mechanisms punctuate the flow of our consciousness awareness in some primitive way. Although some inadequate measure of the subconscious impulses that drive the classification will appear imprinted within the coin of conscious activity, the ultimate springs of these activities will remain largely hidden from review. All we may consciously grasp is simply some hazy intimation of hidden depths, in the way that:

> Great thinkers have vast premonitory glimpses of schemes of relation between terms, which hardly even as verbal images enter the mind, so rapid is the process. We all of us have this permanent consciousness of where our thought is going.[60]

Or:

> The sense of our meaning is an entirely peculiar element of the thought. It is one of those evanescent and "transitive" facts of mind which introspection cannot turn round upon, and isolate and hold up for examination, as an entomologist passes round an insect on a pin. In the (somewhat clumsy) terminology I have used, it pertains to the "fringe" of the subjective state, and is a "feeling of tendency," whose neural counterpart is undoubtedly a lot of dawning and dying processes too faint and complex to be traced.[61]

---

[59] James, Psychology, 436–7.   [60] Ibid., 247.   [61] Ibid., 446.

And finally:

> *And has the reader never asked himself what kind of mental fact is his intention of saying a*
> *thing before he has said it? It is an entirely definite intention, distinct from all other*
> *intentions, an absolutely distinct state of consciousness, therefore; and yet how much of it*
> *consists of definite sensorial images, either of words or things? Hardly anything! Linger,*
> *and words and things come into the mind; the anticipatory intention, the divination is*
> *there no more. But as the words that replace it arrive, it welcomes them successively and*
> *calls them right if they agree with it, it rejects them and calls them wrong if they do not.*[62]

In this contrast between Locke and James, we witness another reflection of the sense data/concept dualism highlighted in 7,xi, for Locke's sharply presented concepts correlate nicely with the similarly articulated sense data of traditional thinking about perception, whereas James' fringed discriminations resemble the evanescent and unanalyzable gestalts of the percept theorists.

Plainly, with respect to phenomenological accuracy, the Jamesian portrait of what we do when we attempt to think clearly is more nearly correct than Locke's utopian assessment. Rarely can we evaluate the *worthiness* of an idea simply from some bare, dictionary-like catalog of its components: we have to "know our way around with the notion" and that capacity often demands that we cobble together our intellectual capacities in complex and rather unruly ways. Our face plot example nicely illustrates the reasons: we can't tell whether a proffered collection of statistical parameters possess genuine descriptive merit until we can harness them somehow to other processes of deep cerebration. And so Richard Feynman is right that, in real life, the quest for "conceptual clarity" often assumes the guise:

> *What I am really trying to do is bring birth to clarity, which is really a half-assedly*
> *thought-out pictorial semi-visional thing.*[63]

"Evaluating the worthiness of ideas"—that is our teenager's intended forte, and we now find ourselves in a position where we can't seem to place any concrete demands on how that is to be accomplished, and must allow him to indulge, as he prefers, in half-assedly thought-out pictorial semi-visional things.

Indeed, although our youth has argued us to our present point of *ur*-philosophical defeat by continually reminding us of the importance of submitting predicates to unexpected forms of intellectual review, it isn't clear, on a Jamesian portrait of veiled judgmental standards, how we manage to achieve any *precision* or *objectivity* in the process, for we now lack Lockean assurances that crisp ingredients lie available within our mushy thoughts. Instead, "conceptual clarity" appears as if it merely represents a new form of diffuse judgment: as conceptual contents pop in our mind, we can either "welcome them successively and call them right if they agree with our expectations and reject them and call them wrong if they do not." This *ur*-philosophical state of affairs should alarm us, for we recall the abundance of crackpots (e.g., the aforementioned

---

[62] James, Psychology, 247. Wittgenstein, Investigations, §245.   [63] Gleick, Genius, 244.

L. Ron Hubbard or Alfred Lawson) who have convinced themselves that they represent paragons of rigor ne plus ultra. But it is hard to find the tracing of conceptual authenticity that discriminates the good from the bad and ugly within our conceptual musings; we require a mark of "true content" and can locate nothing plausible. And thus we find ourselves in the unhappy position of, to quote Hilbert again:

> *looking for something [we] can never find because there is nothing there; and everything gets lost and becomes vague and tangled and degenerates into a game of hide-and-seek.*

Our teenager, however, is quite unperturbed by our difficulties, for he is content to loaf about his room, play his records too loudly, and to welcome great ideas successively and call them right if he agrees with them.

(vii)

Correlational pictures. What has gone wrong? How has our teenager procured this preposterous victory over homework duties? Essentially, the trick is turned through a clever program of "bait and switch," relying upon a tacit appeal to our classical heart strings. Our youth invites us to see misty but invariant conceptual contents persisting through the developmental histories of our predicates and then directs our attention to those favorable moments where the other directivities of routine calculation seem clearly marginal to our focus of evaluative attention, having been displaced by epiphany or other evolutionary development. Thus in section (ii) he highlights those reorientational moments when the old calculus notion of **derivative** ("dy/dt") spawns its "generalized derivative" prolongations, in the course of which previous algorithmic verities are set aside. "See!," our charge exclaims, "your bothersome homework assignments can't embody what is vital in conceptual attainment, so why should I bother with such peripherals?"

The proper reply to such sophism is to observe that directivity judgments of the type "the conceptual content of C is . . ." or "C's proper application is . . . ." are *seasonal* in their evaluative focus: the answer we supply on day A may be quite unlike that supplied on day B. This thesis should hardly be regarded as surprising, given that our previous chapters argued that many common predicates of macroscopic evaluation (e.g., "hardness") behave in essentially similar ways and can obtain a measure of efficiency in linguistic engineering thereby. Phrases like "conceptual content" serve as our chief instruments for highlighting the salient directive factors operative upon a wandering predicate and isolating the future path we find most conducive to effective linguistic function (in 7,ix, I called this their "spotting function"). Given that these salient directivities can shift dramatically from predicate to predicate, as well as altering considerably over the courses of their developmental histories, it isn't surprising that the evaluative term "concept" imitates "hardness" and "redness" and likewise acquires a contextual sensitivity to developmental state. Similar observations apply to the wider family of phrases that monitor linguistic condition in related ways. So we shouldn't be misled into presuming that *any* passing episode reveals the "essence of proper conceptual grasp."

The mechanisms of patch-based prolongation examined in earlier chapters suggest that these seasonalities in evaluative focus will often reflect the *maturity* of a usage's developmental state. A facade structure commonly enlarges when certain strands of practical advantage $\mathcal{R}$ utilized in some established patch A are found to extend profitably beyond its natural boundary ∂A. Often, like the escapees in <u>Grand Illusion</u>, the linguistic agents involved may be utterly oblivious to the border crossing. Once the initial incursion into a fresh patch B has been assayed, trial and error retooling can articulate a rough and ready system of controls that prevents unhelpful information and/or inferential rules from being imported from the original homeland A. Since these nascent controls can assume the innocent guise of salient "facts" in the manner of 7,viii, our wandering employers may be left with little inkling of how far their usage has strayed from patch A. As confidence in the B-situated usage grows, free employments of predicate "P" outside the limiting contours of $\mathcal{R}$-rule routines begin to flower, allowing Uncle Fred ample opportunities to pontificate freely over patch B as well. In short, fairly radical changes can transpire within the transition from A to B even though the society believes that "they are simply using 'P' in the old-fashioned way" (in this innocence, they resemble the Druids of 1,ix).

Eventually, however, the disharmonies between old and new usage can become so pronounced that some more systematic resolution is required (*ur*-philosophical puzzlement with respect to "P" may create a crisis of genuine practical import, for example). This necessity brings forward a new aspect within our discussion: the need to *detoxify old pictures* of usage. I have mentioned "correlational pictures" several times before, *e.g.*, in considering the role that projection rules play in the preface of a geographical atlas (indeed, 6,iii employed the terms "preface" and "picture" interchangeably). The general idea is that pictures embody the generic stories that speakers tell themselves with respect to how their predicate's usage matches to worldly support within normal circumstances of application. The simplest picture of this type is that which we have dubbed "is a dog"/*being a dog* registration (2,ii): over all its applications, "P" is invariably associated with the attribute φ. A more complicated picture claims that "P" lies situated upon a facade: "Within patch A, 'P' is associated with attribute φ; but over patch B, 'P' correlates with attribute ψ." Our discussion has suggested that far more complicated strategic arrangements are possible as well, although I generally utilize facade-like arrangements as our chief foil with respect to unadorned "P"/φ circumstances. When we decide that "is hard" cannot cover its applications in simple "P"/φ manner, but instead represents a quasi-attribute covering, we have *altered our picture* of how "hardness" correlationally behaves and thereby *detoxified* the usage of an erroneous former conception. A major objective in the remainder of this chapter is to investigate how we normally talk about "conceptual contents" during these episodes of detoxification and how the import of the changes in picture is weighed. To this end we will survey a celebrated episode from the history of science (Oliver Heaviside's operational calculus) that displays the central ingredients I have in mind in vivid detail. Heaviside was roundly criticized in his time for his strange methodological attitudes, but time has amply demonstrated that he was right and his critics wrong. Modern philosophy of language can learn much from the rugged common sense that Heaviside evinces during a period of especially vexing linguistic circumstance.

The term "picture" could be employed more widely than I do (and is so utilized by Wittgensteinians). I restrict "pictures" to *generic representations* of the manner in which "P"'s sentential employments *correlate* with worldly features such as physical attributes φ (or more complicated packages of information like quasi-attributes), in possibly complex and contextually sensitive ways. The qualifier "generic" is added because such pictures are always schematic in the manner of the abstract rule for a Mercator projection: no equivalent of concrete geographical fact is contained in the preface itself. Sometimes practitioners struggling with a predicate cite *philosophical excuses* for maintaining that such terms do *not* require any correlations to prove linguistically useful: indeed, the syntactic instrumentalists of Chapter 4 maintain precisely that position. If we wish, we might reasonably say that such thinkers entertain a *non-correlational picture* of language's workings.

However, I do not take this point of view very seriously because I believe that non-correlational explanations of a predicate's successes are rarely appropriate and largely appear in the historical record as temporizing excuses. Thus in 4,i we briefly surveyed Karl Pearson's invocation of neo-Kantian regulative ideals as a means to rationalize certain oddities of derivation in the mechanics of elastic bodies (the term "material particle" needn't display any true correlation to external things, Pearson claims, because it is *we* who impose this predicate upon the world as an organizational framework). But such anti-correlational musings are now seen as simply a stopgap to cover some missing mathematics that Pearson was unable to supply.

My own working assumption is that successful linguistic routines work well only if they covertly code supportive physical information in their component sentences and that, after rough improvements have been made, a usage can be adjusted to a higher stage of improved performance only if its users are able to frame a picture that more accurately captures its actual correlational condition. The best way to render this point of view plausible is simply to investigate salient historical cases in some detail. In consequence, when I write of right and wrong "pictures" in the rest of this chapter, I shall be mainly concerned with a competition between distinct correlational stories, having set aside instrumentalist and other "philosophical" responses as mere inhibitors of progress. But in Chapter 10, I will return to a discussion of opinions like Pearson's in a more direct manner.

In fact, rarely do pictures succeed one another immediately: often an intervening period of *semantic agnosticism* intervenes where no correlative portrait is greatly trusted. The Heaviside case will nicely illustrate the delicate decision that we often confront in steering a useful language along an improving trajectory: when should we construct and trust semantic pictures and when should we leave such endeavors alone? Here we should be guided by the seasonalities described in Ecclesiastes: *there is a time to cast away*

Heaviside

stones and a time to gather stones together. In fact, Heaviside advances a powerful case for waxing agnostic for certain intervals in a predicate's career, as it struggles to regain its sea legs after a rocky prolongation into fresh applicational waters.

In these respects, I will call a developmental history *canonical* if it follows the pattern: reach beyond the patch boundary ∂A through $\mathcal{R}$-based incursion $\rightarrow$ consolidate a wider range of practical and impractical usage over $\mathcal{B}$ $\rightarrow$ detoxify erroneous old picture $\mathcal{P}$ and articulate replacement semantic portrait $\mathcal{P}^{*}$ $\rightarrow$ correct B-patch (and possibly A-patch) usage through the use of $\mathcal{P}^{*}$. In dubbing this pattern "canonical," I merely intend that it roughly qualifies the shape of most of the linguistic developments we shall study, *not* that every predicate's progress fits its pattern. In the history of mechanics, it is not uncommon to find predicates that have suffered a longer succession of these patterns.

The relevance of all this to our teenager's thinking lies in the fact that a sound understanding of a canonical development allows us to appreciate that the operative directivities that affect our shifting evaluations of "correctly uses predicate P" or "understands concept C" do not stem from one concentrated source or are wholly grasped within any single mental act, but interweave with one another in natural developmental pattern, much, in Quine's fashion, like the boat that gets rebuilt at sea. The role that correlational pictures play in all this is crucial, because a correct assessment of their contributions allows us to consign an important improving role to *rigorous procedure*, without succumbing thereby to the alluring flatteries of classical thinking.

(viii)

I heard the voice of an algorithm. Oliver Heaviside was a self-educated Victorian engineer who developed a number of unusual innovations within applied mathematics, including the "operational calculus" method of approaching linear differential equations. He mainly applied these novelties in electrical work, such as the problem of transmitting signals without unacceptable distortion over long telephone lines. He based his discoveries on the suggestions of earlier mathematicians, as outlined below, but pushed their algebraic directivities into far reaching territories through essentially trial-and-error methods. So contrary were these procedures to prevailing conceptions of rigor that the Royal Society declined to publish some of his most valuable results, leaving Heaviside indignant at what he called "the wet blanket" cast by "Cambridge mathematicians." A great individualist,[64] Heaviside responded to these criticisms with great vigor and a commonsensical appreciation for the seasonal necessities of agnosticism. Indeed, it is largely from reading Heaviside that I have come to adopt the semantic opinions articulated here. Although this episode can be portrayed as merely an anomaly within the history of mathematics, I believe that, in fact, Heaviside articulated, in his own vivid and self-created terms, a rather modern attitude with respect to differential equations. As such, a review of its history can bring the typical issues involved in semantic epiphany into admirably sharp focus.

If the reader has only taken a standard calculus course and never witnessed such techniques before, the maneuvers of Heaviside's calculus will seem utterly whimsical. Specifically, its general propensity is to manipulate differential *operators* as if they were *numbers*. That is, beginning with the equation

(1) $dy/dt + y = t^2$,

Heaviside will "factor" it

(2) $[(d/dt) + 1]y = t^2$,

then "divide" it

(3) $y = t^2/(d/dt + 1)$,

and finally "expand" it (in analogy to $1/1 + x = \sum (-1)^{n+1} 1/x^n$, valid if $|x| > 1$)

(4) $y = \left\{ \dfrac{t^2}{d/dt} - \dfrac{t^2}{d^2/dt^2} + \dfrac{t^2}{d^3/dt^3} - \cdots \right\}$.

Prima facie, these are extraordinary liberties to take, for, on the face of it, such procedures are about as sensible as dividing both sides of the movie star equation

Cary Grant = Archie Leach

by "Cary" to derive a conclusion about our eighteenth president:

Grant = Archie (Leach/Cary).

Heaviside then maneuvers the gobbledegook on the right side of his equation to a form he could interpret by one of his self-styled "algebratizing rules". For example, the rule "Algebratize $1/d^n/dt^n$ as n-fold integration" would convert the nonsense above to:

(5) $y = \dfrac{t^3}{3} - \dfrac{t^4}{3.4} + \dfrac{t^5}{3.4.5} - \cdots$

[64] His biographer Nahin observes that Heaviside replaced most of his household furniture with large boulders in his declining years.

which reduces, by standard results on series, to

$$y = 2 - 2t + t^2 - 2e^{-t}.$$

This, <u>mirabile dictu</u>, is the problem's correct answer for an important form of starting state. The odd "algebratizing" rules may lead one to suspect that this particular result has been inserted by hand but, in fact, Heaviside invariably obtained correct answers through such apparent lunacy. Moreover, his algorithm generally found the right answer more quickly than orthodox methods (when the latter could be made to work at all).

In terms of the abstract description provided in the previous section, what has occurred is this. Heaviside begins in the patch of ordinary calculus language and technique **A** but he borrows algorithmic rules $\mathcal{R}$ from numerical algebra that carry him to some odd location **B** lying beyond ∂A (as witnessed by the appearance, not seen in the orthodox calculus, of previously nonsensical expressions like "$t^2/d/dt$"). Through trial and error experimentation (often involving comparison with results known to be correct within **A**), Heaviside revises $\mathcal{R}$ to an improved $\mathcal{R}^*$ with various supplements (<u>e.g.</u>, his "algebratizing" procedures) and curbs. Eventually he becomes sufficiently comfortable within the **B** arena that he often finds it a preferable venue for thinking about the behavior of his electrical circuits, rather than adhering solely to old-fashioned **A**-patch usage. Many anomalies emerge in the course of employing this calculus in comparison to traditional technique, sometimes leading to a complete disagreement in answers (I'll later supply a striking example in the fine print). Nonetheless, when final results are compared to the values measured within a suitable electrical circuit, Heaviside's discordant answers generally prove correct. When the causes of these disparities are finally diagnosed fifty years later, it is discovered that Heaviside's methodology has dragged the regular calculus term "dx/dt" into alignment with a mathematical operation considerably more exotic than that supplied in the orthodox δ/ε treatments of freshman calculus courses. In a similar way, the "functions" Heaviside discusses undergo a related shift in semantic alignment as well (to "generalized functions of "Dirac's δ-function" as we shall discuss in section (ix)). In this manner, Heaviside's **B**-patch extension has doubled back to become multi-valued over the original domain of **A**-coverage, for linear differential equations themselves obtain a revised reading under these natural prolongations.

Heaviside was astonishingly direct in defense of the empirical methodology he followed. Thus he explains how an attempt to conquer new descriptive territory is prosecuted:

> It is by the gradual fitting together of the parts of a distinctly connected theory that we get to understand it, and by the revelation of its consistency. We may begin anywhere, and go over the ground in any way . . . It may be more interesting and instructive not to go by the shortest logical course from one point to another. It may be better to wander about, and be guided by circumstance in the choice of paths, and keep our eyes open to the side prospects, and vary the route later to obtain different views of the same country. Now it is plain enough when the question is that of guiding another over a well-known country, already

well explored, that certain distinct routes may be followed with advantage. But it is somewhat different when it is a case of exploring a comparatively unknown region, containing trackless jungles, mountains and precipices. To attempt to follow a logical course from one to another would then perhaps be absurd. . . . You have first to find out what there is to find out.[65]

If seemingly reliable rules lead to mysterious expressions like "$t^2/\sqrt{d/dt}$," we must "make the best use of them that we can":

[I]n mathematics, the fundamental notions are so simple that one might expect that unlimited developments could be made without coming to anything unintelligible. But we do, and in various directions . . . [W]hen [such] things turn up in the mathematics of physics, the physicist is bound to consider them and make the best use of them that he can.[66]

Did he "understand" what he was doing? Well, certainly in a distributed normativity kind of manner, but he was unable to supply orthodox mathematical underpinnings for his procedures in terms of an adequate associated picture.

But then the rigorous logic of the matter is not plain! Well, what of that? Shall I refuse my dinner because I do not fully understand the processes of digestion? No, not if I am satisfied with the result . . . First, get on, in any way possible, and let the logic be left for later work.[67]

But does this lack of foundation represent a failure that should lead to the utter rejection of his methods? No, someday someone should be able to figure suitable underpinnings, but the time may not be ripe for such an undertaking now:

As the subject opens out, so does the theory improve. But it can only become logical when the subject is very well known indeed, and even then it is bound to be only imperfectly logical.[68]

In every one of these contentions, Heaviside has been absolutely vindicated by subsequent research and displayed uncommon common sense in defending his explorations in the teeth of quite harsh criticism by academic mathematicians.

In being thus pulled onto a new sheet of calculus vocabulary usage, Heaviside is coaxed forward by the beckoning arms of the easy inferential techniques that we learn in high school algebra, now borrowed for the sake of an unaccustomed context. In fact, many mathematicians before Heaviside had responded to these same algorithmic allures. Indeed, Leibniz himself made a few proposals of this type, almost immediately upon inventing the "dx/dt" notation that calls these yearnings forth. Writing to John Bernoulli:

Many things thus far lurking in the summations and progressive differentials will gradually come forth . . . I believe I do not know what is hidden there.[69]

[65] Heaviside, Electromagnetic ii. 3.   [66] Ibid., 9.   [67] Ibid., 8.   [68] Ibid., 123.
[69] Eugene Stephens, The Elementary Theory of Operational Mathematics (New York: McGraw-Hill, 1937), 265–6.

The subsequent history of operational methods has been well documented in recent years[70] and we know that Heaviside acquired the rudiments of his techniques from the Treatise on Differential Equations by the well-known Irish mathematician George Boole. In truth, Heaviside's real genius lay in the ways that he used these venerable techniques to extract truly surprising "hidden information" from differential equations in a manner we'll outline later. However, I will first offer my readers (especially those with a little experience in elementary logic) an opportunity to "wander about, guided by circumstance in the choice of paths" themselves, for it is philosophically edifying to experience first hand the call of a delectable but mysterious algorithm, and to engage in the cut-and-paste work required to adjust such routines to new surroundings. For these purposes, we are happily aided by the fact that Heaviside's muse George Boole, having completed his studies in the operational calculus, then turned his attention to symbolic logic where he practiced a very similar methodology (Boole is best remembered today for his The Laws of Thought[71] of 1854). By approximately retracing some of Boole's logical peregrinations, the reader can participate in a reasonable facsimile of Heaviside's own gambols, without requiring experience in differential equations or electrical engineering. The main difference is that Boole's results turned out rather crabbed and awkward, whereas Heaviside's proved gloriously successful. Their essential affinity in methodology is well worth recalling, for Boole is commonly excoriated for his defective philosophical methodology (by Frege and Michael Dummett inter alia[72]). But these are the wrong lessons to extract from Boole. Indeed, in some aspects, the entire argument in this book represents a lengthy modus tollens upon such conclusions, for the glories of Heaviside's parallel endeavors indicate that something is deeply wrong in the philosophical presumptions of those who criticize Boole for essentially doing the same thing. But I shall come back to these curious issues later.

....

What now ensues properly represents a reconstructed Boole, adequate only to my illustrative purposes (it is not a completely specified method). The real Boole engages in a wide variety of often ill-specified techniques, most of which are motivated by class-focused applications and address purposes other than the "whodunits" emphasized here. He also commonly employs an "expansion theorem" method that is more like Quine's fell swoop resolution technique[73] than the algebraic borrowings I highlight.

....

Let us direct our attention to a particular species of problem that we might call "whodunits": they represent a common staple of the "Challenging Puzzles in

---

[70] Elaine Koppelman, "The Calculus of Operations and the Rise of Abstract Algebra," Arch. Hist. Exact Sci. 8 (1971). I. Grattan-Guinness, The Search for Mathematical Roots (Princeton: Princeton University Press, 2000).

[71] George Boole, An Investigation of the Laws of Thought (New York: Dover, n.d.).

[72] Gottlob Frege, "Boole's Method and my Own" in Posthumous Papers. Michael Dummett, Frege and other Philosophers (Oxford: Oxford University Press, 1996).

[73] W. V. Quine, Methods of Logic (New York: Holt, Reinhart and Winston, 1959).

Logical Reasoning" oeuvre. Here's a simple case (unworthy of the cleverness of real life specimens):

> Somebody has killed poor Mr. Weatherbee and it is known that either Archie and Betty committed the crime together or Veronica did it without the two of them. Likewise, there's convincing evidence that either Jughead did it or Archie did it while Betty remained innocent. Question: under what conditions was Archie involved in the plot?

The answer can be seen to be: just in case he worked in collusion with Betty and Jughead but not Veronica or his conspiracy was formed with Veronica and not the other two.

Although a proper answer is easy to discern in this simple instance, a casual student of modern elementary logic may experience some difficulty in figuring out how to resolve problems of this breed in systematic terms. This is because the techniques usually taught in modern universities are not directive in the right kind of way: they either impose no requirements on *goal* at all (e.g., truth tables) or they attempt to establish some already articulated conclusion (e.g., natural deduction techniques of the sort surveyed in the previous chapter). But our whodunits demand a different category of search: we must hunt for an unknown sentence of the general form "Archie committed the murder just in case . . . ? . . . " Nowadays, we commonly describe what's wanted as a *heuristic* for solving whodunits: a set of instructions that channels logically permitted conclusions along an algorithmic pathway leading to the solution desired. But to Boole—and here he has ordinary usage on his side—, the very phrase "logical method" implies a "heuristic", not the less structured forms of non-algorithmic specification (like axiom schemes or natural deduction systems) that are usually called "logics" today. He writes:

> If we may judge from the mathematical sciences, which are the most perfect examples of the method known, this directive function of Method constitutes its chief office and distinction. The fundamental processes of arithmetic, for instance, are in themselves but the elements of a possible science. To assign their nature is the first business of its method, but to arrange their succession is its subsequent and higher function. In the more complex examples of logical deduction . . . , the aid of a directive method, such as a Calculus alone can supply, is indispensable.[74]

On this old-fashioned way of thinking, "logic" has not yet achieved its proper offices until it supplies "directive methods" like the one we shall now supply.

Let us first restate our data in unambiguous terms.

> Either Archie and Betty committed the crime or Veronica did it, but not both.

> Either Archie did it and Betty remained innocent or Jughead did it, but not both.

Note that some of the "or" 's in the original puzzle have been interpreted *exclusively* (i.e., "A or B" is *not* true if both A and B are true). Let us sketch how a Boolean treatment

[74] Boole, Thought, 11.

of our Archie problem can proceed. The data can be expressed in equational form as:

(1) $(A \bullet B) + V = 1$
(2) $(A \bullet -B) + J = 1$

Subtracting from each side:

(3) $A \bullet B = 1 - V$
(4) $A \bullet -B = 1 - J$

Divide each side and replace $1 - X$ by $- X$:

(5) $A = -V/B$
(6) $A = -J/ - B$

Expand each term by division (analogous to a power series):

(7) $A = (- V \bullet B) + (x \bullet V \bullet -B)$
(8) $A = (- J \bullet -B) + (y \bullet J \bullet B)$

Here the x and y are "indeterminates" that represent unknown multiplicative factors. Cross-multiply these two equations:

(9) $A \bullet A = (-V \bullet B \bullet y \bullet J \bullet B) + (- J \bullet -B \bullet x \bullet V \bullet -J)$

Here terms have been dropped that contain both X and $- X$. Using the associativity and commutativity of $\bullet$, as well as the principle $A \bullet A = A$:

(10) $A = (y \bullet -V \bullet B \bullet J) + (x \bullet -J \bullet -B \bullet V)$.

Comparing (10)'s coefficients with those of (7) and (8) (the algebraic analog is called "the method of undetermined coefficients"), we find

(11) $x \bullet V = 1$
(12) $y \bullet J = 1$

Hence:

(13) $A = (-V \bullet B) + (- J \bullet -B)$.

Oddly enough, we have worked our way to a correct answer: Archie will have participated in the crime only if he works in collusion with Betty but not Veronica or without either Betty or Jughead (and other forms of equivalent answer can be constructed from this). But what a strange route we employed to get there! Let us list some of the questions we might reasonably ask about what we have done.

(a) *What on earth can the "division" of one sentence by another ($- V/B$) mean?*
(b) *Why should any inferential transition in logic be considered like a power series expansion?*

(c) *Likewise, why are we allowed to employ the "method of undetermined coefficients" here? What on earth can the "undetermined variables" x and y signify?*

(d) *Hasn't " + " clearly shifted its meaning between lines (1) and (2) and (9)?*

(e) *Likewise, "1" between lines (1) and (12) (or (2) and (13))?*

And most generally,

(f) *Why do we have any reason to believe that this process heads towards a correct solution of our problem?*

In fact, later critics such as Frege and Schröder criticized Boole severely on all of these accounts. If I had attempted to imitate other aspects of Boole's methods—e.g., his frequent use of the "indeterminate" "0/0" evaluated by what he called a "version of l' Hospital's rule," the weirdness of his procedures would have become even more palpable.

On the other hand, this technique systematically provides answers to a type of problem that the reader might not know how to tackle handily otherwise. Plainly, some coherent informational support must underwrite this method, for, as the British mathematician Richard Woodhouse remarked:

[A] *method which leads to true results must have its logic.*[75]

Later logicians eventually corrected most of the anomalies listed above (insisting that " + " only signify inclusive "or"; avoiding use of " − V/B," etc.), eventually producing "Boolean Algebra" as we know it today, where every step in the allowed derivations follows in a pellucid way from its predecessors (and thus conforms nicely to the "true thought" demands of section (v)). Oddly enough, in introducing these "improvements," we produce a calculus that, by some measures, is less efficient than Boole's original. For such reasons, Boole's champions defended his original techniques as late as 1906; thus, A. T. Shearman:

[Boole's critics] *suggest that all the intermediate processes in a solution ought to be intelligible; but this is not so, because a calculus is a means of reaching correct conclusions by means of the mechanical application of a few logical rules, and it is quite possible that in the application of such rules unintelligible elements may temporarily appear.*[76]

I'll come back to the odd philosophy articulated here (which echoes Boole's own attitude as well as many of his contemporaries like George Peacock and Augustus deMorgan).

I hope the reader, after fooling around with several problems of this ilk, has gained some conviction, through trial and error, that a practical directive method lies here

---

[75] H. J. S. Smith, "On Some of the Methods at Present in Use in Pure Geometry" in Collected Mathematical Papers, i (Bronx: Chelsea, 1979), 6.

[76] A. T. Shearman, The Development of Symbolic Logic (Dubuque, Ia: William C. Brown, n.d.), 68. In his introduction to Couturat's helpful little pamphlet, Philip Jourdain remarks that Boole's school emphasizes the calculus ratiocinator aspects of logic whereas Frege's, its lingua characteristica lineage: Louis Couturat, The Algebra of Logic (Chicago: Open Court, 1914), p. viii.

whose workings are presently veiled in unhelpful mists. In fact, high school algebra
displays a largely accidental (and clearly partial) affinity to the matters now under
consideration and Boole's desire to retain these distracting trappings makes the true
underpinnings of his techniques hard to fathom. We find ourselves in the position of
partially enlightened "rainbow" employers who know that the phenomena we discuss
are not at all like arches, but don't understand their positive nature of the data at all.
What we would like, in fact, is a revised *semantic picture* of our Boolean techniques: an
account that can apportion coherent informational content to each step in our (1)–(13)
derivation and explain why its inferential flow supplies sound results. We want a story
that maps expressions like " $+$ " and "/" to meaningful requirements. In fact, such an
account is easy to provide and it illustrates several of the oddities of linguistic engineering that we have discussed earlier (7,viii). In particular, the routine splits into two
stages within which the same symbols are locally assigned different readings and it
narrows in on the desired solution through a variation upon successive approximations
(4,x). The pseudo-mathematical steps ("power series expansions, " etc.) are merely
irrelevant window dressing for commonplace logical considerations.

....

Let me first sketch the basic idea of our reconstructed reading. A famous result of logic is the
Beth Definability theorem which declares that if a set of data D implicitly defines a letter A, then
an explicit definition of the form " $\equiv$ " will logically follow from D where A does not appear
within B (" $\equiv$ " means "if and only if"). In solving our whodunit, we seek an explicit definition of
this type and it is easy to provide an algorithm for its production (if it exists) as follows.

   Form the conjunction of all the available data $D = S_1 \& S_2 \& S_3 \& \ldots$ and reexpress D's content
in *alternative normal form*: $D_1 \vee D_2 \vee D_3 \vee \ldots$ (that is, each D_i is a conjunction of atomic letters or
their negations). Segregate these D_i terms into three groups:

   (i) terms of the form A & B where no other term takes the form $\sim$A & B
   (ii) terms of the form A & B where $\sim$A & B is also a term
   (iii) terms of the form $\sim$A & B where no other term takes the form A & B

If D establishes an explicit definition for A, it will take the form $E_1 \vee E_2 \vee E_3 \vee \ldots$
where the E_i are all and only the terms of class (i). On the other hand, if any D_i term falls in class
(ii), then no explicit definition of A from D is possible (because D can be made true in the B
manner, forcing the truth of neither A nor $\sim$A). Once $E_1 \vee E_2 \vee E_3 \vee \ldots$ is found, it can be
shortened by various logical equivalences if desired.

   The basic strategy utilized in our Boolean procedure works towards $E_1 \vee E_2 \vee E_3 \vee \ldots$ by
essentially working with alternative normal forms for the *component sentences* S_i of D, rather than
immediately calculating the lengthy $D_1 \vee D_2 \vee D_3 \vee \ldots$ The latter is of course equivalent to the
cross-product of these sentential alternative forms, but we shall gradually multiply our sentential
normal forms together as we go along, discarding as many terms as possible. To this end, we
employ the following trick: if a term T_i within some alternative normal form based upon a subset
of the S_i appears in some danger of falling in the type (ii) class in the final equivalent to D
multiplication, we write that term as "W(S_i)" ("watch out for S_i"). Our computational routine
will produce the required definition of A only if all of these W(S_i) terms eventually disappear
under cross-multiplication.

To further shorten our initial calculations, in the first stages of our calculation, we express our data utilizing *exclusive disjunction* $\tilde{v}$ rather than the usual *inclusive* v. This indicates that, if "B $\tilde{v}$ C" is given, we are effectively supplied with *two* component alternations: (B v C) & ($\sim$B v $\sim$C). In the second stage of our computations, we switch to relying upon inclusive "v" in our working formula, but we remember that we have extra $\sim$B v $\sim$C data from stage 1 to utilize as needed. In particular, we can throw out any cross-multiplied term that contains an "B & C" component.

Based upon these indications, the reader can probably now follow the drift of our Boolean procedures, but here is a line-by-line reading of our example.

Stage 1:

(1) $(A \bullet B) + V = 1$ $(A \& B) \tilde{v} V \equiv D$

(2) $(A \bullet -B) + J = 1$ $(A \& \sim B) \tilde{v} J \equiv D$

(3) $A \bullet B = 1 - V$ $A \& B \equiv D| \sim V$ where "$D| \sim V$" means the terms of D that

survive cross-multiplication by $\sim$V. Given D, $\sim V \equiv (D| \sim V)$

(4) $A \bullet -B = 1 - J$ $A \& \sim B \equiv D| \sim J$

(5) $A = -V/B$ $A \& B \equiv \sim V$

(6) $A = -J/ - B$ $A \& \sim B \equiv \sim J$

We view these "divisions" as simply syntactic indications that we intend to move to the steps (7) and (8) following.

Stage 2:

(7) $A = (- V \bullet B) + (x \bullet V \bullet -B)$ $A \equiv (\sim V \& B) v W(V \& \sim B)$

(8) $A = (- J \bullet -B) + (y \bullet J \bullet B)$ $A \equiv (\sim J \& \sim B) v W(J \& B)$

In these calculations we only write down the terms that will either produce a definition of A or potentially prevent its explicit definability. Thus we know that "$\sim V \& B$" can serve as an E_i in the final explicit definition whereas W(V & $\sim$B) may form part of a type (ii) term that spoils the definition (that is, we aren't sure that we have enough data to settle A's behavior in situations where V & $\sim$B holds). Note that we now understand "+" as V, not $\tilde{v}$.

(9) $A \bullet A = (- V \bullet B \bullet y \bullet J \bullet B) + (- J \bullet -B \bullet x \bullet V \bullet -B)$

(A&A) $\equiv (\sim V \& B \& W(J \& B)) v (\sim J \& B \& W(V \& \sim B))$

This is just the cross-multiplication of (7) and (8), dropping inconsistent factors. If "+" still meant exclusive $\tilde{v}$, this step would be invalid.

(10) $A = (y \bullet -V \bullet B \bullet J) + (x \bullet -J \bullet -B \bullet V)$.

A $\equiv (W(J \& B) \& \sim V \& B) v (W(V \& \sim B) \& \sim J \& \sim B)$

(11) $x \bullet V = 1$

Line (8) tells us that any clause containing the piece $\sim J$ & $\sim$ B belongs to the safe class (i) group, hence the risk term W(V & $\sim$ B) can be dropped. Note that "1" does not mean D as in stage 1, but simply encodes the instruction to drop the "W" term.

(12) $y \bullet J = 1$

Likewise, from line (7).

(13) $A = (-V \bullet B) + (-J \bullet -B)$

$\quad\quad A \equiv (\sim V\,\&B)\,v\,(\sim J\,\&\sim B)$

Having removed all potential type (ii) threats, our problem accepts a solution which we have constructed. It is curious that if a solution isn't forthcoming, our technique quantifies the amount of additional data required to solve for A in its remaining W terms.

Incidently, I have always found the strong inclination of some scholars (and most beginning logic students) to insist that English "or" must carry an exclusive reading curious. Thus F. H. Bradley:

> I cannot admit any possible instance in which alternatives are not exclusive. I confess I should despair of human language, if such distinctions as separate "and" from "or" could be broken down.[77]

Perhaps (this is just idle speculation on my part) Bradley has concentrated upon the opening stages in whodunits, where, indeed, we do like to frame the data in exclusive terms.

. . . .

We have thus provided an alternative *picture* for our Boolean procedure that generically indicates why all derivations of this particular type will lead to sound results and thus the mysteries of its strange workings disappear. On the other hand, once its workings are unraveled in this way, we may reasonably wonder why Boole frames the problem in this algebraic guise, for as the Kneales rightly observe,

> Obviously the nut from which this kernel has been obtained could have been cracked with a lighter hammer.[78]

Although we have obtained limited computational gains over an orthodox Beth's theorem procedure, it is surely not optimal for this kind of question. In fact, we are likely to remain enamored of Boole's techniques only as long we continue to see logic, as Shearman does, only through the prism of high school algebraic technique. Once those irrelevant trappings are removed, we appreciate the informational basis of the method, but it also seems gimcrack and contrived, and we should renounce, with Shakespeare, our former Boolean affections:

> O me! what eyes hath Love put in my head
> Which have no correspondence with true sight?[79]

But Heaviside, through virtually identical policies of discovery, articulates an operational calculus that is quite beautiful and genuinely extracts hidden secrets from the mathematical woodwork. Why? In the final analysis, because he was lucky.

An examination of the sort sketched in the fine print, where we study the unfolding steps of a formalism to insure that they always lead to proper results with respect to the settings for which they are intended, is called a *soundness proof* in a logical context. We have already witnessed another verification of the same type in 4,x, in the guise of

---

[77] F. H. Bradley, The Principles of Logic, i (Oxford: Oxford University Press, 1963), 134.

[78] William and Martha Kneale, The Development of Logic (Oxford: Oxford University Press, 1962), 420.

[79] William Shakespeare, Complete Works (Garden City, NY: Doubleday, 1936), 1432.

probing Euler's method for its *correctness* (the more usual term than *soundness* in such contexts). Through those means we discovered that a naïve employment of Euler's rule is *not* correct and supplementation by a Lipschitz proviso is required (given that Euler's method traffics in step size approximations and rounded off real numbers, further considerations of numerical stability <u>etc.</u> must be checked before results can be fully trusted, but we needn't worry about those complications here[80]). I consider both cases to illustrate studies of predicate behavior in relation to proposed pictures of their intended settings. As we'll see in 10,v, philosophically inclined authors often fuss greatly about logic's parochial forms of correctness proof but completely ignore those appropriate to our more *powerful strands of reasoning*. With respect to the second category, I have already remarked upon the ill-understood rules of rapid geometrical reasoning that we exploit at virtually every moment in our dealings with the world: if we could codify their procedures formally, they would almost certainly not prove sound, raising the even harder question of why they *usually* work well. As we'll see in the next chapter, Euler's method represents a standard way in which we reason about causal processes in our ordinary thinking, which will accordingly fail if a violation of Lipschitz condition inadvertently arises. Both of these methods play larger roles in conducting our thought to useful conclusions than does logical principle alone.

I mention these basic affinities in correctness proof because I strongly believe that the inclination to approach logical principle in quarantine from its more powerful cousins in reasoning represents a great philosophical error, a mistake that has been encouraged by the faulty invariance posited in classical concept thinking. These issues will become prominent in our final chapter.

(ix)

Putting a picture to it. Before proceeding further, let me summarize the main tenets suggested in the remainder of this chapter. At certain moments within the developmental history of a predicate, it becomes evident that its usage has become extended in ways that we do not fully understand. In such seasons, it becomes advantageous to replace our former picture of how our predicates convey physical information with a new account that schematically indicates a different rationale for their operations. In this context we often find that our old partial successes have been reached through a descriptive strategy whose proper contours we failed to appreciate. Such reorientations in picture usually cannot be achieved easily and often an intervening period of *semantic agnosticism* must intercede. But once the directionalities of such policies have been rendered articulate, we can readjust our practices to accord better with their dictates and thereby improve the quality of our future linguistic performance. In this end, we frame correlational pictures of usage that, in generic fashion, sketch how strategically informed

[80] Lawrence F. Shampire, <u>Numerical Solution of Ordinary Differential Equations</u> (New York: Chapman and Hall, 1994).

sentences manage to carry correct information with respect to the physical environments to which the reasoning pattern is pertinent. It is within this localized context, that
correctness proofs—if their details can be worked out—emerge in seasonal importance.

Essentially classical patterns of thinking have led many prominent philosophers (for
reasons we shall consider in 10,v) to mistake these localized "diagnoses of picture
pertinent to predicate P" for "complete accounts of the conceptual contents of P."
Doing so accords our seasonal resettings of picture with an importance far beyond their
proper due and leads to great philosophical confusion in the bargain. In an allied, but
different, vein, the inclination of many philosopher/scientists to succumb to formalist
temptations (canvassed in 4,iii) traces to closely related inclination to exaggerate the
previous evaluative stage in our canonical developments beyond *their* proper due. In
these two forms of monocularly focused enthusiasm, our *ur*-philosophical thinking
comes to imitate our teenager: mistaking a seasonal evaluative episode for a continuing
conceptual invariant.

In a more generalized form, these issues will be reconsidered in Chapter 10, in the
context of a broader survey of lessons learned. In the remainder of the present chapter, I
would like to pursue the Heaviside case in a grittier way, so that we can benefit from its
very salient details as well as Heaviside's practical insights into developmental processes.
Once again I will endeavor to outline the scientific issues in accessible terms, but some
wearied readers may prefer to advance directly to Chapter 10 (bypassing Chapter 9 as
well, which deepens our issues in a context of historical relevance, but is not otherwise
required for our Chapter 10 summation either).

To begin, let us start with the brute fact, nicely evoked by H. T. H. Piaggo:

> *Yet Heaviside's results were always correct! Could a tree be really corrupt if it always
> brought forth good fruit?*[81]

Effective computational routines don't work by accident: somehow their component
sentences must carry, under some possibly rococo coding, valid information about their
target systems. "A method which leads to true results must have its logic," Richard
Waterhouse advised us and we agree with him. The proper way to diagnose that "logic"
is to provide a picture of how Heaviside's manipulations manage to carry content
correctly, just as we investigated Boole's method above.

By the 1920s Heaviside's methods became so popular within electrical engineering
that a number of attempts, of variable rigor themselves, were assayed to place the
operational calculus on a firmer, and less mysterious, foundation (both due to the errors
to which its employment sometimes leads and simple curiosity). Most of these early
attempts invoked the Laplace transform (or one of its cousins) and this interpretative
policy still represents the approach found in most engineering primers to this day (as
such, it was largely perfected by the mathematician Gustav Doetsch[82]). However, appeal

---

[81] H. J. Josephs, "Some Unpublished Notes of Oliver Heaviside" in Heaviside, Electromagnetic, iii. 575.

[82] Gustav Doetsch, Introduction to the Theory and Application of the Laplace Transformation (New York: Springer-
Verlag, 1974). This later account (which is a model of clarity) employs distributions: the 1937 version did not. Earlier
efforts in this line were offered by T. A. Bromwich and John Carson—for details, see Jesper Lützen, "Heaviside's

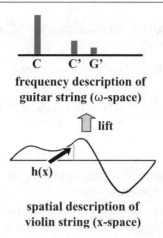

frequency description of
guitar string (ω-space)

⬆ lift

h(x)

spatial description of
violin string (x-space)

to this transform alone fails to address some of Heaviside's most striking innovations, which were not happily provided until the French mathematician Laurent Schwartz articulated his celebrated "generalized functions" or "distributions" in the late 1940s (the Russian mathematician L. Sobolev had pioneered similar ideas somewhat earlier, but this work was less well known and their salience to Heaviside is less obvious). Schwartz's distributions are often popularly dubbed as "the mathematician's rigorization of the Dirac δ-function," referring to the well-known use of generalized functions in the context of quantum mechanics. In truth, δ-function-like ideas appear deeply embedded within Heaviside's work (under the heading of "impulse (or spotting) functions"[83]) and Dirac (who was educated as an electrical engineer) probably borrowed from that source. Once the Laplace transform is combined with distributions, we obtain a fairly nice scheme for parsing Heaviside's syntax (I'll discuss several alternatives in a moment).

Essentially, a Laplace transform offers advantages akin to the Fourier decomposition of vibrating strings into their component modes described earlier (5,vi). Suppose we first think of a string "described in spatial format": that is, we articulate its condition in terms of its transverse height h(x, t) along the string at time t. Its allied velocity (dh(x,t)/dt) is naturally conjugate to h(x, t) and the two functions together capture the string's current state. However, when we alternatively think of the string's state in Fourier's manner (which is, roughly, that of a musician as well), we focus instead upon the quantity of energy E_i stored within each of the string's component harmonics ω (fundamental, octave, twelfth partial, etc.). The spectral function $E(\omega)$, together with its phase distribution $\theta(\omega,t)$, captures our string's condition just as ably as the spatial description. Our Fourier redescription can be pictured as a lift (or "transform") into frequency-based "ω-space" which provides a better setting for calculating many important features of

Operational Calculus and the Efforts to Rigorize It," Arch. Hist. Exact Sci. 21 (1979). The Laplace transformation itself had an earlier career unconnected to Heaviside-like issues: Michael Deakin, "The Development of the Laplace Transform", I and II, Arch. Hist. Exact Sci. 25, 26 (1981–2).

[83] H. J. Josephs, Heaviside's Electric Circuit Theory (London: Methuen and Co., 1950), 51–5.

$$\mathcal{L}(\mathcal{D})x = v(t) \xrightarrow{\textit{easy}} \mathcal{L}x = v/c$$

s-space

$$\text{lift} \uparrow \mathcal{L} \qquad\qquad \text{lower} \downarrow \mathcal{L}^{-1}$$

$$\mathcal{D}(x) = v(t) \dashrightarrow^{\textit{hard}} x = \mathcal{L}^{-1}(v/c)$$

t-space

our string's behavior—musicians generally approach the performance of their instruments in frequency-based terms rather than spatial ones (I should add that, properly speaking, I here discuss Fourier *series*; a true Fourier transform applies only when our string is infinitely long). The Laplace transform story approaches the manipulations of the operational calculus in a similar way: when Heaviside performs his odd "divisions by $1/dt$," he effectively lifts the temporal description of the changing voltage V(x,t) in an electrical circuit into a new descriptive arena, where the same signal is decomposed into "imaginary frequency components" $\mathcal{V}(i\omega)$ or $\mathcal{V}(s)$ (the combination "$i\omega$" is commonly rendered as simply "s"). That is, the Laplace transform $\mathcal{L}$ represents a map that translates our old description into the language of s-space. In performing this lift, we gain advantages of simplicity much like those that attend to guitar strings: the behavior of the target system becomes easier to fathom when treated in s-space terms (in the Heaviside circumstances, intertwined "convolution integrals" within the temporal description unravel into simple multiplications in s-space, rather as the logarithm operation lifts multiplications into additions through a transform). Rather than attempting to solve equations directly in their time-space appearance, we first lift the formulae into s-space by the Laplace transformation $\mathcal{L}$, work various computations up there, and then lower the final results back into time-space (this is often the hardest part). Our chart of this process (mathematicians call it a "commuting diagram") should seem familiar, for we have already witnessed several other examples of this basic "lift-and-conquer" gambit.

. . . .

To be more specific, the Laplace transformation interpretation treats Heaviside's formulae as constituting integral equations in *camouflaged form* (Jean Dieudonné calls such truncations, which were extremely common in nineteenth century mathematical work, "crypto-integral equations"[84]). Consider the simple example provided above:

$$dy/dt + y = t^2,$$

Taking the Laplace transform of both sides, we move into s-space and get

$$\int_0^\infty e^{-st}\, dy/dt\ dt + \int_0^\infty e^{-st} y\ dt = \int_0^\infty e^{-st}\, t^2\ dt$$

[84] Jean Dieudonné, A History of Functional Analysis (Amsterdam: North-Holland, 1981), 4.

Integrating by parts, the first term $\int_0^\infty e^{-st} dy/dt\_dt$ equals:

$$[e^{-st}y]|_0^\infty + s \int_0^\infty e^{-st}y\ dt$$

But, since in Heaviside' electrical context, we only consider output signals y *assumed to be 0 at time t_0 and that also grow slowly at infinity*, the first term drops out, leaving us with

$$\int_0^\infty e^{-st}y\ dt$$

But plugging in this result, our transformed equation legitimately factors into

$$(s + 1) \int_0^\infty e^{-st}y\ dt = \int_0^\infty e^{-st}\ t^2\ dt$$

in neat correspondence to Heaviside's version

$$[(d + 1)/dt]y = t^2.$$

When Heaviside proceeds to "divide" by $(d + 1)/dt$

$$y = t^2/(d/dt + 1),$$

he is actually claiming that

$$\int_0^\infty e^{-st}y\ dt = 1/(s + 1) \int_0^\infty e^{-st}\ t^2\ dt \quad \text{ie.,}\ \mathcal{L}(y) = 1/(s + 1)\mathcal{L}(t^2).$$

He then expands the latter into partial fractions and recovers their individual inverse transforms (which is what he calls "algebratizing the function"). The final result is a correct representation of y(t).

. . . .

These techniques alone do not ratify many of the strange results that, in the electrical context, supply Heaviside's techniques with much of their power. In particular, starting from an "initial condition" where a one-volt potential is applied to a circuit already charged at two volts, Heaviside works his way to an answer that claims the voltage at the starting time was, in fact, five volts! In his electrical context, such results are not so strange, because many circuits display an immediate impulsive response when excited by an applied voltage. So Heaviside's "five volt" answer reflects this internal response mixed together with the applied input and its earlier state. But the orthodox mathematics of "initial conditions" does not know how to formulate Heaviside's expected conditions—from its point of view, Heaviside's calculus appears to have fallen into a contradiction of a "$1 = 2 = 5$" ilk.

. . . .

More specifically, let the circuit obey the equation $dy/dt + 2y = 3du/dt + 2u$ where y(t) is a voltage at some sample point within the circuit and u(t) is an input voltage that is suddenly turned on to a constant value of one volt at t_0 and thereafter. Presumably, our sample point $y(t_0)$ could start with any initial voltage; let us assume it happens to be 2 volts. But when we solve the equation by Heaviside's methods, we obtain an answer of $y(t) = 4e^{-2t} + 1$ which is plainly $= 5$

534 Song of Master Idea

at $t_0 = 0$. In the non-rigorous text from which I extracted this example,[85] this discrepancy is explained away by claiming that 2 volts really represents the circuit's state "at time $t_0 = 0 -$" whereas five volts allegedly represents its "time $0 +$" condition. Extending the Laplace transform to operate over distributions evades these weird appeals to the "two sides" of time t_0.

By the way, apparent "initial conditions" that really demand more complex requirements arise commonly in mathematical physics and often confuse casual observers. The airplane wing and razor blade examples of Chapter 7 represent cases in point.

....

Crudely speaking, our generalized distributions must be able to carry more information at a concentrated point t_0 than a regular function can support. The basic trick of the technique is to relate our distributional gizmo f to nearby functions g that straightforwardly display the behaviors that f needs to exemplify in a spread out manner. The Dirac distribution $\delta(t)$ must behave as if its weight were entirely concentrated at $t = 0$: we appeal to "test functions" to spread this behavior out so that it can be approximately seen over a wider area.

We might observe that there are other, not entirely equivalent, rigorizations of the operational calculus that rely upon other devices to carve out a landscape suitable for Heaviside's manipulation. Among these is the well-known construction of Jan Mikusiński,[86] who extends a base convolution algebra in a manner not unlike that of the manner in which we supplement the integers with rationals through algebraic construction. In many ways, this approach sticks closer to the spirit of Heaviside's original procedures than the Laplace transform treatment, but the latter is easier to explain quickly. Since Heaviside's own explorations are somewhat undisciplined, it is not surprising that his own results sometimes borrow a bit from the special virtues of both formalisms.

As observed before, an important purpose of establishing a semantic picture of this type is that it allows us to check various inference rules $\mathcal{R}$ (e.g., Heaviside's well-known "shift" procedure) for their *generic soundness*: that is, within every mathematical structure $\mathcal{S}$ compatible with the Laplace + distributions reading, $\mathcal{R}$ carries true claims to true claims. Assigning a fresh picture to a term can therefore serve as a novel source of *corrective directivities* upon its usage, as well as needing to be responsible to those already recognized. We must now investigate the question: how should the claims of these possibly competing standards of predicate correctness be weighed against one another?

(x)

Retooling at sea. For convenience, let us recapitulate our discussion in facade-like terms. Earlier we called predicate P's developmental history *canonical* if starting in established patch A with associated picture $\mathcal{P}$, "P" migrates beyond ∂A following some suggestive $\mathcal{R}$-based incursion → wider forms of "P"'s usage over B then nucleate

[85] Chen, <u>Design</u>, 573–4. [86] Jan Mikusiński, <u>Operational Calculus</u>, i (Oxford: Pergamon Press, 1984).

around that entering wedge → "P"'s usage is detoxified by abandoning the old $\mathscr{P}$ in favor of a replacement semantic picture $\mathscr{P}^\star$ → "P"'s usage over both A and B-patches is refined according to the corrections suggested by $\mathscr{P}^\star$. In the case at hand, the linguistic phenomenon unfolds over the kinds of behavior found within electrical circuits (call this $\mathcal{D}$) and we are especially interested in what happens over its linear sector (i.e., when the applied loads and currents remain fairly moderate). In performing the lifts characteristic of the operational calculus, Heaviside develops, by borrowing from algebra, inferential pathways $\mathcal{R}$ that cycle through patch B that utilizes an extended language L* full of "1\dt" expressions and the like absent in the conventional calculus language L native to A. As his confidence in L* grows, Heaviside finds it useful to characterize his circuits directly in its terms without necessarily completing an instrumental cycle back into A. This faith then leads Heaviside to embrace variant cycles that return to A with strange results: viz., our one-two- and five-volt circuit "solutions" that seemingly don't satisfy the initial conditions they were assigned. To make supportive sense of these new claims, applied mathematicians must frame a new picture $\mathscr{P}^\star$ of how Heaviside's expressions (even within the linear part of the original language L) correlate with circuit behavior. That is, we now *detoxify* our circuit equations of their original freshman calculus picture (δ/ε epsilonics and all that) in favor of a distributional replacement $\mathscr{P}^\star$ that allows Heaviside's peculiar "solutions" to genuinely qualify as *solutions* by its new semantical lights. But this revisionary activity occurs mainly over the *linear corner* of the full domain of circuit behavior—non-linear behaviors can't be treated in Heaviside's manner and so we must stick by our old δ/ε picture elsewhere on patch A. We eventually find ourselves in a situation where the significance of electrical terminology has been dragged to assume a new correlational reading over a region that is comprised from patch B and the linear part of old A.

However, supplying such a correlational picture of how Heaviside's procedures manage to carry valid information about circuits does not really explain the *drives* or *directivities* that led his employments to settle upon such fancy supports as their natural moorings. In general, we not only want to know *where* the ark of language has finally landed, but the nature of the flood waters that have deposited it there. In our talk about "concepts" within everyday evaluation, we do take into genuine consideration *all* of the threads required for an understanding of why a predicate behaves as it does over the long term. However, we usually do this only in a seasonal manner—we rarely consider all directivities on equal fours at the same time. In the intervals where detoxification of some old semantic picture is especially wanted, crucial aspects of a predicate's overall personality often drop from view as not especially pertinent to the needs of the moment, which require an alternative correlational account of the information carried within the usage. The mere fact that other directive considerations do not materially aid the specific project of picture reconstruction does *not* show that such factors are not vital to an adequate understanding of how the word behaves in other respects.

The fact that we emphasize (or "spot" in the jargon of 7,ix) certain subgroups of directivities in an episodic fashion can occasion serious confusion if we stoutly insist upon viewing these activities through the invariant eyes of *ur*-philosophical proclivity.

For the latter fancies it can discern "core contents" running constantly beneath the surface of our linguistic activity and believes that this same package of subterranean content will be brought to review whenever we actively appraise the "semantics" of our language in a critical way. This "everything vitally semantic will be on display" presumption automatically fuels the disputes between classicists and pre-pragmatists surveyed in earlier chapters, where each school argues that the packets of improving directivities they favor reflect the true fundament upon which predicative significance rests. But, no, the concrete activities to which each side points typically *alternate* in salience over the history of an improving term and neither selection manages to put "everything semantic on display." Indeed, it is impossible to understand how many evolutionary progressions unfold unless we attend to a considerably wider circle of directive influences than either of our warring philosophical parties concede.

The Heaviside situation exemplifies the lessons I am after quite beautifully, for it happens that the base descriptive rationale for his extended vocabulary is fueled chiefly by hidden *design-oriented* considerations, rather as allied elements contributed to *gear wheel*'s characteristic personality back in Chapter 4. However, there are sound reasons why the richer directivities of design can permissibly drop from view if we are narrowly concerned with finding a better correlational narrative for his procedures. For those focused purposes we only need to determine what sorts of *physical information* are carried along within his successful manipulations. And to this end we might reasonably announce, "It really doesn't matter *why* Heaviside developed this language: somehow it encrypts valid information and we want to crack that code." Indeed, the Laplace transform reading of the operational calculus does not address its motivating concerns in any manifest way, as I shall detail in a moment. If we happen to be men from Mars (or simply armchair-bound philosophers) who have never before witnessed a real life engineer in computational action, we will be unlikely to reconstruct her purposes merely from the correlational data provided within a Laplace transform picture alone. After all, the bare rules that govern the Mercator projection scarcely alert us to the navigational advantages that brought such maps into prominence in the first place. Many a landlubber could pore over its strange correlations endlessly, yet fail to reconstruct those critical directivities.

The remainder of this section will explain more fully why this narrowing of evaluative focus arises in the Heaviside case and why such apparent semantic myopia does not compromise the goals of descriptive improvement sought in rigorization. Some readers may prefer to pass over these details and skip ahead to section xii, which will amplify further upon the theme that classical/neo-pragmatist clashes generally trace to an inflation of seasonal flavors of predicate evaluation into timeless semantic invariants.

I claimed that the directivities of *circuit design* are paramount in Heaviside's endeavors. But how do these manifest themselves? Here are the apologetics that he supplies on his own behalf:

> In physical mathematics the quantities concerned are not arbitrary, but are controlled
> by the special relations involved in certain laws, involving, for instance, the necessary

positiveness and singleness of certain quantities . . . So we have a definite march of events from one state to another, without that common multiplicity so common in pure mathematics. It is these general characteristics that seem to give reality to the mathematics, and serve to guide one along safe paths to useful results . . . [W]hen it is clear that when one is led to ideas and processes which are not understood, and when one has to find ways of attack, the physical guidance becomes more important still. If it be wanting, we are left nearly in the dark. The Euclidean logical way of development is out of the question. That would mean to stand still. First get on, in any way possible, and leave the logic to the later work. . . . When intuition breaks down, something more rudimentary must take its place. This is groping, and it is experimental work; with of course some induction and deduction going along with it.[87]

Throughout the book I have emphasized the fact that distinct practical concerns often force us to attack the same base equations with different inferential tools and, sometimes, novel terminologies. It happens that Heaviside's chief concerns with his circuits are not quite "prediction questions" in the conventional sense, but reflect an opportunity for "understanding" electrical behavior in broader terms that proves available within the realm of linear behavior (an "understanding" that Heaviside uncovers by obeying the beckoning arms of design opportunity). Within philosophy, we tend to be very loose about what "prediction" encompasses, but every applied mathematician knows that a slight shift in problematic (from, in their jargon, a problem of *prediction* to one of *control*) is apt to alter the character and difficulty of the relevant mathematics considerably. Is someone who is attempting to design a mousetrap trying to *predict behavior*? Well, possibly the word "predict" can be stretched that far in everyday use, but a query of a mathematically different order lies before us (when such problems are posed in formal terms, their underlying differences become readily apparent). The relevance of these observations to Heaviside is this: the "space of solutions S" we accept as pertinent to fixed equations E can alter in the face of the background projects at hand. Heaviside's *design interests* favor a space of solutions S* different from those S pertinent to prediction per se. It is precisely the disparity between S* and S that serves as the amatory magnet that draws Heaviside's linguistic experimentations into unexplored terrain. But none of this fuller story of motivation will be laid plain if we are merely informed that the operational calculus can be correlationally supported in a Laplace transform + distributions manner.

. . . .

On several earlier occasions I have complained about the unhelpful "possible world" vocabulary in which many contemporary philosophers like to pose their issues. To me, such lingo tends to shelter theory T syndrome proclivities behind murky veils. Indulgence in such loose veins of characterization does not encourage the close scrutiny required to detect the critical distinctions between S and S∗. In our discussion, the central pertinent equation E is the telegrapher's equation, which many writers will immediately identify as a "theory." E, in turn, arises from

[87] Heaviside, Electromagnetic, ii. 461.

some wider framework T such as Maxwell's equations. Possibly no harm is engendered by these suppositions alone, but such authors swiftly link a range of philosophically charged notions in a tight circle that runs: the theory E delineates a set of acceptable solutions $\mathcal{S}$ which constitutes the set of possible worlds $\mathcal{W}$ permissible under E. If a group of theoretical concepts happens to be implicitly defined by T, then each concept $\mathcal{C}$ in that collection can be understood as a map from each possible world w in $\mathcal{W}$ to an extension within w (if $\mathcal{C}$ is a quantity term, $\mathcal{C}$ should be a function from w into numerical values). More generally, if our concept $\mathcal{C}$ finds its implicit definability home over a range of theories T, T′, etc., then $\mathcal{C}$ is expected to pick out extensions within the union of all pertinent worlds (and will choose the null set in worlds belonging to theories that fail to embrace $\mathcal{C}$). Indeed, one commonly finds "concept" defined in exactly this way in many modern works of a metaphysical bent.[88]

The problem with this assimilation is that, as we shall now see, the "solutions" contained in Heaviside's favored S∗ are not intended as "possible histories" of anything: to presume otherwise is to misunderstand the design rationale for this choice of space (as I also emphasize, S∗, considered in its formal aspects only, does not wear the background motivations for its selection on its sleeve—these stem from other directivities that motivate Heaviside).

More generally, we should be leery of rashly identifying an orthodox "solution set" with "the possible ways the world might be according to E," for many a slip intervenes between the mathematician's "solution space" and the philosopher's "possible world." Allied morals were underscored in Chapter 4 where it was pointed out that the "black sheep phenomenon" often prevents the members of an official "solution set" from aligning tidily with any "range of expected physical possibilities." Some commentators attempt to approach the spaces of variation natural to virtual work and other forms of variational principle in a "possible world" manner as well, but, once again, careful canvass is required, for these spaces are generally chosen because they are naturally tied to possibilities of *manipulation*: how will a structure respond if we wiggle its parts? As such, they do not correspond to the manners in which the system might behave *if left to its own devices*.

....

Here is a useful clue to understanding Heaviside's motivations better. In his fine history of the events that eventually led to Schwartz' delineation of distributions, Jesper Lützen remarked of its characteristic ingredient, the "impulse function" $\delta(t)$:

> The δ-function must have had a very sad childhood since neither mathematicians nor physicists recognized it as belonging to their domain. If mathematicians used it, it was as an intuitive physical notion with no mathematical reality ... On the other hand, physicists usually considered the δ-function, or the point mass, as a pure mathematical idealization which did not exist in nature.[89]

In fact, Heaviside was one of the first who understood the underlying issues with great clarity. But it is also easy to understand why the δ-function's other parents (as Lützen catalogs, there were many) disowned their progeny. We only need to inspect several of the odd properties that such "solutions" commonly embody.

[88] Robert C. Stalnaker, Inquiry (Cambridge, Mass.: M.I.T. Press, 1984), ch. 3.
[89] Jesper Lützen, The Prehistory of the Theory of Distributions (New York: Springer-Verlag, 1982), 110.

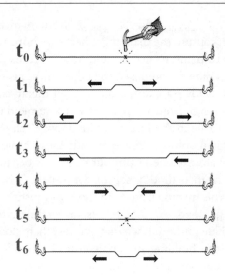

Much of Heaviside's great work was applied to the so-called telegrapher's equation. If we drop one of its terms, we obtain the wave equation ($d^2y/dt^2 = k\, d^2y/dx^2$) and we can use it to demonstrate the physical oddity of impulse solutions in a very simple manner. The wave equation was virtually the first partial differential equation discovered (by d'Alembert) and, almost immediately, a famous dispute between Euler and himself broke out over its proper set of solutions S, because Euler thought that a perfectly plucked string should belong to S, whereas d'Alembert objected that the term "d^2y/dx^2" isn't defined at the plucked point.[90] But the wave equation's impulsive solutions are even more curious than the plucked case. Here we are invited to consider a string initially at rest, with a sharp hammer blow delivered to its midpoint at time t_0. At that instant, the strike's effects will be invisible, but immediately afterward, two raised corners emerge from the center and travel towards the endpoints (where they reflect upside down and backwards, again momentarily vanishing when they collide once again). Even if we are willing to accept this "motion" as coherent, we quickly ascertain that an infinite amount of bending energy needs to be stored within its sharp corners (we get the same divergence if we calculate the total energy stored in its Fourier spectrum as well). So it's understandable why neither physicists nor mathematicians wished to acknowledge such a prodigal infant as kin. The fact that such funny "solutions" can be rigorized in distributional terms doesn't remove their physical oddity one bit.

However, let us now look upon the wave equation's space of solutions S in a somewhat different manner, by attending to their *linear structure*: <u>viz.</u>, if S_1 and S_2 are both solutions, then there is always a third solution S_3 that corresponds to the super-position of S_1+S_2. We furthermore realize that there are several points at the edge

⁹⁰ Clifford Truesdell, "Introduction," to Leonhard Euler, <u>Opera Omnia</u>, ii. 12 (Zurich: Orell Fussli Verlag, 1960).

of this space ($S_{impulse}$, $S_{transient}$ and $S_{long\text{-}term}$) from which the others can be superimposed and which capture the main ingredients we hope to control in our circuit: that is, how it responds to being turned on and how it settles down over time and so forth (4,vii). In fact, Heaviside decomposed the behavior of his telegraph equation into impulsive, transient and steady state terms and by their means saw how to correct for a circuit's lapses through design improvements (by recommending inductances, Heaviside played an important role in the history of long range telephone transmission). By supplementing our original solution space S with these building block solutions as "points at infinity" (they are excluded from our original S because they don't obey the telegrapher's equation in an orthodox sense), we obtain a richer collection S^{*} that contains the design controls missing from S. In short, our justification for considering S^{*} is not that $S_{impulse}$, $S_{transient}$ and $S_{long\text{-}term}$ represent genuine circuit behaviors that we expect to observe in real life (although we may witness their close approximations), but because they comprise the building materials crucial to understanding our circuit's properties under *all* conditions. Indeed, our new "solutions" are better regarded as coding basic properties of response that lie latent *within* the telephone wire itself, as opposed to any impossible history that our wire might display. In short, the physical significance of "solution" shifts from *representing a possible history of the circuit* to covering more abstract inclinations to reshape signals fed into the wire.

Heaviside understood this rationale quite clearly:

> The object of using impulses, involving infinite forces acting for infinitely short periods of time, is to be able to represent with comparative simplicity effects which, considered finitely, might be nearly the same in character, but vastly more complicated in expression.[91]

Also:

> For it is not from formulae representing the expansion of an arbitrary initial state that we can most readily learn the general course of events in the physical problem concerned. We should rather prefer to examine the result of some special initial state, or the result of a disturbance initiated at a single spot, such as an impulsive or continued source. Now it is just in these cases that the expansion theorem shows to best advantage. We obtain our formulae in a very ready manner, without the circumbendibus connected with arbitrary initial states.[92]

By "the result of a disturbance initiated by an impulsive or continued source," he was thinking of what I have labeled as $S_{impulse}$ and $S_{transient}$. Indeed, in avoiding "the circumbendibus of arbitrary initial states," Heaviside means to highlight the advantages for circuit design of his special "solutions" in a manner closely parallel to the considerations that supply rays and caustics with their vital role in the design of practical optical systems.

By articulating matters in terms of a natural completion of an originally unsatisfactory incomplete space, I follow modern ways of thinking (which is the same path that

---

[91] Heaviside, Electromagnetic, ii. 353.   [92] Ibid., 153.

Mikusiński pursues in his rigorization of Heaviside). But Heaviside was unfamiliar with such ideas and instead found his inspirations in the syntactic explorations of the English algebraist school, particularly, Boole's text on differential equations. As we saw, such formalist thinkers were prone to "solve" equations by *inversion* (= taking terms to the other side of an equation by "division") even if no apparent inverse exists. In a *static* electrical context, the equation $V = ZC$ reports that the current (C) supplies a voltage through a factor Z that represents the overall impedance of the circuit. Often we need to utilize this equation in inverted form: $C = 1/ZV$. This inversion represents no difficulty within an algebraic context, but, in Heaviside's time sensitive applications, the terms analogous to C, V and Z contain differential operators. Pursuing an "inversion" here leads to all those peculiar divisions by "1/dt," as well as their even stranger factorizations into "$1\sqrt{dt}$." In his little book on the Laplace transform written for engineers, W. D. Day comments :

> *Equations $V = ZC$ and its [inverse] $C = YZ$ are sometimes called generalizations of Ohms's Law, the term Z (called the* impedance of a circuit*) being a generalization of resistance and the term Y (called the* admittance of the circuit *and equal to the reciprocal of the impedance for the same C and V) being a generalization of conductance. . . . It is not always realized that [here] we have been using an example of what a mathematician would call a "transformation" from the real time variable t into the imaginary variable iω. The problem is posed as one involving currents and voltages, which are functions of time, but we decide to transform it into one involving the same currents and voltages as functions of a new variable, in order to simplify the mathematics . . . In spite of the fact that impedance is a convenient and easily pictured property (its physical nature is usually interpreted as a kind of opposition to current flow), we should not lose sight of the mathematical significance of the step we have taken, because the latter is still there when we extend the concept, even though the physical interpretation is obscure or absent.[93]*

In fact, through his "divisions," Heaviside found a formulaic expression for what is now called the *transfer function* of a circuit and by supplying simple impulsive "voltages" V to the combination "1/Z V" extracted the desired $S_{impulse}$, $S_{transient}$ and $S_{long\text{-}term}$. By these policies, he also found simple ways to picture how a circuit would behave if we try to *improve* it. We witnessed a primitive example of such a methodology in the problem of 4,i where our design task is to move a motorized telescope to a desired position by setting a dial. The critical question for good performance is how much amplification is desired in its feedback circuit c_2? We observed that this design problem can be solved by moving the roots of the formula $k/(x^2 + 2x + k)$ about on the complex plane by adjusting k. In fact, $k/(x^2 + 2x + k)$ represents the transfer function for this problem and we are simply plotting its roots within the s-space of the Laplace transform. But that representation gives an immediate portrait of how different choices of k lead to

[93] W. D. Day, Introduction to Laplace Transforms for Radio and Electronic Engineers (London: Interscience Publishers, 1960), 6. I have altered some of Day's choices of symbols—"s" for his "p"; "i" for his "j", etc.—to fit more usual standards.

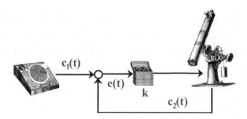

different virtues and vices in our telescope's behavior. This humble illustration supplies a preliminary sense of the vast increase in understanding we can obtain, for design-oriented projects, from a simple lift into s-space. Plainly, we will want $S_{impulse}$, $S_{transient}$ and $S_{long\text{-}term}$ within our "solution space" $S^\star$, for these represent the control space dials we tweak in attempting to improve our circuit's performance.

Here I have described the situation in essentially semantic terms: over how large a space of "solutions" do we want an effective descriptive language for electrical circuits to operate? Heaviside, of course, addresses these same concerns in syntactic garb: "What inferential pathways of connection should I follow to become able to think about circuits with genuine understanding?" Since the desired information must lie latent within the basic equations for the circuit, Heaviside attempts to massage them into a revealing form through trial and error manipulation. In fact, no one in Heaviside's era was prepared to reach his worthy goals in any other manner, given the difficulties of articulating a supportive mathematical structure $S^\star$ with the appropriate number of positions. But a policy of syntactic exploration mandates two vital procedural consequences. (1) Do not automatically follow *any* manipulation officially sanctioned by mathematical rigorists, for many of those will prove of no heuristic worth whatsoever (their "multiplicity" is too great, in Heaviside's expression). (2) At the same time, if a profitable route from A to B seems to be blocked because of a missing transitional step C, find some previously unutilized grammatical combination able to fill that gap (or, more rarely, make up something entirely new). Note that policies (1) and (2) go together: we must considerably winnow the multiplicity of paths before we will be able to discern those useful A to B routes that lack fill-in intermediaries. Through this methodology, Heaviside was led to his sentences involving "$1\sqrt{dt}$" and so forth. Such methods can lead to apparently unintelligible formulae, but the physicist should expect that strange syntactic combinations may actually bring to the surface the very information we need to complete some profitable line of practical reasoning:

> [I]n mathematics, the fundamental notions are so simple that one might expect that unlimited developments could be made without coming to anything unintelligible. But we do, and in various directions ... [W]hen [such] things turn up in the mathematics of physics, the physicist is bound to consider them and make the best use of them that he can.[94]

---

[94] Heaviside, Electromagnetic, ii. 8–9.

This is why a physical mathematician should experimentally attack a starting expression with any form of manipulative procedure that comes to mind, hoping to produce the missing pieces required to complete some profitable procedure. And we certainly shouldn't be deterred by the fact that prior practice has found no need for a facilitating piece of syntax:

> A man would never get anything done if he had to worry over all the niceties of logical mathematics under severe restrictions; say, for instance, that you are bound to go through a gate, but must on no account jump over it or get through the hedge, although that action would bring you at once to your goal.[95]

In all these matters Heaviside is impressed by the strange behaviors of series expansions discussed in 6, vii, especially the divergent series in which Stokes trafficked. It is an astonishing fact about such expressions that they hide all sorts of valuable information that they suddenly disgorge when manipulated in an unexpected manner. As we observed, Michael Berry and others have recently demonstrated that valuable information about borderline regions have been hiding in these expressions unsuspected for many years.

In fact, the earlier English algebraists encouraged such free play with syntactic innovation. Thus John Venn explains that we should regard Boole's inverse combinations ("$A/(1 - B)$") as "interrogative hints": "Would you, perhaps, care to use us?":

> We might conceive the symbols conveying the following hint to us: Look out and satisfy yourselves on logical grounds whether or not there be not an inverse operation to the above. If you can ascertain its existence, then there is one of our number at your service to express it. In fact, having chosen one of us to represent your logical analogue to multiplication, there is another which you are bound in consistency to employ as representative of your logical analogue to its inverse, division,—supposing such an operation to exist.[96]

Or, in Boole's own words with respect to operational methods for differential equations:

> It is the office of the inverse symbol to propose a question, not to describe an operation. It is, in its primary meaning, interrogative, not directive.[97]

However, those authorities explained their policies by misty appeal to the "generality of algebra," whereas Heaviside is merely expounding the virtues of gaining a sense of a new supportive structure by experimentally piecing together routes of practical advantage that crisscross its surface:

> It is of some importance to distinguish between a function in the physical sense, and its mode of expression in symbols standing for numbers . . . It is characteristic of rigorous

[95] Ibid., 122.   [96] John Venn, Symbolic Logic (London: MacMillan and Co., 1894), 74.
[97] George Boole, A Treatise on Differential Equations (Cambridge: Cambridge University Press, 1859), 377.

mathematicians, I think, that they think too much of the formula, and consider that it is the function. No, it is only the dress, and need not be a convenient fit. It is generally too large; or, may be, several dresses are needed, disconnected, for what would be simply a single function in its physical meaning.[98]

Here by "formula" Heaviside seems to mean "formula and its deductive field of viable expansions": he is advising us to attend to the chains of valuable reasoning that initiate from the original circuit equation as head. It is in these terms that he maps out the linear structure of the space $S^\star$, as well as the desirability of including the normally forbidden $S_{impulse}$, $S_{transient}$ and $S_{long\text{-}term}$ as occupiable nodes. Viewed sympathetically, Heaviside's thinking seems quite modern in its contours (akin to the motives that drive functional analysis), rather than merely representing a historical oddity.

However, we can't entirely trust a technique that has been entirely hammered out in this quasi-experimental way, because there large gaps of untested territory where such techniques might (and, in fact, do) go wrong. If some critical engineering project depends upon their trustworthiness, we can't simply adopt an "experimental" attitude: "Oh, let's just build an electrical grid based upon our calculations and see if it works," for such unmonitored faith can bring the economic ruin of a country in its wake. Heaviside has built a chart over electric circuitry that is like Ernest's hub-and-spokes maps: it doesn't supply enough information about what happens when we follow trails that Heaviside hasn't already blazed. And that is why we eventually need to find a mathematical picture of the solutions to support manipulations of Heaviside type, ensuring that it is rich enough to support his $S_{impulse}$, $S_{transient}$ and $S_{long\text{-}term}$ expressions. The Laplace/distribution and Mikusiński proposals both do a good job in capturing, albeit in different ways, the basic "usefully decompose the manner in which a circuit acts upon an impressed signal" structure within Heaviside's thinking.

In such circumstances, an Ecclesiastesian season of semantic agnosticism is clearly mandated, for sometimes a semantically opaque path represents the only route of desirable improvement open to us. In Heaviside's circumstances, we again witness the same contrast in modes of "understanding" that our face plots bring forth: Heaviside gains a better understanding of his circuits precisely through a prolongation policy that insures that he understands them less well than does the wet blanket rigorist who hews firmly to an old fashioned δ/ε picture.

In the next section, I shall attempt to locate the salience of semantic pictures within the fuller story of a predicate's usage. Before we turn to these matters, it is important to observe that, although we might dearly *like* to check rules $\mathcal{R}$ against the settings of $\mathcal{P}$, that task may lie considerably beyond our mathematical means. For example, Augustus Cauchy articulated his famous standards for what infinite series and calculus expressions signify (essentially the δ/ε business familiar from freshman calculus, although Weierstrass later contributed some key elements) because he hoped to understand some

---

[98] Heaviside, Electromagnetic, ii. 462

serious derivational difficulties that arise within celestial mechanics. Despite the enormous progress Cauchy himself made and everything else that has been learned subsequently, we still lack a reliable analysis of whether many of the reasoning patterns $\mathscr{R}$ upon which he originally set his sights are genuinely sound or not.[99] In the meantime, widely employed calculations of planetary motions have been forced to fly ahead on a wing and a prayer, as Brown and Shook explain in their classic exposition of lunar theory:

> While mathematical rigor is desirable when it can be obtained, nearly all progress in knowledge of the effects of these laws would be stopped if complete justification of every step in the process were demanded. The use of formal processes is justified whenever experience shows that the results, not otherwise obtainable, are useful for the prediction of physical phenomena. Thus when calculating with an infinite series whose convergence properties are not known, one has to be guided by the results obtained; if the series appears to be converging with sufficient rapidity to yield the needed degree of accuracy, there is no choice save that of using the numerical values which it gives. We have not attempted to deal with convergence questions, but have retained throughout the practical point of view.[100]

Of course, such uncheckable patches in our methodologies offer all sort of convenient cubby holes in which unexpected semantic surprises can readily hide.

(xi)

Semantic detoxification. A new semantic picture can thus serve as a fresh source of directivities with respect to a predicate, for significant improvements in our linguistic skills can be achieved through determining whether our established inferential principles are sound with respect to the newly proposed settings or not. Through these correlational checks, we discover that Euler's method is not always valid—that attention to the heretofore unnoticed Lipschitz condition is also required (4,x). Heaviside's procedures are also full of analogous lapses, from which the fine grain of the Laplacean setting can protect us (although in a somewhat overly cautious manner). For these purposes, we do *not* need to understand *everything* about a predicate's personality, but only the *mathematical structure* over which the inferential rules we study unfold. The soundness of the procedures of the operational calculus can be adjudicated on the structural basis provided by Laplace transforms and distributions alone, without any wider appreciation of how it happened that Oliver Heaviside's linguistic craft came to park itself over this particular patch of ocean.

One of the problems that make life difficult for the would-be rigorist lies in the fact that many of the background issues that shape Heaviside's approach hover rather

---

[99] Laurence G. Taff, <u>Celestial Mechanics: A Computational Guide for the Practitioner</u> (New York: Wiley-Interscience, 1985), 52, 274, 321.   [100] Ernest Brown and Clarence Shook, <u>Planetary Theory</u> (New York: Dover, 1964), p. x.

amorphously in the air and do not appear registered within the equations themselves. As the analyst J. Schwartz observes in a lovely article:

> *The physicist, looking at [a specific] equation, learns to sense in it the presence of many invisible terms, integral, integrodifferential, perhaps even more complicated kinds of operators, in addition to the differential terms visible, and this sense inspires an entirely appropriate disregard for the purely technical features of the equations he sees. This very healthy self-skepticism is foreign to the mathematical approach . . . Give a mathematician a situation which is the least bit ill-defined—he will first of all make it well defined. Perhaps appropriately but perhaps also inappropriately . . . The mathematician turns the scientist's theoretical assumptions, <u>i.e.</u>, convenient points of analytical emphasis, into axioms and then takes these axioms literally. This brings with it the danger that he may also persuade the scientist to take those axioms literally. The question, central to the scientific invest-igation but intensely disturbing in the mathematical context—what happens to all this if the axioms are relaxed?—, is thereby put into shadow.*[101]

When we resettle Heaviside's techniques upon the platform of Laplace transforms and distributions, this is precisely what we do: we surround Heaviside's barer notation with several forms of previously unattested integral and integro-differential operators ("crypto-integral equations," in Dieudonné's phrase). Of course (and this is what Schwartz intends), Heaviside himself is alive only to a hazy *multiplicity* that is demanded by the practicalities of his descriptive situation—e.g., that his special new "solutions" need to be tolerated in companionship with the regular solutions accepted in conven-tional δ/ε epsilonics. It is the rigorist's task to find surrogates that can occupy the required positions in a manner such that their mathematical underpinnings become clear. The device of dressing up the old equations in fancy operators represents one method for achieving that goal.

But Schwartz makes a deeper observation as well. Heaviside's sense of notational multiplicity derives from an appreciation of the *variation* native to his background problematic: the cloud of alterations and possibilities that scientists must keep in mind before they can trust that their calculations apply to real life circumstances with any measure of certifiable *safety*. Schwartz emphasizes *structural stability* in this regard: will our results hold true if the modeling assumptions within the equations are allowed to vary in the manner we might encounter within real applications? An explicit study of these tolerances sets our problem within what is often called a *control space*: a large arena full of the smaller phase spaces in which we usually study physical equations when we hold their structural parameters fixed. Often surprising complications emerge when otherwise tame equations and procedures are scrutinized from this more expansive point of view. Control spaces are also natural to *planning* and *design* and often inspire us to accept "solutions" to equations that won't seem natural if we look at the formulas merely on phase space level. Indeed, this is exactly where

---

[101] J. Schwartz, "The Pernicious Influence of Mathematics on Science" in E. Nagel, P. Suppes, A. Tarski, eds., <u>Logic, Methodology and Philosophy of Science</u> (Stanford, Calif.: Stanford University Press, 1962), 356–7.

Heaviside's motive for tolerating the somewhat "unphysical" solutions S_{impulse} and $S_{\text{transient}}$ lies.

More generally, every modern applied mathematician recognizes that the *questions* we ask of a target equation can lead to quite different branches of mathematical inquiry: a control problem with respect to a given equation places us in different mathematical territory—usually, a more formidable terrain—than a straightforward prediction problem (a nice example is 6,xiii's distinction between predicting how a frictionless skater will move on an inclined plane and devising the most efficient path to a desired destination). Even asking a question of simple *postdiction* can carry us into a quite different landscape than a *predictive* query. For example, the mathematics of ascertaining the past thermal history of the earth mandates a complex host of auxiliary considerations (because the task represents an ill-posed problem[102]) that can be blissfully ignored when we chart our planet's thermal future (Heaviside, by the way, contributed much to the former problem). That fact that quite distinct branches of mathematics can spring from a common formula according to the demands scientists require of it led Poincaré to remark, in a famous aperçu: "*Without* [physics] *we should not know partial differential equations.*"[103] Indeed, we have already observed that whodunit concerns in logic give rise to Boole-like predicates ("A/B") that are rather different in their conceptual personalities from those emphasized within a conventional modern logic course. Disparities in personality that derive from different forms of background problematic appear quite commonly across classical mechanic's rambunctious family of descriptive notions.

Heaviside always considers his circuits from a perspective of hoping to improve their performance. Throughout this book, the unacknowledged role that *design's* teleological interests often play in wrapping a characteristic personality around a descriptive predicate has been much emphasized, partially as a means of combating narrow stereotypes of "what scientific concepts are like." As observed in 5,viii, much argumentation in recent philosophical discussion essentially turns around faulty presumptions on this score. This situation could be considerably improved if such writers more adequately realized that suitable background concerns can cloak a "predicate of physics"—or any other subject—in virtually any kind of conceptual personality that one might desire.

Nevertheless, the somewhat limited task of the would-be constructor of semantic picture needn't take all of these complexities of motivation and directivity into account: she mainly needs to map out the *structure* of the informational supports upon which the language rests and she needn't attend closely to the richer niceties of motivation to perform her task adequately. But she must get the *multiplicity* of states required right and, as Schwartz emphasizes, the trained mathematician often experiences a good deal of difficulty in extracting what's required from the physicist, who often attends to the required variations appropriately in the *patterns* whereby he utilizes the results of his

[102]   D. N. Ghosh Roy and L. S. Couchman, Inverse Problems and Inverse Scattering of Plane Waves (San Diego, Calif.: Academic Press, 2003), chs.1–3.   [103]   Poincaré, "Analysis," 81.

calculations, but may not otherwise mark the discriminations he observes in explicit notation.

Accordingly, we can't really understand why Heaviside's usage gravitated to its operational calculus "multiplicity" without appreciating the background considerations of circuit design (this represents the arena in which the discipline "sees its opportunities and takes'em," recalling the rapacity of the redoubtable Plunkitt[104]). But no telltale signature of these objectives appears within the semantic constructions offered either by Laplace transforms or Mikusiński's alternative, which merely mark out the *structure* required in the relevant "space of solutions." This neglect of motivational context is clearly demonstrated by the fact that virtually identical tools of rigorization can be applied to settings that manifest not the faintest whiff of design concern in their own circumstances. In adding sentences about "impulsive solutions" to the naïve solution space $\mathcal{S}$ for our circuit equations, we frame an enlargement $\mathcal{S}^*$ akin to amplifying Euclidean geometry with points at infinity. Although it is customary to call these supplements "weak solutions" ("weak" because they relate to the original equations E only in a roundabout way), they often signalize completely different states of affairs in their home locales. Thus, in Heaviside's circumstances the weak solutions signalize "internal dispositions of our circuit to act upon input currents according to certain patterns," rather than true "histories of current flow in a circuit" as the "regular solutions" do. In contrast, within many other forms of continuum mechanics context, the comparable weak solutions do represent "physical histories" just as ably as the regular solutions (for reasons mentioned in the fine print). In other applications yet, they convey a rather surprising *warning content*, "You can't handle *this* kind of case without bringing in more physics" (in the jargon of 4,ix, they provide "lousy encyclopedia" instructions: "For better methods for this particular problem, visit descriptive patch B"). So decoded, it seems odd to designate a sentence that carries the message "I can't solve this kind of problem" as a *solution*, but such answers are invariably designated as "weak solutions" in mathematical practice, just as Heaviside's non-solution "solutions" are likewise so denominated. All that is intended by the "weak solution" label is a structural position within an extended space $\mathcal{S}^*$, allowing correspondent claims to indicate completely different arrangements within their individual settings.

....

In the roughest terms, the basic idea behind a weak solution construction is that we persuade an equation E to issue instructions to a (generalized) function f(x,t) through intermediaries. As in the Euler/d'Alembert case, E doesn't make direct sense at the sharp corner of f(x,t), so we construct a crowd of nearby g(x,t)'s for which the equation's instructions at least make sense (we employ test functions to smear out f's sharp corner into curvy g's). We don't really care about these g's in their own right: we merely want them to speak to f on E's behalf (f says to E, "I can't understand what you're asking me to do, but tell my g-function entourage what you wish and I'll imitate them"). In this way, we fill out the space of regular solutions S to E with f supplements, reaching a nicer S∗ as a result.

[104] William L. Riordon, <u>Plunkitt of Tammany Hall</u> (Boston: Bedford, 1994), 49.

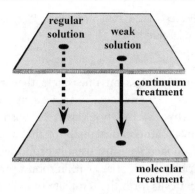

As a general policy for investigating the behavior of equations, it is useful to investigate a sequence of improving guesses to a problem (say, the broken line approximations drawn by Euler's method to a smooth cannon ball motion) and see what their final limit s is like. Without further information, we can often prove only that s qualifies as a weak solution to the E under consideration. With extra work, if we're lucky, s can be shown to comprise a regular solution and hence its physical significance proves unproblematic. Sometimes s merely lacks an upper layer of derivatives demanded by E, a fact that we can often blame on the fact that E's conditions should have been properly articulated in integral equation form (so it is "E" that carries the misleading content, not "s"). Sometimes, as with shock waves, "s" conveys the funny halfway meaning discussed in 4,vii: "Complicated physics that isn't really described by E becomes relevant along my shock front, but we can still stay within E's descriptive patch by adopting the Riemann-Hugoniot trick." But sometimes our weak solutions don't carry any clear meaning except *trouble*: we need to abandon our current approach in this case (in many parts of continuum mechanics we haven't been able to rule out the threat of these unwanted guests, which may spoil the utility of our theory in the very cases where it most matters). As Poston and Stewart remark:

> We have gone beyond the casual assumption of "configurations pretty much like smooth functions if we don't look too closely" . . . We are making very delicate hypotheses (like square integrable second distributional derivatives) about exactly what we would see if we could look arbitrarily close: hypotheses we know to be false. In what sense these "infinitesimal" hypotheses can be said to model the local average nature of elastic solids is a dark and gleaming mystery. The results of the model, in macroscopic predictions, are very successful indeed; but it is hard to say in what sense the theory is more exact than a computer treatment of a finite element model . . . , but with many more rods and strings . . . [L]ong calculations that you do not understand, whether by number crunching or Banach space methods, will land you with bridges that fall down. A theory is only as good as your understanding of it.[105]

In the appended diagram, the *dashed arrow* from the regular solution indicates that the answer it presents to a problem posed in patch A terms can be regarded as satisfactory, even though the same problem can be *optionally* approached in patch B terms if desired. In contrast, the *filled arrow*

---

[105] Tim Poston and Ian Stewart, Catastrophe Theory and its Applications (San Francisco: Pitman, 1978), 300.

from the weak solution indicates that a shift into patch B is *mandatary* for the conditions to which the weak solution corresponds.

....

These considerations may address an uneasiness that many readers of modern mathematics may have secretly experienced: "Gee, I can appreciate how the mechanics of distributions works, but why, exactly, should we ever wish to fool with such things?" It is hard to find an answer in a mathematical textbook that doesn't seem disappointingly unsatisfactory on this score. In fact, we must usually probe deeper into the motivational background of a specific problem, as it arises within robust practice, before we learn the richer setting of purpose that calls forth the template of a distributional treatment. Our mathematics primer has only reported on the bare structure common to applications that are motivationally quite distinct, leaving its explication of "why do this?" rather uncomfortably abstract.

Nonetheless, the fact that a weak solution sometimes signalizes a true current history; sometimes, an internal state; sometimes, a "get out of this patch" warning matters not one whit to the basic task of *picture reformulation,* for its concerns center upon the mathematical correlations that exist between the target language and some generic family $\mathcal{S}$ of supporting landscapes. Thus the Laplace transform story relates Heaviside's peculiar argot to the $\mathcal{S}^*$, without really explaining why $\mathcal{S}^*$ is the relevant collection to consider. Rather than attempting to codify every factor detectible within the conceptual personalities of the predicates under scrutiny, some of the *merit* of a new picture may trace precisely to the fact that it has *ignored or discounted* otherwise central ingredients in the developmental history of the target language. This occurs because a common virtue of a good preface or picture lies in the fact that it offers an effective way of terminating— or, to use my old word, *detoxifying*—some unhelpful thread that has heretofore lain tangled up within a word's elaborate heritage. A nice example is provided in the modern reconsideration of the terms "light ray" and "caustic" discussed in 6,vii. As we noted, unexpected prolongations of these venerable notions help modern designers of optical systems (e.g., a complicated telescope) understand the complex effects that arise within their creations and improve their performance. J. B. Keller (one of the pioneers of this recognition) writes:

> [W]e note that an older, displaced theory—the ray theory—was used to solve asymptotically the problems of a new theory which displaced it—the wave theory. It must be true in general that any outmoded theory which is superseded by an old one is asymptotically correct in some limiting case. Otherwise, it would not have been accepted as a satisfactory theory in the first place. Therefore, it should provide a basis for solving asymptotically the problems of the new theory. This methodological principle, which can be illustrated by other cases as well as the present one, may be helpful in guiding us to the solution of other problems.[106]

---

[106] J. B. Keller, "Rays, Waves and Asymptotics," American Mathematical Monthly, 84, 5 (Sept. 1978).

To this end Keller and his associates first needed to frame the asymptotics-over-wave-pattern picture of how old-fashioned talk about "rays" and "caustics" actually conveys physical information in settings like our razor blade case. From their new portrait, unexpected directivities for "ray" and "caustic" automatically open up: extraordinary new kinds of "ray" and "caustic" enter the modern designer's toolkit, including "rays" that run through wholly imaginary locations. Regarded from this new point of view, Keller's prolongations of terminology are utterly natural and supported in exactly the same manner as the old-fashioned "rays" and "caustics" of traditional optical thinking. But no one laboring under the old image of "rays" as the paths traced by light particles could have possibly foreseen such wild extrapolations into the "imaginary." Through its detoxifying merits, a shift in supportive picture can suddenly open new vistas for old and shopworn vocabulary.

This capacity for radically reorienting an old language highlights one of the characteristic beauties of a new semantic picture: once we decide that we have correctly mapped out the *informational terrain* over which our language sits, then we can settle lots of important questions with respect to future practice without worrying much about other determinants of our usage (we can apprize the naturalness of designating Keller's imaginary extensions as "rays" even if we can't see how a designer might profit from their invocation). Framing a new semantic picture can act as an effective filter against unhelpful inherited prejudices: we can now judge an old inferential rule or recipe solely according to its capacity to perform ably when tested against the range of settings contemplated by our new picture. Such correlational examinations often permit a great winnowing of wheat from chaff in the harvest of directive elements we have acquired from our forebears, a fact that our teenager accurately trumpets. The cruel blacksmith of 5,ii might abandon his reliance upon human victims after a studious attempt to align the macabre steps in his recipe with ascertainable metallurgical processes. The fact that the terms "light ray" and "caustic" originally staggered to their linguistic feet during an era when folks believed that light particles travel along one-dimensional trajectories should not bind our own use of those terms in the future. It's *our* language, after all— why shouldn't we allow "ray" and "caustic" to proceed onward with their previously unrecognized utilities? Much good and little harm will come from this policy as long as we can successfully rid the terms of unwanted former associations through detoxification. But this is easy to do: we merely have to warn the potentially gullible that they must set aside that old malarkey about particle paths. Ancient shackles now severed, "ray" and "caustic" can claim their rightful patrimony of edge diffraction, imaginary beams and all that. To uncover these redirective prolongations, it may even prove necessary to liberate our usage from the prior entanglements of design intent: a good picture diagnoses the structural nature of the information coded within a set of classifications and suggests further natural extensions on that basis. Generally, the precise *purposes* for which we employ a terminology supported in this manner are left to us. In this spirit, we might decide (and, indeed, we often do) to utilize Heaviside-like "weak solutions" in connection with circumstances that do not possess a smidgen of design interest.

All of this is music to Uncle Fred's ears. Nothing he says ever has any purpose, but his aimless assertions can be evaluated for truth-value if an adequate correlational picture stands available. However, the semantic worm turns in his direction only in certain seasons, as we shall see in the next section.

Let me reiterate that a chief manner in which semantic pictures assist us lies in their ability to collaborate or reject established inferential rules $\mathcal{R}$ through generic *investigations of soundness* in the mode of 4,xi. That is, we use our picture $\mathcal{P}$ to carve out a generic range of mathematical settings which we hope will accurately model all of the potential settings in which $\mathcal{R}$ might be applied. $\mathcal{P}$ then certifies $\mathcal{R}$ as *sound* if $\mathcal{R}$ always carries true assertions to true assertions over every setting tolerated by $\mathcal{P}$. Through a generic survey of this correlational kind, we then discover that Euler's method is not always valid—that the heretofore unnoticed Lipschitz condition represents an additional requirement upon applicable setting. Heaviside's procedures likewise turn out to be full of analogous lapses, from which the fine grain of the Laplacean setting can protect us (as we noted, in an overly cautious manner). But for these fine purposes, we needn't understand *everything* relevant to a predicate's personality, but merely the *mathematical settings* over which its pertinent inferential policies unfold.

As we shall later learn, many philosophers have adopted extremist attitudes with respect to the merits of soundness proofs: some claiming that they pierce to the very heart of classical gluing whereas others deny that such investigations offer any significant supportive import at all. Such points of view generally derive from a narrow attention to logical forms of inference only, an inclination that overlooks the fact that the inferential directivities that most shape language tend not to be logical in character at all. We shall return to all of these issues in Chapter 10.

(xii)

Through conceptual thick and thin. But these advantages of a picture are entirely seasonal and moments will arrive when we are well advised to set our pictures by. A just consideration of the Heaviside affair shows that we can't always judge an inferential principle reliably according to any available picture, for such portraits may all prove shortsighted in the manner of Heaviside's δ/ε critics. True, the articulation of sympathetic portraiture, when it is uncovered, can materially advance a usage such as Heaviside's along an improving path—indeed, there invariably comes a time when his sorts of quasi-experimental exploration will flag in yield and a fresh picture is required before further territory is gainsaid. So we should very much appreciate good correlational accounts when we can find them, but we mustn't plump for poor portraits simply through semantic haste. Nor should we expect good pictures to afford the solution to every inferential woe, for it must be recognized that, even with worthy preface theoretically in hand, we often fail to obtain answers to our soundness concerns simply because such investigations are frequently very difficult from a mathematical point of view. Indeed, this situation is very common in science, as we have elsewhere noted. If so, the picture sits comparatively idle, unable to contribute many robust directivities to our investigative endeavors. But when correctness checks can be obtained, they can materially improve our confidence in old rules: we trust computer simulations founded upon a solid set of correctness results more firmly than those whose pictorial underpinnings remain shaky. We still must follow the guidance of the latter in real life, for scientific advance can rarely tarry for poky rigor, but an apt soundness proof assures us that fewer nasty surprises lie ahead ("fewer," however, does not guarantee "none" for our current picture may be unsuited to our real circumstances).

Thus crisp articulation of a picture can help *detoxify* a usage of inherited encumbrances it might wisely shed (we shall consider another recipe for effective detoxification in a minute). Nevertheless, I reiterate that we can't properly understand the directive engines that fuel a predicate's developmental career without attending to factors that a dutiful rigorist can justly neglect. The imperatives of design never vanish from the concerns of the electrical engineer and upon such a basis some future Heaviside (or Martian genius, to recall our teenager's muse) may someday take our circuit classifications along "a new direction of purpose that will aim toward a goal we heretofore had not suspected," casting aside the relevance of the Laplace transform + distributions picture in the process. Even if this never occurs, we rarely abandon the semantic portraits of our forebears entirely—their directive strands remain available as *latent DNA*, to which future prolongations sometimes usefully but unexpectedly attach (8,viii). Most importantly, as Heaviside's travails at the hands of wet blanket rigorists show, it is easy to develop an inadequate picture prematurely and denounce valuable inferential pathways as unworthy upon that basis. He explains that, accordingly, he writes for:

> *practical physicists and electricians; not mathematicians of the Cambridge or conservatory kind, who look the gift-horse in the mouth and shake their heads with solemn smile, or go from Dan to Beersheba and say that all is barren; but of the common field variety, who take the seasons as they come and go, with grateful appreciation.*[107]

Of course, design considerations loom large for this "field variety" audience.

And so seasons in which *picture* and *design* (or allied directive focus) alternatively wax and wane are to be expected. At any given moment, while entangled in the unchartable coils of linguistic development, it is prudent to recall that ascertaining a terminology's proper semantic season can prove enormously difficult: we can't be certain whether we should warmly embrace a proposed picture or remain agnostic a little longer. But these considerations of uncertain ambient directivities represent the ever-present background noise against which the would-be rigorizer necessarily works and, as such, she needn't attempt to capture their additional flavors of conceptual influence within the correlational pictures that she newly mints. Nonetheless, as philosophers observing her efforts in articulating rigorous underpinnings, we mustn't forget those wider circles of inhomogeneous guidance, for they remain vital to understanding the characteristics of developing predicate personalities along their fuller extent. To be sure, it might

---

[107] Heaviside, Electromagnetic, ii. 12.

represent a small improvement in civilized discourse if our high school teachers could persuade us to attach "weighs 5 pounds" unswervingly to *impressed gravitational force*, for that stipulation would obey a transparent "is a dog"/*being a dog* pattern. But everyday usage swims in deeper and harder to chart currents and we soon find ourselves prattling of "weightless astronauts" regardless of semantic vow. As we amply witnessed, unnoticed environmental influences can easily overwhelm the most worthy of simple correlational schemes.

It is important to appreciate that the information that we concretely supply in a reformed picture is comparatively *thin* in its intrinsic qualities, in the same sense that the preface of an atlas is slim in geographical fact (which gets registered within the body of the book). The preface gains its utility through permitting improvements in our skills in working with the maps themselves. For this reason, we should beware of enveloping our worthy endeavors in picture construction in the "true thought" wrappings of classical thinking: when we articulate a generic picture, we scarcely lay bare the "full semantic contents" of our predicates in all their naked glory. No, not if by "full contents", we intend the shifting cloud of directive influences that supply a predicate with its recognizable personality, for many of its critical ingredients will have been omitted from our correlational story *on purpose*. And this is why the merits of any semantic portrait should be approached with a considerable degree of mitigated salt: not everything vital about a language's wherewithal is registered within its gaunt and schematic frame.

....

Appraising the work of Heaviside and Dirac in his pleasant autobiography, Laurent Schwartz comments:

> Thus, [quantum] *physicists lived in a fantastic universe which they knew how to manipulate admirably, and almost flawlessly, without ever being able to justify anything. In this way they made great advances in theoretical physics. They deserved the same reproach as Heaviside: the computations were insane by standards of mathematical rigor, but they gave absolutely correct results, so one could think that a proper mathematical justification must exist. Almost no one looked for one, though ... [I]t's a good thing that theoretical physicists do not wait for mathematical justification before going ahead with their theories!*[108]

However, glosses like this can mislead, because the "theory of distributions" does not *supply a missing content* for Heaviside's calculus, so much as *map out a generic picture* as to how linguistic acts and their supportive contents strategically fit together. As we saw, the rigorization does not address the central questions as to why Heaviside's calculus evolves as it does.

On a related score, it should be observed that, although we often hear loose glosses of the ilk "Schwartz' theory of distributions helps make classical mechanics rigorous," such claims are potentially misleading, in that it is usually only the *local setting* for some small family of equations which has been precisified, not any unit that reaches to the length scale of "classical mechanics." We must be careful to distinguish between the claim "distributions are often useful in rendering the local problems of classical mechanics rigorous" and the more sweeping contention that the

---

[108] Laurent Schwartz, A Mathematician Grappling with His Century, Leila Schneps, trans. (Basel: Nirkhäuser, 2001), 217–18.

entire frame of classical thought has found a happy axiomatic home (not true, argue the considerations of 4,viii). As the Heaviside case displays, the reasons for filling out the "solution space" of a problem can be quite localized in their impertinent particularities and two schemes drawn from different circumstances may not fit together harmoniously.

. . . .

In some form or other, the sundry formalists and pre-pragmatists surveyed in earlier chapters are aware of the inherent thinness of the correlational pictures that we attach to usage and have therefore attempted to argue, through a variety of strategies, that such appendages aren't essential to sound predicative employment at all. Here they typically overreach, because they confuse the inherent value of correlational studies with the "true thought" interpretations that classicists assign them. However, they usually oppose the undue classical emphasis by rightly stressing the importance of a second flavor of semantic detoxification, which I'll dub *pre-pragmatist scrutiny*. In such circumstances, we discard old pictures that we no longer want, not by proposing replacements, but through emphasizing the operational capacities of rules and recipes to which prevailing pictures prove inadequate. Through such means Heaviside endeavors to cast off the stultifying mantle of the Cambridge rigorists and is perfectly correct to do so. If we review the patterns in which our canonical developments unfold, we will witness our two detoxifying policies for eliminating unwanted intensional heritages in seasonal action. Thus, a stereotypical history runs: (1) the predicate "P" gets up and running under the nurturing shelter of picture $\mathcal{P} \rightarrow$ (2) strands of practical advantage $\mathcal{R}$ are developed under $\mathcal{P}$'s aegis, usually regarded as somewhat fallible by its lights $\rightarrow$ (3) other directivities suggest that $\mathcal{R}$ should be expanded and corrected to $\mathcal{R}^\star$ in a manner that $\mathcal{P}$ no longer accommodates ably, initiating an agnostic interval where the indications of $\mathcal{R}^\star$ are trusted more than those supplied by $\mathcal{P} \rightarrow$ (4) a replacement picture $\mathcal{P}^\star$ for "P" is articulated that once again dominates and corrects the indications supplied within rules $\mathcal{R}^\star$. Here the era of pre-pragmatist scrutiny is natural to stage (3), and that of new picture to (4). In the Heaviside history, all of these epoches are sharply defined historically, but, elsewhere, such reversals in trust may prove less stark, ameliorated by basic "never trust any picture or rule too far" considerations. In these cases, we form our ongoing estimates of how "P" should be correctly employed based upon some sort of judicious weighing of what $\mathcal{R}^\star$ and $\mathcal{P}^\star$ each portend. And sometimes oscillations between (3) and (4) occur a number of times over an unfolding career. Both forms of appeal represent important techniques for filtering or detoxification, but neither should be regarded as fully explicating the directive engines that drive a target predicate onward.

In Chapter 4 and elsewhere, I have sketched some of the substantive barriers to progress that scientists in the late nineteenth century sought to overcome through some variation upon instrumentalism or formalism. And they were certainly right to oppose prevailing "true thought" assumptions about the contents of "rigid body" and "force," as well as suspecting that various forms of semantical mimicry underlay many of the alleged "evidences" in favor of atoms (see 10,viii for more). However, in claiming that instrumentalism or formalism can provide the full story of a scientific predicate's personality, such critics open the lid to a full chest of nattering imps that sail in on the winds of holism. Must we really accept these plagues simply to repulse the exaggerations of "true thought" classicism? To be sure, classical thinking insists that language comes tightly glued to the world from the semantic outset and, accordingly, does not readily grant agnostic detoxifications their proper moments in the sun as able cleansers of unwanted directivities. "If you wish to replace old concepts with new," the traditionalist sternly enjoins our pre-pragmatists, "you may do so, but only if you insure that they possess 'true thought' underpinnings as secure as those that attached to the discarded readings. Conceptually unmoored language often leads to scientific mistake, as we recognize from the sometimes terrible errors that failures in rigor have encouraged in the past." Such accusations of "lack of rigor" terrify our pre-pragmatists (one could hardly insult Duhem with a more deeply resented epithet), so they regularly leap into the holistic arms of strict axiomatics, which can at least claim *syntactic rigor* on its behalf. This absolutist choice forces them to argue that scientific endeavor is *never* beholden, in any season, to any form of correlational picture. "The advancement of science," they proclaim, "never requires that its vocabulary be held up by supports we truly understand' in the warm and fuzzy manner that *gear wheel* deceptively evinces. Instead, science's chief duty is to articulate predictive recipes accurately, not to search for some correlative story of underlying support that we especially fancy." Or, as Mach expresses the doctrine in a famous and influential formulation:

> The goal which [science] *has set itself is the* simplest *and* most economical *abstract expression of facts.*[109]

Here "abstract" carries the significance of "follows rules that we understand through effective manipulation," in contrast to the unhelpful "gives rise to warm and fuzzy feelings of contentment" of the sort that attaches, without scientific importance, to *gear wheel*. In truth, it would have been far preferable if our pre-pragmatists had mildly set aside classical demands in Heaviside's manner of seasonally calling for a temporary moratorium on associated pictures.

> There is no self-contained theory possible, even of geometry considered merely as a logical science, apart from practical meaning. For a language is used in its enunciation, which implies developed ideas and processes already in existence, besides the general experience associated therewith. We define a thing in a phrase, using words. These words have to be explained in other words, and so on, for ever, in a complicated maze. There is no bottom to anything. We are all antipodeans and upside down.[110]

But this can be true, while recognizing—as Heaviside did, although often with less than enthusiastic interest—that there are times in which we should build up pictures and times in which we should set pictures aside.

---

[109] Ernst Mach, "The Economical Nature of Physical Inquiry" in Popular Lectures, 207.
[110] Heaviside, Electromagnetic, ii. 2–3.

The characteristic symptomatology displayed in this wavering between classical and formalist extremes traces to the fact that we are vainly attempting to preserve more control over wandering language than we should rightly expect. It is our central theme that language's progressive advances are importantly propelled by interfacial factors that lie outside our active awareness and it is often better for us that they do, because the adaptive capacities of natural linguistic development can easily prove superior to the botched jobs we would provide if we did completely engineer the rails of language. However, if we don't let our managerial ambitions run away with our common sense, we can tinker with language mildly in productive ways. We can supply a pet rabbit with a convenient name; a brawl over folklore can be ameliorated with a definition or two, and so forth. Within this motley assortment of "little ways in which we can help language along" falls construction of semantic pictures and the utility of pre-pragmatist scrutiny: in the right season, each can provide significant guidance for genuine linguistic advance.

So we should neither underestimate nor overestimate the redirective capacities offered by our two forms of detoxification, remembering that significant engines of syntactic advancement lie scattered through language in lots of hidden pockets we may scarcely notice. Significant improvements will always slip in place without our noticing the structural changes incurred: word and world gradually work out an effective bargain between themselves and may not bother calling our attention to their adjustments. In our self-conscious efforts at linguistic management, we act as the captain of a great ocean freighter: the prevailing winds and currents, the thrust of the engines in the boiler room and its own mighty inertia largely determine our vessel's heading, but, as skippers, we can supply a fine tuning to the rudder that, with sufficient luck, will eventually bring our cargo to a desirable port. But sometimes commanding officers prove navigationally inept and need to replaced at sea by wiser heads and better maps. As we know from Mutiny on the Bounty and a dozen other nautical yarns, the captain who adheres too strictly to rules or geographical picture often steers his ship into difficult straits.

Therefore we should beware of any semantic monotheism that insists that we choose between the conceptual primacy of rules or pictures and also recognize that our primary purpose in attending devotedly to either stems from a desire to uncover clues with respect to improved directions of usage, not to report upon the full conceptual personality that has conveyed our target predicate to its current developmental condition. Much of the momentum that carries "gear wheel" or Heaviside's terminology to better moorings traces to design's beckoning call, but this factor represents a dimension of predicative personality that hides in the misty background and commonly proves resistant to sharp diagnosis until long after the fact. Fortunately, we can blithely ignore such wider considerations when we tinker with the little seasonal improvements that lie within our control.

The fact that the tasks of picture articulation or pre-pragmatist scrutiny are rather limited in their capacities has not been adequately recognized, leaving us liable to puff up such detoxification endeavors into something larger than they really are. In all this, the hidden hand of the classical picture of concepts is much to blame. "What is the *mechanism of reference* that glues language to the world in a manner that allows us to communicate with one another effectively?"—that is the single-minded question that classicism, following the prevailing *ur*-philosophical undertow, thrusts before us, offering no apparent choice but to acquiesce in its adhesive presumptions, as does Russell, or reject word/world correlation altogether, in company with the instrumentalists. In fact, there isn't a single "mechanism of reference" and there needn't be: a goodly swirl of distinct directive influences act in concert to park a terminology over some fairly determinate stretch of ocean floor. Our wavering barge may lie more or less securely in place, but its anchoring can always be improved and there remain lots of gaps and patches of nearby topography that, due to unavoidable limitations of complacency and skill, we have scarcely begun to address or anticipate. Even staying steadily in place is often hard work, for we have been fitted out with rather puny motors and paddles and must employ them with considerable strategic sagacity. As aids to the mariner, the cleansing powers of pre-pragmatic scrutiny and picture articulation are both very helpful, but so are the coarse and fault-ridden procedures of everyday life, where we learn that the hardness of a plastic is best appraised with a quick flick of the fingernail without noticing the material-optimized particularities of the routine we adopt. For maintaining our skiff in its proper place, the rough rule of thumb plays as important a role as the axiomatic enclosure or the attached picture.

Nonetheless, insofar as we are able to take a conscious hand in improving the descriptive lot of our predicates, focusing intently upon soundness across generic settings or isolating salient procedures as guiding principles serve us very well in the proper seasons. This specialized attention explains why the "conceptual contents" that we place upon the examining table during a typical effort at detoxification are apt to seem rather *thin*: "How could anyone profitably use a word that only means *that*?," we ask in puzzlement if presented only with a bare correlational picture or some passel of rules emphasized by pre-pragmatists. To which the proper reply is: "We're not attempting to set forth every directive wind relevant to our predicate, but only looking over the factors where we might profitably find clues for improvement." On other forms of evaluative occasion, we may need to attend (if we are able) to component strands (such as background design objectives) that we may safely ignore now.

We might also observe that meritorious correlational accounts are usually accurate only to a certain *degree of diagnostic depth*, which, when further opened up, sometimes reveal unexpected ways in which familiar predicates can be usefully employed. I have already cited the surprising fortunes of "ray" and "caustic" to this effect. Commentators on optical phenomena frequently note that caustics (like the patterns in a teacup or a rainbow) represent the most observable traits in their subject: these spots of intensity represent the aspects of light we can most readily *see*. At no point in the fluctuating fortunes of our sundry theories of light should we ever imagine that the predicate "is a caustic" doesn't correlate with certifiable informational packets *of some kind*. But *exactly* what kind?—there's the rub. The old ray picture from which "caustic" derives its original inspiration—a curve where the light rays pile up—does not correctly report the fine structure of the data we actually report when we speak of "caustics" in optical

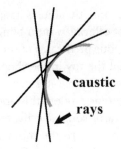

observation. As we observed, Keller and his school have articulated a quite different correlational story in terms of the asymptotic support by the underlying electromagnetic wave distributions. This sharper diagnosis leaves most of the old reasoning patterns in place, but also expands the set of directivities guiding "caustic" and "ray" in valuable but unanticipated ways. Uncovering these hidden layers of finer grain in optics is no different, I submit, than learning that "is red" sits over a complicated quasi-attribute, rather than a simple, single-valued quality. Accordingly, we needn't fall into the pre-pragmatist trap of fearing talk of word/world correlations altogether, lest the seasonal value of pre-pragmatist rule-centered scrutiny be denied in classicism's absolutist manner. No: a claim that "P" correlates with informational structures of type **S** can prove essentially correct *insofar as it goes*, but, within that "insofar as it goes" concession, a whole world of finer filagree may open out in a manner that mandates a healthy season of "first finding out what there is to find out."

These reflections support my general observation that philosophical maxims often rush in when detailed strategic analysis remains unavailable. Critics like Mach and Duhem were right that the tropisms of mechanists are driven by wrongly conceived pictures of how their favored traits operate, but such issues can be satisfactorily resolved only through the close scrutiny we find in applied mathematics. If we elevate our legitimate pre-pragmatist concerns to an inappropriately sweeping "philosophical" level, we will convert a potentially liberating investigation into a dogma likely to occasion harm further along the line, for any conception of scientific endeavor that focuses exclusively upon matters of recipe, syntax or practice will retard its longer term need to develop, when the moment proves ripe, a better correlative account of how its terminology engages with the world. Reverting to Chapter 4's terminology, the fact that direct and distributed normativities wrestle over the steering of our predicates in seesaw ascendancy reflects the essential fact that some measure of foundational uncertainty must be tolerated within the mechanics of linguistic management so that the varied indications of wise strategic improvement can sometimes creep in with their unexpected improvements.

But the everyday manners in which we employ "concept" and its allies baffle us, for we speak in connotative tones that suggest absolutes, when seasonal refinements are actually on our minds. Shifting evaluative behaviors are scarcely uncommon—we have observed how humble "hardness" quietly adjusts its informational assay as it switches from steels to ceramics, while not offering a mumbling word of complaint or warning. In their momentary enthusiasms, our predicates of semantic evaluation behave like the child who is utterly convinced that the toy in the shop window that presently occupies her attention represents the most desirable item she has ever seen, or could possibly contemplate. "The consideration that renders this application of 'P' correct is X," we state, and gesture confidently towards some specific picture or rule. No doubt we have thereby offered sage advice appropriate to the moment, but the single-mindedness in which we advance our appraisal renders us prey for classicism's sucker questions with respect to "the mechanism of reference," for our absolutist forms of speech encourage the belief that the same invariants are at work in whatever evaluative concerns occupy our notice. If, like Heaviside, we are concerned to sever the lingering tentacles of unhelpful pictures, it will seem as if the heart of semantic process must surely lie within the bundles of pragmatically proven procedure that he properly highlights. Twenty years later, these same promising beacons of advancement show signs of strain and decrepitude, leading reformers like Doetsch or Mikusiński to declare, "Oh, a truly effective terminology must locate its moorings within a well-developed correlational picture, with which Heaviside's recipes will only imperfectly cohere. 'Tis a pity that he didn't appreciate this key semantic moral more adequately." Yet their day of comeuppance may arrive as well, when some fresh phalanx of pre-pragmatist advantage carries the brigade of language beyond the reach of their carefully mounted semantic pictures. In truth, we must thin out and scatter the assorted directivities of predicate correctness so that we can justly regard both forms of semantic detoxification as useful activities that seasonally assume an importance against the backdrop of a much richer dynamics in predicative flow.

Because our *ur*-philosophical inclinations so markedly favor a picture of strong semantic invariance, we become readily buffaloed by "concept"'s variegated enterprises. And that opening makes it easy for our teenagers to get us completely befuddled about the conceptual value of homework.

. . . .

A brief additional "excursion to the borders of duplicity and fearful rigor" (in Heaviside's phrase once again[111]) may prove helpful. As J. Schwartz correctly observes, mathematicians generally carve out their studies in rigor at a smaller scale than allows us to adequately appreciate the full set of directive influences that act upon an equation or modeling. It is a shame that this commonsensical advice is not more widely appreciated. For there are certainly mathematicians and philosophers who lack Schwartz' mild perspective and sally forth as unchastened snobs of rigor (a category overlooked by Thackeray). In this vein, consider J. E. Littlewood's complaints about the practices of a prominent Cambridge mathematician of the generation previous (A. R. Forsyth):

> Nowadays . . . [i]f we want to consider well-behaved functions, *e.g.* "continuous" ones of a real variable, or Forsyth's *f(z)*, we define what being such a function means (2 lines for Forsyth's

---

[111] Heaviside, Electromagnetic, ii, 122.

function), and "consider" the class of functions so restricted. That is all. This clear daylight is now a matter of course, but it replaces an obscurity as of midnight.[112]

In fact, in the passage criticized, while not a paragon of pellucid prose, poor Forsyth is attempting to *motivate* the consideration of differential equations that take complex numbers as values, a task of intellectual placement that any reasonable teacher ought to address. Littlewood's "clear daylight" corresponds to a fashion of mathematical tutelage wherein any linkage to motivating concern is intentionally purged from the textbooks, requiring the novice to find friendly mentors who will inculcate them in the Guild's secrets by talking about electric circuits and drawing doodles upon coffee shop napkins. I believe that such unfortunate attitudes arose as a consequence of the bonds that were framed between mathematics and philosophy in the Edwardian period (Russell, Hardy and Littlewood were friends), but such a story would take us too far afield.

That Littlewood paints a misleading picture is easily shown by reflecting on the fact that, even within pure mathematics, the winds of change can overturn the relevance of his "two lines" if mathematicians decide that the problem "has been put in the wrong setting." Here Philip Davis muses on the likely relevance of proofs of "conclusion C is mathematically impossible!" in the face of such shifts:

> *When placed within abstract deductive mathematical structures, impossibility statements are simply definitions, axioms or theorems . . . [But] in mathematics there is a long and vitally important record of impossibilities being broken by the introduction of structural changes. Meaning in mathematics derives not from naked symbols but from the relationship between these symbols and the exterior world. . . . Insofar as structures are added to primitive ideas to make them precise, flexibility is lost in the process. In a number of ways, then, the closer one comes to an assertion of an absolute "no," the less is the meaning that can be assigned to this "no."*[113]

I would rephrase the penultimate sentence as, "Insofar as structures are assigned to terminology to render them precise, there is a danger that critical directive elements hiding in their makeup will have been overlooked." As such, the statement supplies a pretty good summary of the chief morals of the Heaviside affair.

In a manner complimentary to the main text, it should be observed that investigations of correctness relative to picture are commonly most useful when they *don't* attempt to plumb a term to the deepest semantic depths possible. After all, the practical value of a soundness investigation resides in the fact that it increases *our trust* in the results of its procedures when employed in a real life situation, as well as potentially suggesting improvements to our old techniques. Here considerations of *safety* play a large role in fixing the variations in setting over which we investigate our rules. For example, although we *know* that a steel girder is built up from atoms and then framed into higher organizational patterns such as dislocations and crystalline epitaxy, we generally consider none of this fine structure in the settings against which we evaluate the worthiness of the computer routines commonly employed within structural engineering. Instead, we look at how our rules behave over some appropriate space **S** of smoothed out distributions lacking grain or molecular structure. Why? Because the molecular modeling of steel remains highly inaccurate at present, whereas our long familiar continuum models are well tested in their first-order accuracy. As a check on rule reliability, it is more

[112] J. E. Littlewood, A Mathematician's Miscellany (London: Methuen, 1953), 64–5.
[113] Philip Davis, "When Mathematics Says No" in Philip Davis and David Park, eds., No Way: The Nature of the Impossible (New York: W. H. Freeman, 1987), 176–7.

prudent to determine how our rules behave with respect to S's averaged out presentations of data, bracketing off questions of exactly why S's smoothed approximations manage to carry information so successfully at the macroscopic level. The latter concerns require answers, to be sure, and should be vigorously pursued, but it would be unwise to substitute any of their presently suppositional details for S at this point in time.

....

(xiii)

Design imperatives. Throughout this book I have been reluctant to identify "concepts" with any specific worldly feature because I regard the personality characteristic of a predicate as twisted together from a rich variety of directive fibers, as in Wittgenstein's celebrated metaphor:

> And we extend our concept of number as in spinning a thread we twist fiber upon fiber. And the strength of the thread does not reside in the fact that some one fiber runs through its whole length, but in the overlapping of many fibers.[114]

Thus, the encouragement of a dominating picture $\mathcal{P}$ will first supply directive encouragement to various practical recipes $\mathcal{R}$ that entwine themselves about $\mathcal{P}$ as an anchoring base. But these $\mathcal{R}$ may then enlarge in strength and multitude after Heavisidian experimentation and allow our rope to maintain itself in robust worthiness, even though $\mathcal{P}$ itself has become enfeebled and eventually drops almost entirely from view. But some new $\mathcal{P}^*$ will start to wrap its tendrils tentatively about $\mathcal{R}$'s fibers and allow our cord to reach beyond the limited opportunities that the $\mathcal{R}$ rules can themselves offer. And so on, to greater lengths of prolongation. Furthermore, many other forms of sustaining filament thread betwixt these two species of lamination, some of which we have managed to diagnose and many others which we have not. In the former category falls the latent DNA of 7,viii —the discarded or inactive pictures to which later practical strands sometimes attach—, as well as the rarified intimations of intensionality (6,iv) that can provide our rope with a long-term trend around which the more noticeable strands twist, while remaining nearly invisible itself.

The chief exemplars provided in this book for such "intimations" lie in the general arena of *design purpose*: the manner in which the overall contours of a descriptive vocabulary can be subtly shaped by the directive imperatives of *invention*. Not all words carry such directivities, but many do, even within arenas of descriptive mechanics where we may utterly fail to detect their background presence. This is why I consider Franz Reuleaux's insightful diagnosis of *gear wheel*'s hidden design-oriented personality (7,iv) to be so revealing of semantic circumstance in general.

---

[114] Wittgenstein, <u>Investigations</u>, §67.

In the methodological introduction to <u>The Kinematics of Mechanism</u>, Reuleaux supplies a wonderful description of how the allure of an obtainable class of desirable inventions gradually converts a hazily sensed and misty "art" into a concrete and algorithmically regimented "science" (here "science" is employed in the nineteenth century sense of "set of rules"—I have an instruction manual of the period that promises "the only scientific approach to correct banjo playing"). Reuleaux observes that the splendid designs of a prodigious inventor like James Watt (who brought so many astonishing improvements to the steam engine) are explicable only if some hidden set of procedures can capture the means by which such repeated successes are achieved. Accordingly, in the vein of "every successful method must have its logic," Reuleaux searches for the "design science" latent in Watt's achievements. A striking feature of Watt's innovations lies in their *optimality* and *perfected beauty*: he commonly improves his rough designs until their component parts wind up sized in exactly the manner best calculated to produce the motion wanted. He achieves these victories through nebulous "art," without apparent method, but such a large number of delicate improvements cannot have been uncovered through accidental whimsy alone. As noted in 7,iv, Reuleaux is quite successful in his dedicated efforts to codify a set of "machinal ideas" useful to design, which are still employed, with subsequent improvements, by all mechanical engineers today. To sort these out properly, we must first recognize that the personality traits characteristic of machinal classifications are informed by a different directive drummer—namely, effective design—than is evident in other parts of standard physics (recall Reuleaux's complaints about misguided attempts "to thin away machine-problems into those of pure mechanics").

But to Watt himself, these intimations of future "science" are not evident at all: inspiration comes to him merely as "a new hare" to chase, to cite an evocative letter he wrote to a friend:

> I have started a new hare. I have gotten a glimpse of a method of causing a piston rod to move [in a desirable way] . . . I have only tried it in a slight model yet, so cannot build upon it, though I think it is a very probable thing to succeed, and one of the most ingenious simple pieces of mechanism I have contrived.[115]

Indeed, Watt appends a rude sketch of his "hare": the basic plan for what later became his celebrated "parallel motion" (an arrangement of rods that can covert the (almost) back-and-forth thrust movements of the piston into the circular motion wanted for grinding corn or turning wheels—the clumsier forms of earlier linkage policy induced vibrations that often shook their housing apart). Reuleaux observes that Watt often misunderstands the foundations of his own discoveries:

> The links that connect isolated thoughts seem indeed to be almost entirely destroyed,—we have to reconstruct them. We see the whole picture before us only like a faintly outlined or half-washed-out picture, and the painter himself can hardly furnish us with any better

---

[115] Eugene Ferguson, <u>Kinematics of Mechanisms from the Time of Watt</u>, United States National Museum Bulletin 228 (Washington, DC: Smithsonian, 1962), 194–5.

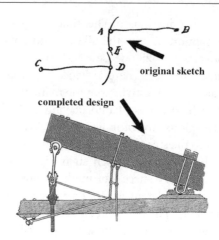

explanation of it than we can discover for ourselves. Indeed, the comparison holds good in
more than one point. In each new region of intellectual creation the inventor works as does
the artist. His genius steps lightly over the airy masonry of reasoning which it has thrown
across to the new standpoint. It is useless to demand from either artist or inventor an
account of his steps.[116]

And this is how an "intimation of intensionality" often appears: as an airy masonry of
reasoning that may not wear its directive foci upon its sleeve. But it is precisely through
bringing into explicitness the pathway of calculation implicit in Watts' crude sketch that
Reuleaux founds the modern "science of mechanism."

Another clever invention of Watt's is his *governor* that regulates, through feedback
control, the amount of fuel admitted to the engine, lest the steam pressure climb too
high (poor control was a frequent cause of steamship calamity[117]).

This kind of design problem does not fall within Reuleaux's orbit, but it represents
exactly the kind of issue that Heaviside's operational calculus addresses so admirably
(our telescope problem is of this type, but transfer function techniques can address far
more difficult cases than this[118]). It is the love call of these design capacities that
eventually lures the predicate "impedance," starting life as "a convenient and easily
pictured property of opposition to current flow," into assuming the role of an "extended
concept whose physical interpretation is obscure or absent" (this phraseology derives
from the W. D. Day passage cited earlier). Obscure, yes, but not absent: like the hidden
modes within Chlandi's plates (5,vii), the transfer function represents as substantial a
feature of an electrical circuit as any quantity it possesses.

---

[116] Reuleaux, <u>Kinematics</u>, 6.

[117] *The b'iler's busted and the whistle's done squalled/ The head captain's gone through the hole in the wall*— Uncle Dave
Macon, "Rockabout my Saro Jane," Vocalion 5152.

[118] During Heaviside's era, simple governors were handled by techniques developed by Maxwell and Routh; the
unification of such techniques came in the 1940s. Otto Mayr, "Maxwell and the Origins of Cybernetics" in Otto Mayr, ed.,
<u>Philosophers and Machines</u> (New York: Science History Publications, 1976).

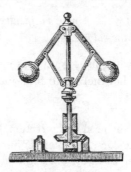

Watt's governor

In dealing with our teenager, we hope to discourage his vision of himself as a conceptual seeker who awaits that "new direction of purpose that aims toward a goal we have heretofore not suspected." "But if you look upon conceptual advance in that naïve way," we fumed, "your efforts are likely to get lost, become vague and tangled, and degenerate into a game of hide-and-seek" (to quote Hilbert again). But we stumbled in our attempts to provide a convincing philosophical counterweight to our ward's ethereal leanings, because sometimes useful predicates *do* embody "new directions of purpose" as wispy intimations whose first apostles, such as Watt or Heaviside, cannot delineate their contours in any clear form at all. So we must properly warn our teenager that useful language is seasonally woven together like a rope: different kinds of capacities and strivings must entangle with one another, sometimes inconsistently and sometimes as mere intimations, to provide the whole with any worth or strength. Any monotheism of conceptual focus is apt to lead to nothing of value whatsoever.

In these regards, we must also caution our youth that temporary mystification is not a cause for mysticism: that he should turn a cold eye upon the hagiographers who describe the "intuitive insight" of some esteemed scientist as a form of supernatural communication with Mother Nature (she won't answer our calls if they are posed like that). Instead, our teenager should heed the flowery, but sage, advice of Reuleaux:

> The peculiar condition consequently presents itself throughout the whole region of investigation into the nature of the machine that the most perfect means have been employed to work upon the results of human thought—without anything being known of the processes of thought which have furnished these results. Terms have been made with this inconsistency, which would not readily be submitted to in any other of the exact sciences, by considering Invention either avowedly or tacitly as a kind of revelation, as the consequences of some higher inspiration.[119]

Lord Kelvin was thinking of mechanics, rather than semantical matters, when he wrote:

> I cannot doubt but that these things, which now seem to us so mysterious, will be no mysteries at all; that the scales will fall from our eyes; that we shall learn to look on things

---

[119] Reuleaux, <u>Kinematics</u>, 3.

in a different way—when that which is now a difficulty will be the only common sense and intelligible way of looking at the subject.[120]

But this, I think, is also how our linguistic mysteries will eventually resolve, if we only take the patience "to find out what there is to find out."

. . . .

Although I quoted Wittgenstein at the head of this section, I believe I understand the tropes in the rope analogy differently than he does. As I interpret him, the component threads represent individually complete "language games": ways of using words that enjoy some obscure form of human completeness that lets them stand as sustainable in their integral right, although, in practice, each thread quickly encourages neighboring attachments. In his estimation, the discerning philosophical eye should be able to unravel these windings of usage and isolate the component games in their individual purity, without engaging in a scrap of research beyond "reflecting upon what we already know as speakers." In contrast, my own "fibers" are the various centers of ascertainable directivity that guide us in employing a predicate in a specific fashion, but these *cannot* stand as sustainable practices in isolation. Nor can we frame an accurate diagnosis of the nature of these strands without investigative work of a straightforwardly empirical and mathematical character.

. . . .

[120] Williams, Hardness, 344.

9*

SEMANTIC MIMICRY

Though analogy is often misleading, it's the least misleading thing we have.
—Samuel Butler[1]

(i)

The varieties of linguistic strategy. Words commonly display quite mercurial personalities: on a given day they can display a flash of temperament that may not be seen again for a long time. A term may have been shaped by factors in which we presently retain little interest, and would prefer to detoxify away, yet their relevance can return to befuddle us in ways we cannot readily anticipate. Normally we would prefer that our descriptive vocabulary not find itself tightly chained to specific strands of practical advantage, but sometimes complete escape from strategy's confines is not easy. Sometimes the shaping winds of opportunity will have placed our predicates over territory from which no viable prolongations extend, however much we might believe that they do and wish that they might.

In this chapter I will extend these themes by drawing upon applied mathematics in two ways: through exploring a manner in which reductive strategies can create *semantic mimics* and by outlining characteristic *signatures* of their presence. We will apply these tools to standard philosophical confusions about causation and learn something further about mitigated skepticism along the way. Although I believe this investigation adds depth to our earlier discussion, it does not introduce any substantially new considerations. Some readers may prefer to pass on to the final chapter.

One of my chief regrets with respect to this book is that, despite its considerable massiveness, I have not been able to explore the varieties of strategic experience in the depth they deserve. To keep other aspects of our discussion in focus, I have largely relied upon facade-like situations where each localized patch behaves normally from a logical

---

[1] Samuel Butler, <u>The Note-books of Samuel Butler</u> (Lantham, Md: University Press of America, 1984).

point of view. However, interesting cases of mimicry arise in which a local patch A looks as if it is semantically supported in manner $\mathcal{M}$, although it is actually propped up in fashion $\mathcal{N}$. Grammatical sentences that would be meaningful if $\mathcal{M}$ represented their proper support do not gather any reading under $\mathcal{N}$ (I have elsewhere[2] dubbed this a situation where the *working grammar* of "P" differs from its *apparent grammar*). Such situations are especially prone to fooling their employers when patch A happens to sit next to, or above, a neighboring B genuinely supported in manner $\mathcal{M}$ (from which "P"'s use over A may have nucleated through prolongation). The "rainbow" example of 7,viii can be regarded as fitting this pattern. However, there is a specific template for variable reduction whereby potentially confusing mimics are commonly formed, for which many fine paradigms are available within applied mathematics. Many celebrated puzzles with respect to causation can be aligned with these patterns, I believe.

It is very common in physical practice to follow a policy of concentrating upon *equilibrium* or *steady-state circumstances*, where we either assume that our system has come to rest or that its relevant velocities are no longer changing. Such assumptions, when viable, offer great simplifications in their governing equations. For example, a simple equation for the movements of a string pulled by gravitation is $\rho\partial^2 y/\partial t^2 = k\partial^2 y/\partial t^2 + c\partial y/\partial t - g\rho(x)$ (i.e., the force pushing a portion of string up or down is the sum of its resistence to bending plus the frictional retarding force plus the gravitational pull). If the density $\rho(x)$ is variable, this behavior can prove fairly complex. However, we also know that any movement initially contained in the string will eventually die away due to the friction. When that happens, the two movement-related terms $\rho\partial^2 y/\partial t^2$ and $c\partial y/\partial t$ become zero. Accordingly, a string in equilibrium obeys a simplified "hanging string" equation $k\partial^2 y/\partial t^2 + g\rho(x) = 0$ that is both easier to treat and may encapsulate the chief concerns we hope to address anyway (we may not care about the details of how Jack and Jill fell down the hill; we only want to know where they finally land).

In adopting this strategy for reducing complexity, all mention of *time* is dropped from our equation and we lose the original equation's capacity to predict how movement is generated within the cord. The longer equation genuinely describes how *causal processes* unfold within the string, but its equilibrium replacement can only report on what the *outcome* of those processes will be after they have brought the string to rest. That is, an active description of the relevant causal processes has been purged from our reduced equation description.

. . . .

When we first encounter them in a freshman calculus course, differential equations tend to all look alike. However, under the sage guidance of Jacques Hadamard,[3] we have gradually learned that this first impression is quite misleading and that a number of basic mathematical

[2] Wilson, "Logical Form."
[3] Hadamard, Cauchy's Problem. Vladimir Maz'ya and Tatyana Shaposhnikova, Jacques Hadamard, A Universal Mathematician (Providence, RI: American Mathematical Society, 1998), ch. 15.

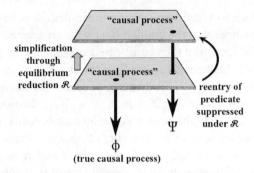

simplification
through
equilibrium
reduction $\mathcal{R}$

"causal process"

"causal process"

reentry of
predicate
suppressed
under $\mathcal{R}$

ψ

ϕ

(true causal process)

considerations group such equations into natural classes: What is the type of its principal operator? What side conditions does it accept? What forms of reductive policy are involved in its coming to our attention? (it is this last question that I especially emphasize here). Following these guidelines, Hadamard points out that our two equations display different mathematical signatures of a technical type (the original is *hyperbolic*; its reduction is generally *elliptic*, although in the simple circumstances illustrated it becomes an o.d.e.).

I shall not depend upon these technical distinctions here, but they certainly clarify the features we should expect to see in a "causal process."

. . . .

It is a curious fact that, although the predicate "causal process" has been officially banished from our reduced arena through our simplifying strategy, it nonetheless displays a perverse tendency to sneak back into our sights again, sometimes returning with a greatly altered significance. Here is a nice example that often confuses physics students, although its underlying strategic logic is simple. The forces that bind together a gravitational planetary system $\mathcal{S}$ can be expressed in terms of an overall potential energy function $V(\mathcal{S})$ that they frame. As such, this V is properly defined only over their high-dimensional joint configuration space (infinite dimensions, in fact, if we allow our planets to display flexible shapes). But if we can somehow assume that our celestial bodies remain locked in place, we can derive an equilibrium formula (Poisson's equation) for a distribution $\mathcal{F}$ that spreads itself out over ordinary, three-dimensional space. $\mathcal{F}$ is often called the "the gravitational field around the planets," but this label is misleading in various ways ($\mathcal{F}$ does not have the same physical status as the electromagnetic field, for example). Why should we want to set up such an $\mathcal{F}$? Well, if we introduce a small comet a into the scene whose mass is so puny it won't wiggle the big planets $\mathcal{S}$ much, $\mathcal{F}$ can be used to conveniently calculate how a will itself move under their influence. "We are calculating how the planets *cause* a to move," we will normally report and thus "cause" has reentered our reduced descriptive arena. And we can push this reductive strategy a bit further: if we know from other sources how the large $\mathcal{S}$ will move, we can use Poisson's equation in progressive time slices to determine how a will move *given* that the $\mathcal{S}$ move in the way that they do (this ploy is commonly called *quasi-static approximation*). Indeed, this treatment handles the behavior of $\mathcal{S}$ in the same "here versus there" manner that the crank movements are treated within a mechanism (6,iv): we turn the $\mathcal{S}$ by hand as if they constitute a mechanical orrery and then observe how puny $\boldsymbol{a}$ reacts to their changing configurations. Plainly, in carrying this policy through, we nowhere describe the full causal process that makes the *entirety* $\mathcal{S} + \boldsymbol{a}$ behave as it does, because in our V to $\mathcal{F}$ reduction, we have thrown out the physics that governs the movements of $\mathcal{S}$ (although a facsimile to it is reinserted later, treating its unfolding stages as already known for the sake of our quasi-static treatment of $\boldsymbol{a}$). In short, the causal processes that genuinely influence $\mathcal{S}$ and $\boldsymbol{a}$ as a totality have been divided up into fractured fragments, with a good part of the story removed from our Poisson's equation treatment altogether. Unfortunately, freshman physics instructors being what they are, many students come away with the false impression that Poisson's equation fully embodies "the causal processes that make the gravitational field behave as it does." Here they are confused—it is quite unlikely they have ever seen the rather complex equations that can lay reasonable claim to *that* descriptive task ("They are utterly intractable, so why drag them forth?," our misleading instructor might retort, by way of excuse). But a simple clue should tip us off that something funny has happened to "cause" along the way: Poisson's equation doesn't mention *time* at all (this is one of Hadamard's signatures of strategic alteration, better articulated in terms of its elliptic character).

In this example, it might be fairly claimed that the term "causal process" hasn't endured radical *property dragging* in the sense that, within its new $\mathcal{F}$-field home, the term still covers *some* of the causal processes that affect $\boldsymbol{a}$ (as opposed to those that affect $\mathcal{S}$).

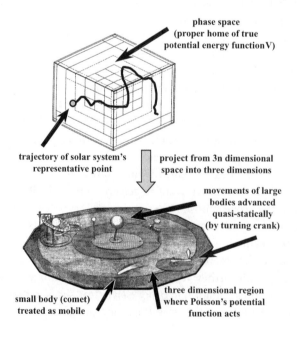

phase space
(proper home of true
potential energy function V)

trajectory of solar system's
representative point

project from 3n dimensional
space into three dimensions

movements of large
bodies advanced
quasi-statically
(by turning crank)

small body (comet)
treated as mobile

three dimensional region
where Poisson's potential
function acts

However, we will soon investigate a quite similar example where the strategic reduction involved carries "causal process" onto quite strange moorings. If so, our usage finds itself in a situation like that pictured, where "causal process" has reentered a covering sheet from which its old supports have been strategically severed. Many confusions in philosophy about "causation" result from not recognizing when this reentry process occurs. Quite often, we expect that the *working grammar* of the upper sheet will be the same as that of the lower, but this isn't true: our reductive policy leaves a lot of little holes that the upper sheet usage can't cover and these explanatory glitches provide a characteristic signature of the fact that a covert reduction strategy lurks in the background (such Swiss cheese-like omissions comprise part of what I call the *fine grain* within the usage).

(ii)

Marching methods. Before we examine several real life transgressions of these backdoor "causal" recidivists, let's shift our attention to Hume's celebrated discussion of causation, which embodies, in my diagnosis, a nice blend of semantic wisdom coupled with serious victimization at the hands of several causal imposters. That Hume could be misled by semantic mimicries was fully understandable in the context of his time and will offer fertile ground for reflecting upon our own place within the unexpected currents of linguistic process. Although Hume is often portrayed as having offered an "analysis" of "causation," he actually denies that such an account can be provided. Instead, he articulates two patently distinct "definitions" of "cause" and comments that, although neither "definition" offers an acceptable rendering of what we expect a "cause" to be, we have no means of improving our account beyond the bounds set by these limiting perspectives. His first account is the familiar criterion of "constant conjunction" of the evidence:

> [W]e may define a cause to be an object followed by another, and where all the objects, similar to the first, are followed by objects similar to the second.[4]

The second "definition" speaks instead of the impression that observing the specific events make upon a rational observer:

> We may . . . form another definition of cause and call it an object followed by another, and whose appearance always conveys the thought of that other.

He maintains that neither account can be regarded as capturing the expected conceptual core of "causation," but instead reports upon its *external effects*: in the first treatment, upon a conjunctive regularity encountered in the *affected behaviors* of the physical objects under investigation and, in the second, allied regularity in our *mental reactions* to such

---

[4] Both passages: Hume, Enquiry, 87. Also: Hume, A Treatise of Human Nature (Oxford: Oxford University Press, 1965), 170. Graciela dePierris, "Hume's Pyrrhonian Skepticism and the Belief in Causal Laws," Journal of the History of Philosophy 39, 3 (2001).

behaviors. Each account seems to involve elements "extraneous and foreign" to what we expect to witness in a causal relationship proper. In particular, mere conjunctive regularity seems too thin a relationship to qualify as causation-as-we-expect-it-to be, because there will surely be accidental correlations between events where no true causal process acts. And the fact that we psychologically link events A and B together seems even more distant from what we want (I will eventually argue that, suitably recast, a fair amount of semantic sagacity is embodied in these claims). Hume then concludes that we are incapable of grasping any proper idea of "cause" and so we must muddle by with his two unsatisfactory surrogates, considered as containing the term "cause" within some kind of conceptual corral, wherein our use of the word can pace around freely, with no better directivities available to guide its employment. As Hume puts the thesis:

> [T]hough both these definitions be drawn from circumstances foreign to the cause, we cannot remedy this inconvenience or obtain any more perfect definition which may point out that circumstance in the cause which gives it a connection with its effect. We have no idea of this connection, nor any distinct notion what it is we desire to know when we endeavor at a conception of it. . . . [I]t is impossible to give any just definition of cause, except what is drawn from something extraneous and foreign to it.[5]

But if "cause" cannot be assigned straightforward conceptual content, an unhappy dilemma arises:

> the necessary conclusion seems to be that we have no idea of connection or power at all, and that these words are absolutely without any meaning when employed either in philosophical reasonings or common life.[6]

According, when we talk of "causes" in ordinary life, we must remain content with some hazy contention lying somewhere within the bounds set by his two weaker "definitions." As I noted, commentators often ignore all this and consider the "mere conjunctive regularity of objects" to represent "Hume's analysis of causation," when, in fact, he *declines* to provide such an "analysis."

In his final assessment, of course, Hume succumbs to dark Pyhrronian skepticism with respect to "cause"—that is, doubts about its underpinnings that are too strong to prove acceptable as a guide to real life activity. However, we can extract a properly *mitigated caution* from the considerations of the previous section. It develops as follows. If we survey our everyday talk of "causal process" in its fullest extent, the landscape will be filled with many returning intruders, whose semantic significance will be largely fueled with *procedural content* ("First do this, then do that"). Such employments often come very near to Hume's "regularity in our mental reactions" end of the spectrum. On the other hand, there are plenty of cases where the phrase "causal process" sits over physically generative relationships that can qualify as the real article, such as the gravitationally driven evolutions described in our *unreduced* planetary equations. Through correlational detective work, the true processes can be eventually segregated

---

[5] Hume, Enquiry, 87.   [6] Ibid., 85.

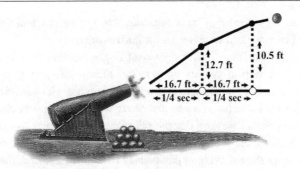

from the imposters. However, sometimes the time for such discriminations is not ripe and we must simply wait, with Heavisidean agnosticism, until the clues required choose to reveal themselves. If so, we should entertain a mild semantic patience with respect to the circumstances before us. At present we lack a "sufficiently distinct" conception of the variety of "causation" on display: whether a genuine causal connection is signified, or merely some procedural imitator. Unlike Hume, we should not regard this ignorance as representing an irremediable limitation of the abiding human condition, but simply a current lack of detail with respect to semantic underpinnings: right now we cannot determine exactly where within Hume's roomy corral our present specimen properly sits. Such *mitigated skepticism*, I contend, represents a practical attitude that can help us avoid *ur*-philosophical blunders in a world where causal pretenders abound.

I will first describe a few ways in which such semantic mimics can arise and then describe how Hume himself was driven to his bleaker, Pyhrronist skepticism about "cause" due to several masqueraders that confused him. Let us first acquire a better feel for how we normally talk about proper causal processes in untutored life (that is, without benefit of the calculus). Although we could investigate our string or planet examples from this point of view, it is most convenient to return to the simple cannon ball circumstances we have discussed several times before, most relevantly in 4,iii. We consider a cannon ball of one pound that is fired into a frictionless air with a muzzle velocity of 94 ft/sec at a 60° angle. The downward force exerted by gravity is -32 kg-m^2/sec^2. If asked to reason the situation through, we may proceed as follows (this represents what engineers call a "back of the envelope calculation").

If we decompose the velocity into horizontal and vertical components, we find that the ball's initial velocity will be 66.7 ft/sec in the horizontal direction. This speed will remain constant because no air resistance or other factors will cause it to slow down. By the same decomposition, the ball's upwards velocity will begin at 50 ft/sec at the moment of firing but gravity will continually cause this skyward velocity to alter. After the first quarter of a second, the ball's velocity will be caused to decrease to approximately 42 ft/sec. During the same interval the ball can be expected to reach a height of roughly 12.5 feet, because that is roughly how far its velocity will cause it to travel, utilizing its 1/4 second averaged velocity of 46 ft/sec as an estimate of its magnitude. Meanwhile, its undiminished horizontal velocity will cause the ball to travel 16.7 ft horizontally.

Reiterating this reasoning, during the second 1/4 second time the projectile will be caused to decelerate to a vertical velocity of 34 ft/sec, thereby carrying it to a total height of approximately 35.5 ft. In the meantime it will have traveled 33.4 feet relative to the ground. Continuing this same schedule of causes and effects, we can draw up the chart supplied in 4,iii.

Studying these results, we learn that the pull of gravity will eventually cause the ball to reach a maximum height of about 38 feet after 1.3 seconds and that it can be expected to fall back to the ground after three seconds, approximately to hundred feet from the cannon.

A quick review of 4,iii reveals that this monolog represents nothing more than the computation procedure called "Euler's method" cast into the ordinary vernacular. As such, it explains why commentators on the history of algorithms often observe that the reasoning pattern was in widespread use long before differential equations, let alone Euler himself, came along.[7]

We should observe that we don't expect the computation just laid out to prove completely accurate, for we recognize that approximations to the velocity are made at each step: "We do that only to obtain some baseline numbers with which to compute," we explain. "If we shorten the interval over which we make those estimates, the graph we draw will capture the *true causal progress* more accurately." And this claim is entirely reasonable.

In these regards, let us pause to consider some misunderstandings that have plagued causation that trace to Ernst Mach and Bertrand Russell. Such commentators declare that, because the differential equations governing our cannon ball do not indicate any time step Δt, only a bare "formula" governs each temporal moment, not a proper relationship of cause and effect (a fact that they regard as exposing causation as an outdated mode of thought). In this vein, Russell writes:

> *I wish, first, to maintain that the word "cause" is so inextricably bound up with misleading associations as to make its complete extrusion from the philosophical vocabulary desirable . . . [Causal] laws, . . . though useful in daily life and in the infancy of a science, tend to be displaced by quite different laws as soon as a science is successful . . . In the motions of mutually gravitating bodies, there is nothing that can be called a cause, and nothing that can be called an effect; there is merely a formula. Certain differential equations can be found, which hold at every instant for every particle in the system, and which, given the configuration and velocities at an instant . . . render the configuration at any other earlier or later instant theoretically calculable.[8]*

But this evaluation is simply a mistake: it is plain that we can express the exact content of our cannon ball equations ($d^2y/dt^2 = -32$ and $d^2x/dt^2 = 0$) in Euler's method terms, realizing as we do that time steps needed to be shortened for increased accuracy (and

---

[7] Jean-Luc Chabert et al., A History of Algorithms (Berlin: Springer-Verlag, 1999).

[8] Bertrand Russell, "On the Notion of Cause" in Mysticism and Logic (Garden City, NY: Doubleday, 1957), 188. Russell inserts the supplement "or the configurations at two instants," but this represents a mathematical misunderstanding. Cf., Sheldon Smith, "Resolving Bertrand Russell's Anti-Realism about Causation," The Monist 83, 2 (2000).

would also recognize the problem with a Lipschitz condition violation were it to arise). In other words, our ordinary thinking about "causal process" captures the same requirements as the differential equations themselves, albeit expressed in terms of "one damn thing after another" language that squeeze in on the central core of the process in approximating terms. As to the "mere formula" part, Russell mistakenly attempts to evaluate Newton's gravitational law in isolation from the manner in which it fits together with classical dynamical principles. Considered properly as a unit, they cleanly delineate how a chosen state of our gravitating universe causally induces its successors (true, signal speed is infinite in these circumstances but that doesn't render the casual succession of events incoherent). Poisson's law, in contrast, *is* a "mere formula" in Russell's sense, because all mention of "time" has been drained from it. But we shouldn't expect to see any record of causal process there, but merely of their outcomes at equilibrium. Clearly, Russell has fallen prey to the "all differential equations look alike" fallacy. If we look into the right kind of equations, we'll find proper causal processes nicely registered.

. . . .

Occasionally, similar sentiments are encountered in textbooks. This comes from W. D. MacMillan's primer of the 1930s:

> It will be observed that Newton's second law says nothing about causation. Since the force and the acceleration are simultaneous, there is no more reason for asserting that force is the cause of acceleration than for asserting that acceleration is the cause of force . . . In the philosophical sense, nothing is known about causation.[9]

Insofar as I can see, such opinions trace to Mach's concern to expunge "animist attitudes" from scientific thinking, a program discussed in 4,ii. The objective is laudable enough in many ways, but causal process needn't suffer on its account.

Incidently, it is rather common for the contents of physical laws to be expressed in what should be regarded as approximation space terms: by setting firm enough demands there, the desired differential conditions can be induced upon the target space. In much nineteenth century writing, spaces of finite elements often serve this utility.

. . . .

Observe that such Euler's method/"one damn thing after another" forms of reasoning *march along* with the cannon ball's unfolding causal processes in jerky facsimile, just as we observed in connection with Hertz' Harpo-imitates-Groucho analogy in 4,iii. That is why computational routines of this ilk are called *marching methods*.

(iii)

Algorithmic borrowing. Let us now take an equation that directly describes causal processes and purge these by considering an equilibrium situation, just as we did with our string in section (i). For sake of vividness, I'll employ a treatment made famous by

---

[9] W. D. MacMillan, Statics and Dynamics of a Particle (New York: Dover, 1958), 37.

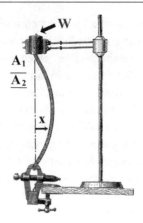

Euler for these purposes (although the string case would provide the same moral). Consider a thin metal strip pinned between the plates of a huge press, its thin end facing the page. Place a heavy load W upon the upper plate. Here's a simple story of how causal processes then play out within our strut. As the metal in the upper portion of the band A_1 becomes bent under W, local stresses will attempt to restraighten the affected section. This movement in A_1 supplies a bending reaction against both W (where its effects will be ineffectual) and upon its lower neighboring section A_2 (here the effect will induce a bending in A_2). A_2 will then cause similar changes in both A_1 and its lower neighbor A_3. All the while these interactions occur, frictional damping will resist movement within each section. Combing these two effects, we expect to see waves rippling up and down the strut, depending upon the exact details of how W was originally pressed upon our strut. Eventually, however, frictional causes will cause our strut to come to rest. If we are supplied with the details of how the bending and damping works, we can write down a dynamical equation very similar to our string example.

....

To wit, $EI(\partial^2 x/\partial y^2) - c(\partial x/\partial t) + \rho \partial^2 x/\partial t^2 = -Wx$ where "EI" reflects material parameters pertinent to our strut. Most likely, more motion will be lost through the cap and base than to the air, but I won't worry with the modeling complications that would entail.

....

If we are architects, we will be uninterested in all this detail (unless we work in an earthquake-prone region). Following Euler (and the strategy suggested in section (i)), let's simply drop the terms that contain *time* in any fashion, resulting in $EI(\partial^2 x/\partial y^2) = -Wx$. As we do this, all of Hadamard's alarm bells go off, for our equation has shifted its classification type (we've now switched to an o.d.e. from a hyperbolic starting equation) and requires different side conditions. But we've antici-pated this change: our shortened equation is designed to describe an equilibrium state only. It happens that, for smallish top end loads W, our equation is only satisfied by the

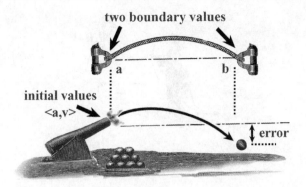

two boundary values

initial values
<a,v>

a   b

error

"trivial solution" representing a straight band. In these circumstances, our strut stays upright under the pressure of a light load; it accommodates the weight through vertical compression alone. But if we load our strut with too big a burden (W greater than $\pi^2 EI/L^2$ if $EI(y)$ is constant), a *bifurcation*[10] arises in the solution structure and two more (non-trivial) solutions emerge for our equation, representing posts that sag either to the left or to the right. The old straight line solution now becomes unstable, in the sense that a small gust of wind will immediately cause the strut to start wobbling all over again. And this means that, under a sufficiently heavy load, we can anticipate that our strut will not be able to support the W burden placed upon it in a straight line, but will buckle with architecturally ruinous consequences. <u>Ergo</u>: design your buildings with W's below the critical value where sag initiates.

If $EI(y)$ varies up and down the band due to changes in thickness, it will require numerical techniques to find the value of the *critical load* W that marks the line between safe and unsafe support. How might we do this? Well, we can experiment with various values of W until we find the lowest one that gives rise to a buckled solution (mathematically, we are looking for the lowest non-trivial eigenfunction/eigenvalue pair, which is two ways degenerate in this case). Swell, but how do we find the solution corresponding to W? Although our strut equation ($EI(\partial^2 x/\partial y^2) = -Wx$) looks very much like our cannon ball formula ($md^2y/dt^2 = -32$), we find that we are asking them different kinds of questions. This is easiest to see if we flip our strut on its side and compare it to the cannon ball. In the latter case, we are wondering, "What happens if we fire a projectile from the *left side* with height **0** *and* starting velocity v?," whereas, with the strut, we enquire, "What happens if your *left and right sides* both hold at **0**?" The former is called an *initial value problem* and the latter, a *two-point boundary condition*.[11] And readers might pause to figure out how they might attempt to address the latter kind of problem in the same "back of the envelope" manner that we applied to the cannon ball.

[10] S. Timoshenko, <u>Theory of Elastic Stability</u> (New York: McGraw-Hill, 1936).

[11] L. Fox and D. F. Mayers, <u>Numerical Solution of Ordinary Differential Equations</u> (London: Chapman and Hall, 1987).

One of the policies that may have come to mind is called the *shooting method*. Let's simply *guess* a strut slope s starting on the left side to complement its **0** and see if that **0,s** combination will induce enough bending to land our strut properly at 0 on the right hand side. Probably not, but we can then adjust our s guess lower or higher depending on the outcome of our first trial. Eventually, we will uncover a **0,s** pair that works properly and we can conclude that the shape it supplies to our strut represents the form of equilibrium sagging sought. Indeed, we've already witnessed a computational situation like this from 4,x: our archer keeps shooting at the target through different angles while his assistant on the other side yells back whether he's "getting hotter or colder" in his ballistic attempts (more formally, this general strategy is called *calculation through successive approximations* <u>per</u> the discussion in 4,xiii). But how do we calculate how our strut should stretch after it leaves **0,s**? "Why not borrow Euler's method from the cannon ball folks?," we ask. "It's pretty good at that kind of thing."

In fact, this is how standard shooting method calculations proceed: they appropriate some marching method package (usually of better quality than Euler's method) and embed it within a fuller computational program. It is useful to picture the workings of this combination as a flow diagram. Beginning with the data that the top end of the strut lies at **0** with an arbitrary slope s (this is just a guess), we feed these values into a

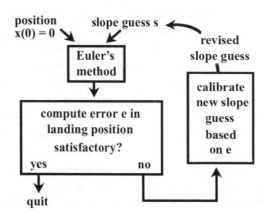

marching method calculation for $EI(\partial^2 x/\partial y^2) = -Wx$. Its result will generate an error e representing the distance that our calculation lands away from **0**. Based upon e, we interpolate a new guess s_1 which we feed back into our marching method hopper. We travel 'round and 'round this recursive loop until we obtain a satisfactory result.

The fact that a marching method routine can be borrowed for this purpose should be regarded as a kind of *computational accident*—it supplies a useful subroutine that helps another kind of strategic project achieve its objectives. As I've tried to stress throughout the book, a lot of our thinking follows such patterns: we can often obtain excellent results by utilizing unsuitable methods in clever ways.

That is how a *proper semantic picture* of a shooting method calculation should be expounded. "But what might happen," I once asked myself, "if someone became confused by the presence of the marching method subroutine and misread the entire computation as the *history* of how 'causal processes' operate within our strut?" To do so would be plainly absurd, for the sequence of the approximate shapes we calculate by our shooting method bear no relationship at all to the real shapes that our strut will pass through as it settles into rest after an initial loading (a rapid exposure camera will reveal a lot of rippling waves traveling up and down our post, quite unlike the shapes our shooting method calculates—after all, we never find our strut with its feet placed two inches to the right of where it's pinned!). "But struts come to equilibrium quite rapidly," I mused, "so perhaps someone might not notice the strangeness of the 'causal history' they believe they've calculated."

As an experiment, I simply took several standard routines for solving our two-point boundary problems and translated their "one damned thing after another" *procedural steps* into causal lingo, to see how plausible the results might sound. Working with the shooting method as I have described it, we obtain a narrative that sounds fairly plausible until we reach the stage where our initial guess of slope requires correction. Its "causal" transliteration is then forced to claim that some mysterious process ensues where a strain arising at the foot of our strut mysteriously "causes" a change in the bending at its top pinning. It would require fancy diversional tactics to slip *that* episode past an audience without their noticing its oddity (in the next section we'll examine a real life case where, through a simple change in the topology of the application, practitioners are fooled in exactly this manner). However, that long distance jump in our computation can be avoided by simply replacing Euler's scheme by some implicit marching scheme such as a central difference approximation (which usually supplies better results than Euler's anyway, although they ask us to solve equations in a lot of unknowns). Commonly, its gaggle of equations are solved by employing a more subtle form of successive approximations and, if so, the ersatz "causal" narrative we construct will sound middling fair, for it appears to tell a story of how "primary causal processes" in the strut become adjusted by subsequent waves of "secondary effects." It runs like this.

> We know that our post is pinned at 0 on the top plate but its compressed condition will result in some small deflection within the initial 1/4 ft section of the strut that lies adjacent to the upper plate. Now we aren't interested in the trivial solution, so we should assume

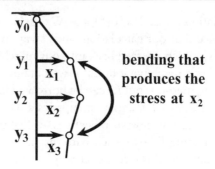

that at the 1/4 foot point y_1, our strut has been **caused** to show a displacement of some finite amount which we can register as, say, $\frac{1}{16}'' \pm \Delta x_1$ where Δx_1 represents a secondary correction factor that we will calculate later after the primary chain of causes and effects in the strut has been traced out. Proceeding onto the $\frac{1}{2}$ foot position y_2, its horizontal displacement will be caused to move to x_2 because of the internal resistence y_2 shows to the bending created by its end point positions y_1 and y_3, which happens to be expressed by the formula $8(x_1 + x_3)/(16 - W)$ (to see why, consult the fine print below). Here we will use $\frac{1}{16}''$ as our x_1 value, but we realize that we may have to consider the secondary effects δx_1 later on. Likewise, the displacement at y_3 will be caused by the bending action of its end points x_2 and x_4, and so on the rest of the way to the bottom of the strut. But at the bottom, the final displacement x_n is caused to be 0 because of its pinning. Putting all of these constraints together, we attempt to calculate what each of the x_1, x_2, x_3 ... need to be. Probably we won't get a consistent result due to neglecting the secondary effect Δx_1, so we should engage in further rounds of causal reasoning to figure out how large its magnitude must be. Once we know that we will have solved our problem (if W is above critical load).

This narrative, to my ears at least, sounds fairly normal, but the history of "events" it credits to the strut are just as spurious as those we obtained from our first shooting method scheme. If we picture this perfectly valid inferential technique as providing a story of how "causal processes" unfold in our strut, we will have fallen victim to *semantic mimicry*.

. . . .

Here I employ the so-called "centered difference scheme" $EI(x_{n+1} - 2x_n + x_{n-1})/\, 2(\Delta y)^2 = -Wx_n$ as our approximating equation with EI set to 1 for simplicity and employing a step Δy of 1/4. This is an "implicit scheme" because it relates the x_n we hope to find to the values that lay on both of its sides. Then, instead of finding concrete x_n values as we march forward along the strut, an implicit scheme collects together all of the local algebra relationships it finds and then solves the whole bunch as a class of simultaneous equations, allowing us to utilize the bottom pinning condition in equality with the upper pinning. To solve this batch of equation I then utilize a Gauss-Seidel routine that begins with a guess of $\frac{1}{16}''$. Our "chains of secondary effects" merely mock the steps in this second procedure.

I might remark that the relaxation method (and, more generally, finite element) schemes mentioned briefly in 4,x also provide nice alternative sources for mimics as well.

....

Such experiments supply prima facie reason to suppose that "cause" migrates rather easily from its core "marching method" home into surrounding territories, where its informational import generally assumes a more abstract and "procedural" quality (such claims *do* report valid facts about our strut, because any sound successive approximation routine progressively puts its target system into ever narrowing descriptive boxes, but they're not of causal process type). In fact, we often use "cause" and "effect" quite commonly in equilibrium or steady state situations, usually without much awareness that their supportive significance has shifted rather considerably from genuine "causal process" contexts (like our cannon ball). I wouldn't say that "causal process" has "shifted its meaning" when this occurs, anymore than it's completely happy to say that "weight" shifts its meaning with respect to space station or "hardness" with respect to plastics (although we sometimes *do* evaluate such matters in "shift of meaning" mode). However, I will say that we sometimes need to *watch out* for such behaviors, lest we talk ourselves into some ill-conceived project otherwise (I'll supply a stunning example in the next section).

If this assumption is right, then Hume displays great, if perhaps accidental, perspicacity in locating general "cause" only within a roomy corral, rather than attempting to "analyze" its contents definitively in the mode of his successors. Within its specializations, "cause" picks up robust local contents and *sometimes* covers the events described in exactly the manner we expect to witness in a true causal process. But sometimes "cause"'s informational significance shifts to one largely of *procedure*: "The next claim to consider after S is $S^\star$." Accordingly, Hume errs in maintaining that "cause" doesn't carry a thick significance: it does, but only locally. But if we try to articulate a common denominator able to reach across a wider spectrum of "cause"'s everyday employments, we are forced to drift towards the procedural side of the corral: "It's useful to think about Y after you think about X."

But we can't think of this as simply a situation where "cause" proves to be "contextually sensitive" in the usual way, because often *we have no way of knowing* what significance "cause" presently carries. That is, we often learn useful ways for dealing with the events around us that we describe in "causal" argot and which we initially picture in a "causal process" manner, yet later turn out, through no fault of our own, to constitute unexpected mimics. Indeed, there have been plenty of occasions in the history of science when some assumed chain of cause and effect has turned out, upon closer inspection, to be misconceived, even though we often leave our old "cause and effect" talk in place afterwards anyway. My favorite example (mentioned in 1,iii) is L. G. Gouy's and Lord Raleigh's rethinking of the "causal processes" that occur when a prism or a finely scored metal plate forms a spectrum. Most of us have been taught that with his prism Newton "untwisted all the shining robe of day"[12] (in James Thomson's

---

[12] James Thomson, "To the Memory of Sir Isaac Newton" in Joan Digby and Bob Brier, eds., Permutations (New York: Quill, 1985), 178.

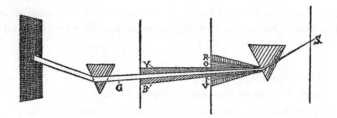

description) or, more prosaically: that sunlight is composed of different wavelengths of homogeneous light blended together, an arrangement that *causes* a prism to fan them out into a spectrum as the different monochromatic waves are carried along different pathways through the glass. In truth, virtually the opposite process occurs: the approximately monochromatic light that leaves our prism or grating is actually *created* upon its exiting surface. As J. R. Partington explains the matter:

> [N]atural light consists of a great number of quite irregular pulses . . . [W]hen white light is resolved into a spectrum of monochromatic waves, the prism or diffraction grating actually produces these from the irregular wave trains or pulses. The monochromatic rays are not actually present as such in white light, as was generally believed since Newton's time.[13]

It then turns out, according to an elaborate (and rather startling) theory developed by Norbert Weiner and A. I. Khinchen[14] (based upon Raleigh and Arthur Schuster's earlier suggestions), that the spectral "decomposition" of "mixed light" actually comprises a surprising *statistical mimicry* of true Fourier decomposition (of the sort encountered in the guitar string modes of 5,viii). It is as if (to modify a comparison offered by Schuster[15]) the dulcet tones of Winston Fitzgerald's violin prove to be created by little imps beating very rapidly with hammers (in fact, that's essentially what happens when we listen to Fitzgerald on a CD). On the new picture, monochromatic light's status as "cause" rather than "effect" becomes inverted topsy-turvy, yet most of us continue to talk in the old, backwards way courtesy of the statistical imitation of decomposition that white light displays.

There is little we can do to ward off mistaken assumptions about "cause" in circumstances such as this—Nature has simply decided to play one of her little jokes of unexpected underpinnings upon us. However, it would be conducive to clearer thinking if we could imitate applied mathematicians and learn to watch more intently for the characteristic symptomatology of "cause" talk that has wandered into a neighboring form of strategic patch.

As we'll see, other types of "causal process" mimicry create the characteristic problems that trouble Hume's own discussion of "cause." It is largely because these

---

[13] Partington, <u>Advanced Treatise</u>, iv. 162.

[14] Norbert Weiner, "Generalized Harmonic Analysis" and "The Historical Background of Harmonic Analysis" in <u>Collected Works</u>, ii (Cambridge, Mass.: MIT Press, 1979).

[15] Arthur Schuster, <u>An Introduction to the Theory of Optics</u> (London: Edward Arnold, 1909), 117.

pretenders compete for his attention that Hume cannot detect a genuine causal process (as displayed within a projectile's flight) even when it sits plainly before his eyes. I would like to complete our study of equilibrium-style mimics before I turn to the details of the masquerades I have in mind, but will observe that it is virtually inconceivable that anyone alive in Hume's era could have possibly discriminated the imposters from the genuine articles. Thanks to the subsequent scrutiny of researchers such as Hadamard, we are far more sagacious about the particular deceptions that plagued Hume, but we must anticipate that allied misapprehensions lie in our futures as well, for we lack the handles upon language that might fully prevent them from arising. It represents an abiding aspect of the human estate that we must sometimes adopt a temporary agnosticism with respect to some of our discourse about "cause and effect": we can be quite certain that there's much *good* in our chatter, yet not be very clear about *what kind* of good it represents.

(iv)

Struggling with a word. Rather to my surprise, I accidently ran across a misunderstanding of "cause" that seems essentially of my "misunderstanding the shooting method" type. Here the faulty picture has serious repercussions, because it sets its victims scurrying off on an ill-conceived pursuit. The case I have in mind shows up in some of the early work on "qualitative reasoning" within the field of artificial intelligence. These researchers set themselves the task of duplicating how an engineer might intuitively reason about the behavior in, say, a plumbing valve without resorting to brute number crunching. They reasonably stressed that we often make do with rough, qualitative estimates such as "the velocity at A is a little bit bigger than the velocity at B" rather than "the velocity at A is 1.6 m/sec while the velocity at B is 1.5 m/sec" in everyday reasoning (I suspect that they underestimate our capacity to keep track of comparative magnitudes, however). Unfortunately, these research efforts were often plagued by a "let's reinvent the wheel" hubris that led their victims into unhelpful confusion. Specifically, they fancied that, because they trafficked in qualitative estimates rather than numbers, they might thereby ignore with impunity all of the strategic distinctions emphasized in this chapter (in particular, the difference between a "marching method" and a "shooting method" calculation). But a basic law of Conservation of Difficulties plainly attends any form of descriptive problem like this: the mere fact that you wish to ignore important distinctions doesn't mean that they'll be willing to go away. Replacing ".1 m/sec faster" by the phrase "a little bit faster" can't possibly make our problems of computational strategy any easier.

As noted, we sometimes utilize the word "cause" in long calculations, even if it merely carries a procedural significance there (as in a successive approximations procedure). Insofar as I can see, this simple foible convinced our AI researchers that they were on the brink of articulating some wondrous calculus for "causal reasoning" that had heretofore alluded the orthodox physicist. In this vein, John S. Brown and Johan de Kleer[16] boldly declare that causality is "something [that] modern physics provides no formalism for treating" and de Kleer, in conjunction with Daniel Weld, brags:

> Qualitative physics is interested in constructing explanations for how things work. This perspective establishes the context in which we are interested in causality. . . . There is a great deal of tension in the qualitative physics community about the question whether causality is solely an artifact of human reasoning. Whether this is true or not, human beings expect to be provided explanations in causal terms . . . Many qualitative physics researchers have adopted the far stronger thesis that causality is fundamental and plays a central role in reasoning about physical systems. But where can qualitative physics look for insights into how causality is used? Physics is of little help here because it long ago discarded the notion of cause in favor of empirical laws expressed as constraints.[17]

This last remark about "constraints" apparently echoes the views of Russell cited above and is wrong for the same reasons. And it is rather common to find analogous claims erroneously advanced in both the philosophical and statistical "causation" literature: science may eschew causes, but we ordinary folks still like'em.[18] There are certainly reasons to be interested in notions of "causal factor" that do not neatly comport with the narrow notion of "causal process" we are investigating, but many assertions of this class arise from Russell-like muddles about how differential equations contain their data.

In any case, let's examine a typical example from de Kleer and Brown[19] to see how their thinking gets fooled by a fairly simple mimicry. They invite us to consider a regulatory water valve as pictured. They hope that their self-styled qualitative calculus will analyze the valve's workings. I don't quite see how they expect to answer this question, given the paucity of data they supply (e.g., they say nothing about the stiffness of the spring that pulls the gate out of the fluid stream). In fact, I believe they need to tacitly assume that the valve has been *designed* to stabilize around a constant flow of a prescribed amount. But then their claim that "physics can't deal with this sort of problem" is patently false, because we are looking at a simple control problem like the telescope of 8,viii and, as such, is readily adjudicated by Heaviside's operational calculus. De Kleer and Brown would likely protest that they want their qualitative calculus to be able to diagnosis potential malfunction in the manner of a good plumber, but surely it is easy to pinpoint deviation from optimal performance once the parameters of the latter have been determined. Insofar as I can see, their extraordinary claims about "physics' limitations" stem from stereotypes with respect to physical method quite comparable to those entertained by Sir William Hamilton (the Scots philosopher again,

[16] Johan de Kleer and John Seely Brown, "A Qualitative Physics Based on Confluences" in Daniel S. Weld and Johan de Kleer, Readings in Qualitative Reasoning about Physical Systems (San Mateo, Calif.: Morgan Kaufman, 1990), 88. Steven Brady, "Artificial Intelligence and Robotics" in Masoud Yazdani, ed., Artificial Intelligence (New York: Chapman and Hall, 1986), 140.

[17] Daniel S. Weld and Johan de Kleer, "Causal Explanations of Behavior" in Weld and de Kleer, Readings, 611–12.

[18] Judea Pearl, Causation: Models, Reasoning and Inference (New York: Cambridge University Press, 2000). Nancy Cartwright, Nature's Capacities and their Measurement (Oxford: Oxford University Press, 1989).

[19] De Kleer and Brown, "Qualitative Physics," 89.

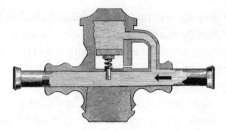

not the Irish mathematician), despite the fact that they seem to have enjoyed some measure of engineering training.

For whatever reason, they believe that "physics" can only track how the water will rush through the valve, without any regard for teleological purpose. Theoretically, it is possible to calculate how water arriving from the right might flow through our plumbing, but this project will require tracking a horrifically complicated causal process involving a good deal of turbulence, mixing of layers and back flow: a project very much at the far edge of current day computing technique. Instead, it is most common in situations like this, if we merely want to know the magnitude of the flow after its complicated transients have died away, to assume a condition of *steady state flow*. That is, we assume the fluid is in motion, but unchanging in its flow properties everywhere. This condition allows us to drop (or replace with constants) many of the problematic terms in our fluid equations, just as we did in equilibrium cases earlier. Indeed, two allied forms of steady state reduction were on view in Chapter 6: the Kutta/Joukowsky calculation of air foil lift and the Sommerfield calculation of razor blade diffraction. As occurred in the equilibrium approaches, when we drop terms in this manner, we have thereby thrown away the parts of the governing equations that determine how *causal processes* act in the valve (by the way: we must assume that the spring constant and input flow lie in the right ranges, for otherwise steady state flow will not be possible). However, if we're merely interested in the stabilized flow, we won't care about these details (which, as we remarked, are nearly impossible to follow anyway).

However, the fact that we are dealing with a self-regulating feedback system prompts de Kleer and Brown to ask a different set of questions about their device: how will its parts adjust if we set the inlet pressure higher or lower? Addressing this question provides a "diagnosis of the machine's functionality" in their estimation. Here is the manner in which they typically articulate their objectives:

> One common source of insight in qualitative physics is to examine methods engineers use to reason about devices ... Causal explanations are important to the engineer because they are an explicit representation of how a device achieves its behavior. This explanation itself forms the basis for subsequent reasoning. In design tasks, it is important to reason backwards to pinpoint what could have caused the symptoms.[20]

[20] Weld and de Kleer, "Causal Explanations of Behavior" in Weld and de Kleer, Readings, 611–12.

In fact, several potential projects are muddied together here, but let us stick with the task of ascertaining how our plumbing arrangement responds to alterations in inlet pressure. In fact, this is a question we can resolve by utilizing a shooting method-like technique: starting with a new pressure on the right (and assuming that the left hand pressure stays constant), we progressively adjust our solution so that a variant of two-point boundary problem is solved (note that, although we say the "inlet pressure" has increased to 5 ft-lbs/sec, the final pressure we accord to the right hand side of the valve can be expected to differ in its totality due to back flow considerations). So far, so good, but our authors presume that we have thereby traced the *history of the causal events* that restabilize our valve at the higher pressure. This is an error of interpretation exactly equivalent to reading our strut calculations in a "causal process" manner (however, the topology of the plumbing mollifies the peculiar top to bottom causal jump that betrayed the mimicry of our strut calculations). As before, none of the intermediate steps computed in our successive approximations scheme bear any relationship to the aqueous events that actually cause the steady flow to emerge, for all information pertinent to that purpose has been discarded.

Somewhat imperfectly, the "qualitative physics" account that Brown and de Kleer construct attempts to follow what would otherwise represent a perfectly viable computational plan familiar to physicists. However, because they instruct their program to write "X causes Y to happen" rather than "after X, Y is calculated," they wind up with a descriptive printout that talks about a large number of fictitious "causal episodes." They are dimly aware that peculiar elements have crept into their story, but they cheerfully conclude that these oddities actually represent a *virtue* of their methodology, in the following astonishing passage:

> [T]hese complications and impediments concerning causality come as a result of asking the question "How does change come about?" Modern physics tends to sidestep this question by adopting a modeling perspective which cannot, in principle, account for change. The central thermodynamic principle that underlies the construction of almost every model is that of quasi-static approximation: the device is presumed always to be infinitesimally near equilibrium ... Our solution is to leave the original models unchanged, but define a new kind of causality (which we call mythical causality) that describes the trajectory of non-equilibrium "states" the device goes through before it reachieves a situation in which the quasi-statical models are valid ... [We also add] additional criteria [that] help to reconstruct what the behavior below the quasi-static level must have been if the world were causal.[21]

Needless to say, this is complete bosh. The belief that they are "reconstructing the behavior between quasi-static levels" represents a clear display of mistaking a successive approximation calculation for a causal history. The most minimal attention to the real life effects that arise when the water pressure in a plumbing system is altered should have warned them that their "behavior below the quasi-static level" story bears no

---

[21] DeKleer and Brown, "Qualitative Physics," 115.

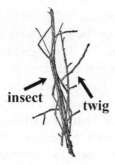

insect   twig

Biological mimicry

direct relation to reality (according to their picture: I continue to stress that successive approximations carry information but not of the type *pictured*).

On the other hand, except for their dumbfounding arrogance, we should pardon our AI researchers for falling prey to a rather universal human foible: it takes a good deal of experimental cleverness to determine how water actually flows through a valve[22] and so we become suckers for fill-in narratives that bear little resemblance to the events that actually occur. Even the wisest among us will be occasionally fooled by semantic masquerade in thinking about "causal processes," for Nature is simply too good at constructing plausible mimics for matters to prove otherwise (this observation, we might note, stands at the core of many of Mach and Duhem's misapprehensions with respect to atomist hypotheses (4,i)—proponents of the latter often argued upon the basis of masquerades in their time, as we shall see in 10,viii). Closer attention to detail always has the potential to unmask what we have assumed to be "histories" as merely convenient way stations within a calculation. And so we discover that Newton did *not* "unweave the rainbow": an incoming bundle of component light beams does *not* cause a prism to transmit them along diverging paths; rather, the glass causes the creation of the monochromatic rays along its lateral edge. The basic utilities of our *talk* of "spectral decomposition" can proceed ahead more or less unscathed, but our portraiture of their cause-and-effect relationships has been turned upside down.

I have not followed the adventures of "qualitative physics" reasoning amongst AI researchers since these initial attempts, but if improvements have been made, they must surely reflect the basic strategic distinctions of applied mathematics in a fitter manner (as my principle of the Conservation of Difficulties would suggest). It would be curious if these improvements have crept in unnoticed, as "further factual corrections" in the mode of 7,viii.

I have spent some time on this example, because it beautifully illustrates a rather common linguistic process: a linguistic methodology $\mathcal{R}$ (Euler's method) well suited

---

[22] Milton Van Dyke, An Album of Fluid Motion (Stanford, Calif.: The Parabolic Press, 1982). M. Saminy, K. S. Breuer, L. G. Leal and P. H. Steen, A Gallery of Fluid Motion (Cambridge: Cambridge University Press, 2003).

to phenomena A described by predicate "P" ("cause") gets borrowed as a subroutine within some methodology $\mathscr{R}^*$ (shooting method) natural to a distinct variety of physical situation B. In the process, "P" is dragged in the borrowing as well, pulling "cause" away from a patch where it genuinely conveys the significance of *bringing about as a subsequent causal effect* into an adjoining sheet where it merely signifies *being the next thing to consider in a steady state computation*. The ensuing *ur*-philosophical confusion then inspires our authors' wild claims that the qualitative calculus captures the "general meaning of 'cause'" found in everyday thinking, which physics fails to represent (such boastful claims are structurally equivalent to contending that the Securities Exchange hasn't been monitoring banks correctly because it has failed to attend to the shores of rivers). For allied reasons, I suspect that many philosophical discussions of "cause" suffer similar debilities, albeit of a less dramatically manifest sort.[23] It seems scarcely credible that a writer can airily ignore all distinction between dynamic, equilibrium, steady state, quasi-static, etc. strategies, yet be left with anything intelligible to say about "cause" or "causal process." None of this shows, of course, that "cause" represents an abnormally misbehaved word, for we have seen that many words string their global usages together through similar borrowings and continuations. However, we must beware of positing without examination that "cause," or the many other predicates which are prone to rove in wayward fashion, rests upon uniform pillars. Indeed, very strange confusions can hide under the common coverlet of "cause" if we allow it to range too freely in its prolongations. In this fashion, I believe that Hume has proven deeply prescient in his reluctance to analyze "cause" as we find it in its everyday, rambling and extended glory. The corral is set at too large a diameter for that project to prove of any profit.

Incidently, our AI researchers are right to observe that background design or diagnostic purpose can alter the conceptual personality of a predicate considerably. Such ingredients form an undeniable component of *cause*'s multiplex character within its everyday dimensions. But they are quite wrong to presume that the intrinsic nature of physics or differential equations prevent those qualities from being manifested within their presence: such misapprehensions arise from the toxic blend of scientific stereotype and classical invariance of which I have often complained over the course of this book's many pages. A predicate might start life upon the narrow road to prediction, yet soon be pulled aside by the tempting wiles of design. Such deviations in projected course occur all the time, to all sorts of words, whether in dedicated scientific use or not, and language usually winds up better off in the end result. But those reorientations in direction do not indicate that we can't, as the indulgent but genuine owners of our language, filter away unwanted alternative courses through detoxifying techniques when we must. Indeed, it is not wise to begin programming a computer to reason about "cause" unless we have first established a clear picture of exactly what sort of worldly relationship we want our predicate to follow.

---

[23] In fact, the ill-effects of underestimating "cause"'s many variations reach through many regions of philosophy: for a critique, see George M. Wilson, The Intentionality of Human Action (Stanford, Calif.: Stanford University Press, 1989).

Newton

(v)

Newtonian counterfeits. As we observed in section (ii), a cannon ball falling under the influence of gravity provides an admirable paradigm of a causal process whose scientific treatment had become fully familiar by Hume's time. Why then did he claim that we have "no idea of this connection, nor any distinct notion what it is we desire to know when we endeavor at a conception of it," given that a range of entirely adequate examples lay close to hand? The answer lies in the fact that a number of causal imposters also clamored for his attention and these distracted him from attending to cases where our understanding of "causal process" proves uncompromised by mimicry. Even in the best of circumstances (4,viii), classical mechanical doctrine organizes itself as an uneven facade and, during Hume's era, matters were considerably worse, for many distinct but manageable topics were indiscriminately jumbled together due to a lack of suitable mathematical discriminations. In addition, Newton's glorious accomplishments were popularly described in terms that discouraged judicious criticism. Thus James Thomson scarcely hints in his heroic ode that the British Lion had prevaricated mildly on mechanism and slightly cheated with respect to the billiard table:

> Have ye not listened while he bound the suns
> And planets to their spheres! the unequal task
> Of humankind till then. Oft had they rolled
> O'er erring man the year, and oft disgraced
> The pride of schools, before their course was known
> Full in its causes and effects to him,
> All-piercing sage! who sat not down and dreamed
> Romantic schemes, defended by the din
> Of specious words, and tyranny of names[24]

---

[24] Thomson, "Ode to Newton," 176. Nicolson, <u>Newton Demands</u> has a lovely discussion of these issues.

Oddly enough, attitudes of "Newton got the fundamentals of classical mechanics entirely right at their inception" persist in many intellectual circles even today (surely, it does not demean his achievements to claim somewhat less).

In mentioning mechanism and billiard balls, I thereby reference our earlier discussions of the oddities that attach to those specific topics (4,ix; 6,xii; 7,iv). But poor Hume, who knew little mathematics, had every reason to presume that Newton had treated these twin subjects in unified and impeccable fashion. But this isn't true—standard Newtonian approaches in both instances incorporate rather severe policies for descriptive reduction. Overlooking our cannon ball circumstances, Hume thereby elects to scrutinize several methodological imposters as "paradigmatic examples of the scientific treatment of cause." He fails to find any non-attenuated form of "causal process" revealed therein, but this desultory result is not surprising—the critical ingredients had already been leached away by the very policies adopted for their strategic treatment.

In fact, the full story is a bit more complicated than this, for machinal ideas and impactive collisions (i.e., billiard balls) engage in a variety of good cop/bad cop routine that is admirably calculated to befuddle Hume (and, for that matter, most of the rest of us). Most forms of traditional mechanics (I set aside the Boscovitchean point particle lineage) allow material bodies to engage with one another through direct *contact* and *action at a distance* (such as gravitation) interactions. As we observed in our earlier discussion of Boyle's strictures on mechanical understandability (3,iii), action at a distance accounts were often regarded as explanatorily inferior to contact interaction accounts, possibly even by Newton himself. However, insofar as illustrations of genuine causation are wanted, a restriction to contact action needn't pose a hindrance, for completely unproblematic causal processes operate mainly in this manner (e.g., the propagation of waves through deep water or a piece of steel). Unfortunately for Hume, the mathematics required to formulate even the simplest of these processes—partial differential equations[25]—had not yet been invented and so physicists of his era were forced to incorporate some measure of severe reductive strategy in their attempts to treat any form of contact problem (steady state or equilibrium reductions comprising very standard fare). Typically, the mechanisms of true causal process get cast overboard as unmanageable ballast along the way.

Another common method for circumventing the difficulties of contact action is to exploit the fact that systems of hard bodies which smoothly roll or slide across each other's surface can be approximately captured in algebraic or differential terms (physicists usually call these *constraints*). If these conditions are not over-constrained (i.e., the contacting bodies display freedom of movement), then we find ourselves approximately within the realm of *mechanism* as it was later formulated by Reuleaux, whose investigative efforts were described at some length in 7,iv. It is a basic fact of our conceptual phenomenology that, even to this day, we intuitively feel that we *understand* the workings of devices that suit these limitations better, in the warm and fuzzy manner I have highlighted before (as 4,i noted, a prime motive of the semantic instrumentalists

---

[25] S. B. Engelsman, Families of Curves and the Origins of Partial Differentiation (Amsterdam: North-Holland, 1984).

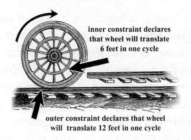

inner constraint declares
that wheel will translate
6 feet in one cycle

outer constraint declares that wheel
will translate 12 feet in one cycle

was precisely to undercut the influence of this continuing traditional bias). Following Reuleaux, we have observed that this strong faith conceals a substantial element of tropospheric misapprehension, for machinal ideas operate happily only within narrow niches and are greatly prone to the lousy encyclopedia phenomenon, where we must drop into other arenas of physical description to handle circumstances in which our mechanisms often find themselves (to provide yet another example, consider a wheel with two concentric hubs—"Aristotle's wheel" of antiquity—: when its center rolls onto the upper shelf, the system becomes over-constrained and its governing physics must be shuttled off to an altogether different patch of descriptive treatment). In fact, our strong sense of "understanding" seems to arise largely from subliminally associated algorithmic routines for design and diagnosis rather then mechanical prediction <u>per se</u> (I'll comment a bit more on these issues in a moment).

In 7,iv when we discussed these same topics, we noted that Descartes attempted to found a complete world view centered upon the local patch of mechanism. Here he was deceived, for the realm of coherent machinal thinking represents, when properly viewed against the context of Nature's far more abundant varieties, a quite narrowly supported and largely man-made plot of descriptive opportunity. However, it is a patch that readily encourages tropospheric complacency: with a few natural supplements its doctrines look as if they might very well pass—in the dark with a light behind them—for a complete theory of the inorganic world. It is this faith that makes folks entertain the hunch that perhaps the inanimate world functions as a gigantic piece of clockwork. As Hume expressed the thesis in Cleanthes' voice:

> Look round the world: Contemplate the whole and every part of it: You will find it to be nothing but one great machine, subdivided into an infinite number of lesser machines, which again admit of subdivisions, to a degree beyond what human senses and faculties can trace and explain. All of these various machines, and even their most intimate parts, are adjusted to each other with an accuracy, which ravishes into admiration all men, who have ever contemplated them.[26]

Captivated by this vision, advocates of the "mechanical hypothesis" perch themselves upon machinery's little outcropping and imagine that everywhere else the universe looks much the same, if only the blueprints of their microscopic hardware could be

---

[26] David Hume, <u>Dialogues on Natural Religion</u> (Indianapolis: Bobbs-Merrill, 1947), 143.

deciphered. In fact, a lot of little prolongations move off the patch of mechanism far enough to create a <u>prima facie</u> *hope* that the whole affair can be eventually prolonged into boundless coverage. These deceitful intimations of mechanistic completion convince enthusiasts that, surely, a complete physics of clockwork can be rationally articulated, even if they have not managed to turn the trick effectively thus far: the proper formalism must be just around the corner (faith in a benevolent deity can make these fond wishes burn brighter, of course). In this fashion, Descartes and his confrères succumbed to tropospheric complacency of a completely natural and often unavoidable kind, for, in many cases, our sense of how our predicates might prolong into unexamined waters is no better than that of fuzzy intimation. But in mechanism's case, Reuleaux has laid bare the underlying strategic mathematics of machine presupposition and thereby revealed its many descriptive holes. Benefitting from this perspective, it can seem odd that anyone ever fancied that machinal ideas might be prolonged to cover every form of natural event, so palpable seem its lapses to us. In our clearer apprehension of mechanism's unbreachable tropospheric confinement, we echo the theme that was beautifully articulated by Kelvin in the previous chapter: *the scales have fallen from our eyes and we have learned to look upon things in a different way, so that what were once difficulties are now the only commonsensical and intelligible way of looking at the subject.*

....

In rough terms, Descartes requires a set of a dynamical principles that will control the time evolution of a mechanism internally. There are a number of popular forms of generalized inertia and conservation of work capacity that give intimation that such an articulation might be possible (it isn't, but the reasons are rather subtle). I also see Descartes as providing a primitive mechanism for fracture, which would begin to address his problems with over-constraint. I believe that his notorious rules of collision represent a somewhat inept attempt to articulate principles along these lines. More generally, Descartes' physics has been rather unsympathetically judged by commentators, for we moderns (if we haven't benefitted from Reuleaux's insights) commonly fall prey to forms of mechanical complacency closely allied to those that animate Descartes' researches.[27]

....

However, our primary concern is not with Cartesianism, admirable as it is in many respects, but with the prevailing Newtonianism of Hume's era. By then it was recognized, albeit in a less than satisfactory manner, that conventional machinal ideas could not adequately accommodate abrupt and apparently *percussive contact interactions* such as those involved in billiard ball impact. Nor could mechanism deal adequately with motive forces of virtually any kind, whether in gravity's guise or coiled within a watch's mainspring. However, it was commonly presumed that Newton had found a way to treat billiard ball collisions in a fully satisfactory manner. If so, these same impulsive interactions, now regarded as acting as rapid and infinitesimal hammer blows over

[27] Mark Wilson, "Mechanism and Cartesian Physics," <u>Topoi</u> 14 (1997). I plan to improve the account provided there at some point.

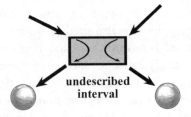

undescribed
interval

an extended span of time, might be treated as foundational to a understanding of the processes that occur in watch springs and ordinary rolling and sliding. In fact, unfortunate inclinations to decompose contact actions into impactive infinitesimals persist strongly throughout the entire nineteenth century.

Such thinking is an error. In terms of actual accomplishment, Newton's approach to billiard ball collisions *blocks out the temporal interval* in which the collision occurs and patches over this lacuna through the crude "coefficient of restitution" policy discussed in 4,vi (a direct treatment of the causal processes involved will need to track the formation of the stress waves that form and then pulsate through the interiors of the two balls, but the very difficult mathematics required for this project lay far beyond the capacities of Hume's time—and, perhaps, our own). But if this lapse is not noticed, one can be easily betrayed by a clever handoff from the arena of mechanism into that of impact. The "warm and fuzzy feelings of understanding" that we entertain with respect to everyday mechanical occurrence—that is, in situations of rolling and sliding—convince us that such circumstances represent paradigms cases of what "causal processes" are like (this is the "good cop" side of the story, although it is misleading). But then we are informed that, in point of fact, the physical processes that underlie the events of rolling and sliding can be further explicated according to a billiard ball impact model, as successfully provided by the great Newton (this is the "bad cop" side of the story). If we fall for this fast shuffle—as Hume surely did—, then we find upon our plate as the "paradigmatic form of causal interaction" a treatment that practices a particularly blatant form of *causal process avoidance*. So blatant, in fact, that its funny features would have been earlier noticed by critics except for the fact that it gains an alibi for its oddities by leaning upon the prestige of clockwork "causality" (which itself comprises a mimic of section (i)'s here-versus-there variety).

I would like to stress again, because such concerns accord so nicely with this book's overarching themes, that our "warm and fuzzy feelings" towards mechanism trace most directly to subconscious capacities for *design and diagnosis*, rather than to prediction proper. We show a great capacity to fiddle around with our watches—that is, manipulate their parts to ascertain their possibilities of movement—and then deduce how they must move if the piece is to work correctly. As we learned from our earlier discussion, nice algorithms exist for the optimal sizing and performance of a planar mechanism and I believe that our conviction of "understanding the watch" derives from complex sources allied to these (as we noted, the commendable project of our AI researchers was to uncover those routines and program them on a computer). Because of the manner in which our time piece is designed (i.e., engineered to implement one degree of here-versus-there freedom), we can also predict—*as long as the watch works more or less properly*—how it will overall behave without knowing virtually anything substantive about the processes that actually cause its successive states to occur: why its spring unwinds; why its contacting gears move as approximate rigid bodies, rather than distorting, and so forth. But this is merely an artifact of narrow tropospheric confinement. As soon as we consider circumstances that lie off the narrow shelf of machinal opportunity—our clockwork is relocated to some unaccustomed climate such as the surface of Pluto, say—, we shall quickly discover how shallow our "predictive understanding" has been, for we now must shift to quite different patches of physical description to gain any sense of how our watch is likely to behave. In short, most of us don't *understand* the "warm and fuzzy understanding" we bring to mechanism very well, for the latter rests upon an unexpected and somewhat subterranean set of facade-governed skills.

In any event, encouraged by the popular, but misleading, description of Newtonian physics as "billiard ball mechanics," Hume understandably fastens upon such circumstances as representing the "typical physical interaction." Indeed, Hume characterizes impact as "perfect an instance of the relation of cause and effect as any which we know":

> Here is a billiard ball lying on the table, and another ball is moving towards it with great rapidity. They strike; and the ball which was formerly at rest acquires a motion. This is as perfect an instance of the relation of cause and effect as any which we know either by sensation or reflection. Let us therefore examine it. It is evident that the two balls touched one another before the motion was communicated, and that there was no interval between the shock and the motion . . .
>
> Were a man such as Adam created in the full vigor of understanding, without experience, he would never be able to infer motion in the second ball from the motion and impulse of the first. It is not anything that reason sees in the cause which makes us infer the effect . . . But no inference from cause to effect amounts to a demonstration. Of which there is this evident proof. The mind can always conceive any effect to follow from any cause, and indeed any event to follow upon another: whatever we conceive is possible, at least in a metaphysical sense; but wherever a demonstration takes place the contrary is impossible and implies a contradiction.[28]

But if we inspect the rigid qualities we assign to our billiard balls, none of them has any evident bearing upon what occurs in a collision:

> Solidity, extension, motion—these qualities are all complete in themselves and never point out any other event which may result from them. The scenes of the universe are continually shifting, and one object follows another in an uninterrupted succession; but the power or force which actuates the whole machine is entirely concealed from us and never discovers itself in any of the sensible qualities of body.[29]

---

[28] David Hume, "An Abstract of a Treatise of Human Nature" in Enquiry, 188.   [29] Hume, Enquiry, 75.

But it is scarcely surprising that the "power or force which actuates the whole machine" seem "entirely concealed from us," because it is we, through our misplaced trust in Newton, who have left the critical events out of our accounting by a rather drastic strategy of reductive interpolation. Believing that he has confronted the "typical causal interaction" in as intimate a manner as man can, Hume concludes that we lack the requisite concepts to render what occurs in the causal breach intelligible. Indeed, if science could tell no better tale than Newton's coefficient of restitution story for billiard balls, he would be right.

> But as to the causes of these general causes, we should in vain attempt their discovery, nor shall we ever be able to satisfy ourselves by any particular explication of them. These ultimate springs and principles are totally shut up from human curiosity and inquiry. Elasticity, gravity, cohesion of parts, communication of motion by parts—these are probably the ultimate causes and principles which we shall ever discover in nature; and we may esteem ourselves sufficiently happy if, by accurate inquiry and reasoning, we can trace up the particular phenomena to, or near to, these general principles . . . Thus the observation of human blindness and weakness is the result of all philosophy, and meets us, at every turn, in spite of our endeavors to elude or avoid it.[30]

In the fullness of time and mathematical development, classical mechanics articulates better models for billiard ball impact where internal processes of rapidly altering stress and strain are addressed in a manner that seems far more "understandable" than the gappy story Newton provides. But these descriptions took a long time to materialize and Hume can be pardoned for believing that Newton's non-distorting treatment represents the last word on the subject.

Unfortunately lost in all this confusion is gravity's straightforward causal involvement with our cannon ball. But, as we have seen, its plain virtues of honest causal process were disparaged by the complaints of traditionalists that its causal processes are not "understandable" in a warm and fuzzy way. Here we must wait until Reuleaux before we find a convincing story of why such standards of "understandability" are unreasonable (Mach and Duhem appreciated that they were ill-considered as well, but they substitute instrumentalist philosophizing for the strategic identification that Reuleaux provides). Once those obstacles are properly clarified—and this diagnostic task also lies far beyond the capacities of Hume's time—, it is simply false that "the power or force which actuates the cannon ball's flight is entirely concealed from us." It lies plainly on view in the cannon ball's differential equations; Hume has missed its expression through expecting that something "warmer and fuzzier" needs to be found. Although focused upon practical concerns of a somewhat different nature, I see the sundry formalists and instrumentalists we discussed in Chapter 4 as hoping to establish, in allied spirit, that our understanding of scientific predicates can be much "thinner" than traditionalists demand. However as those parties articulated the doctrine, it cast physics' understandings of the world behind eerie veils, supplying only a

---

[30] Ibid., 45.

"residue of similarity" in Helmholtz' phrase. In this book's opinion, the proper manner for dealing with the swarm of intensional directivities that crowd around a predicate is not to thin them away into syntactic shallowness, but to *sort them out* according to the descriptive missions they advance (or impede). Unfortunately—as Hume's struggles with "cause" amply illustrate—that "sorting out" cannot be accomplished readily, through some sweeping brush stroke of "philosophizing about concepts." More often unraveling the underlying confusions will require the idio-syncratic study of the affected predicate's impertinent peculiarities. And when we can't yet accomplish that, we are left with little recourse except hunch and the council of past example.

. . . .

I might remark, as a fuller account of our story, that non-distorting billiard balls, shorn of "coefficient of restitution" mollification, should presumably qualify as "absolutely hard." That some paradox attaches to the perfectly inflexible body has struck many observers besides Hume, ranging from the prize set by the St. Petersburg Academy in 1720[31] to Johnny Mercer's "Something's Gotta Give." In the Optics Newton himself posited that atoms, could they be denuded of their attendant action-at-a-distance forces, would prove perfectly hard in this sense, a fact that he believed would eventually cause the energy budget of the universe to run down.[32] As earlier remarks may have made clear, the assumption of truly rigid bodies lies in considerable tension with other forms of classical doctrine, but it was long assumed that either they, or mass points, needed to be assumed as "foundational" to the physics of flexible bodies (vide the comments in 4, i).

. . . .

Our discussion can be summarized as follows. The mechanists favor a picture of "clockwork physics" based upon what is, in fact, a design opportunity that eschews genuine causal process altogether. Descartes attempts to extend this portrait to all nature, including in his Principles a rather unsuccessful attempt to cover mild cases of impact. Newton later provides the first adequate (differential equation) modeling of genuine causality in gravitational contexts, which is promptly dismissed as inad-equate by the mechanists in light of their faulty paradigm of "intelligibility." The Newtonians evade this philosophical criticism by cowardly conceding that gravita-tion's true workings remain opaque, but science isn't obliged to press to such a deeper core of understanding in any case. At the same time, Newtonians advance greater claims for their master's treatment of impact than it properly deserves, claiming in the bargain that Newtonian impact somehow does satisfy higher stan-dards of "intelligibility," frequently contrasting their (incomplete) achievements to Descartes' proposals, now uncharitably interpreted in a colliding body sort of man-ner. The resulting sum constitutes a rather formidable body of scientific folklore that

[31] Wilson L. Scott, The Conflict between Atomism and Conservation Theory 1644–1869 (London: MacDonald, 1970). J. Clerk Maxwell composed a poem on the subject: Wells, Curious Mathematics, 255.
[32] Isaac Newton, Optics (New York: Dover, 1952), query 31.

assures Hume: the unified successes of Newtonian physics rest upon interactions that it treats, in the final analysis, always in a brute force "this happens after that" manner, whether the events in question involve gravitation, cohesion or impact. Hume, attempting to survey the greatest accomplishments of Newtonian thinking, now popularized as "billiard ball mechanics," understandably focuses his attention upon impact and correctly finds virtually nothing in their Newtonian treatment beyond constant conjunction. In turn, Hume's Spartan conclusions later influence Mach in his anti-intentionalist views with respect to the proper character of scientific predicates.

Hume's musing on billiard balls led him to his celebrated dark pessimism about causation:

> We never can, by our utmost scrutiny, discover anything but one event following another, without being able to comprehend any force or power by which the cause operates or any connection between it and its supposed effects.[33]

As such, it conjures up all of the veil of predication worries we examined in Chapter 3: that we often confront quantities within the workings of Nature for which we have been supplied with no appropriate conceptual representative (causation may represent a true physical relationship, but it does not lie within the orbit of concepts we can grasp). With respect to Nature's basic animating traits we find ourselves permanently trapped in a distanced relationship like that we bear to Bismarck (we can grasp descriptive concepts that, more or less adequately, hedge the old statesman in, but we can never become directly acquainted with him). And so we return to the eerie structural hypotheses of Chapter 3, where we can never grasp the internal content of whatever real world attribute directly supports the predicate "is under a gravitational attraction"; we can merely describe it as "the I-know-not-what that induces masses to congregate towards one another according to Newton's inverse square law."

Besides the specific forms of mimicry that we have traced in this chapter, which can eventually be unraveled given sufficient time and talent, underlying Hume's thinking is the classical assumption that, whatever the true personality of "cause" is like, we grasp its qualities, insofar as such understandings lie within our reach at all. But here he is fooled by "understanding" 's seasonal shifts in focus. When we claim that we "understand a predicate fully" in everyday evaluation, we make substantial claims about our capacities, but not in the absolutist frame that Hume imagines. For the factors we highlight along one dimension of "understanding" (e.g., design) can easily obscure considerations more important for other purposes (such tradeoffs comprised a chief moral of our face plot example (8, iii): "understanding" legal tender from a family resemblance perspective runs up considerably different gradients than "understanding" it in statistical terms). In time, such differential qualities of development can be overcome: we can gradually improve our appreciation of all the dimensions of "grasp" and

---

[33] Hume, Enquiry, 84–5.

"understanding" that envelop a predicate, but not all at once and not in any single gulp of classical grasp.

In the meantime, I submit, a gentle dash of mitigated skepticism with respect to the directions in which our words will wend may help to diminish our *ur*-philosophical disappointments.

10

THE CRITIC OF NATURE
AND GENIUS

*[N]ature does not follow the rule, but suggests it. Reason is the interpreter and critic of
nature and genius, not their lawgiver and judge. He must be a poor creature indeed
whose practical considerations do not in almost all cases outrun his deliberate
understanding, or who does not feel much more than he can give a reason for . . . But it
does not follow that the dumb and silent pleading of the former (though sometimes,
nay often mistaken) is less true than that of its babbling interpreter, or that we are
never to trust its dictates without consulting the express authority of reason.*

William Hazlitt[1]

(i)

Mitigated expectations. In summary: I have suggested that we commonly fall into
conceptual puzzlement through our native inclination to presume that invariant and
wholly controllable packets of recoverable content stand behind our everyday predicates of macroscopic classification. In actual fact, the mildly divergent directivities that
pull a term into fresh arenas can stem from widely scattered sources, including shaping
factors that merely lie latent as environmental opportunities yet to be exploited. Our
exuberant but rash faith in simple semantical arrangements often leads us to confuse
layered facades for simple "is a dog" / *being a dog* alignments and to mistake discordant
networks governed by delicate boundary controls for well-integrated bodies of harmonious opinion. Given these proclivities, it is small wonder that we become befuddled
when suppressed fissures or multi-valued schizophrenias pop into view, often at the
very moments when descriptive language appears to be accommodating our linguistic
objectives most admirably. Academic philosophy, by and large, has not sufficiently
prepared us for these emergent wrinkles. True, we have been amply warned that old

---

[1] William Hazlitt, "On Genius and Common Sense" in <u>Table Talk</u>, 31, 35.

ways of speaking must sometimes be abandoned when descriptive hardships loom. We expect to abandon old terminological friends if they falter: "We gave those old concepts of classical mechanics a good run for their money, but they eventually grew old and tired and so we needed to replace them with the fresh stable of notions that quantum scientists have cultivated." But in the muddles with which we have been concerned, it is not these sorts of descriptive difficulties that have troubled us, but a puzzling symptomatology characteristic of linguistic *success*. And this is where our book's main moral lies: a well-adjusted terminology is likely to display odd quirks as its standards of performance improve. The varieties of restless usage we have canvassed should not be regarded as undesirable or otherwise "conceptually inadequate," for the odd grain they manifest generally represents the implementation of unanticipated informational controls that are required in better recipes for practical planning and prediction, providing tighter tolerances against anticipated mistake. Thus a failure to meet "is a dog"/*being a dog* expectations needn't signalize poor linguistic manufacture. Quite the contrary, such arrangements merely reflect the trifling cost to be paid for strategies otherwise admirably suited to agents of limited capacity who find themselves situated within challenging environments. The only fault of which we can reasonably complain is that these complexly supported arrangements sometimes confuse us, having generally slipped into place rather silently, through the disorderly but steady hammering of little directive corrections here and there. Often the processes that compile these funds of adroitly curbed usage have been so intent about their business that they have neglected to inform us, their beneficiaries, of the supportive rationales behind their engineering (as Reuleaux would put it, our predicative usage has been assembled with *great art,* but we are unable to discern the *science* within it). But if we are susceptible to the intellectual hubris encouraged by classicism's whispering blandishments, we will fancy that our "spying attention" can surely discern the contents that supply our predicates with their bewildering personalities, if we only squint in their direction hard enough. But, when put to the test, these promised powers of discernment prove mistier than we anticipate, for the supportive directives that shape our predicates don't lie in one place, but spread themselves along the entire ebb and flow of natural linguistic development.

Nonetheless, our faith in our intellectual acumen is often so secure that we scarcely find this simple, skeptical answer credible and search for more extraordinary apologetics to explain our linguistic puzzlements, to the degree that we soon find ourselves denying that apples are truly red or that musical performance can be ably preserved on a tape recorder.

Why do we do this? Why are we so inclined to fancy that, whatever the unhappy facts of the universe may be like, we qualify as complete masters of our own thoughts—that we can act, as Descartes insists, as "little gods within our minds," if nowhere else? Why do we presume that, if we merely think hard enough, we can fully know the contents of our predicates and all their possibilities, even when their physical actualities lie utterly out of reach? Richard Steele observes in The Tatler:

> When I run back in my imagination all the men I have ever known and conversed with in my whole life, there are but a few who have not used their faculties in the pursuit of what it

is impossible to acquire; or left the possession of what they might have been, at setting out,
masters, to search for it where it was out of their reach.[2]

Seeking assurances in circumstances where satisfactory remedies cannot be supplied
represents an abiding aspect of human frailty. Inventing gods in hopes of controlling the
weather provides an obvious case; the behavior of most of us within a gambling casino
provides an abundant source of parallel example. We may pass muster as sober souls in
our daily affairs, but grant us an hour's span within a gaming hall and we prove geysers
of spurious hope and worthless incantation. Sitting before an obdurate slot machine, we
frame an unending stream of evanescent hypotheses: *If I can only repeat that same special*
twist against the lever, I'm sure I will win again; It's now time for this device to show a win;
I think I've figured out how to plan my wagers to insure a long term gain. The mathematician
Richard Epstein trenchantly observes:

> *The number of "guaranteed" betting systems, the proliferation of myths and fallacies con-*
> *cerning such systems, and the countless people believing, propagating, venerating, protecting,*
> *and swearing by such systems are legion. Betting systems constitute one of the oldest delusions*
> *of gambling history. Betting system votaries are spiritually akin to the proponents of perpetual*
> *motion machines, butting their heads against the second law of thermodynamics.[3]*

In the final analysis, I wonder if our *ur*-philosophical favoritism towards the fully
graspable, if elusive, conceptual packages promised in classical thinking does not trace to
this same primitive desire to control the uncontrollable: in the case at hand, the wandering propensities of language in its developmental interactions with the world, rather
than the climate or the slot machine.

Be this as it may, we should adopt a more chastened appraisal of our capacities to
foresee where the currents of our thoughts head, without thereby falling into an
exaggerated pessimism that these tasks lie beyond all human competence. With
patience, we can puzzle out the true support for a usage quite ably, but semantic
mimicry and allied obstacles represent a common enough occurrence that it often takes
longer to reach our goal than we generally presume. In conceptual matters, we do well
to assume a greater proportion of the attitude that Hume calls *mitigated skepticism*:

> *There is one mistake to which* [philosophers] *seem liable, almost without exception; they*
> *confine too much their principles, and make no account of the vast variety which nature has*
> *so much affected in all her operations. When a philosopher has once laid hold of a favorite*
> *principle, which perhaps accounts for many natural effects, he extends the same principle*
> *over the whole creation, and reduces to it every phenomenon, though by the most violent*
> *and absurd reasoning. Our own mind being narrow and constricted, we cannot extend our*
> *imagination to the variety and extent of nature, but imagine that she is as much bounded*
> *in her operation as we are in our speculation.[4]*

---

[2] Richard Steele, "Ambition," The Tatler, no. 202 in English Humorists of the Eighteenth Century (New York: Century, 1906), 111.
[3] Richard Epstein, The Theory of Gambling and Statistical Logic (New York: Academic Press, 1977), 53.
[4] Hume, "Sceptic," 93.

As such, this passage provides a nice description of the tropospheric complacency of which I have also warned. However, Hume errs in his estimate of how we might avoid these pitfalls: he recommends that we should henceforth "avoid all distant and high inquiries" and confine ourselves "to common life and to such subjects as fall under daily practice and experience." In truth, the far-reaching claims of basic physics stand a better chance of remaining relatively impervious to the surprises emerging from the cracks I have described, whereas commonplace macroscopic classification ("is hard," "is red," "is a gear wheel") is apt to prove positively riddled with them. There is an old maxim that "physics is simpler in the small," a claim that does not contend that fundamental principle is mathematically simple in any obvious sense, but that small scale physics can be excepted from the leaps and fissures that typically emerge when a huge set of descriptive variables is reduced to a tractable set. Thus it is not the exotic predicates of fundamental physics that can be expected to behave oddly and to reside upon complex supports, but the subtly adjusted classifications of common life: the vocabulary that allows us to frame fruitful recipes for complex macroscopic systems after a quick peek or a poke. Swiftness and largeness of scale have their price: compensatory forms of ad hoc rule must patrol unexpected boundaries and the natural propensities of otherwise admirable inferential schemes need to be sharply trimmed. Macroscopic objects are too complex to be discussed straightforwardly and a healthy toleration of irregularity allows localized arenas to be framed in which more roundabout forms of descriptive strategy can flourish.

The net effect of these adjustments to practicality is not to diminish the *truth* of what we ordinarily say, but to increase it. Our everyday claims with respect to "red," "weight" and "hardness" should qualify, by and large, as fully correct, although the *pictures* or *prefaces* we frame with respect to their supports may sometimes prove wanting. The modern engineer may still chatter of classical "force," "strain" and "mass" with respect to her projects as if the year were still 1890 (although with computers available), but she claims nothing false thereby. To be sure, the Victorians themselves fell victim to sundry misconceptions about the real world with respect to these same predicates, for they believed their descriptive reach prolonged to small and large scale applications that we no longer accept. But their modern descendants suffer no illusions: they know how their usages lie founded in quantum mechanical behaviors or, like Heaviside, they have lapsed into simple agnosticism about such matters. In today's use, the vocabulary of classical physics has become detoxified of the previous erroneous pictures of older times, leaving the assertions of contemporary engineers unblemished by any tincture of faulty opinion. In a related vein, when we declare an apple to be red, we easily may—or may not—suffer from venerable misconceptions as to "is red"'s proper footing, but such marginal ur-philosophical opinions rarely reach to the informational core contained within our everyday claims and should not be regarded as seriously impugning their truth.

Accordingly, we needn't fall into the extremities of the following position, extracted from an article by two contemporary philosophers, Paul Boghossian and David Velleman. They weigh our Chapter 2 worries as to whether the objects of the external world are "really colored" or not and, in fellowship with Joseph Addison, they decide not:

> We therefore conclude that when someone calls some [external object] red, in an everyday context, he is asserting a falsehood. Indeed, our account of color experience, when joined with the plausible hypothesis that color discourse reports the contents of color experience, yields the consequence that all statements attributing colors to external objects are false.[5]

They then provide the following extraordinary "apology' for the palpable oddness of their claim:

> Consider one of the many harmless falsehoods that we tolerate in everyday discourse: the statement that the sun rises. When somebody says that the sun rises, his remark has the same content as the visual experience that one has when watching the horizon at an appropriately early hour. That is, the sun actually looks like it's moving, and that the sun moves in this manner is what most people mean when talking about sunrise. When somebody says that the sun rises, he is wrong; and he usually knows he is wrong, but he says it anyway. Why?
>
> When one understands why talk about sunrise is false, one also understands that its falsity makes no difference in ordinary life ... [Such a] belief will not mislead him about any of the phenomenon he normally encounters; and it will in fact give him correct guidance about many such phenomena. His judgements about the time of day, the weather, the best placement of crops, the location of glare and of shadows at noon, will all be correct despite being derived from premises about a stationary earth and a revolving sun . . . . Only an undue fascination with the truth could lead someone to reform ordinary discourse about the sun.
>
> Talk about colors is just like talk about sunrise in these respects. That is, life goes on as if objects are colored in the way they seem to be. Experience refutes few if any of the conclusions derived from beliefs about objects' colors and many true conclusions are derived from such beliefs.[6]

But surely my readers will not readily agree that it is *false* that the sun rises! Don't detoxifying claims such as the following make perfectly good sense?

> Years ago, folks used to believe that, when the sun rises, it actually hid below the horizon before dawn and later climbed into the sky propelled by some form of locomotion.

Quite generally, terms of robust descriptive utility accept a good deal of detoxifying revision of associated picture without incurring the charge of having altered their meanings (or, to restate the observation more accurately, "from anyone *except philosophers*," for classically inclined thinkers frequently evoke covert shifts in meaning to explain how their invariant conception of predicative content accords with everyday linguistic experience (3,vii)).

---

[5] Paul Boghossian and David Velleman, "Color as a Secondary Quality" in Alex Byrne and David Hilbert, Readings on Color, i (Cambridge, Mass.: MIT Press, 1997), 98.   [6] Ibid., 99.

Boghossian and Velleman's radical conclusions apparently stem from commonplace forms of classical assumption, i.e., the belief that the only "common content" invariant capable of underwriting the spectrum of assertions we make with respect to "is red" directly suits sensory experience only, preventing *redness* from accurately qualifying as the quality of a rose or signpost. Following this trail of traditional opinion ("Redness is a simple, indescribable trait"), they fail to anticipate either the virtues of quasi-attribute support or the fact that traditional semantic pictures, however deeply ingrained, can be detoxified by observations of the type offered by Ralph Evans (7,x). Their mistake suggests a general moral: whenever we find ourselves denouncing the "strict truth" of commonplace assertion, we should first wonder, "Are we sure that we have pictured the semantic support for this terminology correctly?," rather than succumbing to a philosophical vision where we purposively say false things at every moment, like the South Sea islanders who substitute inappropriate classifiers for correct words (e.g., "rock" for "mango") during taboo seasons, lest the gods become offended at their blunt truthfulness.[7]

And so the proper conclusion we should accept, as citizens of limited capacity within a complex world, is that we should *not want* the firm control over linguistic unfolding that classical grasp and gluing promise: we must allow a measure of *open texture* within our procedures so that our language can take its opportunities when it sees them, even if we, as its official masters, are too obtuse to notice the improvements. In claiming that we possess limited capacities for controlling our language, I should clarify that I really mean "for controlling our language *wisely*" because we *can*, in fact, lay down stern methodological percepts as to when a sentence ought to be employed and when not. Indeed, Quine, in his dream of a final "web of belief" monitored with beauteous "theory T" pattern and regularity, expects essentially that. But the web of language becomes dangerously brittle if we string its wires too tightly: viable descriptive techniques require a healthy allowance for tinkering if the roundabout adjustments that suit their subjects are to be found. Or so the hard-won lessons of applied mathematics warn us.

Accepting this open texture requires that we abandon cherished ambitions with respect to unchallenged linguistic management and accept the fact that, in large part, language wanders where it listeth (and is better off for doing so). This is not to say that we cannot participate in its ongoing management in a host of extremely useful ways—indeed, "concept" and its evaluative cousins represent the vocabulary in which we typically express our redirective assessments. But we must avoid becoming the prigs of inflexible methodology that classicism and theory T syndrome thinking encourage, for the Heaviside affair (and many like histories) advise us that, within the hidden crevices of the familiar and shopworn terminology, commonly sit seeds that, when conditions are ripe, suddenly blossom and carry our old words into extraordinary fresh adventures, in the manner of the great prospects that suddenly descend upon humble "$\sqrt{-1}$" when its time for flourishing arrives. We should not diminish the wonder of these unfolding processes by clinging grimly to classical-leaning excuses that preserve the

---

[7] For a mild version, see R. N. H. Bulmer, "Karam Colour Categories," Kivung 1, 3 (1968).

illusion that we remain masters of linguistic process but confer no evident advantage except to flatter our sense of semantic entitlement. By "excuse," I intend glosses such as, "Gee, look what happens when we *change the meaning* of '$\sqrt{x}$' ". No: the changes we witness in "$\sqrt{x}$" are entirely due to changes in season, not evidence that mathematicians made an inadvertent slip in holding firmly to predicative content.

(ii)

Sublime imagination. An odd feature of academic philosophy in the times I write is that it is characterized by a general willingness to reject the classical picture as excessively strong or rigid, yet is likewise loath to relinquish any *privilege* attached to the old picture (rather like the progeny of wealth who has given away all of his money yet still expects to keep servants). In particular, many writers still presume that it remains their primary duty to serve as *regulators of conceptual possibility*, in the manner outlined in 3,iv, where a philosopher should be able, through keen scrutiny, to legislate sternly whether some proposed employment for a predicate is truly compatible with its invariant contents. If we retain an essentially classical vision of "clear thinking," such a job description will seem wholly in order. We can mentally transport ourselves back to 1900 and participate once again in those stirrings of Great Expectations, when Russell and his allies believed they had uncovered the key to permanent and unchallenged rigor in intellectual endeavor. But today's philosophers have largely rejected the foundation of classical gluing upon which such claims originally rested, and few of us struggle with substantive scientific difficulties in the old manner. Yet we continue to advertize our conceptual wares in essentially the same way. Why? Doctrinal inertia supplies the only credible answer I know.

In this regard, we should be acutely aware that, of all that the "possibilities" that hover around a term in some fashion or other, some are more revealing of how the vocabulary functions practically than others. Yes, the arch-like capacities of fairy support still cling to "rainbow," and comprise a non-negligible portion of its latent DNA. But worrying excessively about this old baggage can distract our attention away from the more important question of how we manage to carry real world information within a format as odd as this. In this regard, recall Nathaniel Hawthorne's contempt (2,iii) for those narrow-minded persons who overlook the "possibilities of things," where the contrary "possibility" that his story illustrates simply represents another riff on the familiar "the human soul encased in an inanimate object" whimsy that has proved the stuff of scary stories since the dawn of campfires.[8] As latent DNA, original pictures of semantic support, whether for "human soul" or "rainbow," are apt to linger in a language indefinitely as potentially active ingredients, no matter how dramatically our estimates of the real life underpinnings of these terms alter subsequently. As such, it is sometimes useful to muse upon the possibilities that feed upon these fossilized

---

[8] William Steig, <u>Sylvester and the Magic Pebble</u> (New York: Simon and Schuster, 1969).

directivities: they reflect ingredients that still lie genuinely buried within our concepts of *soul* and *rainbow*. The problem comes—and I believe this occurs with Hawthorne—when our skills in trafficking in such nugatory fantasies get mistaken for a general capacity to know "all the possibilities of things." After all, when genuine *ur*-philosophical perplexity attacks, it often creeps from the cracks within some real life facade that we have mistaken for integral doctrine, not because we have overlooked animate snow children or the fairies that frolic upon the spines of rainbows. Likewise, when we strive to improve the usage's performance through a better picture as in 8,ix's circumstances, we must positively exclude the flotsam of irrelevant semantic accretion (we observed that the ability to *detoxify* usage is the capacity that drew attention to philosophy of language in the late nineteenth century). In these contexts, a liberal brandishing of Hawthornian possibilities can function like a magician's misdirection: we fumble clumsily with our right hand so that our audience will ponder the possibilities of where it has stashed the silk (in the meantime, we have secretly secured the scarf up our left sleeve with a vanishing tube). If we have grown accustomed (and most academic philosophers have) to citing the "self-consistent worlds of classical physics" as an illustration of the kinds of "conceptual possibilities" in which the philosopher trades, then we are less likely to be on the vigilant lookout for the covert property draggings that represented the true reason why that subject puzzled the Victorians so. The impulse to lecture neuroscientists upon the possibilities of snow children ought to be chastened, not encouraged (raw prejudice frequently masquerades itself as "broad-mindedness" of a Hawthornian manner). But contemporary philosophical training, which loves all of its "possibilities" equally, often promotes unwise practices, rather as in the Colleges of Erewhon:

> The main feature in their system is the prominence which they give to a study which I can only translate by the word "hypothetics." They argue thus—that to teach a boy merely the nature of the things that exist in the world round him, and about which he will have to be conversant during his whole life, would be giving him but a narrow and shallow conception of the universe, which it is urged might contain all manner of possibilities which are not now to be found therein . . . To imagine a set of utterly strange and impossible contingencies, and require the youths to give intelligent answers to the questions that arise therefrom, is reckoned the fittest conceivable way of preparing them for the actual conduct of their affairs in after life.[9]

With respect to the considerations raised in the previous chapter, consider Ernest Sosa and Michael Tooley's report on prevailing contemporary attitudes with respect to analyzing "cause":

> [M]ight it not be plausibly argued that causal relations possess an intrinsic nature, so that causation must be one and the same relation in all possible worlds? . . . But if this is right, then one can appeal to the possibility of worlds that involve causation, but do not contain the physicalist relation in question—or, more radically, that contain no physicalist states

---

[9] Samuel Butler, <u>Erewhon</u> (New York: Random House, 1927), 206–7.

at all—in order to draw the conclusion that causation cannot, even in this world, be
identical with any physicalist relation.[10]

The concern expressed is that any explication of "cause" in terms of the concrete processes described in physics might not touch upon "the true nature of cause" because spooks in Hawthornian stories might cause events to occur by ectoplasmic means. I can't see how fussing abundantly about the causal capacities of ghosts is likely to reveal the basic dichotomies between marching method and steady state strategies that seem essential to appreciating the rather different forms of information that claims of "cause" and "leads to" convey within those respective contexts (9,ii). But it is precisely a failure to distinguish the two (because of their mutual mimicry) that appears to be central within many of the conceptual puzzles to which "cause" is heir, including Hume's own celebrated worries. And mistakes along this same line can readily generate ill-conceived scientific projects as well, as our AI example attests (9,iii).

For allied reasons, we should cast a cold eye upon invocations such as the following (from Frank Jackson):

> *There is an important sense in which we know the live possibilities as far as color is concerned. We know that objects have dispositions to look one or another color, that they have dispositions to modify incident and transmitted light in ways that underlie their dispositions to look one or another color, that they have physical properties responsible for both these dispositions, and that subjects have experiences as of things looking one or another color. We also know that this list includes all the possibly relevant properties . . . Color thus presents a classical case of the location problem. The colors must, if they are to be instantiated anywhere, be findable somehow, somewhere in accounts that mention dispositions to look colored and affect light, the physical bases of these dispositions, and color experiences.*[11]

By "the location problem," Jackson intends the spectrum of unyielding options offered in Chapter 2, where we seemed unable to house the trait *realizing the Symphony in G Minor* happily in either mental or physical dominions. The effective import of a passage such as this (whether its author so intends or not) is to encourage the student to set aside her books on practical color technology and initiate a program of armchair musing. But, plainly, Jackson's list of "live possibilities" overlooks the facade-based factors highlighted in 7,x, all of which were suggested by inspecting the surface grain that emerges vividly only when everyday color talk is pressed to higher grades of performance, as in industrial application. Restricting our attention to the orbit of "live possibilities" that occur to us in the courses of untutored meditation is likely to preserve unsuitable semantic pictures past their expiration date, rather than helping us reach a better understanding of the framework of localized controls that keeps our complicated macroscopic employment of "is red" viable.

A factor that creates considerable confusion in these regards is the fact that the pictures we construct of word usage are almost invariably *generic* in character: we

[10] Ernest Sosa and Michael Tooley, Causation (Oxford: Oxford University Press, 1993), 3.
[11] Jackson, Metaphysics to Ethics, 87.

examine how our terminology behaves across a variety of models for its possible application. This circumstance feeds the false impression that we are thereby considering "possible worlds" akin to our own to discern how the "intension" of the term behaves (5, viii). In ascertaining the soundness of a recipe or reasoning rule, we usually seek some (partial) guarantee of safety for our procedures and it becomes natural to consider salient variation within this context. But, as we observed in our consideration of why we examine electrical equations over the "solution spaces" we do (8,x), the modelings considered may not represent "possible worlds" in anything like the philosopher's sense. To consider another example, the use of variational principles is quite common in mechanics, for they allow an engineer to write down reduced variable equations with much greater reliability than if the forces upon the structure were completely modeled in a bottom-up manner. This advantage traces to the fact that the variations considered represent a fairly direct report on how the structure will *wiggle when manipulated* (often physicists call such a schedule "a varied history for the mechanism"). However, if we don't manipulate the mechanism, it is unlikely that it can possibly move in those ways under its own locomotion. Indeed, the reduced equations formed on this basis may not prove especially valuable for *prediction* (because frictional effects are likely to intrude too quickly), but can prove exactly the ticket for calculating *optimal steering* (6,xiv sketches the standard illustration of a skater sliding upon a sloped plane). Once again, the family of "possibilities" registered within the formalism of orthodox variational principle are somewhat *design-oriented* in their salience, a revealing fact we are unlikely to notice if we view the matter entirely through the foggy spectacles of the "possible worlds of classical physics." In attempting to understand the shaping directivities of language, not all possibilities are created equal.

....

Some of our specific examples reveal deeper reasons to distrust the "concepts as maps from possible worlds" point of view. Under the influence of the theory T syndrome, philosophers presume that "theories" lay out their "possibilities" as models in the logician's sense, which are then identified with the solution space natural to a given equational form. But we have seen that, in real life, considerable strategic delicacy enters into the framing of what qualifies as a relevant "solution" to a localized problematic. At various points in my career, I have been interested in issues such as "Is Newtonian mechanics truly deterministic?" Having come from an educational background thoroughly infused in theory T lore, I was mystified, as were a number of careful philosophers of like-minded concern,[12] by the fact that the existence and uniqueness results actually found in the mathematical literature typically focus upon quite narrow classes of equations, entirely insufficient to address the sorts of questions that we philosophers asked. The proper resolution of our puzzlement lies in the fact that the equations we parochially find in practice almost invariably reflect subtle and strategically framed policies of variable reduction, with the net effect that it can be rather hard to frame an accurate assessment of the overall matrix in which they appear. Accordingly, we cannot form the alleged "set of Newtonian possible worlds" simply by lumping together the local solutions we encounter. In fact, these

[12] Richard Montague, "Deterministic Theories" in Formal Philosophy (New Haven: Yale University Press, 1974). Earman, Primer on Determinism.

considerations prevent us from easily determining whether an uneven facade stands before us (as seems to be the case in the classical context) or a flat structure doctrine amendable to axiomatization et al. I believe that contemporary metaphysicians have been far too cavalier about what are, in fact, quite delicate issues.

. . . .

On the other hand, when we wish to understand the full story of why a predicate displays the "conceptual personality" it does, we must plainly attend to a considerably more extensive set of directive factors than are manifested within the thin considerations of information-bearing capacity alone. But, in the context of that fuller narrative, the directive factors that push words to and fro don't appear to us as worlds, but as *little nudges*: "That thingamajig in the sky sure looks like an *arch*" or "Gee, look what happens when I divide the other side of the equation by dx." Most of the time when we evaluate a usage using the term "concept," we do *not* wish to attend to a predicate's full personality, in all its inconsistent and meandering glory, but only to those component threads that, at the given moment, promise to show us a path to improvement. In these regards, rich billows of armchair smoke represent one of the chief ways in which we can lull our observant faculties to sleep, for we must beware of the soporific vapors of "I know all the possibilities that concept C tolerates" (Uncle Fred is fond of disquisitions upon that topic). Unless we can first chase away the fairies that climb on rainbows and stop confusing steady state arrangements for dynamical circumstances, we will not concentrate effectively on how our target language conveys data within a real world setting. In those moments where we need to delimit the information-bearing capacities of a mysterious inferential routine, we must push away the clouds of ersatz "possibilities" that hover about it and cultivate instead the real world exemplars where such reasonings have been pushed to a heightened quality of performance. Only there are we likely to find the surface clues indicative of the subterranean strategies that allow such vocabularies to work, including the variations important to their secure functioning. Oftentimes our general knowledge has not advanced to a stage where this is possible.

But if this somewhat dour assessment of our linguistic circumstances is correct, neoclassical expectations that we can augur our semantic futures fully in the near term and unravel every *ur*-philosophical muddle must be slackened—their underlying operations may simply lie beyond our current scientific or strategic ken (such matters will eventually become clear, but not now). As philosophers, we should become wary of those temptations of overconfidence that Hume describes so well:

> *The imagination of man is naturally sublime, delighted with whatever is remote and extraordinary, and running without control, into the most distant parts of space and time in order to avoid the objects which custom has rendered too familiar to it.*[13]

But Hume himself frequently vacillated between tempered and inflated doubt and, in his philosophical practices, he often violates the cautions I consider most desirable, for the unmoored possibility represents one of his chief philosophical weapons. This is not

[13] Hume, Enquiry, 170.

surprising, for he is also a firm adherent of the classical picture of concepts, to the degree that he expects that the contents of acceptable traits can be analyzed as categorizations of private impressions. And that point of view quickly leads to the most radical forms of Pyhrronist skepticism, with respect to cause, geometry and the external world.

But, famously, he acknowledged that qualms of this ilk are scarcely beneficial to life or science:

> [Such skeptic] *must acknowledge, if he will acknowledge anything, that all human life must perish were his principles universally and steadily to prevail. All discourse, all action would immediately cease, and men remain in a total lethargy till the necessities of nature, unsatisfied, put an end to their miserable existence.*[14]

But a more moderated circumspection might improve our lot in life to measurable degree:

> *In general, there is a degree of doubt and caution and modesty which, in all kinds of scrutiny and decision, ought forever to accompany a just reasoner. There is, indeed, a more mitigated skepticism or academical philosophy which may be both durable and useful, and which may, in part, be the result of this Pyrrhonism or excessive skepticism when its undistinguished doubts are, in some measure, corrected by common sense and reflection.*[15]

For the reasons surveyed in 1,x, I am dubious that any useful mitigation is likely to descend from undistinguished doubts, but Hume also observes that there is another fount of wisdom upon which we can more reliably, if imperfectly, rely: the human histories of past follies and successes. At least that is the only guidance I know to supply with respect to wandering words, having foresworn the Archimedean advantages promised within the classical picture. Cultivating an appreciation of *ur*-philosophical muddles as they have arisen in the prior career of everyday descriptive endeavor can help us better anticipate future pitfall, rather as a historian's exemplars might serve to temper the follies of the politician (true: in practice, such advice is generally ignored, with the excuse that every fresh situation looks a little different than the past). In any event, that is why this book is largely stuffed with linguistic curiosities: the queer little turns that show how varied the puzzling currents of language can be.

(iii)

The philosophical investigation of concepts. The open texture aspects required in a profitable predicate's directivities can help us resolve a faulty dilemma has become prominent in contemporary debate (similar issues were posed in other terms in Sir William Hamilton's era). Within the framework of classical assumption, the proper contents of concepts are held to lie potentially open to *conscious review*, although opinions differ as to its exact phenomenology (Locke is contrasted with William James on

---

[14] Hume, Enquiry, 168.   [15] Ibid., 169.

this score in 8,iv). Those aspects of practical classification that transpire at a wholly unconscious level—the discrimination mechanisms whereby we categorize objects as looking like prototypical birds, say—are not regarded as "forming part of the concept" according to a classical traditionalist. But the latter's appeals to our "spying attention" make us uneasy today, so recent discussion has become inclined to preserve the flavor of this old-fashioned division between conscious deliberation and subpersonal computation by evoking the *rational norms* of interpersonal discussion in lieu of classical "presentations to consciousness." Prima facie, making a "cut" in terms of rational standards seems commonsensical: if I ask, "Is it *correct* to classify a kiwi under the concept *bird*?," I normally evoke taxonomic considerations that have little evident bearing upon the perceptual mechanisms that allow us to recognize most birds on the wing quickly.

However, a contrary school of philosophers influenced by modern cognitive science seems quite insensitive to these basic concerns. For example, Ned Block cheerfully suggests that we think of "concepts" thus:

> [C]onceptual role is a matter of the causal role of the expression in reasoning and deliberation and, in general, in the way the expression combines and interacts with other expressions so as to mediate between sensory inputs and behavioral outputs.[16]

But, insofar as such a "causal role" is well defined (and it plainly is not), Block has made no attempt to distinguish between those aspects that reflect the proper content of the concept *bird* and those that merely facilitate its recognition in a quite approximate way. Or, to vary the example, all advanced animals (including humans) reason geometrically about the world very ably, but the inferential routines they unconsciously employ are surely *heuristic* in nature: measured according to the strict standards of a Euclid or Hilbert, the policies will sometimes lead to faulty conclusions. Here the aperçu of C. S. Peirce aptly applies: we often do not understand the workings of our most powerful patterns of reasoning. Indeed, if some cognitive scientist were to explicitly lay out the mute workman routines whereby our brains anticipate the appearances of objects under shifts in position, we would likely be baffled by their transitions: "Why do these odd rules conclude that B holds from A," we will wonder, "for that step is both strange and patently fallacious?" In framing this judgment, we employ the proper notion of *triangle* as a *norm* whereby the workings of our unconscious routines are scrutinized for their *correctness*. Such scrutiny is likely to reveal the underlying algorithm to represent one of those heuristic policies, well known to computer scientists and often resistant to rational unraveling, that locate valuable results with high probability only.

This contrast between *norm of correctness* and *effective thinking routine* should not surprise us, for recall the face plots of 8,iii. In a natural sense of "understand," we perfectly understand what dimensional data compiled with respect to a dollar bill signifies after we have taken an orthodox class in multi-variant statistics. On the other

---

[16] Ned Block, "Advertisement for a Semantics for Psychology" in Ned Block, ed., Readings in Philosophy of Psychology (Cambridge, Mass.: Harvard University Press, 1980), 93

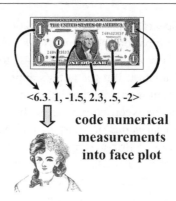

<6.3. 1, -1.5, 2.3, .5, -2>

⇩ code numerical
measurements
into face plot

hand—and this is the rationale for recalibrating our money data in the guise of cartoon faces—, our brains can better "understand" the significance of our information in transformed form (in the sense of forming hypotheses readily), for our mute routines of unconscious search are more effectively specialized in their abilities to rank faces according to similitude. And this disparity induces a natural contextuality in how we employ the evaluative term "understand." Typically we highlight the grouping advantages of face plots thus:

> I didn't really understand the properties of these bills until I saw them as faces.

On the other hand, a criminologist accustomed to such techniques but newly educated in statistics might well exclaim:

> I didn't really understand the face-derived properties of these bills until I learned that they code for multi-variant dimensional measurements.

Indeed, this contrast very much recalls the seasonalities of Chapter 8: having already framed a correct *picture* of the correlational significance of the phrase "bills that belong in the same family," we have learned how to prolong its practical employment significantly through a lift into unexpected inferential turf. It will be this increase in intellectual bounty that elicits our attention when we speak of "understanding." A party long familiar with the lift technique, but having not appreciated its correlational underpinnings heretofore, will likely employ "understand" with a much different emphasis.

Almost certainly, our native skills in intuitive geometry work via routines of a face plot character: inferential procedures that swiftly supply practically valuable answers, but do not crisply embody what we intuitively regard as the "true contents" of our geometrical concepts. For that second purpose, classroom Euclidian doctrine delineates the necessary ingredients far more ably, just as a statistics course expounds the proper content of a *multi-variant set of dimension measurements* more directly than a computer routine that sorts faces into rough families through oddball measures. But we utilize "understand" to highlight both aspects of the situation, depending upon the occasion.

On a wholly traditional picture, it is the "spying intelligence" of the conscious observer that extracts the true concept from its supporting miasma of heuristics. It is hard not to believe, as we did with respect to our teenager's aspirations in Chapter 8, that some rather occult portrayal of human capacity is here assumed (rather like the rarified intellectual capacities that the Cartesian soul will preserve after it severs itself from its body-contaminated passions). As we just noted, many philosophers in corrective response maintain that the key to such discriminations lies, not in mystic peeping, but in acquiescence to public intellectual norms shared within a society. That is, amongst ourselves in community we lay down standards of rational inquiry to which geometrical assertions must comply, whatever deviant reasoning patterns happen to be sounded within our antechamberal whisperings. Such social assumptions seem implicit in Michael Dummett's advice as to how the topic of linguistic meaning (including predicative expressions) should be properly approached:

> *The fact that the use of language is a conscious rational activity—we might say the rational activity—of intelligent agents must be incorporated into any . . . description* [of what it is that we learn when we learn to speak], *because it is integral to the phenomenon of the use of human language . . . When we converse with others, we are continually concerned to discern the point of what they say, that is, their reasons or motives for saying what they do . . .* [W]e *can estimate somebody's purpose, motive, or intention only against the background of what we presume him to know . . . A* [theory of linguistic meaning] *should not therefore aspire to be a theory giving a causal account of linguistic utterances, in which human beings figure as natural objects, making and reacting to vocal sounds and marks on paper in accordance with certain natural laws . . . We have no need of such a theory . . . We can, in general, make some unfamiliar human activity . . . intelligible . . .* [by] *describ*[ing] *the practice and the institutions that surround the practice, and then it becomes intelligible as an activity of rational agents.*[17]

From this point of view, Block has made the exact mistake of identifying concepts with the mental hardware as it might be mentioned in "a causal account of linguistic utterances," thereby failing to do justice to the manner in which we rationally adjudicate the conceptual capacities of others in everyday interaction, e.g., when we ask whether Sonya's employment of "triangle" lies in accordance with Euclid or not. We commonly "bring concepts to mind" to scrutinize their contents, but we can hardly haul forth a complete Blockian "conceptual role" in this fashion.

It has therefore struck many contemporary writers that normative evaluations of a communal ilk represent an altogether different kettle of fish than we encounter in normal scientific inquiry. Thus John McDowell writes:

> [T]*here is a familiar and impressive tradition of reflection about common-sense psychology, according to which the point of its concepts lies in their providing a kind of understanding of persons and their doings that is radically unlike the understanding that the natural sciences can yield. This tradition's insights are never taken sufficiently seriously by people*

[17] Michael Dummett, <u>The Logical Basis of Metaphysics</u> (Cambridge, Mass.: Harvard University Press, 1991), 91–2.

> who suggest "folk psychology" is a proto-theory of the operations of those internal
> mechanisms, to be refined and perhaps wholly superseded as we learn more about what goes
> on inside our heads. For instance, natural-scientific investigations of how what is literally
> internal controls behavior would seek theories whose power to explain would be propor-
> tional to their power to predict. But folk-psychological concepts can express a kind of
> understanding of a person that seems to have little or no relation to predictive power . . . If
> the understanding that common-sense psychology yields is sui generis, there is no reason to
> regard it as a primitive version of the understanding promised by a theory of inner
> mechanisms. The two sorts of understanding need not compete for room to occupy.[18]

Here McDowell primarily emphasizes the "commonsense psychology" status of
understanding the concept triangle, but he will presumably also reject the Block-like claim
that the proper contents of *being a triangle* itself will be supplied within a "theory of
inner mechanisms." Borrowing a phrase from Wilfred Sellars, McDowell frequently
contends that concepts "live in a space of reasons, not causes" and I believe that this
slogan demands that whatever transpires on a neurophysiological (or otherwise subpersonal) level should be dismissed as irrelevant to the normative aspects of "concept"
talk that McDowell wishes to highlight.

To be sure, McDowell allows that "natural scientific" explanations might prove
feasible in their own right:

> This leaves it open that investigations of an "engineering" sort might be fine for other
> purposes.[19]

But within "engineering" is *not* where the true home of "concept" in its primary usage
lies and it is not the vein that a philosopher should primarily pursue. What the psychologist attempts to study is simply not the same gizmo as the student of self-critical
human nature seeks.

In the other camp, Block cheerfully allows that alternative approaches to "concept"
may prove viable "for other purposes" and his liberal dispensation (which is not otherwise specified) may extend so far as to embrace the suggestions of McDowell and
Dummett. But if such munificence is so liberally extended, then Block has essentially
conceded the battle to establish that "conceptual roles" (or anything much like them)
have much direct relevance to the traditional problems of philosophy. For in every
instance that I can think of, the conceptual questions that emerge as central in such
disputes are invariably of the type, "In these circumstances, what is the *proper way* to
employ *redness* (or *force* or *pain*)?" Jolly capitulation on Block's part to the public norm
school (or to classical traditionalists) is tantamount to a confession, "What do I know
about standard philosophical problems? I'm just an engineer; you'd better ask my
colleagues across the hall."

---

[18] John McDowell, "Knowledge and the Internal" in <u>Meaning, Knowledge and Reality</u> (Cambridge, Mass.: Harvard
University Press, 1998), 412–13.

[19] John McDowell, <u>Mind and World</u> (Cambridge, Mass.: Harvard University Press, 1996), p. xxi. Although I redeem
some of these claims below, I certainly do not endorse the "limited purposes of science" doctrine implicit here. To
paraphrase Jafee's gofer in <u>Twentieth Century</u>, I yield the lamp of Scientism to no one.

For such reasons, the psychologically oriented author Jerry Fodor (although no defender of "cognitive roles" per se) is less inclined to be accommodating to McDowellish opinion. He believes that philosophical attempts to separate the "properly human" manipulation of concepts crisply from every subpersonal consideration impede a proper understanding of real human capacity through a sentimental fussiness. He rhetorically asks:

> Does any of [these worries about concepts] really matter except to philosophers over sherry?

He continues:

> Oddly enough, I think perhaps [they do]. We are in the midst of a major interdisciplinary attempt to understand the mental process by which human behavior accommodates to the world's demands—an attempt to understand human rationality, in short. Concepts are the pivot that this project turns upon since they are what mediate between the mind and the world. Concepts connect with the world by representing it, and they connect with the mind by being the constituents of beliefs. If you get it wrong about what concepts are, almost certainly you'll get the rest of it wrong too.[20]

That is, unlike investigations in the vein of McDowell (or, for that matter, Chapter 1's gentleman with the "the"'s in a box), cognitive science seeks a general understanding of *concept* that will qualify as truly important scientifically. And this, Fodor tacitly presumes, will provide the skeleton key to the old-fashioned puzzles about concepts such as we find in Locke or Descartes:

> I'm afraid the bottom line is there is no room where McDowell wants to wiggle . . . If that's right, then epistemology needs to bend and McDowell will have to cool it a little about justification.[21]

Although I prefer to be genially tolerant of the projects of others, I'm afraid that, in these regards, Fodor is more likely to prove correct than conceptual segregationists of the public norm school. However, I also believe that Fodor is quite mistaken in his view of our philosophical priorities as well.

However, my own reasons for rejecting the notion that the standards we apply in conceptual evaluation have much to do with societal engagement stem from different quarters. With respect to the terms of macroscopic classification we have investigated, public norms, however formulated, are implausible as a final magistrate of conceptual correctness, simply because our chief objective is to make practical headway within an often uncooperative external world. Getting matters right with respect to those externals usually matters more to us than the opinions of our chums. Recalling Heaviside's operational calculus from Chapter 8, to what final authority do we appeal in attempting to settle whether moving from "$df(x)/dt = g(x)$" to "$f(x) = g(x)/1/dt$" represents a *correct use* of the concept *derivative*? In such journeys "beyond the railhead,"

---

[20] Jerry Fodor, In Critical Condition (Cambridge, Mass.: MIT Press, 1998), 33.   [21] Ibid., 8.

Heaviside follows his own drummer, out of step with every societal trend. Nonetheless, it is he who has utilized the notion rightly, correctly extracting its latent potential, not his meeker, "wet blanket" critics. Likewise, he is right to stretch *impedance* into its strangely prolonged application, although every prevailing norm *except* that of "getting the right answer" would argue against such policy. Once we are concerned with building a skyscraper or laying down long extensions of telegraphic cable, the standard "get the right answer" represents our avatar, not any norms promulgated by civic rigorists. Indeed, these are the exact concerns that led us to conclude, long ago (2,vi), that environmentally determined standards of correct use apply even to the lonely mumblings of a solitary Robinson Crusoe, at least when he is busy counting his goats or planning a mousetrap. Restored to the embrace of society, our Crusoe will be fully justified, like Heaviside, in sticking by his terminological guns if his privately evolved calculations allow him to design better mousetraps than his critics.

Both of our opposing schools have overlooked the important fund of basic directivity emphasized throughout this book: the reshaping impulses that arise from *interfacial accommodation*, where available linguistic capacity gradually adjusts itself to suit beneficial environmental opportunity. We have amply observed that our shifting employment of "concept" commonly serves to highlight those aspects of predicate management that prove salient within the seasonal moment: if we are exploring a prolongation into virgin territory, the indications of rewarded distributed normativity usually provide our prevailing indicators of correct use ("'$f(x) = g(x)/1/dt$' must be a correct equation because it leads to valid conclusions"), whereas, at a later date, the same issues may be directed to a newly minted picture instead ("That sentence proves correct when evaluated according to the Laplace transform"). Both forms of focus stand relatively detached from either social norms or the psychological vicissitudes of how our mute workmen happen to push "$1/dt$" about. We are mainly concerned to bring our descriptive vocabulary into improved correlation with the workings of electrical circuits: this is the guiding star which sets the "norm" to which our predicate is chiefly responsible. Beside it, the demands of society and psychological implementation pale. We are grateful, of course, to the subconscious searching that first suggested the basic transitions of the operational calculus to us, but we owe such considerations little conceptual fealty. We want to follow the *thread of correctness* that lies buried in the routine's partial successes; we don't expect our mute workmen (who are rather dumb, after all) to provide further edification on this score.

Forcing an artificial invariance upon these natural processes of shifting focus renders their natural etiology incoherent: we can hunt for lions generally without having some beast in sight from the outset. Since the days of Kronecker and Dedekind, it has been realized that very deep affinities exist between algebraic number theory and algebraic fields, enough to motivate a coherent attempt to articulate in what it might consist (many mathematicians claim the goal has been fulfilled in the exertions of Grothendiek's school). It is easy to appreciate in what such "intimations of structure" consist, but it would be silly to maintain that such a framework was "fully grasped" by the earliest mathematicians to whom the affinity first dimly appeared. This is why I compared a "concept" to a long rope twisted together from shorter threads of concrete directivities: if we follow its length, an improving predicate will help us to increasingly tighten our grip upon the physical world, but our guiding cord will not be comprised of just one fiber.

Of course, in stressing the interfacial factors that influence correct application, I did not mean to endorse McDowell's claim that the "engineering aspects" of language are irrelevant to proper conceptual behavior. Quite the contrary, virtually everything we have here considered in respect to conceptual behavior qualifies as an engineering concern of a linguistic ilk. And certainly the nitty gritty oddities of our psychological makeups play a significant, if not insurmountable, role in determining which solutions will prove viable with respect to a desired end (vide the complex way in our usage of "red" needs to be adjusted and controlled). As our earlier discussion of machine design noted (7,iv), it is wisest to approach an engineering project by first mapping out a basic skeleton of optimal design, for that study outlines the strategic options open to us. We can then worry whether our plan can be implemented within the steel, brass and grease of the available world. I likewise believe that a similar emphasis on *basic design* best explains the schedule of shifting evaluations of conceptual correctness upon which we focus during a typical course of improving our predicates along a canonical history. For these purposes, the interfacial directivities ignored by both McDowell and Fodor emerge as the factors of greatest salience, whereas social norms and psychological implementations seem largely complicating irritants.

Much of the trouble I find in Fodor and McDowell's presentations traces to the fact that they consider predicates only in the context of relatively static circumstances, whereas the richer utilities of "concept" talk I emphasize emerge most vividly when we are struggling to steer balky or misbehaving words in better directions.

Within present day cognitive science, much of the research on "concepts" is concerned with the mechanics of storage and computation whereby basic recognitional capacities are achieved, a very important task, albeit somewhat oblique to the issues considered in this book. In these regards, I believe that overblown exhortations such as Fodor's ("Concepts are the pivot," etc.) offer neither philosophy nor psychology much benefit: in the former case, because traditional philosophical problems about concepts largely concern issues that Fodor leaves unaddressed ("How should we understand our capacities for rigorous methodology?") and, in the latter, because Fodor's rhetoric encourages a sweeping theory T syndrome picture of the doctrines a good psychologist should be setting forth. In contrary allegiance with the "the little details are often what matters most" viewpoint that has served as a prime subtext throughout this book, I predict that searching for some one-size-fits-all approach to "concepts" is likely to prove sterile. In truth, I believe that Fodor can summons cognitive scientists to his piping call only through confusing their endeavors with the standard concerns of philosophical focus.

(iv)

Pursuits of "truth". Although this book has largely focused upon the manners in which *predicate*-centered terms such as "concept" operate within our fluctuating evaluative affairs, philosophy of language over the last fifty years has mainly concerned itself, in one manner or another, with the notions of "true" and "false," as they serve to characterize the condition of *sentential* acts in allied ways. The historical reasons that explain this shift in emphasis are too complex to recount here, but the displacement has engendered a fair number of unfortunate misunderstandings. Once again hidden presumptions with respect to conceptual invariance have converted a small but useful way station along the winding trail of everyday linguistic evaluation into a permanent edifice both Grand and Mythic. In reaction, the philosophical community divides into two camps: enthusiasts who defend the grandeur and critics who denounce its superstition. A more balanced assessment of the situation falls between these extremes, but, unless we keep a fairly sharp eye on the idiosyncrasies of linguistic evaluation, we won't be able to locate the proper ground on which to stand.

Indeed, this "truth" business illustrates a basic motif that I have sounded since the opening chapter: an excellent scheme for getting rid of an unwanted difficulty in philosophy is to invite the topic for a drive through the Realm of Universals. Many a philosophical awkwardness has gotten itself conveniently lost in this manner, for the landscape seems so drab and unexciting that it rarely invites suspicious scrutiny. It is through such out-of-sight erasures that an alluring portrayal of "truth" as a resplendent Shangri-la we should seek is built up, as we find the fantasy recounted in Frege's writings:

> Just as "beautiful" points the ways for aesthetics and "good" for ethics, so do words like "true" for logic. All sciences have truth as their goal; but logic is also concerned with it in a quite different way: logic has much the same relation to truth as physics has to weight or heat. To discover truth is the task of all sciences; it falls to logic to discern the laws of truth.[22]

Today Frege's rhetorical exuberance is generally tamed into more tepid academicese, but his project still dominates philosophical discourse, whether as enthusiastic embrace or equally charged rejection. But the entire overheated affair, in my diagnosis, is but the aftermath of classical concept's furtive whisperings: its picture of firm semantic gluing goads us into addressing a seasonal evaluative emphasis as if it were a steely concern both overwhelming and all-consuming.

On the basis of Chapter 8's ruminations upon developmental pattern, Frege's contention that "discovering truth is the task of all sciences" can be supplied with a bland and unexceptionable reading, although hardly the one he intends. We have observed that, within the unfolding seasons of our canonical histories, there generally come moments when the semantic picture prevailing heretofore requires adjustment, detoxification or scrutinized reinstatement after an intervening interval of Heaviside-style agnosticism. At such times, the basic (distributed) correctness of a large stretch of established routine will not lie in serious doubt, but the margins of its applicability will seem uncertain and we will not know how many of Uncle Fred's unfocused assertions can be credited with ascertainable truth-values. In framing a new *picture*, we articulate

[22] Gottlob Frege, "Thoughts" in <u>Collected Papers</u>, Peter Geach and R. H. Stoothoff, trans. (Oxford: Basil Blackwell, 1984), 351.

generic models that supply us with a a new diagnosis of how our old sentences garner truth-values from their surroundings. As we do so, we engage in both a *description* and a *repair* of past usage, for the foremost utility of a fresh picture lies in its capacities to move our usage to improved performance. But such diagnosis represents a fallible enterprise: like the wet blanket rigorists of Heaviside's day, we may mistakenly reject a valuable linguistic extension as unsound or, as with the Roaring Twenties crowd of applied mathematicians who struggled to interpret Heaviside's lingo on the basis of Laplace transforms without distributions, succeed in improving the potential of some of it, yet fail to redeem other worthy aspects. It is typical of semantic picture constructions that they assign truth-values within their attached models to sentences via recursions upon their component parts. That is, in simple "is a dog"/*being a dog* circumstances, we utilize the easy construction pattern familiar from Tarski's logical work: beginning with a base correlation between primitive vocabulary and model (<u>e.g.</u>, *Domain*: {Tramp, Lady, Unnamed Siamese}; *t*: Tramp; *l*: Lady; *p*: Peggy; *is a dog*: {Tramp, Lady}, we recursively assign stage-by-stage truth-values to claims such as "Tramp and Lady are dogs but something else isn't" (<u>i.e.</u>, Dt & Dl & ($\exists$x)~Dx). But if we picture the language's support in a more complex way—such as the simple facades of 8,i—, the evaluation rules become more elaborate with different domains correspondent to our localized patches (so we might have: *Domain$_1$*: {Tramp, Lady, Unnamed Siamese}; *Domain$_2$*: {Peter, Wendy, Nana}; *t*: Tramp; *l*: Lady; *n*: Nana; *is a dog*: {Tramp, Lady}$_1$, {Nana}$_2$—under this picture, crosspatch claims such as "Dt & Dn & ($\exists$x)~Dx" will not qualify as well formed or supported). And if we decide the support of our usage proves as complexly supported as "red," "force" or "rainbow," our inductive specification of truth-relative-to-model may need to be very complex indeed (7,v observed that simple surface syntax is often achieved at the cost of extremely convoluted supportive readings within modern mathematics). If the effort required to construct a recursive picture of this type will not result in much beneficial improvement (as is undoubtedly the case with "rainbow"), we might very well decide to abandon our project of constructing a step-by-step picture of its workings with a simple dismissal: "Oh, just remember that this kind of talk is oddly and incompletely grounded." In so doing, we fail to bring "rainbow" talk to the higher level of accurate performance that it might conceivably achieve, but, in such circumstances, who really cares about that? (on this score we might cite Keats on the follies of "unweaving the rainbow"). But we would be very foolish to adopt such a cavalier attitude to the vocabulary utilized within Euler's method ("step size Δt," <u>etc.</u>), for the buildings we design sometimes fall down when we follow Euler's routine blindly into regions where its claims stand unsupported.

I stress again that these picture-focused endeavors are generally worthwhile only after a reasonable body of trustworthy technique has been collected, Heaviside style. But, after a large degree of mapping out "the practical go of things," further improvement in task performance usually mandates that we map out a consistent scheme of generic correlation wherein grammatical parts are systematically supported by physical traits, quasi-attributes, probabilistically averaged qualities or some allied variety of informational package (this book has stressed that the family of available

options is very wide). If we manage to reach such a sharpened reassessment happily, we will be blest with a well-understood language, at least for the time being. As such, the picture can provide valuable fresh directivities—"check for violation of a Lipschitz condition"—that can tighten our future employments into improved word/world alignment. Based upon the correlational picture we have now developed, we can ask whether a sentence proffered is true, false or ungrounded with respect to its supportive arena, whether the sentence appears within the courses of practical application or simply during one of Uncle Fred's garrulous rambles. However, we must always remember, as good mitigated skeptics, that the seeds of shaping directivities unfold in covert ways, and it can easily turn out that our splendid new semantical picture has missed some of the worthy capacities that lie latent within the usage. At such point, we must either return to the semantic drawing board or renounce our pictorial ambitions for the time being and declare with Heaviside, "It looks as if there's a lot more finding out what there is to find out to be done, so let's let the correlational picture business lapse for awhile."

Through this narrative of seasonal occupation, we might concede to Frege that "all sciences have truth as their goal," although we should quickly add that model rail-roading, home gardening et al. suit our playbill equally well, for those worthy topics display optimizing propensities every bit the equal of theoretical physics (hobbyists also wish their linguistic recipes to work out well). But now let us insert a dash of classical invariance into the story just told and observe how rapidly modest human project hardens into the enterprise of occult majesty that Frege describes. Immediately swept aside is all of our cautious recognition that a diagnosis in terms of semantic picture represents but a localized tool that we utilize, often with imperfect results, to reorient a well-established usage along a somewhat improved course. Once classical invariance has cast its rigid glue over our proceedings, poor wandering words become solidly indentured to ghost attribute masters and the sentences that qualify as "true" or "false" become locked in place as well. The judicious tinkering and exploration that language requires to gradually tighten its hold upon the world is now denounced as craven disloyalty: words have their proper meanings and we evade our duties when we engage in Heavisidean dwiddling ("If you want to change the subject and pursue some other kind of truth, that may be okay but don't pretend that you are settling the truth-values that we originally specified").

Such rigidly classical thinking raises proper alarms within the pre-pragmatist ranks of Chapter 5 (as well as other forms of philosophical school), who rightly doubt that human beings have a capacity to weld words so firmly to world, at least given the comparatively small degree of initial effort that we typically consign to the task (recall the engineering travails of the interplanetary explorer from 5,iii). The most radical rejoinders—typical of pragmatists properly so-called—claim that the very notion of word/world correlation is somehow deeply incoherent. Thus William James insists:

> [T]he knower is not simply a mirror floating with no foot-hold anywhere and passively reflecting an order that he comes upon and finds simply existing. The knower is an actor and co-efficient of the truth. Mental interests, hypotheses, postulates, so far as they are bases for human action—action which to a large extent transforms the world—help make the truth they declare. In other words, there belongs to mind, from birth upward, a spontaneity, a vote.[23]

Of course, it is not easy to explain in what manner we represent "coefficients" of the truth that dinosaurs existed, but I do not propose to probe those wooly jungles further. In more recent times—following Quine and now led by Stephen Leeds and Hartry Field (although many versions abound)[24]—, a less extreme rejection of "true" as signalizing word/world correlation has emerged that downgrades the utility of "true" without cheerfully discarding the objective physical world along with the bath water of Fregean crusade. This view is commonly dubbed *deflationism* and will be discussed further below. In my appraisal, such efforts at "truth" diminishment go too far, for they neglect the term's role in the vital projects of semantic detoxification that our study of Heaviside has brought to the fore.

In fact, the rotten element within Frege's exaggerated quest is not the notion of word/world correspondence <u>per se</u>, but the hidden assumption of *classical invariance*: the idea that when we learn the meanings of our words, we have thereby settled how all matters of correct use ought to be addressed. But this assumption is both wrong and foolish. It is *wrong* because our capacities to set our macroscopic vocabulary running along entirely foreseeable paths are limited: we can baptize rabbits in backyard hutches, but not in faraway locales; we can sharpen our terminology by constructing semantic pictures, but their value is always limited. It is *foolish* because the complexities of the world require that usage display an openness to idiosyncratic adaption, as I have emphasized throughout. So the moral we should properly draw is that we can't achieve solid and useful word/world correspondence *easily*: we can't simply gesture at a bunch of colored samples and declare, "No matter where in the universe you roam, *being red* will be like this" (we engage in unwarranted tropospheric complacency when we make such asseverations). Nonetheless, we can gradually improve our semantic lot considerably if we put a lot of hard (and variegated) work into the project over a long stretch of time. Constructing a better picture

---

[23] William James, "Remarks on Spencer's Definition of Mind as Correspondence," <u>Collected Essays and Reviews</u> (New York: Longmans, Green and Co, 1920), 67.

[24] Stephen Leeds, "Theories of Reference and Truth," <u>Erkenntis</u> 13 (1978). Hartry Field, <u>Truth and the Absence of Fact</u> (Oxford: Oxford University Press, 2001).

of our language helps, but only at the margins, for the more significant engines of our advancement lie within our successful recipes and routines.

On the other hand, we should not wallow in excessive skepticism with respect to our semantic capacities. We can recognize dogs quite capably and know enough about their worldly distribution that we can confidently declare that "is a dog," in its core patch employments, has become nicely attached to a biological attribute of considerable firmness: *belonging to Canis familiaris*. But there are other common classificatory phrases—" is in full turbulence," say, and, probably, "temperature"—, whose worldly correspondents have been *roughly* mapped out, but with many uncertain spots remaining, to the degree that we can anticipate, to the point of moral certainty, that many surprising wrinkles of inadequate attachment await us with respect to those predicates, for which we have virtually no means of reliable anticipation.

In 1,x and 5,viii, I complained that exaggerated doubts with respect to word/world alignment are often driven by demands that firm evidence be shown of the "reference relationship" that binds a speaker's use of "rabbit" to *being a rabbit* (Quine's discussion of the "indeterminacy of reference" represents a well-known case in point). But this is a funny way to look at our concerns. We know a lot about the world and its contents but we don't always know as much about the workings of our words as we sometimes fancy, especially with respect to their finer details. We may know a lot of certifiable facts concerning the sundry hardnesses of rocks, steels and plastics, but not understand adequately how the evaluative term "is hard" itself manages to codify physical information with the benefit of an almost invisible honeycombing of boundary line cracks (just as we can know a lot about viruses without adroitly grasping the workings of an electron scanning microscope). Through the dedicated efforts of the Samuel Williams and David Tabors of the world, we have learned detailed facts about each side of the semantic equation for "hardness," through whose benefit we can now frame sharper recipes for choosing materials within engineering applications. Tabor and Williams made their discoveries by directly studying how our linguistic determinations of "hardness" correlate with supportive physical structures. Likewise, if we have learned of Heaviside's procedures within Edwardian times, we can still be certain that a broad swath of his strange sentences must be *somehow true* because their distributed correctness is so evident. This conviction will motivate us to call upon the gaggle of applied mathematicians whose devoted labors, after forty years of struggle, eventually uncovered several forms of directly supported "truth" for Heaviside's syntactic excursions. Plainly, we can sensibly wonder, "What makes his assertions true?" without presuming that some prior semantic act has welded their grammatical parts firmly to the world everywhere or that there must necessarily be a unique way in which their supportive correspondence can be diagnosed (as the alternative proposals of Mikusiński and others demonstrate (8,ix)).

In other words, the word "true" displays tergiversations of appraisal that are largely consequent upon the same vicissitudes that macroscopic predicates suffer: "true" adjusts its evaluative focus with respect to sentences according to the same directive winds as blow their component predicates hither and yon. During the more agnostic stages of its career, "truth"'s appraisals will be adjudicated largely on the basis of the distributed correctness of practical routines, but these determinations can be expected to shift eventually to picture-based assessments as the semantic moment seems ripe. Ill-advisedly compressing these natural, but mercurial, moods into an invariant sentential personality only conjures up strange semantic marvels: the floated overbearing tyrant of the Fregeans or the starved wraith cherished by deflationists.

(v)

A logical chicken or egg. To render flesh on these bare bones, let us rehearse a dilemma that brings forth dichotomist attitudes nicely. Since the 1930s, a self-styled "semantical approach" to logic has become popular that follows the lead of the logician Alfred Tarski (building, in turn, upon concerns implicit within the work of Kurt Gödel). Here a range of permissible mathematical structures are framed in the basic "is a dog"/ *being a dog* manner displayed in the previous section (i.e., the structure with the dogs and cats in it) and, upon their basis, standard forms of logical argumentation are evaluated for their generic *soundness* (that is, conclusion S legitimately follows from premises A, B, C . . . just in case no permissible structure makes all of A, B, C . . . true, yet renders S false). As a simple illustration of how an inferential rule can be validated by such methods, consider the common device of "reasoning by cases":

> Premises : (1) A or B holds; (2) A implies S; (3) B implies S. Conclusion:
> S must hold.

Here is the kind of "soundness argument" commonly supplied on its behalf in logic primers:

> *Why? Assume, to the contrary, that 𝓢 is not forced to be true, but that, instead, a structure S is possible where A or B and not S are both in 𝓢. But we know that if A holds in any situation, then S must hold there as well. So A cannot be true in 𝓢, because, otherwise, S would need to be true there too. But since A or B holds in 𝓢, then the truth of not A forces B to be true in 𝓢. But B's truth then forces, by premise 2, S to be true in 𝓢, which contradicts our assumption. Hence, we have no choice but to conclude that S follows from (1)–(3).*

To a writer like Michael Dummett, such argumentation forms a central part of a "theory of meaning" for a language, which he explains as follows:

> *[A] theory of meaning must provide for every component of our practice in the use of language an understanding of the way it works: we seek, not merely a description of our practice, but a grasp of how it functions. A semantics in terms of which a given fragment of logical theory can be proved to be sound . . . supplies an answer to the question: How must our language be conceived to work—what model must we have for the meanings of our sentences—if the practice of deductive inference in which we engage is to be justified?*[25]

---

[25] Michael Dummett, "The Justification of Deduction" in Truth and Other Enigmas (Cambridge, Mass.: Harvard University Press, 1978), 311–12.

And further:

> One particular aspect [of meaning] will be taken as central, as constitutive of the meaning
> of a given sentence . . . ; all other features of the use of the sentence will then be explained by
> a uniform account of their derivation from that feature taken as central.[26]

In the case of Frege's picture, Dummett believes that the story we provided with respect
to argument by cases constitutes a part of how a *truth-functional* theory of meaning for
the language would work, because at the center of our structural description lie
recursive rules that stipulate how the truth-values of compound sentences depend upon
the truth-values of simpler parts (or associates, if quantifications are involved). As such,
if we know the truth-conditions that determine when the atomic parts of speech (e.g., a
predicate like "is a dog") applies to a target object or not, then we will understand the
truth-conditional requirements for every sentence in our language. Such a perspective is
entirely satisfactory to Russell,[27] for he would merely add: "Yes, and we understand
those base conditions through grasping the universal *being a dog*." Classical gluing, as it
was explicated in Chapter 3, provides a nice exemplar of what Dummett has in mind by
a "core conception of meaning": it tells us a crisp story of what we must learn to
"become competent in the use of a language." The subsequent justification of inferential
principles through soundness proof then serves as Dummett's central illustration of how
we should expect "to derive a subsidiary feature of use from a core conception." To lay
out the stages in this "derivation" explicitly, it will follow the following stages. (1)
Predicate "D" is assigned, as core associate, the truth-conditions implicit within the trait
being square. (2) From (1), we can extract, as a minimal bare husk, the fact that "P"
possesses an associated extension α, it matters not what it is. (3) Based upon the fact P is
aligned with a set α and similar data with respect to other vocabulary (i.e. that "t" aligns
with an object), we frame our recursive rules for evaluating the truth-conditions of
arbitrary sentences S with respect to generically selected structures $\mathcal{S}$ (where $\mathcal{S}$ is
unpacked as a Tarskian domain and interpretation structure $<\mathcal{D}, \mathcal{J}>$ in the manner of
our dog and cat domain example). (4) A proposed inferential rule $\mathcal{R}$ (such as argument
by cases) is then examined as to whether it is generically truth-preserving over all
allowable $\mathcal{S}$ in the manner of the soundness argument provided. If $\mathcal{R}$ passes this generic
test, we conclude that it represents an acceptable logical principle (on the grounds that,
no matter what the facts about dogs and Tramp are like in the real world, we can safely
trust the conclusions reached by $\mathcal{R}$). In short, Dummett expects that a characteristic
derivational chain will canonically descend from a "theory of meaning": "core meaning"
assignments are provided for all predicates and names relative to allowed structures
$\mathcal{S} \rightarrow$ truth-value for all sentences are defined relative to $\mathcal{S} \rightarrow \mathcal{R}$-rule inferential validity
is induced by generic soundness over all $\mathcal{S}$.

Let me hasten to add that, according to Dummett, philosophers might favor a
truth-functional theory of meaning even if they do not subscribe to full classical

---

[26] Michael Dummett, Frege: Philosophy of Language (New York: Harper, 1973), 456–7.

[27] It is unclear whether Frege himself looked upon linguistic process in quite this way, possibly favoring an approach
closer to that of the neo-Kantians. However, I will follow Dummett's interpretation of Frege's intent here.

gluing of a Russellian stripe. For his own part, Dummett believes that such truth-functional stories make excessive demands upon the capacities of human employers and favors a replacement "core theory," a point to which I shall briefly return. Let us first observe, before we turn to the deflationists, that Dummett's requirements automatically install 1,vi's *semantic finality* within any acceptable "theory of meaning." Few "truth-functional theories of meaning," as they appear within the philosophical literature, provide many details as to exactly what we manage to grasp when we "understand the truth-conditions for *being a dog*," preferring to devote most of their attention to the structural considerations that relate compound sentences to simpler forms. The prevailing attitude is generally: "I will lay out the schematic framework into which 'is a dog' falls and leave it to my readers to amplify its specific contents a little bit." This omission of task is not an oversight; most neo-classicists would blanch if pressed to supply substantive details as to what might comprise the a priori "truth-conditions" relevant to "Tramp is dog" (vide 3,vi). In any case, the notion that a semantic picture represents a framework to be fattened up with a few particularities arises, in my opinion, from a fundamental misunderstanding of the good works that semantic pictures provide us. Under our more modest and seasonal diagnosis, *skeletal details* are precisely what a semantic picture hopes to provide, for through their means we diagnose the *generic patterns* whereby words meet their worldly supports de facto. And often we can achieve quite a bit along this line without managing to codify at the same time all of the directive factors responsible for bringing such alignments into position. Indeed, we can plainly locate an adequate rigorizing support for Heaviside's calculus without understanding much about the stray wisps of directive whimsy that have gradually prodded Leibniz' original "dy/dx" notation into assuming the great electrical accomplishments it reveals under Heaviside's hand (like the meek tailor who eventually slays giants). And we should view the full story of "is a dog"'s "content" in the same way: its accounting requires a hefty narrative that shouldn't be squeezed into the wee box of the base step of a Tarski-style recursion.[28]

Indeed, deflationist criticisms of Dummettian demands often begin with the blunt observation that it is silly to suppose that we have provided *any* useful information about a predicate by intoning, in sonorous voice, "To understand the predicate 'is a dog,' we must grasp the truth-conditions for *being a dog*." And they continue this "the Emperor has no clothes" critique of truth-functional accounts of meaning by complaining about the vacuousness apparent in the "soundness proof" we offered earlier. "That argument you advanced on behalf of argument by cases is patently circular. It employs logic to argue in favor of logic in the same vein as we might 'prove' that $2 + 2 = 4$ by:

> Clearly $2 - 2 = 0$. Adding $+ 2$ to each side, $2 = 2$. Adding $+ 2$ once again, $2 + 2 = 2 + 2$. But the right hand side is clearly 4, so $2 + 2 = 4$.

[28] My complaint bears some relation to those registered by Hartry Field in his well-known "Tarski's Theory of Truth" (reprinted in Field's Absence of Fact), although I resolve the tensions in a rather different manner.

"But plainly," our critics continue, "we've assumed *as a rule* what we intended to prove in the last step; the rest is merely window dressing. By equal rights, you might as well have shown argument by cases sound in the following manner:

> *Why? Consider any possible situation 𝔖 in which A or B holds. That fact supplies us with two basic cases to consider. (case 1): A is true in 𝔖. By premise (2), S will be true in 𝔖 as well. (case 2): B is true in 𝔖. Premise (3) then tells us that S is also true in 𝔖. Either way, S has to be true in 𝔖 and that exhausts all the situations we need to consider.*

"But this 'proof' patently appeals to argument by cases to justify its own worth. What can the utility of *that* exercise be?"

So what form of evaluation do we offer when we declare a sentence to be "true"? Deflationists provide a novel answer: adding "is true" after a claim S is essentially equipollent to decorating S within supplementary filagree or rephrasing an active assertion as a passive. Such notational bric-a-brac is not entirely useless because it can be employed for various purposes. If I have branded all my cows with numerals, I can instruct the foreman to round up only the dogies that bear prime numbers. Just so, I can exploit the "is true" filagree to make assertions like "Everything Nixon believes is true," for the implausible contention that I agree with Nixon on everything is not conveniently rendered otherwise. From this point of view, the so-called Tarski biconditionals (sentences that match the schema "φ" is true if and only if φ) seem as if they must be largely constitutive of "true"'s utility, for they tell us when to add the filagree and when to take it away.

More generally, deflationists scoff at Frege's "pursuit of truth" mission as gaseous rhetoric: "Yes, we change our minds about sentences all the time as we meander through life, generally obedient to accepted principles of scientific methodology and the like. But we will have taken to chasing our own shadows if we fancy, following the truth-functional vision, that we are thereby discovering which sentences were pinned to the world as correspondingly 'true' long ago, from the first moments when the inventors of English hammered out our native tongue around the dining tables of Camelot." If asked about the status of argument by cases, our dissenting sect usually maintains that it, and the other principles of logic, simply constitute syntactic manipulations that we follow as part of scientific methodology, to which any rational discourse must submit if it is to be judged as coherent (if our deflationists are Kantian-inclined, argument by cases is then described as a norm or regulative principle). "The emptiness of the soundness proof you provided demonstrates," they continue, "that our lot in life is that of being irredeemably trapped within a *logocentric predicament*[29] wherein we must embrace logical rules as valid before any consideration can be given to their supportive underpinnings."

Both parties to this dispute have misunderstood the real, but limited, value that soundness proof evaluations play within the unfolding histories of predicative

---

[29] The phrase apparently derives from the American logician Henry Sheffer. See also Lewis Carroll, "What the Tortoise Said to Achilles" in James R. Newman, ed., The World of Mathematics, iv (New York: Simon and Schuster, 1956). Not all modern deflationists take such radical stand, however.

vocabulary. This oversight is the consequence of considering logical specimens in isolation from their cousin projects in *rule ratification*, such as scrutinizing Euler's rule to learn its proper range of applications (4,xi) or propping up the operational calculus through Laplace transforms and distributions (8,ix). As I have continuously stressed, such studies also represent examinations of how inferential rules $\mathcal{R}$ behave with respect to generic associated models of their word/world correlations. Nobody would fancy that such probes are trivial or uninformative, yet they utilize "true in $\mathcal{S}$" determinations in exactly the same manner as our logical examination. The "soundness proofs" beloved of the logician are entirely of the same character as the "examinations of correctness" that applied mathematicians provide in a wide range of contexts.

As we observed earlier, the relevancies of such evaluations are not independent of one another. A simple "is a dog"/*being a dog* picture of how the logical connectives operate may need to be abandoned when our portrait of how the stronger forms of inferential principle work within the language requires adjustment. For example, having unmasked the semantic mimicry described in 9,iii we are now obliged to recognize that narratives formerly interpreted in marching method mode must be regarded as implementing a successive approximation approach to a steady state problem instead. This shift in picture requires that previously freely interpretable statements of the form "A leads to B" (or even "A causes B") need to be contextually interpreted according to sequential position within the discourse. If so, Uncle Fred can no longer be allowed the liberty of extracting claims S_1 and S_2 from different stages of our narrative and combining them thus: "S_1 and S_2." Under the old marching method picture, Fred's freely exported combination qualifies as permissible, if pointless, but the procedure must be now renounced as incoherent under our new recognition. This discovery is quite analogous to learning (from the $+2 = -2$ paradox of 6,vi) that Uncle Fred cannot freely combine claims situated upon different sheets of the Riemann surface for $\sqrt{z}$. In this sense, the trustworthiness of simple logical rules such as "from statements A and B, conclude the compound 'A and B'" stands intrinsically hostage to the pictures we provide for the non-logical aspects of our usage.

Contrary to the globalist picture of methodology that Quine evokes (and wholeheartedly accepted by most logocentrists), the familiar principles of standard logic do *not* represent directive principles never to be abandoned except in exigencies of last resort. Instead, they represent rather feeble junior partners whose fortitude immediately gives way at the least indication of semantic trouble elsewhere. Such flabbiness does not indicate that logical principle has been proved *wrong* exactly, but it does show that such dictates are easily rendered *irrelevant* (the situation is as if logic has been assigned a fancy badge, "Supreme Potentate of Scientific Inquiry," yet none of its ministers pay a lick of attention to its instructions if they don't suit their own ambitions).

It is in this context that the proper salience of our soundness proof for reasoning by cases should be addressed: our examination lays down the prerequisite conditions to be sought within the more detailed pictures that we provide for our more powerful reasoning rules. If we can't align "D" and "t" with simple sets and objects inside those models, we must be wary of following logical impulse willy-nilly across this language (which is not to say that it isn't often desirable to *reformulate* our surface usage so that logical principle can be freely employed once again, as the fine print of 7,v observed). To articulate this same point in another fashion, our soundness proof sketches the *minimal structure* that we must locate within a broader scale picture to be able to announce, "Whew! At least we won't have to worry about failures of logical reasoning here."

Obviously, classical conceptual invariance is primarily responsible for the mis-understandings we have surveyed. Dummett decides, "The attitudes we adopt towards logical reasoning must be based upon what we truly grasp from our semantic begin-nings, lest logical principle seem unreliable afterward" (I will return to this fear that "logic might crumble" in a minute). But these prerequisites seem rather strong; indeed, Dummett himself doubts that we can truly lay down firm truth-conditions for all our predicates <u>ab initio</u>, roughly, because of misgivings similar to those expressed under "pre-pragmatism" in Chapter 5. Unfortunately, the unhelpful imp of classical invariance whispers in Dummett's ear that he should thereby scale back his faith in standard inferential principle until it rests upon a safer "theory of meaning" more congruent with our actual capacities, such as understanding procedures for verification. And by this ill-starred route, sturdy inferential helpmates such as argument by cases soon fall under house quarantine, except in feeble situations.

In wandering down this unhappy path, Dummett is much influenced by Brouwerian intuitionism, which may *possibly* have its merits within "pure mathematics" (tho' I doubt it), but is plainly disastrous within applied science. In the course of building up his favored techniques, Heaviside employs argument by cases extensively in his explora-tions. True, he will later need to prune some of logic's potential excursions over his language, but, at the present moment, he can't determine where they lie. He is sur-veying an ill-understood landscape and mustn't allow his explorations to be hamstrung by drastic restrictions upon argument by cases (it is unlikely that the prohibitions that will later emerge can be articulated in purely logical terms). For now, Heaviside needs to pursue lines of thought $\mathcal{L}_1$ quite strenuously, while eschewing the apparently similar $\mathcal{L}_2$, for no better reason than the $\mathcal{L}_1$ seem to lead to better results. But if he abandons every logical rule known to lead to trouble on occasion, he will be scarcely left any inferential machinery with which to make his forays (given his opinions with respect to the strictures of "Cambridge mathematicians," imagine what he would have made of those exacted by the intuitionist!). Indeed, in most of the relevant cases examined in this book, it is the simple rule, "from A and B, conclude the combination 'A and B'", that even-tually proves unsustainable on a global scale. But we discover that fact by first allowing the conjunction principle its full head of steam, and then discovering where the cross-cut prohibitions that tame the ensuing "$+2 = -2$"-style paradoxes must be placed. The notion that we must somehow set up logic so that application of its principles will never "crumble" in this manner is utterly unrealistic. But that is the cul-de-sac into which classical invariance has tacitly lured us.

Indeed, after many scary stories heard around the classical campfire, the threat that "logic might crumble" has become an imaginary bogeyman that has chased innumer-able philosophers into positions far beyond the circle of common sense: "Why, if you hold *that* doctrine, then logic will not prove inviolate." Well, "inviolate" can be here supplied with different readings, but we plainly prime ourselves with poor advice if we fancy that we must sternly hew to the conjunction rule's syntactic imperatives, damn whatever "$+2 = -2$" paradoxes come along. Such unwise admonitions trace, I believe, to the root fear that the possibilities of "rigorous science" will be lost if we do not steer the vessel of language firmly onward according to a predetermined schedule. But our observations with respect to open texture argue that such injunctions do not represent wise navigational policies with respect to the linguistic river in which we find ourselves; we should expect to control the wigglings of our barges only at the margins, and allow the currents of natural improvement to do the rest. "But how can we know where we are going?," demands the philosopher frightened of the bogeyman. Well, we may not know precisely, for that is largely an arrangement that Mother Nature and our evolving language work out between themselves, and they may not alert us to their decisions until we are far along to sea. However, our trip will be much happier if we heed Oliver Heaviside's advice and relax our navigational preconceptions somewhat. From this vantage point, Dummett and his critics have both adopted the topsy-turvy emphasis of which our muse rightly complains:

> And no doubt the logic of it all will have to be found out experimentally. And then, finally, I suppose "rigorous" mathematicians will put the logic at the beginning, and pretend they knew all about it before they began.[30]

....

The criticism offered of Dummett can be expressed in an alternative way. Like it or not, the chief structures found in the physical world appear to embody the features of real-valued manifolds in some way or other. In scrutinizing reasoning rules within physics for soundness, we hope to learn whether the rules will fail with respect to manifolds of *that type*, no matter how unacceptably Platonistic their structures may appear to a doubting intuitionist. "I don't care about safety in *your* sense," we should retort. "I want to know if my buildings can possibly fall down and I want to rely upon models that capture the potentially threatening *physical situations* as directly as as possible."

....

(vi)

The critical role of "truth-condition." Within our canonical developments, directive influences that have hovered over a predicate heretofore can be screened, rejected or supplanted in two different fashions. (1) We decide, on the basis of Heavisidean *utility*, that old-fashioned strictures represent unnecessary constraints upon the profitable employment of our target phrase. (2) We determine, following the guidance of a reformulated *picture*, that its usage can be extended, but at the price of more delicate

---

[30] Heaviside, Electromagnetic, iii. 370. Heaviside is discussing the strange manner in which different varieties of divergent expansion cover a domain in the sectoral fashion described in 6,vii.

restrictions upon inferential principle. In both cases, we report these discoveries within the argot of "correct concept use": "A proper understanding of the concept of *derivative* must regard the manipulations of the operational calculus as *generally correct*"; "A proper understanding of the concept of *derivative* must regard some of Heaviside's inferences as sometimes *incorrect*." These two claims are compatible, of course, but we commonly establish the first conclusion during an interval of semantic agnosticism, whereas the latter generally stems from a developed correlative picture as its source. In both circumstances, we rid our usage of accumulated burdens that have compromised its utility in its prior stages of development.

As noted earlier, the practical motive that drove Hertz and his comrades into the arms of syntactic instrumentalism (4,iii) is precisely that of ridding mechanics of constraining heritages of this breed (I have cited Boylean requests for "mechanical understandability" as an example, but 10,viii will outline the concerns more directly pertinent to the late nineteenth century). But an instrumentalist approach to word semantics will prove excessively confining, for no stern rules of a syntactically precise ilk are likely to suit macroscopic predicates ably. Indeed, we often discover those failures in rule-governed behavior precisely through an investigation of their performance over a class of anticipated structures (as Euler's method was scrutinized in 4,x).

Thus Hertz <u>et al.</u> err in mistaking a seasonally important activity of directive winnowing for a revelation of the "true meanings" that support scientific predicates. "No, they require substantive worldly underpinnings to display an appreciable measure of practical success," we aver, in accordance with the determination that any profitable method must have its logic. It seems to me that Dummett's vision of a "theory of meaning" suffers from the same narrowed focus as plagues our instrumentalists, but this time in application to our second category of semantic evaluation, those centering upon conformity to a semantic picture. This new emphasis also makes the "contents" of our predicates seem strangely thin, for most of the recognitional and inferential abilities that allow us to employ "dog" or "hardness" with everyday profit have been dropped from view as not especially relevant to their proper correlational picture (from its stern point of view, our practical capacities are generally somewhat inaccurate).

In short, both forms of evaluative sifting (1) and (2) serve important purposes, but merely as seasonal strands within the fuller mixture of fibers that supply our terminology with strength and allow it to stretch into an improving future. But we should not expect to reconstitute the complete cable by dilating into artificial fatness any particular thread that we encounter along its composite entirety.

Our inclination to favor semantic doctrines of a monotheist cast traces, I believe, to our *ur*-philosophical loyalty to classical invariance, partially because of the false linguistic controllability it promises. But "concept" and "truth"'s own behaviors play a role in generating such spurious semantic projects as well, because, at different junctures within a predicate's career, concerns about syntactic rules and truth-conditions are likely to become locally dominating, to the exclusion of any other interest. This leads "concept" and "true" to *speak* in strong absolutes, but the standards they actually *apply* are seasonal and shifting in their foci. During those occasions when framing a better correlational picture of the information encoded within a usage is imperative, our attention concentrates upon diagnosing the alignments evident within our profitable patterns of speech, as well as recommending, as best we can, a course of better implementation and improvement for the rough correlations we discern. Here our primary loyalty is to greater success, not to the complex factors that have brought us to this brink. Applied mathematicians often come to the aid of, e.g., gunners who have worked out some imperfect Euler-like algorithm through trial-and-error tinkering and then ask for improvements upon their routines. To provide this help, the reforming mathematician rarely needs to learn a lot about ballistics, but simply needs to be shown the old routine and provided with a rough sense of the basic factors salient within the environment. In contrast, we almost certainly won't be able to appreciate how our crew *discovered* their routine without substantial knowledge of the practical vicissitudes of artillery gunnery, but these richer considerations do not impinge greatly upon our mathematician's narrower task. Accordingly, as we seek to *improve* linguistic performance, the formative currents that have advanced our usage to the stage of meriting such appraisal commonly drop from view as either unimportant or worthy of jettison.

In short, the generic accounts of correlation that emerge from a typical investigation of rule soundness will seem oddly thin and incomplete if they are mistaken for a complete delineation of the "semantic contents" of their component predicates, for the central engines that drive their extending applications have been largely set aside (innocent observers are rightly puzzled when philosophers inform them that the "semantics" of "is a dog" can be wholly captured by a *set*). And the missing factors cannot be restored to our skeletal account simply by fattening up the information provided at the base vocabulary level a bit, for such attempts fundamentally misunderstand the extremely varied ways in which directive influences can act upon evolving words. Indeed, the interfacial factors of environmental accord can hardly be expected to fit on any recursive frame, for these shaping influences commonly act upon an evolving language at the supra-sentential level of what 5,ii calls *strands of practical advantage*. Our study of the hidden hand of design imperatives (8,xiii) reveals that we are often oblivious to such shaping influences, even though their directivities affect our employments with considerable salience. So we commonly encounter situations where linguistic agents believe, quite fervently, that their usage stoutly adheres to simple "is a dog"/*being a dog* alignment, when, in fact, the stronger undertow of practical adaptation has long since carried their craft to some far different section of river. Is it therefore proper to claim, "We understand the *truth-conditions* of 'is red'?" Or "is hard"? Or "is weightless"? The answers we supply will generally depend upon seasonal considerations: in some evaluative moods we will cheerfully announce, "Yes" and, sometimes, "No, not really." A "theory of meaning" in Dummett's sense will scarcely satisfy us, because it omits many of the central factors that contribute to overall "predicate personality," as we encounter it along the boulevards of everyday chatter. And, really, we don't frame a semantic picture in order to *capture* the full directive richness that surrounds a term; we use it as a tool for *eliminating* some of the portions we no longer want.

Accordingly, we can concur with our radically skeptical friends, the deflationists, that visions of "truth" cast as the Holy Grail of rational inquiry are chimerical and that "grasp of truth-conditions" cannot represent the chalice in which the fullness of language can be completely poured. However, our deflationist allies seem strangely unappreciative of the limited, yet real, benefits that reconsideration of semantic picture can bring, wherein we genuinely manage to tease out the correlational supports that underwrite the ascriptions of "true" and "false" we assign to a usage, in a diagnostic manner that reaches far beyond the "attaching an 'is true' filagree to S" terms that they tolerate. There are seasons in the life of a predicate when working out a correlational account of its underpinnings becomes vital and it is in these moments that our evaluative token "true" truly emerges into its natural element.

We observed that deflationists typically insist that claims of the form (it is usually called the Tarski T-schema) are wholly and <u>a priori</u> inviolate for any sentence φ:

"φ" is true if and only if φ.

But this policy is either unwise or irrelevant during the epochs in which an old picture need to be detoxified and rebuilt—we no longer trust any old sentence φ that we can frame in our old vocabulary. Instead, we usually attempt to first quarantine those stretches of usage in which a problematic predicate appears with indubitable value (e.g., within 5,ii's strands of practical advantage) and then gingerly investigate which patches and lines of connection seem reliable outside these narrow walkways. Thus, in pondering Heaviside's mysterious "1/dt," we first compile a catalog of "1/dt" facilitated computations and declare, "Okay, this is the target class we must be able to redeem; how far can we safely wander away from its proscribed patterns?" As we do this, we usually select "S is true" as our natural means of announcing that S appears to possess sound supportive foundations: "Okay, it's clear that '$f(1/dt)\varepsilon^{at} = f(a)\varepsilon^{at}$' always needs to be true." And we might reach such decisions on various grounds. Thus:

(1) Given that "$dy/dt = t^2 - y$" holds, then the claim "$y = t^2/(d/dt + 1)$" must qualify as true as well, because it is central to all of Heaviside's successes.

as well as:

(1) Given that "$dy/dt = t^2 - y$" holds, then the claim "$y = t^2/(d/dt + 1)$" must qualify as true as well, because it means that $\mathcal{L}(y) = 1/(s + 1)\mathcal{L}(t^2)$ where $\mathcal{L}$ is the Laplace transform and where Heaviside's initial conditions are assumed to apply.

In (1), we have put "$y = t^2/(d/dt + 1)$" into our "true" column because of its *distributed correctness*, whereas, in (2) we have begun to explore which Heavisidean sentences obtain a *direct support* when evaluated within a semantic picture based upon the Laplace transform (this use of "distributed" and "direct" support is introduced in 4,v). As I've often emphasized, these two criteria for truth naturally shade into one another, because a calculational routine is unlikely to produce reliable results unless its component sentences carry valid physical information under some (possibly elaborate) coding. Our direct story (as in (2)) simply tries to tease out a coding that can support this algorithmic correctness (as we've seen, there needn't be a unique method).

In using "true" and "false" thus, we certainly *seem* to be sorting out sentences according to their capacity to correlate with supportive fact, under the assumption that their component parts will generally correlate with genuine information as well (there can be spare wheels in the mix as well, analogs to the grisly accretions in which the Japanese blacksmith of 5,ii envelopes his practical recipes for sword manufacture). True, we don't expect to see the tight correlations everywhere promised by classical gluing; on the contrary, it is likely that large patches of electrically oriented sentences expressible in Heavisidean phraseology are completely unmoored in any consideration presently available to us (large stretches of non-linear expression, for example). Whether those lapses can be eventually filled in or whether their employment should be barred in the future remains to be seen. In short, we appear to be investigating correlational circumstances that lie somewhere between classical promise and deflationist denial. And this is the territory in which, in my opinion, the evaluative term "true" really earns its oats.

At such moments of critical assessment, if asked, "Do you wholeheartedly embrace the Tarski biconditional:

$$\text{`}y = t^2/(d/dt + 1)\text{'} \text{ is true if and only if } y = t^2/(d/dt + 1)?,$$

I would say, "Gee, I don't know because I'm not yet certain whether its right hand side makes any sense or not." And I might plausibly say that even if I had utilized that very sentence in a calculation ten minutes before. If challenged on this score, I might plausibly reply, "That was then and this is now. It is now time to scrutinize that old usage carefully and so I must bracket my former employments carefully, as I try to probe which of their workings are genuinely supported. Surely you don't expect me to remain loyal to every scrap of language I have uttered at any time in the past?"

In allied spirit, if we ask, "What are the conditions required to make Heaviside's $y = t^2/(d/dt + 1)$' *correct* within a derivation?," we anticipate an answer of substantive character—e.g., "Just in case $\mathcal{L}(y) = 1/(s + 1)\mathcal{L}(t^2)$." To be fobbed off with the deflationist rejoinder—"Just in case $y = t^2/(d/dt + 1)$"—will strike us as either a lame joke or as displaying utter incomprehension with respect to the ills to which potentially unmoored language is prone. As we noted, deflationists insist that " $y = t^2/(d/dt + 1)$' is true if and only if $y = t^2/(d/dt + 1)$" qualifies as some form of *truism* about language, but I'm not familiar with truisms that run great risk of proving senseless. Might "As*^vh try7&" represent a truism?

The feature that obscures this simple observation lies in the fact that contemporary philosophers of language generally concentrate upon language during its most unruffled moments, where no suspicion of anything rotten in its semantic underpinnings disturbs the pastoral scene. I might possibly allow that, if language usage could linger long in this cheery and static condition, most of the employments of "true" we might witness therein would conform docilely obedient to deflationist pattern (I have not investigated this claim seriously, as it strikes me as unverifiably counterfactual in its asseverations). Nonetheless, our studies suggest that macroscopic descriptive language rarely remains so dormantly placid, although we detect this turmoil most transparently when descriptive predicates are put to sharper performance challenges than are commonly demanded within philosophy's lecture halls. It is during these higher pressure junctures that words like "concept" and "truth-condition" emerge from their deflationist slumbers and supply the natural argot we utilize to express, in a contextually sensitive manner, the discriminations of linguistic support required to cure our muddles and advance our precisions.

When I have asked deflationists how they regard such enterprises of critical evaluation, they sometimes reply, "Oh, but those sorts of concerns don't reflect the proper meaning of 'true,' which comes completely encapsulated within the Tarski truisms we have provided." How they ascertain this crisp demarcation in descriptive task eludes me, although they sometimes gesture hazily towards the "ways in which we originally learned the term 'true'" (these suggestions avail me little, for I don't remember my parents once saying, "'S' is true if and only if S," although I *do* recall being informed at a tender age: "It's actually the sunlight reflected in raindrops that makes statements about rainbows true"). In fact, such off-handed apriorism rests upon tacit assumptions of invariance and semantic finality which, I have argued, represent exactly the sorts of unhelpful doctrine we have inherited from classical tradition and ought to discard if we hope to understand natural linguistic process better.

To be sure, more tolerant deflationists might make room in their accounts of language for studies of the sort I highlight here, although, for reasons that I don't understand, they are rarely willing to accord them the philosophical honorific of "semantics." Similar forbearance, of course, is not offered by the larger population of "veil of predication" deflationists who regard talk of word/world correlation as inherently incoherent in the manner sketched in 2,vi. However, both schools undervalue the main observation advanced here: if we hope to bring an established use of language to a higher standard of task performance, it is virtually inevitable that we must traffic in non-trivial correlational investigations at opportune moments. Complex and incomplete word/world alignments come with the territory in which "true" and "concept" live and I don't see the purpose of pretending otherwise. We may swaddle infant "true" in comforting truisms, but, sooner or later, it must face the harsh adult world of correlational complication.

. . . .

Just as this book was to be shipped to the typesetter, Penelope Maddy kindly showed me a draft section of a manuscript (tentatively entitled Second Philosophy) where she discusses many of the ambient flavors of deflationism in useful detail (to prevent her own projects from inflating into enormity imitative of ours, she seeks a rationale for *sidestepping* many of our key issues). From this survey she extracts a purged variant of deflationism that is not at all hostile to the correlational studies defended here. Indeed, this more genial creed is willing to embrace them all, only not under the headings of "truth" or "truth-value" (Hartry Field has sometimes employed the phrase "indication relations" to roughly cover the kinds of correlational considerations I have emphasized, although I believe this term captures their purposes rather poorly). Insofar as I can determine, this genial doctrine represents more Maddy's own concoction than anything to be found in her nearest avatars (Field and Stephen Leeds, who occupy holist positions closer to Quine's). I am pleased with this propitiation in project, although I see no especial reason why the generalization-forming virtues of "true" should enjoy some higher semantic status than our correlational tasks. In fact, the considerations that historically motivate a contrary partiality unfold roughly as follows: "If 'truth' and 'truth-value' are to prove useful terms in delineating the architecture of language, their contents must be *implicitly defined* by some background framework of theoretical assumption. But this tacit 'theory' may advance strong presumptions about human capacity that are incompatible with any naturalistic account of our behavior (indeed, contrary sentiments in this vein are quite common). But such anti-scientific pretensions must be defanged if we are to maintain a consistently naturalistic portrait of our universe." For a period, figures like Hartry Field nurtured the hope that our problematic notions might be successfully tamed through some kind of causal recounting, but, as the prospects for this enterprise dimmed, it seemed more prudent to scale our everyday evaluations of linguistic "truth" into a quite minimal activity, leaving room for some wiser science of "indication relations" to grow up in its place.

But our facade models for "hardness," "force," *et al.* demonstrate that many descriptive terms adjust their applications to local circumstances quite adeptly, without benefit of sweeping "theories" as backdrops. The same observation holds for "truth" and its kinfolk. Any errant inclination on their part to outstrip their naturalistic upbringings needn't be opposed by restricting their dominions of valid application so drastically; we should instead reject the tacit preconceptions about "theory" that tick ominously within the innocent looking wrappings of "a theory of truth." More generally, the would be "naturalist" is well advised to walk softly and carry a lot of facts, rather than brandishing dubious weapons of an Ockham's razor ilk.

. . . .

As we noted, the road that leads to the framing of a satisfactory picture can begin by pursuing clues found virtually anywhere, ranging from a direct microscopic examination of the physical events that occur within indentation impression to an almost purely syntactic survey of coherence across a field of grammatical sentences. Before moving onto other topics, I'd like to mention a rather pure case of the latter, as it sheds considerable light on how "truth" gets utilized within non-trivial evaluative discourse.

The case I have in mind is that of William Rowan Hamilton (the Irish mathematician, not the philosopher discussed previously) and his efforts to invent "triples." Among their many virtues, regular complex numbers such as $3 + 2\sqrt{-1}$ (which we will henceforth write as $3 + 2i$) allow us to express two-dimensional coordinates on the plane

	1	i	j
1	1	i	j
i	i	-1	?
j	j	?	-1

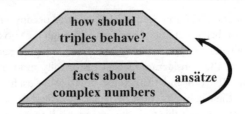

(thus $3 + 2i$ can be viewed as designating the location $<3, 2>$). It was natural to hope for a three-dimensional analog relying upon a second root of -1, which Hamilton designated as j, expecting to deal with quantities such as $3 + 2i - 6j$. However, it is not immediately clear how such numerals should be multiplied and, for this, Hamilton needed to figure out how to complete the unit multiplication table for 1, i and j. If we simply guess at the ?'s marked, we will soon be disappointed. So Hamilton attempted a more sophisticated approach. He begins with certain desirata $\mathcal{D}$ he expected his numbers to obey, such as the natural *addition rule* $(aj + bj = (a = b)j)$, the absence of *zero divisors* (no $a.b = 0$ unless either $a = 0$ or $b = 0$) and the *law of norms* (the product of two numbers of the form $a^2 + b^2$ must be of the same form). To these he would experiment with sundry trial assumptions (mathematicians call these ansätze), hoping to force out a completion of his multiplication table (its entries would then provide a *diagram* in the logician's sense of the mathematical structure he sought). For example, an experiment might begin with "$ij = -1$":

> Let's see, if "$ij = -1$" is true, then "$ij - i^2 = 0$" will need to be true as well and if that happens, then "$i(j - i) = 0$" will be true ... Oh drat, this isn't going to work out, because that gives us a divisor of zero unless $i = j$, in which case we're back to the regular complex numbers.

Hankins' excellent book on Hamilton[31] (from which our discussion primarily benefits) gives a nice account of the inconclusive results obtained from the alternative ansatz $ij = -ji$ (Hamilton did *not* presume commutativity). We can diagram these exploratory processes with a diagram in the manner of Chapters 6 and 7. We begin in the lower patch of the usual facts about the complex numbers $\mathcal{C}$, throw in an ansatz or two, and see what inductively grows in the upper patch according to the principles in $\mathcal{D}$. Several outcomes can occur: (1) we achieve complete success and find workable values for our table; (2) we encounter inconsistent clashes with what we knew before or trivializations (i.e., our ansatz forces $i = j$); (3) we only derive partial results. If the latter, our results may suggest further ansätze to try, but perhaps not (if the former, we add another leaf to our diagram). Restated in the language of "truth," Hamilton would have said to himself, "I know that $\mathcal{C}$ and $\mathcal{D}$ must be true of my triples, but I don't know what multiplication results are correct. Let me assume that '$ij = -ji$' is true, and see what other triplet facts are forced to be true as well."

---

[31] Thomas L. Hankins, <u>Sir William Rowan Hamilton</u> (Baltimore: Johns Hopkins Press, 1980).

Unfortunately, there simply is no structure that satisfies Hamilton's demands for his triples (as was later proved by G. Frobenius). One fine day it occurred to Hamilton that perhaps he should add a third unit k into his mix. It didn't take long to force the enlarged multiplication table and thus were his *quaternions* born. He described the incident in a letter to his son Archibald in 1865:

> *Every morning in the early part of* [October, 1843], *on my coming down to breakfast, your (then) little brother William Edwin, and yourself, used to ask me, "Well, Papa, can you multiply triplets?" Whereto I was always obliged to reply, with a sad shake of the head: "No, I can only add and subtract them." But on the 16th day of that month . . . I was walking along the Royal Canal. . . . and although* [your mother] *talked with me now and then, yet an under-current of thought was going on in my mind, which gave at last a result, whereof it is not too much to say that I felt at once the importance. An electric circuit seemed to close; and a spark flashed forth, the herald (as I foresaw immediately) of many long years to come of definitely directed thought and work.*[32]

Unfortunately, quaternions never produced a bountiful mathematical yield comparable to that of the complex numbers and they have generally languished in obscurity since (although a minority sect has kept them alive within mechanism kinematics and a more recent enthusiasm has blossomed within theoretical physics).

It is very easy to construct other examples where towers of induced truths build up from a set of established facts *c*, growth principles $\mathcal{D}$ and ansätze *a* (recall Annie and Frank from Annie Get your Gun each announcing, "Anything you can do, I can do better," add "Frank can drink liquor faster than a flicker" as an ansatz and consider how the tower of claimed skills climbs). Within philosophical circles, the most celebrated example of such recursive hierarchies arises when the predicate "is true" itself occurs in $\mathcal{D}$ and *a*, sometimes leading to inconsistencies; sometimes not. In certain natural cases, the tower climbs into the transfinite.

I mention these considerations because I believe they indicate that "truth"'s role as an evaluator of language organically leads it into circumstances of this layered tower type, simply because the predicate serves as our natural linguistic vehicle for keeping track of how possible support spreads through a language under critical investigation. As such, we come very close to the "revision rule" picture of "truth"'s workings that has been articulated by Anil Gupta and Nuel Belnap.[33] Gupta also suggests that many predicates gain their foothold in language only through a combination of weak $\mathcal{D}$ and *c*-type specifications (he calls certain analogs to my $\mathcal{D}$ "circular definitions"), a point of view that is underwritten by the prevailing need for open texture that I have emphasized.

. . . .

In their formal work, Gupta and Belnap require that their ansatz assumption *a* represent a syntactically complete extension of *c*, which typically induces many inconsistencies between

---

[32] Michael J. Crowe, A History of Vector Analysis (New York: Dover, 1985), 29.

[33] Anil Gupta and Nuel Belnap, The Revision Theory of Truth (Cambridge, Mass.: MIT Press, 1999).

	1	i	j	k
1	1	i	j	k
i	i	-1	k	-j
j	j	-k	-1	i
k	k	j	-i	-1

layers. I find that my manner of setting up the circumstances fits the historical cases of critical scrutiny utilizing ansätze better. And there seems to be no universal recipe that prescribes how conflicts between layers should be ameliorated when they are found to arise. My thinking about philosophy of language in general owes much to many discussions with Gupta over the years.

....

There is a curious coda to our history. Hamilton was originally motived by the desire to both introduce a three-dimensional "number" akin to an imaginary number *and* reproduce formal results somewhat like those found in complex analysis (rather as Boole strived to keep technique in logic close to that borrowed from linear equations). It is these dual desires that led him to include the ban on zero divisors as a necessary prerequisite to his explorations. When the physically minded Oliver Heaviside encountered quaternions in the context of Maxwell's Treatise on Electromagnetism, he saw no reason for retaining the zero divisor restriction (not heeding the siren voices of complex analysis' beauties) and thus were the modern three-dimensional vectors quickly engendered, now commonly exploited within virtually every scientific topic studied in the modern university (to Heaviside "$a·b = 0$" merely indicates that a and b are orthogonal).

(vii)

Understanding others. Classical thought promises a simple, and beguiling, picture of what occurs when we understand one another and the nature of our endeavors when we translate a foreign tongue: we bring concepts to notice that match, as closely as we manage, the precise conceptions that our subjects attach to their own predicates (mutatis mutandis for the other parts of speech). This matching task is not always easy, the traditionalist concedes, because other people frequently entertain universals with which we enjoy little truck. Extracting the right items from the far hinterlands of the Realm of Concepts can sometimes represent a considerable struggle.

Of course, there is no prospect of diminishing the exaggerations of classicism if such opinions are left to stand, because all of its central assumptions with respect to presentation and invariance are contained therein. For this reason, W. V. Quine has proposed (as we fleetingly noted in 5,vii) that issues of translation and interpretation should be set instead within a framework of constructing a large scale alignment between the target language and our own, meeting certain criteria for an "adequate translation manual" (following Quine, I characterize this ploy as that of "mapping into a home language"). As observed in our earlier discussion (5,ix), Quine is mortally afraid that any toleration of locally supportive *attributes* might serve to reinstate all the machinery of classical thinking, so he demands that his translation mapping be constructed according to the holistic principles delineated in Word and Object. This tale makes it likely that rather different maps might accommodate his desirata equitably, and thus is Quine's specter of "indeterminancy of translation" precipitated, along with the rest of the accouterments characteristic of his developed point of view.

Behind all of this bustle is Quine's recognition that he must somehow reconstruct, within his scheme, the good works that "concept," "property" and "truth-value" perform on our behalf in run-of-the-mill linguistic evaluation, without thereby collapsing meekly into classicism. The "mapping to a home language" story supplies such an answer, tho' a very unfortunate one, in my opinion. At root, its unhappy features trace to Quine's decision to harness (with a few allowances for disorganized learning) the implicit definability normativities claimed within the old theory T syndrome as his chief means for evading the rigid clasp of classical gluing. By now, we have trudged through enough scientific exemplars to recognize that Quine's globalist methodologies—the ties that try to bind everything up into a colossal "web of belief"—are insufficiently sensitive to the world's prickly demands, which commonly require effective descriptive language to fracture into little patches, at least within macroscopic application. In matters of translation and interpretation, we should attend to these localized structures if we can, not waft away into the mists of holistic language-to-language mappings.

Accordingly, Quine's portrait approaches the typical problems in translation in a completely wrong-headed spirit. After all, in evaluating our own usage, we find that our focal interests diverge rather significantly over time, so we should hardly expect less in dealing with a foreign tongue or agent who happens to be affected by linguistic currents significantly different than our own. Rather, a given native utterance may naturally evoke simultaneous layers of evaluation that seem inconsistent from a classical perspective (or by the lights of a unitary "mapping to English" scheme), yet will, quite innocently and appropriately, capture different measures of the set of directive influences acting upon its component words. That is, to properly appreciate what someone, in a different time and with a different language, intends to convey in a text, we strive to bring forth, as best we can, the key directive strands that would have been operative upon what she said. To do this may require separate attention to a fair number of pertinent factors. In assaying a translation, we seek a rendering that, in tolerably brief compass, conveys the relevant shaping factors as effectively as we can manage. It is a truism that good translations should be comprehensible to ourselves, but that does not prevent us from assigning to the predicates in the text conglomerations of shaping elements far different from any that affect our own words.

It may fix our ideas if we peruse a few toy examples, which are specifically con-structed to embody simple prolongations from an originating patch in the manners of Chapters 6 and 7 (I will consider some real life circumstances below, but they are inevitably rather messy in their contours). Suppose that we have a people—the Golddiggers—who formerly believed that the sun is a big bird that flies to its nest at sundown, but have now acculturated into the modern world. However, they still retain their old word for sun ("oneymay") but now regard it as designating a gaseous star just as we do. So how we might we translate a specimen text from their language? Let our target discourse be:

> Ereway inhay the oneymay.
> Ethay iesksays are unnysay.

Looking up "oney" and "may" in a Golddigger–English dictionary we find the entries *bird* and *big* respectively. Taken in conjunction, it may seem wisest overall to render our specimen text as:

> It's time for lunch. Did you realize that solar flares are common this time of year?

Why did I adopt the not-perfectly-correspondent, "It's time for lunch"? Well, the following wouldn't do very well, absent a lot of supplementary explanation:

> The big bird is presently overhead. Did you realize that solar flares are common this time of year?

This second rendering makes the speaker sound schizophrenically confused and is likely to evoke the puzzled response, "What does this guy actually believe about the sun?" On the other hand, if our first sentence appears imbedded in an venerable children's folk song, we are likely to prefer a translation that inclines oppositely:

> The big bird is flying high overhead/ Soon it will return to its goslings,

rather than:

> It's time for lunch/Soon it will return to its goslings.

Why do we make these choices? Clearly, a brace of directive elements attach to the term "oney," some of which (its ancient avian associations) still linger as latent DNA. Clearly, "oney"'s bundle of directivities is not perfectly matched by any English equivalent. In confronting our text, we must choose those ingredients that appear most salient within the original and harness English equivalents to our translation task as we can. Thus we select "It's time for lunch" because its directive summons to practical action is most central to our subject's concerns in (1), whereas utilizing *"The big bird is flying high overhead"* would ineptly highlight directive ingredients that are not strongly active in (1)'s speech act and obscure the more vital operational import of the call to supper. In (2), on the other hand, the latent DNA of inactive directivities must now be brought forward, lest the rhyme's <u>modus operandi</u> make no sense. And there are plainly mixed passages where the translator's ploy of utilizing pidgin coinages is required to suggest the right connections:

> *The oney-sun is presently overhead. By the way, did you know that once upon a time people used to believe that a literal "oney" (= bird) climbed in the sky?*

Here we insert "oney" itself into the translation to mark a historical etymology that possesses no English analog. If all else fails, there is always the footnote!

Or, to vary the example, suppose some translated Golddigger text startles us by reading:

> *How now, orange cow?*

As background, it might emerge that the color specialists of the society have convinced everyone that the "true colors" of material goods are best registered via discrimination through a reduction tube, as explained in 7,x, and that "brown" can be adequately replaced by "dim orange" (we observed that our own researchers sometimes hint at the same recommendation). To remove the mistaken impression that the Golddiggers keep gaudy cows, we might alter our gloss to:

> *How now, aperture orange hue cow?*

However, its manifest poetic difficulties may force a retreat to

> *How now, brown cow?*

In this case, as translators, we must hope that the fact that the Golddiggers find a stronger affinity between citrus fruits and cattle than we do will not emerge as a significant element elsewhere in our text.

Another revealing case is provided by the Druids of 1,ix, whose early employment of "bird" has not yet been subjected to any directive factors that incline them to consider B-29's as "birds" or not. Psychologically, the Druids are much like us (in the movie, they dress remarkably like Hollywood starlets <u>circa</u> 1952). Indeed, an algebraist might designate their active set of directivities as $Bird/\mathcal{D}_1$, <u>i.e.</u>, our conception of *being a bird* restricted to the Druid domain $\mathcal{D}_1$. In this wise, there is really nothing we fail to understand about our subjects: their linguistic activities are as transparent to us as our own. Nonetheless, particular specimens of native text might require special care if we are unaccustomed to compensating for "scaled back to $|\mathbfit{D}_1$," effects:

> *"I had a dream," the seer reported. "A great metal object with wheels and bombs hanging beneath its wings landed in the field. It might have been a bird."*

"Why does he say, 'It might have been a bird?' when it's obviously a bomber?," we wonder. If we are so obtuse about earlier stages of civilization, we may require a neologism or a footnote.

I scarcely pretend to have advanced a full "theory of translation" in these sketchy ruminations, anymore than I have supplied a complete accounting of every directive influence that might wash a wandering word to and fro (even within my narrow target class of macroscopic classificatory predicates). What I *am* suggesting is that, if we pay sufficient attention, we discover that our evaluations of foreign speech are often contextually adjusted to spotlight the patterns of directive influence that have been discussed in earlier chapters (as well as many others). So a translation should not be viewed as a "map into English" in the proper sense, for our policies conceal a complex battery of layered evaluations. In many respects, the directive factors I emphasize (e.g., extending the distributed correctness of an effective routine; imitating manners of usage borrowed from adjacent patches; consulting a semantic picture for guidance) play a role in my own thinking roughly comparable to Quine's own appeals to "sentential conditioning" (although I hope my proposals reflect a more realistic appraisal of viable linguistic engineering). As such, I see no deep doctrinal reason why Quine, liberated from unnecessary assumptions of univocalism and invariance, shouldn't *prefer* a treatment of the translational enterprise that pays close attention to localized structure, rather than washing away its detail in excessively averaged holism. But I'll return to this issue in a few pages.

In musing on these matters, it occurred to me that an excellent arena for testing these suggestions might lie in alchemical language, whose descriptive vocabulary stretches over a goodly number of strands of practical advantage involving chemical manufacture, yet is simultaneously driven by bizarre winds that blow from the scattered corners of Platonism and assorted religious traditions, amply laced with borrowings extracted analogically from peculiar conceptions of celestial and animal fact. So I attempted some survey of how commentators on alchemy strive to make their subjects comprehensible to us. Consider the term "mercury" (or, often, "quicksilver"), as it would have been employed in writings of the early 1600s. If we consider the tally sheets that detail the shipments from a Elizabethan cinnabar mine or some kindred text, we will feel inclined to evaluate most of their sundry "mercury" assertions more or less as we might a contemporary speaking about the standard liquid metal. Thus, with little cavil, (1) will be reckoned as true and (2) false:

> (1) [P]ots [should be] *molded from the best potter's clay, for if there are defects the quicksilver flies out in the fumes.*[34]

---

[34] Georgius Agricola, <u>De Re Metallica</u>, trans. Herbert and Lou Hoover (New York: Dover, 1950), 428 (yes: *that* Herbert Hoover).

(2) *In the mineral holes and dens . . . of the earth are found Vegetables which in long*
     success of time . . . do put off the Vegetable nature, and put on a Mineral.[35]

On the other hand, within alchemical tradition the term "mercury" undergoes a vast
and bizarre extension in usage, to an extent that it severs almost all connection with
liquid metal-centered employments (alchemist adepts often wrote of "*our* mercury" to
distinguish its extraordinary qualities from those of the impure stuff to which it was, in
some dim way, connected). The general picture is summarized by John Read as follows:

> [T]*he sulphur-mercury theory appears basically as a derivative of the theory of the Four*
> *Elements. The apposition of the two opposed, or contrary elements, fire and water, now*
> *assumed a new guise. "Fire" became "sulphur," and "water" became "mercury." These*
> *names must not be identified with material substances, sulphur (brimstone) and mercury*
> *(quicksilver). They represented abstract principles, consisting of hot and dry (sulphur) and*
> *cold and moist natures (Mercury) "natures." In alchemical writings they were often called*
> *"sophic" (or philosopher's) sulphur and "sophic" mercury, or "our" sulphur and "our"*
> *mercury, in order to distinguish them from the material substances bearing the same name.*
> *In the main, sophic mercury stood for the property of combustibility or the spirit of fire,*
> *and sophic mercury for that of fusibility or the mineral spirit of metals.*[36]

However, note the studied vagueness of Read's claim that "our mercury" "stood for the
property of combustibility or the spirit of fire," as if a "property" and a "spirit" could
possibly represent the "same thing." In fact, this awkwardness comes with the territory,
for the alchemists often reveled in nearly contradictory—or, as they sometimes put it,
"hermaphrodite"—accounts of "our mercury"'s special qualities. Thus the Polish
alchemist Michael Sendivogius:

> *He is all things, who was but one; he is nothing, and his number is entire; . . . he is a spirit,*
> *and yet he hath a Body; . . . he is a Beast, and yet hath the Wings of a Bird; . . . he is Life,*

---

[35] Paracelsus, from Stanton J. Linden, <u>The Alchemy Reader</u> (Cambridge: Cambridge University Press, 2003), 164.
[36] John Read, <u>From Alchemy through Chemistry</u> (New York: Dover, 1995), 17–18.

> *yet kills all things; . . . he flyeth from the Fire, yet Fire is made of him; he is Water, yet wets not . . . .*[37]

In another work Read comments:

> *Philalethes in an exuberant mood let loose the following deluge of synonyms for mercury: "Mercury is our doorkeeper, our balm, our honey, our oil, urine, may-dew, mother, egg, secret furnace, oven true fire, venomous dragon, Theriac, ardent wine, Green Lion, Bird of Hermes, Goose of Hermogenes, two-edged sword in the hand of the Cherub that guards the Tree of Life . . . ." The unrestrained use by alchemists of symbols, emblems, allegory, and other forms of cryptic and mystical expression, led to a unique literature, laden with a mass of enigma, metaphoric expression, and jargon which defies description.*[38]

Plainly for passages such as these it would be silly to attempt to match these employments of "mercury" closely with any predicate we might utilize in everyday life or any notion we are likely to have entertained before we have investigated alchemical writings. Instead, we attempt, through scare quotes, capitalizations and borrowed modifiers, to induce a flow of rhetorical connection that allows a modern reader to roughly appreciate how "mercury" gets blown hither and yon, once its usage becomes detached from the practicalities of the cinnabar mine and assumes the earnest trappings of alchemical quest. To follow these winds, we must gather some appreciation of how sundry borrowings from Scripture, mythology and astrology serve to fatten up the discourse and lead it to spread over vast applicational dominions to which it seems very speciously attached (more on this in a moment). Accordingly, a passage such as the following (from Gerber) merits an explanatory footnote from its editor (Stanton Linden):

> (A) [Mercury] *easily adheres to three minerals, <u>viz.</u> to Saturn, and Jupiter, and Sol, but to Luna more difficultly.*[39]

Compare this with:

> (B) [Mercury] *easily adheres to three minerals, <u>viz.</u> to lead, and tin, and gold, but to silver more difficultly*

(B)'s more prosaic rendering foregrounds the correct chemical information that Gerber provides, yet that version plainly misses (A)'s simultaneous stretch into otherworldly contention with respect to how astrological influences affect "our mercury." And that is why Linden provides a footnote: he must superimpose (B) over (A) to provide his readers with a warmer sense of how Gerber's original claim behaves over its patches of greater and lesser practicality.

Charles Nicholl observes in his <u>The Chemical Theatre</u>:

> *"As it were in a riddle and cloudie voyce," says the preface to the <u>Mirror of Alchimy</u>, "they have left unto us a certaine and most excellent science." In a sense, the bizarre and*

---

[37] Charles Nicholl, <u>The Chemical Theatre</u> (Pleasantville, NY: Akadine Press, 1997), 95.
[38] John Read, <u>Prelude to Chemistry</u> (Cambridge, Mass.: MIT Press, 1966), 92–3.   [39] Linden, <u>Reader</u>, 85.

nebulous language of alchemy [represents] *... the archaeological remains of a buried philosophy. But ... even while it lived and prospered alchemy was as much a literary phenomenon as a practiced "science." The texts of this "alchemical renaissance" fully exploited the rich dramatic and poetic qualities of alchemical language, and we can hear in them today the tones that might have excited a receptive ear in 1605 ... There is in all this an element of code, the "cloudie voyce" as a veil against the impure and uninitiated*[40]

Indeed, one of the hurdles required in "understanding" alchemical claims suitably lies in recognizing that the rhetorical *tricks of evasion* characteristic of "orgone" society (5,i) are commonly employed throughout.

> *It must need be taught from mouth to mouth*
> *And also he shall (be he never so loath)*
> *Regard it with sacred and most dreadful oath,*
> *So blood and blood maie have no parte*
> *But only vertue wynneth this holi art.*[41]

Any guru worthy of their salt knows the benefits of never being definite about any issue of substance or consequence of unhappy prospect (it's "somewhat like tricks o' the cards, to cheat a man with charming," Surly complains in The Alchemist[42]). Vagaries as to how "our mercury" relates to the familiar silvery stuff abound: the latter is claimed to represent an "imperfect" condition of the former, but only in a manner that is utterly impervious to confutation. The net effect of all this "charming" produces a nicely quarantined patchwork: local usages admirably adapted to assay and the manufacture of perfumes (a popular skill of the adepts) border upon abstruse sprawls of the same vocabulary moored

---

[40] Nicholl, Theatre, 90–2.

[41] Thomas Norton quoted in C. J. S. Thompson, Alchemy and Alchemists (New York: Dover, 2002), 96.

[42] Ben Jonson, Three Plays (New York: Hill and Wang, n.d.), 247.

to virtually nothing except myth and astrology (I'll qualify this claim in a minute, how-ever). Great prodigalities of decoupling envelop the "recipes" offered for the celebrated "philosopher's stone," the chief item of miraculous capacities that these parties sought. No correspondent substance, we can assume, was ever produced nor were many adepts deluded into thinking that they had done so, so the sketchy formulae for the "Stone" found in surviving manuscripts usually represent blends of opaque instructions passed along through tradition, reports of pregnant hopes ("I think I came very close to making the Stone on the day when I did Y") and straightforward bluffing ("Folks won't accept my other recipes as valuable unless I throw in hazy hints that such manufacture contributes to making the Stone"). These vagaries were facilitated by the standard guru device of insisting that a formula so sacred can only be articulated in a "riddle and cloudie voyce." In combination, such policies allowed many worthies to fancy that "the Stone must be just around the corner," despite its many years of non-appearance.

Now we can't really "understand alchemical language" in a meaningful way unless we enjoy some awareness of the mechanisms that hold this oddly jointed patchwork together, yet this is a sense of "understanding" that no right thinking alchemist would have possessed. Quite the contrary, Sendivogius claims of his usages:

> Let me therefore admonish the gentle reader that my meaning is to be apprehended not so much from the outward husk of my words, as from the inward spirit of Nature.[43]

It should tartly remind us of our own linguistic horizons that a parallel semantic con-fidence informs our trust that our own "mercury" bonds solidly to an underlying natural kind attribute.

All the same, if we merely pay attention to the engines of doctrinal borrowing that stitch the old texts together, we are likely to remain at sea with respect to alchemical thinking, for it is hard to understand why so many serious folks would have devoted so much sooty labor over several centuries to unrewarded syntactic outpourings. Considering our orgonist community again, we can easily appreciate how, under the dominion of some mesmerizing guru, a populace might subscribe to the most foolhardy projects for a decent span of time, but such arrangements become more puzzling as they wear on endlessly to no evident advantage.[44] "What *rewards* did these people extract from their pursuits?," we ask in pre-pragmatic mood. "What were the spurs that kept the alchemists going?" Here we can be grateful to the scientists who have patiently deciphered the chemical events that these experimenters likely observed when they carried out some of their characteristic recipes. For example, William H. Brock writes:

> In Michael Maier's <u>Atalanta fugiens</u> (1618), we read "the gray wolf devours the King, after which it is buried on a pyre, consuming the wolf and restoring the King to life." All becomes

---

[43] Nicholl, <u>Theatre</u>, 90–2.

[44] "In this time of the Consciousness Revolution, the alchemists are teaching us to re-examine the world and ourselves through our own observations and experiences and to take on faith nothing that we've been told"—back jacket blurb to Arthur E. Waite, <u>Alchemists Through the Ages</u> (Blauvelt: Rudolf Steiner, 1970). Apparently, you can fool *some* of the people *all* of the time.

clear when it is realized that this refers to an extraction of gold from its alloys by skimming off lesser metal sulfides formed by a reaction with antimony sulphide and the roasting of the resultant gold-antimony alloy until only gold remains.[45]

Such an underlying practical routine helps us better understand why the wild syntactic effusions about "wolves" and "kings" that decorate its assertional surface hold in place as firmly as they do: the whole is like the Japanese blacksmith's appalling recipe (5,ii). A good sword emerges at the end, and that success makes our smithy conservatively retain human victims within his procedures, despite the fact that more vigorous experimentation would reveal that ammonia baths could serve his purposes just as well.[46] Brock's decoding of the real world underpinnings for Maier's strange parable thereby provides us with a palpable sense of, "Oh, now I can appreciate how much of the alchemical point of view might have naturally grown from such recipes as a seed." Once again, we see how a given alchemical sentence often needs to be parsed from several vantage points, if we hope to gain a proper appreciation of its raison d'être within its original context. This perspectival multiplicity stems, I submit, from essentially the same foundations as our "oneymay" example.

Here is another example that likewise calls for a doubled reading, but with an additional twist. Consider this recipe from Joan Baptista von Helmont's De Febribus, as presented in William Newman and Lawrence Principe's Alchemy Tried in the Fire. This story can be likewise wedded firmly to a chemical substratum:

> *If you distill oil of vitriol from running mercury, the oil is coagulated with the mercury, and they both remain in the bottom in the form of a snow. Whatever you distill thence is mere water. But this snow, if it is washed, becomes a yellow powder, which is easily reduced into running mercury in just the same weight as before. But if you distill off the wash water, you have a pure alum in the bottom, from the acid salt of vitriol. Therefore dissolvents are mutated even if the dissolved lose nothing of their substance or matter.*[47]

Von Helmont's purpose is to argue that "oil of vitriol" (sulfuric acid) can lose its chemical potency (become "mutated") in a sequence of chemical reactions upon mercury without losing its identity, that it "gradually becomes enfeebled in the same way as an animal might become tired after exerting itself." Newman and Principe outline the chemical setting over which this recipe unfolds as follows (they provide interesting glosses on many other aspects of the passage as well):

> *What he has observed (in modern terms) is the hydrolysis of the white mercuric sulphate ($HgSO_4$) into a yellow basic sulphate ($HgSO_4.2H_2O$) . . . When the wash water is distilled off, Van Helmont finds a fine crystalline residue, which he calls "alum," but which is in reality [a portion of] redissolved mercuric sulphate. . . . [If] he washed the mercuric sulphate "snow" with a very large amount of water, . . . the crystals of mercuric sulphate*

[45] William H. Brock, The Norton History of Chemistry (New York: W. W. Norton, 1992), 19.

[46] Rolf E. Hummel, Understanding Materials Science (New York: Springer, 2004), 132. He also explains the rationale behind the smithy who prepared excellent sword steel by feeding it in pellets to a chicken.

[47] William Newman and Lawrence Principe, Alchemy Tried in the Fire (Chicago: University of Chicago Press, 2002), 81.

will be large and will actually resemble those of alum (potassium aluminum sulphate) and thus be very unlike the powdery "snow" initially formed.

Observe how helpful this gloss proves for "understanding Van Helmont's thinking": it allows us to understand how the stages in his linguistic recipe correlate with the laboratory events he witnessed. On the other hand, to comprehend the flow of the recipe, we also need to recognize that he did not apply "alum" and similar words as we might. In particular, it is crucial to the logic of his reasoning that "snow" (finely crystallized $HgSO_4$) is not recognized as the same stuff as his "alum" and that he did not actually regain all of the "running mercury" with which he had begun. By observing how the two layers of supported recipe procedure and alchemical picture work together to frame van Helmont's line of thought, we can gain a quite reasonable appreciation of why he says what he does. Any monotone evaluation of either "support" or "picture" would not allow us to understand the text nearly as well. In order to convey this appreciation to their readers, Newman and Principe comment upon the text from several vantage points; they do not attempt to find any phrase that translates "alum" into a modern English equivalent (what might it be?: our "alum" (potash sulfate)—too strong; "looks like alum"—too weak). Instead, they leave van Helmont's "alum" in place, rather as we utilized "oney" and "oney-sun" in our "English" glosses above.

In these passages, I have concentrated upon *translation* or—what might be a better name for the process—*textual gloss*, and argued that, in real life, we do not insist upon the single-valued mapping into English that Quine assumes to represent the sine qua non of "understanding an alien speech." I believe that if we scrutinize how we employ our usual evaluative terms—"concept," "content," "truth-condition," "truth," "refers"—within alchemy-like circumstances, we will find analogous layers within these appraisals as well. For example, we often employ the satellite term "conception" to induce a desirable bifurcation with respect to "alum": "Van Helmont operated with the same basic *concept* of *being alum* as we do, but he had a different *conception* of what it was like." "Did he use the term 'alum' wrongly then?" "No, he was loyal to his own conception as he understood it." Plainly (to drop into my own argot), we are trying to express that alchemical "alum" begins over the same patch of related chemicals as our own (in fact, that patch was so baptized in alchemical times), but has extended differently than our own ("conception" often serves to mark a branch prolongation).

The same multiplicity applies to "true," "truth-condition" and "refers." Consider Van Helmont's claim:

But if you distill off the wash water, you have a pure alum in the bottom, from the acid salt of vitriol.

Is this *true*? Demanded so baldly, we will likely refuse to answer and beg instead for greater contextuality in our reply: "Well, 'yes' in the sense that he is correctly reporting the appearance of a substance that would have been considered an "alum" at the time. But 'no' in the sense that he believes that this "alum" is of a different chemical nature than his 'snow.'" To what does "alum" refer? Again, we demur: "In one sense, it refers to a wide assortment of chemical forms that includes the specimen of mercuric sulphate before him. On the other hand, it really doesn't refer to a definite property at all, because there's no sustainable way to distinguish, in principle, the forms of mercuric sulphate he designates as 'alum' from those he designates as 'snow,' as chemists will eventually realize once they investigate those forms further" (philosophers will sometimes articulate the latter claim as follows:[48] "Various forms of speaker's reference can be attributed to alchemical 'alum,' but no firm form of semantic reference seems viable"). In this case, evaluative terms like "true" and "refers" gain some foothold, but in allied circumstances, their very intervention is apt to seem forced ("reference" has become largely a philosopher's notion; it is not naturally applied within many cases of real life textual appraisal).

Returning to some of our oldest themes: "concept" and family might have served us a better turn if they had alerted us to their shifting appraisals in a more blatant fashion, rather than shuffling ahead with low key modifiers such as "conception" and "in one sense." In truth, however, these terms have acquired their evaluative complexities through adaption to circumstances, and it's our fault that we've not kept up with them in our accompanying *ur*-philosophical pictures.

(viii)

The schedules of our time. In my assessment, Quine simply made a wrong turn in the course of pursuing pre-pragmatist hunch in a vein that commenced similarly to my own, through choosing to *imitate* classical invariance and univocalism, rather than resolutely shunning their allures. True: if he had not selected the "mapping" fork along the interpretative road, he would have never been led to the doctrines that made him famous; but that consideration should not spur us to follow him yonder, for his celebrated opinions represent themes that only a radical skeptic could love. To be sure, I agree with Quine that significant looseness and dislocation appears scattered through our language's present attachments to the world, although never in the tremendous expanses he alleges. And I would not dub this lack of firmness an "indeterminacy of reference," if only because of the large amount of classical baggage that cleaves to the term "reference" nowadays.

However, many of Quine's strongest admirers have detected in his "mapping into a home language" musings confirmation of profound truths about the human condition that run deeper, perhaps, than any thesis that Quine himself intends. Such heightened strains strike me as closely allied to the neo-idealist themes charted in 2,vi. I have particularly in mind the melancholy conceit that the characteristics we attribute to other people—indeed, to every worldly event that passes before our gaze—come irrevocably tinctured with the contributions of our own point of view—that our interpreting gaze

[48] Keith Donnellan, "Reference and Definite Descriptions," Philosophical Review 75, 3 (1966).

forever locks our conclusions into an ego-centered orbit from which they can never escape. In this vein, the English logician John Venn wrote in 1889:

> [O]ur logical scheme is avowedly constructed from the present point of view. It does not, or should not, profess to be anything else than an interpretation of remote times by the schedules and forms of our own time.[49]

But the popularity of this unsettling fantasy, it seems to me, is much the consequence of not observing closely the furtive behaviors of our dowdy friends, "concept" and "understanding," imagining them to hold frozen in place when they actually scuttle busily about.

Throughout this book, I've frequently complained about philosophical impulses that advise us to frame sweeping, "big picture" accounts of issues that are better resolved, I think, if we act oppositely and attend to the small details that are likely to have escaped our attention. And this consideration brings us back to those hazy doctrines of holistic understanding that we briefly canvassed in 5,xii.

To this end, let us briefly survey the neo-Quinean views of Donald Davidson. He articulates a sweeping account of "human interpretation" (= the policies whereby we understand the activities of others) which amplifies Quine's "mapping into English" story into a prodigious vision of how we bring target subjects into equilibration with our own proclivities as best we can:

> The possibility of understanding the speech or actions of an agent depends upon the existence of a fundamentally rational pattern, a pattern that must, in general outline, be shared by all rational creatures. We have no choice, then, but to project our own logic on to the language and beliefs of another. This means it is a constraint on possible interpretations of sentences held true that they are (within reason) logically consistent with another.[50]

Here Davidson argues against the notion that alien people can meaningfully entertain "conceptual schemes" different than our own. I happen to find most invocations of "variant conceptual schemes" problematic, but not for Davidson's reasons. In most cases, attributions of strange scheme (e.g., Evans-Prichard's contention that the Neur possess a different scheme for time reckoning than our own) covertly rest upon a conception of classical conceptual grasp much like Russell's: the Neur have somehow located novel notions within the Land of Universals and clasped them to their bosom as a bundled "scheme." It can well be that this people pursue unexpected but viable strategies in time-reckoning that we should tease out and appreciate, but not through collapsing their variant mechanisms into compact, but allusive, ghost attributes "of different scheme."

But Davidson's concerns are entirely different. He believes that the very act of "understanding the Neur" requires that we must map their beliefs to ours as best we

---

[49] John Venn, The Principles of Empirical or Inductive Logic, (London: MacMillan and Co., 1889), 16.
[50] Donald Davidson, "The Structure and Content of Truth," Journal of Philosophy 87, 6 (1990), 320.

can, perhaps univocally forcing (to revert to our earlier example) "Ereway inhay the oneymay" onto "It's time for lunch" (for that's what *we* would usually signify with the utterance if we worked as anthropologists within their midst). Why does Davidson think such a thing? Oddly enough, his official arguments upon their behalf commonly rest upon assumptions deeply reminiscent of theory T syndrome days. As such, his views prove automatic grist for the mills of Chapters 6 and 7. For example, in generalization of Quine's coordinate-system-like sets of "analytical hypotheses," Davidson claims that any sensible application of a concept such as *hardness* to another agent requires that we find in her behavior allegiance to general requirements such as a belief in *transitivity*: if a̲ scratches b̲ and b̲ scratches c̲, then a̲ must scratch c̲. Under the influence of the Stanford school of measurement, he writes:

> Unless this law (or some sophisticated variant) holds, we cannot easily make sense of the concept of [hardness].[51]

The claim is that, without such behavioral regularities, our assignments of *measured hardness values* to physical objects can't be justified. He then assumes that our evaluations of psychological state must be obedient to some comparable schedule of general "norms":

> Just as we cannot intelligibly assign a [hardness] to any object unless a comprehensive theory holds of objects of that sort, we cannot attribute any propositional attitude to an agent except within the framework of a viable theory of his beliefs, desires, intentions, and decisions.

In very short order, these electrifying consequences spill forth:

(a) In the case of psychological notions, allegiance to these "norms" will pull their enslaved terminology away from any fixed physicalist base and cause them to classify events according to non-physical standards.

(b) The psychological interpretation of alien or distant cultures requires that these same general norms continue to be applied, with only minor tinkering tolerated.

These should seem like heady consequences to extract from such a slender reed of operationalist dogma. Of course, in 6,ix we observed that *hardness* does not obey Davidson's strictures globally, any more than "$\sqrt{z}$" does. Working backwards in modus tollens from such considerations, basic reflection on the available varieties of linguistic engineering advises us that other people might easily pursue quite different descriptive policies than we do and that any reasonable portrait of "human interpretation" should take account of that divergence. We can ably understand someone else's strategic policies, I have argued, by considering them *in pieces*. We ask: How do the basic underlying strategies work out relative to expected environmental setting? How are the

---

[51] Davidson, "Mental Events", 221. Davidson's own example is *length* which is confusing, as it is properly a vector quantity.

required lifts and Stokes-line curbs implemented within the language? Which factors induce patch-to-patch prolongation? What are the parent patches to these processes, from which later stages borrow? What accompanying pictures do they embrace? And so forth, running through a lengthy list of salient shaping considerations (I reiterate that I do not in the least pretend to have diagnosed them all in this book). And our brief examination of alchemy suggests that *piecework answers* of this type are exactly what scholars of the subject have slowly and painfully assembled for our benefit, allowing us to genuinely understand how an alien people of long ago behaved. I see scarcely a trace of "making them seem like us" in any of it.

Why are so many able thinkers attracted to holistic opinions like Davidson's?[52] Let us set aside silly arguments of the criterial flavor just recounted. I often get the impression that many of Davidson's admirers, in their deepest hearts, entertain opinions in the same family as those surveyed earlier in the book (e.g., those of William James or Coleridge). At the risk of articulating the underlying point of view too crudely, it runs something like this. The discriminations we make with respect to the deepest issues—e.g., whether a moral act is right or wrong—stem from inaccessible cords buried deep within us (lodged, perhaps, only within the soul and not the brain). As such, their classificatory urges draw from the full wealth of our personal experience and character, rather than any localizable source. In working with others within a linguistic community, we have learned to employ terms such as "conceptual content" and "truth-value" as a means of forging these individual impulses into an accepted framework of public objectivity, so that we can reach our common ends more ably. Within this accommodation between individual impulse and public adjudication, our usual evaluative notion of a predicate's "proper meaning" is engendered. If so, we can meaningfully attribute allied conceptual contents to folks outside our community only if we can see their behaviors as similar to ours across a wide ranging scale of reactive inclinations. Insofar as a community displays alien patterns within their behaviors, we lack means of assigning their words contents, because that would require capturing, per impossible, their hidden springs of experience and character. They display a "form of life" that we cannot discuss in "conceptual content" terms, as a Wittgensteinian might put it.

Pursued in this mode, our Davidsonian doctrines take on a cast of which the Lake Poets would have approved. Thus thesis (a) reflects a vital aspect of the human condition: in judging whether someone is in pain or not, we draw upon a rich system of public/individual discrimination that simply cannot be calibrated with any point of view found in physical science. Even more mystically, every classification whatsoever ipso facto partakes of the general features of a moral point of view and, by these means, the features of the entire physical world become "humanized" again, for every conceptual discrimination (even with respect to rocks and quarks) inherently filters through our public methods for objectifying holistic points of view.

Accordingly, not only is man the measure of all things on this view, but perhaps even man-in-our-little corner-of-society-that-reads-academic-books. And I find that such

---

[52] Gary Ebbs, "The Very Idea of Sameness of Extension Across Time," American Philosophical Quarterly 37, 3 (2000).

a perspective often serves as an excuse for smug and short-sighted opinions akin to those of Chapter 2's critic. To refashion Hume, such thinkers fancy that other peoples are "as much bounded in their operations as we are in our speculations." "Not so fast," I complain in reply. "I have little to offer with respect to the 'moral point of view,' but I know something about rocks, and you are rapidly walling yourselves behind an unpleasant veil of predication with these lines of thought. Insofar as I can determine, your problems in 'understanding' the alien tribe stem largely from a grim insistence, driven by unexamined *ur*-philosophical currents, to capture everything about your subjects in one breath, when we should properly address their activities by patiently outlining how a large variety of perfectly communicable factors work together in tandem. To understand the words of others does not entail that we should be able to live their lives as they would, but that we should adequately appreciate the welter of shifting directivities that nudge their terminological employments forward. These are not ineffable processes, but they are hard to chart, for their shaping winds can arise from any corner of the compass and we must expect to employ a large arsenal of evaluative tools to capture their complexities, even to first order adequacy. Potentially, we can understand the Druids perfectly, but we will never be able to react to circumstance exactly as they might, for that would require that we forget about airplanes and the other everyday paraphernalia of which they are blissfully unaware."

....

In his <u>Henry James and the Modern Moral Life</u>,[53] Robert Pippin articulates the opinion (which he attributes to James) that confidence in our judgments with respect to the obligations owed to others often rests upon a widely distributed set of expectations about "what is to be done" within familiar circumstances. When, because of changes in mode of living forced by alterations in commerce or allied cause, the erstwhile coherence of those exemplars sometimes splinters, thereby leaving the most perceptive and well-intentioned observers within a society—those from whom we normally expect the wisest judgments—unable to find their moral compass, because they best perceive the hairline cracks that have now penetrated the clear glass of old verities. Pippin tells me that such venerable doctrines of historicity are often dismissed by zealous Davidsonians as "misunderstanding the basis of moral attribution." That opinion exactly reveals the viewpoint of which I complain. In mild analogy, those nineteenth century physicists who were most acutely aware of patch-based tensions in the way that "force" was employed, but could find no replacement foundation for its manifest utilities, often lost the confidence required to press "force" into its more advanced schedules of descriptive capacity.

....

Returning to distinct, yet related, issues (canvassed in preliminary fashion in 5,xi), I rarely find it helpful to regard the divisions that rend two scientists asunder as arising from fundamental differences in gestalt or paradigm, in the manner of Thomas Kuhn. An invocation to cloak disagreements within the amorphous doughiness of a "paradigm" directs our attention towards a blank and unhelpful sky, when we should instead

[53] Robert Pippin, <u>Henry James and the Modern Moral Life</u> (Cambridge: Cambridge University Press, 2000), 27.

scour the ground for the little things that better explain the intractability of the issues at hand.

An instructive case in point is provided by the mode in which the late nineteenth century disputes over atomism are commonly discussed today, often in the mode of a warfare between scientists locked in loyalty to distinct visions, one reactive and one progressive. Such characterizations often encourage "heroes versus villains" approaches to the histories of science and philosophy. I invariably find these narratives uncharitable, lacking an adequate appreciation of the nearly invisible turns of the screw that some-times lead the best of us, while exquisitely practicing the most judicious methodologies, charging up alleyways from which a shamefaced retreat will later be required. It hap-pens that our current inclination to look upon the old atomist debates in gestalt-washed terms has been much encouraged by the philosophical writings of one of the central participants within that debate, Pierre Duhem. There is a certain irony in this, because Duhem's scientific opinions fell on the wrong side of the atomism divide, because he argued in favor of a macroscopic mechanics that incorporates phenomenalist level thermodynamics as an unreduced component. As a side consequence of portraying disagreements as he did, Duhem damaged his own reputation, because, from a "little detail" point of view, his remarks often prove prescient of modern opinion whereas those of many of the atomists do not. But the salience of his observations have been largely washed away in the flood waters of "heroes and villains" adjudication, much encouraged by Duhem's own polemics.

. . . .

To be more specific about his theoretical preferences, Duhem favors a unified formalism where thermodynamic and chemical quantities enter on equal terms with purely mechanical notions (he also tolerates certain measures of micro-structure in the manner of his colleagues, the Cosserat brothers). For a good summary of the principles to which Duhem's philosophical speculations stood as prolegomena, see his "On Some Recent Extensions of Statics and Dynamics" of 1901.[54] Modern engineering practice generally follows Duhem in this tolerant acceptance of non-mechanical notions within "mechanics."

. . . .

I have already related part of this story in 6,xii (although laden with technicalities that some readers might have elected to skip; they will not be needed, as other factors will be emphasized here). The competing gestalts that Duhem emphasizes are (a) the form of thermodynamics-based physical theory he favors ("narrow but abstract minds," he claims, gravitate to these), and (b) atomist or narrowly mechanical conceptions attractive to those who think "in broad but shallow visualizing terms." He objects to the latter because he believes that such "minds" are prejudicially chained to "metaphysical conceptions of matter" that descend from traditions such as Boyle's (3,vi). To scientists afflicted with such biases, *heat* and *chemical potential* cannot serve as primitive elements

---

[54] Pierre Duhem,"On Some Recent Extensions of Statics and Dynamics" in <u>Mixture and Chemical Composition</u>, Paul Needham, trans. (Dordrecht: Kluwer, 2002).

within physical theory because the mechanical underpinnings that would make such notions *intelligible* have not yet been articulated—which is why they often gravitate towards atomic hypotheses despite the paltry empirical evidence that could then be offered in their favor (Ernst Mach is much exercised in his own writings by this same "prejudice"). Duhem also believes that his foes are adventitiously satisfied by local modelings of a mechanical cast, even though the details of those constructions rarely cohere with each other (here Duhem correctly detects some of the facade-like character that comprises a genuine part of classical mechanical practice). To Duhem, these twin propensities arise together because, to mechanism-addled individuals, the warm sense of "full understanding" that washes over them whenever they perceive a mechanical model utterly trumps any worries about the petty incongruencies that render their applications mutually incompatible (writing of their "visualizing tendencies" serves as Duhem's manner of expressing the notion of *fully presented content* that forms an important part of Boylean preference—see 3,vii). Duhem, like Mach, rejects as an unhelpful burden upon scientific progress any dedicated allegiance to (quoting now from Mach):

> the physico-mechanical view of the world [that is] . . . bequeathed to us as an heirloom from our forebears.[55]

An "abstract and logically organized mind" like Duhem's prefers a rigorous formalization of physical doctrine amenable to ready axiomatization, unconcerned with any deeper "understanding" of the predicates chosen but devotedly alive to inconsistencies in how such doctrines are applied. Here is how Duhem himself frames these issues:

> For abstract minds the reduction of facts to laws and the reduction of laws to theories will truly constitute intellectual economies; each of these operations will diminish to a very large degree the trouble their minds will have to take in order to acquire a knowledge of physics. . . . They have no difficulty in conceiving of an idea which abstraction has stripped of everything that would stimulate the sensuous memory; they grasp clearly and completely the meaning of a judgment connecting such ideas; they are skillful in following, untiringly and unwaveringly, down to its final consequences, the reasoning which adapts such judgments for its principles.
>
> [On the other hand,] there are some minds that have a wonderful aptitude for holding in their imaginations a complicated collection of disparate objects. . . . The minds possessing this power need the help of sensuous memory in order to have conceptions; the abstract idea stripped of everything to which this memory can give shape seems to vanish like an impalpable mist. . . . Will such visualizing minds regard an abstract physical theory as an intellectual economy? Surely not. . . . [They will maintain that] a physicist will logically have the right first to regard matter as continuous and then to consider it as formed of separate atoms . . . [56]

---

[55] Mach, "Economical," 187, 190.   [56] Duhem, <u>Aim</u>, 56, 103 (rearranged)

Duhem's "official" view is that both tendencies are legitimate insofar as the aims of science are concerned and only "metaphysical faith" can choose between them (however, the opposition is described in such a stacked manner that only a nitwit would fail to side with the "abstract"—i.e., French—"minds").

Mach adopts a simple instrumentalist position with respect to any descriptive predicate that lies outside psychic experience and regards efforts to find external support for any such terms as "metaphysical"—viz., utterly beyond the pale. Duhem, however, takes a somewhat different position and embraces "metaphysics" as a topic that lies outside scientific deliberation, but isn't inherently to be despised. In the peculiar article "The Physics of a Believer" appended to The Aim and Structure of Physical Theory, he draws the astonishing conclusion that, although purely scientific considerations cannot favor his own methods over those of a Kelvin, a "metaphysical faith" of a religious origin may:

> What is this metaphysical affirmation that a physicist will make, despite the nearly forced restraint imposed upon his [scientific methodology]? He will affirm that underneath the observational data, the only data accessible to his methods of study, are hidden realities whose essence cannot be grasped by the same methods, and that these realities are arranged in a certain order which physical science cannot directly contemplate.[57]

And, lo!, it turns out that, as a practicing Catholic, Duhem's favored thermomechanics turns out to be in greater "metaphysical" harmony with the Thomistic traditions of his church than its mechanical rivals (apparently, British citizenship, Protestantism, mechanism and "weak minds" run together).

But if we set aside all of this unhelpful talk of "gestalts," "paradigms" and "metaphysical mind sets," we will discover some interesting ways in which Duhem's atomist opponents were sometimes misled by misdiagnosed semantic pictures and that Duhem and Mach were right to object to their practices, although not because of their "allegiance to faulty metaphysics." The real nub of the problem lay in much tinier, although, in some cases, rather tricky, concerns. In particular, consider the main argument that Ludwig Boltzmann advances in his "On the Indispensability of Atomism in Natural Science":

> Do not imagine that by means of the word continuum or the writing down of a differential equation, you have acquired a clear conception of the continuum. On closer scrutiny the differential equation is merely the expression of the fact that one must first imagine a finite number, ... only then is the number to grow until its further growth has no significance. ... [T]he basic equations of elasticity can be generally solved only if one first imagines a finite number of elementary particles that act on each other according to certain simple laws and then once again looks for the limit as this number increases. This limit is thus once again the real definition of the basic equations and the picture that from the outset assumes a large but finite number seems once more simpler.[58]

---

[57] Duhem, Aim, 56, 103 (rearranged).
[58] Ludwig Boltzmann, "On the Indispensability of Atomism in Natural Science" in Theoretical Physics and Philosophical Problems, Paul Foulkes, trans. (Dordrecht: D. Reidel, 1974), 43–4.

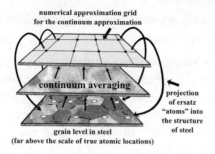

numerical approximation grid
for the continuum approximation

continuum averaging

projection
of ersatz
"atoms" into
the structure
of steel

grain level in steel
(far above the scale of true atomic locations)

Not to mince words, these claims are complete bosh and stem from a faulty picturing of differential equations more primitive than any considered in Chapter 9. A common way to approximate formulas of this type is to employ *finite differences*, whereby a continuously distributed object such as a steel plate of metal is covered with a grid of approximation points algebraically tied together in rough approximation to their parent differential equations (the method is merely a version of Euler's method adapted to higher dimensional processes). In Boltzmann's time, it was common to "derive" the differential equations of physics by starting with an allied grid of points and then squeezing them together through taking a limit (in the fine print to 4,i we observed that Karl Pearson was driven into the arms of neo-Kantianism in an effort to justify this practice, which good engineers studiously avoid nowadays). An inspection of the Boltzmann passage reveals that he is arguing for "atoms" on the grounds that they are needed as worldly support for the nodes on such a grid.

But consider how absurd this argument is. Steel, as we've already noted, is comprised of a very complex hierarchy of complicated grain, and the usual equations utilized in engineering for such plates reflect a homogenizing blurring over its top level of structure (only then do we witness isotropy, for example). Boltzmann's finite difference grid arises from an approximation to *that* averaged equation and the grid point "atoms" he projects into reality from their checkerboard have absolutely no relationship to any microscopic structure found within the steel (its true atoms will be located elsewhere, with completely different properties). In short, he is offering a brief for projected *ghost atoms*, just as surely as classical thought argues for ghost attributes.

I've not found Duhem responding directly to this kind of argument, but Mach does:

> When a geometer wishes to understand the form of a curve, he first resolves it into small rectilinear elements. In doing this, however, he is fully aware that these elements are only provisional and arbitrary devices for comprehending in parts what we cannot comprehend as a whole. When the law of the curve is found he no longer thinks of the elements. Similarly, it would not become physical science to see in its self-created, changeable, economical tools, molecules and atoms, realities behind phenomena . . . The atom must remain a tool for representing phenomena, like the functions of mathematics. Gradually, however, as the intellect, by contact with its subject matter grows in discipline, physical science will give up its mosaic play with stones and seek out, the boundaries and the bed in

658 Nature and Genius

which the living stream of phenomena flows. The goal which it has set itself is the simplest and most economical abstract expression of facts.[59]

This response is curious in two ways. First of all, Mach correctly senses that Boltzmann has his foundational priorities backwards: the continuous differential equation should be regarded as corresponding more closely to reality than the finite difference grid that approximates it. On the other hand, Mach still suffers from confusion as to how differential equations in physics should be justified, and mistakenly concedes that "the atom must remain a tool for representing phenomena" where "atom" has plainly assumed the significance of "grid point in a derivation," not anything that can properly be considered as an atom. Note—and this feature is typical of the nineteenth century origins of so many methodological themes widely accepted in philosophical circles even today—that Mach is attempting to combat what is essentially a *mathematical misdiagnosis* on Boltzmann's part with a *philosophical maxim* (as remarked in 4,i, this is true of many of Karl Pearson's endeavors as well). Once again, we should take warning from this: just because we presently lack the tools to resolve a problem, we shouldn't attempt to bridge the gap with slogans.

It so happens that a more sophisticated confusion of the same general type lies at the base of the methodological inconsistencies that Duhem detects in Kelvin and Maxwell, but it would take us too far afield to diagnose its details here. In this case delicate considerations speak in favor of Duhem and the British lions in different respects, although no one then alive could have then sorted out what they are, because they hinge upon discoveries in functional analysis and mechanics dating to the mid-twentieth century. In the meantime, before such matters became plain, Duhem and Kelvin could only forge ahead on semantic hunch, each encouraged to some degree by the funds of practical success to which their differing policies gave rise. But neither party labored under gestalt handicap, inescapable conceptual scheme or national mind—they were simply caught in the position of betting in a horse race whose winner won't cross the finish line until 1965. For the steed in question happens to be the supportive logic that stands behind the equational derivations common within nineteenth century continuum physics and, as Heaviside advises us, such nags often show up late:

> *Logic has nothing to do with it, either with the fact, the discovery, or its use. At the very same time it must be said that a sufficiently profound study of the subject would ultimately lead to the logic of its laws, as a final result. What I do strongly object to is the idea that the logic should come first, or else you prove nothing. Yet perhaps the majority of academical mathematical books are written under this idea. In reality the logic is the very last thing, and that is not final.*[60]

There is nothing like an unresolved wager to induce fist fights, but the parties don't battle because they fail to *understand* one another.

---

[59] Mach, "Economical," 206–7.   [60] Heaviside, Electromagnetic iii. 370.

"Come, let us hunt the —"

. . . .

In brief, what I have in mind is this. In the mid-twentieth century it was realized that a large class of differential equations can be usefully approximated, after they are reexpressed in variational form, by so-called *finite elements* (which are related to the *finite differences* discussed above, but are of more sophisticated capacity). If we dress up these schemata in strings and guiding rods, we obtain "models" remarkably like those in which Kelvin commonly traffics and of which Duhem complains. However, because of their intimate ties to variational principles and the geometrical constitutive hypotheses required to extract differential equations of viable simplicity, Kelvin often hews a path of *safer* derivational policy for real materials than Duhem himself, even though the modelings he employs do not *appear* to be consistent with one another.

. . . .

Through the writings of Duhem, Kuhn, Quine and a thousand others, we have grown accustomed to parsing such disputes as great clashes over paradigm and philosophy, often cheering the winner as a Prometheus of right thinking within a sea of uncomprehending colleagues. But do we really wish to view our capacities for interpersonal communication in such a gloomy light? In truth, a large amount of rhetorical smoke can spring from a very small seed if neither disputant can locate it accurately, simply because it is tiny and complex in its operations. When this happens, we are likely to witness one of those unavoidable episodes of "strong men struggling with a word," in the manner of Hilaire Belloc's fierce hunters who couldn't resolve how "gnu" should be pronounced.[61] If Duhem and Kelvin could be seated around the table with a modern functional analyst, I wager that each would come away feeling half vindicated in their methodologies but half embarrassed to have been taken in by the semantic mimicries of such a wee word as "dy/dx." Their dispute could have been ironed out in ten minutes, but only by utilizing diagnostic tools then unavailable.

In the same vein, let us again review those extraordinary inclinations to dub the most innocuous facts of word/world correlation as "metaphysical" ("in the pejorative sense," adds Mark Johnston). In this mode, Crispin Wright contends:

> *If . . . certain expressions in a branch of our language function as singular terms, and descriptive and identity contexts containing them are true by ordinary criteria, there is no room for any ulterior failure of "fit" between those contexts and the structure of states of*

---

[61] Hilaire Belloc, "The Gnu", <u>A Moral Alphabet</u> in <u>Cautionary Tales</u>, 312–13. The illustration is by "B.T.B."

affairs which make them true. So there can be no philosophical science of ontology, no well founded attempt to see past our categories of expression and glimpse the way the world is truly furnished.[62]

Here Wright primarily has "singular terms" (= names) in view, but his reservations presumably extend to predicates as well. I suppose I might be dubious of a "philosophical science of ontology" if I knew what it was, but I am quite certain that we all do quite well at "glimpsing the way the world is truly furnished," although commonly only in mid-range detail. Underlying Wright's thinking unfurls some strange carpet of exaggerated assumption akin to Mach's, Quine's or Duhem's: that we can somehow observe our words functioning on our behalf with perfect clarity, yet we can't, at the same time, perceive the rabbits that cavort before us, or the rocks or the noses upon our face. But why should we wish to believe such extraordinary claims? We know quite a bit about the *redness* and *hardness* that we encounter in the world, but each trait hides its little secrets as well, whose evidences first emerge as little fissures in usage whose presence we can't quite explain. Perhaps, before we embrace prodigal opinions such as Wright's, a suspicious glance should be first cast in the direction of those Uriah Heaps of *ur*-philosophical chicanery, "concept" and company, as they toil at the drudgery of our semantic bookkeeping?

While we're on such topics, surely we should reopen the case of Helen Keller, whose claims to understand *redness* adequately had, at last evaluation (3,viii), been brusquely denied? However, if she becomes as sophisticated about the real world supports behind "is red"'s employments as Ralph Evans (7,x), she might be properly claimed to *understand* the term *better* than most of us. We have observed that "understand" can switch its evaluative focus as promiscuously as does "concept." If we so choose, we can employ "understand" to appraise the fact that Keller cannot classify red objects in ordinary settings as swiftly as you or I. But what exactly does that prove? By the same lights, a Treasury agent ignorant of statistics can claim that she "understands" the markings of a counterfeit bill better than we, although statistically instructed folk can fairly counter with an equally founded, "Yes, but we understand them better than *you*, in another sense of 'understand.'" And so our claims of "understanding" with respect to Keller scarcely carry the absolutist significance that we *ur*-philosophically believed them to have, back in Chapter 3 when we haughtily dismissed her claims to conceptual grasp as ridiculous.

These observations, of course, merely reiterate my contention that a complex set of skills can be adequately appreciated *in pieces*, without our needing to be able to imitate the overall bundle in question. Of course, the classical picture of concepts vehemently opposes this scattered approach to "conceptual understanding," demanding that Keller must be able to grasp the same concentrated content that allegedly hovers just before our mind's eye. In framing this false portrait of our intellectual skills, classical doctrine displays its deep affinities with the assumptions of crisp presentation familiar within traditional sense data accounts of perception (7,xi).

---

[62] Crispin Wright, <u>Frege's Conception of Numbers as Objects</u> (Aberdeen: Aberdeen University Press, 1983), 52.

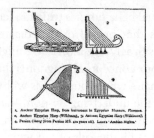

Ancient harps

Finally, let us return glancingly to the musical pastures of Chapter 2. The ethno-musicologist Bruno Nettl recalls:

> *I was about to leave my lesson of Persian music in the spacious old house in south Teheran when my teacher suddenly fixed me with his forefinger: "You will never understand this music. There are things that every Persian on the street understands instinctively which you will never understand no matter how hard you try."* . . . *I blurted out, "I don't really expect to understand it that way, I am just trying to figure out how it is put together." "Oh well, that is something you can probably learn, but it's not really very important."*[63]

The instructor underestimated what we might potentially learn about his music, but he may very well be right: if our understanding becomes too dispersed, it might not seem very important. And if we can no longer hear the lyres of Homer's age with relish, that may, or may not, represent a loss, given that we might need to abandon the harmonic pleasures of Mozart to do so. But that scattering within our comprehension doesn't mean that we can't, with a lot of hard work, gain a pretty solid understanding of why Achilles once wept to hear those strains.

(ix)

An isthmus of a middle state. To recapitulate a very long book in a very short verse, our semantic lot in life is that described in Pope's <u>Essay on Man</u>:

> *Plac'd on this isthmus of a middle state,*
> *A Being darkly wise, and rudely great:*
> *With too much knowledge for the Skeptic's side,*
> *With too much weakness for the Stoic's side.*
> *He hangs between; in doubt to act, or rest;*
> *In doubt to deem himself a God, or Beast:* . . .
> *Sole judge of Truth, in endless Error hurl'd:*
> *The glory, jest, and riddle of the world!*[64]

---

[63] Bruno Nettl, <u>The Study of Ethnomusicology</u> (Urbana: University of Illinois Press, 1983), 259.
[64] Alexander Pope, <u>Selected Works</u> (New York: Random House, 1948), 107.

Nothing latent in our conceptual behavior lies truly hidden from us—the rationality and the errors of the directivities that propel our predicates onward can eventually be unraveled, but not always in as timely a fashion as we might desire. We certainly cannot rule as the little gods within our conceptual kingdoms that classical thinking promises, able to augur from our armchairs where our words will wend. With patience and toil, we can eventually puzzle out the underlying wisdom—and the unavoidable follies—of linguistic leaps of faith executed long before. But, in the final analysis, we must acknowledge that the winds that fill the sails of our words are not primarily of our own manufacture or determination, but arise from mischievous Nature, for, as Samuel Johnson prefaces his Dictionary:

> Total and sudden transformations of language seldom happen; conquests and migrations are now very rare: but there are other causes of change, which, though slow in their operation, and invisible in their progress, are perhaps as much superior to human resistance, as the revolutions of the sky, or intumescence of the tide.[65]

[65] Samuel Johnson, "Preface," Dictionary, 138.

INDEX OF TOPICS

INDEX OF AUTHORS